Handbook of Approximate Bayesian Computation

Chapman & Hall/CRC

Handbooks of Modern Statistical Methods

Series Editor

Garrett Fitzmaurice, *Department of Biostatistic, Harvard School of Public Health, Boston, MA, U.S.A.*

The objective of the series is to provide high-quality volumes covering the state-of-the-art in the theory and applications of statistical methodology. The books in the series are thoroughly edited and present comprehensive, coherent, and unified summaries of specific methodological topics from statistics. The chapters are written by the leading researchers in the field, and present a good balance of theory and application through a synthesis of the key methodological developments and examples and case studies using real data.

Longitudinal Data Analysis
Edited by Garrett Fitzmaurice, Marie Davidian, Geert Verbeke, and Geert Molenberghs
Handbook of Spatial Statistics
Edited by Alan E. Gelfand, Peter J. Diggle, Montserrat Fuentes, and Peter Guttorp
Handbook of Markov Chain Monte Carlo
Edited by Steve Brooks, Andrew Gelman, Galin L. Jones, and Xiao-Li Meng
Handbook of Survival Analysis
Edited by John P. Klein, Hans C. van Houwelingen, Joseph G. Ibrahim, and Thomas H. Scheike
Handbook of Mixed Membership Models and Their Applications
Edited by Edoardo M. Airoldi, David M. Blei, Elena A. Erosheva, and Stephen E. Fienberg
Handbook of Missing Data Methodology
Edited by Geert Molenberghs, Garrett Fitzmaurice, Michael G. Kenward, Anastasios Tsiatis, and Geert Verbeke
Handbook of Design and Analysis of Experiments
Edited by Angela Dean, Max Morris, John Stufken, and Derek Bingham
Handbook of Cluster Analysis
Edited by Christian Hennig, Marina Meila, Fionn Murtagh, and Roberto Rocci
Handbook of Discrete-Valued Time Series
Edited by Richard A. Davis, Scott H. Holan, Robert Lund, and Nalini Ravishanker
Handbook of Big Data
Edited by Peter Bühlmann, Petros Drineas, Michael Kane, and Mark van der Laan
Handbook of Spatial Epidemiology
Edited by Andrew B. Lawson, Sudipto Banerjee, Robert P. Haining, and María Dolores Ugarte
Handbook of Neuroimaging Data Analysis
Edited by Hernando Ombao, Martin Lindquist, Wesley Thompson, and John Aston
Handbook of Statistical Methods and Analyses in Sports
Edited by Jim Albert, Mark E. Glickman, Tim B. Swartz, Ruud H. Koning
Handbook of Methods for Designing, Monitoring, and Analyzing Dose-Finding Trials
Edited by John O'Quigley, Alexia Iasonos, Björn Bornkamp
Handbook of Quantile Regression
Edited by Roger Koenker, Victor Chernozhukov, Xuming He, and Limin Peng
Handbook of Environmental and Ecological Statistics
Edited by Alan E. Gelfand, Montserrat Fuentes, Jennifer A. Hoeting, Richard Lyttleton Smith

For more information about this series, please visit: https://www.crcpress.com/go/handbooks

Handbook of Approximate Bayesian Computation

Edited by

S. A. Sisson

Y. Fan

M. A. Beaumont

CRC Press
Taylor & Francis Group
Boca Raton London New York

CRC Press is an imprint of the
Taylor & Francis Group, an **informa** business
A CHAPMAN & HALL BOOK

CRC Press
Taylor & Francis Group
6000 Broken Sound Parkway NW, Suite 300
Boca Raton, FL 33487-2742

Printed on acid-free paper

International Standard Book Number-13: 978-1-4398-8150-7 (Hardback)

Library of Congress Cataloging-in-Publication Data

Names: Sisson, S. A. (Scott A), editor. | Fan, Y. (Yanan), editor. |
Beaumont, M. A. (Mark A.), editor.
Title: Handbook of approximate Bayesian computation / edited by S.A. Sisson,
Y. Fan, M.A. Beaumont.
Description: Boca Raton, Florida : CRC Press, [2019] | Includes
bibliographical references and index.
Identifiers: LCCN 2018010970 | ISBN 9781439881507 (hardback : alk. paper) |
ISBN 9781315117195 (e-book) | ISBN 9781439881514 (web pdf) | ISBN
9781351643467 (epub) | ISBN 9781351633963 (mobi/kindle)
Subjects: LCSH: Bayesian statistical decision theory. | Mathematical analysis.
Classification: LCC QA279.5 .H36 2019 | DDC 519.5/42--dc 3
LC record available at https://lccn.loc.gov/2018010970

Visit the Taylor & Francis Web site at
http://www.taylorandfrancis.com

and the CRC Press Web site at
http://www.crcpress.com

Contents

Preface ix

Editors xi

Contributors xiii

I Methods 1

1 Overview of ABC 3
S. A. Sisson, Y. Fan, and M. A. Beaumont

2 On the History of ABC 55
Simon Tavaré

3 Regression Approaches for ABC 71
Michael G.B. Blum

4 ABC Samplers 87
S. A. Sisson and Y. Fan

5 Summary Statistics 125
Dennis Prangle

6 Likelihood-Free Model Choice 153
Jean-Michel Marin, Pierre Pudlo, Arnaud Estoup, and Christian Robert

7 ABC and Indirect Inference 179
Christopher C. Drovandi

8 High-Dimensional ABC 211
David J. Nott, Victor M.-H. Ong, Y. Fan, and S. A. Sisson

9 Theoretical and Methodological Aspects of Markov Chain Monte Carlo Computations with Noisy Likelihoods 243
Christophe Andrieu, Anthony Lee, and Matti Vihola

10 Asymptotics of ABC **269**
Paul Fearnhead

11 Informed Choices: How to Calibrate ABC with Hypothesis Testing **289**
Oliver Ratmann, Anton Camacho, Sen Hu, and Caroline Colijn

12 Approximating the Likelihood in ABC **321**
Christopher C. Drovandi, Clara Grazian, Kerrie Mengersen, and Christian Robert

13 A Guide to General-Purpose ABC Software **369**
Athanasios Kousathanas, Pablo Duchen, and Daniel Wegmann

14 Divide and Conquer in ABC: Expectation-Propagation Algorithms for Likelihood-Free Inference **415**
Simon Barthelmé, Nicolas Chopin, and Vincent Cottet

II Applications **435**

15 Sequential Monte Carlo-ABC Methods for Estimation of Stochastic Simulation Models of the Limit Order Book **437**
Gareth W. Peters, Efstathios Panayi, and Francois Septier

16 Inferences on the Acquisition of Multi-Drug Resistance in *Mycobacterium Tuberculosis* Using Molecular Epidemiological Data **481**
Guilherme S. Rodrigues, Andrew R. Francis, S. A. Sisson, and Mark M. Tanaka

17 ABC in Systems Biology **513**
Juliane Liepe and Michael P.H. Stumpf

18 Application of ABC to Infer the Genetic History of Pygmy Hunter-Gatherer Populations from Western Central Africa **541**
Arnaud Estoup, Paul Verdu, Jean-Michel Marin, Christian Robert, Alex Dehne-Garcia, Jean-Marie Cornuet, and Pierre Pudlo

19 ABC for Climate: Dealing with Expensive Simulators **569**
Philip B. Holden, Neil R. Edwards, James Hensman, and Richard D. Wilkinson

20 ABC in Ecological Modelling **597**
Matteo Fasiolo and Simon N. Wood

21 ABC in Nuclear Imaging **623**
Y. Fan, Steven R. Meikle, Georgios I. Angelis,
and Arkadiusz Sitek

Index **649**

Preface

Approximate Bayesian computation (ABC) is now recognised as the first member of the class of 'likelihood-free' methods, which have been instrumental in driving research in Monte Carlo methods over the past decade. ABC originated from the need to address challenging inferential problems in population genetics, where the complexity of a model meant that the associated likelihood function was computationally intractable and could not be evaluated numerically in any practical amount of time. At its heart, ABC is a very simple method: numerical evaluation of the likelihood function is replaced with an assessment of how likely it is the model could have produced the observed data, based on simulating pseudo-data from the model and comparing it to the observed data.

The idea is remarkably simple, but this simplicity also means that ABC methods are highly accessible as analysis tools in a manner similar to the way in which the accessibility of the Metropolis–Hastings algorithm was responsible for the propagation of Bayesian inferential methods in the 1990s. The last ten years have seen a surge in interest from the statistical community, as researchers have realised how powerful ABC can be as a computational technique. Major advances have been made in terms of computation and theory. In addition, ABC methods have been extensively applied to wide ranging and diverse applications in many disciplines. With such rapid developments, it seemed to us that now was a good time to put together an overview of ABC research and its applications. This will be the first book that synthesises the most important aspects of ABC research within a single volume.

This handbook is intended to be a comprehensive reference for both developers and users of ABC methodology. While ABC methods are still relatively new, in this Handbook, we have attempted to include a substantial amount of the fundamental ideas so that the material covered will be relevant for some time. Graduate students and researchers new to ABC wishing to become acquainted with the field will be able to find instruction on the basic theory and algorithms. Many chapters are written in tutorial style with detailed illustrations and examples. Part I first provides an overview of the basic ABC method and then some history on it's developmental origins. This is followed by detailed expositions on various aspects of ABC techniques, including algorithm development, construction and choice of summary statistics, model choice, asymptotic theory, and variants and extensions of ABC algorithms that deal with particular modelling situations. A purpose-written chapter on software available for implementing ABC methods is also provided.

For those who are interested in seeing how ABC is implemented in practice, Part II contains a number of applied chapters written by discipline experts, detailing the use of ABC in particular discipline applications. These are intended to be exemplar ABC implementations in a range of fields, with topics including financial market dynamics, multi-drug resistance, systems biology, population genetics, climate simulators, dynamic ecological models, and medical imaging.

When planning this book, we soon realised that no single source can give a truly comprehensive overview of ABC research and application—there is just too much of it and its development is moving too fast. Instead, the editorial goal was to obtain high quality contributions that may stand the test of time. We have enlisted a group of contributors who are at the very forefront of ABC research and application. To ensure a uniformly high quality, all handbook contributions (including those written by members of the editorial panel) were submitted to a rigorous peer review process and many underwent several revisions. We thank the anonymous referees for their thoughtful and focused reviews. We also thank the editorial team from Chapman & Hall/CRC Press for their advice and patience throughout the production process.

We are proud to be able to bring together this handbook on an exciting and novel statistical technique that has greatly expanded the modelling horizons of applied research across the board. We hope that ABC will have as much impact in the next ten years as it has had in the previous decade.

S. A. Sisson

Y. Fan

M. A. Beaumont

Editors

S. A. Sisson is a professor of statistics at the University of New South Wales, Sydney. He is a past president of the Statistical Society of Australia and a deputy director of the Australian Research Centre of Excellence in Mathematical and Statistical Frontiers. He has received a 2017 ARC Future Fellowship and the 2011 Moran Medal from the Australian Academy of Science. Scott is interested in computational statistics and methods, particularly in computationally challenging modelling scenarios, extreme value theory, and the application of Bayesian statistics to problems in a wide range of disciplines.

Y. Fan is a senior lecturer in statistics at the University of New South Wales, Sydney. She obtained her PhD in statistics, on Markov chain Monte Carlo methods, from the University of Bristol. Currently, her research interests include development of computational statistical methods, Bayesian modelling, image analysis, and climate modelling.

M. A. Beaumont is a professor of statistics at the University of Bristol. His research interests focus mainly on problems in population genetics and the statistical methodology associated with them.

Contributors

Christophe Andrieu
School of Mathematics
University of Bristol
Bristol, United Kingdom

Georgios I. Angelis
Faculty of Health Sciences
The University of Sydney
Sydney, Australia

Simon Barthelmé
CNRS
Gipsa-lab
Grenoble INP
Grenoble, France

M. A. Beaumont
School of Mathematics
University of Bristol
Bristol, United Kingdom

Michael G.B. Blum
Laboratoire TIMC-IMAG, UMR 5525
Université Grenoble Alpes, CNRS
Grenoble, France

Anton Camacho
Centre for the Mathematical Modelling of Infectious Diseases
London School of Hygiene and Tropical Medicine
London, United Kingdom

Nicolas Chopin
ENSAE-CREST
Paris, France

Caroline Colijn
Department of Mathematics
Imperial College London
London, United Kingdom

Jean-Marie Cornuet
CBGP, INRA, CIRAD, IRD, Montpellier SupAgro
University of Montpellier
Montpellier, France

Vincent Cottet
ENSAE-CREST
Paris, France

Alex Dehne-Garcia
CBGP, INRA, CIRAD, IRD, Montpellier SupAgro
University of Montpellier
Montpellier, France

Christopher C. Drovandi
School of Mathematical Sciences
Queensland University of Technology
Brisbane, Australia

Pablo Duchen
Department of Biology
University of Fribourg
Fribourg, Switzerland

Neil R. Edwards
School of Environment, Earth and Ecosystem Sciences
Open University
Milton Keynes, United Kingdom

Arnaud Estoup
CBGP, INRA, CIRAD, IRD, Montpellier SupAgro
University of Montpellier
Montpellier, France

Y. Fan
School of Mathematics and Statistics
University of New South Wales
Sydney, Australia

Matteo Fasiolo
School of Mathematics
University of Bristol
Bristol, United Kingdom

Paul Fearnhead
Department of Mathematics and Statistics
Lancaster University
Lancaster, United Kingdom

Andrew R. Francis
Centre for Research in Mathematics
Western Sydney University
Sydney, Australia

Clara Grazian
Nuffield Department of Medicine
University of Oxford
Oxford, United Kingdom

James Hensman
PROWLER.io
Cambridge, United Kingdom

Philip B. Holden
School of Environment, Earth and Ecosystem Sciences
Open University
Milton Keynes, United Kingdom

Sen Hu
School of Mathematics and Statistics
University College Dublin
Dublin, Ireland

Athanasios Kousathanas
Department of Biology
University of Fribourg
Fribourg, Switzerland

Anthony Lee
School of Mathematics
University of Bristol
Bristol, United Kingdom

Juliane Liepe
Quantitative and Systems Biology
Max-Planck-Institute for Biophysical Chemistry
Göttingen, Germany

Jean-Michel Marin
IMAG
University of Montpellier, CNRS
Montpellier, France

Steven R. Meikle
Faculty of Health Sciences and Brain and Mind Centre
The University of Sydney
Sydney, Australia

Kerrie Mengersen
School of Mathematical Sciences
Queensland University of Technology
Brisbane, Australia

David J. Nott
Department of Statistics and Applied Probability
National University of Singapore
Singapore

Victor M.-H. Ong
Department of Statistics and Applied Probability
National University of Singapore
Singapore

Efstathios Panayi
Department of Actuarial Mathematics and Statistics
Heriot-Watt University
Edinburgh, United Kingdom

Gareth W. Peters
Department of Actuarial Mathematics and Statistics
Heriot-Watt University
Edinburgh, United Kingdom

Dennis Prangle
School of Mathematics Statistics and Physics
Newcastle University
Newcastle upon Tyne
United Kingdom

Pierre Pudlo
Institut de Mathématiques de Marseille
Aix-Marseille Université
Marseille, France

Oliver Ratmann
Department of Mathematics
Imperial College London
London, United Kingdom

Christian Robert
CEREMADE
Université Paris Dauphine
Paris, France

Guilherme S. Rodrigues
CAPES Foundation
Ministry of Education of Brazil
Brasília, Brazil

Francois Septier
Institute Mines-Telecom Lille
CRIStAL UMR CNRS 9189
Lille, France

S. A. Sisson
School of Mathematics and Statistics
University of New South Wales
Sydney, Australia

Arkadiusz Sitek
Radiology Department
Massachusetts General Hospital and Harvard Medical School
Boston, Massachusetts

Michael P.H. Stumpf
Melbourne Integrative Genomics
School of BioSciences and School of Mathematics and Statistics
University of Melbourne
Parkville, Australia

Mark M. Tanaka
School of Biotechnology and Biological Sciences
University of New South Wales
Sydney, Australia

Simon Tavaré
Department of Applied Mathematics and Theoretical Physics
University of Cambridge
Cambridge, United Kingdom

Paul Verdu
UMR 7206, Ecoanthropology and Ethnobiology
CNRS-MNHN-Université Paris Diderot-Sorbonne Paris Cité
Paris, France

Matti Vihola
Department of Mathematics and Statistics
University of Jyväskylä
Jyväskylä, Finland

Daniel Wegmann
Department of Biology
University of Fribourg
Fribourg, Switzerland

Simon N. Wood
School of Mathematics
University of Bristol
Bristol, United Kingdom

Richard D. Wilkinson
School of Mathematics and Statistics
University of Sheffield
Sheffield, United Kingdom

Part I

Methods

1

Overview of ABC

S. A. Sisson, Y. Fan, and M. A. Beaumont

CONTENTS

1.1	Introduction	3
1.2	Likelihood-Free Intuition	6
1.3	A Practical Illustration: Stereological Extremes	7
	1.3.1 Background and model	8
	1.3.2 Analysis	8
1.4	A g-and-k Distribution Analysis	11
1.5	Likelihood-Free Methods or ABC?	14
1.6	The Approximate Posterior Distribution	15
	1.6.1 Simple examples	19
1.7	The Use of Summary Statistics	23
	1.7.1 Summary statistic basics	23
	1.7.2 Some practical issues with summary statistics	28
1.8	An ABC Analysis in Population Genetics	34
1.9	Levels of Approximation in ABC	41
1.10	Interpretations of ABC	43
1.11	Further Reading	44
1.12	Conclusion	45
Acknowledgements		45
References		46

1.1 Introduction

In Bayesian inference, complete knowledge about a vector of model parameters, $\theta \in \Theta$, obtained by fitting a model $\mathcal{M}$, is contained in the posterior distribution. Here, prior beliefs about the model parameters, as expressed through the prior distribution, $\pi(\theta)$, are updated by observing data $y_{obs} \in \mathcal{Y}$ through the likelihood function $p(y_{obs}|\theta)$ of the model. Using Bayes' theorem, the resulting posterior distribution:

$$\pi(\theta|y_{obs}) = \frac{p(y_{obs}|\theta)\pi(\theta)}{\int_\Theta p(y_{obs}|\theta)\pi(\theta)d\theta},$$

contains all necessary information required for analysis of the model, including model checking and validation, predictive inference, and decision making. Typically, the complexity of the model and/or prior means that the posterior distribution, $\pi(\theta|y_{obs})$, is not available in closed form, and so numerical methods are needed to proceed with the inference. A common approach makes use of Monte Carlo integration to enumerate the necessary integrals. This relies on the ability to draw samples $\theta^{(1)}, \theta^{(2)}, \ldots, \theta^{(N)} \sim \pi(\theta|y_{obs})$ from the posterior distribution so that a finite sample approximation to the posterior is given by the empirical measure:

$$\pi(\theta|y_{obs}) \approx \frac{1}{N}\sum_{i=1}^{N} \delta_{\theta^{(i)}}(\theta),$$

where $\delta_Z(z)$ denotes the Dirac measure, defined as $\delta_Z(z) = 1$ if $z \in Z$ and $\delta_Z(z) = 0$ otherwise. As the size of the sample from the posterior gets large, then the finite sample approximation better approximates the true posterior so that $\lim_{N\to\infty} \frac{1}{N}\sum_{i=1}^{N} \delta_{\theta^{(i)}}(\theta) \to \pi(\theta|y_{obs})$, by the law of large numbers. As a result, the expectation of a function $a(\theta)$ under $\pi(\theta|y_{obs})$ can be estimated as:

$$\begin{aligned}\mathbb{E}_\pi[a(\theta)] &= \int_\Theta a(\theta)\pi(\theta|y_{obs})d\theta \\ &\approx \int_\Theta a(\theta)\frac{1}{N}\sum_{i=1}^{N}\delta_{\theta^{(i)}}(\theta)d\theta = \frac{1}{N}\sum_{i=1}^{N} a(\theta^{(i)}).\end{aligned}$$

There are a number of popular algorithms available for generating samples from posterior distributions, such as importance sampling, Markov chain Monte Carlo (MCMC) and sequential Monte Carlo (SMC) (Chen et al. 2000; Doucet et al. 2001; Del Moral et al. 2006; Brooks et al. 2011).

Inherent in such Monte Carlo algorithms is the need to numerically evaluate the posterior distribution, $\pi(\theta|y_{obs})$, up to a normalisation constant, commonly many thousands or millions of times. For example, in the Metropolis–Hastings algorithm, an MCMC algorithm, this arises through computing the probability that the Markov chain accepts the proposed move from a current point θ to a proposed point $\theta' \sim q(\theta, \theta')$, where q is some proposal density, given by $\alpha(\theta, \theta') = \min\left\{1, \frac{\pi(\theta'|y_{obs})q(\theta',\theta)}{\pi(\theta|y_{obs})q(\theta,\theta')}\right\}$. Similarly in SMC algorithms, the incremental particle weight is given by $w_t(\theta_t) = \frac{\pi_t(\theta_t|y_{obs})L_{t-1}(\theta_t,\theta_{t-1})}{\pi_{t-1}(\theta_{t-1}|y_{obs})M_t(\theta_{t-1},\theta_t)}$, where M_t and L_{t-1} are transition kernels, and π_t

denotes a function strongly related to the posterior distribution, such as $\pi_t(\theta_t|y_{obs}) = [\pi(\theta_t|y_{obs})]^{t/T}\pi(\theta_t)^{1-t/T}$. Evaluating acceptance probabilities or particle weights clearly requires evaluation of the likelihood function.

However, for an increasing range of scientific problems – see Section 1.11 for a selection – numerical evaluation of the likelihood function, $\pi(y_{obs}|\theta)$, is either computationally prohibitive, or simply not possible. Examples of the former can occur where the size of the observed dataset, y_{obs}, is sufficiently large that, in the absence of low dimensional sufficient statistics, evaluating the likelihood function even once is impracticable. This can easily occur in the era of Big Data, for example, through large genomic datsets. Partial likelihood intractability can arise, for instance, in models for Markov random fields. Here, the likelihood function can be written as $p(y_{obs}|\theta) = \frac{1}{Z_\theta}\tilde{p}(y_{obs}|\theta)$, where $\tilde{p}(y_{obs}|\theta)$ is a function that can be evaluated, and where the normalisation constant, $Z_\theta = \sum_{\mathcal{Y}} \tilde{p}(y|\theta)$, depends on the parameter vector θ. Except for trivial datasets, the number of possible data configurations in the set $\mathcal{Y}$ means that brute-force enumeration of Z_θ is typically infeasible (Møller et al. 2006; Grelaud et al. 2009). While there are algorithmic techniques available that arrange for the intractable normalising constants to cancel out within, for example, Metropolis–Hastings acceptance probabilities (Møller et al. 2006), or that numerically approximate Z_θ through, for example, path sampling or thermodynamic integration, these are not viable when $\tilde{p}(y|\theta)$ itself is also computationally intractable. Instances when the complete likelihood function is unavailable can also occur when the model density function is only implicitly defined, for example, through quantile or characteristic functions (Drovandi and Pettitt 2011; Peters et al. 2012). Similarly, the likelihood function may only be implicitly defined as a data generation process.

In these scenarios, if the preferred model is computationally intractable, the need to repeatedly evaluate the posterior distribution to draw samples from the posterior makes the implementation of standard Bayesian simulation techniques impractical. Faced with this challenge, one option is simply to fit a different model that is more amenable to statistical computations. The disadvantage of this approach is that the model could then be less realistic, and not permit inference on the particular questions of interest for the given analysis. A more attractive alternative, may be to consider an approximation to the preferred model, so that modelling realism is maintained at the expense of some approximation error. While various posterior approximation methods are available, 'likelihood-free' Bayesian methods, of which approximate Bayesian computation (ABC) is a particular case, have emerged as an effective and intuitively accessible way of performing an approximate Bayesian analysis.

In this chapter, we aim to give an intuitive exploration of the basics of ABC methods, illustrated wherever possible by simple examples. The scope of this exploration is deliberately limited, for example, we focus only on the use of simple rejection sampling-based ABC samplers, in order that this chapter will provide an accessible introduction to a subject which is given more detailed and advanced treatments in the rest of this handbook.

1.2 Likelihood-Free Intuition

The basic mechanism of likelihood-free methods can be fairly easily understood at an intuitive level. For the moment, we assume that data generated under the model, $y \sim p(y|\theta)$, are discrete. Consider the standard rejection sampling algorithm for sampling from a density $f(\theta)$:

Standard Rejection Sampling Algorithm

Inputs:

- A target density $f(\theta)$.
- A sampling density $g(\theta)$, with $g(\theta) > 0$ if $f(\theta) > 0$.
- An integer $N > 0$.

Sampling:
For $i = 1, \ldots, N$:

1. Generate $\theta^{(i)} \sim g(\theta)$ from sampling density g.
2. Accept $\theta^{(i)}$ with probability $\frac{f(\theta^{(i)})}{Kg(\theta^{(i)})}$ where $K \geq \max_\theta \frac{f(\theta)}{g(\theta)}$.
   Else go to 1.

Output:
A set of parameter vectors $\theta^{(1)}, \ldots, \theta^{(N)}$ which are samples from $f(\theta)$.

If we specify $f(\theta) = \pi(\theta|y_{obs})$, and suppose that the prior is used as the sampling distribution, then the acceptance probability is proportional to the likelihood, as then $f(\theta)/Kg(\theta) \propto p(y_{obs}|\theta)$. While direct evaluation of this acceptance probability is not available if the likelihood is computationally intractable, it is possible to stochastically determine whether or not to accept or reject a draw from the sampling density, *without* numerical evaluation of the acceptance probability. The following discussion assumes that the data y are discrete (this will be relaxed later).

This can be achieved by noting that the acceptance probability is proportional to the probability of generating the observed data, y_{obs}, under the model $p(y|\theta)$ for a fixed parameter vector, θ. That is, suitably normalised, the likelihood function $p(y|\theta)$ can be considered as a probability mass function for the data. Put another way, for fixed θ, if we generate a dataset from the model $y \sim p(y|\theta)$, then the probability of generating our observed dataset exactly, so that $y = y_{obs}$, is precisely $p(y_{obs}|\theta)$. From this observation, we can use the Bernoulli event of generating $y = y_{obs}$ (or not) to determine whether to accept (or reject) a draw from the sampling distribution, in lieu of directly evaluating the probability $p(y_{obs}|\theta)$.

This insight permits a rewriting of the simple rejection sampling algorithm, as given below. A critical aspect of this modified algorithm is that it does not require numerical evaluation of the acceptance probability (i.e. the likelihood function). Note that if sampling is from $g(\theta)$ rather than the prior $\pi(\theta)$, then the acceptance probability is proportional to $p(y_{obs}|\theta)\pi(\theta)/g(\theta)$. In this case, deciding whether to accept a draw from $g(\theta)$ can be split into two stages: firstly, as before, if we generate $y \sim p(y|\theta)$, such that $y \neq y_{obs}$, then we reject the draw from $g(\theta)$. If, however, $y = y_{obs}$, then we accept the draw from $g(\theta)$ with probability $\pi(\theta)/[Kg(\theta)]$, where $K \geq \max_\theta f(\theta)/g(\theta)$. (These two steps may be interchanged so that the step with the least computational overheads is performed first.) Importance sampling versions of this and later algorithms are examined in chapter 4.

Likelihood-Free Rejection Sampling Algorithm

Inputs:

- A target posterior density $\pi(\theta|y_{obs}) \propto p(y_{obs}|\theta)\pi(\theta)$, consisting of a prior distribution $\pi(\theta)$ and a procedure for generating data under the model $p(y_{obs}|\theta)$.
- A proposal density $g(\theta)$, with $g(\theta) > 0$ if $\pi(\theta|y_{obs}) > 0$.
- An integer $N > 0$.

Sampling:
For $i = 1, \ldots, N$:

1. Generate $\theta^{(i)} \sim g(\theta)$ from sampling density g.
2. Generate $y \sim p(y|\theta^{(i)})$ from the likelihood.
3. If $y = y_{obs}$, then accept $\theta^{(i)}$ with probability $\frac{\pi(\theta^{(i)})}{Kg(\theta^{(i)})}$, where $K \geq \max_\theta \frac{\pi(\theta)}{g(\theta)}$. Else go to 1.

Output:
A set of parameter vectors $\theta^{(1)}, \ldots, \theta^{(N)}$ which are samples from $\pi(\theta|y_{obs})$.

1.3 A Practical Illustration: Stereological Extremes

In order to illustrate the performance of the likelihood-free rejection sampling algorithm, we perform a re-analysis of a stereological dataset with a computationally intractable model first developed by Bortot et al. (2007).

1.3.1 Background and model

Interest is in the distribution of the size of *inclusions*, microscopically small particles introduced during the production of steel. The steel strength is thought to be directly related to the size of the largest inclusion. Commonly, the sampling of inclusions involves measuring the maximum cross-sectional diameter of each observed inclusion, $y_{obs} = (y_{obs,1}, \ldots, y_{obs,n})^\top$, obtained from a two-dimensional planar slice through the steel block. Each cross-sectional inclusion size is greater than some measurement threshold, $y_{obs,i} > u$. The inferential problem is to analyse the unobserved distribution of the largest inclusion in the block, based on the information in the cross-sectional slice, y_{obs}. The focus on the size of the largest inclusion means that this is an extreme value variation on the standard stereological problem (Baddeley and Jensen 2004).

Each observed cross-sectional inclusion diameter, $y_{obs,i}$, is associated with an unobserved true inclusion diameter V_i. Anderson and Coles (2002) proposed a mathematical model assuming that the inclusions were spherical with diameters V, and that their centres followed a homogeneous Poisson process with rate $\lambda > 0$ in the volume of steel. The distribution of the largest inclusion diameters, $V|V > v_0$ was assumed to follow a generalised Pareto distribution, with distribution function:

$$\Pr(V \leq v|V > v_0) = 1 - \left[1 + \frac{\xi(v - v_0)}{\sigma}\right]_+^{-1/\xi}, \tag{1.1}$$

for $v > v_0$, where $[a]_+ = \max\{0, a\}$, following standard extreme value theory arguments (Coles 2001). However, the probability of observing the cross-sectional diameter $y_{obs,i}$ (where $y_{obs,i} \leq V_i$) is dependent on the value of V_i, as larger inclusion diameters give a greater chance that the inclusion will be observed in the two-dimensional planar cross-section. This means that the number of observed inclusions, n, is also a random variable. Accordingly, the parameters of the full spherical inclusion model are $\theta = (\lambda, \sigma, \xi)^\top$.

Anderson and Coles (2002) were able to construct a tractable likelihood function for this model by adapting the solution to Wicksell's corpuscle problem (Wicksell 1925). However, while their model assumptions of a Poisson process are not unreasonable, the assumption that the inclusions are spherical is not plausible in practice.

Bortot et al. (2007) generalised this model to a family of ellipsoidal inclusions. While this model is more realistic than the spherical inclusion model, there are analytic and computational difficulties in extending likelihood-based inference to more general families of inclusion (Baddeley and Jensen 2004; Bortot et al. 2007). As a result, ABC methods are a good candidate procedure to approximate the posterior distribution in this case.

1.3.2 Analysis

For simplicity, suppose that we are interested in the spherical inclusions model, so that the true posterior distribution can be estimated directly. Suppose also

$y = y_{obs}$, it may now be possible to satisfy the more relaxed criterion. This will accordingly result in a greater range of θ values that will be accepted and thereby increase the variability of the posterior approximation. The more relaxed the criterion (i.e. the larger the value of h), the greater the resulting variability.

It is possible to be more precise about the exact form of the posterior obtained through this adjusted procedure – this will be discussed in detail in the next section. However, for this particular analysis, based on samples $\lambda^{(1)}, \ldots, \lambda^{(N)}$ and datasets $n^{(1)}, \ldots, n^{(N)}$ obtained from the likelihood-free rejection algorithm, it can be seen that as the posterior approximation is constructed from those values of $\lambda = \lambda^{(i)}$ such that $\|n^{(i)} - n_{obs}\| \leq h$, then the posterior approximation can firstly be expressed as:

$$\begin{aligned}\hat{\pi}(\lambda|n_{obs}) = \frac{1}{N}\sum_{i=1}^{N} \delta_{\lambda^{(i)}}(\lambda) &= \frac{1}{N} \sum_{\lambda^{(i)}:\|n^{(i)}-n_{obs}\|\leq h} \delta_{\lambda^{(i)}}(\lambda) \\ &= \sum_{h'=-h^*}^{h^*} \left(\frac{1}{N} \sum_{\lambda^{(i)}:(n^{(i)}-n_{obs})=h'} \delta_{\lambda^{(i)}}(\lambda) \right),\end{aligned}$$

where h^* is the largest integer such that $\|h^*\| \leq h$. It then follows that:

$$\lim_{N\to\infty} \hat{\pi}(\lambda|n_{obs}) = \sum_{h'=-h^*}^{h^*} \Pr(n = n_{obs} + h')\pi(\lambda|n_{obs} + h'). \tag{1.2}$$

That is, the 'likelihood-free' approximation of the posterior, $\pi(\theta|y_{obs})$, is precisely an average of the individual posterior distributions $\pi(\lambda|n_{obs} + h')$ for $h' = -h^*, \ldots, h^*$, weighted according to $\Pr(n = n_{obs} + h')$, the probability of observing the dataset, $n_{obs} + h'$, based on samples drawn from the (prior predictive) distribution $p(n|\lambda)\pi(\lambda)$. This can be loosely observed from Figure 1.1, in which the approximations for $h = 10$ and $h = 20$ in panel (b) respectively correspond to rough visual averages of the centre three and all five displayed posteriors in panel (a). For $h = 0$ we obtain $\lim_{N\to\infty} \hat{\pi}(\lambda|n_{obs}) = \pi(\lambda|n_{obs})$ as for standard Monte Carlo algorithms.

Similar interpretations and conclusions arise when the data y are continuous, as we examine for a different model in the following subsection. This also allows us to introduce a fundamental concept in ABC methods, the use of summary statistics.

1.4 A g-and-k Distribution Analysis

The univariate g-and-k distribution is a flexible unimodal distribution that is able to describe data with significant amounts of skewness and kurtosis. Originally developed by Tukey (1977) (see also Martinez and Iglewicz 1984;

Hoaglin 1985; and Rayner and MacGillivray 2002), the g-and-k and related distributions have been analysed in the ABC setting by Peters and Sisson (2006), Allingham et al. (2009), Drovandi and Pettitt (2011), and Fearnhead and Prangle (2012) among others. Its density function has no closed form, but is alternatively defined through its quantile function as:

$$Q(q|A,B,g,k) = A + B\left[1 + c\frac{1-\exp\{-gz(q)\}}{1+\exp\{-gz(q)\}}\right](1+z(q)^2)^k z(q) \tag{1.3}$$

for $B > 0, k > -1/2$, where $z(q) = \Phi^{-1}(q)$ is the q-th quantile of the standard normal distribution function. The parameter c measures overall asymmetry, and is conventionally fixed at $c = 0.8$ (resulting in $k > -1/2$) (Rayner and MacGillivray 2002). This distribution is very flexible, with many common distributions obtained or well approximated by particular parameter settings, such as the normal distribution when $g = k = 0$. Given $\theta = (A, B, g, k)^\top$, simulations $z(q) \sim N(0, 1)$ drawn from a standard normal distribution can be transformed into samples from the g-and-k distribution through equation (1.3).

Figure 1.2 shows a scatterplot of samples from the likelihood-free approximation of the posterior $\pi(\theta|y_{obs})$ (grey dots), based on a simulated dataset y_{obs} of length $n = 1,000$ generated from the g-and-k distribution

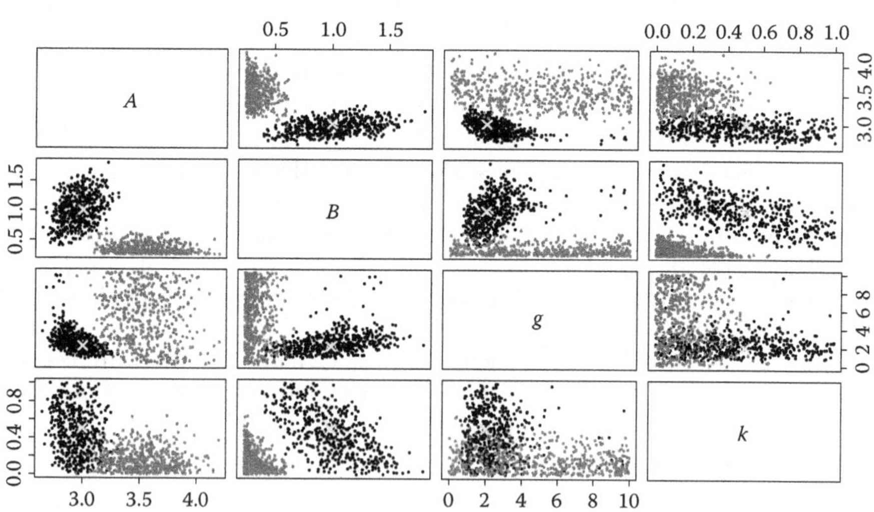

FIGURE 1.2
Pairwise scatterplots of samples from the likelihood-free approximation to the posterior using the full dataset (grey dots) and four summary statistics (black dots). True parameter values $(A, B, g, k) = (3, 1, 2, 0.5)$ are indicated by the cross $\times$.

with parameter vector $\theta_0 = (3, 1, 2, 0.5)^\top$. This analysis was based on defining $\|y - y_{obs}\| = (y - y_{obs})^\top \hat{\Sigma}^{-1}(y - y_{obs}) \leq h$ as Mahalanobis distance, with h given by the 0.005 quantile of the differences $\|y - y_{obs}\|$ for $i = 1, \ldots, N = 100{,}000$ Monte Carlo samples from the joint prior $\pi(\theta) = \pi(A)\pi(B)\pi(g)\pi(k) = N(1, 5) \times N(0.25, 2) \times U(0, 10) \times U(0, 1)$. The matrix $\hat{\Sigma}$ was determined as the sample covariance matrix of y using 2,000 samples generated under the model $y|\theta_0$ with $\theta = \theta_0$ fixed at its true value.

As is apparent from Figure 1.2, the likelihood-free approximation to $\pi(\theta|y_{obs})$ (grey dots) is particularly poor, the true parameter vector θ_0 is not even close to the estimated posterior samples. This outcome is a direct result of the dimension of the comparison $y - y_{obs}$. The chance of generating an $n = 1{,}000$-dimensional vector y that is close to y_{obs}, even if $\theta = \theta_0$, is vanishingly small. The odds of matching y with y_{obs} can be increased by redefining both in terms of their order statistics, although the chances still remain extremely low (see Example 3 in Section 1.7.1 for an illustration). This means that h must be relatively large, which results in accepting samples $\theta^{(i)}$ that generate data $y^{(i)}$ that are not actually close to y_{obs}, and thereby producing a poor approximation to $\pi(\theta|y_{obs})$.

The obvious way to avoid this problem is to reduce the dimension of the data comparison $y - y_{obs}$. Suppose that lower dimensional statistics $s = S(y)$ and $s_{obs} = S(y_{obs})$ are available, such that $S(y)$ is sufficient for, or highly informative for θ under the model, but where $\dim(S(y)) \ll \dim(y)$. Then the comparison $\|y - y_{obs}\|$ might be replaced by $\|s - s_{obs}\|$ without too much loss of information, but with the advantage that the dimension of $S(y)$ is now much lower. That is, step 3 in the likelihood-free rejection sampling algorithm could be further replaced by:

Likelihood-Free Rejection Sampling Algorithm

3. Compute $s = S(y)$.
   If $\|s - s_{obs}\| \leq h$, then accept $\theta^{(i)}$ with probability, $\frac{\pi(\theta^{(i)})}{Kg(\theta^{(i)})}$ where $K \geq \max_\theta \frac{\pi(\theta)}{g(\theta)}$. Else go to 1.

Using this idea, Drovandi and Pettitt (2011) suggested the statistics

$$S_A = E_4, \quad S_B = E_6 - E_2, \quad S_g = (E_6 + E_2 - 2E_4)/S_B,$$
$$\text{and } S_k = (E_7 - E_5 + E_3 - E_1)/S_B$$

as informative for A, B, g, and k, respectively, so that $S(y) = (S_A, S_B, S_g, S_k)^\top$, where $E_1 \leq E_2 \leq \ldots \leq E_8$ are the octiles of y. Repeating the above g-and-k analysis but using the four-dimensional comparison $\|s - s_{obs}\|$ rather than $\|y - y_{obs}\|$ (and recomputing $\hat{\Sigma}$ and h under the same conditions), the resulting posterior samples are shown in Figure 1.2 (black dots).

The difference in the quality of the approximation to $\pi(\theta|y_{obs})$ when using $S(y)$ rather than y, is immediately apparent. The true parameter value θ_0 is now located firmly in the centre of each pairwise posterior sample, several parameters (particularly A and g) are more precisely estimated, and evidence of dependence between parameters (as is to be expected) is now clearly seen.

While it is unreasonable to expect that there has been no loss of information in moving from y to $S(y)$, clearly the overall gain in the quality of the approximation to the likelihood-free posterior has been worth it in this case. This suggests that the use of summary statistics $S(y)$ is a useful tool more generally in approximate Bayesian computational techniques.

1.5 Likelihood-Free Methods or ABC?

The terms *likelihood-free* methods and *approximate Bayesian computation* are both commonly used to describe Bayesian computational methods developed for when the likelihood function is computationally intractable or otherwise unavailable. Of course, 'likelihood-free' is arguably a misnomer, in no sense is the likelihood function not involved in the analysis. It is the function used to generate the data $y \sim p(y|\theta)$, and it accordingly must exist, whether or not it can be numerically evaluated or written down. Rather, in this context, 'likelihood-free' refers to any likelihood-based analysis that proceeds without direct numerical evaluation of the likelihood function. There are several techniques that could be classified according to this description.

'Approximate Bayesian computation,' commonly abbreviated to 'ABC', was first coined by Beaumont et al. (2002) in the context of Bayesian statistical techniques in population genetics (although see Tavaré 2019, Chapter 2, this volume) and refers to the specific type of likelihood-free methods considered in this book. In particular, given the 'approximate' in ABC, it refers to those likelihood-free methods that produce an approximation to the posterior distribution resulting from the imperfect matching of data $\|y - y_{obs}\|$ or summary statistics $\|s - s_{obs}\|$.

Thus, the likelihood-free rejection algorithm described above with $h = 0$, which only accepts samples, θ, which have exactly reproduced the observed data y_{obs}, is not an ABC algorithm, as the method produces exact samples from the posterior distribution – there is no approximation. (The Monte Carlo approximation of the posterior is not considered an approximation in this sense.) It is, however, a likelihood-free method. Whereas, the likelihood-free rejection algorithm which may accept samples if $\|y - y_{obs}\| \leq h$, for $h > 0$, is an ABC algorithm, as the samples will be drawn from an approximation to the posterior distribution. Similarly, when the sampler may alternatively accept samples if $\|s - s_{obs}\| \leq h$, for any $h \geq 0$ (including $h = 0$), the resulting samples are also drawn from an approximate posterior distribution. As such, this is also an ABC algorithm. The only exception to this is the case where $h = 0$ and the summary statistics are sufficient: here there is no posterior approximation, the algorithm is then likelihood-free, but not an ABC method.

With a few exceptions (such as indirect inference, see Chapter 7), all of the methods considered in this book are both ABC and (by definition) likelihood-free methods. The aim of any ABC analysis is to find a practical way of performing the Bayesian analysis, while keeping the approximation and the computation to a minimum.

1.6 The Approximate Posterior Distribution

In contrast to the intuitive development of likelihood-free methods in the previous Sections, we now describe the exact form of the ABC approximation to the posterior distribution that is produced from the likelihood-free rejection algorithm. The procedure of (i) generating θ from the sampling distribution, $g(\theta)$, (ii) generating data, y, from the likelihood, $p(y|\theta)$, conditional on θ, and (iii) rejecting θ if $\|y - y_{obs}\| \leq h$, is equivalent to drawing a sample (θ, y) from the joint distribution proportional to:

$$I(\|y - y_{obs}\| \leq h)p(y|\theta)g(\theta),$$

where I is the indicator function, with $I(Z) = 1$ if Z is true, and $I(Z) = 0$ otherwise. If this sample (θ, y) is then further accepted with probability proportional to $\pi(\theta)/g(\theta)$, this implies that the likelihood-free rejection algorithm is sampling from the joint distribution proportional to:

$$I(\|y - y_{obs}\| \leq h)p(y|\theta)g(\theta)\frac{\pi(\theta)}{g(\theta)} = I(\|y - y_{obs}\| \leq h)p(y|\theta)\pi(\theta). \qquad (1.4)$$

Note that if $h = 0$, then the θ marginal of (1.4) equals the true posterior distribution, as:

$$\begin{aligned}\lim_{h\to 0}\int I(\|y - y_{obs}\| \leq h)p(y|\theta)\pi(\theta)dy &= \int \delta_{y_{obs}}(y)p(y|\theta)\pi(\theta)dy \\ &= p(y_{obs}|\theta)\pi(\theta).\end{aligned}$$

That is, for $h = 0$, the likelihood-free rejection algorithm draws samples, (θ, y), for which the marginal distribution of the parameter vector is the true posterior, $\pi(\theta|y_{obs})$. (The marginal distribution of the auxiliary dataset y is a point mass at $\{y = y_{obs}\}$ in this case.)

It is useful in the following to generalise the above formulation slightly. In (1.4), the indicator term $I(\|y - y_{obs}\| \leq h)$ only takes the values 0 or 1.

This is useful in the sense that it allows clear '*If* $\|y - y_{obs}\| \leq h$, *then* ...' statements to be made in any algorithm, which can simplify implementation. However, it is intuitively wasteful of information, as it does not discriminate between those samples, θ, for which the associated dataset y exactly equals the observed dataset y_{obs}, and those samples, θ, for which the associated dataset is the furthest away from y_{obs}, (i.e. $\|y - y_{obs}\| = h$). As the former case produces samples that are exact draws from the true posterior distribution, whereas the latter case does not, this produces a motivation for a more continuous scaling from 1 (when $y = y_{obs}$) to 0 (when $\|y - y_{obs}\|$ is large).

This can be achieved by replacing the indicator function, $I(\|y - y_{obs}\| \leq h)$, with a standard smoothing kernel function, $K_h(u)$, with $u = \|y - y_{obs}\|$, where:

$$K_h(u) = \frac{1}{h} K\left(\frac{u}{h}\right).$$

Kernels are symmetric functions such that $K(u) \geq 0$ for all u, $\int K(u)du = 1$, $\int uK(u)du = 0$, and $\int u^2 K(u)du < \infty$. Here, $h > 0$ corresponds to the scale parameter, or 'bandwidth' of the kernel function. Several common forms for kernel functions are given in Table 1.1, and these are illustrated in Figure 1.3. Following convention, we define $\lim_{h\to 0} K_h(u)$ as a point mass at the origin ($u = 0$).

An alternative specification of a smoothing kernel for multivariate datasets is obtained by writing $u = y - y_{obs}$, where $u = (u_1, \ldots, u_n)^\top$, $y = (y_1, \ldots, y_n)^\top$, and $y_{obs} = (y_{obs,1}, \ldots, y_{obs,n})^\top$, so that $u_i = y_i - y_{obs,i}$. Then we can write $K_h(u) = \prod_{i=1}^n K_{h_i}(u_i)$, where the scale parameter of each individual kernel function, $K_{h_i}(u_i)$, may vary. A further, more general specification may determine $K_h(u)$ as a fully multivariate, smooth and symmetric function, satisfying the above moment constraints. One such example is a multivariate $N(0, \Sigma)$ distribution, for some fixed covariance matrix Σ.

Substituting the kernel function, $K_h(u)$, into the likelihood-free rejection algorithm results in the ABC rejection sampling algorithm:

TABLE 1.1
The Functional Forms of Several Common Kernel Functions

Kernel	$K(u)$
Uniform	$\frac{1}{2}I(\lvert u\rvert \leq 1)$
Triangular	$(1 - \lvert u\rvert)I(\lvert u\rvert \leq 1)$
Epanechnikov	$\frac{3}{4}(1 - u^2)I(\lvert u\rvert \leq 1)$
Biweight	$\frac{15}{16}(1 - u^2)^3 I(\lvert u\rvert \leq 1)$
Gaussian	$\frac{1}{\sqrt{2\pi}}e^{-\frac{1}{2}u^2}$

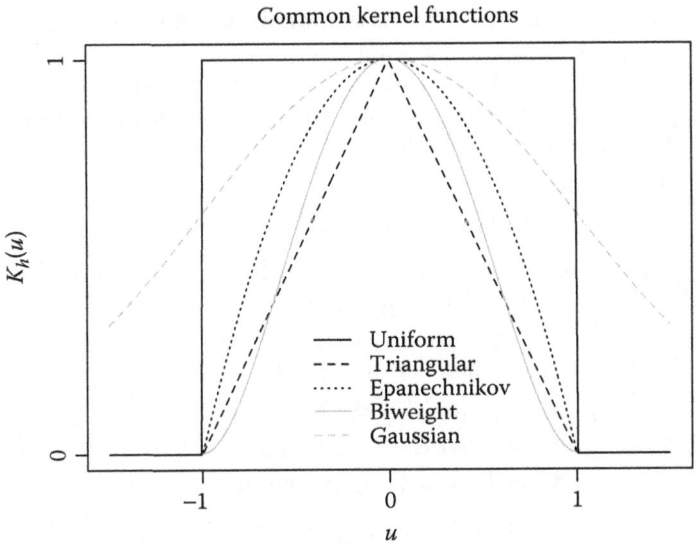

FIGURE 1.3
Standard kernel functions, $K(u)$, listed in Table 1.1 plotted on a common scale (with maximum at 1).

ABC Rejection Sampling Algorithm

Inputs:

- A target posterior density $\pi(\theta|y_{obs}) \propto p(y_{obs}|\theta)\pi(\theta)$, consisting of a prior distribution $\pi(\theta)$ and a procedure for generating data under the model $p(y_{obs}|\theta)$.
- A proposal density $g(\theta)$, with $g(\theta) > 0$ if $\pi(\theta|y_{obs}) > 0$.
- An integer $N > 0$.
- A kernel function $K_h(u)$ and scale parameter $h > 0$.

Sampling:
For $i = 1, \ldots, N$:

1. Generate $\theta^{(i)} \sim g(\theta)$ from sampling density g.
2. Generate $y \sim p(y|\theta^{(i)})$ from the likelihood.
3. Accept $\theta^{(i)}$ with probability $\frac{K_h(\|y-y_{obs}\|)\pi(\theta^{(i)})}{Kg(\theta^{(i)})}$, where $K \geq K_h(0)\max_\theta \frac{\pi(\theta)}{g(\theta)}$. Else go to 1.

Output:
A set of parameter vectors $\theta^{(1)}, \ldots, \theta^{(N)} \sim \pi_{ABC}(\theta|y_{obs})$.

In order to determine the form of the target distribution, $\pi_{ABC}(\theta|y_{obs})$, of this algorithm, we can follow the same argument as before. By (i) generating θ from the importance distribution, $g(\theta)$, (ii) generating data, y, from the likelihood, $p(y|\theta)$, conditional on θ, and then (iii) accepting the sample (θ, y) with probability proportional to $K_h(\|y - y_{obs}\|)\pi(\theta^{(i)})/g(\theta^{(i)})$, this results in samples from the joint distribution:

$$\pi_{ABC}(\theta, y|y_{obs}) \propto K_h(\|y - y_{obs}\|)p(y|\theta)\pi(\theta). \tag{1.5}$$

When $K_h(u)$ is the uniform kernel (see Table 1.1), then (1.5) reduces to (1.4). Accordingly, we define the ABC approximation to the true posterior distribution as:

$$\pi_{ABC}(\theta|y_{obs}) = \int \pi_{ABC}(\theta, y|y_{obs})dy, \tag{1.6}$$

where $\pi_{ABC}(\theta, y|y_{obs})$ is given by (1.5).

As before, as $h \to 0$, so that only those samples, θ, that generate data for which $y = y_{obs}$ are retained, then (1.5) becomes:

$$\begin{aligned}\lim_{h\to 0} \pi_{ABC}(\theta, y|y_{obs}) &\propto \lim_{h\to 0} K_h(\|y - y_{obs}\|)p(y|\theta)\pi(\theta)\\ &= \delta_{y_{obs}}(y)p(y|\theta)\pi(\theta),\end{aligned}$$

and so $\lim_{h\to 0} \pi_{ABC}(\theta|y_{obs}) \propto \int \delta_{y_{obs}}(y)p(y|\theta)\pi(\theta)dy = p(y_{obs}|\theta)\pi(\theta)$. That is, samples from the true posterior distribution are obtained as $h \to 0$. However, $h = 0$ is not a viable choice in practice, as for continuous y_{obs}, it corresponds to an algorithm with an acceptance rate of zero.

To see what marginal distribution the ABC rejection algorithm is sampling from for $h > 0$, we can integrate $\pi_{ABC}(\theta, y|y_{obs})$ over the auxiliary data margin, y.

A natural question to ask is, how accurate is this approximation? Re-writing the right hand side of (1.5) without the prior distribution, $\pi(\theta)$, we can similarly define the ABC approximation to the true likelihood, $p(y|\theta)$, for a fixed value of θ, as:

$$p_{ABC}(y_{obs}|\theta) = \int K_h(\|y - y_{obs}\|)p(y|\theta)dy. \tag{1.7}$$

In this manner, ABC can be interpreted as a regular Bayesian analysis, but with an approximated likelihood function.

Working in the univariate case for simplicity of illustration, so that $y, y_{obs} \in \mathcal{Y} = \mathbb{R}$ and $\|u\| = |u|$, we can obtain:

$$\begin{aligned}p_{ABC}(y_{obs}|\theta) &= \int K_h(|y - y_{obs}|)p(y|\theta)dy\\ &= \int K(u)p(y_{obs} - uh|\theta)du\\ &= \int K(u)\left[p(y_{obs}|\theta) - uhp'(y_{obs}|\theta) + \frac{u^2h^2}{2}p''(y_{obs}|\theta) - \dots\right]du\\ &= p(y_{obs}|\theta) + \frac{1}{2}h^2p''(y_{obs}|\theta)\int u^2K(u)du - \dots,\end{aligned} \tag{1.8}$$

using the substitution $u = (y_{obs} - y)/h$, a Taylor expansion of $p(y_{obs} - uh|\theta)$ around the point y_{obs}, and the kernel function properties of $K_h(u) = K(u/h)/h$, $\int K(u)du = 1$, $\int uK(u)du = 0$ and $K(u) = K(-u)$. The above is a standard smoothing kernel density estimation expansion, and assumes that the likelihood, $p(y|\theta)$, is infinitely differentiable. As with kernel density estimation, the choice of scale parameter is more important than the choice of kernel function in terms of the quality of the approximation.

Then, the pointwise bias in the likelihood approximation for fixed θ can be expressed as:

$$b_h(y|\theta) := p_{ABC}(y|\theta) - p(y|\theta), \tag{1.9}$$

as a function of y, which to second order can be written as

$$\hat{b}_h(y|\theta) = \frac{1}{2}h^2\sigma_K^2 p''(y|\theta),$$

where $\sigma_K^2 = \int u^2 K(u)du$ is the variance of the kernel function. Accordingly, the magnitude of the bias is reduced if h is small, corresponding to better approximations. Clearly, the second derivative of the likelihood function, $p''(y|\theta)$, is typically also unavailable if the likelihood function itself is computationally intractable. When $y, y_{obs} \in \mathcal{Y}$ is multivariate, a similar derivation to the above is available. In either case, the ABC approximation to the true posterior is defined through (1.6).

In a similar manner, we can determine the pointwise bias in the resulting ABC posterior approximation. From (1.9) we can write:

$$\begin{aligned} b_h(y_{obs}|\theta)\pi(\theta) &= p_{ABC}(y_{obs}|\theta)\pi(\theta) - p(y_{obs}|\theta)\pi(\theta) \\ &= \pi_{ABC}(\theta|y_{obs})c_{ABC} - \pi(\theta|y_{obs})c, \end{aligned} \tag{1.10}$$

where $c_{ABC} = \int p_{ABC}(y_{obs}|\theta)\pi(\theta)d\theta > 0$ and $c = \int p(y_{obs}|\theta)\pi(\theta)d\theta > 0$. Rearranging (1.10), we obtain:

$$\begin{aligned} a_h(\theta|y_{obs}) &:= \pi_{ABC}(\theta|y_{obs}) - \pi(\theta|y_{obs}) \\ &= \frac{b_h(y_{obs}|\theta)\pi(\theta) + \pi(\theta|y_{obs})c}{c_{ABC}} - \pi(\theta|y_{obs}) \\ &= \frac{b_h(y_{obs}|\theta)\pi(\theta)}{c_{ABC}} + \pi(\theta|y_{obs})\left(\frac{c}{c_{ABC}} - 1\right), \end{aligned} \tag{1.11}$$

as a function of θ. As $h \to 0$, then $b_h(y_{obs}|\theta) \to 0$ from (1.9), and so $p_{ABC}(y_{obs}|\theta) \to p(y_{obs}|\theta)$ pointwise, for fixed θ. Further, $c/c_{ABC} \to 1$ as h gets small, so that $a_h(\theta|y_{obs}) \to 0$.

1.6.1 Simple examples

In many simple cases, the ABC approximation to the posterior distribution can be derived exactly.

Example 1:
Suppose that the observed data, y_{obs}, is a single draw from a continuous density function $p(y|\theta)$, and that θ is a scalar. If we consider the particular case where $K_h(\|u\|)$ is the uniform kernel on $[-h, h]$ (see Table 1.1), and $\|u\| = |u|$, then we have:

$$\begin{aligned} \pi_{ABC}(\theta|y_{obs}) &\propto \pi(\theta) \int_{-\infty}^{\infty} K_h(|y - y_{obs}|) p(y|\theta) dy \\ &= \frac{\pi(\theta)}{2h} \int_{y_{obs}-h}^{y_{obs}+h} p(y|\theta) dy \\ &= \pi(\theta) \frac{[P(y_{obs} + h|\theta) - P(y_{obs} - h|\theta)]}{2h}, \end{aligned} \tag{1.12}$$

where $P(y|\theta) = \int_{-\infty}^{y} p(z|\theta)dz$ is the cumulative distribution function of $y|\theta$. Noting that as $\lim_{h \to 0}[P(y_{obs} + h|\theta) - P(y_{obs} - h|\theta)]/2h = p(y_{obs}|\theta)$ via L'Hopital's rule, then $\pi_{ABC}(\theta|y_{obs}) \to \pi(\theta|y_{obs})$ as $h \to 0$, as required. Also, $[P(y_{obs} + h|\theta) - P(y_{obs} - h|\theta)]/2h \approx 1/2h$ for large h, and so $\pi_{ABC}(\theta|y_{obs}) \to \pi(\theta)$ as $h \to \infty$.

Suppose now that $p(y|\theta) = \theta e^{-\theta y}$, for $\theta, y \geq 0$, is the density function of an *Exp*(θ) random variable, and that the prior $\pi(\theta) \propto \theta^{\alpha-1}e^{-\beta\theta}$ is given by a *Gamma*(α, β) distribution with shape and rate parameters $\alpha > 0$ and $\beta > 0$. Then from (1.12), and for $0 < h < y_{obs} + \beta$, we can directly obtain:

$$\begin{aligned} p_{ABC}(y_{obs}|\theta) &= \frac{1}{2h} e^{-\theta y_{obs}} (e^{\theta h} - e^{-\theta h}) \\ \hat{b}_h(y_{obs}|\theta) &= \frac{1}{6} h^2 \theta^3 e^{-\theta y_{obs}} \\ \pi_{ABC}(\theta|y_{obs}) &= \frac{\theta^{\alpha-1} e^{-\theta(y_{obs}+\beta)} \left(e^{\theta h} - e^{-\theta h}\right)}{\frac{\Gamma(\alpha)}{(y_{obs}+\beta-h)^\alpha} - \frac{\Gamma(\alpha)}{(y_{obs}+\beta+h)^\alpha}}, \end{aligned}$$

where $\Gamma(\alpha) = \int_0^\infty z^{\alpha-1}e^{-z}dz$ is the gamma function.

Figure 1.4a illustrates the true likelihood function, $p(y|\theta)$, (black dashed line) and the ABC approximation to the true likelihood function, $p_{ABC}(y|\theta)$, (solid grey line) as a function of y for $h = 0.91$ and $\theta = 2$. Also shown (grey dashed line) is the second order approximation to the ABC likelihood function, $p(y|\theta) + \hat{b}_h(y|\theta)$. In this case, the second order approximation provides a reasonable representation of the ABC likelihood, $p_{ABC}(y|\theta)$. For other choices of h and θ, the quality of this representation will vary.

The ABC approximation, $\pi_{ABC}(\theta|y_{obs})$, to the true posterior $\pi(y_{obs}|\theta)$, given $y_{obs} = 2$ and $\alpha = \beta = 1.2$ is shown in Figure 1.4b for various values of $h = 0.01, \ldots, 2.7$ (grey lines). The true posterior is illustrated by the black dashed line. For small h ($h = 0.01$), $\pi_{ABC}(\theta|y_{obs})$ is indistinguishable from the true posterior. As h increases, so does the scale of the approximate posterior, which begins to exhibit a large loss of precision compared to the true posterior. Both mean and mode of $\pi_{ABC}(\theta|y_{obs})$ increase with h.

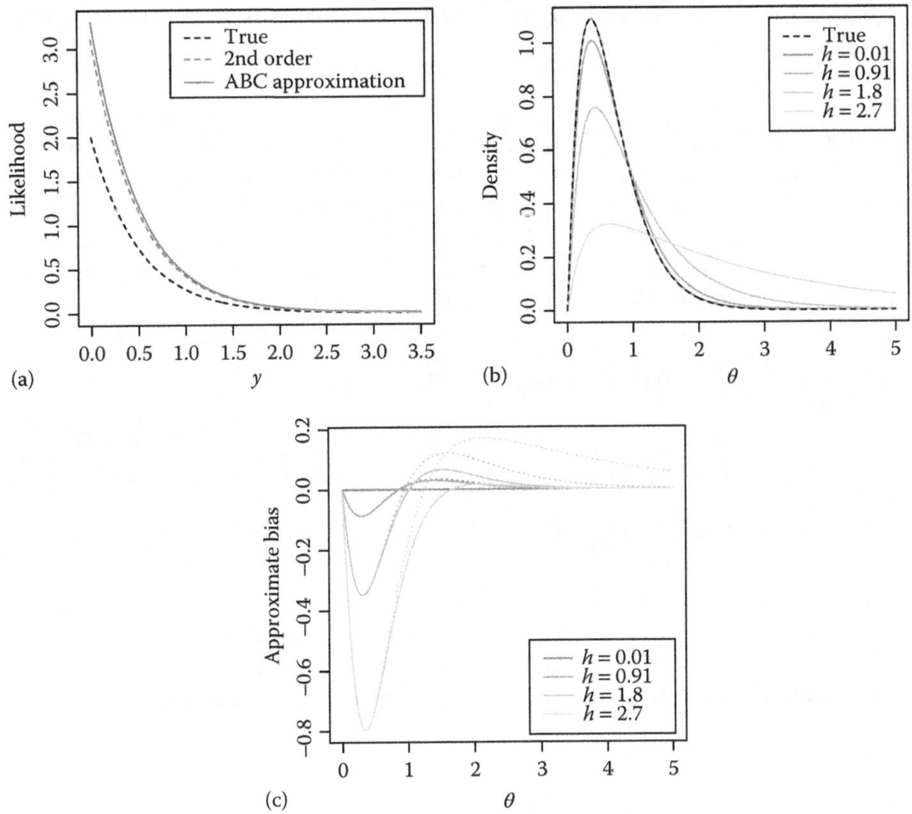

FIGURE 1.4
Approximations involved in the ABC analysis of the Exponential-Gamma example. (a) Various likelihood functions with $h = 0.91$ and $\theta = 2$. The true likelihood function, $p(y|\theta)$, and the ABC approximation to the likelihood, $p_{ABC}(y|\theta)$, are denoted by black-dashed and solid grey lines respectively. The second order approximation to $p_{ABC}(y|\theta)$, given by $p(y|\theta) + \hat{b}_h(y|\theta)$, is illustrated by the grey-dashed line. (b) The ABC posterior approximation, $\pi_{ABC}(\theta|y_{obs})$ with $y_{obs} = 2$ for various values of $h = 0.01, 0.91, 1.80, 2.70$. (c) Approximation bias in the ABC posterior as a function of h for $y_{obs} = 2$. Dashed lines indicate the exact bias $a(\theta|y_{obs})$ for each h, whereas solid lines denote the second order bias $\hat{a}(\theta|y_{obs})$.

Finally, Figure 1.4c shows the resulting bias, $a_h(\theta|y_{obs})$, in the ABC posterior approximation as a function of θ and h. Dashed and solid lines, respectively, show the exact bias $a_h(\theta|y_{obs})$ and the second order bias $\hat{a}_h(\theta|y_{obs})$ (defined as $a_h(\theta|y_{obs})$ in (1.11), but with $\hat{b}_h(y|\theta)$ substituted for $b_h(y|\theta)$). Clearly, the bias in the main body of the distribution, particularly in the region around the mode, is well described by the second order approximation,

$\hat{a}_h(\theta|y_{obs})$, whereas the bias in the distributional tails is more heavily influenced by terms of higher order than two.

Example 2:
Suppose that the observed data, $y_{obs} = (y_{obs,1}, \ldots, y_{obs,n})^\top$, are n independent draws from a univariate $N(\theta, \sigma_0^2)$ distribution, where the standard deviation, $\sigma_0 > 0$, is known. For this model, we know that $p(y_{obs}|\theta) \propto p(\bar{y}_{obs}|\theta)$, where $\bar{y}_{obs} = \frac{1}{n}\sum_i y_{obs,i}$, as the sample mean is a sufficient statistic for θ. If we specify $K_h(u)$ as a Gaussian $N(0, h^2)$ kernel (see Table 1.1), then the ABC approximation to the likelihood, $p(\bar{y}_{obs}|\theta)$ is given by

$$\begin{aligned}
p_{ABC}(\bar{y}_{obs}|\theta) &= \int_{-\infty}^{\infty} K_h(|\bar{y} - \bar{y}_{obs}|)p(\bar{y}|\theta)d\bar{y} \\
&= \int_{-\infty}^{\infty} \frac{1}{\sqrt{2\pi}h} \exp\left\{-\frac{(\bar{y} - \bar{y}_{obs})^2}{2h^2}\right\} \frac{\sqrt{n}}{\sqrt{2\pi}\sigma_0} \exp\left\{-\frac{n(\bar{y} - \theta)^2}{2\sigma_0^2}\right\} d\bar{y} \\
&\propto \exp\left\{-\frac{(\theta - \bar{y}_{obs})^2}{2(\sigma_0^2/n + h^2)}\right\}
\end{aligned}$$

for $h \geq 0$. That is, $\bar{y}_{obs} \sim N(\theta, \sigma_0^2/n{+}h^2)$ under the ABC approximation to the likelihood. In comparison to the true likelihood, for which $\bar{y}_{obs} \sim N(\theta, \sigma_0^2/n)$, the variance is inflated by h^2, the variance of the Gaussian kernel. Accordingly, if the prior for θ is given by a $N(m_0, s_0^2)$ distribution, where m_0 and $s_0 > 0$ are known, then:

$$\pi_{ABC}(\theta|y_{obs}) = \phi\left(\frac{m_0 s_0^{-2} + \bar{y}_{obs}(\sigma_0^2/n + h^2)^{-1}}{s_0^{-2} + (\sigma_0^2/n + h^2)^{-1}}, \frac{1}{s_0^{-2} + (\sigma_0^2/n + h^2)^{-1}}\right),$$

where $\phi(a, b^2)$ denotes the density of a $N(a, b^2)$ distributed random variable.

Clearly $\pi_{ABC}(\theta|y_{obs}) \to \pi(\theta|y_{obs})$ as $h \to 0$. However, the approximation will be quite reasonable if σ^2/n is the dominating component of the variance so that h is small in comparison (Drovandi 2012). A similar result to the above is available in the case of a multivariate parameter vector, θ.

Figure 1.5a illustrates the resulting ABC posterior approximation $\pi_{ABC}(\theta|y_{obs})$ with $\bar{y}_{obs} = 0$ when $\sigma^2/n = 1$ for the improper prior given by $m_0 = 0, s_0^2 \to \infty$, so that the true posterior distribution is $N(0, 1)$ (dashed line). The approximation is clearly quite reasonable for $h = 0.5$ and $h = 0.1$, as then $h^2 < \sigma_0^2/n$. Figure 1.5b shows the same posterior approximations, but based on a uniform kernel over $[-h, h]$ for $K_h(u)$, rather than the Gaussian $N(0, h^2)$ kernel. This ABC posterior is derived from (1.12). The resulting forms for $\pi_{ABC}(\theta|y_{obs})$ are no longer within the Gaussian family for $h > 0$, exhibit a flatter behaviour around the mean, and are more concentrated around the mean due to the compact support of the uniform kernel. The approximations with either kernel perform well for small h.

This example additionally provides some insight into the asymptotic behaviour of the ABC posterior approximation. Following standard likelihood

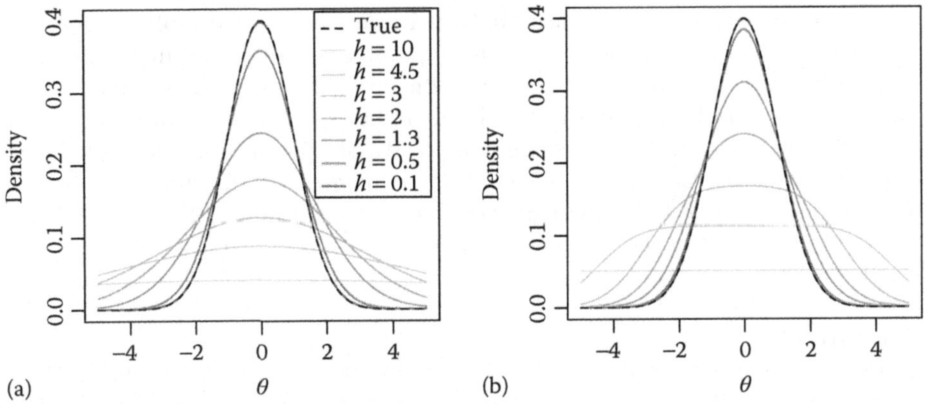

FIGURE 1.5
ABC posterior approximations, $\pi(\theta|y_{obs})$, for a $N(0,1)$ target distribution (dashed lines) for various values of kernel scale parameter h. The posterior approximations are based on (a) $N(0,h^2)$ and (b) uniform over $[-h,h]$ kernel functions, $K_h(u)$.

asymptotic results, when the amount of data, n, becomes large, the true likelihood function, $p(y|\theta)$, will approximately behave as a Gaussian distribution. As most prior distributions will have little impact in this setting (they will be approximately constant over the region of high posterior density), it follows that the ABC posterior approximation, $\pi_{ABC}(\theta|y_{obs})$, will follow a Gaussian distribution with a variance that is inflated by an h^2 term. Consequently, the ABC posterior approximation, $\pi_{ABC}(\theta|y_{obs})$, may then in principle be improved simply by rescaling the posterior variance to remove this term (Drovandi 2012).

1.7 The Use of Summary Statistics

1.7.1 Summary statistic basics

Despite the development in the previous Section, the ABC posterior approximation $\pi_{ABC}(\theta|y_{obs}) \propto \int K_h(\|y-y_{obs}\|)p(y|\theta)p(\theta)dy$ is rarely used in practice. This is because, except in very specific scenarios (such as when y_{obs} is very low dimensional, or when the likelihood function $p(y|\theta)$ factorises into very low dimensional components), it is highly unlikely that $y \approx y_{obs}$ can be generated from $p(y|\theta)$ for any choice of θ for realistic datasets. This results in the need to use a large value of the kernel scale parameter h in order to achieve viable rejection sampling algorithm acceptance rates (or a similar loss of performance in other algorithms), and in doing so produce poorer ABC posterior approximations.

In the stereological extremes analysis in Section 1.3 we replaced the full dataset y_{obs} with a sufficient statistic n_{obs} for the model parameter λ when estimating $\pi(\theta|y_{obs}) = \pi(\lambda|n_{obs})$. As sufficient statistics can be much lower dimensional than the full dataset, it is clear that greater approximation accuracy can be achieved for the same computational overheads when using low dimensional statistics (which is hinted at in the g-and-k distribution analysis in Section 1.4). The following example, based on Drovandi (2012), highlights the computational benefits in using lower dimensional, and less variable sufficient statistics.

Example 3:
Suppose that $y = (y_1, y_2)^\top$, where $y_i \sim$*Binomial*(n, θ) with $\theta \sim U(0, 1)$. Consider three possible vectors of sufficient statistics: $s^1 = (y_1, y_2)^\top$ is the full dataset, $s^2 = (y_{(1)}, y_{(2)})^\top$ are the order statistics $y_{(1)} \leq y_{(2)}$, and $s^3 = y_1 + y_2$ is the sum of the two individual values. All three vectors of statistics are sufficient for this simple model.

It is easy to compute the marginal distribution of each summary statistic $p_i(s^i) = \int_0^1 p(s^i|\theta)\pi(\theta)d\theta$ as follows:

$$\begin{aligned}
p_1(s^1) &= \int_0^1 \prod_{i=1}^{2} \binom{n}{y_i} \theta^{y_i}(1-\theta)^{n-y_i} d\theta \\
&= \binom{n}{y_1}\binom{n}{y_2} B(y_1 + y_2 + 1, 2n - y_1 - y_2 + 1), \\
p_2(s^2) &= [2 - I(y_{(1)} = y_{(2)})] \int_0^1 \prod_{i=1}^{2} \binom{n}{y_i} \theta^{y_i}(1-\theta)^{n-y_i} d\theta \\
&= [2 - I(y_{(1)} = y_{(2)})] \binom{n}{y_1}\binom{n}{y_2} B(y_1 + y_2 + 1, 2n - y_1 - y_2 + 1), \\
p_3(s^3) &= \int_0^1 \binom{2n}{s^3} \theta^{s^3}(1-\theta)^{2n-s^3} d\theta \\
&= \binom{2n}{s^3} B(s^3 + 1, 2n - s^3 + 1) \\
&= 1/(2n+1),
\end{aligned}$$

where $\mathrm{B}(a, b) = \int_0^1 z^{a-1}(1-z)^{b-1}dz$ is the beta function. Here, $p_i(s^i)$ is the probability of generating the vector s^i under an ABC rejection sampling algorithm with sampling distribution given by the prior, $g(\theta) = \pi(\theta)$. That is, $p_i(s_i)$ is the acceptance probability of the algorithm if we only accept those sufficient statistics that exactly match the observed sufficient statistics.

Suppose that we observe $y_{obs} = (y_{obs,1}, y_{obs,2})^\top = (1, 2)^\top$ from $n = 5$ experiments. From the above, we have algorithm acceptance rates of:

$$p_1(s^1_{obs}) = \frac{5}{132} \approx 0.038, \quad p_2(s^2_{obs}) = \frac{5}{66} \approx 0.076, \quad \text{and} \quad p_3(s^3_{obs}) = \frac{1}{11} \approx 0.091,$$

where s^i_{obs} denotes the statistic s^i derived from y_{obs}. The probability $p_1(s^1)$ is the probability of generating first $y_1 = 1$, and then $y_2 = 2$. As a result, $p_1(s^1)$ will decrease rapidly as the length of the observed dataset y_{obs} increases. The probability $p_2(s^2)$ corresponds to the probability of generating either $y = (1,2)^\top$ or $y = (2,1)^\top$, which are equivalent under the binomial model. Hence, s^2 has twice the probability of s^1 of occurring. Finally, the probability $p_3(s^3)$ is the probability of generating $y = (1,2)^\top, (2,1)^\top, (0,3)^\top$, or $(3,0)^\top$. Each of these cases is indistinguishable under the assumed model, and so the event s^3 occurs with the largest probability of all.

Quite clearly, while still producing samples from the true target distribution, $\pi(\theta|y_{obs})$, the impact on the efficiency of the sampler of the choice of sufficient statistics is considerable, even for an analysis with only two observations, y_1 and y_2. The most efficient choice is the minimal sufficient statistic. The differences in the acceptance rates of the samplers would become even greater for larger numbers of observations, n.

While the optimally informative choice of statistic for an ABC analysis is a minimal sufficient statistic, this may still be non-viable in practice. For example, if the minimal sufficient statistic is the full dataset y_{obs}, sampling from $\pi_{ABC}(\theta|y_{obs})$ will be highly inefficient even for moderately sized datasets. Similarly, in a scenario where the likelihood function may not be known beyond a data generation procedure, identification of any low-dimensional sufficient statistic (beyond, trivially, the full dataset y_{obs}) may be impossible. Further, low dimensional sufficient statistics may not even exist, depending on the model.

In general, a typical ABC analysis will involve specification of a vector of summary statistics $s = S(y)$, where $\dim(s) \ll \dim(y)$. The rejection sampling algorithm with then contrast s with $s_{obs} = S(y_{obs})$, rather than y with y_{obs}. As a result, this procedure will produce samples from the distribution $\pi_{ABC}(\theta|s_{obs})$ as follows:

ABC Rejection Sampling Algorithm

Inputs:

- A target posterior density $\pi(\theta|y_{obs}) \propto p(y_{obs}|\theta)\pi(\theta)$, consisting of a prior distribution $\pi(\theta)$ and a procedure for generating data under the model $p(y_{obs}|\theta)$.
- A proposal density $g(\theta)$, with $g(\theta) > 0$ if $\pi(\theta|y_{obs}) > 0$.
- An integer $N > 0$.
- A kernel function $K_h(u)$ and scale parameter $h > 0$.
- A low dimensional vector of summary statistics $s = S(y)$.

Sampling:
For $i = 1, \ldots, N$:

1. Generate $\theta^{(i)} \sim g(\theta)$ from sampling density g.
2. Generate $y \sim p(y|\theta^{(i)})$ from the likelihood.
3. Compute summary statistic $s = S(y)$.
4. Accept $\theta^{(i)}$ with probability $\frac{K_h(\|s-s_{obs}\|)\pi(\theta^{(i)})}{Kg(\theta^{(i)})}$ where $K \geq K_h(0) \max_\theta \frac{\pi(\theta)}{g(\theta)}$. Else go to 1.

Output:
A set of parameter vectors $\theta^{(1)}, \ldots, \theta^{(N)} \sim \pi_{ABC}(\theta|s_{obs})$.

Similar to the discussion in Section 1.6, it can be seen that the ABC posterior approximation now has the form:

$$\pi_{ABC}(\theta|s_{obs}) \propto \int K_h(\|s - s_{obs}\|)p(s|\theta)\pi(\theta)ds, \tag{1.13}$$

where $p(s|\theta)$ denotes the likelihood function of the summary statistic $s = S(y)$ implied by $p(y|\theta)$. (That is, $p(s|\theta) = \int_{\mathcal{Y}} \delta_s(S(y))p(y|\theta)dy$.) If we let $h \to 0$, so that only those samples, θ, that generate data for which $s = s_{obs}$ are retained, then:

$$\begin{aligned}
\lim_{h\to 0} \pi_{ABC}(\theta|s_{obs}) &\propto \int \lim_{h\to 0} K_h(\|s - s_{obs}\|)p(s|\theta)\pi(\theta)ds \\
&= \int \delta_{s_{obs}(s)}p(s|\theta)\pi(\theta)ds \\
&= p(\theta|s_{obs})\pi(\theta).
\end{aligned}$$

Hence, samples from the distribution $\pi(\theta|s_{obs})$ are obtained as $h \to 0$. If the vector of summary statistics, $s = S(y)$, is sufficient for the model parameters, then $\pi(\theta|s_{obs}) \equiv \pi(\theta|y_{obs})$, and so samples are produced from the true posterior distribution. However, if $S(y)$ is not sufficient – and this is typically the case in practice – then the ABC posterior approximation is given by (1.13), where in the best scenario (i.e. as $h \to 0$) the approximation is given by $\pi(\theta|s_{obs})$.

The following example illustrates the effect of using a non-sufficient summary statistic.

Example 4:
Consider again the univariate Gaussian model in Example 2. Suppose that we modify this example (Drovandi 2012), so that the model still assumes that the observed data $y_{obs} = (y_{obs,1}, \ldots, y_{obs,n})^\top$ are random draws from

a univariate $N(\theta, \sigma_0^2)$ distribution, but where we now specify an insufficient summary statistic, $s = \bar{y}_{1:n'} = \frac{1}{n'}\sum_{i=1}^{n'} y_i$ with $n' < n$.

Writing $s_{obs} = S(y_{obs})$, the resulting ABC approximation to the likelihood function becomes:

$$p_{ABC}(s_{obs}|\theta) = \int K_h(s - s_{obs})p(s|\theta)ds$$

$$\propto \int_{-\infty}^{\infty} \frac{1}{\sqrt{2\pi}h} \exp\left\{-\frac{(s - s_{obs})^2}{2h^2}\right\} \frac{\sqrt{n'}}{\sqrt{2\pi}\sigma_0} \exp\left\{-\frac{n'(s-\theta)^2}{2\sigma_0^2}\right\} ds$$

$$\propto \exp\left\{-\frac{(\theta - s_{obs})^2}{2(\sigma_0^2/\omega n + h^2)}\right\},$$

where $\omega = n'/n$ is the proportion of the n observations used in the vector of summary statistics. That is, $s_{obs} \sim N(\theta, \sigma^2/\omega n + h^2)$. When $\omega = 1$, then $s_{obs} = \bar{y}_{obs}$ is sufficient for θ and so $s_{obs} \sim N(\theta, \sigma^2/n + h^2)$ recovers the same result as Example 2.

When $n' < n$, so that s is no longer sufficient for θ, the mean of the Gaussian likelihood function is centred on the mean $\bar{y}_{obs,1:n'}$ rather than $\bar{y}_{obs,1:n}$, but more critically, the variance of the Gaussian likelihood is $\sigma^2/\omega n + h^2$. It is evident that there are now two sources of error, both of which inflate the variance of the likelihood. The first, h^2, arises through the matching of the simulated and observed data through the Gaussian kernel. The second source of error comes from the $0 < \omega < 1$ term, which can be interpreted as the degree of inefficiency of replacing y by $s = S(y)$. That is, the use of non-sufficient statistics reduces the precision of the likelihood (and by turn, the posterior distribution) in this case.

From Example 2, it follows that when n is large and the posterior is asymptotically Gaussian, the ABC posterior approximation, $\pi_{ABC}(\theta|s_{obs})$, can be improved by rescaling to remove h^2 from the posterior variance. However, correcting for the lack of sufficiency in the summary statistic, s, would require knowledge of the relative inefficiency of s over y, which may be difficult to obtain in practice.

The choice of summary statistics for an ABC analysis is a critical decision that directly affects the quality of the posterior approximation. Many approaches for determining these statistics are available, and these are reviewed in Blum et al. (2013) and Prangle (2019), Chapters 3 and 5, this volume. These methods seek to trade off two aspects of the ABC posterior approximation that directly result from the choice of summary statistics. The first is that $\pi(\theta|y_{obs})$ is approximated by $\pi(\theta|s_{obs})$. As this represents an irrevocable potential information loss, the information content in s_{obs} should be high. The second aspect of the ABC posterior approximation is that the simulated and observed summary statistics are compared within a smoothing kernel $K_h(\|s - s_{obs}\|)$ as part of the form of $\pi_{ABC}(\theta|s_{obs})$ (1.13). As stochastically matching s and s_{obs} becomes increasingly difficult as the dimension of the summary statistics increases, the dimension of s should be low.

As such, the dimension of the summary statistic should be large enough so that it contains as much information about the observed data as possible, but also low enough so that the curse-of-dimensionality of matching s and s_{obs} is avoided. For illustration, in Example 3, the optimum choice of summary statistic is a minimal sufficient statistic. However, for other models it may be the case that the dimension of the minimal sufficient statistic is equal to that of the original dataset. As this will cause curse-of-dimensionality problems in matching s with s_{obs}, it is likely that a more accurate ABC posterior approximation can be achieved by using a lower-dimensional non-sufficient statistic, rather than remaining within the class of sufficient statistics. This was indeed the case in the g-and-k distribution analysis in Section 1.4.

1.7.2 Some practical issues with summary statistics

Even with the above principles in mind, summary statistic choice remains one of the most challenging aspects of implementing ABC in practice. For instance, it is not always viable to continue to add summary statistics to s until the resulting ABC posterior approximation does not change for the worse, as is illustrated by the following example.

Example 5:
Suppose that $y = (y_1, \ldots, y_n)^\top$ with $y_i \sim Poisson(\lambda)$. Combined with conjugate prior beliefs $\lambda \sim Gamma(\alpha, \beta)$, this gives $\lambda|y \sim Gamma(\alpha + n\bar{y}, \beta + n)$. For this model, we know that the sample mean $\bar{y}$ is a sufficient statistic. However, we also know that the mean and variance of a $Poisson(\lambda)$ model are both equal to λ, and so we might also expect the sample variance v^2 to also be informative for λ, although it is not sufficient. Suppose that we observe $y_{obs} = (0, 0, 0, 0, 5)^\top$, which gives $(\bar{y}_{obs}, v^2_{obs}) = (1, 5)$. Here, as the sample mean and variance are quite different from each other, we might expect that the Poisson model is not appropriate for these data.

Figure 1.6 illustrates various ABC posterior approximations to the true target distribution (solid lines) based on a prior with $\alpha = \beta = 1$ (dashed lines), with $K_h(u)$ specified as a uniform kernel over $[-h, h]$ and $\|u\|$ representing Euclidean distance. The top row illustrates the resulting posterior approximations, $\pi(\lambda|s_{obs})$, when the summary statistics s are given as the sample mean $\bar{y}$ (a), the sample standard deviation v (b), or both (c) when the kernel scale parameter is $h = 0$. Using $s = \bar{y}$ recovers the true posterior exactly, which is no surprise as $\bar{y}$ is a sufficient statistic. Using $s = v$ produces an informed ABC approximation, but one which is based on a variance that is consistent with a larger mean under the Poisson model. When $s = (\bar{y}, v)^\top$, then we again obtain the true posterior distribution as $\pi(\lambda|\bar{y}_{obs}, v_{obs}) \equiv \pi(\lambda|\bar{y}_{obs})$ through sufficiency, and the additional information that v brings about the sample y has no effect on the ABC estimated posterior.

The bottom row in Figure 1.6 shows the same information as the top row, except that the kernel scale parameter is now non-zero ($h = 0.3$).

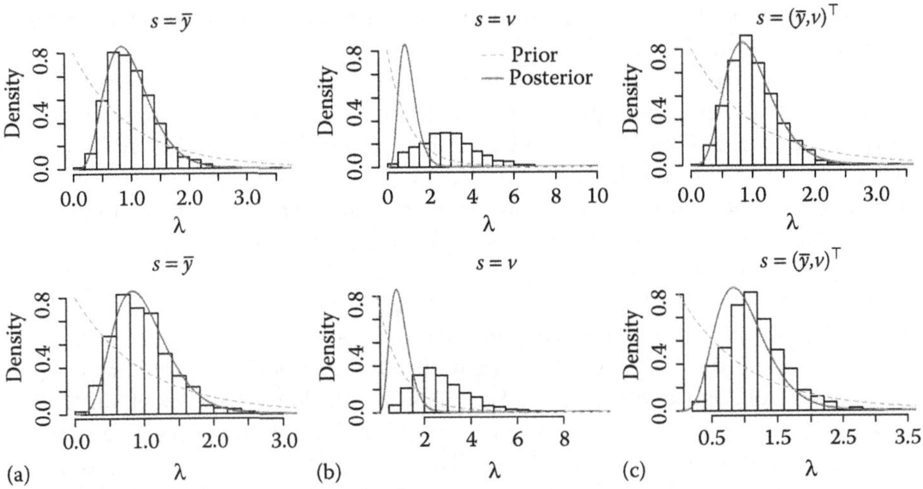

FIGURE 1.6
Various ABC posterior approximations (histograms) for a $Gamma(\alpha+\bar{y}, \beta+n)$ target distribution (solid line) with a $Gamma(\alpha,\beta)$ prior (dashed lines). Columns illustrate posterior estimates based on (a) sample mean $s = \bar{y}$, (b) standard deviation $s = v$, and (c) $s = (\bar{y}, v)^\top$ as summary statistics. Top row shows results with $h = 0$ and the bottom row with $h = 0.3$.

The posterior approximations based on $s = \bar{y}$ and $s = v$ are minor deviations away from those in the top row when $h = 0$. This occurs as the values of λ that are able to reproduce the observed summary statistics within a non-zero tolerance $h = 0.3$ are slightly different to those that can reproduce the summary statistics exactly. However, the third panel with $s = (\bar{y}, v)^\top$ is clearly biased to the right, with the resulting ABC posterior approximation visually appearing to be a loose average of those distributions with $s = \bar{y}$ and $s = v$.

This behaviour is different from when $h = 0$. In that case, when adding more information in the vector of summary statistics in going from $s = \bar{y}$ to $s = (\bar{y}, v)^\top$, the posterior approximation does not change as the summary statistic $s = \bar{y}$ is sufficient, and it is being matched exactly. However, when $h > 0$, because the ABC algorithm allows a non perfect matching of the sufficient statistic $\bar{y}$, it additionally allows the extra information in the sample standard deviation v to also contribute to the approximation. In this case, because the observed summary statistics $\bar{y}_{obs}$ and v_{obs} are inconsistent with respect to the model, this then results in a strongly biased fit when moving from $s = \bar{y}$ to $s = (\bar{y}, v)^\top$.

As such, while it may be tempting to include progressively more summary statistics into s_{obs} until the ABC posterior approximation does not change appreciably, the assumption that this will provide the most accurate posterior approximation is clearly incorrect. Even if s_{obs} contains sufficient statistics for the model, the inclusion of further statistics can still bias the posterior approximation, particularly in the case where the observed data are inconsistent with the model.

The identification of suitable summary statistics is clearly a critical part of any analysis. Accordingly, many techniques have been developed for this purpose. See e.g. Chapters 3 and 5, this volume for a detailed review and comparison of these methods. While the choice of summary statistics is itself of primary importance, it is less appreciated that the distance measure $\|\cdot\|$ can also have a substantial impact on ABC algorithm efficiency, and therefore the quality of the posterior approximation.

Consider the distance measure $\|s - s_{obs}\| = (s - s_{obs})^\top \Sigma^{-1}(s - s_{obs})$. Here, we can specify the covariance matrix Σ as the identity matrix to produce Euclidean distance, or as a diagonal matrix of non-zero weights to give weighted Euclidean distance (e.g. Hamilton et al. 2005; Luciani et al. 2009), or as a full covariance matrix to produce Mahalanobis distance (e.g. Peters et al. 2012; Erhardt and Sisson 2016). To see why standard and weighted Euclidean distance can be a poor choice, consider the setting in Figure 1.7, where candidate parameter values, θ, generating continuous bivariate statistics, $s|\theta$, $s = (s_1, s_2)^\top$, are accepted as draws from $\pi_{ABC}(\theta|s_{obs})$ if s lies within a ball of radius h, centered on s_{obs}. That is, K_h is the uniform kernel on $[-h, h]$, and $\|\cdot\|$ denotes Euclidean distance.

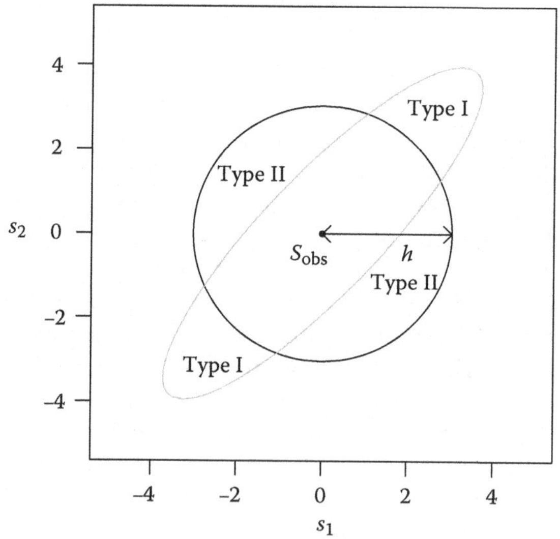

FIGURE 1.7
The concept of type I and II errors for accept/reject decisions in ABC samplers under a uniform kernel, $K_h(u)$, over $[-h, h]$ and Euclidean distance, $\|\cdot\|$. The circle represents the acceptance region for a simulated summary statistic $s = (s_1, s_2)^\top$, centred on s_{obs}. The ellipse represents the possible dependence between s_1 and s_2.

If we reasonably suppose that the elements of s may be dependent and on different scales, their true distribution under the model may be better represented by an ellipse (grey lines). As such, an efficient ABC algorithm should accept candidate draws from $\pi_{ABC}(\theta|s_{obs})$ if $s|\theta$ lies within this ellipse. Consequently, implementing a circular acceptance region (implying independence and identical scales) induces both type I (i.e. candidate samples are rejected when they should be accepted) and type II (i.e. candidate samples are accepted when they should be rejected) errors.

Work linking the ABC posterior with non-parametric density estimation methods (Blum 2010; see Section 1.10) provides support for this argument. Here, for a multivariate kernel $K_H(u) = \det(H)^{-1}K(H^{-1}u)$, where K is a symmetric multivariate density function with zero mean and finite variance, a general rule of thumb is to specify the bandwidth matrix as $H \propto \Sigma^{-1/2}$, where Σ is the covariance matrix of the data (e.g. Scott 1992; Wand and Jones 1995). In the ABC context, this is equivalent to defining $\|\cdot\|$ as Mahalanobis distance, where Σ is the covariance matrix of s (or $s|\theta$).

Note that the above argument assumes that the summaries s_1 and s_2 are both informative for the model parameter θ. For example, in the case where $s_1 + s_2$ is uninformative, but $s_1 - s_2$ is informative, then it is credible that the circular acceptance region could result in a more accurate ABC posterior approximation than that resulting from the elliptical region. In general, the best acceptance region is tied up with the choice of the summary statistics in a more complicated way than that presented here (see e.g. Prangle 2017 for a discussion).

The following example illustrates the effect that different covariance matrices Σ can have on the ABC posterior approximation.

Example 6:
Suppose that the model is specified as $y_1, \ldots, y_{50} \sim N(\theta, 1)$, with a uniform prior $\theta \sim U(-5, 5)$. Various sufficient statistics are available for this model. We consider two alternatives: $s^1 = (\bar{y}_{1:40}, \bar{y}_{41:50})^\top$ and $s^2 = (\bar{y}_{1:25} - \bar{y}_{26:50}, \bar{y}_{26:50})^\top$, where $\bar{y}_{a:b} = (b - a + 1)^{-1}\sum_{i=a}^{b} y_i$. In each case, given the observed sufficient statistics $s_{obs} = (0, 0)^\top$, the exact posterior distribution $\pi(\theta|y_{obs})$ is $N(0, 1/50)$ truncated to $(-5, 5)$. However, the covariance matrices of s^1 and s^2 for fixed θ are quite different (though they do not depend on the exact value of θ), namely

$$\text{Cov}(s^1|\theta) = \begin{pmatrix} 1/40 & 0 \\ 0 & 1/10 \end{pmatrix}, \quad \text{Cov}(s^2|\theta) = \begin{pmatrix} 2/25 & -1/25 \\ -1/25 & 1/25 \end{pmatrix}, \tag{1.14}$$

with a negative correlation between the elements of s^2 of $-1/\sqrt{2} \approx -0.71$. We implement ABC using the distance measure as $\|s - s_{obs}\| = (s - s_{obs})\Sigma^{-1}(s - s_{obs})'$ and consider the impact of the choice of Σ.

We use a version of the ABC rejection sampling algorithm (see box) that maintains a sample $\theta^{(1)}, \ldots, \theta^{(N)}$ of size N from the ABC posterior approximation, which progressively lowers the kernel scale parameter h until a stopping rule is satisfied. On algorithm termination, the samples are identical to those samples that would have been obtained under the standard ABC rejection sampling algorithm if it was implemented with the lowest value of h achieved under the stopping rule. This allows us to implement a rejection sampling algorithm that will terminate when a pre-specified degree of accuracy has been achieved. The (random) number of iterations obtained before algorithm termination will accordingly be an indicator of the efficiency of the model specification – in this case, the effect of different covariance matrices Σ.

ABC Rejection Sampling Algorithm (with Stopping Rule)

Initialise:

For each particle $i = 1, \ldots, N$:

- Generate $\theta^{(i)} \sim \pi(\theta)$ from the prior, $y^{(i)} \sim p(y|\theta^{(i)})$ from the likelihood.
- Compute summary statistics $s^{(i)} = S(y^{(i)})$ and distance $\rho^{(i)} = \|s^{(i)} - s_{obs}\|$.
- Generate $u^{(i)} \sim \mathrm{U}(0,1)$ that determines whether to accept the particle. (i.e. accept if $u^{(i)} \leq K_h(\rho^{(i)})/K_h(0)$.)
- Determine the *smallest* h that results in the acceptance of all N particles. For example,

$$h = \sqrt{\max_i\{-[\rho^{(i)}]^2/(2\log(u^{(i)}))\}} \qquad \text{or} \qquad h = \max_i\{\rho^{(i)}\}$$

  if (respectively)

$$K_h(\rho) \propto \exp\{-\rho^2/(2h^2)\} \qquad \text{or} \qquad K_h(\rho) \propto 1 \text{ on } [-h, h].$$

- Calculate the acceptance probabilities $W^{(i)} = K_h(\rho^{(i)})/K_h(0)$, $i = 1, \ldots, N$.

Simulation:

While the stopping rule is not satisfied, repeat:

1. Identify the index of the particle that will first be rejected if h is reduced: $r = \arg_i \min\{W^{(i)} - u^{(i)}\}$.
2. Set the new value of h to be the lowest value which would result in the acceptance of all particles, except particle r.
3. Recompute acceptance probabilities $W^{(i)}$ given the new value of h.

4. Replace particle r by repeating:
   (a) Generate $\theta^{(r)} \sim \pi(\theta)$, $y^{(r)} \sim p(y|\theta^{(i)})$, $u^{(r)} \sim U(0,1)$.
   (b) Compute $s^{(r)} = S(y^{(r)})$, $\rho^{(r)} = \|s^{(r)} - s_{obs}\|$, $W^{(r)} = K_h(\rho^{(r)})/K_h(0)$

   Until $u^{(r)} \leq W^{(r)}$.

Output:
A set of parameter vectors $\theta^{(1)}, \ldots, \theta^{(N)} \sim \pi_{ABC}(\theta|s_{obs})$, with h determined as the largest achieved value that satisfies the stopping rule.

Table 1.2 displays the average number of data generation steps [i.e. generating $y \sim p(y|\theta)$] in each algorithm implementation, per final accepted particle, as a function of smoothing kernel type and the form of Σ, based on 100 replicate simulations of $N = 500$ samples. The stopping rule continued algorithm execution until an estimate of the absolute difference between empirical ($F_N(\theta)$) and true ($F(\theta)$) model cumulative distribution functions was below a given level. Specifically, when $\sum_{i=1}^{N} |F_N(\theta^{(i)}) - F(\theta^{(i)})| < 0.01825$. In Table 1.2, the true form of Σ is given by $\text{Cov}(s^1|\theta)$ and $\text{Cov}(s^2|\theta)$ (1.14), and the diagonal form refers to the matrix constructed from the diagonal elements of $\text{Cov}(s^2|\theta)$.

The summary statistics for $s = s^1$ are independent, but are on different scales. Accordingly, when this difference of scale is accounted for ($\Sigma =$ true), algorithm efficiency, and therefore ABC posterior approximation accuracy, is greatly improved compared to when the difference in scale is ignored

TABLE 1.2
Mean Number of Summary Statistic Generations per Final Accepted Particle (with Standard Errors in Parentheses), as a Function of the Form of Covariance Matrix, Σ, and Smoothing Kernel K_h, and for Two Different Sets of Sufficient Statistics $s^1 = (\bar{y}_{1:40}, \bar{y}_{41:50})^\top$ and $s^2 = (\bar{y}_{1:25} - \bar{y}_{26:50}, \bar{x}_{26:50})^\top$. Results are Based on 100 Replicates of Posterior Samples of Size $N = 500$

Summary Statistic	Kernel	Form of Σ: Identity		Diagonal		True	
$s = s^1$	Uniform	134.7	(5.8)			84.5	(2.4)
	Epanechnikov	171.6	(4.7)			111.1	(3.8)
	Triangle	232.3	(7.1)			153.0	(5.1)
	Gaussian	242.4	(6.5)			153.6	(4.9)
$s = s^2$	Uniform	182.5	(5.6)	161.0	(4.1)	84.4	(2.4)
	Epanechnikov	245.5	(6.6)	209.2	(7.2)	111.1	(3.8)
	Triangle	336.3	(8.9)	277.2	(6.9)	144.2	(3.8)
	Gaussian	368.2	(12.6)	289.7	(9.7)	157.7	(4.3)

(Σ = identity). The summary statistics s^2 are both negatively correlated and on different scales. As for s^1, when summary statistic scale is taken into consideration (Σ = diagonal), an improvement in algorithm efficiency and ABC posterior approximation accuracy is achieved compared to when it is ignored. However, in this case, further improvements are made when the correlation between the summary statistics is also accounted for (Σ = true). These results are consistent regardless of the form of the smoothing kernel K_h. Note that the uniform kernel produces the most efficient algorithm and most accurate ABC posterior approximation, and that this steadily worsens as the form of the kernel deviates away from the uniform density, with the worst performance obtained under the Gaussian kernel.

This approach has been implemented in practice by, for example, Luciani et al. (2009) and Erhardt and Sisson (2016), who identify some value of $\theta = \theta^*$ in a high posterior density region via a pilot analysis and then estimate $\text{Cov}(s|\theta^*)$ based on repeated draws from $p(s|\theta^*)$.

1.8 An ABC Analysis in Population Genetics

To illustrate some of the points concerning summary statistics, we consider here a population genetic example, very similar to that considered in the paper by Pritchard et al. (1999), a key paper in the development of ABC methods. In population genetics, we are often confronted with sequence data (as illustrated in Table 1.3), and we wish to infer demographic parameters that may be associated with such data. The standard modelling framework that is used is Kingman's coalescent (Hein, Schierup, and Wiuf 2004), which describes the genealogical relationship of DNA sequences in a sample. The general likelihood problem that we wish to solve then can be represented as:

$$p(y_{obs}|\phi) = \int_H p(y_{obs}|H)p(H|\phi)dH,$$

where y_{obs} represents the observed set of sequences in a sample, ϕ is an unobserved vector of parameters, and H represents the unobserved genealogy history, including mutations. A common mutation model, used here, is the infinite-sites model, in which every mutation that occurs in a genealogy is unique. Typically H is high dimensional, represented as a variable-length vector of times of events in the genealogical history, and the types of events. Although the likelihood can be computed exactly for simple demographic models and small datasets (Hein et al. 2004) it is generally more flexible to resort to Monte Carlo methods (Marjoram and Tavaré 2006).

One approach is through importance sampling. Here, an instrumental distribution $q_{\phi,y}(H)$ is available that describes the distribution of all genealogical histories H that are consistent with the data y, as a function of the model parameters ϕ. The distribution $q_{\phi,y}(H)$ is easy to simulate from and has a

TABLE 1.3
Infinite Sides Data Simulated with *ms* in a Format Suitable for the *Genetree* Program. The Left Hand Column Gives the Number of Times the Sequence on the Right is Observed in the Sample (of Size 30 in this Case). The Ancestral Type is denoted by 0, and the Mutant (Derived) Type is Denoted by 1. The Length of the Sequence is Equal to the Number of Segregating Sites S and is Equal to the Number of Mutations that Occurred int he Genealogy. All Sequences that Share a Mutation at a Given Position are Descendent (and Possibly Further Mutated) Copies of the Sequence in which that Mutation First Occurred. The Sequences are Ordered Lexicographically

1	:	00000000000000000000000010001000000000000000000
1	:	00000000000000000000001010001000000000000101001
1	:	00000000000000010000000010001000100010000101001
5	:	00000010000001000000000000000000000000000000000
1	:	00000010000001000000000000000000000001000000000
2	:	00000010000001000000000000000000001000000000000
1	:	00000010000001000000000000000000100000000000000
2	:	00000010000001000001000000000000000000000000000
2	:	00000010000001000100000000000000000000000000000
1	:	00000010000110000010000000000000000000000000000
1	:	00000010010010000000100000000000000000001000000
1	:	00000010010010000000110000000000000000000000000
1	:	00000010100010000000000100100000000000000000110
2	:	00000110001001000000000000000000000000000000000
2	:	00001001000000000000000010100000000000000010000
2	:	00010000000000000000000000000001000100000001000
1	:	00100000000000010000000000001000000000110101000
1	:	01000010000001000000000000000000000000000000000
2	:	10000000000000000000000010001000000000000101001

known functional form that can be directly evaluated. It also has the property that $p(y|H') = 1$ for $H' \sim q_{\phi,y}(H)$. Hence, $p(y_{obs}|\phi)$ can be estimated by:

$$\hat{p}(y_{obs}|\phi) = \frac{1}{N}\sum_{i=0}^{N}\frac{p(H^{(i)}|\phi)}{q_{\phi,y_{obs}}(H^{(i)})}$$

where $H^{(i)} \sim q_{\phi,y_{obs}}(H)$ for $i = 1, \ldots, N$.

In this analysis, we compare an ABC approach to the above importance sampling method that targets the true likelihood. The aim is to investigate the performance of different summary statistics on ABC inferences, using the importance sampling-based inferences as a (noisy) ground truth.

The demographic model that generates the data is one of smooth exponential expansion. In this model, the current population size N_0 contracts backwards in time as $N_0(t) = N_0 \exp(-\beta t)$, where time t is expressed in units of $2N_0$ and $\beta = 2N_0 b$ is the growth rate in this scaled time. An additional parameter in the model is the scaled mutation rate $\theta_0 = 4N_0\mu$.

In the ABC analysis, simulations are carried out using the *ms* program of Hudson (2002). A technical complication that needs to be accounted for when using *ms* is that time in this program is scaled in units of $4N_0$ rather than $2N_0$ that appears standardly in most treatments (e.g. Hein et al. 2004), and, more importantly, in the *Genetree* importance sampling program (Griffiths and Tavare 1994) that is used for the ground truth. The data in Table 1.3 were generated using the *ms* command:

```
ms 20 1 -t 50 -G 30,
```

which simulates one instance of 20 sequences with $\theta = 50$ and $\alpha = 30$, where $\alpha = \beta/2$ (because of the different scaling of time, noted above). Assuming independent uniform priors $U(0, 200)$ for each parameter $\phi = (\theta_0, \alpha)^\top$, it is straightforward to generate particles by sampling parameter values from the prior and then compute an importance weight for each particle using an algorithm suggested by Stephens and Donnelly (2000). The implementation here (described in Maciuca 2003) is a modification of the *Genetree* program to include the Stephens and Donnelly algorithm, following De Iorio and Griffiths (2004). Although the particles could be used directly for weighted density estimation, it is computationally easier to first re-sample them in proportion to their weights $w^{(i)}$, because the distribution of weights is typically very skewed (they have high variability). For the data in Table 1.3, $N = 10^8$ generated particles yielded an effective sample size (estimated by $(\sum_i w^{(i)})^2 / \sum_i w^{(i)2})$ of around 300. The following analyses are based on re-sampling 1000 particles.

For the ABC analysis, parameter values $\phi = (\theta_0, \alpha)^\top$ are simulated from the prior, data sets are simulated using *ms*, and summary statistics computed. The four summary statistics examined comprise the number of segregating sites, S_0, which corresponds to the number of mutations in the genealogy under the infinite sites mutation model, the average pairwise Hamming distance between all pairs of sequences in the sample, π_0, Tajima's D, and Fay and Wu's H_0. These latter two statistics express the difference in estimates of the scaled mutation parameter θ_0, assuming a standard coalescent model (i.e. with no population growth), based on two different unbiased estimators, one of which is π_0. The average pairwise distance, π_0, is directly an estimate of θ_0 because in the standard constant size model the expected time to coalescence for a pair of sequences is $2N_0$, and therefore the expected number of mutations occurring down both branches since the common ancestor is $(2N_0 + 2N_0)\mu$. Other estimators have been developed, based on the number of segregating sites (Watterson's estimator, used in Tajima's D), or the number of segregating sites weighted by the number of times the mutant type occurs in the sample (Fu's estimator, used in Fay and Wu's H_0). Only under the standard constant size model will these estimators all have the same expectation, and therefore deviations between them can be used to identify departures from this model. Negative values of D and positive values of H_0 are expected to be found in growing populations. The output of the *ms* program can be piped to a program `sample_stats`, included with *ms*, which computes these four summary statistics. The observed summary statistics are:

$$s_{obs} = (\pi_0, S_0, D, H_0)^\top = (5.90, 42, -1.64, 3.67)^\top.$$

ABC methods were implemented by first simulating $N = 1,000,000$ parameter values from the $U(0, 200)$ prior distributions, storing these in the file `params.txt` (in the order indicated by the key-word `tbs`), and then running the *ms* program with the command:

```
ms 20 1 -t tbs -G tbs < params.txt.
```

The summary statistics corresponding to these simulated data were then obtained and then $\|s - s_{obs}\|$ computed as Euclidean distance. The ABC posterior approximation was obtained by using a uniform kernel K_h over $[-h, h]$ and determining the kernel scale parameter h as the value retaining the 1000 samples for which $s^{(i)}$ is closest to s_{obs}.

The summary statistics are measured on different scales. A common practice is to centre and scale them using the standard deviation for each summary statistic sampled from the prior predictive distribution. [However, some authors argue that the motivations for this are flawed, as an arbitrary change in the prior can change the scaling of a summary statistic within the analysis. Instead, following a similar discussion to that in Example 6, the scaling should be based on $\text{Cov}(s|\theta^*)$ for some value of $\theta = \theta^*$ in the high posterior density region, rather than $\text{Cov}(s)$. See e.g. Erhardt and Sisson 2016.] For the present analysis, the prior predictive sample standard deviations for π_0, S_0, D, and H_0 are 14.3, 69.0, 0.50, and 7.3, respectively. In Figure 1.8, the estimated posterior distributions using both scaled and unscaled summary statistics are shown.

Figure 1.8 compares the resulting ABC posterior approximation using (a) all four summary statistics, (b) D and H_0 only, (c) π_0 and S_0 only, or (d) π_0 or S_0 alone. The first point to note is that the data, although quite informative about θ_0 or α jointly, do not allow us to make a very detailed inference about either parameter individually that is, they are only partially identifiable in the model – at least for these data. This is the case both for the full-likelihood and ABC inferences, although the density for the full-likelihood method, as estimated by importance sampling, tends to be more localised towards the true parameter value (indicated by a +).

When all four summary statistics are used (panel a) the 95% high posterior density envelope for ABC is quite similar to that for importance sampling (black line), but is shifted towards higher values of α and θ_0. Scaled or unscaled summary statistics give similar results. The ABC posterior approximation for π_0 and S_0 together (panel b) is very similar to that for the

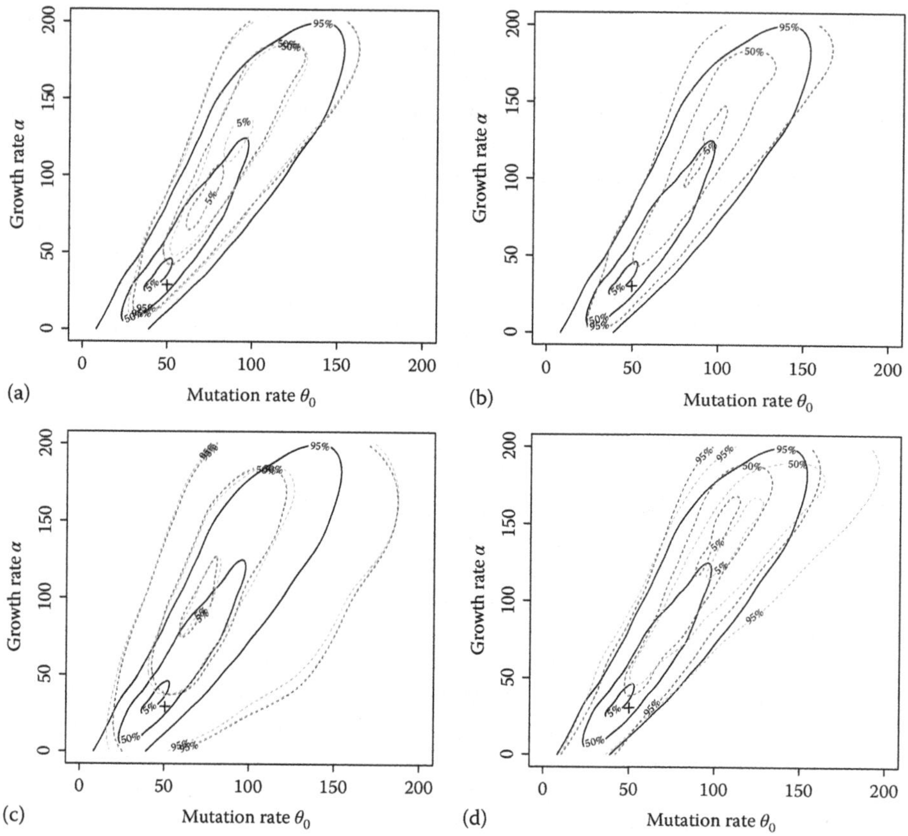

FIGURE 1.8
Various ABC posterior approximations using different summary statistics and scalings, compared to the 'ground truth' importance sampling-based posterior (black lines). The true parameter value is indicated by a +. Estimates show the 95%, 50%, and 5% highest posterior density contours. ABC posteriors are based on (a) all four summary statistics; (b) π_0 and S_0 only; (c) D and H_0 only; and (d) π_0 (light grey dotted) and S_0 (dark grey dotted). For panels (a)–(c) the ABC posterior is based on scaled summary statistics (dark grey dotted line), and unscaled summary statistics (light grey dotted line).

full set of summary statistics. In this case, the distances for scaled and unscaled summaries are the same because S is discrete and matched exactly. This outcome perhaps indicates that one should be cautious of adding summaries such as Tajima's D because it is simply a non-linear function of π_0 and S_0. Whereas H_0 includes additional information from the site frequency spectrum, and would be expected to be informative (positive H_0 indicates a deficit of high-frequency derived mutations compared with that expected under the standard model). Using D and H_0 together (panel c) yields a less

TABLE 1.4
Data from Locus 9pMB8 Surveyed in 11 Biaka Pygmies (Hammer et al. 2010), Using the Same Layout as for Table 1.3

1 :	000000000000000000000000000000000010100001
1 :	000000000000000000000000001000000000000010
1 :	000000000000000000000001010100111001000100
4 :	000000000000000011010000000100000000000000
1 :	000000000000000111010010000100000100000000
4 :	000000000000000111010010000101000100000000
1 :	000000000000010000000000000000000000000000
1 :	000000000000100111010000000100000000000000
1 :	000000000001000000000000010100000000001100
1 :	000000000010000000000000010100000000000100
1 :	000000000100001000000100100000000010000000
1 :	000000010000000000000000000000000010100001
1 :	000100001000000000000000001000000000000100
1 :	001000000001000000101000010100000000010100
1 :	010001100000000000000000001000000000000000
1 :	100010000000000011010000000110000000000000

concentrated posterior approximation. Both statistics are based on the difference of two estimators of mutation rate, and therefore it is unsurprising that θ_0 is not well localised. The posteriors based on π_0 and S_0 individually (panel d), superficially look surprisingly similar to the full-likelihood posterior. However there is much stronger support for larger values of θ_0 and α than in the importance-sampling based posterior.

We conduct a similar analysis with sequence data published in Hammer et al. (2010) from locus 9pMB8 surveyed in 11 Biaka pygmies (resulting in 22 sequences). The data are shown in Table 1.4. Like the simulated data above, there are 42 sites that are segregating within the Biaka sample and which are compatible with the infinite sites model. The ABC simulations were performed as previously, using all four summary statistics. The observed summary statistics for these data are:

$$s_{obs} = (\pi_0, S_0, D, H_0)^\top = (7.52, 42, -1.35, 4.0)^\top.$$

The posterior computed using importance sampling was also computed as before, but required 12×10^8 particles to achieve a similar effective sample size to that for the previous dataset.

It is immediately apparent from Figure 1.9 that the ABC posterior approximation and ground-truth posterior are very similar, unlike the previous analysis. This differing behaviour is not due to Monte Carlo error. The result illustrates a point that outside the exponential family there is no single, low-dimensional set of summary statistics s that will be highly informative for θ, for all observed datasets. Summary statistics that work well for one dataset

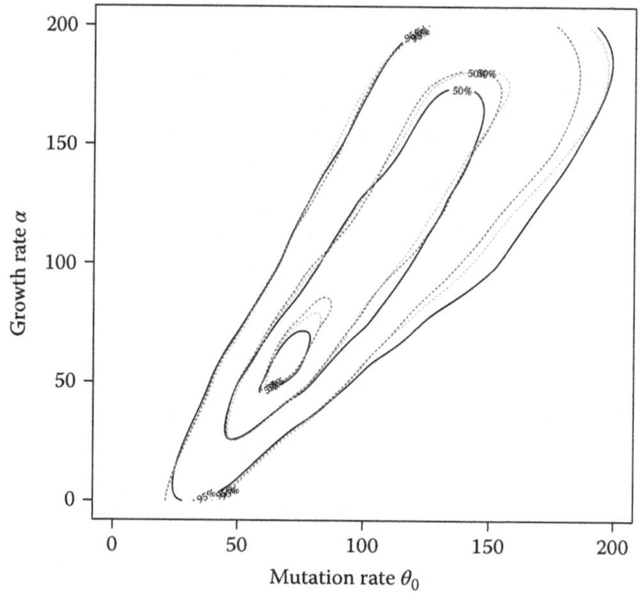

FIGURE 1.9
Comparison of ABC posterior approximations (dotted lines) and full-likelihood (black lines) posterior for the Biaka pygmy data in Table 1.4. ABC posterior approximations are based on all four summary statistics, which are scaled (dark grey dotted line) and unscaled (light grey dotted line).

may perform less well on another. In the case of the two datasets considered here, it may be argued that in the latter, despite the smaller sample size, there is a stronger signal of growth in these data, which is more readily captured by the summary statistics. For the simulated data, the signal is less strong, and information in other summary statistics, such as the site frequency spectrum or higher moments of the distribution of pairwise Hamming distances, may be required for the ABC posterior to better match the true posterior.

From a computational perspective, the 10^6 ABC simulations took about 3 minutes on a desktop computer, whereas 10^8 importance sampling simulations took around 4 hours, i.e. the computational effort per iteration is broadly similar for both approaches. The algorithms used in each are 'similar yet different', in that they both generate genealogical trees, but in one case the tree is constrained by the data, and in the other it is independent of the data. Naively one might think that an importance sampling algorithm should be more efficient because it always generates a tree that is compatible with the data. However, it is typically very difficult to devise an algorithm that samples trees in proportion to their conditional distribution under the model, and therefore genealogical importance sampling tends to be inefficient, as illustrated here, where 10^8 simulations only give an effective sample size of around 300. Of course, it is possible to use sequential methods, or a pseudo-marginal method to improve efficiency (Beaumont 2003; Cornuet et al. 2012; Andrieu et al. 2019a), but similar approaches are available for ABC as well.

1.9 Levels of Approximation in ABC

The primary challenge in implementing an ABC analysis is to reduce the impact of the approximation, while restricting the required computation to acceptable levels. In effect, this is the usual 'more computation for more accuracy' tradeoff. It is therefore worthwhile to briefly summarise the quality and nature of the approximations involved in any ABC analysis. While some of these approximations are common with standard Bayesian analyses, in particular points 1 and 5 below, within the ABC framework these have additional, more subtle implications. In order, from model conception to implementation of the analysis, the ABC approximations are:

1. *All models are approximations to the real data-generation process.*

   While this is true for any statistical analysis, this approximation can produce an ABC-specific issue if the assumed model is not sufficiently flexible to be able to reproduce the observed summary statistics. In this scenario, the kernel scale parameter h will necessarily be large (as all simulated data are far from the observed data), and as a consequence, the quality of the ABC approximation may be low. Further, if, for this inflexible model, the observed summary statistics contain conflicting information for a model parameter, this may cause additional bias in the posterior approximation for this parameter, as is illustrated in Example 5. In summary, this means that the more unlikely a model is to have generated the observed data, the worse the ABC approximation will be. In general, this is problematic, as it implies that routine inspection of the fitted ABC posterior may not in itself be enough to determine model adequacy, as the ABC posterior may be a poor estimate of the true posterior, and poor data generation models may appear more likely (with $h > 0$) than they actually are (with $h = 0$). By extension, this also implies that posterior model probabilities of inadequate models (constructed from the normalising constant of the poorly estimated ABC posterior distribution) may also be affected, although this has yet to be fully explored in the literature. See Chapter 10, in this handbook for an exploration of related ABC asymptotics results to date, and Chapter 6 for particular methods for performing ABC model choice.

2. *Use of summary statistics rather than full datasets.*

   The full posterior distribution $\pi(\theta|y_{obs}) \propto p(y_{obs}|\theta)\pi(\theta)$ is replaced by the partial posterior $\pi(\theta|s_{obs}) \propto p(s_{obs}|\theta)\pi(\theta)$, where $s_{obs} = S(y_{obs})$ is a vector of summary statistics. If S is sufficient for θ, then there is no approximation at this stage. More commonly, for non-sufficient S, there is a loss of information.

3. *Weighting of summary statistics within a region of the observed summary statistics.*

   The partial posterior $\pi(\theta|s_{obs})$ is replaced by the ABC approximation to the partial posterior:

$$\pi_{ABC}(\theta|s_{obs}) \propto \pi(\theta) \int K_h(\|s - s_{obs}\|)p(s|\theta)ds,$$

   where K_h is a standard smoothing kernel with scale parameter $h \geq 0$. If $h = 0$ or in the limit as $h \to 0$ then there is no further approximation at this stage. In most cases, however, $h > 0$ and so ABC makes use of a kernel density estimate as an approximation to the true likelihood function. This aspect of approximation can be a particular problem in ABC when the number of model parameters θ is large, as then the vector of summary statistics, s, must be equivalently large for parameter identifiability, and, hence, the comparison $\|s - s_{obs}\|$ will suffer from the curse of dimensionality.

4. *Approximations due to other ABC techniques.*

   There are a number of other ABC techniques not discussed in this chapter that are optionally implemented in ABC analyses in order to improve some aspect of the approximations in points 1 and 2 or to achieve a greater computational performance. Many of these are discussed in later chapters, but some common methods involve post-processing techniques such as regression and marginal adjustments (e.g. Beaumont et al. 2002; Blum and François 2010; Blum et al. 2013; Blum 2019; Nott et al. 2019), or develop alternative approximations to the intractable likelihood function, while remaining in the ABC framework, such as expectation-propagation ABC, synthetic likelihoods, and copula or regression-density estimation models (e.g. Wood 2010; Fan et al. 2013; Barthelmé and Chopin 2014; Barthelmé et al. 2019; Li et al. 2017; Price et al. 2017).

5. *Monte Carlo error.*

   In common with most Bayesian analyses, performing integrations using Monte Carlo methods introduces Monte Carlo error. Typically, this error may be reduced by using larger numbers of samples from the posterior or by reducing the variability of importance weights. The same is true for an ABC analysis, although with the additional point that more posterior samples effectively allow for a lower kernel scale parameter h and consequently an improved ABC posterior approximation. As a result, for a fixed number of Monte Carlo samples, the choice of kernel scale parameter represents a typical bias-variance tradeoff: if h is large, more posterior draws are available, reducing variance, but at the cost of a poorer ABC approximation; if h is small, the ABC posterior approximation is improved, but Monte Carlo variance is increased.

1.10 Interpretations of ABC

There are a number of closely related ways in which ABC methods may be understood or interpreted. The most common of these is conditional density estimation of the posterior (e.g. Blum 2010; Bonassi et al. 2011) in the sense usually understood in a conventional Bayesian analysis. Before observing the data, the distribution $\pi(\theta, y) = p(y|\theta)\pi(\theta)$ describes prior beliefs about the model parameters and credible datasets under the model. When a dataset y_{obs} is observed, interest is then in the conditional distribution of θ given that $y = y_{obs}$. In the ABC setting, $\pi(\theta, y)$ is represented by the joint sample $(\theta^{(i)}, y^{(i)}) \sim \pi(\theta, y)$, $i = 1, \ldots, N$. Weighting the vectors $\theta^{(i)}$ based on the value of $\|y^{(i)} - y_{obs}\|$ (larger weights for smaller $\|y^{(i)} - y_{obs}\|$), then produces an empirical conditional density estimate of $\pi(\theta|y_{obs})$.

Similarly, we have already discussed that the ABC approximation to the true likelihood, $p_{ABC}(y_{obs}|\theta)$, is a kernel density estimate of $p(y|\theta)$, following (1.7) and (1.8). This allows ABC to be considered as a regular Bayesian analysis with an approximated likelihood function.

Fearnhead and Prangle (2012) noted that the ABC approximation to the posterior can be considered as a continuous mixture of posterior distributions:

$$\begin{aligned}\pi_{ABC}(\theta|y_{obs}) &\propto \int K_h(\|y - y_{obs}\|)p(y|\theta)\pi(\theta)dy \\ &= \int w(y)\pi(\theta|y)dy,\end{aligned}$$

where $\pi(\theta|y) = p(y|\theta)\pi(\theta)/\pi(y)$, with weight function $w(y) \propto K_h(\|y - y_{obs}\|)\pi(y)$. This is the continuous equivalent of equation (1.2) obtained during the analysis of stereological extremes in Section 1.3.2.

While ABC is most often thought of as an approximate method, Wilkinson (2013) pointed out that ABC methods can be considered as exact if $e = y - y_{obs}$ (or $e = \|y - y_{obs}\|$) is considered as the error (either from observation error or model misspecification) obtained in fitting the model $p(y|\theta)$ to the observed data y_{obs}. From this perspective, the smoothing kernel K_h is simply the density function of this error, so that $e \sim K_h$, and h is a scale parameter to be estimated.

Finally, while ABC methods are universally used for the analysis of models with computationally intractable likelihood functions, it is often overlooked that they also provide a useful inferential mechanism for tractable models.

As an illustration, consider a scenario where a standard Bayesian analysis is available for a complex, but incorrect model, given the observed dataset. Under this model, predictions of some particular quantity of interest, $T(y)$, could be precise, but completely implausible due to the limitations in the model. Consider now an ABC analysis based on this model, based on matching summary statistics that include $T(y)$. ABC methods would identify those parameter values θ that are most likely to have produced these statistics under the model. This means that predictions of $T(y)$ under the ABC approximation now have some chance of being accurate (although they may be less precise), as the model may be able to predict the summary statistics, including $T(y)$, even if it can't accurately predict the full dataset. This allows ABC to be interpreted as a mechanism for fitting models based on summary statistics that may in fact be more useful than the exact inference with the full dataset. An explicit example of this in the robust model selection context was given by Li et al. (2017).

Related arguments allow ABC to be thought of as a natural method to fit models when the full dataset (y_{obs}) is only partially observed (s_{obs}) and has missing data (see e.g. Chapter 16, this volume). ABC methods have also been used to determine weakly informative prior distributions in a regular tractable Bayesian analysis, exploiting the mechanism of predictive data matching to identify a priori non-viable regions of the parameter space (Nott et al. 2016).

1.11 Further Reading

ABC methods have been extensively and rapidly developed since their first modern appearance in Tavaré et al. (1997) and Pritchard et al. (1999). Naturally a number of review articles have been written for various discipline audiences to review the techniques available at the time. While with time such reviews can rapidly become dated, they often provide useful perspectives on ABC methods as viewed at the time. See, for example, the reviews by Beaumont (2010), Bertorelle et al. (2010), Blum et al. (2013), Csilléry et al. (2010), Sisson and Fan (2011), Marin et al. (2012), Turner and Zandt (2012), Robert (2016), Erhardt and Sisson (2016), Lintusaari et al. (2016), and Drovandi (2017). Each of the chapters in this handbook also makes for excellent reading and review material on focused aspects of ABC (Andrieu et al. 2019b; Barthelmé et al. 2019; Blum 2019; Drovandi 2019; Drovandi et al. 2019; Fan and Sisson 2019; Fearnhead 2019; Kousathanas et al. 2019; Marin et al. 2019; Nott et al. 2019; Prangle 2019; Ratmann et al. 2019; Tavaré 2019).

Because ABC methods are now recognised as a standard Bayesian tool, their scientific reach has effectively become as extensive as standard Bayesian methods. While it is accordingly futile to exhaustively describe all areas in which ABC has applied, the below selection is provided to provide a flavour of the impact ABC methods have had. Beyond the applications in this handbook,

ABC methods have been successfully applied to applications in α-stable models (Peters et al. 2012), archaeology (Wilkinson and Tavaré 2009), cell biology (Johnston et al. 2014; Vo et al. 2015a,b), coalescent models (Tavaré et al. 1997; Fan and Kubatko 2011), ecology (Jabot and Chave 2009; Wood 2010), evolutionary history of mosquitos (Bennett et al. 2016), filtering (Jasra et al. 2012), extreme value theory (Erhardt and Smith 2012; Erhardt and Sisson 2016), financial modelling (Peters et al. 2012), host-parasite systems (Baudet et al. 2015), HIV contact tracing (Blum and Tran 2010), human evolution (Fagundes et al. 2007), hydrological models (Nott et al. 2014), infectious disease dynamics (Luciani et al. 2009; Aandahl et al. 2012), infinite mixture models for biological signalling pathways (Koutroumpas et al. 2016), image analysis (Nott et al. 2014), long range dependence in stationary processes (Andrade and Rifo 2015), operational risk (Peters and Sisson 2006), quantile distributions (Allingham et al. 2009; Drovandi and Pettitt 2011), pathogen transmission (Tanaka et al. 2006), phylogeography (Beaumont et al. 2010), protein networks (Ratmann et al. 2007, 2009), population genetics (Beaumont et al. 2002), psychology (Turner and Zandt 2012), single cell gene expression (Lenive et al. 2016), spatial point processes (Shirota and Gelfand 2016), species migration (Hamilton et al. 2005), state space models (Vakilzadeh et al. 2017), stochastic claims reserving (Peters et al. 2012), susceptible-infected-removed (SIR) models (Toni et al. 2009), trait evolution (Slater et al. 2012), and wireless communications engineering (Peters et al. 2010). Within this handbook novel analyses can be found in Peters et al. (2019), Rodrigues et al. (2019), Liepe and Stumpf (2019), Estoup et al. (2019), Holden et al. (2019), Fasiolo and Wood (2019) and Fan et al. (2019).

1.12 Conclusion

ABC methods are based on an inherently simple mechanism, simulating data under the model of interest and comparing the output to the observed dataset. While more sophisticated ABC algorithms and techniques have subsequently been developed (and many of these are discussed in more detail in this handbook), this core mechanic remains a constant. It is this methodological simplicity that has made ABC methods highly accessible to researchers in across many disciplines. We anticipate that this will continue in the future.

Acknowledgements

SAS is supported by the Australian Research Council under the Discovery Project scheme (DP160102544) and the Australian Centre of Excellence in Mathematical and Statistical Frontiers (CE140100049).

References

Aandahl, R. Z., J. Reyes, S. A. Sisson, and M. M. Tanaka (2012). A model-based Bayesian estimation of the rate of evolution of VNTR loci in Mycobacterium Tuberculosis. *PLoS Computational Biology 8*, e1002573.

Allingham, D. R., A. R. King, and K. L. Mengersen (2009). Bayesian estimation of quantile distributions. *Statistics and Computing 19*, 189–201.

Anderson, C. W. and S. G. Coles (2002). The largest inclusions in a piece of steel. *Extremes 5*, 237–252.

Andrade, P. and L. Rifo (2015). Long-range dependence and approximate Bayesian computation. *Communications in Statistics: Simulation and Computation*, in press.

Andrieu, C., A. Lee, and M. Vihola (2019a). Theoretical and methodological aspects of MCMC computations with noisy likelihoods. In S. A. Sisson, Y. Fan, and M. A. Beaumont (Eds.), *Handbook of Markov chain Monte Carlo.* Chapman & Hall/CRC Press, Boca Raton, FL.

Andrieu, C., A. Lee, and M. Vihola (2019b). Theoretical and methodological aspects of MCMC computations with noisy likelihoods. In S. A. Sisson, Y. Fan, and M. A. Beaumont (Eds.), *Handbook of Approximate Bayesian Computation.* Chapman & Hall/CRC Press, Boca Raton, FL.

Baddeley, A. and E. B. V. Jensen (2004). *Stereology for Statisticians.* Chapman & Hall/CRC Press, Boca Raton, FL.

Barthelmé, S. and N. Chopin (2014). Expectation propagation for likelihood-free inference. *Journal of the American Statistical Association 109*, 315–333.

Barthelmé, S., N. Chopin, and V. Cottet (2019). Divide and conquer in ABC: Expectation-Propagation algorithms for likelihood-free inference. In S. A. Sisson, Y. Fan, and S. A. Sisson (Eds.), *Handbook of Approximate Bayesian Computation.* Chapman & Hall/CRC Press, Boca Raton, FL.

Baudet, C., B. Donati, C. Sinaimeri, P. Crescenzi, C. Gautier, C. Matias, and M.-F. Sagot (2015). Cophylogeny reconstruction via an approximate Bayesian computation. *Systematic Biology 64*, 416–431.

Beaumont, M. A. (2003). Estimation of population growth or decline in genetically monitored populations. *Genetics 164*(3), 1139–1160.

Beaumont, M. A. (2010). Approximate Bayesian computation in evolution and ecology. *Annual Review of Ecology, Evolution and Systematics 41*, 379–406.

Beaumont, M. A., R. Nielsen, C. P. Robert, J. Hey, O. Gaggiotti, L. Knowles, A. Estoup, et al. (2010). In defence of model-based inference in phylogeography. *Molecular Ecology 19*, 436–466.

Beaumont, M. A., W. Zhang, and D. J. Balding (2002). Approximate Bayesian computation in population genetics. *Genetics 162*, 2025–2035.

Bennett, K. L., F. Shija, Y.-M. Linton, G. Misinzo, M. Kaddumukasa, R. Djouaka, O. Anyaele, et al. (2016). Historical environmental change in Africa drives divergence and admixture of *aedes aegypti* mosquitoes: A precursor to successful worldwide colonization? *Molecular Ecology 25*, 4337–4354.

Bertorelle, G., A. Benazzo, and S. Mona (2010). Abc as a flexible framework to estimate demography over space and time: Some cons, many pros. *Molecular Ecology 19*, 2609–2625.

Blum, M. G. B. (2010). Approximate Bayesian computation: A non-parametric perspective. *Journal of the American Statistical Association 105*, 1178–1187.

Blum, M. G. B. (2019). Regression approaches for ABC. In S. A. Sisson, Y. Fan, and M. A. Beaumont (Eds.), *Handbook of Approximate Bayesian Computation*. Chapman & Hall/CRC Press, Boca Raton, FL.

Blum, M. G. B. and O. François (2010). Non-linear regression models for approximate Bayesian computation. *Statistics and Computing 20*, 63–75.

Blum, M. G. B., M. A. Nunes, D. Prangle, and S. A. Sisson (2013). A comparative review of dimension reduction methods in approximate Bayesian computation. *Statistical Science 28*, 189–208.

Blum, M. G. B. and V. C. Tran (2010). HIV with contact-tracing: A case study in approximate Bayesian computation. *Biostatistics 11*, 644–660.

Bonassi, F. V., L. You, and M. West (2011). Bayesian learning from marginal data in bionetwork models. *Statistical Applications in Genetics and Molecular Biology 10*(1).

Bortot, P., S. G. Coles, and S. A. Sisson (2007). Inference for stereological extremes. *Journal of the American Statistical Association 102*, 84–92.

Brooks, S. P., A. Gelman, G. Jones, and X.-L. Meng (Eds.) (2011). *Handbook of Markov Chain Monte Carlo*. Chapman & Hall/CRC Press, Boca Raton, FL.

Chen, M.-H., Q.-M. Shao, and J. G. Ibrahim (2000). *Monte Carlo Methods in Bayesian Computation*. Springer-Verlag.

Coles, S. G. (2001). *An Introduction to Statistical Modelling of Extreme Values*. Springer-Verlag.

Cornuet, J., J.-M. Marin, A. Mira, and C. P. Robert (2012). Adaptive multiple importance sampling. *Scandinavian Journal of Statistics 39*(4), 798–812.

Csilléry, K., M. G. B. Blum, O. E. Gaggiotti, and O. François (2010). Approximate Bayesian computation in practice. *Trends in Ecology and Evolution 25*, 410–418.

De Iorio, M. and R. C. Griffiths (2004). Importance sampling on coalescent histories. i. *Advances in Applied Probability 36*, 417–433.

Del Moral, P., A. Doucet, and A. Jasra (2006). Sequential Monte Carlo samplers. *Journal of the Royal Statistical Society, Series B 68*, 411–436.

Doucet, A., N. de Freitas, and N. Gordon (2001). *Sequential Monte Carlo Methods in Practice.* Springer-Verlag.

Drovandi, C. C. (2012). *Bayesian Algorithms with Applications.* Ph. D. thesis, Queensland University of Technology.

Drovandi, C. C. (2017). Approximate Bayesian computation. *Wiley StatsRef: Statistics Reference Online*, 1–9.

Drovandi, C. C. (2019). ABC and indirect inference. In S. A. Sisson, Y. Fan, and M. A. Beaumont (Eds.), *Handbook of Approximate Bayesian Computation.* Chapman & Hall/CRC Press, Boca Raton, FL.

Drovandi, C. C., C. Grazian, K. Mengersen, and C. P. Robert (2019). Approximating the likelihood in approximate Bayesian computation. In S. A. Sisson, Y. Fan, and M. A. Beaumont (Eds.), *Handbook of Approximate Bayesian Computation.* Chapman & Hall/CRC Press, Boca Raton, FL.

Drovandi, C. C. and A. N. Pettitt (2011). Likelihood-free Bayesian estimation of multivariate quantile distributions. *Computational Statistics and Data Analysis 55*, 2541–2556.

Erhardt, R. and S. A. Sisson (2016). Modelling extremes using approximate Bayesian computation. In *Extreme Value Modelling and Risk Analysis.* Chapman & Hall/CRC Press.

Erhardt, R. and R. L. Smith (2012). Approximate Bayesian computing for spatial extremes. *Computational Statistics & Data Analysis 56*, 1468–1481.

Estoup, A., P. Verdu, J.-M. Marin, C. P. Robert, A. Dehne-Garcia, J.-M. Corunet, and P. Pudlo (2019). Application of approximate Bayesian computation to infer the genetic history of Pygmy hunter-gatherers populations from West Central Africa. In S. A. Sisson, Y. Fan, and M. A. Beaumont (Eds.), *Handbook of Approximate Bayesian Computation.* Chapman & Hall/CRC Press, Boca Raton, FL.

Fagundes, N. J. R., N. Ray, M. A. Beaumont, S. Neuenschwander, F. M. Salzano, S. L. Bonatto, and L. Excoffier (2007). Statistical evaluation of alternative models of human evolution. *Proceedings of the National Academy of Science of the United States of America 104*, 17614–17619.

Fan, H. H. and L. S. Kubatko (2011). Estimating species trees using approximate Bayesian computation. *Molecular Phylogenetics and Evolution 59*, 354–363.

Fan, Y., S. R. Meikle, G. Angelis, and A. Sitek (2019). ABC in nuclear imaging. In S. A. Sisson, Y. Fan, and M. A. Beaumont (Eds.), *Handbook of Approximate Bayesian Computation.* Chapman & Hall/CRC Press, Boca Raton, FL.

Fan, Y., D. J. Nott, and S. A. Sisson (2013). Approximate Bayesian computation via regression density estimation. *Stat 2*(1), 34–48.

Fan, Y. and S. A. Sisson (2019). ABC samplers. In S. A. Sisson, Y. Fan, and M. A. Beaumont (Eds.), *Handbook of Approximate Bayesian Computation.* Chapman & Hall/CRC Press, Boca Raton, FL.

Fasiolo, M. and S. N. Wood (2019). ABC in ecological modelling. In S. A. Sisson, Y. Fan, and M. A. Beaumont (Eds.), *Handbook of Approximate Bayesian Computation.* Chapman & Hall/CRC Press, Boca Raton, FL.

Fearnhead, P. (2019). Asymptotics of ABC. In S. A. Sisson, Y. Fan, and M. A. Beaumont (Eds.), *Handbook of Approximate Bayesian Computation.* Chapman & Hall/CRC Press, Boca Raton, FL.

Fearnhead, P. and D. Prangle (2012). Constructing summary statistics for approximate Bayesian computation: Semi-automatic approximate Bayesian computation. *Journal of the Royal Statistical Society, Series B 74*, 419–474.

Grelaud, A., C. P. Robert, J.-M. Marin, F. Rodolphe, and J.-F. Taly (2009). ABC likelihood-free methods for model choice in Gibbs random fields. *Bayesian Analysis 4*, 317–336.

Griffiths, R. C. and S. Tavare (1994). Sampling theory for neutral alleles in a varying environment. *Philosophical Transactions of the Royal Society B: Biological Sciences 344*(1310), 403–410.

Hamilton, G., M. Currat, N. Ray, G. Heckel, M. A. Beaumont, and L. Excoffier (2005). Bayesian estimation of recent migration rates after a spatial expansion. *Genetics 170*, 409–417.

Hammer, M. F., A. E. Woerner, F. L. Mendez, J. C. Watkins, M. P. Cox, and J. D. Wall (2010). The ratio of human X chromosome to autosome diversity is positively correlated with genetic distance from genes. *Nature Genetics 42*(10), 830–831.

Hein, J., M. Schierup, and C. Wiuf (2004). *Gene Genealogies, Variation and Evolution: A Primer in Coalescent Theory.* Oxford University Press, Oxford.

Hoaglin, D. C. (1985). Summarizing shape numerically: The g-and-h distributions. In D. C. Hoaglin, F. Mosteller, and J. W. Tukey (Eds.), *Exploring Data Tables, Trends and Shapes.* Wiley, New York.

Holden, P. B., N. R. Edwards, J. Hensman, and R. D. Wilkinson (2019). ABC for climate: dealing with expensive simulators. In S. A. Sisson, Y. Fan, and M. A. Beaumont (Eds.), *Handbook of Approximate Bayesian Computation.* Chapman & Hall/CRC Press, Boca Raton, FL.

Hudson, R. R. (2002). Generating samples under a wright–fisher neutral model of genetic variation. *Bioinformatics 18*(2), 337–338.

Jabot, F. and J. Chave (2009). Inferring the parameters of the netural theory of biodiversity using phylogenetic information and implications for tropical forests. *Ecology Letters 12*, 239–248.

Jasra, A., S. Singh, J. Martin, and E. McCoy (2012). Filtering via ABC. *Statistics and Computing 22*, 1223–1237.

Johnston, S., M. J. Simpson, D. L. S. McEwain, B. J. Binder, and J. V. Ross (2014). Interpreting scratch assays using pair density dynamics and approximate Bayesian computation. *Open Biology 4*(9), 140097.

Koutroumpas, K., P. Ballarini, I. Votsi, and P.-H. Cournede (2016). Bayesian parameter estimation for the Wnt pathway: An infinite mixture models approach. *Bioinformatics 32*, 781–789.

Kousathanas, A., P. Duchen, and D. Wegmann (2019). A guide to general purpose ABC software. In S. A. Sisson, Y. Fan, and M. A. Beaumont (Eds.), *Handbook of Approximate Bayesian Computation.* Chapman & Hall/CRC Press, Boca Raton, FL.

Lenive, O., P. D. W. Kirk, and M. P. H. Stumpf (2016). Inferring extrinsic noise from single-cell gene expression data using approximate Bayesian computation. *BMC Systems Biology 10*, 81.

Li, J., D. J. Nott, Y. Fan, and S. A. Sisson (2017). Extending approximate Bayesian computation methods to high dimensions via Gaussian copula. *Computational Statistics and Data Analysis 106*, 77–89.

Liepe, J. and M. P. H. Stumpf (2019). ABC in systems biology. In S. A. Sisson, Y. Fan, and M. A. Beaumont (Eds.), *Handbook of Approximate Bayesian Computation.* Chapman & Hall/CRC Press, Boca Raton, FL.

Lintusaari, J., M. U. Gutmann, R. Dutta, S. Kaski, and J. Corander (2016). Fundamentals and recent developments in approximate Bayesian computation. *Systematic Biology*, in press.

Luciani, F., S. A. Sisson, H. Jiang, A. R. Francis, and M. M. Tanaka (2009). The epidemiological fitness cost of drug resistance in Mycobacterium tuberculosis. *Proceedings of the National Academy of the Sciences of the United States of America 106*, 14711–14715.

Maciuca, S. (2003). Project report.

Marin, J. M., P. Pudlo, C. P. Robert, and R. J. Ryder (2012). Approximate Bayesian computational methods. *Statistics and Computing 22*, 1167–1180.

Marin, J.-M., P. Pudlo, A. Estoup, and C. P. Robert (2019). Likelhood-free model choice. In S. A. Sisson, Y. Fan, and M. A. Beaumont (Eds.), *Handbook of Approximate Bayesian Computation*. Chapman & Hall/CRC Press, Boca Raton, FL.

Marjoram, P. and S. Tavaré (2006). Modern computational approaches for analysing molecular genetic variation data. *Nature Reviews Genetics 7*(10), 759–770.

Martinez, J. and B. Iglewicz (1984). Some properties of the Tukey g and h family of distributions. *Communications in Statistics: Theory and Methods 13*, 353–369.

Møller, J., A. N. Pettitt, R. Reeves, and K. Berthelsen (2006). An efficient Markov chain Monte Carlo method for distributions with intractable normalising constants. *Biometrika 93*, 451–458.

Nott, D. J., C. C. Drovandi, K. Mengersen, and M. Evans (2016). Approximation of Bayesian predictive p-values with regression ABC. *Bayesian Analysis*, in press.

Nott, D. J., Y. Fan, L. Marshall, and S. A. Sisson (2014). Approximate Bayesian computation and Bayes linear analysis: Towards high-dimensional ABC. *Journal of Computational and Graphical Statistics 23*, 65–86.

Nott, D. J., V. M.-H. Ong, Y. Fan, and S. A. Sisson (2019). High-dimensional ABC. In S. A. Sisson, Y. Fan, and M. A. Beaumont (Eds.), *Handbook of Approximate Bayesian Computation*. Chapman & Hall/CRC Press, Boca Raton, FL.

Peters, G. W., Y. Fan, and S. A. Sisson (2012). On sequential Monte Carlo, partial rejection control and approximate Bayesian computation. *Statistics and Computing 22*, 1209–1222.

Peters, G. W., I. Nevat, S. A. Sisson, Y. Fan, and J. Yuan (2010). Bayesian symbol detection in wireless relay networks via likelihood-free inference. *IEEE Transactions on Signal Processing 56*, 5206–5218.

Peters, G. W., E. Panayi, and F. Septier (2019). SMC-ABC methods for estimation of stochastic simulation models of the limit order book. In S. A. Sisson, Y. Fan, and M. A. Beaumont (Eds.), *Handbook of Approximate Bayesian Computation*. Chapman & Hall/CRC Press, Boca Raton, FL.

Peters, G. W. and S. A. Sisson (2006). Bayesian inference, Monte Carlo sampling and operational risk. *Journal of Operational Risk 1*, 27–50.

Peters, G. W., S. A. Sisson, and Y. Fan (2012). Likelihood-free Bayesian inference for α-stable models. *Computational Statistics and Data Analysis 56*, 3743–3756.

Prangle, D. (2017). Adapting the ABC distance function. *Bayesian Analysis 12*, 289–309.

Prangle, D. (2019). Summary statistics in approximate Bayesian computation. In S. A. Sisson, Y. Fan, and M. A. Beaumont (Eds.), *Handbook of Approximate Bayesian Computation*. Chapman & Hall/CRC Press, Boca Raton, FL.

Price, L. F., C. C. Drovandi, A. Lee, and D. J. Nott (2017). Bayesian synthetic likelihood. *Journal of Computational and Graphical Statistics*, in press.

Pritchard, J. K., M. T. Seielstad, A. Perez-Lezaun, and M. W. Feldman (1999). Population growth of human Y chromosomes: A study of Y chromosome microsatellites. *Molecular Biology and Evolution 16*(12), 1791–1798.

Ratmann, O., C. Andrieu, T. Hinkley, C. Wiuf, and S. Richardson (2009). Model criticism based on likelihood-free inference, with an application to protein network evolution. *Proceedings of the National Academy of Sciences of the United States of America 106*, 10576–10581.

Ratmann, O., A. Camacho, S. Hu, and C. Colijn (2019). Informed choices: How to calibrate ABC with hypothesis testing. In S. A. Sisson, Y. Fan, and M. A. Beaumont (Eds.), *Handbook of Approximate Bayesian Computation*. Chapman & Hall/CRC Press, Boca Raton, FL.

Ratmann, O., O. Jorgensen, T. Hinkley, M. Stumpf, S. Richardson, and C. Wiuf (2007). Using likelihood-free inference to compare evolutionary dynamics of the protien networks of h. pylori and p. falciparum. *PLoS Computational Biology 3*, e230.

Rayner, G. and H. MacGillivray (2002). Weighted quantile-based estimation for a class of transformation distributions. *Computational Statistics & Data Analysis 39*(4), 401–433.

Robert, C. P. (2016). Approximate Bayesian computation: A survey on recent results. In R. Cools and D. Nuyens (Eds.), *Monte Carlo and Quasi-Monte Carlo Methods*, pp. 185–205. Springer.

Rodrigues, G. S., A. R. Francis, S. A. Sisson, and M. M. Tanaka (2019). Inferences on the acquisition of multidrug resistance in mycobacterium tuberculosis using molecular epidemiological data. In S. A. Sisson, Y. Fan, and M. A. Beaumont (Eds.), *Handbook of Approximate Bayesian Computation*. Chapman & Hall/CRC Press, Boca Raton, FL.

Scott, D. W. (1992). *Multivariate Density Estimation: Theory, Practice and Visualisation*. John Wiley & Sons.

Shirota, S. and A. E. Gelfand (2016). Approximate Bayesian computation and model validation for repulsive spatial point processes. *https://arxiv.org/abs/1604.07027*.

Sisson, S. A. and Y. Fan (2011). Likelihood-free Markov chain Monte Carlo. In *Handbook of Markov chain Monte Carlo*, pp. 219–341. Chapman & Hall/CRC Press, Boca Raton, FL.

Slater, G. J., L. J. Harmon, D. Wegmann, P. Joyce, L. J. Revell, and M. E. Alfaro (2012). Fitting models of continuous trait evolution to incompletely sampled comparative data using approximate Bayesian computation. *Evolution 66*, 752–762.

Stephens, M. and P. Donnelly (2000). Inference in molecular population genetics. *Journal of the Royal Statistical Society: Series B (Statistical Methodology) 62*(4), 605–635.

Tanaka, M. M., A. R. Francis, F. Luciani, and S. A. Sisson (2006). Using Approximate Bayesian Computation to estimate tuberculosis transmission parameters from genotype data. *Genetics 173*, 1511–1520.

Tavaré, S. (2019). On the history of ABC. In S. A. Sisson, Y. Fan, and M. A. Beaumont (Eds.), *Handbook of Approximate Bayesian Computation*. Chapman & Hall/CRC Press, Boca Raton, FL.

Tavaré, S., D. J. Balding, R. C. Griffiths, and P. Donnelly (1997). Inferring coalescence times from DNA sequence data. *Genetics 145*, 505–518.

Toni, T., D. Welch, N. Strelkowa, A. Ipsen, and M. P. H. Stumpf (2009). Approximate Bayesian computation scheme for parameter inference and model selection in dynamical systems. *Journal of the Royal Society Interface 6*, 187–202.

Tukey, J. W. (1977). Modern techniques in data analysis. In *NSF-Sponsored Regional Research Conference*. Southeastern Massachusetts University, North Dartmouth, MA.

Turner, B. M. and T. V. Zandt (2012). A tutorial on approximate Bayesian computation. *Journal of Mathematical Psychology 56*, 69–85.

Vakilzadeh, M. K., Y. Huang, J. L. Beck, and T. Abrahamsson (2017). Approximate Bayesian computation by subset simulation using hierarchical state space models. *Mechanical Systems and Signal Processing 84*, 2–20.

Vo, B. N., C. C. Drovandi, A. N. Pettitt, and G. J. Pettet (2015a). Melanoma cell colony expansion parameters revealed by approximate Bayesian computation. *PLoS Computational Biology 11*(12), e1004635.

Vo, B. N., C. C. Drovandi, A. N. Pettitt, and M. J. Simpson (2015b). Quantifying uncertainty in parameter estimates for stochastic models of collective cell spreading using approximate Bayesian computation. *Mathematical Biosciences 263*, 133–142.

Wand, M. P. and M. C. Jones (1995). *Kernel Smoothing.* Chapman & Hall/CRC Press, Boca Raton, FL.

Wicksell, S. D. (1925). The corpsucle problem: A mathematical study of a biometric problem. *Biometrika 17*, 84–99.

Wilkinson, R. D. and S. Tavaré (2009). Estimating primate divergence times by using conditioned birth-and-death processes. *Theoretical Population Biology 75*, 278–285.

Wilkinson, R. L. (2013). Approximate Bayesian computation (ABC) gives exact results under the assumption of model error. *Statistical Applications in Genetics and Molecular Biology 12*, 129–141.

Wood, S. N. (2010). Statistical inference for noisy nonlinear ecological dynamic systems. *Nature 466*, 1102–1104.

2

On the History of ABC

Simon Tavaré

CONTENTS

2.1 Introduction 55
2.2 Coalescent Trees and Mutation 55
2.3 Statistical Inference 59
2.4 Computationally Intensive Methods 60
2.5 A Bayesian Approach 62
2.6 ABC Takes Off 64
2.7 Conclusion 65
2.8 Acknowledgements 66
References 66

2.1 Introduction

What follows is a personal view of the evolution of ABC – approximate Bayesian computation – up to 2003 and it is certainly not intended to be an exhaustive review. ABC arose as an inferential method in population genetics to address estimation of parameters of interest such as mutation rates and demographic parameters in cases where the underlying probability models had intractable likelihoods. To set the scene I will give a very brief introduction to genealogical trees and the effects of mutation, focusing on the simplest case in which a panmictic population is assumed to be very large and of constant size N and within which there is no recombination. The treatment follows that in Tavaré (2004).

2.2 Coalescent Trees and Mutation

The ancestral relationships among n individuals sampled at random from the population can be described by Kingman's coalescent (Kingman, 1982)

Looking back into the past, the sample has n distinct ancestors for time T_n, at which point two individuals are chosen at random to coalesce, the sample then having $n-1$ distinct ancestors. We continue merging random pairs of ancestors in such a way that for time T_j the sample has j distinct ancestors, for $j = n-1, \ldots, 2$. The times T_j are independent exponential random variables with mean:

$$\mathbb{E}[T_j] = \frac{2}{j(j-1)}.$$

In this setting, time is measured in units of N generations. The height of the tree is $T_{\text{MRCA}} = T_n + \cdots + T_2$, the time to the most recent common ancestor (MRCA) of the sample.

This description produces random coalescent trees, as illustrated in Figure 2.1.

It is worth noting that $\mathbb{E}[T_{\text{MRCA}}] = 2(1 - 1/n)$, while the average time for which the sample has just two ancestors is $\mathbb{E}[T_2] = 1$. Thus, the height of the tree is influenced most by T_2, as Figure 2.1 clearly shows.

Conditional on the coalescent tree of the sample, mutations in the genomic region of interest are modelled in two steps. In the first, potential mutations are poured down the branches of the coalescent tree from the MRCA according to independent Poisson processes of rate $\theta_0/2$, where $\theta_0 = 2Nu$ is the compound mutation parameter, u being the chance of a mutation occurring in the genomic region in a given generation. Once the locations of mutations are determined, their effects are modeled by a mutation process that changes the current type.

I will describe three mutation models, the first being the so-called *infinitely many alleles model*, used originally to study the behaviour of allozyme frequencies. Mutations arising on the branches of the coalescent tree are marked by a sequence $U_j, j = 0, 1, \ldots$ of distinct labels, a mutation on a branch replacing the current label with the next available U. An example is given in Figure 2.2, which shows the sample of size $n = 5$ represented by labels U_2, U_2, U_5, U_3, and U_3, respectively.

The particular values of the types observed in a sample are not of interest; rather, it is the number of types $C_j(n)$ represented j times in the sample, for $j = 1, 2, \ldots, n$, that records the information in these data. In the example above, there are $K_5 = 3$ types, with $C_1(5) = 1, C_2(5) = 2$. Note that $C_1(n) + 2C_2(n) + \cdots + nC_n(n) = n$.

In the second example, known as the *infinitely many sites model*, we think of the genomic region of interest as the unit interval (0,1), and assume that each mutation in the coalescent tree arises at a novel position in the genomic region. These positions may be realised as values in a sequence $U_j, j = 0, 1, \ldots$ of distinct labels (generated, e.g., as independent random variables uniformly distributed in $(0, 1)$). Figure 2.2 can be used to illustrate this model too. Each individual in the sample is represented as a sequence of mutations back to

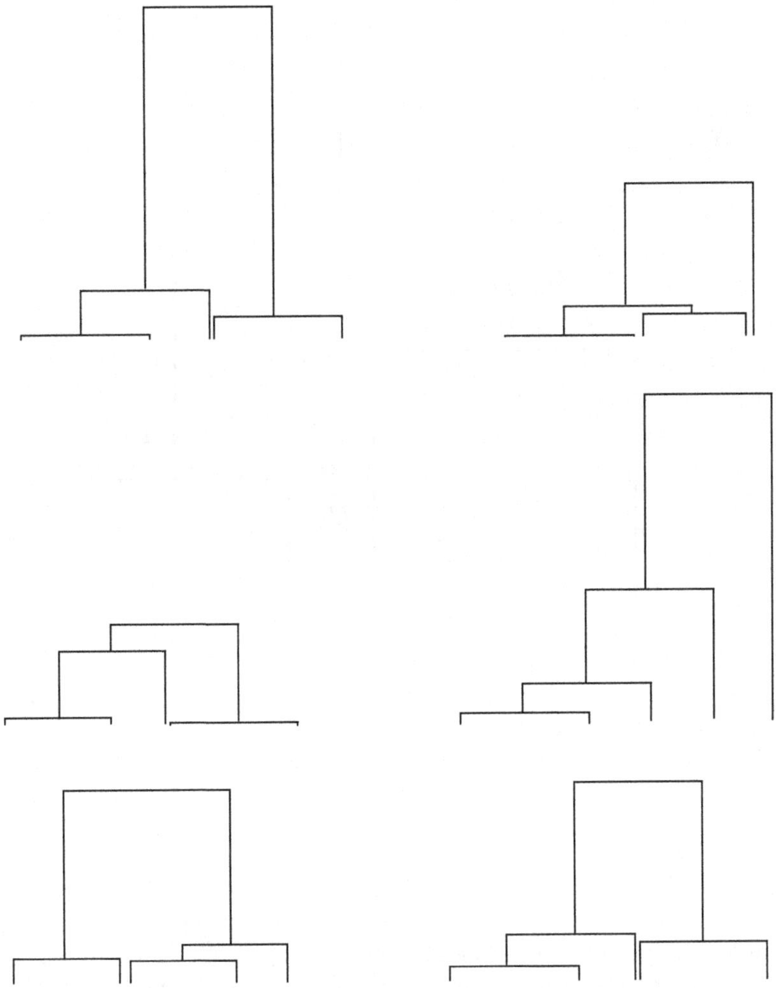

FIGURE 2.1
Six realisations of coalescent trees for a sample of size $n = 5$, drawn on the same scale. (Reprinted by permission from Springer Nature, *Lectures on Probability Theory and Statistics*, Volume 1837, Part I, 2004, Tavaré, S., Copyright 2004.)

the root. Reading from left to right in the figure, we see that the first two individuals in the sample have mutations at locations $\{U_1, U_2\}$, the third at $\{U_1, U_2, U_4, U_5\}$, and the fourth and fifth at $\{U_0, U_3\}$.

We can write these in rather more conventional 'DNA style' by representing the ancestral type as 0, mutants as 1, and recording the mutation status of

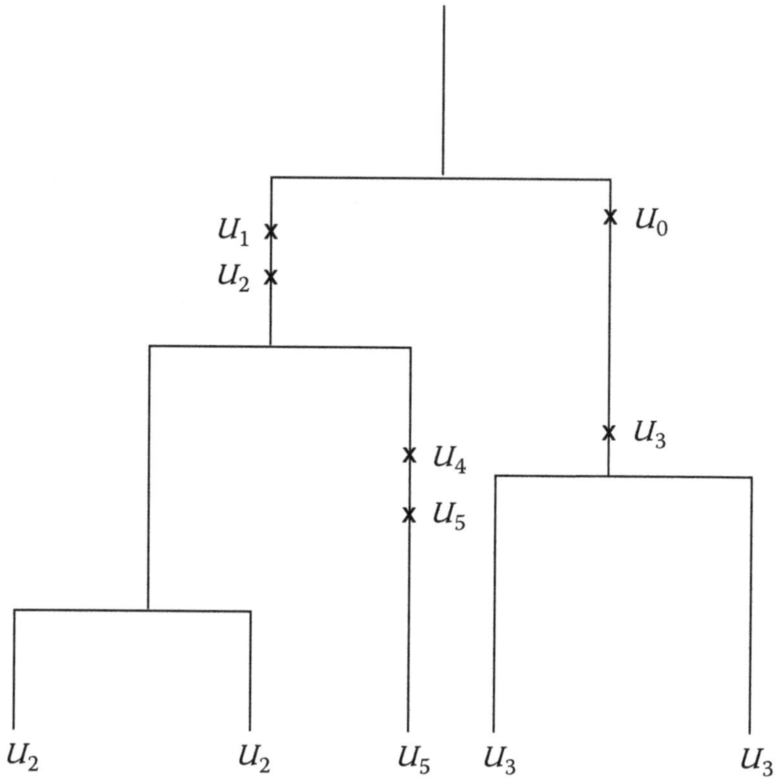

FIGURE 2.2
A coalescent tree for $n = 5$ with mutations $U_0, U_1, \ldots, U_5$ marked on the branches. The labels at the bottom of the tree give the type of each individual, assuming the infinitely many alleles model. Two types (U_2, U_3) are represented twice and one type (U_5) once. (Reprinted by permission from Springer Nature, *Lectures on Probability Theory and Statistics*, Volume 1837, Part I, Tavaré, S., Copyright 2004.)

each individual in an $n \times s$ matrix, where s is the number of mutations, or *segregating sites*, that have occurred in the coalescent tree. For our example, the sequences are:

$$\begin{array}{c} \\ 1 \\ 2 \\ 3 \\ 4 \\ 5 \end{array} \begin{array}{c} \begin{array}{cccccc} U_0 & U_1 & U_2 & U_3 & U_4 & U_5 \end{array} \\ \begin{pmatrix} 0 & 1 & 1 & 0 & 0 & 0 \\ 0 & 1 & 1 & 0 & 0 & 0 \\ 0 & 1 & 1 & 0 & 1 & 1 \\ 1 & 0 & 0 & 1 & 0 & 0 \\ 1 & 0 & 0 & 1 & 0 & 0 \end{pmatrix} \end{array}. \tag{2.1}$$

In practice, of course, the ancestral labeling is often not known, and neither is the time-ordered labeling of the mutations; the sequences would be recorded by ordering the columns according to positions along the genome. Any sequence dataset consistent with the infinitely many sites model, such as that in (2.1), can be represented by a rooted tree if the ancestral labeling is known and as an unrooted tree if it is not. See Chapter 5 of Tavaré (2004) for further details.

In the previous examples mutations have a simple structure, in that they do not allow for back mutations, for example. More detailed models of sequence evolution have also been developed. For example, rather than representing the DNA region as a unit interval, it might be described as a series of m completely linked sites, each site containing one of the letters A, C, G, or T. There are now $M = 4^m$ possible values (or haplotypes) for each sequence. Mutations are laid down on the coalescent tree as before, the results of each mutation being given by an $M \times M$ mutation probability matrix $\mathcal{P}$. The (i, j)th element of $\mathcal{P}$ gives the probability that sequence i mutates to sequence j when a mutation occurs. These models are referred to collectively as *finite sites models.* For historical amusement, when these models were used in the early days of sequence data the sample sizes were $n \approx 60$ and the length of the sequences $m \approx 360$ (Ward et al., 1991). How things have changed!

2.3 Statistical Inference

Statistical inference for the parameter θ_0 for the infinitely many alleles model was the subject of Ewens' celebrated paper (Ewens, 1972). Ewens established that $K_n := C_1(n) + \cdots + C_n(n)$, the number of types observed in the sample, is sufficient for θ_0, and that the (moment and) maximum likelihood estimator $\widehat{\theta}_{\mathrm{E}}$ of θ_0 is the solution of the equation:

$$K_n = \sum_{j=0}^{n-1} \frac{\theta}{\theta + j}.$$

In large samples, $\widehat{\theta}_{\mathrm{E}}$ is asymptotically Normally distributed with mean θ_0 and variance $\theta_0 / \log n$. This last result goes some way to explaining why accurate estimation of θ_0 is hard; even modern-day sample sizes do not make much progress.

For the infinitely many sites mutation model, θ_0 has traditionally been estimated by making use of the summary statistic S_n, the number of segregating sites observed in the sample. Watterson's classic paper (Watterson, 1975) derived the basic results. We note first that:

Conditional on the total length $L_n = 2T_2 + \cdots + nT_n$ of the branches of the coalescent tree, S_n has a Poisson distribution with mean $L_n\theta_0/2$.

Hence, unconditionally:

$$\mathbb{E}[S_n] = \frac{\theta_0}{2}\mathbb{E}[L_n] = \frac{\theta_0}{2}\,2\left(1 + \frac{1}{2} + \cdots + \frac{1}{n-1}\right) = \theta_0 \sum_{j=1}^{n-1} \frac{1}{j}. \tag{2.2}$$

This gives Watterson's unbiased estimator:

$$\widehat{\theta}_{\mathrm{W}} = S_n \Bigg/ \sum_{j=1}^{n-1} \frac{1}{j}.$$

In large samples, $\widehat{\theta}_{\mathrm{W}}$ is approximately normally distributed with mean θ_0 and variance $\theta_0/\log n$. The rate of decay of the variance of $\widehat{\theta}_{\mathrm{W}}$ and $\widehat{\theta}_{\mathrm{E}}$, the reciprocal of the logarithm of the sample size n (rather than of the sample size itself, as might have been anticipated), reflects the dependence among the observations arising from the tree structure of the coalescent.

2.4 Computationally Intensive Methods

Computationally intensive methods have been used to fit stochastic models to data for many years. Among the early examples are Ottestad (1956), Edwards (1967), Trudgill and Ross (1969), Hoel and Mitchell (1971). Edwards noted that:

> Particular emphasis will be placed on the need to formulate sound methods of 'estimation by simulation' on complex models.

Ross (1972) explored approximate likelihood methods to fit models to data, and a systematic treatment was provided in the paper by Diggle and Gratton (1984).

By the early 1990s, the emergence of DNA sequence data led to a number of computational inference methods in population genetics. Among these is Lundstrom et al. (1992), who analysed mitochondrial data by using a finite sites model to describe the behaviour of purine-pyrimidine sites across the region. Their first approach compared the expected number of sites of different types with the observed numbers, and estimated parameters by matching expected to observed numbers as closely as possible. Their second approach was a composite likelihood method that treated the sites as independent. Kuhner et al. (1995) developed an ingenious Metropolis–Hastings Monte Carlo method to estimate the parameter θ in another finite sites model, exploiting the coalescent structure to generate a likelihood curve from which inference could be made.

Griffiths and Tavaré (1994c,b) introduced another approach to full-likelihood-based inference by exploiting a classical result about Markov chains. For a discrete-time Markov chain $\{X_k, k \geq 0\}$ with state space $\mathcal{S}$ and transition matrix $P = (p_{xy}, x, y \in \mathcal{S})$, let $\mathcal{A}$ be a set of states for which the hitting time:

$$\eta = \inf\{k \geq 0 : X_k \in \mathcal{A}\},$$

is finite with probability one starting from any $x \in \mathcal{T} := \mathcal{S} \setminus \mathcal{A}$. Let f be a function on $\mathcal{S}$, and define:

$$u_x(f) = \mathbb{E}\left[\prod_{k=0}^{\eta} f(X_k) \mid X_0 = x\right],$$

for all $X_0 = x \in \mathcal{S}$ (so that $u_x(f) = f(x), x \in \mathcal{A}$). Then for all $x \in \mathcal{T}$:

$$u_x(f) = f(x) \sum_{y \in \mathcal{S}} p_{xy} u_y(f). \tag{2.3}$$

A simulation approach to solve equations such as the one in (2.3) follows: simulate a trajectory of the chain X starting at x until it hits $\mathcal{A}$ at time η, compute the value of product $\prod_{k=0}^{\eta} f(X_k)$, and repeat this many times to obtain an estimate of $u_x(f)$. In the applications in Griffiths and Tavaré (1994c,b), coalescent-based recursions for likelihoods were reduced to this form. The method is essentially a version of von Neumann and Ulam's suggestion for matrix inversion, as described in Forsythe and Leibler (1950), and improved by sequential Monte Carlo by Halton (1962, 1970). Further examples may be found in Chapter 6 of Tavaré (2004). Felsenstein et al. (1999) showed how to exploit importance sampling to design more efficient ways to (in our language) choose the process X, and this resulted in a number of more effective inference methods; see, for example, Stephens and Donnelly (2000) and Griffiths et al. (2008). A Markov chain Monte Carlo (MCMC) approach to inference for the infinitely many sites model appears in Racz (2009).

Summary statistics continued to be used for inference, as illustrated by Weiss and von Haeseler (1998), who described what is essentially the frequentist version of ABC in the context of inference about population history, based on the number of segregating sites and the mean pairwise distance among the sequences. They produced a likelihood surface over a grid of parameter values, approximating the likelihood by repeated simulation of the model and recording the proportion of simulated values of the statistics that were sufficiently close to the observed values.

The distributions of unobservable features of coalescent models, such as T_{MRCA}, conditional on observed values of the data, have also been studied by

Monte Carlo methods. Griffiths and Tavaré (1994a) considered inference for T_{MRCA} under the infinitely many sites model, using data of the form (2.1) and exploiting a version of the approach outlined in (2.3). Fu and Li (1997) studied a similar problem, but using the maximal value of the number of nucleotide differences between any pair of sequences in the dataset as the observed statistic. Their method uses a simple form of density estimation to approximate T_{MRCA}.

2.5 A Bayesian Approach

Bayesian methods provide a natural setting for inference not just about model parameters, but also about unobservables in the underlying model. Tavaré et al. (1997) illustrated this for the infinitely many sites model by developing a rejection algorithm for simulating observations from T_{MRCA} and θ, conditional on the number of segregating sites $S_n = s$ seen in the data. The method is based on the observation made above (2.2) that, conditional on the times $T_2, \ldots, T_n$ and θ, the number of segregating sites in the sample of size n has a Poisson distribution with mean $\mathbb{E}[S_n | T_n, \ldots, T_2, \theta] = \theta\, L_n/2$; we write $S \sim \text{Po}(\theta L_n/2)$.

Suppose then that θ has prior distribution $\pi()$, and let $p(s|\lambda)$ denote the probability that a Poisson random variable with mean λ has value s:

$$p(s|\lambda) = \frac{e^{-\lambda}\lambda^s}{s!}, s = 0, 1, \ldots .$$

The rejection algorithm is:

A1 Generate $\theta \sim \pi(\cdot)$.

A2 Generate $T_n, \ldots, T_2$ from the coalescent model. Calculate $L_n = \sum_{j=2}^{n} jT_j$ and $T_{\text{MRCA}} = \sum_{j=2}^{n} T_j$.

A3 Accept $(\theta, T_{\text{MRCA}})$ with probability proportional to:

$$\alpha = p(s|\theta L_n/2).$$

Accepted values of this algorithm have the required distribution, that of $(\theta, T_{\text{MRCA}})$ given $S_n = s$.

The previous method may be viewed as an application of the rejection algorithm, which proceeds as follows. For discrete data $\mathcal{D}$, probability model $\mathcal{M}$ with parameters θ having prior $\pi()$, we can simulate observations from:

$$f(\theta|\mathcal{D}) \propto \mathbb{P}(\mathcal{D}|\theta)\, \pi(\theta), \tag{2.4}$$

via

B1 Generate $\theta \sim \pi(\cdot)$.

B2 Accept θ with probability proportional to the likelihood $\mathbb{P}(\mathcal{D}|\theta)$.

This method can be extended dramatically in its usefulness using the following, stochastically equivalent, version:

C1 Generate $\theta \sim \pi(\cdot)$.

C2 Simulate an observation $\mathcal{D}'$ from model $\mathcal{M}$ with parameter θ.

C3 Accept θ if $\mathcal{D}' = \mathcal{D}$.

For the example in algorithm A above, C3 takes the form

C3 Simulate an observation $S \sim \text{Po}(\theta L_n/2)$, and accept $(T_{\text{MRCA}}, \theta)$ if $S = s$.

While algorithms B and C are probabilistically identical, C is much more general in that one does not need to compute probabilities explicitly to make it work; only simulation is needed. Version C is due to Rubin (1984). Surprisingly, the result does not seem to be described in text books that focus on simulation.

The drawback in C is clear. It will typically be the case that for a given value of θ, the chance of the outcome $\mathcal{D}' = \mathcal{D}$, namely $\mathbb{P}(\mathcal{D}|\theta)$, is either vanishingly small or very time consuming to compute, resulting in an algorithm that does not work effectively. This is where ABC finally comes into play, in the form of the following scheme. We start with a metric ρ to compare datasets and a tolerance $h \geq 0$, and then:

D1 Generate $\theta \sim \pi(\cdot)$.

D2 Simulate an observation $\mathcal{D}'$ from model $\mathcal{M}$ with parameter θ.

D3 Compute $\rho := \rho(\mathcal{D}', \mathcal{D})$, and accept θ as an approximate draw from $f(\theta|\mathcal{D})$ if $\rho \leq h$.

The parameter h measures the tension between computability and accuracy. If ρ is a metric, then $\rho = 0 \implies \mathcal{D}' = \mathcal{D}$, so that such an accepted θ is indeed an observation from the true posterior.

Pritchard et al. (1999) were the first to describe a version of this scheme, in which the datasets in D3 were compared through a choice of summary statistics. Thus, ρ compares how well a set of simulated summary statistics matches the observed summary statistics. If the statistics are sufficient for θ, then when $h = 0$, the accepted values of θ are still from the true posterior based on the full data. This begs the question of how one might identify 'approximately sufficient' statistics, a topic covered in Chapter 5, this volume. The method is also applicable to continuous data.

2.6 ABC Takes Off

Beaumont et al. (2002) showed that it might be better to soften the hard cutoff suggested in algorithm D, by making use of the all the simulated values. They proposed to weight values of θ by the size of the corresponding distance ρ; smaller values of ρ suggest an observation whose distribution is closer to the required posterior. They made a number of suggestions for how the weights might be chosen, and then used to produce a sample with better sampling properties than the original hard cut-off method.

The development of ABC was predicated on the availability of computational power and the lack of tractable likelihoods. The latter is also an issue for MCMC methods, and this motived Marjoram et al. (2003) to suggest an MCMC method that does not need likelihoods in its implementation.

In the present setting the idea behind classical MCMC (Hastings, 1970) is to construct an ergodic Markov chain that has $f(\theta|\mathcal{D})$ as its stationary distribution. In skeleton form, it works as follows:

E1 The chain is now at θ.

E2 Propose a move to θ' according to a proposal distribution $q(\theta, \theta')$.

E3 Calculate the Hastings ratio:

$$\alpha = \min\left(1, \frac{\mathbb{P}(\mathcal{D}|\theta')\pi(\theta')q(\theta', \theta)}{\mathbb{P}(\mathcal{D}|\theta)\pi(\theta)q(\theta, \theta')}\right). \tag{2.5}$$

E4 Move to θ' with probability α, else remain at θ.

It is the ratio of likelihoods in E3 that might cause problems. Marjoram et al. (2003) proposed the following:

F1 The chain is now at θ.

F2 Propose a move to θ' according to $q(\theta, \theta')$.

F3 Generate $\mathcal{D}'$ using parameter θ'.

F4 If $\mathcal{D}' = \mathcal{D}$, go to F5, else remain at θ.

F5 Calculate:

$$\alpha = \min\left(1, \frac{\pi(\theta')q(\theta', \theta)}{\pi(\theta)q(\theta, \theta')}\right).$$

F6 Move to θ' with probability α, else remain at θ.

This likelihood-free method does indeed have the correct stationary distribution. In practice the rejection step is often replaced by a version of D3:

F4 If $\rho(\mathcal{D}', \mathcal{D}) \leq h$, go to next step, else return θ,

and this too might involve a comparison of summary statistics. Marjoram et al. (2003) were able to assess the effect of summary statistics in a population genetics problem, and Plagnol and Tavaré (2004) used the method in a problem concerning divergence times of primate species.

Marjoram et al. (2003) also suggested that the likelihood terms in (2.5) be approximated by estimates of the form:

$$\hat{\mathbb{P}}(\mathcal{D}|\theta) = \frac{1}{B} \sum_{j=1}^{B} \mathbb{I}(\mathcal{D}'_j = \mathcal{D}),$$

for B independent simulations of the model with parameter θ; algorithm F is the special case $B = 1$. Beaumont (2003) made a similar suggestion, and this motivated the development of the 'pseudo-marginal method' (Andrieu and Roberts, 2009), discussed in Chapter 12 , this volume.

2.7 Conclusion

The term 'Approximate Bayesian computation' has arisen more than once. For example, Sweeting (1996) used it to describe computation based on the asymptotic behaviour of signed roots of log-density ratios. He argued that:

> ...analytic approximation still has an important role to play in Bayesian statistics.

In the setting of the present handbook, and in some sense at the other end of the analytical spectrum, it was Beaumont et al. (2002) who coined the term 'approximate Bayesian computation', in the article that made ABC the popular technique it has become.

Where did the acronym ABC arise? By the time we submitted Marjoram et al. (2003) for publication at the end of 2002, the University of Southern California (USC) group had held many meetings on what we then called ABC. In the submitted version, the term ABC appeared twice:

> ...we have the following approximate Bayesian computation (ABC) scheme for data $\mathcal{D}$ summarised by S

and:

> ...and it is often useful to replace the full data by a number of judiciously chosen summary statistics. The resulting approximate Bayesian computation, which we dub ABC, allows us to explore scenarios which are intractable if the full data are used.

In the published version, ABC does not appear, because of house style at the time in the Proceedings of the National Academy of Sciences. This from the proofs:

> D – AU: Per PNAS style, nonstandard abbreviations are allowed only when used at least 5 times in the main text.

A missed opportunity for the National Academy of Sciences! In 2003, I gave an invited lecture at the Royal Statistical Society entitled (with a certain amount of bravado) 'Who needs likelihoods', in which ABC appeared several times, as the write-up in the RSS News in October 2003 showed. It concluded:

> The lively discussion that followed reinforced our feeling that we were not hearing the last of ABC.

This observation turned out to be true, and ABC has become a standard approach in the statistician's toolbox. New areas of application arise frequently, as the rapidly expanding literature shows. One area that would repay deeper analysis is that of cancer evolution, a field that is producing enormous amounts of DNA sequence and phenotype data and for which there is a dearth of inference methods. For an early application see Tsao et al. (2000). Inference for agent-based models in stem cell biology appears in Sottoriva and Tavaré (2010), which motivated the approach in Sottoriva et al. (2013) for colorectal cancer.

2.8 Acknowledgements

I thank Dr. Andy Lynch and two anonymous reviewers for helpful comments on an earlier version of this article, and Paul Gentry, Conference and Events Manager at the Royal Statistical Society, for tracking down the Royal Statistical Society News report on ABC.

References

Andrieu, C. and G. Roberts (2009). The pseudo-marginal approach for efficient Monte Carlo computations. *Ann Stat 37*, 697–725.

Beaumont, M. A. (2003). Estimation of population growth or decline in genetically monitored populations. *Genetics 164*, 1139–1160.

Beaumont, M. A. (2008). Joint determination of topology, divergence times and immigration in population trees. In S. Matsumara, P. Forster, and C. Renfrew (Eds.), *Simulations, Genetics and Human Prehistory*, pp. 135–154. McDonald Institute for Archaeological Research, Cambridge, UK.

Beaumont, M. A., W. Zhang, and D. J. Balding (2002). Approximate Bayesian computation in population genetics. *Genetics 162*, 2025–2035.

Diggle, P. J. and R. J. Gratton (1984). Monte Carlo methods of inference for implicit statistical models. *J R Statist Soc B 46*, 193–227.

Edwards, A. W. F. (1967). A biometric case history: Studies on the distribution of sexes in families, 1889–1966. Abstract 1286. *Biometrics 23*, 176.

Ewens, W. (1972). The sampling theory of selectively neutral alleles. *Theor Popul Biol 3*, 87–112.

Felsenstein, J., M. Kuhner, J. Yamato, and P. Beerli (1999). Likelihoods on coalescents: A Monte Carlo sampling approach to inferring parameters from population samples of molecular data. In F. Seillier-Moiseiwitsch (Ed.), *Statistics in Molecular Biology and Genetics*, Volume 33 of *IMS Lecture Notes–Monograph Series*, pp. 163–185. Institute of Mathematical Statistics and and American Mathematical Society, Hayward, CA.

Forsythe, G. and R. Leibler (1950). Matrix inversion by the Monte Carlo method. *Math Comput 26*, 127–129.

Fu, Y.-X. and W.-H. Li (1997). Estimating the age of the common ancestor of a sample of DNA sequences. *Mol Biol Evol 14*, 195–199.

Griffiths, R. and S. Tavaré (1994a). Ancestral inference in population genetics. *Stat Sci 9*, 307–319.

Griffiths, R. and S. Tavaré (1994b). Sampling theory for neutral alleles in a varying environment. *Phil Trans R Soc Lond B 344*, 403–410.

Griffiths, R. and S. Tavaré (1994c). Simulating probability distributions in the coalescent. *Theor Popul Biol 46*, 131–159.

Griffiths, R. C., P. A. Jenkins, and Y. S. Song (2008). Importance sampling and the two-locus model with subdivided population structure. *Adv Appl Probab 40*, 473–500.

Halton, J. (1962). Sequential Monte Carlo. *Proc Camb Philos Soc 58*, 57–58.

Halton, J. (1970). A retrospective and prospective study of the Monte Carlo method. *SIAM Rev. 12*, 1–63.

Hastings, W. K. (1970). Monte Carlo sampling methods using Markov chains and their applications. *Biometrika 57*, 97–109.

Hoel, D. G. and T. J. Mitchell (1971). The simulation, fitting, and testing of a stochastic cellular proliferation model. *Biometrics 27*, 191–199.

Kingman, J. (1982). On the genealogy of large populations. *J Appl Probab 19A*, 27–43.

Kuhner, M., J. Yamato, and J. Felsenstein (1995). Estimating effective population size and mutation rate from sequence data using Metropolis-Hastings sampling. *Genetics 140*, 1421–1430.

Lundstrom, R., S. Tavaré, and R. Ward (1992). Estimating mutation rates from molecular data using the coalescent. *Proc Natl Acad Sci USA 89*, 5961–5965.

Marjoram, P., J. Molitor, V. Plagnol, and S. Tavaré (2003). Markov chain Monte Carlo without likelihoods. *Proc Natl Acad Sci USA 100*, 15324–15328.

Ottestad, P. (1956). On the size of the stock of Antarctic fin whales relative to the size of the catches. *The Norwegian Whaling Gazette 45*, 298–308.

Plagnol, V. and S. Tavaré (2004). Approximate Bayesian computation and MCMC. In H. Niederreiter (Ed.), *Monte Carlo and Quasi-Monte Carlo Methods 2002*, pp. 99–114. Springer-Verlag Berlin, Germany.

Pritchard, J. K., M. T. Seielstad, A. Perez-Lezaun, and M. W. Feldman (1999). Population growth of human Y chromosomes: A study of Y chromosome microsatellites. *Mol Biol Evol 16*, 1791–1798.

Racz, M. Z. (2009). MCMC and the infinite sites model. Technical report, Department of Statistics Summer Project, University of Oxford, Oxford, UK.

Ross, G. J. S. (1972). Stochastic model fitting by evolutionary operation. In J. R. N. Jeffers (Ed.), *Mathematical Models in Ecology*, pp. 297–308. Blackwell Scientific Publications, Oxford, UK.

Rubin, D. B. (1984). Bayesianly justifiable and relevant frequency calculations for the applied statistician. *Ann Stat 12*, 1151–1172.

Sottoriva, A., I. Spiteri, D. Shibata, C. Curtis, and S. Tavaré (2013). Single-molecule genomic data delineate patient-specific tumor profiles and cancer stem cell organization. *Cancer Res 73*(1), 41–49.

Sottoriva, A. and S. Tavaré (2010). Integrating approximate Bayesian computation with complex agent-based models for cancer research. In G. Saporta and Y. Lechevallier (Eds.), *COMPSTAT 2010 – Proceedings in Computational Statistics*, pp. 57–66. Physica Verlag Heidelberg.

Stephens, M. and P. Donnelly (2000). Inference in molecular population genetics. *J R Stat Soc B 62*, 605–655.

Sweeting, T. J. (1996). Approximate Bayesian computation based on signed roots of log-density ratios (with discussion). *Bayesian Stat 5*, 427–444.

Tavaré, S. (2004). Ancestral inference in population genetics. In J. Picard (Ed.), *Lectures On Probability Theory And Statistics: Ecole d'Eté dc Probabilités de Saint-Flour XXXI – 2001*, Volume 1837 of *Lecture Notes in Mathematics*, pp. 1–188. Springer-Verlag, New York.

Tavaré, S., D. J. Balding, R. C. Griffiths, and P. Donnelly (1997). Inferring coalescence times for molecular sequence data. *Genetics 145*, 505–518.

Trudgill, D. L. and G. J. S. Ross (1969). The effect of population density on the sex ratio of *Heterodera rostochiensis*: A two dimensional model. *Nematologica 15*, 601–607.

Tsao, J., Y. Yatabe, R. Salovaara, H. J. Jarvinen, J. P. Mecklin, L. A. Aaltonen, S. Tavaré, and D. Shibata (2000). Genetic reconstruction of individual colorectal tumor histories. *Proc Natl Acad Sci USA 97*, 1236–1241.

Ward, R., B. Frazier, K. Dew, and S. Pääbo (1991). Extensive mitochondrial diversity within a single amerindian tribe. *Proc Natl Acad Sci USA* 8720–8724.

Watterson, G. A. (1975). On the number of segregating sites in genetical models without recombination. *Theor Popul Biol 7*, 256–276.

Weiss, G. and A. von Haeseler (1998). Inference of population history using a likelihood approach. *Genetics 149*, 1539–1546.

3

Regression Approaches for ABC

Michael G.B. Blum

CONTENTS

3.1 Introduction 71
3.2 Principle of Regression Adjustment 72
3.2.1 Partial posterior distribution 72
3.2.2 Rejection algorithm followed by adjustment 73
3.2.3 Regression adjustment 73
3.2.4 Fitting regression models 74
3.2.5 Parameter transformations 76
3.2.6 Shrinkage 77
3.3 Theoretical Results about Regression Adjustment 77
3.4 Application of Regression Adjustment to Estimate Admixture Proportions Using Polymorphism Data 79
3.5 Regression Methods Besides Regression Adjustment 81
3.6 Conclusion 82
References 83

3.1 Introduction

In this chapter, we present regression approaches for approximate Bayesian computation (ABC). As for most methodological developments related to ABC, regression approaches originate with coalescent modeling in population genetics [3]. After performing rejection sampling by accepting parameters that generate summary statistics close enough to those observed, parameters are *adjusted* to account for the discrepancy between simulated and observed summary statistics. Because adjustment is based on a regression model, such approaches are coined as *regression adjustment* in the following.

Regression adjustment is a peculiar approach in the landscape of Bayesian approaches where sampling techniques are usually proposed to account for mismatches between simulations and observations [4, 5]. We suggest various reasons explaining why regression adjustment is now a common step in practical applications of ABC. First, it is convenient and generic because the simulation mechanism is used to generate simulated summary statistics as a first step, and it is not used afterwards. For instance, the software *ms* is used to generate DNA sequences or genotypes when performing ABC inference in population genetics [6]. Statistical routines, which account for mismatches, are completely separated from the simulation mechanism and are used in a second step. Regression adjustment can therefore be readily applied in a wide range of contexts without implementation efforts. By contrast, when considering sampling techniques, statistical operations and simulations are embedded within a single algorithm [5, 7], which may require new algorithmic development for each specific statistical problem. Second, regression approaches have been shown to produce reduced statistical errors compared to rejection algorithms in a quite diverse range of statistical problems [3, 8, 9]. Last, regression approaches are implemented in different ABC software including *DIYABC* [10] and the R *abc* package [11].

In this chapter, I introduce regression adjustment using a comprehensive framework that includes linear adjustment [3], as well as more flexible adjustments, such as non-linear models [8]. The first section presents the main concepts underlying regression adjustment. The second section presents a theorem that compares theoretical properties of posterior distributions obtained with and without regression adjustment. The third section presents a practical application of regression adjustment in ABC. It shows that regression adjustment shrinks posterior distributions when compared to a standard rejection approach. The fourth section presents recent regression approaches for ABC that are not based on regression adjustment.

3.2 Principle of Regression Adjustment

3.2.1 Partial posterior distribution

Bayesian inference is based on the posterior distribution defined as:

$$\pi(\theta|y_{\text{obs}}) \propto p(y_{\text{obs}}|\theta)\pi(\theta), \tag{3.1}$$

where $\theta \in \mathbb{R}^p$ is the vector of parameters, and y_{obs} are the data. Up to a renormalising constant, the posterior distribution depends on the prior $\pi(\theta)$ and on the likelihood function $p(y_{\text{obs}}|\theta)$. In the context of ABC, inference is no longer based on the posterior distribution $\pi(\theta|y_{\text{obs}})$, but on the *partial* posterior distribution $\pi(\theta|s_{\text{obs}})$, where s_{obs} is a q-dimensional vector of descriptive statistics. The partial posterior distribution is defined as follows:

$$\pi(\theta|s_{\text{obs}}) \propto p(s_{\text{obs}}|\theta)\pi(\theta). \tag{3.2}$$

Obviously, the partial posterior is equal to the posterior if the descriptive statistics s_{obs} are sufficient for the parameter θ.

3.2.2 Rejection algorithm followed by adjustment

To simulate a sample from the partial posterior $p(\theta|s_{\text{obs}})$, the rejection algorithm followed by adjustment works as follows:

---

1. Simulate n values $\theta^{(i)}$, $i = 1, \ldots, n$, according to the prior distribution π.
2. Simulate descriptive statistics $s^{(i)}$ using the generative model $p(s^{(i)}|\theta^{(i)})$.
3. Associate with each pair $(\theta^{(i)}, s^{(i)})$ a weight $w^{(i)} \propto K_h(\|s^{(i)} - s_{\text{obs}}\|)$, where $\| \cdot - \cdot \|$ is a distance function, $h > 0$ is the bandwidth parameter, and K is an univariate statistical kernel with $K_h(\| \cdot \|) = K(\| \cdot \|/h)$.
4. Fit a regression model where the response is θ and the predictive variables are the summary statistics s [equations (3.3) or (3.5)]. Use a regression model to adjust the $\theta^{(i)}$ in order to produce a weighted sample of adjusted values. Homoscedastic adjustment [equation (3.4)] or heteroscedastic adjustment [equation (3.6)] can be used to produce a weighted sample $(\theta_c^{(i)}, w^{(i)})$, $i = 1, \ldots, n$, which approximates the posterior distribution.

To run the rejection algorithm followed by adjustment, there are several choices to make. The first choice concerns the kernel K. Usual choices for K encompass uniform kernels that give a weight of 1 to all accepted simulations and zero otherwise [12] or the Epanechnikov kernel for a smoother version of the rejection algorithm [3]. However, as for traditional density estimation, the choice of statistical kernel has a weak impact on estimated distribution [13]. The second choice concerns the threshold parameter h. For kernels with a finite support, the threshold h corresponds to (half) the window size within which simulations are accepted. For the theorem presented in Section 3.4, I assume that h is chosen without taking into account the simulations $s^{(1)}, \ldots, s^{(n)}$. This technical assumption does not hold in practice, where we generally choose to accept a given percentage p, typically 1% or 0.1%, of the simulations. This practice amounts at setting h to the first p-quantile of the distances $\|s^{(i)} - s_{\text{obs}}\|$. A theorem where the threshold depends on simulations has been provided [14]. Choice of threshold h corresponds to bias-variance tradeoff. When choosing small values of h, the number of accepted simulations is small and estimators might have a large variance. By contrast, when choosing large values of h, the number of accepted simulations is large and estimators might be biased [1].

3.2.3 Regression adjustment

The principle of regression adjustment is to adjust simulated parameters $\theta^{(i)}$ with non-zero weights $w^{(i)} > 0$ in order to account for the difference between the simulated statistics $s^{(i)}$ and the observed one s_{obs}. To adjust parameter values, a regression model is fitted in the neighbourhood of s_{obs}:

$$\theta^{(i)} = m(s^{(i)}) + \varepsilon, \quad i = 1, \cdots, n, \tag{3.3}$$

where $m(s)$ is the conditional expectation of θ given s, and ε is the residual. The regression model of equation (3.3) assumes *homoscedasticity*, for example, it assumes that the variance of the residuals does not depend on s. To produce samples from the partial posterior distribution, the $\theta^{(i)}$'s are adjusted as follows:

$$\begin{aligned} \theta_c^{(i)} &= \hat{m}(s_{\text{obs}}) + \hat{\varepsilon}^{(i)} \\ &= \hat{m}(s_{\text{obs}}) + (\theta^{(i)} - \hat{m}(s^{(i)})), \end{aligned} \tag{3.4}$$

where $\hat{m}$ represents an estimator of the conditional expectation of θ given s, and $\hat{\varepsilon}^{(i)}$ is the ith empirical residual. In its original formulation, regression adjustment assumes that m is a linear function [3], and it was later extended to non-linear adjustments [8]. Other regression adjustment techniques have also been proposed when parameters and summary statistics are real-valued functions [15].

The homoscedastic assumption of equation (3.3) may not be always valid. When the number of simulations is not very large because of computational constraints, local approximations, such as the homoscedastic assumption, are no longer valid because the neighborhood corresponding to simulations for which $w^{(i)} \neq 0$ is too large. Regression adjustment can account for heteroscedasticity that occurs when the variance of the residuals depend on the summary statistics. When accounting for heteroscedasticity, the regression equation can be written as follows [8]:

$$\theta^{(i)} = m(s^{(i)}) + \sigma(s^{(i)})\zeta, \quad i = 1, \cdots, n, \tag{3.5}$$

where $\sigma(\mathbf{s})$ is the square root of the conditional variance of θ given s, and ζ is the residual. Heteroscedastic adjustment involves an additional scaling step in addition to homoscedastic adjustment (3.4) (Figure 3.1):

$$\begin{aligned} \theta_{c'}^{(i)} &= \hat{m}(s_{\text{obs}}) + \hat{\sigma}(s_{\text{obs}})\hat{\zeta}^{(i)} \\ &= \hat{m}(s_{\text{obs}}) + \frac{\hat{\sigma}(s_{\text{obs}})}{\hat{\sigma}(s^{(i)})}(\theta^{(i)} - \hat{m}(s^{(i)})), \end{aligned} \tag{3.6}$$

where $\hat{m}$ and $\hat{\sigma}$ are estimators of the conditional mean and of the conditional standard deviation.

3.2.4 Fitting regression models

Equations (3.4) and (3.6) of regression adjustment depend on the estimator of the conditional mean $\hat{m}$ and possibly of the conditional variance $\hat{\sigma}$. Model

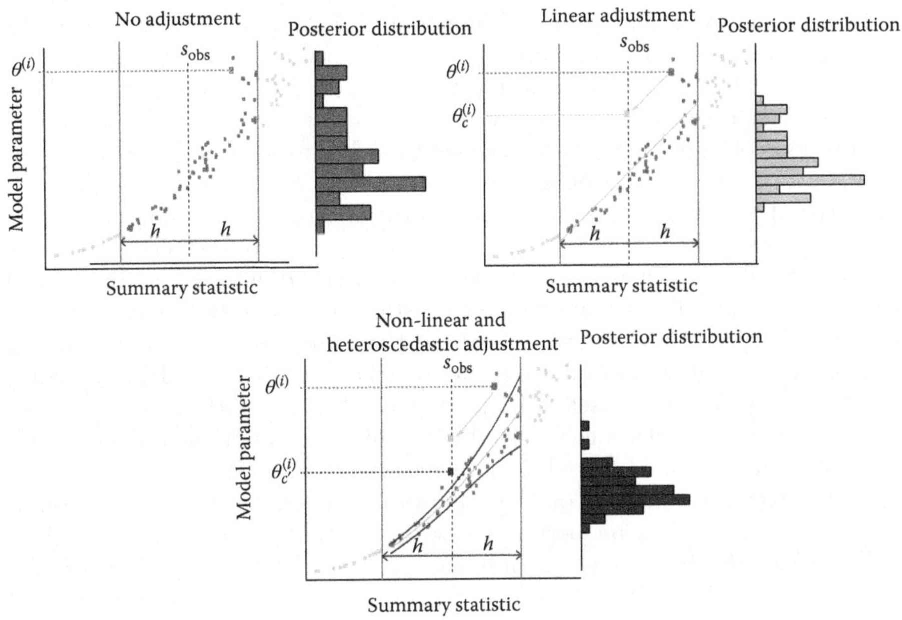

FIGURE 3.1
Posterior variances of the histograms obtained after adjustment are reduced compared to the posterior variance obtained without adjustment.

fitting is performed using weighted least squares. The conditional mean is learned by minimising the following weighted least square criterion:

$$E(m) = \sum_{i=1}^{n} (\theta^{(i)} - m(s^{(i)}))^2 w^{(i)}. \tag{3.7}$$

For linear adjustment, we assume that $m(s) = \alpha + \beta s$ [3]. The parameters α and β are inferred by minimising the weighted least square criterion given in equation (3.7).

For heteroscedastic adjustment [equation (3.4)], the conditional variance should also be inferred. The conditional variance is learned after minimisation of a least square criterion. It is obtained by fitting a regression model where the answer is the logarithm of the squared residuals. The weighted least squares criterion is given as follows:

$$E(\log \sigma^2) = \sum_{i=1}^{n} \left(\log((\hat{\varepsilon}^{(i)})^2) - \log \sigma^2(s) \right)^2 w^{(i)}.$$

Neural networks have been proposed to estimate m and σ^2 [8]. This choice was motivated by the possibility offered by neural networks to reduce the dimension of descriptive statistics via an internal projection on a space of lower dimension [16].

In general, the assumptions of homoscedasticity and linearity [equation (3.3)] are violated when the percentage of accepted simulation is large. By contrast, heteroscedastic and non-linear regression models [equation (3.5)] are more flexible. Because of this additional flexibility, the estimated posterior distributions obtained after heteroscedastic and non-linear adjustment is less sensitive to the percentage of accepted simulations [8]. In a coalescent model, where the objective was to estimate the mutation rate, heteroscedastic adjustment with neural networks was found to be less sensitive to the percentage of accepted simulations than linear and homoscedastic adjustment [8]. In a model of phylodynamics, it was found again that statistical error obtained with neural networks decreases at first – because the regression method requires a large enough training dataset – and then reaches a plateau [9]. However for a larger phylodynamics dataset, statistical error obtained with neural networks increases for higher tolerance values. Poor regularisation or the limited size of neural networks were advanced as putative explanations [9].

In principle, estimation of the conditional mean m and of the conditional variance σ^2 can be performed with different regression approaches. For instance, the `R` *abc* package implements different regression models for regression adjustment including linear regression, ridge regression, and neural networks [11]. Lasso regression is another regression approach that can be considered. Regression adjustment based on lasso (least absolute shrinkage and selection operator) was shown to provide smaller errors than neural network in a phylodynamic model [9]. An advantage of lasso, ridge regression and neural networks compared to standard multiple regression is that they account for the large dimension of the summary statistics using different regularisation techniques. Instead of considering regularised regression, there is an alternative where the initial summary statistics are replaced by a reduced set of summary statistics or a combination of the initial summary statistics [17,18]. The key and practical advantage of regression approaches with regularisation is that they implicitly account for the large number of summary statistics and the additional step of variable selection can be avoided.

3.2.5 Parameter transformations

When the parameters are bounded or positive, parameters can be transformed before regression adjustment. Transformations guarantee that the adjusted parameter values lie in the range of the prior distribution [3]. An additional advantage of the *log* and *logit* transformations is that they stabilise the variance of the regression model and make regression model (3.3) more homoscedastic [1].

Positive parameters are regressed on a logarithm scale $\phi = \log(\theta)$,

$$\phi = m(\mathbf{s}) + \varepsilon.$$

Parameters are then adjusted on the logarithm scale:

$$\phi_c^{(i)} = \hat{m}(s_{\text{obs}}) + (\phi^{(i)} - \hat{m}(s^{(i)})).$$

The final adjusted values are obtained by exponentiation of the adjusted parameter values:

$$\theta_c^{(i)} = \exp(\phi_c^{(i)}).$$

Instead of using a logarithm transformation, bounded parameters are adjusted using a logit transformation. Heteroscedastic adjustment can also be performed after log or logit transformations.

3.2.6 Shrinkage

An important property of regression adjustment concerns posterior shrinkage. When considering linear regression, the empirical variance of the residuals is smaller than the total variance. In addition, residuals are centred for linear regression. These two properties imply that for linear adjustment, the empirical variance of $\theta_c^{(i)}$ is smaller than the empirical variance of the non-adjusted values $\theta^{(i)}$ obtained with the rejection algorithm. Following homoscedastic and linear adjustment, the posterior variance is consequently reduced. For non-linear adjustment, shrinkage property has also been reported, and the additional step generated by heteroscedastic adjustment does not necessarily involve additional shrinkage when comparing $\theta_c^{(i)}$ to $\theta_{c'}^{(i)}$ [1]. However, shrinkage obtained with regression adjustments can be excessive, especially for small tolerance rates, and it can affect posterior calibration. In a model of admixture, 95% credibility intervals obtained with regression adjustments were found to contain only 84% of the true values in less favourable situations [19].

3.3 Theoretical Results about Regression Adjustment

The following theoretical section is technical and can be skipped by readers not interested by mathematical results about ABC estimators based on regression adjustment. In this section, we give the main theorem that describes the statistical properties of posterior distributions obtained with or without regression adjustment. To this end, the estimators of the posterior distribution are defined as follows:

$$\hat{\pi}_j(\theta|s_{\text{obs}}) = \sum_{i=1}^{n} \tilde{K}_{h'}(\theta_j^{(i)} - \theta)w^{(i)},\ j = 0, 1, 2, \tag{3.8}$$

where $\theta_0^{(i)} = \theta^{(i)}$ (no adjustment), $\theta_j^{(i)} = \theta_c^{(i)}$ for $j = 1, 2$ (homoscedastic adjustment), $\tilde{K}$ is an univariate kernel, and $\tilde{K}_{h'}(\cdot) = \tilde{K}(\cdot)/h'$. Linear adjustment corresponds to $j = 1$ and quadratic adjustment corresponds to $j = 2$. In nonparametric statistics, estimators of the conditional density with adjustment have already been proposed [20, 21].

To present the main theorem, we introduce the following notations: if X_n is a sequence of random variables and a_n is a deterministic sequence, the notation $X_n = o_P(a_n)$ means that X_n/a_n converges to zero in probability, and $X_n = O_P(a_n)$ means that the ratio X_n/a_n is bounded in probability when n goes to infinity. The technical assumptions of the theorem are given in the appendix of [1].

Theorem 3.1. *We assume that conditions (A1)–(A5) given in the appendix of [1] hold. The bias and variance of the estimators* $\hat{\pi}_j(\theta|s_{\text{obs}})$, $j = 0, 1, 2$ *are given by:*

$$E[\hat{\pi}_j(\theta|s_{\text{obs}}) - \pi(\theta|s_{\text{obs}})] = C_1 h'^2 + C_{2,j} h^2 + O_P((h^2 + h'^2)^2) + O_P(\frac{1}{nh^q}), \quad (3.9)$$

$$\text{Var}[\hat{\pi}_j(\theta|s_{\text{obs}})] = \frac{C_3}{nh^q h'}(1 + o_P(1)), \quad (3.10)$$

where q *is the dimension of the vector of summary statistics, and the constants* C_1, $C_{2,j}$, *and* C_3 *are given in [1].*

Proof: See [1].

There are other theorems that provide asymptotic biases and variances of ABC estimators but they do not study the properties of estimators arising after regression adjustment. Considering posterior expectation (e.g. posterior moments) instead of the posterior density, Barber et al. [22] provides asymptotic bias and variance of an estimator obtained with rejection algorithm. Biau et al. [14] studied asymptotic properties when the window size h depends on the data instead of being fixed in advance.

Remark 1. Curse of dimensionality The mean square error of the estimators is the sum of the squared bias and of the variance. With elementary algebra, we can show that for the three estimators $\hat{\pi}_j(\theta|s_{\text{obs}})$, $j = 0, 1, 2$, the mean square error is of the order of $n^{-1/(q+5)}$ for an optimal choice of h. The speed with which the error approaches 0 therefore decreases drastically when the dimension of the descriptive statistics increases. This theorem highlights (in an admittedly complicated manner) the importance of reducing the dimension of the statistics. However, the findings from these asymptotic theorems, which are classic in non-parametric statistics, are often much more pessimistic than the results observed in practice. It is especially true because asymptotic theorems in the vein of theorem 3.1 do not take into account correlations between summary statistics [23].

Remark 2. Comparing biases of estimators with and without adjustment It is not possible to compare biases (i.e. the constant $C_{2,j}$, $j = 0, 1, 2$) for any statistical model. However, if we assume that the residual distribution of ε in equation (3.3) does not depend on s, then the constant $C_{2,2}$ is 0. When assuming homoscedasticity, the estimator that achieves asymptotically the smallest mean square error is the estimator with quadratic adjustment $\hat{p}_2(\theta|s_{\text{obs}})$. Assuming additionally that the conditional expectation m is linear in s, then both $\hat{p}_1(\theta|s_{\text{obs}})$ and $\hat{p}_2(\theta|s_{\text{obs}})$ have a mean square error lower than the error obtained without adjustment.

3.4 Application of Regression Adjustment to Estimate Admixture Proportions Using Polymorphism Data

To illustrate regression adjustment, I consider an example of parameter inference in population genetics. Description of coalescent modeling in population genetics is out of the scope of this chapter, and we refer interested readers to dedicated reviews [24, 25]. This example illustrates that ABC can be used to infer evolutionary events, such as admixture between sister species. I assume that two populations (A and B) diverged in the past and admixed with admixture proportions p and $1-p$ to form a new hybrid species C that subsequently splits to form two sister species C_1 and C_2 (Figure 3.2). Simulations are performed using the software DIYABC (Do It Yourself Approximate Bayesian Computation) [10]. The model of Figure 3.2 corresponds to a model of divergence and admixture between species of a complex of species from the butterfly gender *Coenonympha*. We assume that 2 populations of the Darwin's Heath (*Coenonympha darwiniana*) originated through hybridisation between the Pearly Heath (*Coenonympha arcania*) and the Alpine Heath (*Coenonympha gardetta*) [26]. A total of 16 summary statistics based on Single Nucleotide Polymorphisms (SNPs) are used for parameter inference [26]. A total of 10^6 simulations are performed and the percentage of accepted simulations is of 0.5%.

I consider four different forms of regression adjustment: linear and homoscedastic adjustment, non-linear (neural networks) and homoscedastic adjustment, linear and heteroscedastic adjustment, and non-linear and

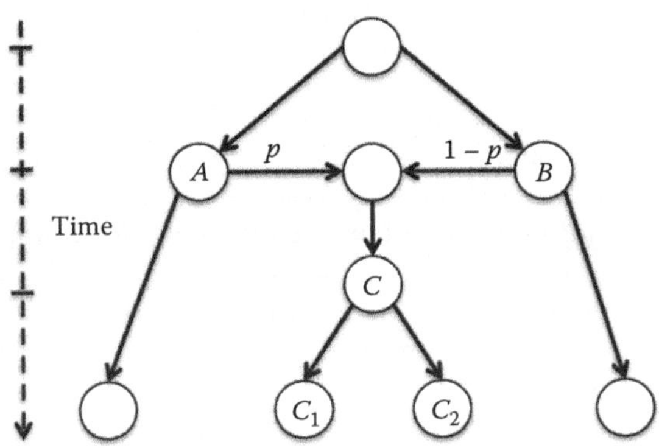

FIGURE 3.2
Graphical description of the model of admixture between sister species. Two populations (A and B) diverge in the past and admixed with admixture proportions p and $1-p$ to form a new hybrid species C that subsequently diverges to form two sister species (C_1 and C_2).

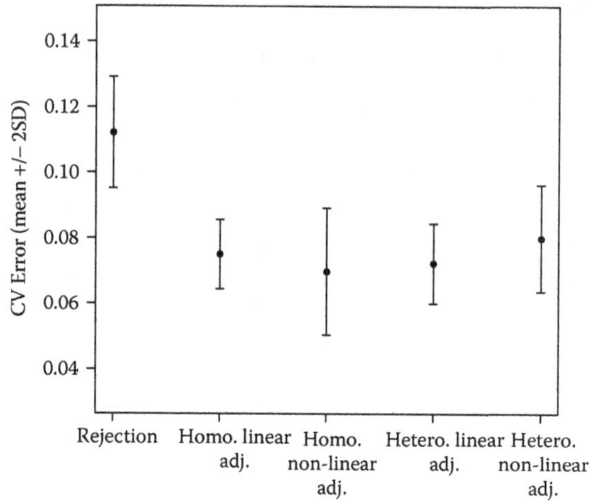

FIGURE 3.3
Errors obtained when estimating the admixture proportion p with different ABC estimators. The errors are obtained with cross-validation and error bars (two standard deviations) are estimated with bootstrap. *adj.* stands for adjustment.

heteroscedastic adjustment. All adjustments were performed with the R package *abc* [11, 27]. I evaluate parameter inference using a cross-validation criterion [11]. The cross-validation error decreases when considering linear adjustment (Figure 3.3). However, considering heteroscedastic instead of homoscedastic adjustment does not provide an additional decrease of the cross-validation error (Figure 3.3).

Then, using real data from a butterfly species complex, we compare the posterior distribution of the admixture proportion p obtained without adjustment, with linear and homoscedastic adjustment, and with non-linear and homoscedastic adjustment (Figure 3.4). For this example, considering regression adjustment considerably changes the shape of the posterior distribution. The posterior mean without adjustment is of $p = 0.51$ [95% $C.I. = (0.12, 0.88)$]. By contrast, when considering linear and homoscedastic adjustment, the posterior mean is of 0.93 [95% $C.I. = (0.86, 0.98)$]. When considering non-linear and homoscedastic adjustment, the posterior mean is 0.84 [95% $C.I. = (0.69, 0.93)$]. Regression adjustment confirms a larger contribution of *C. arcania* to the genetic composition of the ancestral *C. darwiniana* population [26]. This example shows that regression adjustment not only shrinks credibility intervals, but can also shift posterior estimates. Compared to rejection, the posterior shift observed with regression adjustments provides a result that is more consistent with published results [26].

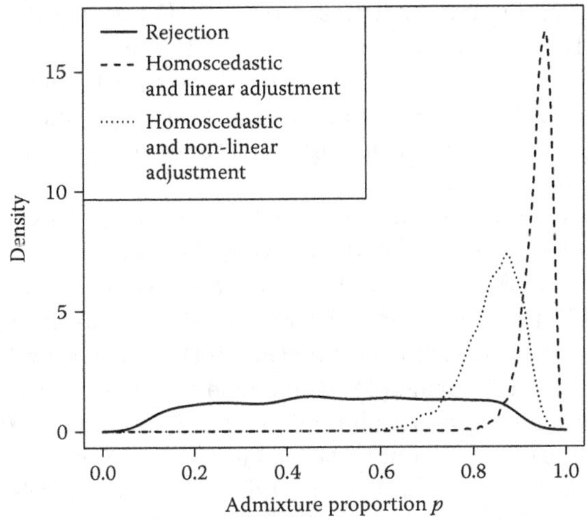

FIGURE 3.4
Posterior distribution of the admixture coefficient p obtained using real data from a butterfly species complex to compute observed summary statistics. This example shows that regression adjustment not only shrinks credibility intervals, but can also shift posterior estimates. The admixture coefficient p measures the relative contribution of *Coenonympha arcania* to the ancestral *C. darwiniana* population.

3.5 Regression Methods Besides Regression Adjustment

There are other regression methods besides regression adjustment that have been proposed to estimate $E[\theta|s_{\text{obs}}]$ and $\pi(\theta|s_{\text{obs}})$ using the simulations as a training set. A first set of methods consider kernel methods to perform regression [28]. The principle is to define a kernel function K to compare observed statistics to simulated summary statistics. Because of the so-called kernel trick, regression with kernel methods amounts at regressing θ with $\Phi(s)$, where $< \Phi(s), \Phi(s') >= K(s, s')$ for two vector of summary statistics s and s'. Then, an estimate of the posterior mean is obtained as follows:

$$E[\theta|s_{\text{obs}}] = \sum_{i=1}^{n} w_i \theta^{(i)}, \tag{3.11}$$

where w_i depends on the inverse the Gram matrix containing the values $K(s^{(i)}, s^{(j)})$ for $i, j = 1 \ldots, n$. A formula to estimate posterior density can also be obtained in the same lines as formula (3.11). Simulations suggest that kernel ABC gives better performance than regression adjustment when high-dimensional summary statistics are used. For a given statistical error, it was reported that fewer simulations should be performed when using kernel ABC instead of regression adjustment [28]. Other kernel approaches have been proposed for ABC where simulated and observed samples or summary statistics are directly compared through a distance measure between empirical probability distributions [29, 30].

Another regression method in ABC that does not use regression adjustment considers quantile regression forest [19]. Generally used to estimate conditional mean, random forests also provide information about the full conditional distribution of the response variable [31]. By inverting the estimated conditional cumulative distribution function of the response variable, quantiles can be inferred [31]. The principle of quantile regression forest is to use random forests in order to give a weight w_i to each simulation $(\theta^{(i)}, s^{(i)})$. These weights are then used to estimate the conditional cumulative posterior distribution function $F(\theta|s_{\text{obs}})$ and to provide posterior quantiles by inversion. An advantage of quantile regression forest is that tolerance rate should not be specified and standard parameters of random forest can be considered instead. A simulation study of coalescent models shows that regression adjustment can shrink posterior excessively by contrast to quantile regression forest [19].

3.6 Conclusion

This chapter introduces regression adjustment for approximate Bayesian computation [3, 8]. We explain why regression adjustment shrinks posterior distribution, which is a desirable feature because credibility intervals obtained with rejection methods can be too wide [1]. When inferring admixture with single nucleotide polymorphisms data in a complex of butterfly species, the posterior distribution obtained with regression adjustment was not only shrunk when compared to standard rejection, but also shifted to larger values, which confirm results obtained for this species complex with other statistical approaches [26]. We have introduced different variants of regression adjustment, and it might be difficult for ABC users to choose which adjustment is appropriate in their context. We argue that there is no best strategy in general. In the admixture example, we found, based on a cross-validation error criterion, that homoscedastic linear adjustment provides considerable improvement compared to rejection. More advanced adjustments provide almost negligible improvements if no improvement at all. However, in a model of phylodynamics, non-linear adjustment was reported to achieve considerable improvement compared to linear adjustment [9]. In practical applications of ABC, we suggest to compute errors, such as cross-validation estimation errors to choose a particular method for regression adjustment.

With the rapid development of complex machine learning approaches, we envision that regression approaches for approximate Bayesian computation can be further improved to provide more reliable inference for complex models in biology and ecology.

References

[1] M G B Blum. Approximate Bayesian computation: A nonparametric perspective. *Journal of the American Statistical Association*, 105(491):1178–1187, 2010.

[2] K Csilléry, M G B Blum, O E Gaggiotti, and O François. Approximate Bayesian computation (ABC) in practice. *Trends in Ecology & Evolution*, 25(7):410–418, 2010.

[3] M A Beaumont, W Zhang, and D J Balding. Approximate Bayesian computation in population genetics. *Genetics*, 162(4):2025–2035, 2002.

[4] P Marjoram, J Molitor, V Plagnol, and S Tavaré. Markov chain Monte Carlo without likelihoods. *Proceedings of the National Academy of Sciences*, 100(26):15324–15328, 2003.

[5] S A Sisson, Y Fan, and M M Tanaka. Sequential Monte Carlo without likelihoods. *Proceedings of the National Academy of Sciences*, 104(6): 1760–1765, 2007.

[6] R R Hudson. Generating samples under a wright–fisher neutral model of genetic variation. *Bioinformatics*, 18(2):337–338, 2002.

[7] M A Beaumont, J-M Cornuet, J-M Marin, and C P Robert. Adaptive approximate Bayesian computation. *Biometrika*, 96(4):983–990, 2009.

[8] M G B Blum and O François. Non-linear regression models for Approximate Bayesian computation. *Statistics and Computing*, 20:63–73, 2010.

[9] E Saulnier, O Gascuel, and S Alizon. Inferring epidemiological parameters from phylogenies using regression-ABC: A comparative study. *PLoS Computational Biology*, 13(3):e1005416, 2017.

[10] J-M Cornuet, P Pudlo, J Veyssier, A Dehne-Garcia, M Gautier, R Leblois, J-M Marin, and A Estoup. DIYABC v2. 0: A software to make approximate Bayesian computation inferences about population history using single nucleotide polymorphism, DNA sequence and microsatellite data. *Bioinformatics*, 30(8):1187–1189, 2014.

[11] K Csilléry, O François, and M G B Blum. ABC: An R package for approximate Bayesian computation (ABC). *Methods in Ecology and Evolution*, 3(3):475–479, 2012.

[12] J K Pritchard, M T Seielstad, A Perez-Lezaun, and M W Feldman. Population growth of human Y chromosomes: A study of Y chromosome microsatellites chromosome microsatellites. *Molecular Biology and Evolution*, 16(12):1791–1798, 1999.

[13] B W Silverman. *Density Estimation for Statistics and Data Analysis*, Volume 26. CRC Press, London, UK, 1986.

[14] G Biau, F Cérou, A Guyader, et al. New insights into approximate Bayesian computation. In *Annales de l'Institut Henri Poincaré, Probabilités et Statistiques*, volume 51, pp. 376–403. Institut Henri Poincaré, Paris, 2015.

[15] G S Rodrigues, D J Nott, and S A Sisson. Functional regression approximate Bayesian computation for Gaussian process density estimation. *Computational Statistics & Data Analysis*, 103:229–241, 2016.

[16] B D Ripley. Neural networks and related methods for classification. *Journal of the Royal Statistical Society. Series B (Methodological)*, volume 56, pp. 409–456, 1994.

[17] P Fearnhead and D Prangle. Constructing summary statistics for approximate Bayesian computation: Semi-automatic approximate Bayesian computation. *Journal of the Royal Statistical Society: Series B (Statistical Methodology)*, 74(3):419–474, 2012.

[18] M G B Blum, D Nunes, D Prangle, S A Sisson, et al. A comparative review of dimension reduction methods in approximate Bayesian computation. *Statistical Science*, 28(2):189–208, 2013.

[19] L Raynal, J-M Marin, P Pudlo, M Ribatet, C P Robert, and A Estoup. ABC random forests for Bayesian parameter inference. *arXiv preprint arXiv:1605.05537*, 2016.

[20] R J Hyndman, D M Bashtannyk, and G K Grunwald. Estimating and visualizing conditional densities. *Journal of Computing and Graphical Statistics*, 5:315–336, 1996.

[21] B E Hansen. Nonparametric conditional density estimation. Working paper available at http://www.ssc.wisc.edu/ bhansen/papers/ncde.pdf, 2004.

[22] S Barber, J Voss, and M Webster. The rate of convergence for approximate Bayesian computation. *Electronic Journal of Statistics Electronic Journal of Statistics Electronic Journal of Statistics*, 9:80–105, 2015.

[23] D W Scott. *Multivariate Density Estimation: Theory, Practice, and Visualization.* John Wiley & Sons, New York, 2009.

[24] R R Hudson. Gene genealogies and the coalescent process. *Oxford Surveys in Evolutionary Biology*, 7(1):44, 1990.

[25] N A Rosenberg and M Nordborg. Genealogical trees, coalescent theory and the analysis of genetic polymorphisms. *Nature Reviews Genetics*, 3(5):380–390, 2002.

[26] T Capblancq, L Després, D Rioux, and J Mavárez. Hybridization promotes speciation in coenonympha butterflies. *Molecular Ecology*, 24(24):6209–6222, 2015.

[27] R Core Team. *R: A Language and Environment for Statistical Computing.* R Foundation for Statistical Computing, Vienna, Austria, 2016.

[28] S Nakagome, K Fukumizu, and S Mano. Kernel approximate Bayesian computation in population genetic inferences. *Statistical Applications in Genetics and Molecular Biology*, 12(6):667–678, 2013.

[29] J Mitrovic, D Sejdinovic, and Y Teh. DR-ABC: Approximate Bayesian computation with kernel-based distribution regression. In *Proceedings of ICML 2016,* volume 48, pp. 1482–1491. JMLR W&CP, New York, 2016.

[30] M Park, W Jitkrittum, and D Sejdinovic. K2-ABC: Approximate Bayesian computation with infinite dimensional summary statistics via kernel embeddings. *Proceedings of AISTATS,* pp. 398–407, 2016.

[31] N Meinshausen. Quantile regression forests. *Journal of Machine Learning Research*, 7(Jun):983–999, 2006.

4

ABC Samplers

S. A. Sisson and Y. Fan

CONTENTS

4.1 Introduction 88
4.2 Rejection and Importance Sampling 89
  4.2.1 Rejection sampling 89
  4.2.2 Importance sampling 91
  4.2.3 Importance/rejection sampler variants 95
    4.2.3.1 Rejection control importance sampling 95
    4.2.3.2 ABC importance sampling 96
    4.2.3.3 ABC rejection sampling with stopping rule 99
    4.2.3.4 Rejection-based ABC algorithms for expensive simulators 99
    4.2.3.5 Marginal ABC samplers 100
4.3 Markov Chain Monte Carlo Methods 102
  4.3.1 ABC-Markov chain Monte Carlo samplers 102
  4.3.2 Augmented space ABC-Markov chain Monte Carlo samplers 105
  4.3.3 Other ABC-Markov chain Monte Carlo samplers 107
4.4 Sequential Monte Carlo Sampling 109
  4.4.1 Sequential importance sampling 110
  4.4.2 Sequential Monte Carlo samplers 113
4.5 Discussion 116
Acknowledgements 117
References 117

4.1 Introduction

Approximate Bayesian computation (ABC) is a phrase that describes a collection of methods and algorithms designed to perform a Bayesian analysis using an approximation to the true posterior distribution, when the likelihood function implied by the data generating process is computationally intractable. For observed data $y_{obs} \in \mathcal{Y}$, the likelihood function $p(y|\theta)$ depends on a vector of model parameters $\theta \in \Theta$, from which prior beliefs $\pi(\theta)$ may be updated into posterior beliefs $\pi(\theta|y_{obs}) \propto p(y_{obs}|\theta)\pi(\theta)$ via Bayes' theorem. In the standard ABC framework (see e.g. Sisson et al. 2018, Chapter 1, this volume), the ABC approximation to $\pi(\theta|y_{obs})$ is given by:

$$\pi_{ABC}(\theta|s_{obs}) \propto \int K_h(\|s - s_{obs}\|)p(s|\theta)\pi(\theta)ds, \tag{4.1}$$

where $K_h(u) = K(u/h)/h$ is a standard kernel density function with scale parameter $h > 0$, $\|\cdot\|$ is an appropriate distance metric (e.g. Euclidean or Mahalanobis distance), $p(s|\theta)$ is the (intractable) likelihood function of the low-dimensional vector of summary statistics $s = S(y)$ implied by $p(y|\theta)$, and $s_{obs} = S(y_{obs})$. Defining $K_h(\|s - s_{obs}\|) \to \delta_{s_{obs}}(s)$ as $h \to 0$, where $\delta_Z(z)$ denotes the Dirac measure, defined as $\delta_Z(z) = 1$ if $z \in Z$ and $\delta_Z(z) = 0$ otherwise, then as a result:

$$\lim_{h\to 0} \pi_{ABC}(\theta|s_{obs}) \propto \int \delta_{s_{obs}}(s)p(s|\theta)\pi(\theta)ds = p(s_{obs}|\theta)\pi(\theta) \propto \pi(\theta|s_{obs}).$$

Accordingly, $\pi_{ABC}(\theta|s_{obs})$ provides an approximation to the partial posterior $\pi(\theta|s_{obs})$, which becomes more accurate as h gets small. If the summary statistics s are sufficient for θ, then $\pi(\theta|s_{obs})$ will equal $\pi(\theta|y_{obs})$, and so, for small h, the ABC posterior approximation $\pi_{ABC}(\theta|s_{obs})$ will be a good approximation of the true posterior. If either s is not sufficient, or h is not small, then the ABC posterior approximation $\pi_{ABC}(\theta|s_{obs})$ will be of the form (4.1).

In terms of drawing samples from the approximate posterior $\pi_{ABC}(\theta|s_{obs})$, the choice of summary statistics $s = S(y)$ is typically considered known, and interest is then in sampling from $\pi_{ABC}(\theta|s_{obs})$, for a specific and low value of h, as efficiently as possible. The more efficient the simulation procedure, the further h can be lowered within the sampling framework, resulting in samples from a more accurate approximation of $\pi(\theta|s_{obs})$.

In this chapter, we survey the various forms of ABC algorithms that have been developed to sample from $\pi_{ABC}(\theta|s_{obs})$. These have broadly followed the familiar Monte Carlo classes of algorithms, including rejection and importance sampling, Markov chain Monte Carlo (MCMC), and sequential Monte Carlo (SMC)-based algorithms. While each of these classes have their ABC-specific implementations and characteristics, in general, they target the joint distribution of parameter vector θ and summary statistic s given by:

$$\pi_{ABC}(\theta, s|s_{obs}) \propto K_h(\|s - s_{obs}\|)p(s|\theta)\pi(\theta). \tag{4.2}$$

By noting that (4.1) is obtained from (4.2) by integrating over s (i.e. $\pi_{ABC}(\theta|s_{obs}) = \int \pi_{ABC}(\theta, s|s_{obs})ds$), samples from $\pi_{ABC}(\theta|s_{obs})$ can be obtained by first drawing samples from (4.2) and then discarding the marginal s values.

An alternative, but related Monte Carlo approach is based on sampling from (4.1) directly, by obtaining an unbiased and non-negative estimate of the ABC posterior distribution function, such as:

$$\hat{\pi}_{ABC}(\theta|s_{obs}) \propto \frac{\pi(\theta)}{T}\sum_{t=1}^{T} K_h(\|s(t) - s_{obs}\|),$$

where $s(1), \ldots, s(T) \sim p(s|\theta)$ are samples from the intractable model given θ, and then using this estimate in place of $\pi_{ABC}(\theta|s_{obs})$ within a standard Monte Carlo algorithm (e.g. Del Moral et al. 2012). This approach falls within the family of *pseudo-marginal* Monte Carlo methods (Beaumont 2003; Andrieu and Roberts 2009), a more general class of likelihood-free samplers that has gained popularity outside of the ABC setting. For a detailed and in-depth discussion of the connections between ABC-MCMC algorithms and pseudo-marginal MCMC methods see Andrieu et al. (2019) (Chapter 9, this volume).

4.2 Rejection and Importance Sampling

4.2.1 Rejection sampling

The earliest ABC samplers (e.g. Tavaré et al. 1997; Pritchard et al. 1999) were basic rejection sampling algorithms. Under the standard rejection sampling framework (e.g. Ripley 1987; Liu 2001), interest is in obtaining samples from some target distribution $f(\theta) = Z^{-1}\tilde{f}(\theta)$, which is known up to a normalising constant $Z = \int_{-\infty}^{\infty} \tilde{f}(\theta)d\theta$. The standard rejection sampling algorithm obtains draws $\theta' \sim g(\theta)$ from a sampling density $g(\theta)$ from which it is trivial to sample, such that:

$$\tilde{f}(\theta) \leq Mg(\theta),$$

for all θ and some positive constant $M > 0$. The draws θ' are then accepted as independent samples from the target density $f(\theta)$ with probability $\frac{\tilde{f}(\theta')}{Mg(\theta')}$. To see that the above procedure is correct, for simplicity, consider the case in which θ is univariate, with the extension to multivariate θ being straightforward. Define A as the event that a sample θ' from $g(\theta)$ is accepted. Then, the overall acceptance rate of the algorithm is:

$$\Pr(A) = \int \frac{\tilde{f}(\theta)}{Mg(\theta)}g(\theta)d\theta = \frac{1}{M}\int \tilde{f}(\theta)d\theta = \frac{Z}{M},$$

and hence the distribution of accepted draws is:

$$\Pr(\theta' \leq \theta | A) = \frac{\Pr(\theta' \leq \theta, A)}{\Pr(A)} = \frac{\int_{-\infty}^{\theta} \frac{\tilde{f}(\theta')}{Mg(\theta')} g(\theta') d\theta'}{P(A)} = F(\theta),$$

as required, where $F(\theta) = \int_{-\infty}^{\theta} f(z)dz$ is the distribution function associated with $f(\theta)$. The efficiency of the algorithm is associated with the value of M, with smaller values of M (subject to $\tilde{f}(\theta) \leq Mg(\theta)$, $\forall\theta$) corresponding to more efficient samplers. That is, for fixed $g(\theta)$, the optimum choice is $M = \max_\theta \frac{\tilde{f}(\theta)}{g(\theta)}$. Good choice of the sampling distribution $g(\theta)$, for example to approximate $f(\theta)$, can result in smaller values of M.

The ABC version of the rejection sampler was discussed in Sisson et al. (2018) (Chapter 1, this volume), which we reproduce here as Algorithm 4.1.

Algorithm 4.1: ABC Rejection Sampling

Inputs:

- A target posterior density $\pi(\theta|y_{obs}) \propto p(y_{obs}|\theta)\pi(\theta)$ consisting of a prior distribution $\pi(\theta)$ and a procedure for generating data under the model $p(y_{obs}|\theta)$.
- A proposal density $g(\theta)$, with $g(\theta) > 0$ if $\pi(\theta|y_{obs}) > 0$.
- An integer $N > 0$.
- A kernel function $K_h(u)$ and scale parameter $h > 0$.
- A low-dimensional vector of summary statistics $s = S(y)$.

Sampling:

For $i = 1, \ldots, N$:

1. Generate $\theta^{(i)} \sim g(\theta)$ from sampling density g.
2. Generate $y^{(i)} \sim p(y|\theta^{(i)})$ from the model.
3. Compute summary statistic $s^{(i)} = S(y^{(i)})$.
4. Accept $\theta^{(i)}$ with probability $\frac{K_h(\|s^{(i)} - s_{obs}\|)\pi(\theta^{(i)})}{Mg(\theta^{(i)})}$ where $M \geq K_h(0) \max_\theta \frac{\pi(\theta)}{g(\theta)}$.
   Else go to 1.

Output:

A set of parameter vectors $\theta^{(1)}, \ldots, \theta^{(N)} \sim \pi_{ABC}(\theta|s_{obs})$.

Originally developed by Pritchard et al. (1999) following earlier ideas by Tavaré et al. (1997), the ABC rejection sampling algorithm is typically described heuristically as follows: for the candidate parameter vector $\theta' \sim g(\theta)$, a dataset y' is generated from the (intractable) generative model and summary statistics $s' = S(y')$ computed. If the simulated and observed datasets are similar (in some manner), so that $s' \approx s_{obs}$, then θ' could credibly have generated the observed data under the given model, and so θ' is retained and forms part of the sample from the ABC posterior distribution $\pi(\theta|s_{obs})$. Conversely, if s' and s_{obs} are dissimilar, then θ' is unlikely to have generated the observed data for this model, and so θ' is discarded. The parameter vectors accepted under this approach offer support for s_{obs} under the model and so may be considered to be drawn approximately from the posterior distribution $\pi(\theta|s_{obs})$. In this manner, the evaluation of the likelihood $p(y_{obs}|\theta')$, essential to most Bayesian posterior simulation methods, is replaced by an evaluation of the proximity of summaries of a simulated dataset s' to the observed summaries s_{obs}.

More precisely, this algorithm targets $\pi_{ABC}(\theta, s|s_{obs})$ given by (4.2), the joint distribution of parameter vector and summary statistic given s_{obs}. Accordingly, the sampling distribution is also defined on this space as $g(\theta, s) = p(s|\theta)g(\theta)$, and the acceptance probability of the vector (θ, s) is then given by:

$$\frac{\pi_{ABC}(\theta, s|s_{obs})}{Mg(\theta, s)} \propto \frac{K_h(\|s - s_{obs}\|)p(s|\theta)\pi(\theta)}{Mp(s|\theta)g(\theta)} = \frac{K_h(\|s - s_{obs}\|)\pi(\theta)}{Mg(\theta)}.$$

The normalising constant M is similarly given by:

$$M \geq \max_{s,\theta} \frac{K_h(\|s - s_{obs}\|)p(s|\theta)\pi(\theta)}{p(s|\theta)g(\theta)} = K_h(0) \max_{\theta} \frac{\pi(\theta)}{g(\theta)},$$

with $\max_s K_h(\|s - s_{obs}\|) = K_h(0)$ resulting from the zero-mean, symmetry and (typically) unimodal characteristics of standard kernel density functions. Accordingly, the construction of the target and sampling distributions on the joint space (θ, s) results in the form of the acceptance probability and normalisation constant M being free of intractable likelihood terms. An example implementation of this algorithm is given in Sisson et al. (2018) (Chapter 1, this volume).

4.2.2 Importance sampling

One down side of rejection sampling is the need to determine a near optimal value for the normalising constant M in order to produce an efficient algorithm. Importance sampling is a procedure that, rather than calculating acceptance probabilities, avoids this by alternatively assigning the draw $\theta' \sim g(\theta)$ an (importance) weight $w(\theta') = f(\theta')/g(\theta')$. The weighted vector θ' is then a draw from $f(\theta)$, and desired expectations under the target distribution f are computed as weighted expectations under the importance sampling density g.

To see this, suppose that we are interested in estimating the expectation:

$$E_f[h(\theta)] = \int h(\theta)f(\theta)d\theta.$$

By defining $w(\theta) = f(\theta)/g(\theta)$ we have:

$$E_g[w(\theta)h(\theta)] = \int w(\theta)h(\theta)g(\theta)d\theta = \int h(\theta)f(\theta)d\theta = E_f[h(\theta)].$$

In this manner we can then estimate the expectation as:

$$E_f[h(\theta)] \approx \frac{1}{N}\sum_{i=1}^{N} w^{(i)}h(\theta^{(i)}),$$

where $w^{(i)} = w(\theta^{(i)})$, and where $\theta^{(i)} \sim g(\theta)$ are draws from g. In the more typical case where the target distribution is unnormalised, so that $f(\theta) = Z^{-1}\tilde{f}(\theta)$, we can work with $\tilde{f}(\theta)$ only by defining $\tilde{w}(\theta) = \tilde{f}(\theta)/g(\theta)$ and then noting that:

$$E_g[\tilde{w}(\theta)] = \int \tilde{w}(\theta)g(\theta)d\theta = \int \tilde{f}(\theta)d\theta = Z \approx \frac{1}{N}\sum_{i=1}^{N}\tilde{w}^{(i)}, \tag{4.3}$$

for $\theta^{(i)} \sim g(\theta)$, where $\tilde{w}^{(i)} = \tilde{w}(\theta^{(i)})$. As a result, the expectation $E_f[h(\theta)]$ may be approximated as:

$$\begin{aligned} E_f[h(\theta)] &= \int h(\theta)f(\theta) = \frac{1}{Z}\int h(\theta)\tilde{f}(\theta) \\ &= \frac{1}{Z}\int \tilde{w}(\theta)h(\theta)g(\theta)d\theta = \frac{E_g[\tilde{w}(\theta)h(\theta)]}{E_g[\tilde{w}(\theta)]} \\ &\approx \frac{\frac{1}{N}\sum_{i=1}^{N}\tilde{w}^{(i)}h(\theta^{(i)})}{\frac{1}{N}\sum_{i=1}^{N}\tilde{w}^{(i)}} = \sum_{i=1}^{N} W^{(i)}h(\theta^{(i)}), \end{aligned} \tag{4.4}$$

for $\theta^{(i)} \sim g(\theta)$, where $W^{(i)} = \tilde{w}^{(i)}/\sum_{j=1}^{N}\tilde{w}^{(j)}$ denotes normalised weights. This approximation is not unbiased due to the biased estimator of $1/Z$, although the bias becomes small as N becomes large.

From an ABC perspective, importance sampling works much the same as rejection sampling. The target distribution is $\pi_{ABC}(\theta, s|s_{obs})$, and the importance distribution on joint parameter value and summary statistics space is $g(\theta, s) = p(s|\theta)g(\theta)$. As a result, the (unnormalised) importance weights are computed as:

$$\frac{\pi_{ABC}(\theta, s|s_{obs})}{g(\theta, s)} \propto \frac{K_h(\|s - s_{obs}\|)p(s|\theta)\pi(\theta)}{p(s|\theta)g(\theta)} = \frac{K_h(\|s - s_{obs}\|)\pi(\theta)}{g(\theta)} := \tilde{w}(\theta),$$

which is again free of intractable likelihood terms. The full ABC importance sampling algorithm is given in Algorithm 4.2.

As with rejection sampling, the choice of the (marginal) importance distribution $g(\theta)$ is crucial to the efficiency of the algorithm. In standard importance sampling, if $g(\theta) \propto \tilde{f}(\theta)$, then $\tilde{w}(\theta) \propto 1$. In this case, there is no variation in

Algorithm 4.2: ABC Importance Sampling

Inputs:

- A target posterior density $\pi(\theta|y_{obs}) \propto p(y_{obs}|\theta)\pi(\theta)$, consisting of a prior distribution $\pi(\theta)$ and a procedure for generating data under the model $p(y_{obs}|\theta)$.
- An importance sampling density $g(\theta)$, with $g(\theta) > 0$ if $\pi(\theta|y_{obs}) > 0$.
- An integer $N > 0$.
- A kernel function $K_h(u)$ and scale parameter $h > 0$.
- A low-dimensional vector of summary statistics $s = S(y)$.

Sampling:
For $i = 1, \ldots, N$:

1. Generate $\theta^{(i)} \sim g(\theta)$ from importance sampling density g.
2. Generate $y^{(i)} \sim p(y|\theta^{(i)})$ from the model.
3. Compute summary statistic $s^{(i)} = S(y^{(i)})$.
4. Compute weight $\tilde{w}^{(i)} = K_h(\|s^{(i)} - s_{obs}\|)\pi(\theta^{(i)})/g(\theta^{(i)})$.

Output:
A set of weighted parameter vectors $(\theta^{(1)}, \tilde{w}^{(1)}), \ldots, (\theta^{(N)}, \tilde{w}^{(N)}) \sim \pi_{ABC}(\theta|s_{obs})$.

the importance weights, and each sample $\theta^{(i)}$ contributes equally when computing posterior expectations via (4.4). However, if $g(\theta)$ is different to $\tilde{f}(\theta)$, then the variability in $\tilde{w}(\theta)$ from θ means that some samples $\theta^{(i)}$ will contribute more than others in this computation. In extreme cases, Monte Carlo estimates of expectations can be highly variable when they are dominated by a small number of $\theta^{(i)}$ with relatively large weights. This is known as sample degeneracy. Accordingly, for importance sampling algorithms, the focus is on reducing the variability of $\tilde{w}(\theta)$ over θ.

A common measure of the degree of sample degeneracy is the effective sample size (ESS) (Liu et al. 1998; Liu 2001), estimated as:

$$ESS = \left(\sum_{i=1}^{N}[W^{(i)}]^2\right)^{-1}, \quad 1 \leq ESS \leq N, \tag{4.5}$$

which is computed using the normalised weights $W^{(i)} = \tilde{w}^{(i)}/\sum_{j=1}^{N}\tilde{w}^{(j)}$. The ESS is an estimate of the effective number of equally weighted $\theta^{(i)}$ in a given weighted sample, which can be loosely interpreted as the information content. When $g(\theta) \propto \tilde{f}(\theta)$ so that we have samples directly from $f(\theta)$, then $W^{(i)} = 1/N$

and $ESS = N$. However, when there is severe particle degeneracy in the extreme case where $W^{(1)} = 1$ and $W^{(i)} = 0$ for $i = 2, \ldots, N$, then $ESS = 1$.

Specifically in the ABC framework where $\tilde{w}(\theta^{(i)}) = K_h(\|s^{(i)} - s_{obs}\|)\pi(\theta^{(i)})/g(\theta^{(i)})$, if g is diffuse compared to the (marginal) target distribution $\pi_{ABC}(\theta|s_{obs})$, samples $\theta^{(i)} \sim g(\theta)$ from regions of low posterior density will tend to generate summary statistics $s^{(i)}$ that are very far from the observed statistics s_{obs}, and this will produce low weights $w^{(i)}$ compared to samples $\theta^{(i)}$ in regions of high posterior density. This is the same as for the standard importance sampling case. However, in ABC importance sampling, an additional factor is that the importance weight $\tilde{w}(\theta)$ is a function of the kernel function $K_h(\|s - s_{obs}\|)$, which contains the stochastic term s. This has some implications, which are also relevant for sequential Monte Carlo-based ABC samplers, discussed in Section 4.4.

When K_h has non-compact support, such as when $K_h(u) = \phi(u; 0, h^2)$, where $\phi(x; \mu, \sigma^2)$ denotes the Gaussian density function with mean μ and variance σ^2, the importance weight $\tilde{w}^{(i)}$ is guaranteed to be non-zero for each i. However, the resulting importance weight can be highly variable, depending on whether $s^{(i)}$ is close to or far from s_{obs}. This typically produces samples $(\theta^{(i)}, \tilde{w}^{(i)})$ with low effective sample sizes.

If K_h has a compact support (and this is typical in most ABC implementations), then $\tilde{w}^{(i)} = 0$ is likely for small h, even when $\theta^{(i)}$ is in a high posterior density region. This means that Algorithm 4.2 will return many $(\theta^{(i)}, \tilde{w}^{(i)})$ for which the weight is exactly zero, resulting in low effective sample sizes, and maybe even complete algorithm failure if $\tilde{w}^{(i)} = 0$ for all $i = 1, \ldots, N$. As a result, a common variation of Algorithm 4.2 is to repeat steps 1–4 for each i, until a non-zero weight has been generated. This effectively introduces a rejection sampling step within the importance sampling algorithm. This idea (c.f. Fernhead and Prangle 2012) can be used to improve the ESS for ABC importance sampling algorithms, regardless of the choice of the kernel, by modifying step 4 in Algorithm 4.2 to be

Algorithm 4.3: ABC Importance/Rejection Sampling

Same as Algorithm 4.2, but replacing step 4 of Sampling with:

4. With probability $K_h(\|s^{(i)} - s_{obs}\|)/K_h(0)$ set $\tilde{w}^{(i)} = \pi(\theta^{(i)})/g(\theta^{(i)})$, else go to 1.

When K_h has compact support, this ensures that steps 1–4 of Algorithm 4.2 are repeated until $K_h(\|s^{(i)} - s_{obs}\|)$ is non-zero. When K_h has non-compact support, this offers some control over the variability of the weights, as only samples for which $s^{(i)}$ is reasonably close to s_{obs} are likely to be accepted.

Under Algorithm 4.3, in the particular case of when K_h is the uniform kernel on $[-h, h]$, if in addition $g(\theta) \propto \pi(\theta)$, so that the importance distribution is proportional to the prior, then $\tilde{w}^{(i)} \propto 1$ for any i. This results in $W^{(i)} = 1/N$ and $ESS = N$, and the ABC importance sampling algorithm effectively reduces to the ABC rejection sampling algorithm, but without the need to compute the normalising constant M. This setup is very common in practice, as it removes the need to compute importance weights, and to worry about algorithm performance with respect to effective sample size, which is always maximised. However, in this case, algorithm performance is dominated by the number of times steps 1–4 are repeated before a sample $\theta^{(i)}$ is accepted. In general, the efficiency of Algorithm 4.3 is a combination of the resulting effective sample size and the number of repetitions of the sampling steps 1–4.

4.2.3 Importance/rejection sampler variants

There are many variants on ABC importance and rejection samplers. A few of these are detailed in the following, chosen either because of their popularity, or because of their links with particular ABC samplers discussed in later sections.

4.2.3.1 Rejection control importance sampling

Liu et al. (1998) developed a general importance-rejection algorithm technique known as rejection control, with the aim of reducing the number of $\theta^{(i)}$ samples that are produced with very small weights in an importance sampler. This method was exploited within an ABC sequential Monte Carlo framework by Sisson et al. (2007) and Peters et al. (2012) (see Section 4.4), however, it may also be implemented directly within an ABC importance sampler as outlined in the following.

Suppose that a weighted sample $(\theta^{(i)}, \tilde{w}^{(i)})$ is drawn from $f(\theta)$ using an importance sampling algorithm. In order to control the size of the importance weight, $\tilde{w}^{(i)}$ is compared to some pre-specified threshold value $c > 0$. If $\tilde{w}^{(i)} > c$ then the weight is considered sufficiently large, and the sample $\theta^{(i)}$ is accepted. However, if $\tilde{w}^{(i)} < c$, then $\theta^{(i)}$ is probabilistically rejected, with a higher rejection rate for lower $\tilde{w}^{(i)}$. In this manner, the variability of the accepted importance weights can be reduced. In particular, each sample $\theta^{(i)}$ is accepted with probability:

$$r^{(i)} = \min\left\{1, \frac{\tilde{w}^{(i)}}{c}\right\},$$

which results in the automatic acceptance of samples for which $\tilde{w}^{(i)} > c$ and an acceptance probability of $\tilde{w}^{(i)}/c$ otherwise. This means that larger c results in less variable weights, although at the price of more rejections. The accepted samples are then draws from the modified importance sampling distribution:

$$g^*(\theta) = M^{-1} \min\left\{1, \frac{\tilde{w}(\theta)}{c}\right\} g(\theta),$$

where $\tilde{w}(\theta) = \tilde{f}(\theta)/g(\theta)$ and with normalising constant $M = \int \min\{1, \tilde{w}(\theta)/c\}$ $g(\theta)d\theta$. As a result, setting:

$$\tilde{w}^*(\theta) = \frac{\tilde{f}(\theta)}{g^*(\theta)} \qquad \tilde{w}^{*(i)} = \frac{\tilde{f}(\theta^{(i)})}{g^*(\theta^{(i)})} = M\frac{\tilde{w}^{(i)}}{r^{(i)}}, \tag{4.6}$$

means that the samples $(\theta^{(i)}, \tilde{w}^{*(i)})$ will be weighted samples from $f(\theta)$, but with the property that:

$$Var_{g^*}\left[\frac{\tilde{f}(\theta)}{g^*(\theta)}\right] \leq Var_g\left[\frac{\tilde{f}(\theta)}{g(\theta)}\right]. \tag{4.7}$$

That is, the rejection control algorithm can reduce the variance of the importance weights (Liu 2001). While it may be difficult to evaluate $M = E_g\left[\min\left\{1, \frac{\tilde{w}(\theta)}{c}\right\}\right]$ analytically, it may be estimated from the samples $(\theta^{(i)}, \tilde{w}^{(i)})$ via:

$$\hat{M} \approx \frac{1}{N}\sum_{i=1}^{N} \min\left\{1, \frac{\tilde{w}^{(i)}}{c}\right\}.$$

If an estimate of M is not required, its computation can be avoided for importance sampling purposes by calculating the normalised weights $W^{*(i)} = \tilde{w}^{*(i)}/\sum_{j=1}^{N}\tilde{w}^{*(j)}$, as the M term then cancels in numerator and denominator.

As with ABC rejection sampling (Algorithm 4.2), the ABC implementation of rejection control importance sampling targets $\pi_{ABC}(\theta, s|s_{obs})$, resulting in a weight calculation of $\tilde{w}^{(i)} = K_h(\|s^{(i)} - s_{obs}\|)\pi(\theta^{(i)})/g(\theta^{(i)})$. The full algorithm is given in Algorithm 4.4.

Note that while Algorithm 4.4 requires pre-specification of the rejection threshold c, a suitable value may be practically difficult to determine in advance. As such, Algorithm 4.4 may be alternatively executed by first implementing steps 1–3 only for $i = 1, \ldots, N$, and then specifying c as some quantile of the resulting empirical distribution of $\tilde{w}^{(1)}, \ldots, \tilde{w}^{(N)}$. Following this, Algorithm 4.4 may then continue implementation from step 4 onwards for each $i = 1, \ldots, N$ (e.g. Peters et al. 2012).

As with ABC importance/rejection sampling (Algorithm 4.3), when K_h has compact support, rejection control will replace those samples for which the simulated and observed summary statistics are too far apart, resulting in $\tilde{w}^{(i)} = 0$. More generally, however, rejection control provides much greater control over the variability of the weights regardless of K_h, producing more uniform weights for larger c. The price for this control is the greater number of rejections induced as c increases (Peters et al. 2012).

4.2.3.2 ABC importance sampling

While most published descriptions of importance and rejection sampling ABC algorithms follow the format given in Algorithms 4.1 through 4.4, in practice it

Algorithm 4.4: ABC Rejection Control Importance Sampling

Inputs:

- A target posterior density $\pi(\theta|y_{obs}) \propto p(y_{obs}|\theta)\pi(\theta)$ consisting of a prior distribution $\pi(\theta)$ and a procedure for generating data under the model $p(y_{obs}|\theta)$.
- An importance sampling density $g(\theta)$, with $g(\theta) > 0$ if $\pi(\theta|y_{obs}) > 0$.
- An integer $N > 0$.
- A kernel function $K_h(u)$ and scale parameter $h > 0$.
- A low-dimensional vector of summary statistics $s = S(y)$.
- A rejection control threshold $c > 0$.

Sampling:
For $i = 1, \ldots, N$:

1. Generate $\theta^{(i)} \sim g(\theta)$ from importance sampling density g.
2. Generate $y^{(i)} \sim p(y|\theta^{(i)})$ from the model and compute summary statistic $s^{(i)} = S(y^{(i)})$.
3. Compute weight $\tilde{w}^{(i)} = K_h(\|s^{(i)} - s_{obs}\|)\pi(\theta^{(i)})/g(\theta^{(i)})$.
4. Reject $\theta^{(i)}$ with probability $1 - r^{(i)} = 1 - \min\{1, \frac{\tilde{w}^{(i)}}{c}\}$, and go to Step 1.
5. Otherwise, accept $\theta^{(i)}$ and set modified weight $\tilde{w}^{*(i)} = \tilde{w}^{(i)}/r^{(i)}$.

Output:
A set of weighted parameter vectors $(\theta^{(1)}, \tilde{w}^{*(1)}), \ldots, (\theta^{(N)}, \tilde{w}^{*(N)}) \sim \pi_{ABC}(\theta|s_{obs})$.

---

is not uncommon to deviate from these and implement a slight variation. The reason for this is that Algorithms 4.1 through 4.4 require pre-specification of the kernel scale parameter $h > 0$, without which, importance weights cannot be calculated and accept/reject decisions cannot be made. In reality, as the scale of the distances $\|s^{(i)} - s_{obs}\|$ is unlikely to be known in advance, it is difficult to pre-determine a suitable value for h.

Algorithm 4.5 presents a variation on the ABC importance sampler of Algorithm 4.3 that avoids pre-specification of h. Here, a large number N' of $(\theta^{(i)}, s^{(i)})$ pairs are generated from the importance sampling distribution $p(s|\theta)g(\theta)$. These are the only samples that will be used in the algorithm, so the computational overheads are fixed at N' draws from the model, unlike Algorithm 4.3, in which the number of draws is random and unknown in advance. The N samples for which $s^{(i)}$ is closest to s_{obs} (as measured by $\|\cdot\|$) are then identified, and h determined to be the smallest possible value so that only these N samples have non-zero weights (assuming a kernel K_h with compact support). Once h is fixed, the importance weights can be calculated

Algorithm 4.5: ABC k-nn Importance Sampling

Inputs:

- A target posterior density $\pi(\theta|y_{obs}) \propto p(y_{obs}|\theta)\pi(\theta)$ consisting of a prior distribution $\pi(\theta)$, and a procedure for generating data under the model $p(y_{obs}|\theta)$.
- An importance sampling density $g(\theta)$, with $g(\theta) > 0$ if $\pi(\theta|y_{obs}) > 0$.
- Integers $N' \gg N > 0$.
- A kernel function $K_h(u)$ with compact support.
- A low-dimensional vector of summary statistics $s = S(y)$.

Sampling:
For $i = 1, \ldots, N'$:

1. Generate $\theta^{(i)} \sim g(\theta)$ from importance sampling density g.
2. Generate $y^{(i)} \sim p(y|\theta^{(i)})$ from the model.
3. Compute summary statistic $s^{(i)} = S(y^{(i)})$.

- Identify the N-nearest neighbours of s_{obs} as measured by $\|s^{(i)} - s_{obs}\|$.
- Index these nearest neighbours by $[1], \ldots, [N]$.
- Set h to be the largest possible value, such that $K_h(\max_i\{\|s^{([i])} - s_{obs}\|\}) = 0$.
- Compute weights $\tilde{w}^{([i])} = K_h(\|s^{([i])} - s_{obs}\|)\pi(\theta^{([i])})/g(\theta^{([i])})$ for $i = 1, \ldots, N$.

Output:
A set of weighted parameter vectors $(\theta^{([1])}, \tilde{w}^{([1])}), \ldots, (\theta^{([N])}, \tilde{w}^{([N])}) \sim \pi_{ABC}(\theta|s_{obs})$.

as before, and the N samples $(\theta^{(i)}, \tilde{w}^{(i)})$ with non-zero $\tilde{w}^{(i)}$ are returned as weighted samples from $\pi_{ABC}(\theta|s_{obs})$.

This approach is explicitly used in for example, Beaumont et al. (2002) and Blum et al. (2013), and implicitly in many other ABC implementations. The differences between Algorithms 4.3 and 4.5 may seem small- if the value of h determined in Algorithm 4.5 was used in Algorithm 4.3, then (assuming the same pseudo-random numbers used in the appropriate places) the resulting draws from $\pi_{ABC}(\theta|s_{obs})$ would be identical. However, Algorithm 4.5 is based on a k-nearest neighbour algorithm for density estimation of the ABC likelihood function, and so possesses very different theoretical properties compared to Algorithm 4.3. This k-nearest neighbour approach is discussed and analysed in detail in the rejection sampling context by Biau et al. (2015).

4.2.3.3 ABC rejection sampling with stopping rule

A version of ABC rejection sampling (Algorithm 4.1) which similarly does not require pre-specification of the kernel scale parameter $h > 0$ is presented in Example 6 (Section 1.7.2) of Sisson et al. (2018, Chapter 1, this volume). We do not reproduce this algorithm here for brevity. The algorithm identifies the smallest value of h needed to accept exactly N samples before some stopping rule is achieved. This stopping rule could be based on an overall computational budget [such as using exactly N' total draws from $p(s|\theta)$] or on some perceived level of accuracy of the resulting ABC posterior approximation. If the stopping rule is based on an overall computational budget of exactly N' draws from $p(s|\theta)$ (and again the same pseudo-random numbers), this algorithm will produce exactly the same final samples from $\pi_{ABC}(\theta|s_{obs})$ as Algorithm 4.1, were the ABC rejection sampler to adopt the identified choice of h. Of course, the advantage here is that the value of h is automatically determined.

4.2.3.4 Rejection-based ABC algorithms for expensive simulators

It is not uncommon for the data generation step $y^{(i)} \sim p(y|\theta^{(i)})$ in ABC algorithms to be expensive and thereby dominate the computational overheads of the algorithms. While there are a few principled ways to mitigate this (see discussion of Prangle et al. 2017 and Everitt and Rowińska 2017 in Section 4.4.2), within rejection-based ABC algorithms, it is sometimes possible to reject a proposed sampler $\theta^{(i)}$ *before* generating the data $y^{(i)} \sim p(y|\theta^{(i)})$. To see this, note that e.g. step 4 of Algorithm 4.3

4. With probability $K_h(\|s^{(i)} - s_{obs}\|)/K_h(0)$ set $\tilde{w}^{(i)} = \pi(\theta^{(i)})/g(\theta^{(i)})$, else go to 1.

can be alternatively implemented as:

4. With probability $\pi(\theta^{(i)})/g(\theta^{(i)})$ set $\tilde{w}^{(i)} = K_h(\|s^{(i)} - s_{obs}\|)/K_h(0)$, else go to 1.

This means that steps 2 and 3 of Algorithm 4.3 (generate $y^{(i)} \sim p(y|\theta^{(i)})$ and compute $s^{(i)} = S(y^{(i)})$) need not be performed until the event in step 4 with probability $\pi(\theta^{(i)})/g(\theta^{(i)})$ has occurred. This allows for a possible early rejection of $\theta^{(i)}$ before any data generation needs to take place. (Note that if $g(\theta) = \pi(\theta)$ there is no benefit to be gained.) This modification trades some computational savings for weights $\tilde{w}^{(i)}$ constructed from different terms, and thereby having different variance properties. This idea, which is a standard technique in standard sequential Monte Carlo samplers (e.g. Del Moral et al. 2006), can be implemented in any rejection-based ABC algorithm, including ABC-MCMC and ABC-SMC samplers (Sections 4.3 and 4.4).

4.2.3.5 Marginal ABC samplers

Until now we have presented ABC algorithms as producing samples $(\theta^{(i)}, s^{(i)})$ exactly from the joint distribution $\pi_{ABC}(\theta, s|s_{obs}) \propto K_h(\|s - s_{obs}\|)p(s|\theta)\pi(\theta)$. As a result, samples $\theta^{(i)}$ from the ABC approximation to the posterior $\pi(\theta|y_{obs})$ given by:

$$\pi_{ABC}(\theta|s_{obs}) \propto \int K_h(\|s - s_{obs}\|)p(s|\theta)\pi(\theta)ds,$$

may be obtained by marginalising over the realised $s^{(i)}$. An alternative approach to construct an ABC algorithm could be to directly target the (marginal) posterior $\pi_{ABC}(\theta|s_{obs})$ rather than the joint posterior $\pi_{ABC}(\theta, s|s_{obs})$. This approach becomes apparent when noting that $\pi_{ABC}(\theta|s_{obs})$ can be estimated pointwise (up to proportionality), for fixed θ, as:

$$\int K_h(\|s - s_{obs}\|)p(s|\theta)\pi(\theta)ds \approx \frac{\pi(\theta)}{T} \sum_{t=1}^{T} K_h(\|s(t) - s_{obs}\|) := \hat{\pi}_{ABC}(\theta|s_{obs}),$$

where $s(1), \ldots, s(T) \sim p(s|\theta)$ are T independent draws of summary statistics from the partial likelihood $p(s|\theta)$ for a given θ. This Monte Carlo estimate of $\pi_{ABC}(\theta|s_{obs})$ is unbiased up to proportionality (in that $E_{s|\theta}[\hat{\pi}_{ABC}(\theta|s_{obs})] \propto \pi_{ABC}(\theta|s_{obs})$), and so $\hat{\pi}_{ABC}(\theta|s_{obs})$ may be used in place of $\pi_{ABC}(\theta|s_{obs})$ in a standard rejection or importance sampler which targets $\pi_{ABC}(\theta|s_{obs})$. Using this substitution will produce a random, estimated acceptance probability or importance weight. However, because it is also unbiased (up to proportionality), the resulting target distribution will remain the same as if the exact weight had been used i.e. $\pi_{ABC}(\theta|s_{obs})$, although the sampler weights/acceptances will become more variable. This is the so-called marginal ABC sampler (e.g. Marjoram et al. 2003; Reeves and Pettitt 2005; Sisson et al. 2007; Ratmann et al. 2009; Toni et al. 2009; Peters et al. 2012, among others).

As with standard Monte Carlo estimates, the number of Monte Carlo draws T affects the variability of the ABC posterior estimator. Bornn et al. (2017) explore the question of how many draws, T, produces the most efficient overall sampler in the context of ABC rejection and Markov chain Monte Carlo algorithms. If T is large, the estimate of $\pi_{ABC}(\theta|s_{obs})$ is accurate, and so the acceptance probability is accurate, but at the cost of many Monte Carlo draws, however, if T is small, the acceptance probability is highly variable, but is much cheaper to evaluate. When using a uniform kernel K_h, Bornn et al. (2017) conclude that in fact $T = 1$ is the most efficient, as (loosely) the combination of T draws used to accept one $\theta^{(i)}$ could be better used to accept up to T different $\theta^{(i)}$'s, each using one Monte Carlo draw per ABC posterior estimate.

The idea of the marginal ABC sampler is closely related to the construction of the more recently developed pseudo-marginal sampler (Beaumont 2003; Andrieu and Roberts 2009), a more general class of likelihood-free sampler that has gained popularity outside of the ABC setting. Here, rather than treating $\hat{\pi}_{ABC}(\theta|s_{obs})$ as an unbiased estimate of $\pi_{ABC}(\theta|s_{obs})$ in an algorithm that targets $\pi_{ABC}(\theta|s_{obs})$, an alternative joint posterior distribution can be constructed:

$$\pi_{ABC}(\theta, s(1), \ldots, s(T)|s_{obs}) \propto \left[\frac{1}{T}\sum_{t=1}^{T} K_h(\|s(t) - s_{obs}\|)\right] \left[\prod_{t=1}^{T} p(s(t)|\theta)\right] \pi(\theta), \tag{4.8}$$

which is defined over the joint posterior of θ and all T summary statistic replicates (e.g. Sisson and Fan 2011; Del Moral et al. 2012), where $T = 1$ gives the usual ABC joint posterior $\pi_{ABC}(\theta, s|s_{obs})$. A useful property of this form of joint posterior is that the θ-marginal distribution is the same for any value of T, and in particular:

$$\int \cdots \int \pi_{ABC}(\theta, s(1), \ldots, s(T)|s_{obs}) ds(1) \ldots ds(T) = \pi_{ABC}(\theta|s_{obs}).$$

This means that any sampler targeting $\pi_{ABC}(\theta, s(1), \ldots, s(T)|s_{obs})$ can produce samples from $\pi_{ABC}(\theta|s_{obs})$. Consider now an importance sampler targeting $\pi_{ABC}(\theta, s(1), \ldots, s(T)|s_{obs})$ with the importance sampling density:

$$g(\theta, s(1), \ldots, s(T)) = g(\theta) \prod_{t=1}^{T} p(s(t)|\theta).$$

The resulting importance weight is:

$$\begin{aligned} \frac{\pi_{ABC}(\theta,s(1),\ldots,s(T)|s_{obs})}{g(\theta,s(1),\ldots,s(T))} &\propto \frac{\left[\frac{1}{T}\sum_{t=1}^{T} K_h(\|s(t)-s_{obs}\|)\right]\left[\prod_{t=1}^{T} p(s(t)|\theta)\right]\pi(\theta)}{g(\theta)\prod_{t=1}^{T} p(s(t)|\theta)} \\ &= \frac{\hat{\pi}_{ABC}(\theta|s_{obs})}{g(\theta)}. \end{aligned}$$

This means that any marginal ABC sampler targeting $\pi_{ABC}(\theta|s_{obs})$ through the unbiased estimate of the ABC posterior given by $\hat{\pi}_{ABC}(\theta|s_{obs})$ is directly equivalent to an exact algorithm targeting $\pi_{ABC}(\theta, s(1), \ldots, s(T)|s_{obs})$. That is, all marginal ABC samplers are justified by their equivalent joint space ABC algorithm.

This idea also extends to using unbiased approximations of posterior distributions within MCMC samplers (see next section), where the technique has expanded beyond ABC algorithms to more general target distributions. Here, it is more generally known as pseudo-marginal Monte Carlo methods. See Andrieu et al. (2019) (Chapter 9, this volume) for a more detailed discussion of the connections between ABC marginal samplers and pseudo-marginal MCMC methods.

4.3 Markov Chain Monte Carlo Methods

Markov chain Monte Carlo (MCMC) methods are a highly accessible class of algorithms for obtaining samples from complex distributions (e.g. Brooks et al. 2011). By constructing a Markov chain with the target distribution of interest as its limiting distribution, following chain convergence, a realised random sample path from this chain will behave like a (serially correlated) sample from the target distribution,. Their strong performance and simplicity of implementation has made MCMC algorithms the dominant Monte Carlo method for the past two decades (Brooks et al. 2011). As such, it is only natural that MCMC-based ABC algorithms have been developed.

4.3.1 ABC-Markov chain Monte Carlo samplers

The Metropolis–Hastings algorithm is the most popular class of MCMC algorithm. Given the current chain state $\theta^{(i)}$, the next value in the sequence is obtain by sampling a candidate value θ' from a proposal distribution $\theta' \sim g(\theta^{(i)}, \theta) = g(\theta|\theta^{(i)})$, which is then accepted with probability $a(\theta, \theta') = \min\{1, \frac{f(\theta')g(\theta', \theta^{(i)})}{f(\theta)g(\theta^{(i)}, \theta')}\}$ so that $\theta^{(i+1)} = \theta'$, or otherwise rejected so that $\theta^{(i+1)} = \theta^{(i)}$. Under this mechanism, the target distribution is $f(\theta)$, and there is great flexibility in the choice of the proposal distribution g. An implementation of this sampler in the ABC setting is given in Algorithm 4.6. ABC MCMC algorithms were originally developed by Marjoram et al. (2003). See for example, Bortot et al. (2007), Wegmann et al. (2009), Ratmann et al. (2009), Sisson and Fan (2011) and Andrieu et al. (2019) (Chapter 9, this volume) for more discussion on ABC-MCMC samplers.

As with ABC importance and rejection samplers, the target distribution of ABC MCMC algorithms is the joint ABC posterior $\pi_{ABC}(\theta, s|s_{obs})$. On this space, the proposal distribution becomes:

$$g[(\theta, s), (\theta', s')] = g(\theta, \theta')p(s'|\theta'),$$

and, as a result the acceptance probability of the proposed move from $(\theta^{(i)}, s^{(i)})$ to $(\theta', s') \sim g[(\theta^{(i)}, s^{(i)}), (\theta', s')]$ becomes $a[(\theta^{(i)}, s^{(i)}), (\theta', s')] = \min\{1, \alpha[(\theta^{(i)}, s^{(i)}), (\theta', s')]\}$, where

$$\begin{aligned}
\alpha[(\theta^{(i)}, s^{(i)}), (\theta', s')] &= \frac{\pi_{ABC}(\theta', s'|s_{obs})g[(\theta', s'), (\theta^{(i)}, s^{(i)})]}{\pi_{ABC}(\theta^{(i)}, s^{(i)}|s_{obs})g[(\theta^{(i)}, s^{(i)}), (\theta', s')]} \\
&= \frac{K_h(\|s' - s_{obs}\|)p(s'|\theta')\pi(\theta')}{K_h(\|s^{(i)} - s_{obs}\|)p(s^{(i)}|\theta^{(i)})\pi(\theta^{(i)})} \frac{g(\theta', \theta^{(i)})p(s^{(i)}|\theta^{(i)})}{g(\theta^{(i)}, \theta')p(s'|\theta')} \\
&= \frac{K_h(\|s' - s_{obs}\|)\pi(\theta')}{K_h(\|s^{(i)} - s_{obs}\|)\pi(\theta^{(i)})} \frac{g(\theta', \theta^{(i)})}{g(\theta^{(i)}, \theta')},
\end{aligned}$$

which is free of intractable likelihood terms, $p(s|\theta)$, and so may be directly evaluated.

Algorithm 4.6: ABC Markov Chain Monte Carlo Algorithm

Inputs:

- A target posterior density $\pi(\theta|y_{obs}) \propto p(y_{obs}|\theta)\pi(\theta)$ consisting of a prior distribution $\pi(\theta)$, and a procedure for generating data under the model $p(y_{obs}|\theta)$.
- A Markov proposal density $g(\theta, \theta') = g(\theta'|\theta)$.
- An integer $N > 0$.
- A kernel function $K_h(u)$ and scale parameter $h > 0$.
- A low-dimensional vector of summary statistics $s = S(y)$.

Initialise:
Repeat:

1. Choose an initial parameter vector $\theta^{(0)}$ from the support of $\pi(\theta)$.
2. Generate $y^{(0)} \sim p(y|\theta^{(0)})$ from the model and compute summary statistics $s^{(0)} = S(y^{(0)})$,

until $K_h(\|s^{(0)} - s_{obs}\|) > 0$.

Sampling:
For $i = 1, \ldots, N$:

1. Generate candidate vector $\theta' \sim g(\theta^{(i-1)}, \theta)$ from the proposal density g.
2. Generate $y' \sim p(y|\theta')$ from the model and compute summary statistics $s' = S(y')$.
3. With probability:0

$$\min\left\{1, \frac{K_h(\|s' - s_{obs}\|)\pi(\theta')g(\theta', \theta^{(i-1)})}{K_h(\|s^{(i-1)} - s_{obs}\|)\pi(\theta^{(i-1)})g(\theta^{(i-1)}, \theta')}\right\}$$

set $(\theta^{(i)}, s^{(i)}) = (\theta', s')$. Otherwise set $(\theta^{(i)}, s^{(i)}) = (\theta^{(i-1)}, s^{(i-1)})$.

Output:
A set of correlated parameter vectors $\theta^{(1)}, \ldots, \theta^{(N)}$ from a Markov chain with stationary distribution $\pi_{ABC}(\theta|s_{obs})$.

Algorithm 4.6 satisfies the detailed balance (time reversibility) condition with respect to $\pi_{ABC}(\theta, s|s_{obs})$, which ensures that $\pi_{ABC}(\theta, s|s_{obs})$ is the stationary distribution of the Markov chain. Detailed balance states that:

$$\pi_{ABC}(\theta, s|s_{obs})P[(\theta, s), (\theta', s')] = \pi_{ABC}(\theta', s'|s_{obs})P[(\theta', s'), (\theta, s)],$$

where the Metropolis–Hastings transition kernel P is given by:

$$P[(\theta, s), (\theta', s')] = g[(\theta, s), (\theta', s')]a[(\theta, s), (\theta', s')].$$

Assuming that (without loss of generality) $a[(\theta', s'), (\theta, s)] = \min\{1, \alpha[(\theta', s'), (\theta, s)]\} = 1$ (and so $a[(\theta, s), (\theta', s')] = \alpha[(\theta, s), (\theta', s')]$), the detailed balance condition is satisfied since:

$$\begin{aligned}
& \pi_{ABC}(\theta, s|s_{obs})P[(\theta, s), (\theta', s')] \\
= \quad & \pi_{ABC}(\theta, s|s_{obs})g[(\theta, s), (\theta', s')]\alpha[(\theta, s), (\theta', s')] \\
= \quad & \frac{K_h(\|s - s_{obs}\|)p(s|\theta)\pi(\theta)}{z} g(\theta, \theta')p(s'|\theta') \frac{K_h(\|s' - s_{obs}\|)\pi(\theta')g(\theta', \theta)}{K_h(\|s - s_{obs}\|)\pi(\theta)g(\theta, \theta')} \\
= \quad & \frac{K_h(\|s' - s_{obs}\|)p(s'|\theta')\pi(\theta')}{z} g(\theta', \theta)p(s|\theta) \\
= \quad & \pi_{ABC}(\theta', s'|s_{obs})P[(\theta', s'), (\theta, s)],
\end{aligned}$$

where $z = \int\int K_h(\|s - s_{obs}\|)p(s|\theta)\pi(\theta)dsd\theta$ is the normalisation constant of $\pi_{ABC}(\theta, s|s_{obs})$ (e.g. Sisson and Fan 2011).

Marjoram et al. (2003) found that the ABC-MCMC algorithm offered an improved acceptance rate over rejection sampling-based ABC algorithms with the same scale parameter h, although at the price of serial correlation in the Markov chain sample path $\theta^{(1)}, \ldots, \theta^{(N)}$. Thus, for kernels K_h with compact support, the same mechanism that causes many rejections or zero weights in ABC rejection and importance samplers, now results in many rejected proposals in the ABC-MCMC algorithm. The difference here is that the chain simply remains at the current state $\theta^{(i)}$ for long periods of time, giving additional posterior weight to $\theta^{(i)}$. Techniques for improving the performance of standard MCMC algorithms may also be applied to ABC-MCMC samplers. However there is one feature of ABC-MCMC that is different to that of the standard algorithm, that is particularly acute when using kernel functions K_h with compact support.

Consider a proposed move from $\theta^{(i)}$ to θ'. In standard MCMC, the acceptance probability is based on the relative density of the posterior evaluated at θ' compared to that evaluated at $\theta^{(i)}$. In ABC-MCMC, the density of the posterior at θ' is determined through the ability of the model to generate a summary statistic $s' \sim p(s|\theta')$ that is close to s_{obs}, as measured through $K_h(\|s' - s_{obs}\|)$. That is, to move to θ', a summary statistic s' must be generated that is close enough to s_{obs}. This is the standard ABC mechanism. However, the result of this for the ABC-MCMC algorithm is that it means that the acceptance rate of the sampler is directly related to the value of the (intractable) likelihood function evaluated at θ'. As a result, the sampler may mix rapidly in regions of high posterior density, but will have much worse mixing in regions of relatively low posterior density (Sisson et al. 2007). For this reason, ABC-MCMC samplers can often get stuck in regions of low posterior density for long periods time, effectively producing convergence issues for the algorithm.

This effect is more pronounced when the kernel K_h has compact support, such as the uniform kernel on $[-h, h]$ which is endemic in ABC implementations, although it is still present for kernels defined on the real line, such as the Gaussian density kernel. In a study of 'sojourn time' within ABC-MCMC samplers (that is, the number of consecutive iterations in the sampler in which a univariate parameter θ remained above some high threshold), Sisson and Fan (2011) found empirically that samplers with uniform kernels had a substantially higher expected sojourn time than samplers with Gaussian kernels, indicating that the latter had superior chain mixing in distributional tails. Despite this, ABC-MCMC samplers are routinely implemented with uniform kernels K_h.

Chain mixing can be improved by alternatively targeting the joint posterior distribution $\pi_{ABC}(\theta, s(1), \ldots, s(T)|s_{obs})$ given by (4.8). Under this (pseudo) marginal sampler framework (Section 4.2.3.5) as $T \to \infty$, the mixing properties of the ABC-MCMC approach that of the equivalent standard MCMC sampler directly targeting $\pi_{ABC}(\theta|s_{obs})$ (if it would be possible to numerically evaluate the density function). Sisson and Fan (2011) empirically demonstrated this improvement, as measured in sojourn times, as T increases. Of course, this improvement of chain mixing comes at the price of overall sampler performance, as the computational overheads of generating $s(1), \ldots, s(T)$ for large T would be extremely high. The results of Bornn et al. (2017), that $T = 1$ is the optimum efficiency choice for uniform kernels K_h, also hold for ABC MCMC samplers.

4.3.2 Augmented space ABC-Markov chain Monte Carlo samplers

The standard ABC MCMC sampler (Algorithm 4.6) requires pre-specification of the kernel scale parameter h. As with ABC rejection and importance samplers, there are a number of ways in which lack of knowledge of a suitable value for the kernel scale parameter can be incorporated into the basic algorithm. Most of these methods also attempt to improve chain mixing over the standard algorithm, which uses a fixed, low value of h. At the very simplest level, this could involve adaptively adjusting h as a function of $\|s - s_{obs}\|$ at the current and proposed states of the chain, and either allow h to slowly reduce to some target value to improve convergence at the start of the sampler (e.g. Ratmann et al. 2007, Sisson and Fan 2011, p. 325) or adaptively choose h to achieve some pre-determined overall sampler acceptance probability.

Augmenting the dimension of the target distribution is a common strategy to improve the performance of Monte Carlo algorithms. In order to help the ABC-MCMC sampler escape from regions of low posterior density, Bortot et al. (2007) proposed augmenting the joint ABC posterior $\pi_{ABC}(\theta, s|s_{obs})$

to additionally include the kernel bandwidth h, treating this as an unknown additional parameter. The resulting joint posterior distribution is given by:

$$\pi_{ABC}(\theta, s, h|s_{obs}) \propto K_h(\|s - s_{obs}\|)p(s|\theta)\pi(\theta)\pi(h),$$

and the resulting ABC approximation to the partial posterior $\pi(\theta|s_{obs})$ is then given by:

$$\dot{\pi}_{ABC}(\theta|s_{obs}) = \int\int \pi_{ABC}(\theta, s, h|s_{obs})dsdh, \qquad (4.9)$$

where $h > 0$. Here, h is treated as a tempering parameter in the manner of simulated tempering (Geyer and Thompson 1995), with larger and smaller values, respectively, corresponding to 'hot' and 'cold' tempered posterior distributions. Larger values of h increase the scale of the kernel density function K_h, under which the sampler is more likely to accept proposed moves and thereby alleviating the sampler's mixing problems, although at the price of a less accurate posterior approximation. Lower values of h produce a more accurate posterior approximation, but will induce slower chain mixing. The density $\pi(h)$ is a pseudo-prior, which serves to influence the mixing of the sampler through the tempered distributions.

Note that the augmented space ABC posterior approximation $\dot{\pi}_{ABC}(\theta|s_{obs})$ given by (4.9) will in general be different to that of $\pi_{ABC}(\theta|s_{obs})$, as the latter contains a fixed value of h, whereas the former integrates over the uncertainty inherent in this parameter. Rather than use (4.9) as the final ABC approximation to $\pi(\theta|s_{obs})$, Bortot et al. (2007) chose to remove those samples $(\theta^{(i)}, s^{(i)}, h^{(i)})$ for which $h^{(i)}$ was considered too large to come from a good approximation to $\pi(\theta|s_{obs})$. In particular, they examined the distribution of $\theta^{(i)}|h^{(i)} \leq h^*$, aiming to choose the largest value of h^*, such that the distribution of $\theta^{(i)}|h^{(i)} \leq h^*$ did not change if h^* was reduced further. The resulting ABC posterior approximation is therefore given by:

$$\ddot{\pi}_{ABC}(\theta|s_{obs}) = \int_0^{h^*} \int \pi_{ABC}(\theta, s, h|s_{obs})dsdh.$$

This approach effectively permits an *a posteriori* evaluation of an appropriate value h^*, such that the approximation $\ddot{\pi}_{ABC}(\theta|s_{obs})$ is as close as possible (subject to Monte Carlo variability) to the true posterior $\pi(\theta|s_{obs})$.

A similar idea was explored by Baragatti et al. (2013) in an ABC version of the parallel tempering algorithm of Geyer and Thompson (1995). Here, $M > 1$ parallel ABC-MCMC chains are implemented with different kernel density scale parameters $h_M < h_{M-1} < \ldots < h_1$, with state transitions allowed between chains so that the states of the more rapidy mixing chains (with higher h values) can propagate down to the more slowly mixing chains (with lower h). The final ABC posterior approximation is the output from the chain with $h = h_M$. A related augmented space ABC sampler based on the equi-energy MCMC sampler of Kou et al. (2006) could similarly be implemented.

Ratmann et al. (2009) take the auxiliary space ABC sampler of Bortot et al. (2007) beyond the solely mechanical question of improving Markov chain mixing and towards estimation of the distribution of $s - s_{obs}$ under the model. This is more in line with the ABC approximation $\dot{\pi}_{ABC}(\theta|s_{obs})$ given by (4.9), and the interpretation of the ABC approximation to $\pi(\theta|s_{obs})$ as an exact model in the presence of model error due to Wilkinson (2013). It additionally allows an assessment of model adequacy. Instead of comparing s to s_{obs} through $K_h(\|s - s_{obs}\|)$ with a single h, Ratmann et al. (2009) alternatively make the comparison independently and univariately for each of the q summary statistics in $s = (s_1, \ldots, s_q)^\top$ via $K_{h_r}(\tau_r - |s_r - s_{obs,r}|)$ for $r = 1, \ldots, q$. Here, τ_r is the parameter denoting the true, but unknown, discrepancy between the r-th summary statistics of s and s_{obs}, i.e. $|s_r - s_{obs,r}|$, and so if $\tau_r = 0$, then the model can adequately explain the observed data as described through the r-th summary statistic. The full model has a joint target distribution of:

$$\begin{aligned}
&\pi_{ABC}(\theta, s(1), \ldots, s(T), \tau|s_{obs}) \\
\propto\quad &\min_r \left[\frac{1}{Th_r}\sum_{t=1}^{T} K_{h_r}\left(\tau_r - |s_r(t) - s_{obs,r}|\right)\right]\left[\prod_{t=1}^{T} p(s(t)|\theta)\right]\pi(\theta)\pi(\tau),
\end{aligned}$$

based on T samples $s(1), \ldots, s(T) \sim p(s|\theta)$, where $s_r(t)$ is the r-th element of $s(t)$, and $\pi(\tau) = \prod_{r=1}^{q} \pi(\tau_r)$. The minimum over the univariate density estimates aims to focus the model on the most conservative estimate of model adequacy, while also reducing computation over τ to its univariate margins. Here, interest is in the posterior distribution of τ in order to determine model adequacy (i.e. if the posterior marginal distribution of $\pi_{ABC}(\tau_r|s_{obs})$ is centred on 0), whereas the margin specific kernel scale parameters h_r are determined via standard kernel density estimation arguments over the observed sample $|s_r(t) - s_{obs,r}|$ for $t = 1, \ldots, T$.

4.3.3 Other ABC-Markov chain Monte Carlo samplers

The field of MCMC research with tractable target distributions is fairly mature, and it is not difficult to imagine that many known techniques can be directly applied to ABC-MCMC algorithms to improve their performance. Different forms of algorithms include Hamiltonian Monte Carlo ABC samplers (Meeds et al. 2015), which use a moderate number of simulations under the intractable model to produce an ABC estimate of the otherwise intractable gradient of the potential energy function, multi-try Metropolis ABC (Aandahl 2012; Kobayashi and Kozumi 2015), which uses multiple proposals to choose from at each stage of the sampler to ensure improved mixing and acceptance rates, in addition to the various augmented space samplers discussed in the previous section (Bortot et al. 2007; Ratmann et al. 2009; Baragatti et al. 2013). Of course, transdimensional ABC-MCMC samplers can also be implemented for multi-model posterior inference.

General improvements in efficiency can be obtained by using quasi Monte Carlo ABC methods to form efficient proposal distributions (Cabras et al. 2015). In a similar manner, Neal (2012) developed a coupled ABC MCMC sampler which uses the same random numbers to generate the summary statistics for different parameter values, and showed this algorithm to be more efficient than the standard ABC MCMC sampler.

Within the standard ABC MCMC sampler, Kousathanas et al. (2016) proposed using a subset $\tilde{s} \subseteq s$ of the vector of summary statistics within the acceptance probability when updating a subset of the model parameters conditional on the rest. Here the idea was to reduce the dimension of the comparison $\|\tilde{s} - \tilde{s}_{obs}\|$ within the kernel K_h to increase the efficiency and mixing of the algorithm. Rodrigues (2017) developed a related algorithm based on the Gibbs sampler.

Lee and Łatuszyński (2014) present an analysis of the variance bounding and geometric ergodicity properties of three reversible kernels used for ABC MCMC, previously suggested by Lee et al. (2012), which are based on the uniform kernel K_h. Given that current state of the chain is $\theta^{(i)}$ and a proposed new state is drawn from $\theta' \sim g(\theta^{(i)}, \theta)$, the following algorithms were examined (where $I(\cdot)$ denotes the indicator function):

- *Method 1:* Draw $s'(1), \ldots, s'(T) \sim p(s|\theta')$.

  Accept the move $\theta^{(i+1)} = \theta'$ (and $s(t) = s'(t)\ \forall t$) with probability

$$\min\left\{1, \frac{\left[\sum_{t=1}^{T} I(\|s'(t) - s_{obs}\| \leq h)\right] \pi(\theta')g(\theta', \theta^{(i)})}{\left[\sum_{t=1}^{T} I(\|s(t) - s_{obs}\| \leq h)\right] \pi(\theta^{(i)})g(\theta^{(i)}, \theta')}\right\}$$

  else reject and set $\theta^{(i+1)} = \theta^{(i)}$.

- *Method 2:* Draw $s(1), \ldots, s(T-1) \sim p(s|\theta^{(i)})$ and $s'(1), \ldots, s'(T) \sim p(s|\theta')$. Accept the move $\theta^{(i+1)} = \theta'$ with probability

$$\min\left\{1, \frac{\left[\sum_{t=1}^{T} I(\|s'(t) - s_{obs}\| \leq h)\right] \pi(\theta')g(\theta', \theta^{(i)})}{\left[1 + \sum_{t=1}^{T-1} I(\|s(t) - s_{obs}\| \leq h)\right] \pi(\theta^{(i)})g(\theta^{(i)}, \theta')}\right\}$$

  else reject and set $\theta^{(i+1)} = \theta^{(i)}$.

- *Method 3:* Reject the move and set $\theta^{(i+1)} = \theta^{(i)}$ with probability

$$1 - \min\left\{1, \frac{\pi(\theta')g(\theta', \theta^{(i)})}{\pi(\theta^{(i)})g(\theta^{(i)}, \theta)}\right\}.$$

  For $T = 1, 2, \ldots$ draw $s(T) \sim p(s|\theta^{(i)})$ and $s'(T) \sim p(s|\theta')$ until $\sum_{t=1}^{T} I(\|s(t) - s_{obs}\| \leq h) + I(\|s'(t) - s_{obs}\| \leq h) \geq 1$.

  If $I(\|s'(T) - s_{obs}\| \leq h) = 1$ then set $\theta^{(i+1)} = \theta'$ else set $\theta^{(i+1)} = \theta^{(i)}$.

Method 1 is the acceptance probability constructed from the standard Monte Carlo estimate of the ABC posterior $\hat{\pi}_{ABC}(\theta|s_{obs})$ using a fixed number, T, of summary statistic draws, as described in Section 4.2.3.5. Method 2 is the same as Method 1, except that $T-1$ of the summary statistics of the current chain state $s|\theta^{(i)}$ are regenerated anew in the denominator of the acceptance probability. The idea here is to help the Markov chain escape regions of low posterior probability more easily than under Method 1, at the cost of higher computation. Method 3 produces a random number of summary statistic generations, with computation increasing until either $s|\theta^{(i)}$ or $s'|\theta'$ is sufficiently close to s_{obs}.

Under some technical conditions, Lee and Łatuszyński (2014) conclude that Methods 1 and 2 cannot be variance bounding, and that Method 3 (as with the standard Metropolis–Hastings algorithm if it were analytically tractable) can be both variance bounding and geometrically ergodic. Overall these results, in addition to other methods for constructing estimates of intractable likelihoods (e.g. Buchholz and Chopin 2017), are very interesting from the perspective of future simulation-based algorithm design.

4.4 Sequential Monte Carlo Sampling

It can be difficult to design an importance sampling density $g(\theta)$ that is able to efficiently place a large number of samples in regions of high posterior density. Sequential Monte Carlo (SMC) and sequential importance sampling (SIS) algorithms are designed to overcome this difficulty by constructing a sequence of slowly changing intermediary distributions $f_m(\theta)$, $m = 0, \ldots, M$, where $f_0(\theta) = g(\theta)$ is the initial importance sampling distribution, and $f_M(\theta) = f(\theta)$ is the target distribution of interest. A population of particles (i.e. samples $\theta^{(i)}$, $i = 1, \ldots, N$) is then propagated between these distributions, in sequence, so that $f_1(\theta), \ldots, f_{M-1}(\theta)$ act as an efficient importance sampling bridge between $g(\theta)$ and $f(\theta)$. There are a number of techniques available for specification of the intermediary distributions (e.g. Geyer and Thompson 1995, Del Moral et al. 2006). There is a rich literature on the construction of efficient SMC and SIS algorithms. See e.g. Liu et al. (1998), Gilks and Berzuini (2001), Neal (2001), Doucet et al. (2001) and Chopin (2002) among others. These algorithms invariably involve some combination of three main ideas.

Given a weighted sample $(\theta_{m-1}^{(1)}, w_{m-1}^{(1)}), \ldots, (\theta_{m-1}^{(N)}, w_{m-1}^{(N)})$ from intermediary distribution $f_{m-1}(\theta)$, the *reweighting* step propagates the particles to the next intermediary distribution $f_m(\theta)$. This could involve a simple importance reweighting, or something more involved if hybrid importance/rejection schemes are employed (e.g. Liu et al. 1998).

Depending on the efficiency of the transitions between $f_{m-1}(\theta)$ and $f_m(\theta)$, the variability of the importance weights $w_m^{(i)}$ could be very high, with some particles having very small weights, and others having very large weights – commonly known as particle degeneracy. This can be measured through the effective sample size (4.5) (Liu et al. 1998, Liu 2001). The *resampling* step is designed to replenish the particle population by resampling the particles from their empirical distribution $(\theta_m^{(1)}, w_m^{(1)}), \ldots, (\theta_m^{(N)}, w_m^{(N)})$. In this manner, particles with low weights in regions of low density will likely be discarded in favour of particles with higher weights in regions of higher density. Following resampling, the effective sample size will be reset to N as each weight will then be set to $w_m^{(i)} = 1/N$. Resampling should not occur too frequently. A common criterion is to resample when the effective sample size falls below a pre-specified threshold, typically $E = N/2$. See e.g. Douc et al. (2005) for a review and comparison of various resampling methods.

Finally, the *move* step aims to both move the particles to regions of high probability, and increase the particle diversity in the population. The latter is important since, particularly after resampling, particles with high weights can be replicated in the sample. Any transition kernel $F_m(\theta, \theta')$ can be used for the move step, although an MCMC kernel is a common choice (e.g. Gilks and Berzuini 2001) as it results in the importance weight being unchanged, although there is also the chance that the proposed move is rejected. Other kernels, such as $F_m(\theta, \theta') = \phi(\theta'; \theta, \sigma_m^2)$ to add a random normal scatter to the particles, will require the importance weights to be modified. See e.g. Del Moral et al. (2006) for discussion on different forms of the move kernel.

4.4.1 Sequential importance sampling

In the ABC framework a natural choice for the sequence of intermediary distributions is

$$f_m(\theta) = \pi_{ABC,h_m}(\theta, s|s_{obs}) \propto K_{h_m}(\|s - s_{obs}\|)p(s|\theta)\pi(\theta),$$

for $m = 0, \ldots, M$, indexed by the kernel scale parameter, where the sequence $h_0 \geq h_1 \geq \ldots \geq h_M$ is a monotonic decreasing sequence. Accordingly, each successive distribution with decreasing h_m, will be less diffuse and a closer approximation to $\pi(\theta|s_{obs})$ (Sisson et al. 2007). A sequential importance sampling version of the ABC rejection control importance sampler (Algorithm 4.4) is given in Algorithm 4.7.

This algorithm is a particular version of the sampler proposed by Peters et al. (2012) (see also Sisson et al. 2007) who incorporated the partial rejection control mechanism of Liu et al. (1998) and Liu (2001) into the SMC sampler framework. When applied in the ABC setting, rejection control provides one means of controlling the otherwise highly variable particle weights. As with the ABC rejection control importance sampler (Algorithm 4.4), samples from an importance sampling distribution $g_m(\theta)$, constructed from the samples from the previous population targeting $f_{m-1}(\theta)$, are combined with the rejection control mechanism in order to target $f_m(\theta)$.

Algorithm 4.7: ABC Sequential Rejection Control Importance Sampling Algorithm

Inputs:

- A target posterior density $\pi(\theta|y_{obs}) \propto p(y_{obs}|\theta)\pi(\theta)$, consisting of a prior distribution $\pi(\theta)$ and a procedure for generating data under the model $p(y_{obs}|\theta)$.
- A kernel function $K_h(u)$ and a sequence of scale parameters $h_0 \geq h_1 \geq \ldots \geq h_M$.
- An initial sampling distribution $g(\theta)$, and a method of constructing subsequent sampling distributions $g_m(\theta)$, $m = 1, \ldots, M$.
- An integer $N > 0$.
- A sequence of rejection control thresholds values c_m, $m = 1, \ldots, M$.
- A low dimensional vector of summary statistics $s = S(y)$.

Initialise:
For $i = 1, \ldots, N$:

- Generate $\theta_0^{(i)} \sim g(\theta)$ from initial sampling distribution g.
- Generate $y_0^{(i)} \sim p(y|\theta_0^{(i)})$ and compute summary statistics $s_0^{(i)} = S(y_0^{(i)})$.
- Compute weights $w_0^{(i)} = K_{h_0}(\|s_0^{(i)} - s_{obs}\|)\pi(\theta_0^{(i)})/g(\theta_0^{(i)})$.

Sampling:
For $m = 1, \ldots, M$:

1. Construct sampling distribution $g_m(\theta)$.
2. For $i = 1, \ldots, N$:
   (a) Generate $\theta_m^{(i)} \sim g_m(\theta)$, $y_m^{(i)} \sim p(y|\theta_m^{(i)})$ and compute $s_m^{(i)} = S(y_m^{(i)})$.
   (b) Compute weight $w_m^{(i)} = K_{h_m}(\|s_m^{(i)} - s_{obs}\|)\pi(\theta_m^{(i)})/g_m(\theta_m^{(i)})$.
   (c) Reject $\theta_m^{(i)}$ with probability $1 - r_m^{(i)} = 1 - \min\{1, \frac{w_m^{(i)}}{c_m}\}$, and go to step 2a.
   (d) Otherwise, accept $\theta_m^{(i)}$ and set modified weight $w_m^{*(i)} = w_m^{(i)}/r_m^{(i)}$.

Output:
A set of weighted parameter vectors $(\theta_M^{(1)}, w_M^{*(1)}), \ldots, (\theta_M^{(N)}, w_M^{*(N)})$ drawn from $\pi_{ABC}(\theta|s_{obs}) \propto \int K_{h_M}(\|s - s_{obs}\|)p(s|\theta)\pi(\theta)ds$.

The initial sampling distribution $g(\theta)$ can be any importance sampling density, as with standard importance sampling algorithms. There are a number of adaptive ways to construct the subsequent importance distributions

$g_m(\theta)$ for $m = 1, \ldots, M$, based on the population of samples from the previous intermediary distribution $(\theta_{m-1}^{(1)}, w_{m-1}^{*(1)}), \ldots, (\theta_{m-1}^{(N)}, w_{m-1}^{*(N)})$. The simplest of these is to specify $g_m(\theta)$ as some standard parametric family, such as the multivariate normal distribution, with parameters estimated from the previous particle population (e.g. Chopin 2002). Another option is to construct a kernel density estimate of the distribution of the previous particle population $g_m(\theta) = \sum_{i=1}^{N} W_{m-1}^{*(i)} F_m(\theta_{m-1}^{(i)}, \theta)$ where $W_{m-1}^{*(i)} = w_{m-1}^{*(i)} / \sum_{j=1}^{N} w_{m-1}^{*(j)}$, and $F_m(\theta, \theta')$ is some forward mutation kernel describing the probability of moving from θ to θ', such as $F_m(\theta, \theta') = \phi(\theta'; \theta, \Sigma_m)$, the multivariate normal density function centred at θ and with covariance matrix Σ_m (Del Moral et al. 2006; Beaumont et al. 2009; Peters et al. 2012). These and other possibilities may also be constructed by first reweighting the draws from the previous population $f_{m-1}(\theta)$ so that they target $f_m(\theta)$.

If the kernel K_h has compact support then step 2c of Algorithm 4.7 will automatically reject any $\theta_m^{(i)}$ for which $K_{h_m}(\|s_m^{(i)} - s_{obs}\|) = 0$. (This also happens for Algorithm 4.4.) This practical outcome occurs for most ABC SIS and SMC algorithms used in practice, as use of the uniform kernel is predominant (e.g. Sisson et al. 2007, Toni et al. 2009, Beaumont et al. 2009, Del Moral et al. 2012), although the rejection of $\theta^{(i)}$ is sometimes hard coded as in Algorithm 4.3, rather than being part of a more sophisticated importance weight variance control mechanism, such as rejection control.

In the limit as rejection thresholds $c_m \to 0$ for $m = 1, \ldots, M$ (and defining $0/0 := 1$), the rejection control mechanism will allow all particles to proceed to the next stage of the algorithm. Therefore $c_m \to 0$ represents a standard sequential importance sampler that will likely result in the collapse of the particle population (i.e. all weights $w_m^{*(i)} = 0$) in the ABC setting, for low h_m. However, non-zero rejection control thresholds c_m permit a finer scale control over the importance weights $w_m^{(i)}$ beyond distinguishing between zero and non-zero weights, with larger c_m resulting in more similar weights with less variability, though at the price of higher computation through more rejections. In this manner, rejection control provides one way in which ABC SMC algorithms may be implemented with kernels K_h that are non-uniform, or have non-compact support, without which the effective sample size of the sampler would deteriorate almost immediately for low h_m (Peters et al. 2012).

As with the ABC rejection control importance sampler (Algorithm 4.4), suitable rejection thresholds may be dynamically determined during algorithm run-time by, for each m, first implementing steps 2a and 2b for $i = 1, \ldots, N$, specifying c_m as some function (such as a quantile) of the empirical distribution of the realised $w_m^{(1)}, \ldots, w_m^{(N)}$, and then continuing Algorithm 4.7 from step 2c onwards for each $i = 1, \ldots, N$ (Peters et al. 2012).

The sequence of scale parameters $h_0 \geq h_1 \geq \ldots \geq h_M$ in Algorithm 4.7 has been presented as requiring pre-specification in order to implement the sampler. However, as with any annealing-type algorithm, identifying an efficient sequence is a challenging problem. Fortunately, as with the automatic determination of the rejection control thresholds c_m, choice of the scale parameters can also be automated, and one such method to achieve this is discussed in the next section. To initialise the algorithm efficiently, setting $h_0 = \infty$ would result in all particles $\theta_0^{(1)}, \ldots, \theta_0^{(N)}$ having relatively similar weights $w_0^{(i)} = \pi(\theta_0^{(i)})/g(\theta_0^{(i)})$, as a function of the prior and initial sampling distributions.

4.4.2 Sequential Monte Carlo samplers

An alternative representation of population based algorithms is the sequential Monte Carlo sampler (Del Moral et al. 2006). Here, the particles are defined on the space of the path that each particle will take through the sequence of distributions $f_0(\theta), \ldots, f_M(\theta)$. Hence, if $\theta_m^{(i)} \in \Theta$, then the path of particle i through the first m distributions is given by $\theta_{1:m}^{(i)} = (\theta_1^{(i)}, \ldots, \theta_m^{(i)}) \in \Theta^m$ for $m = 1, \ldots, M$. SMC samplers explicitly implement each of the reweighting, resampling and move steps, and at their most general level have sophisticated implementations (e.g. Del Moral et al. 2006). A number of SMC samplers have been developed in the ABC framework (see Sisson et al. 2007, Toni et al. 2009, Beaumont et al. 2009, Drovandi and Pettitt 2011a, Del Moral et al. 2012). Algorithm 4.8 presents a generalisation (to general kernels K_h) of the adaptive ABC SMC sampler of Del Moral et al. (2012).

This algorithm provides an alternative method to rejection control to avoid the collapse of the particle population, for an arbitrary choice of kernel K_h, by making particular sampler design choices. Firstly, the probability of generating particles $\theta_m^{(i)}$ with identically zero weights $w_m^{(i)} = 0$ is reduced by increasing the number of summary statistics drawn to T, thereby targeting the joint distribution $\pi_{ABC}(\theta, s(1), \ldots, s(T)|s_{obs})$ as described in Section 4.2.3.5, although at the price of greater computation. Within the scope of an ABC SMC sampler that makes use of MCMC kernels within the move step (as with Algorithm 4.8), the alternative algorithms analysed by Lee and Łatuszyński (2014) (see Section 4.3.3) could also be implemented (e.g. Bernton et al. 2017).

In combination with the increased number of summary statistic replicates, Algorithm 4.8 directly controls the degree of particle degeneracy in moving from distribution $f_{m-1}(\theta)$ to $f_m(\theta)$. In particular, the next kernel scale parameter $h_m < h_{m-1}$ is chosen as the value which results in the effective sample size following the reweighting step, being reduced by a user specified proportion, α. In this manner, the sample degeneracy will reduce in a controlled manner at each iteration, and the sequence of h_m will adaptively reduce at exactly the rate needed to achieve this. When the effective sample size is reduced below some value E, resampling occurs and resets the effective sample size back to N, and the process repeats. As a result, resampling repeatedly occurs automatically after a fixed number of reweighting steps, as determined by α.

Algorithm 4.8: ABC Sequential Monte Carlo Algorithm

Inputs:

- A target posterior density $\pi(\theta|y_{obs}) \propto p(y_{obs}|\theta)\pi(\theta)$ consisting of a prior distribution $\pi(\theta)$ and a procedure for generating data under the model $p(y_{obs}|\theta)$.
- A kernel function $K_h(u)$, and an integer $N > 0$.
- An initial sampling density $g(\theta)$ and sequence of proposal densities $g_m(\theta, \theta')$, $m = 1, \ldots, M$.
- A value $\alpha \in [0, 1]$ to control the effective sample size.
- A low dimensional vector of summary statistics $s = S(y)$.

Initialise:
For $i = 1, \ldots, N$:

- Generate $\theta_0^{(i)} \sim g(\theta)$ from initial sampling distribution g.
- Generate $y_0^{(i)}(t) \sim p(y|\theta_0^{(i)})$ and compute summary statistics $s_0^{(i)}(t) = S(y_0^{(i)})$ for $t = 1, \ldots, T$.
- Compute weights $w_0^{(i)} = \pi(\theta_i^0)/g(\theta_0^{(i)})$, and set $m = 1$.

Sampling:

1. Reweight: Determine h_m such that $ESS(w_m^{(1)}, \ldots, w_m^{(N)}) = \alpha ESS(w_{m-1}^{(1)}, \ldots, w_{m-1}^{(N)})$ where

$$w_m^{(i)} = w_{m-1}^{(i)} \frac{\sum_{t=1}^{T} K_{h_m}(\|s_{m-1}^{(i)}(t) - s_{obs}\|)\pi(\theta_m^{(i)})}{\sum_{t=1}^{T} K_{h_{m-1}}(\|s_{m-1}^{(i)}(t) - s_{obs}\|)\pi(\theta_{m-1}^{(i)})},$$

   and then compute new particle weights and set $\theta_m^{(i)} = \theta_{m-1}^{(i)}$ and $s_m^{(i)}(t) = s_{m-1}^{(i)}(t)$ for $i = 1, \ldots, N$, and $t = 1, \ldots, T$.

2. Resample: If $ESS(w_m^{(1)}, \ldots, w_m^{(N)}) < E$ then resample N particles from the empirical distribution function $\{\theta_m^{(i)}, s_m^{(i)}(1), \ldots, s_m^{(i)}(T), W_m^{(i)}\}$ where $W_m^{(i)} = w_m^{(i)} / \sum_{j=1}^{N} w_m^{(j)}$ and set $w_m^{(i)} = 1/N$.

3. Move: For $i = 1, \ldots, N$: If $w_m^{(i)} > 0$:

   - Generate $\theta' \sim g_m(\theta_m^{(i)}, \theta)$, $y'(t) \sim p(y|\theta_m^{(i)})$ and compute $s'(t) = S(y'(t))$ for $t = 1, \ldots, T$.

- Accept θ' with probability

$$\min\left\{1, \frac{\sum_{t=1}^{T} K_{h_m}(\|s'(t) - s_{obs}\|)\pi(\theta')g(\theta', \theta_m^{(i)})}{\sum_{t=1}^{T} K_{h_m}(\|s_m^{(i)}(t) - s_{obs}\|)\pi(\theta_m^{(i)})g(\theta_m^{(i)}, \theta')}\right\}$$

  and set $\theta_m^{(i)} = \theta'$, $s_m^{(i)}(t) = s'(t)$ for $t = 1, \ldots, T$.

4. Increment $m = m + 1$. If stopping rule is not satisfied, go to 1.

Output:
A set of weighted parameter vectors $(\theta_M^{(1)}, w_M^{(1)}), \ldots, (\theta_M^{(N)}, w_M^{(N)})$ drawn from $\pi_{ABC}(\theta|s_{obs}) \propto \int K_{h_M}(\|s - s_{obs}\|)p(s|\theta)\pi(\theta)ds$.

This algorithm requires a stopping rule to terminate. If left to continue, h_m would eventually reduce very slowly, which is an indication that the sampler can no longer efficiently move the particles around the parameter space. Del Moral et al. (2012) argue that this identifies natural values of h_M that should then be adopted. In particular, they terminate their algorithm when the MCMC move rate drops below 1.5%, which then determines the final value of h_M. Alternative strategies to adaptively choose the kernel scale parameter sequence have been proposed by Drovandi and Pettitt (2011a), Drovandi and Pettitt (2011b), Silk et al. (2013) and Daly et al. (2017).

SMC algorithms provide many easy opportunities for sampler adaptation, unlike MCMC samplers which are constrained by the need to maintain the target distribution of the chain. For example, within ABC SMC algorithms, Prangle (2017) adaptively learns the relative weightings of the summary statistics within the distance function $\|s - s_{obs}\|$ to improve efficiency, Bonassi and West (2015) construct adaptive move proposal kernels $g_m(\theta_m^{(i)}, \theta) = \sum_{i=1}^{N} \nu_{m-1}^{(i)} F_m(\theta_{m-1}^{(i)}, \theta)$ based on weighting components of g_m, via $\nu_{m-1}^{(i)}$, based on the proximity of $s_{m-1}^{(i)}$ to s_{obs}, and Filippi et al. (2013) develop a different method of adaptively constructing the sequence of intermediary distributions, $f_m(\theta)$, based on Kullback–Leibler divergences between successive distributions.

Other ideas can be incorporated within ABC SMC algorithms in particular settings, or can use the ideas from ABC SMC algorithms to tackle problems related to posterior simulation. For example, Prangle et al. (2017) use ideas from rare event modelling to improve sampler efficiency within ABC SMC algorithms. When simulation from the model $p(s|\theta)$ is expensive, Everitt and Rowińska (2017) first use a cheap approximate simulator within an ABC SMC algorithm to rule out unlikely areas of the parameter space, so that expensive computation with the full simulator is avoided until absolutely necessary. Jasra et al. (2012) implement an ABC approximation within an SMC algorithm to perform filtering for a hidden Markov model. Dean et al. (2014) and Yildirim et al. (2015) use ABC SMC methods for optimisation purposes (with a different sequence of intermediary distributions), so as to derive maximum (intractable) likelihood estimators for hidden Markov models.

4.5 Discussion

ABC samplers have proved to be highly accessible and simple to implement, and it is this that has driven the popularity and spread of ABC methods more generally. Multi-model versions of each of these algorithms are available (e.g. Toni et al. 2009, Chkrebtii et al. 2015) or can be easily constructed, with ABC posterior model probabilities and Bayes factors being determined by the relative values of $\frac{1}{N}\sum_{i=1}^{N} K_h(\|s^{(i)} - s_{obs}\|)$ under each model. Although here the user needs to clearly understand the ideas of summary statistic informativeness for model choice (e.g. Marin et al. 2014) and the problems involved in computing Bayes factors as $h \to 0$ (Martin et al. 2017).

Improvements to general ABC samplers include increasing algorithmic efficiency by using quasi Monte Carlo methods (Buchholz and Chopin 2017), and the use of multi-level rejection sampling (Warne et al. 2017) (see Jasra et al. 2017 for the SMC version) for variance reduction. The lazy ABC method of Prangle (2016) states that it may be possible to terminate expensive simulations $s \sim p(s|\theta)$ early, if it is also possible to calculate the probability that the full simulation when run to completion would have been rejected. Diagnostics to determine whether the kernel scale parameter h_m is sufficiently low that $\pi_{ABC}(\theta|s_{obs})$ is indistinguishable from $\pi(\theta|s_{obs})$ were developed by Prangle et al. (2014). There are many related results on the rate of convergence of ABC algorithms as measured through the mean squared error of point estimates (Blum 2010; Fernhead and Prangle 2012; Calvet and Czellar 2015; Biau et al. 2015; Barber et al. 2015). ABC samplers have also allowed previously unclear links to other algorithms to become better understood – for example, Nott et al. (2012) have reinterpreted the Kalman filter as an ABC algorithm, and Drovandi (2019) (Chapter 7, this volume) has comprehensively described the links between ABC and indirect inference.

A number of algorithms related to ABC methods have emerged, including Bayesian empirical likelihoods (Mengersen et al. 2013) and bootstrap likelihoods (Zhu et al. 2016), the synthetic likelihood (Wood 2010, Drovandi et al. 2019, Chapter 12, this volume), the expectation-propagation ABC algorithm (Barthelmé and Chopin 2014; Barthelmé et al. 2019, Chapter 14, this volume), Albert et al. (2015)'s particle-based simulated annealing algorithm, and

Forneron and Ng (2016)'s optimisation-based likelihood free importance sampling algorithm. Perhaps the biggest offshoot of ABC samplers is the more general pseudo-marginal Monte Carlo method (Beaumont 2003; Andrieu and Roberts 2009), which implements exact Monte Carlo simulation with an unbiased estimate of the target distribution, of which ABC is a particular case. See Andrieu et al. (2019) (Chapter 9, this volume) for an ABC-centred exploration of these methods.

Acknowledgements

SAS is supported by the Australian Research Council under the Discovery Project scheme (DP160102544), and the Australian Centre of Excellence in Mathematical and Statistical Frontiers (CE140100049).

References

Aandahl, R. Z. (2012). Likelihood-free Bayesian methods for inference using stochastic evolutionary models of Mycobacterium tuberculosis. Ph. D. thesis, University of New South Wales, Sydney.

Albert, C., H. R. Keunsch, and A. Scheidegger (2015). A simulated annealing approach to approximate Bayesian computation. *Statistics and Computing 25*, 1217–1232.

Andrieu, C., A. Lee, and M. Vihola (2019). Theoretical and methodological aspects of MCMC computations with noisy likelihoods. In S. A. Sisson, Y. Fan, and M. A. Beaumont (Eds.), *Handbook of Approximate Bayesian Computation*. Boca Raton, FL, Chapman & Hall/CRC Press.

Andrieu, C. and G. O. Roberts (2009). The pseudo-marginal approach for efficient Monte Carlo computations. *Annals of Statistics 37*, 697–725.

Baragatti, M., A. Grimaud, and D. Pommeret (2013). Likelihood-free parallel tempering. *Statistics and Computing 23*, 535–549.

Barber, S., J. Voss, and M. Webster (2015). The rate of convergence of approximate Bayesian computation. *Electronic Journal of Statistics 9*, 80–105.

Barthelmé, S. and N. Chopin (2014). Expectation-propagation for likelihood-free inference. *Journal of the American Statistical Association 109*, 315–333.

Barthelmé, S., N. Chopin, and V. Cottet (2019). Divide and conquer in ABC: Expectation-propagation algorithms for likelihood-free inference. In S. A. Sisson, Y. Fan, and M. A. Beaumont (Eds.), *Handbook of Approximate Bayesian Computation*. Boca Raton, FL, Chapman & Hall/CRC Press.

Beaumont, M. A. (2003). Estimation of population growth or decline in genetically monitored populations. *Genetics 164*, 1139–1160.

Beaumont, M. A., J.-M. Cornuet, J.-M. Marin, and C. P. Robert (2009). Adaptive approximate Bayesian computation. *Biometrika 96*, 983–990.

Beaumont, M. A., W. Zhang, and D. J. Balding (2002). Approximate Bayesian computation in population genetics. *Genetics 162*, 2025–2035.

Bernton, E., P. E. Jacob, M. Gerber, and C. P. Robert (2017). Inference in generative models using the Wasserstein distance. *arXiv:1701.05146*.

Biau, G., F. Cérou, and A. Guyader (2015). New insights into approximate Bayesian computation. *Annales de l'Institut Henri Poincare (B) Probability and Statistics 51*, 376–403.

Blum, M. G. B. (2010). Approximate Bayesian computation: A nonparametric perspective. *Journal of the American Statistical Association 105*, 1178–1187.

Blum, M. G. B., M. A. Nunes, D. Prangle, and S. A. Sisson (2013). A comparative review of dimension reduction methods in approximate Bayesian computation. *Statistical Science 28*, 189–208.

Bonassi, F. V. and M. West (2015). Sequential Monte Carlo with adaptive weights for approximate Bayesian computation. *Bayesian Analysis 10*, 171–187.

Bornn, L., N. Pillai, A. Smith, and D. Woodward (2017). The use of a single pseudo-sample in approximate Bayesian computation. *Statistics and Computing 27*, 583–590.

Bortot, P., S. G. Coles, and S. A. Sisson (2007). Inference for stereological extremes. *Journal of the American Statistical Association 102*, 84–92.

Brooks, S. P., A. Gelman, G. L. Jones, and X.-L. Meng (Eds.) (2011). *Handbook of Markov Chain Monte Carlo*. CRC Press.

Buchholz, A. and N. Chopin (2017). Improving approximate Bayesian computation via quasi Monte Carlo. https://arxiv.org/abs/1710.01057.

Cabras, S., M. E. C. Nueda, and E. Ruli (2015). Approximate Bayesian computation by modelling summary statistics in a quasi-likelihood framework. *Bayesian Analysis 10*, 411–439.

Calvet, L. E. and V. Czellar (2015). Accurate methods for approximate Bayesian computation filtering. *Journal of Financial Econometrics 13*, 798–838.

Chkrebtii, O. A., E. K. Cameron, S. A. Campbell, and E. M. Bayne (2015). Transdimensional approximate Bayesian computation for inference in invasive species models with latent variables of unknown dimension. *Computational Statistics & Data Analysis 86*, 97–110.

Chopin, N. (2002). A sequential particle filter method for static models. *Biometrika 89*, 539–551.

Daly, A. C., D. J. Gavaghan, C. Holmes, and J. Cooper (2017). Hodgkin-Huxley revisited: Reparametrization and identifiability analysis of the classic action potential model with approximate Bayesian methods. *Royal Society Open Science 2*, 150499.

Dean, T. A., S. S. Singh, A. Jasra, and G. W. Peters (2014). Parameter estimation for hidden Markov models with intractable likelihoods. *Scandinavian Journal of Statistics 41*, 970–987.

Del Moral, P., A. Doucet, and A. Jasra (2006). Sequential Monte Carlo samplers. *Journal of the Royal Statistical Society, Series B 68*, 411–436.

Del Moral, P., A. Doucet, and A. Jasra (2012). An adaptive sequential Monte Carlo method for approximate Bayesian computation. *Statistics and Computing 22*, 1009–1020.

Douc, R., O. Cappe, and E. Moulines (2005). Comparison of resampling schemes for particle filtering. *Proceedings of the 4th International Symposium on Image and Signal Processing and Analysis*, 64–69.

Doucet, A., N. de Freitas, and N. Gordon (Eds.) (2001). *Sequential Monte Carlo Methods in Practice.* Springer.

Drovandi, C. and A. Pettitt (2011a). Estimation of parameters for macroparasite population evolution using approximate Bayesian computation. *Biometrics 67*, 225–233.

Drovandi, C. and A. Pettitt (2011b). Likelihood-free Bayesian estimation of multivariate quantile distributions. *Computational Statistics & Data Analysis 55*, 2541–2556.

Drovandi, C. C. (2019). ABC and indirect inference. In S. A. Sisson, Y. Fan, and M. A. Beaumont (Eds.), *Handbook of Approximate Bayesian Computation.* Boca Raton, FL, Chapman & Hall/CRC Press.

Drovandi, C. C., C. Grazian, K. Mengersen, and C. P. Robert (2019). Approximating the likelihood in approximate Bayesian computation. In S. A. Sisson, Y. Fan, and M. A. Beaumont (Eds.), *Handbook of Approximate Bayesian Computation.* Boca Raton, FL, Chapman & Hall/CRC Press.

Everitt, R. G. and P. A. Rowińska (2017). Delayed acceptance ABC-SMC. https://arxiv.org/abs/1708.02230.

Fernhead, P. and D. Prangle (2012). Constructing summary statistics for approximate Bayesian computation: Semi-automatic approximate Bayesian computation (with discussion). *Journal of the Royal Statistical Society, Series B 74*, 419–474.

Filippi, S., C. P. Barnes, J. Cornebise, and M. P. H. Stumpf (2013). On optimality of kernels for approximate Bayesian computation using sequential Monte Carlo. *Statistical Applications in Genetics and Molecular Biology 12*, 1–12.

Forneron, J.-J. and S. Ng (2016). A likelihood-free reverse sampler of the posterior distribution. *Advances in Econometrics 36*, 389–415.

Geyer, C. J. and E. A. Thompson (1995). Annealing Markov chain Monte Carlo with applications to ancestral inference. *Journal of the American Statistical Association 90*, 909–920.

Gilks, W. R. and C. Berzuini (2001). Following a moving target: Monte Carlo inference for dynamic Bayesian models. *Journal of the Royal Statistical Society, Series B 63*, 127–146.

Jasra, A., S. Jo, D. J. Nott, C. Shoemaker, and R. Tempone (2017). Multilevel Monte Carlo in approximate Bayesian computation. https://arxiv.org/abs/1702.03628.

Jasra, A., S. S. Singh, J. Martin, and E. McCoy (2012). Filtering via ABC. *Statistics and Computing 22*, 1223–1237.

Kobayashi, G. and H. Kozumi (2015). Generalized multiple-point Metropolis algorithms for approximate Bayesian computation. *Journal of Statistical Computation and Simulation 85*, 675–692.

Kou, S. C., Q. Zhou, and W. H. Wong (2006). Equi-energy sampler with applications in statistical inference and statistical mechanics. *Annals of Statistics 34*, 1581–1619.

Kousathanas, A., C. Leuenberger, J. Helfer, M. Quinodoz, M. Foll, and D. Wegmann (2016). Likelihood-free inference in high-dimensional models. *Genetics 203*, 893–904.

Lee, A., C. Andrieu, and A. Doucet (2012). Discussion of a paper by P. Fearnhead and D. Prangle. *Journal of the Royal Statistical Society, Series B 74*, 419–474.

Lee, A. and Łatuszyński (2014). Monte Carlo methods for approximate Bayesian computation. *Biometrika 101*, 655–671.

Liu, J. S. (2001). *Monte Carlo Strategies in Scientific Computing.* Springer-Verlag, New York.

Liu, J. S., R. Chen, and W. H. Wong (1998). Rejection control and sequential importance sampling. *Journal of the American Statistical Association 93*, 1022–1031.

Marin, J.-M., N. Pillai, C. P. Robert, and J. Rousseau (2014). Relevant statistics for Bayesian model choice. *Journal of the Royal Statistical Society, Series B 76*, 833–859.

Marjoram, P., J. Molitor, V. Plagnol, and S. Tavaré (2003). Markov chain Monte Carlo without likelihoods. *Proceedings of the National Academy of Sciences of the United States of America 100*, 15324–15328.

Martin, G. M., D. T. Frazier, E. M. R. Renault, and C. P. Robert (2017). The validation of approximate Bayesian computation: Theory and practice. Technical report, Dept. of Econometrics and Business Statistics, Monash University.

Meeds, E., R. Leenders, and M. Welling (2015). Hamiltonian ABC. *Uncertainty in Artificial Intelligence 31*, 582–591.

Mengersen, K., P. Pudlo, and C. P. Robert (2013). Bayesian computation via empirical likelihood. *Proceedings of the National Academy of Sciences of the United States of America 110*, 1321–1326.

Neal, P. (2012). Efficient likelihood-free Bayesian computation for household epidemics. *Statistics and Computing 22*, 1239–1256.

Neal, R. (2001). Annealed importance sampling. *Statistics and Computing 11*, 125–139.

Nott, D. J., L. Marshall, and M. N. Tran (2012). The ensemble Kalman filter is an ABC algorithm. *Statistics and Computing 22*, 1273–1276.

Peters, G. W., Y. Fan, and S. A. Sisson (2012). On sequential Monte Carlo, partial rejection control and approximate Bayesian computation. *Statistics and Computing 22*, 1209–1222.

Prangle, D. (2016). Lazy ABC. *Statistics and Computing 26*, 171–185.

Prangle, D. (2017). Adapting the ABC distance function. *Bayesian Analysis 12*, 289–309.

Prangle, D., M. G. B. Blum, G. Popovic, and S. A. Sisson (2014). Diagnostic tools for approximate Bayesian computation using the coverage property. *Australia and New Zealand Journal of Statistics 56*, 309–329.

Prangle, D., R. G. Everitt, and T. Kypraios (2017). A rare event approach to high-dimensional approximate Bayesian computation. *Statistics and Computing*, in press.

Pritchard, J. K., M. T. Seielstad, A. Perez-Lezaun, and M. W. Feldman (1999). Population growth of human Y chromosomes: A study of Y chromosome microsatellites. *Molecular Biology and Evolution 16*, 1791–1798.

Ratmann, O., C. Andrieu, T. Hinkley, C. Wiuf, and S. Richardson (2009). Model criticism based on likelihood-free inference, with an application to protein network evolution. *Proceedings of the National Academy of Sciences of the United States of America 106*, 10576–10581.

Ratmann, O., O. Jorgensen, T. Hinkley, M. Stumpf, S. Richardson, and C. Wiuf (2007). Using likelihood-free inference to compare evolutionary dynamics of the protein networks of h. pylori and p. falciparum. *PLoS Computational Biology 3*, e230.

Reeves, R. W. and A. N. Pettitt (2005). A theoretical framework for approximate Bayesian computation. In A. R. Francis, K. M. Matawie, A. Oshlack, and G. K. Smyth (Eds.), *Proceedings of the 20th International Workshop for Statistical Modelling, Sydney Australia, July 10–15, 2005*, pp. 393–396.

Ripley, B. D. (1987). *Stochastic Simulation.* John Wiley & Sons.

Rodrigues, G. S. (2017). New methods for infinite and high-dimensional approximate Bayesian computation. Ph. D. thesis, University of New South Wales, Sydney.

Silk, D., S. Filippi, and M. P. H. Stumpf (2013). Optimising threshold schedules for approximate Bayesian computation sequential Monte Carlo samplers: Applications to molecular systems. *Statistical Applications in Genetics and Molecular Biology 12.*

Sisson, S. A. and Y. Fan (2011). Likelihood-free MCMC. In S. Brooks, A. Gelman, G. L. Jones, and X.-L. Meng (Eds.), *Handbook of Markov Chain Monte Carlo*, pp. 313–335. Chapman & Hall/CRC Press.

Sisson, S. A., Y. Fan, and M. A. Beaumont (2019). Overview of approximate Bayesian computation. In S. A. Sisson, Y. Fan, and M. A. Beaumont (Eds.), *Handbook of Approximate Bayesian Computation.* Chapman & Hall/CRC Press.

Sisson, S. A., Y. Fan, and M. M. Tanaka (2007). Sequential Monte Carlo without likelihoods. *Proceedings of the National Academy of Sciences of the United States of America 104*, 1760–1765. Errata (2009), 106:16889.

Tavaré, S., D. J. Balding, R. C. Griffiths, and P. Donnelly (1997). Inferring coalescence times from DNA sequence data. *Genetics 145*, 505–518.

Toni, T., D. Welch, N. Strelkowa, A. Ipsen, and M. P. H. Stumpf (2009). Approximate Bayesian computation scheme for parameter inference and model selection in dynamical systems. *Journal of the Royal Society Interface 6*, 187–202.

Warne, D. J., R. E. Baker, and M. J. Simpson (2017). Multilevel rejection sampling for approximate Bayesian computation. https://arxiv.org/abs/1702.03126.

Wegmann, D., C. Leuenberger, and L. Excoffier (2009). Efficient approximate Bayesian computation coupled with Markov chain Monte Carlo without likelihood. *Genetics 182*, 1207–1218.

Wilkinson, R. L. (2013). Approximate Bayesian computation (ABC) gives exact results under the assumption of model error. *Statistical Applications in Genetics and Molecular Biology 12*, 129–141.

Wood, S. N. (2010). Statistical inference for noisy nonlinear ecological dynamic systems. *Nature 466*, 1102–1104.

Yildirim, S., S. S. Singh, T. A. Dean, and A. Jasra (2015). Parameter estimation in hidden Markov models with intractable likelihoods using sequential Monte Carlo. *Journal of Computational and Graphical Statistics 24*, 846–865.

Zhu, W., J. M. Marin, and F. Leisen (2016). A bootstrap likelihood approach to Bayesian computation. *Australia and New Zealand Journal of Statistics 58*, 227–224.

5

Summary Statistics

Dennis Prangle

CONTENTS

5.1 Theory 127
5.1.1 The curse of dimensionality 127
5.1.2 Sufficiency 128
5.2 Strategies for Summary Statistic Selection 129
5.3 Subset Selection Methods 129
5.3.1 Methods 130
Approximate sufficiency 130
Entropy/loss minimisation 131
Mutual information 131
Regularisation approaches 132
Related methods 132
5.3.2 Discussion 133
Comparison of subset selection methods 133
Advantages and disadvantages of subset selection 133
5.4 Projection Methods 134
5.4.1 Dimension reduction techniques 134
Partial least squares 134
Linear regression 135
Boosting 135
5.4.2 Generating training data 136
5.4.3 Discussion 137
Comparison of projection methods 137
Advantages and disadvantages of projection methods 137
5.5 Auxiliary Likelihood Methods 137
5.5.1 Methods 138
Maximum likelihood estimators 138
Likelihood distance 139
Scores 139
5.5.2 Discussion 139
Comparison of auxiliary likelihood methods 139
Which auxiliary likelihood? 140
Advantages and disadvantages of auxiliary likelihood methods 140

5.6 Model Choice ... 141
5.6.1 Theory ... 141
Curse of dimensionality ... 141
Sufficiency and consistency ... 142
5.6.2 Methods ... 143
Using an encompassing model ... 143
Mutual information ... 143
Projection/classification methods ... 143
Local error rates ... 144
5.6.3 Discussion ... 145
Comparison of methods ... 145
Prospects ... 145
5.7 Discussion ... 145
Empirical performance ... 145
Which method to use ... 146
Prospects ... 147
References ... 148

To deal with high dimensional data, ABC algorithms typically reduce them to lower dimensional *summary statistics* and accept when simulated summaries $S(y)$ are close to the observed summaries $S(y_{\text{obs}})$. This has been an essential part of ABC methodology since the first publications in the population genetics literature. Overviewing this work Beaumont et al. (2002) wrote: 'A crucial limitation of the...method is that only a small number of summary statistics can usually be handled. Otherwise, either acceptance rates become prohibitively low or the tolerance...must be increased, which can distort the approximation', and related the problem to the general issue of the *curse of dimensionality*: many statistical tasks are substantially more difficult in high dimensional settings. In ABC, the dimension in question is the number of summary statistics used.

To illustrate the issue, consider an ABC rejection sampling algorithm. As more summary statistics are used, there are more opportunities for random discrepancies between $S(y)$ and $S(y_{\text{obs}})$. To achieve a reasonable number of acceptances, it is necessary to use a large threshold and accept many poor matches. Therefore, as noted in the earlier quote, using too many summary statistics distorts the approximation of the posterior. On the other hand, if too few summary statistics are used, some fine details of the data can be lost. This allows parameter values to be accepted which are unlikely to reproduce these details. Again, a poor posterior approximation is often obtained.

As a result of the considerations earlier, a good choice of ABC summary statistics must strike a balance between low dimension and informativeness. Many methods have been proposed aiming to select such summary statistics, and the main aim of this chapter is to review these. There is some overlap between this material and the previous review paper of Blum et al. (2013). This chapter adds coverage of recent developments, particularly on auxiliary likelihood methods and ABC model choice. However, less detail is provided on each method here, due to the larger number now available. The chapter focuses on summary statistic selection methods which can be used with standard ABC algorithms. Summary statistic methods for more specialised recent algorithms (see Chapters 8, 11, and 21 of this book for example are discussed only briefly. Secondary aims of the chapter are to collate relevant theoretical results and discuss issues which are common to many summary statistic selection methods.

An overview of the chapter is as follows. Section 5.1 is a review of theoretical results motivating the use of summary statistics in ABC. Section 5.2 describes three strategies for summary statistic selection and introduces some general terminology. Sections 5.3–5.5 describe particular methods from each strategy in turn. Up to this point, the chapter concentrates on ABC for parameter inference. Section 5.6 instead considers summary statistics for ABC model choice, covering both theory and methods. Section 5.7 concludes by summarising empirical comparisons between methods, discussing which method to use in practice, and looking at prospects for future developments.

5.1 Theory

This section describes why methods for selecting summary statistics are necessary in more theoretical detail. Section 5.1.1 discusses the curse of dimensionality, showing that low dimensional informative summaries are needed. The concept of *sufficient statistics*, reviewed in Section 5.1.2, would appear to provide an ideal choice of summaries. However, it is shown that low dimensional sufficient statistics are typically unavailable. Hence, methods for selecting low dimensional insufficient summaries are required.

Note that from here to Section 5.5, the subject is ABC for parameter inference, understood to mean inference of *continuous* parameters. Theoretical results on ABC for model choice will be discussed in Section 5.6. These are also relevant for inference of discrete parameters.

5.1.1 The curse of dimensionality

A formal approach to the curse of dimensionality is to consider how the error in an ABC approximation is related to the number of simulated datasets produced, n. It can be shown that, at least asymptotically, the rate at which the error decays becomes worse, as the dimension of the data increases. For example, Barber et al. (2015) consider mean squared error of a Monte Carlo estimate produced by ABC rejection sampling. Under optimal ABC tuning and some regularity conditions, this is shown to be $O_p(n^{-4/(q+4)})$, where q denotes $\dim S(y)$. This is an asymptotic result in a regime where n is large and the ABC bandwidth h is close to zero. Several authors (Blum, 2010a; Fearnhead and Prangle, 2012; Biau et al., 2015) consider different definitions of error and different ABC algorithms and prove qualitatively similar results. That is, similar asymptotic expressions for error are found with slightly different terms in the exponent of n. While these asymptotic results may not exactly capture behaviour for larger h, they strongly suggest that high dimensional summaries typically give poor results.

Note that it is sometimes possible to avoid the curse of dimensionality for models whose likelihood factorises, such as state space models for time series data. This can be done by performing ABC in stages for each factor. This allows summary statistics to be chosen for each stage, rather than requiring high dimensional summaries of the entire model. Jasra (2015) reviews this approach for time series data. See also the related method in Chapter 21 of this book.

5.1.2 Sufficiency

Two common definitions of sufficiency of a statistic $s = S(y)$ under a model parameterised by θ are as follows. See Cox and Hinkley (1979) for full details of this and all other aspects of sufficiency covered in this section. The classical definition is that the conditional density $\pi(y|s, \theta)$ is invariant to θ. Alternatively, the statistic is said to be *Bayes sufficient* for θ if $\theta|s$ and $\theta|y$ have the same distribution for any prior distribution and almost all y. The two definitions are equivalent for finite dimensional θ. Bayes sufficiency is a natural definition of sufficiency to use for ABC, as it shows that in an ideal ABC algorithm with sufficient S and $h \to 0$, the ABC target distribution equals the correct posterior. It can also be used to consider sufficiency for a subset of the parameters, which is useful later when ABC model choice is considered.

For independent identically distributed data, the Pitman–Koopman–Darmois theorem states that under appropriate assumptions, only models in the exponential family possess a sufficient statistic with dimension equal to the dimension of θ. For other models, the dimension of any sufficient statistic increases with the sample size. Exponential family models generally have tractable likelihoods so that ABC is not required. This result strongly suggests that for other models, low dimensional sufficient statistics do not exist.

Despite this result there are several ways in which notions of sufficiency can be useful in ABC. First, stronger sufficiency results are possible for ABC model choice, which are outlined in Section 5.6. Second, notions of approximate and asymptotic sufficiency are used to motivate some methods described later (i.e., those of Joyce and Marjoram, 2008 and Martin et al., 2014).

5.2 Strategies for Summary Statistic Selection

This chapter splits summary statistic selection methods into three strategies: *subset selection*, *projection*, and *auxiliary likelihood*. This section gives an overview of each and introduces some useful general terminology. The categories are based on those used in Blum et al. (2013) with slight changes: auxiliary likelihood is added and 'regularisation' is placed under subset selection. In practice, the categories overlap, with some methods applying a combination of the strategies.

Subset selection and projection methods require a preliminary step of choosing a set of *data features* $z(y)$. For subset selection these can be thought of as *candidate summary statistics*. For convenience, $z(y)$ is often written simply as z below and is a vector of scalar transformations of y, $(z_1, z_2, \dots, z_k)$. The *feature selection* step of choosing z is discussed further below. Both methods also require *training data* $(\theta_i, y_i)_{1 \le i \le n_{\text{train}}}$ to be created by simulation.

Subset selection methods select a subset of z, typically that which optimises some criterion on the training data. Projection methods instead use the training data to choose a projection of z, for example, a linear transformation $Az + b$, which performs dimension reduction.

Auxiliary likelihood methods take a different approach which does not need data features or training data. Instead, they make use of an approximating model whose likelihood (the 'auxiliary' likelihood) is more tractable than the model of interest. This may be chosen from subject area knowledge or make use of a general approach, such as composite likelihood. Summary statistics are derived from this approximating model, for example, its maximum likelihood estimators.

All these methods rely on some subjective input from the user. In subset selection methods, candidate summaries $z(y)$ must be supplied. A typical choice will be a reasonably small set of summaries which are believed to be informative based on subject area knowledge. (Large sets become too expensive for most methods to work with.) Projection methods also require a subjective choice of $z(y)$. A typical choice will be many interesting features of the data, and various non-linear transformations. These may not be believed to be informative individually, but permit a wide class of potential projections. There is less requirement for $\dim z$ to be small than for subset selection. Auxiliary likelihood methods instead require the subjective choice of an approximate likelihood (discussed in Section 5.5.2.)

5.3 Subset Selection Methods

Subset selection methods start with candidate summary statistics $z = (z_1, z_2, \dots, z_k)$ and attempt to select an informative subset. The following methods fall into two groups. The first three run ABC for many possible subsets and choose the best based on information theoretic or other summaries of the output. This requires ABC to be run many times, which is only computationally feasible for rejection or importance sampling ABC algorithms, since these allow simulated datasets to be re-used. The final method, regularisation, takes a different approach. All these methods are described in Section 5.3.1. Section 5.3.2 compares the methods and discusses the strengths and weaknesses of this strategy.

5.3.1 Methods

Approximate sufficiency (Joyce and Marjoram, 2008)

Joyce and Marjoram (2008) propose a stepwise selection approach. They add/ remove candidate summary statistics to/from a subset one at a time and assess whether this significantly affects the resulting ABC posterior. The motivation is that given sufficient statistics $S(\cdot)$ of minimal dimension, adding further summaries will not change $\pi(\theta|S(y_{\text{obs}}))$, but removing any summary will. This would be a test for sufficiency, but requires perfect knowledge of $\pi(\theta|S(y_{\text{obs}}))$. Joyce and Marjoram propose a version of this test based on having only a density estimate and argue it is a test of *approximate sufficiency*. A further approximation is due to using $\pi_{\text{ABC}}(\theta|S(y_{\text{obs}}))$ in place of $\pi(\theta|S(y_{\text{obs}}))$.

The approach involves testing whether the change from using summaries $S(y)$ to $S'(y)$ has a significant effect on the ABC posterior. As various subsets are compared, this test will be repeated under many choices of $S(y)$ and $S'(y)$. A change is deemed significant if:

$$\left| \frac{\hat{\pi}_{\text{ABC}}(\theta|S'(y_{\text{obs}}))}{\hat{\pi}_{\text{ABC}}(\theta|S(y_{\text{obs}}))} - 1 \right| > T(\theta), \tag{5.1}$$

where $\hat{\pi}_{\text{ABC}}(\theta)$ is an estimated posterior density based on ABC rejection sampling output, detailed shortly. The threshold $T(\cdot)$ is defined to test the null hypothesis that ABC targets the same distribution under both sets of summary statistics and control for multiple comparison issues arising from testing (5.1) for several θ values. See the appendix of Joyce and Marjoram for precise details of how this is defined.

Only the case of scalar θ is considered by Joyce and Marjoram. Here they propose letting $\hat{\pi}_{\text{ABC}}(\theta)$ be a histogram estimator. That is, the support of θ is split into bins $B_1, B_2, \ldots, B_b$, and the proportion of the accepted sample in bin B_i gives $\hat{\pi}_{\text{ABC}}(\theta)$ for $\theta \in B_i$. In practice, (5.1) is evaluated at a finite set of parameters $\theta_1 \in B_1, \theta_2 \in B_2, \ldots, \theta_b \in B_b$.

Joyce and Marjoram note it is not obvious how to implement their method for higher dimensional parameters. This is because the sample size required to use the histogram estimator of $\hat{\pi}_{\text{ABC}}(\theta)$ becomes infeasibly large, and the paper's choice of $T(\cdot)$ is specific to this estimator.

Entropy/loss minimisation (Nunes and Balding, 2010)

Nunes and Balding (2010) propose two approaches. The first aims to find the subset of z which minimises the entropy of the resulting ABC posterior. The motivation is that entropy measures how concentrated, and thus informative, the posterior is, with lowest entropy being most informative. In practice, an estimate of the entropy is used which is based on a finite sample from the ABC posterior: an extension of the estimate of Singh et al. (2003).

One criticism of the entropy criterion (Blum et al., 2013) is that in some circumstances an ABC posterior having smaller entropy does not correspond to more accurate inference. For example it is possible that given a particularly precise prior the correct posterior may be more diffuse. (See the next page for a further comment on this.)

The second approach aims to find the subset of z which optimises the performance of ABC on datasets similar to y_{obs}, by minimising the average of the following loss function (root mean squared error):

$$\left[t^{-1}\sum_{i=1}^{t}||\theta_i - \theta'||_2\right]^{1/2}.$$

Here, θ' is the parameter value which generated data y', and $(\theta_i)_{1\le i\le t}$ is the ABC output sample when y' is used as the observations. Performing this method requires generating (θ', y') pairs such that y' is close to y_{obs}. Nunes and Balding recommend doing so using a separate ABC analysis whose summary statistics are chosen by entropy minimisation. To summarise, this is a two-stage approach. First, select summaries by entropy minimisation and perform ABC to generate (θ', y') pairs. Second, select summaries by minimising root mean squared error.

Mutual information (Barnes et al., 2012; Filippi et al., 2012)

Barnes et al. (2012) and Filippi et al. (2012) discuss how sufficiency can be restated in terms of the concept of *mutual information*. In particular, sufficient statistics maximise the mutual information between $S(y)$ and θ. From this, they derive a necessary condition for sufficiency of $S(y)$: the Kullback–Leibler (KL) divergence of $\pi(\theta|S(y_{\text{obs}}))$ from $\pi(\theta|y_{\text{obs}})$ is zero for example:

$$\int \pi(\theta|y_{\text{obs}}) \log \frac{\pi(\theta|y_{\text{obs}})}{\pi(\theta|S(y_{\text{obs}}))} d\theta = 0.$$

This motivates a stepwise selection method to choose a subset of z. A statistic z_i is added to the existing subset $S(y)$ to create a new subset $S'(y)$ if the estimated KL divergence of $\pi_{\text{ABC}}(\theta|S(y_{\text{obs}}))$ from $\pi_{\text{ABC}}(\theta|S'(y_{\text{obs}}))$ is above a threshold. One proposed algorithm chooses z_i to maximise this divergence (a 'greedy' approach). Another attempts to save time by selecting any acceptable z_i. Steps are also provided to remove statistics that have become unnecessary. Two approaches are given for estimating KL divergence between two ABC output samples. The stepwise selection algorithm terminates when the improvement in KL divergence is below a threshold. To determine a suitable threshold, it is suggested to perform ABC several times with fixed summaries and evaluate the KL divergences between samples. From this, a threshold can be found indicating an insignificant change.

The mutual information method is closely related to the previous methods in this section. Like the method of Joyce and Marjoram (2008), it seeks a subset $S(y)$, such that adding further statistics does not significantly change the ABC posterior. However, the KL criterion has the advantage that it can be applied when $\dim \theta > 1$. It does share the disadvantage that $\pi_{\text{ABC}}(\theta|S(y_{\text{obs}}))$ is used in place of $\pi(\theta|S(y_{\text{obs}}))$, which is a poor estimate unless $h \approx 0$. Maximising mutual information as in Barnes et al., can be shown to be equivalent to minimising the *expected* entropy of $\pi(\theta|S(y))$ (taking expectation over y). This provides some information theoretic support for the entropy minimisation approach of Nunes and Balding (2010), but also gives more insight into its limitations: the entropy of $\pi(\theta|S(y_{\text{obs}}))$, which they consider may not be representative of expected entropy. Barnes et al. also argue their method is more robust than Nunes and Balding's to the danger of selecting more statistics than is necessary for sufficiency.

Barnes et al. extend their method to model choice. This is discussed in Section 5.6.

Regularisation approaches (Blum, 2010b; Sedki and Pudlo, 2012; Blum et al., 2013)

This method was proposed by Sedki and Pudlo (2012) and Blum et al. (2013). The idea is to fit a linear regression with response θ and covariates z based on training data and perform variable selection to find an informative subset of z. Variable selection can be performed by minimising AIC or BIC (see Blum et al., 2013, for details of these in this setting including how to calculate.) This typically requires a stepwise selection approach, although the cost of this could be avoided by using the lasso (Hastie et al., 2009) rather than ordinary regression. The papers propose also using the fitted regression for ABC regression post-processing, as well as summary statistic selection. A related earlier approach (Blum, 2010b) proposed using a local linear regression model and performing variable selection by an empirical Bayes approach: maximising the likelihood of the training data after integrating out the distribution of the regression parameters.

Related methods

Heggland and Frigessi (2004) provide asymptotic theory on summary statistic selection in another likelihood-free method, indirect inference. Roughly speaking, the most useful summary statistics are those with low variance and expectation that changes rapidly with the parameters. This theory is used to select summary statistics from a set of candidates by numerically estimating their variance and the derivative of their expectation based on a large number of simulations from promising parameter values. It would be useful to develop ABC versions of this theory and methodology. Ratmann et al. (2007) go some way towards providing the latter for a particular application.

5.3.2 Discussion

Comparison of subset selection methods

Joyce and Marjoram (2008) is based on rough ideas of sufficiency and can be implemented only in the limited setting of a scalar parameter. The entropy minimisation method of Nunes and Balding (2010) and the approach of Barnes et al. (2012) are successively more sophisticated information theoretic approaches. The latter has the best theoretical support of methods based on such arguments. However, all these methods are motivated by properties of $\pi(\theta|S(y_{\text{obs}}))$, but then use $\pi_{\text{ABC}}(\theta|S(y_{\text{obs}}))$ in its place. For sufficiently large h these may be quite different, and it is unclear what effect this will have on the results. The loss minimisation method of Nunes and Balding (2010) avoids this problem. It chooses S so that ABC results on simulated data optimise a specified loss function given a particular choice of h. The question of robustness to the choice of which simulated datasets are used to assess this is still somewhat open, but otherwise this method seems to provide a gold standard approach. The drawback of all the previous methods is that they can be extremely expensive (see below). Regularisation methods are cheaper, but have received little study, so their properties are not well understood.

Advantages and disadvantages of subset selection

An advantage of subset selection methods is interpretability. If the summaries in z have intuitive interpretations, then so will the resulting subset. This is especially useful for model criticism by the method of Ratmann et al. (2009). Here, one investigates whether the simulated $S(y)$ values accepted by ABC are consistent with s_{obs}. If some particular component of s_{obs} cannot be matched well by the simulations, this suggests model mis-specification. Understanding the interpretation of this summary can hopefully suggest model improvements.

A disadvantage is the implicit assumption that a low dimensional informative subset of z exists. Subset selection can be thought of as a projection method which is restricted to projections to subsets. However, it may be the case that the best choice is outside this restriction, for example, the mean of the candidate summaries.

Further disadvantages are cost and scalability problems. For small k (denoting $\dim z$), it may be possible to evaluate the performance of all subsets of z and find the global optimum. For large k, this is prohibitively expensive and more sophisticated search procedures such as stepwise selection must be used. However, such methods may only converge to a local optimum, and the computational cost still grows rapidly with k.

Computational cost is particularly large for the methods which require ABC to be re-run for each subset of z considered. The computing requirement is typically made more manageable by using the same simulations for each ABC analysis. However, this restricts the algorithm used in this stage to ABC rejection or importance sampling. Therefore, the resulting summary statistics have not been tested at the lower values of h, which could be achieved using ABC-MCMC or ABC-SMC and may not be good choices under these algorithms.

Finally, as discussed in Section 5.2, subset selection methods require a feature selection stage to choose the set of potential summaries z. In all the papers cited earlier, this step is based on subjective choice and is crucial to good performance. Comparable subjective choices are also required by the strategies described later. However a particular constraint here is that $\dim z$ cannot be too large or the cost of subset selection becomes infeasible.

5.4 Projection Methods

Projection methods start with a vector of data features $z(y) = (z_1, z_2, \ldots, z_k)$ and attempts to find an informative lower dimensional projection, often a linear transformation. To choose a projection, training data $(\theta_i, y_i)_{1 \le i \le n_{\text{train}}}$ are created by simulation from the prior and model and some dimension reduction technique is applied. Section 5.4.1 presents various dimension reduction methods which have been proposed for use in ABC. Section 5.4.2 describes variations in how the training data are generated. It is generally possible to match up the two parts of the methodology as desired. Section 5.4.3 compares the methods and discusses the strengths and weaknesses of this strategy.

5.4.1 Dimension reduction techniques

Partial least squares (Wegmann et al., 2009)

Partial least squares (PLS) aim to produce linear combinations of covariates which have high covariance with responses and are uncorrelated with each other. In the ABC setting, the covariates are $z_1, \ldots, z_k$ and the responses are $\theta_1, \ldots, \theta_p$. The ith PLS component $u_i = \alpha_i^T z$ maximises:

$$\sum_{j=1}^{p} \text{Cov}(u_i, \theta_j)^2,$$

subject to $\text{Cov}(u_i, u_j) = 0$ for $j < i$ and a normalisation constraint on α_i, such as $\alpha_i^T \alpha_i = 1$. In practice, empirical covariances based on training data are used. Several algorithms to compute PLS components are available.

These can produce different results, as they use slightly different normalisation constraints. For an overview of PLS, see Boulesteix and Strimmer (2007).

PLS produces $\min(k, n_{\text{train}} - 1)$ components. Wegmann et al. (2009) use the first c components as ABC summary statistics, with c chosen by a cross-validation procedure. This aims to minimise the root mean squared error in a linear regression of θ on $u_1, \ldots, u_c$. This approach is similar to the regularisation subset selection methods of Section 5.3.

Linear regression (Fearnhead and Prangle, 2012)

Fearnhead and Prangle (2012) fit a linear model to the training data: $\theta \sim N(Az + b, \Sigma)$. The resulting vector of parameter estimators $\hat{\theta}(y) = Az + b$ is used as ABC summary statistics. This is a low dimensional choice: $\dim \hat{\theta}(y) = \dim \theta = p$.

Motivation for this approach comes from the following result. Consider $\pi(\theta | S(y_{\text{obs}}))$, which is the ABC target distribution for $h = 0$. Then $S(y) = \text{E}(\theta | y)$ can be shown to be the optimal choice in terms of minimising quadratic loss of the parameter means in this target distribution. Fitting a linear model produces an estimator of these ideal statistics.

A linear regression estimate of $\text{E}(\theta | y)$ is crude, but can be improved by selecting good $z(y)$ features. Fearnhead and Prangle propose comparing $z(y)$ choices by looking at the goodness of fit of the linear model, in particular the BIC values. Another way to improve the estimator is to train it on a local region of the parameter space. This is discussed in Section 5.4.2. Fearnhead and Prangle use the name 'semi-automatic ABC' to refer to summary statistic selection by linear regression with these improvements. Good performance is shown for a range of examples and is particularly notable when $z(y)$ is high dimensional (Fearnhead and Prangle, 2012; Blum et al., 2013).

As Robert (2012) points out, the theoretical support for this method is only heuristic, as it focuses on the unrealistic case of $h = 0$. Another limitation is that these summaries focus on producing good point estimates of θ, and not on accurate uncertainty quantification. Fearnhead and Prangle propose a modified ABC algorithm ('noisy ABC'), which tackles this problem to some extent.

The earlier discussion also motivates using more advanced regression-like methods. Fearnhead and Prangle investigate the lasso (Hastie et al., 2009), canonical correlation analysis (Mardia et al., 1979), and sliced inverse regression (Li, 1991) (see Prangle, 2011 for details). The former two do not produce significant improvements over linear regression in the applications considered. The latter produces large improvements in one particular example, but requires significant manual tuning. Many further suggestions can be found in the discussions published alongside Fearnhead and Prangle (2012).

Boosting (Aeschbacher et al., 2012)

Boosting is a non-linear regression method. Like linear regression, it requires training data and outputs predictors $\hat{\theta}(y)$ of $\text{E}(\theta | y)$, which can be used as ABC summary statistics. Boosting is now sketched for scalar θ. For multi-variate θ, the whole procedure can be repeated for each component. The approach begins by fitting a 'weak' estimator to the training data. For this, Aeschbacher et al. (2012) use linear regression with response θ and a single covariate: whichever feature in $z(y)$ maximises reduction in error (e.g. mean squared error). The training data are then weighted according to the error under this estimator. A second weak estimator is fitted to this weighted training data and a weighted average of the first two estimators is formed. The training data are weighted according to its error under this, and a third weak estimator is formed, and so on. Eventually the process is terminated, and the final weighted average of weak estimators is output as a 'strong' estimator. The idea is that each weak estimator attempts to concentrate on data which have been estimated poorly in previous steps. See Bühlmann and Hothorn (2007) for a full description.

5.4.2 Generating training data

A straightforward approach to draw training data pairs (θ, y) is to sample θ from the prior and y from the model conditional on θ. This approach is used by Wegmann et al. (2009), for example. In rejection or importance sampling ABC algorithms, this training data can be re-used to implement the ABC analysis. Hence, there is no computational overhead in producing training data. For other ABC algorithms, this is not the case.

Fearnhead and Prangle (2012) and Aeschbacher et al. (2012) use different approaches to generate training data which aim to make the projection methods more effective. The idea is that the global relationship between θ and y is likely to be highly complicated and hard to learn. Learning about the relationship close to y_{obs} may be easier. This motivates sampling training pairs from a more concentrated distribution.

Aeschbacher et al. (2012) implement this by performing a pilot ABC analysis using $S(y) = z$ (i.e. all the data features). The accepted simulations are used as training data for their boosting procedure. The resulting summary statistics are then used in an ABC analysis.

Fearnhead and Prangle (2012) argue that such an approach might be dangerous. This is because $S(y)$ is only trained to perform well on a concentrated region of (θ, y) values and could perform poorly outside this region. In particular it is possible that $S(y) \approx S(y_{\text{obs}})$ in regions excluded by the pilot analysis, producing artefact posterior modes. Fearnhead and Prangle instead recommend performing a pilot ABC analysis using ad-hoc summary statistics. This is used to find a *training region* of parameter space, R, containing most of the posterior mass. Training θ values are drawn from the prior truncated to R, and y values from the model. Summary statistics are fitted and used in a main ABC analysis, which also truncates the prior to R. This ensures that θ regions excluded by the pilot remain excluded in the main analysis. Note that this truncation approach was introduced by Blum and François (2010) in a regression post-processing context.

5.4.3 Discussion

Comparison of projection methods

Partial least squares is a well established dimensional reduction method. However, it does not have any theoretical support for use in ABC and sometimes performs poorly in practice (Blum et al., 2013). Fearnhead and Prangle (2012) provide heuristic theoretical support to the approach of constructing parameter estimators for use as ABC summary statistics, and show empirically implementing this by the simple method of linear regression can perform well in practice. It is likely that more sophisticated regression approaches will perform even better. Boosting is one example of this. A particularly desirable goal would be a regression method which can incorporate the feature selection step and estimate $E(\theta|y)$ directly from the raw data y. This is discussed further in Section 5.7.

Advantages and disadvantages of projection methods

Projection methods avoid some of the disadvantages of subset selection methods. In particular, the high computational costs associated with repeating calculations for many possible subsets are avoided. Also, a wider space of potential summaries is searched, not just subsets of z. However, this means that the results may be less interpretable and thus harder to use for model criticism. (For further discussion of all these points see Section 5.3.2.)

Another advantage of projection methods is that they can be implemented on almost any problem. This is in contrast to auxiliary likelihood methods, which require the specification of a reasonable approximate likelihood.

Projection methods require a subjective choice of features $z(y)$, as do subset selection methods. However, projection methods have more freedom to choose a large set of features and still have a feasible computational cost, and some methods provide heuristic tools to select between feature sets (i.e. Fearnhead and Prangle, 2012 use BIC.).

5.5 Auxiliary Likelihood Methods

An intuitively appealing approach to choosing summary statistics for ABC inference of a complicated model is to use statistics which are known to be informative for a simpler related model. This has been done since the earliest precursors to ABC in population genetics (e.g. Fu and Li, 1997; Pritchard et al., 1999). Recently, there has been much interest in formalising this approach. The idea is to specify an approximate and tractable likelihood for the data, referred to as an *auxiliary likelihood*, and derive summary statistics from this. Several methods to do this have been proposed, which are summarised in Section 5.5.1 and discussed in Section 5.5.2. There have also been related proposals for new likelihood-free methods based on auxiliary likelihoods which are covered elsewhere in this volume (see Chapters 8 and 12).

First some notation and terminology is introduced. The auxiliary likelihood is represented as $p_A(y|\phi)$. This can be thought of as defining an *auxiliary model* for y. This differs from the model whose posterior is sought, which is referred to here as the *generative model*. The auxiliary model parameters, ϕ, need not correspond to those of the generative model, θ. A general question is which auxiliary likelihood to use. This is discussed in Section 5.5.2, including a description of some possible choices. To assist in reading Section 5.5.1, it may be worth keeping in mind the simplest choice: let the auxiliary model be a tractable simplified version of the generative model.

5.5.1 Methods

Maximum likelihood estimators

Here, $S(y) = \hat{\phi}(y) = argmax_\phi p_A(y|\phi)$. That is, the summary statistic vector is the maximum likelihood estimator (MLE) of ϕ given data y under the auxiliary model. To use this method, this MLE must be unique for any y. Typically, $S(y)$ must be calculated numerically, which is sometimes computationally costly.

This approach was proposed by Drovandi et al. (2011), although Wilson et al. (2009) use a similar approach in a particular application. It was motivated by a similar choice of summaries in another likelihood-free method, *indirect inference* (Gourieroux et al., 1993). The terminology **ABC-IP** for this approach was introduced by Gleim and Pigorsch (2013): 'I' represents indirect and 'P' using parameter estimators as summaries.

Some theoretical support for ABC-IP is available. Gleim and Pigorsch (2013) note that classical statistical theory shows that $\hat{\phi}(y)$ is typically *asymptotically sufficient* for the auxiliary model (see Chapter 9 of Cox and Hinkley, 1979 for full details). This assumes an asymptotic setting, where $n \to \infty$ as the data becomes more informative. As a simple example, y could consist of n independent identically distributed observations. Asymptotic sufficiency implies $\hat{\phi}(y)$ asymptotically summarises all the information about the auxiliary model parameters.

Ideally, $\hat{\phi}(y)$ would also be asymptotically sufficient for the generative model. Gleim and Pigorsch show this is the case if the generative model is nested within the auxiliary model. However, having a tractable auxiliary model of this form is rare. Martin et al. (2014) note that even without asymptotic sufficiency for the generative model, *Bayesian consistency* can be attained. That is, the distribution $\pi(\theta|\hat{\phi}(y))$ asymptotically shrinks to a point mass on the true parameter value. They give necessary conditions for consistency: essentially that in this limit, $\hat{\phi}(y)$ perfectly discriminates between data generated by different values of θ.

Likelihood distance

Gleim and Pigorsch (2013) suggest a variation on ABC-IP which uses the distance function:

$$||\hat{\phi}(y), \hat{\phi}(y_{\text{obs}})|| = \log p_A(y_{\text{obs}}|\hat{\phi}(y_{\text{obs}})) - \log p_A(y_{\text{obs}}|\hat{\phi}(y)). \tag{5.2}$$

This is the log likelihood ratio for the auxiliary model between the MLEs under the observed and simulated datasets. They refer to this as **ABC-IL:** 'L' represents using a likelihood distance.

It is desirable that $||\hat{\phi}(y), \hat{\phi}(y_{\text{obs}})|| = 0$ if and only if $\hat{\phi}(y) = \hat{\phi}(y_{\text{obs}})$. This requires p_A to be well behaved. For example, it suffices that $y \mapsto \hat{\phi}(y)$ is a one-to-one mapping. However, weaker conditions can sometimes be used: see Drovandi et al. (2015) Section 7.3, for example.

Scores

Gleim and Pigorsch (2013) suggest taking:

$$S(y) = \left(\frac{\partial}{\partial \phi_i} \log p_A(y|\phi) \Big|_{\phi=\hat{\phi}(y_{\text{obs}})} \right)_{1 \leq i \leq p}. \tag{5.3}$$

This is the score of the auxiliary likelihood evaluated under parameters $\hat{\phi}(y_{\text{obs}})$. As earlier, $\hat{\phi}(y_{\text{obs}})$ is the MLE of y_{obs} under the auxiliary likelihood. Gleim and Pigorsch refer to this approach as **ABC-IS:** 'S' refers to using score summaries.

The motivation is that the score has similar asymptotic properties to those described earlier for the MLE (Gleim and Pigorsch, 2013; Martin et al., 2014), but is cheaper to calculate. This is because numerical optimisation is required once only, to find $\hat{\phi}(y_{\text{obs}})$, rather than every time $S(y)$ is computed. Drovandi et al. (2015) also note that ABC-IS is more widely applicable than ABC-IP, as it does not require existence of a unique MLE for $p_A(y|\phi)$ under all y, only under y_{obs}.

Some recent variations of ABC-IS include: application to state-space models by using a variation on Kalman filtering to provide the auxiliary likelihood and use of a *marginal score* (Martin et al., 2014); alternatives to the score function (5.3) based on *proper scoring rules* (Ruli et al., 2014); and using a *rescaled score* when the auxiliary model is a composite likelihood (Ruli et al., 2016).

5.5.2 Discussion

Comparison of auxiliary likelihood methods

ABC-IS has the advantage that it is not based on calculating the MLE repeatedly. This can be computationally costly, may be prone to numerical errors, and, indeed, a unique MLE may not even exist. Furthermore, the asymptotic properties discussed earlier suggest the score-based summaries used by ABC-IS encapsulate similar information about the auxiliary likelihood as the MLE. This recommends use of ABC-IS. However, empirical comparisons by Drovandi et al. (2015) suggest the best auxiliary likelihood method in practice is problem specific (see Section 5.7 for more details).

Which auxiliary likelihood?

Various choices of auxiliary likelihood have been used in the literature. Examples include the likelihood of a tractable alternative model for the data or of a flexible general model, such as a Gaussian mixture. Another is to use a tractable approximation to the likelihood of the generative model, such as composite likelihood (Varin et al., 2011). There is a need to decide which choice to use.

An auxiliary likelihood should ideally have several properties. To produce low dimensional summary statistics, it should have a reasonably small number of parameters. Also, it is desirable that it permits fast and accurate computation of the MLE or score. These two requirements are easy to assess. A third requirement is that the auxiliary likelihood should produce summary statistics which are informative about the generative model. This seems harder to quantify.

Drovandi et al. (2015) recommend performing various goodness-of-fit tests to see how well the auxiliary likelihood matches the data. Similarly, Gleim and Pigorsch (2013) use the BIC to choose between several possible auxiliary likelihoods. Such tests are computationally cheap and give insight into the ability of the model to summarise important features of the data. However, it is not clear that performing better in a goodness-of-fit test necessarily results in a better ABC approximation. Ideally, what is needed is a test of how well an auxiliary likelihood discriminates between datasets drawn from the generative model under different parameter values. How to test this is an open problem.

Advantages and disadvantages of auxiliary likelihood methods

Auxiliary likelihood methods avoid the necessity of choosing informative data features required by subset selection and projection methods. This is replaced by the somewhat analogous need to choose an informative auxiliary likelihood. However, such a choice may often be substantially easier, particularly if well-developed tractable approximations to the generative model are already available. In other situations, both tasks may be equally challenging.

Another advantage of auxiliary likelihood methods is that they avoid the computational cost of generating training data, as required by preceding methods. Instead, they make use of subject area knowledge to propose auxiliary likelihoods. In the absence of such knowledge, one could try to construct an auxiliary likelihood from training data. This is one viewpoint of how projection methods based on regression operate.

5.6 Model Choice

ABC can be applied to inference when there are several available models $M_1, M_2, \ldots, M_r$. See Chapter 7 in this volume or Didelot et al. (2011) for details of algorithms. This section is on the problem of choosing which summary statistics to use here. The aim of most work to date is to choose summaries suitable for inferring the posterior model weights. The more challenging problem of also inferring model parameters is mentioned only briefly.

A natural approach used by some early practical work is to use summary statistics which are informative for parameter inference in each model. Unfortunately, except in a few special cases, this can give extremely poor model choice results, as highlighted by Robert et al. (2011). The issue is that summary statistics, which are good for parameter inference within models, are not necessarily informative for choosing between models. This section summarises more recent theoretical and practical work which shows that informative summary statistics can be found for ABC model choice. Therefore, ABC model choice can now be trusted to a similar degree to ABC parameter inference.

The remainder of this section is organised as follows. Section 5.6.1 re-examines sufficiency and other theoretical issues for ABC model choice, as there are some surprisingly different results to those described earlier in the chapter for the parameter inference case. Section 5.6.2 reviews practical methods of summary statistic selection, and Section 5.6.3 gives a brief discussion.

Note that model choice can be viewed as inference of a discrete parameter $m \in \{1, 2, \ldots, r\}$ indexing the available models. Therefore, the following material would also apply to ABC inference of a discrete parameter. However, as elsewhere in this chapter, the phrase 'parameter inference' is generally used in the section as shorthand to refer to inference of *continuous* parameters.

5.6.1 Theory

Curse of dimensionality

As described earlier ABC suffers a curse of dimensionality when dealing with high dimensional data. Theoretical work on this, summarised in Section 5.1.1, has focused on the parameter inference case. However, the technical arguments involved focus on properties of the summary statistics, rather than of the parameters. Therefore, it seems likely that the arguments can be adapted to give unchanged results for model choice simply by considering the case of discrete parameters.

This means it remains important to use low dimensional summary statistics for ABC model choice.

Sufficiency and consistency

As for parameter inference case, the ideal summaries for ABC model choice would be low dimensional sufficient statistics. Unlike the case of parameter inference, such statistics do exist for ABC model choice, and results are also available on links to consistency and sufficiency for parameter inference. These theoretical results are now summarised and will motivate some of the methods for summary statistic choice described in the next section.

First, some terminology is defined, based on the definitions of sufficiency in Section 5.1.2. Let θ_i represent the parameters associated with model M_i. Statistics $S(y)$ that are sufficient for θ_i under model M_i will be referred to below as *sufficient for parameter inference* in that model. Now consider the problem of jointly inferring $\theta_1, \theta_2, \ldots, \theta_r, m$, where m is a model index. This is equivalent to inference on an encompassing model in which the data are generated from M_i conditional on θ_i when $m = i$. Statistics will be referred to as *sufficient for model choice* if they are Bayes sufficient for m in this encompassing model.

Didelot et al. (2011) show that sufficient statistics for model choice between models M_1 and M_2 can be found by taking parameter inference sufficient statistics of a model in which both are nested. This result is of limited use, as such parameter inference sufficient statistics rarely exist in low dimensional form (see discussion in Section 5.1.2). However, it has useful consequences in the special case where M_1 and M_2 are both exponential family distributions for example:

$$\pi(y|\theta_i, M_i) \propto \exp\left[s_i(x)^T \theta_i + t_i(x)\right],$$

for $i = 1, 2$. In this case, $s_i(x)$ is a vector of parameter inference sufficient statistics for model M_i and $\exp[t_i(x)]$ is known as the base measure. Didelot et al. (2011) show that sufficient statistics for model choice are the concatenation of $s_1(x), s_2(x), t_1(x)$, and $t_2(x)$.

Prangle et al. (2014) prove that the following vector of statistics is sufficient for model choice:

$$T(y) = (T_1(y), T_2(y), \ldots, T_{r-1}(y)),$$
$$\text{where} \quad T_i(y) = \pi(y|M_i) / \sum_{j=1}^{r} \pi(y|M_j).$$

Here, $T_i(y)$ is the evidence under M_i divided by the sum of model evidences. Furthermore, any other vector of statistics is sufficient for model choice if and only if it can be transformed to $T(y)$.

Thus low dimensional sufficient statistics exist if the number of models r is reasonably small. This results may seem at first to contradict the arguments of Section 5.1, that these are only available for exponential family models.

A contradiction is avoided because model choice is equivalent to inferring the discrete parameter m, and a model with a discrete parameter can be expressed as an exponential family.

A related result is proved by Marin et al. (2014). They give necessary conditions on summary statistics $S(y)$ for $\Pr(m|S(y))$ to be consistent in an asymptotic regime corresponding to highly informative data. That is, these conditions allow for perfect discrimination between models in this limiting case. In addition to several technical conditions, the essential requirement is that the limiting expected value of the summary statistic vector should differ under each model.

5.6.2 Methods

Using an encompassing model (Didelot et al., 2011)

As described earlier, Didelot et al. (2011) prove that sufficient statistics for model choice between exponential family models can be found by concatenating the parameter sufficient statistics and base measures of each model. Situations where this can be used are rare, but one is the Gibbs random field application considered by Grelaud et al. (2009). In this case, the base measures are constants and so can be ignored.

Mutual information (Barnes et al., 2012)

This method was described earlier for the case of parameter inference. To recap briefly, it is a subset selection method motivated by the concept of mutual information, which sequentially adds or removes candidate summary statistics to or from a set. Each time, a Kullback–Leibler divergence between the ABC posterior distributions under the previous and new sets is estimated. The process terminates when the largest achievable divergence falls below a threshold.

Barnes et al. (2012) adapt this method to find summary statistics for joint model and parameter inference as follows. First, they estimate sufficient statistics for parameter inference under each model, and concatenate these. Next, they add further statistics until model sufficiency is also achieved. Alternatively, the method could easily be adapted to search for statistics which are sufficient for model choice only.

Projection/classification methods (Estoup et al., 2012; Prangle et al., 2014)

The idea here is to use training data to construct a classifier which attempts to discriminate between the models given data y. Informative statistics are taken from the fitted classifier and used as summary statistics in ABC. This can be thought of as a projection approach mapping high dimensional data y to low dimensional summaries.

Two published approaches of this form are now described in more detail. Training data $(\theta_i, m_i, y_i)_{1 \leq i \leq n_{\text{train}}}$ are created, where y_i is drawn from $\pi(y|\theta_i, m_i)$ (generating θ_i, m_i pairs is discussed below). A vector of data features $z(y)$ must be specified. A classification method is then used to find linear combinations $\alpha^T z(y)$ which are informative about m, the model index. Estoup et al. (2012) use linear discriminant analysis, and Prangle et al. (2014) use logistic regression (for the case of two models). For a review of both, see Hastie et al. (2009), who note they typically give very similar results. As motivation, Prangle et al. observe that, in the two model case, logistic regression produces a crude estimate of logit$[\Pr(M_1|y)]$, which would be sufficient for model choice as discussed earlier (extending this argument to more than two models is also discussed).

The simplest approach to drawing θ_i, m_i pairs is simply to draw m_i from its prior (or with equal weights if this is too unbalanced) and θ_i from the appropriate parameter prior. Prangle et al. (2014) observe that it can sometimes be hard for their classifier to fit the resulting training data. They propose instead producing training data which focus on the most likely θ regions under each model (as in the similar approach for parameter inference in Section 5.4.2). The resulting summary statistics are only trained to discriminate well in these regions, so a modified ABC algorithm is required to use them. This involves estimation of some posterior quantities and so may not be possible in some applications.

Alternatively, Estoup et al. (2012) first perform ABC with a large number of summary statistics. The accepted output is used as training data to fit model choice summary statistics. These are then used in regression post-processing. This avoids the need for a modified ABC algorithm, but the first stage of the analysis may still suffer from errors due to the curse of dimensionality.

Local error rates (Stoehr et al., 2014)

Stoehr et al. (2014) compare three sets of summary statistics for ABC model choice in the particular setting of Gibbs random fields. The idea is to pick the choice which minimises the *local error rate*, defined shortly. This method could easily be used more generally, for example, as the basis of a subset selection method similar to the loss minimisation method of Nunes and Balding (2010).

The local error rate is $\Pr(\hat{M}(y_{\text{obs}}) \neq M|y_{\text{obs}})$, where $\hat{M}(y)$ is the model with greatest weight under the target distribution of the ABC algorithm given data y. (This can be interpretted as using a 0-1 loss function). In practice, this quantity is unavailable, but it can be estimated. Suppose a large number of $(y_i, M_i)_{1 \leq i \leq n_{\text{val}}}$ *validation* pairs have been generated. Stoehr et al. suggest running ABC using each y_i in turn as the observations and evaluating an indicator variable δ_i, which equals 1 when ABC assigns most weight to model M_i. Non-parametric regression is then used to estimate $\Pr(\delta = 1|y_{\text{obs}})$.

This is challenging if $\dim y$ is large, so dimension reduction is employed. Stoehr et al. use linear discriminant analysis for this (as in the Estoup et al., 2012 approach described earlier.) To reduce costs, a cross-validation scheme is used so that the same simulations can be used for ABC analyses and validation.

5.6.3 Discussion

Comparison of methods

The approach of Didelot et al. (2011) – choosing sufficient statistics for an encompassing model – is only useful in specialised circumstances, such as choice between exponential family models. The other methods listed earlier are more generally applicable. Their advantages and disadvantages are similar to those discussed earlier for corresponding parameter inference methods. In particular, the two subset selection methods have the disadvantage that they have high computational costs if there are many candidate summary statistics.

Prospects

Comparatively few summary statistic selection methods have been proposed for the model choice setting. Thus, there is potential to adapt other existing approaches from the parameter inference case for use here. In particular, it would be interesting to see whether regularisation or auxiliary likelihood approaches can be developed.

Another promising future direction is to construct model choice summary statistics using more sophisticated classification methods than those described earlier, for example, random forests or deep neural networks. As an alternative to using these methods to produce summary statistics, some of them can directly output likelihood-free inference results (Pudlo et al., 2015).

Finally, choosing summary statistics for joint ABC inference of model and parameters is a desirable goal. One approach is to separately find summaries for model choice and for parameter inference in each model and concatenate these. However, it may be possible to produce lower dimensional informative summaries by utilising summaries which are informative for several of these goals. Finding methods to do this is an open problem.

5.7 Discussion

Empirical performance

Most papers proposing methods of summary statistic choice report some empirical results on performance. These show some merits to all the proposed methods. However, it is difficult to compare these results to each other as there are many differences between the applications, algorithms, and tuning choices used. Two studies are reported here which compare several methods on multiple parameter inference applications. Little comparable work exists for model choice methods.

Blum et al. (2013) compare several subset selection and projection methods on three applications using ABC rejection sampling. They conclude: 'What is very apparent from this study is that there is no single "best" method of dimension reduction for ABC.' The best performing methods for each application are: the two stage method of Nunes and Balding (2010) on the smallest example ($k = 6$). the AIC and BIC regularisation methods on a larger example ($k = 11$), and the linear regression method of Fearnhead and Prangle (2012) on the largest example ($k = 113$). (Recall $k = \dim z$ i.e. the number of data features.)

Drovandi et al. (2015) compare auxiliary likelihood methods on several applications using ABC MCMC. They conclude: 'Across applications considered in this paper, ABC IS was the most computationally efficient and led to good posterior approximations'. However, they note that its posterior approximation was not always better than ABC-IP and ABC-IL, so that again, the best approach seems problem specific.

Which method to use

Although many methods for choosing summary statistics have been proposed, there are no strong theoretical or empirical results about which to use in practice for a particular problem. Also, the area is developing rapidly, and many new approaches can be expected to appear soon. Therefore, only very general advice is given here.

Each of the strategies discussed has its particular strengths. When a small set of potential summaries can be listed, subset selection performs a thorough search of possible subsets. When a good tractable approximate likelihood is available, auxiliary likelihood methods can make use of it to produce informative parameter inference summaries, although they are not yet available for model choice. Projection methods are highly flexible and can be applied to almost any problem.

It seems advisable to consider subset selection or auxiliary likelihood methods in situations that suit their strengths and projection methods otherwise. The question of which methods are most appealing within each strategy is discussed within their respective sections. For parameter inference, the empirical comparisons described earlier can also provide some guidance.

Ideally, if resources are available, the performance of different methods should be assessed on the problem of interest. This requires repeating some or all of the analysis for many simulated datasets. To reduce the computation required, Bertorelle et al. (2010) and Blum et al. (2013) advocate performing a large set of simulations and re-using them to perform the required ABC analyses. This restricts the algorithm to ABC rejection or importance sampling.

Finally, note that there is considerable scope to modify the summary statistics generated by the methods in this chapter. For example, the user may decide to choose a combination of statistics produced by several different methods or add further summary statistics based on subject area insights.

Prospects

This chapter has shown how, amongst other approaches, classification and regression methods can be used to provide ABC summary statistics from training data. There are many sophisticated tools for these in the statistics and machine learning literature, which may produce more powerful ABC summary statistic selection methods in future. It would be particularly desirable to find methods which do not require a preliminary subjective feature selection stage. One promising approach to this is regression using deep neural networks (Goodfellow et al, 2016), although it is unclear whether the amount of training data required to fit these well would be prohibitively expensive to simulate. Another possibility is to come up with dictionaries of generally informative features for particular application areas. Fulcher et al. (2013) and Stocks et al. (2014) implement ideas along these lines for time series analysis and population genetics, respectively.

A topic of recent interest in ABC is the case where a dataset y is made up of many observations which are either independent and identically distributed or have some weak dependence structure. Several approaches to judging the distance between such datasets have been recently proposed. These fall somewhat outside the framework of this chapter, as they bypass producing conventional summary statistics and instead simply define a distance. An interesting question is the extent to which these alleviate the curse of dimensionality. The methods base distances on: classical statistical tests (Ratmann et al., 2013); the output of classifiers (Gutmann et al., 2018); and kernel embeddings (Park et al., 2016).

This chapter has concentrated on using summary statistics to reduce the ABC curse of dimensionality and approximate the true posterior $\pi(\theta|y_{\text{obs}})$. However, other aspects of summary statistics are also worth investigating. First, it is possible that dimension-preserving transformations of $S(y)$ may also improve ABC performance. This is exploited by Yıldırım et al. (2015) in the context of a specific ABC algorithm for example. Second, several authors (Wood, 2010; Girolami and Cornebise, 2012; Fasiolo et al., 2016) discuss cases where the true posterior is extremely hard to explore, for example, because it is very rough with many local modes. They argue that using appropriate summary statistics can produce a better behaved $\pi(\theta|s_{\text{obs}})$ distribution, which is still informative about model properties of interest.

References

Aeschbacher, S., M. A. Beaumont, and A. Futschik (2012). A novel approach for choosing summary statistics in approximate Bayesian computation. *Genetics 192*(3), 1027–1047.

Barber, S., J. Voss, and M. Webster (2015). The rate of convergence for approximate Bayesian computation. *Electronic Journal of Statistics 9*, 80–105.

Barnes, C. P., S. Filippi, M. P. H. Stumpf, and T. Thorne (2012). Considerate approaches to constructing summary statistics for ABC model selection. *Statistics and Computing 22*, 1181–1197.

Beaumont, M. A., W. Zhang, and D. J. Balding (2002). Approximate Bayesian computation in population genetics. *Genetics 162*, 2025–2035.

Goodfellow, I., Y. Bengio, and A. Courville (2016). *Deep Learning.* Cambridge, MA: MIT press

Bertorelle, G., A. Benazzo, and S. Mona (2010). ABC as a flexible framework to estimate demography over space and time: Some cons, many pros. *Molecular Ecology 19*(13), 2609–2625.

Biau, G., F. Cérou, and A. Guyader (2015). New insights into approximate Bayesian computation. *Annales de l'Institut Henri Poincaré (B) Probabilités et Statistiques 51*(1), 376–403.

Blum, M. G. B. (2010a). Approximate Bayesian computation: A nonparametric perspective. *Journal of the American Statistical Association 105*(491), 1178–1187.

Blum, M. G. B. (2010b). Choosing the summary statistics and the acceptance rate in approximate Bayesian computation. In *Proceedings of COMPSTAT'2010*, pp. 47–56. New York: Springer.

Blum, M. G. B. and O. François (2010). Non-linear regression models for approximate Bayesian computation. *Statistics and Computing 20*(1), 63–73.

Blum, M. G. B., M. A. Nunes, D. Prangle, and S. A. Sisson (2013). A comparative review of dimension reduction methods in approximate Bayesian computation. *Statistical Science 28*, 189–208.

Boulesteix, A.-L. and K. Strimmer (2007). Partial least squares: A versatile tool for the analysis of high-dimensional genomic data. *Briefings in Bioinformatics 8*(1), 32–44.

Bühlmann, P. and T. Hothorn (2007). Boosting algorithms: Regularization, prediction and model fitting. *Statistical Science*, 477–505.

Cox, D. R. and D. V. Hinkley (1979). *Theoretical Statistics.* CRC Press: Boca Raton, FL.

Didelot, X., R. G. Everitt, A. M. Johansen, and D. J. Lawson (2011). Likelihood-free estimation of model evidence. *Bayesian Analysis 6*(1), 49–76.

Drovandi, C. C., A. N. Pettitt, and M. J. Faddy (2011). Approximate Bayesian computation using indirect inference. *Journal of the Royal Statistical Society: Series C 60*(3), 317–337.

Drovandi, C. C., A. N. Pettitt, and A. Lee (2015). Bayesian indirect inference using a parametric auxiliary model. *Statistical Science 30*(1), 72–95.

Estoup, A., E. Lombaert, J.-M. Marin, T. Guillemaud, P. Pudlo, C. P. Robert, and J.-M. Cornuet (2012). Estimation of demo-genetic model probabilities with approximate Bayesian computation using linear discriminant analysis on summary statistics. *Molecular Ecology Resources 12*(5), 846–855.

Fasiolo, M., N. Pya, and S. N. Wood (2016). A comparison of inferential methods for highly nonlinear state space models in ecology and epidemiology. *Statistical Science 31*(1), 96–118.

Fearnhead, P. and D. Prangle (2012). Constructing summary statistics for approximate Bayesian computation: Semi-automatic ABC. *Journal of the Royal Statistical Society: Series B 74*, 419–474.

Filippi, S., C. P. Barnes, and M. P. H. Stumpf (2012). Contribution to the discussion of Fearnhead and Prangle (2012). *Journal of the Royal Statistical Society: Series B 74*, 459–460.

Fu, Y.-X. and W.-H. Li (1997). Estimating the age of the common ancestor of a sample of DNA sequences. *Molecular Biology and Evolution 14*(2), 195–199.

Fulcher, B. D., M. A. Little, and N. S. Jones (2013). Highly comparative time-series analysis: The empirical structure of time series and their methods. *Journal of the Royal Society Interface 10*(83), 0048.

Girolami, M. and J. Cornebise (2012). Contribution to the discussion of Fearnhead and Prangle (2012). *Journal of the Royal Statistical Society: Series B 74*, 460–461.

Gleim, A. and C. Pigorsch (2013). Approximate Bayesian computation with indirect summary statistics. Technical report, University of Bonn, Bonn, Germany.

Gourieroux, C., A. Monfort, and E. Renault (1993). Indirect inference. *Journal of Applied Econometrics 8*(S1), S85–S118.

Grelaud, A., C. P. Robert, J.-M. Marin, F. Rodolphe, and J. F. Taly (2009). ABC likelihood-free methods for model choice in Gibbs random fields. *Bayesian Analysis 4*(2), 317–336.

Gutmann, M. U., R. Dutta, S. Kaski, and J. Corander. (2018). Likelihood-free inference via classification. *Statistics and Computing, 28*(2), 411–425.

Hastie, T., R. Tibshirani, and J. Friedman (2009). *The Elements of Statistical Learning.* Springer.

Heggland, K. and A. Frigessi (2004). Estimating functions in indirect inference. *Journal of the Royal Statistical Society: Series B 66*(2), 447–462.

Jasra, A. (2015). Approximate Bayesian computation for a class of time series models. *International Statistical Review 83*, 405–435.

Joyce, P. and P. Marjoram (2008). Approximately sufficient statistics and Bayesian computation. *Statistical Applications in Genetics and Molecular Biology 7.* Article 26.

Li, K.-C. (1991). Sliced inverse regression for dimension reduction. *Journal of the American Statistical Association 86*(414), 316–327.

Mardia, K. V., J. T. Kent, and J. M. Bibby (1979). *Multivariate Analysis.* Academic Press: London, UK.

Marin, J.-M., N. S. Pillai, C. P. Robert, and J. Rousseau (2014). Relevant statistics for Bayesian model choice. *Journal of the Royal Statistical Society: Series B 76*(5), 833–859.

Martin, G. M., B. P. M. McCabe, W. Maneesoonthorn, and C. P. Robert (2014). Approximate Bayesian computation in state space models. *arXiv preprint arXiv:1409.8363.*

Nunes, M. A. and D. J. Balding (2010). On optimal selection of summary statistics for approximate Bayesian computation. *Statistical Applications in Genetics and Molecular Biology 9*(1). Article 34.

Park, M., W. Jitkrittum, and D. Sejdinovic (2016). K2-ABC: Approximate Bayesian computation with kernel embeddings. In *Artificial Intelligence and Statistics*, 398–407.

Prangle, D. (2011). Summary statistics and sequential methods for approximate Bayesian computation. Ph.D. thesis, Lancaster University, Lancaster, UK.

Prangle, D., P. Fearnhead, M. P. Cox, P. J. Biggs, and N. P. French (2014). Semi-automatic selection of summary statistics for ABC model choice. *Statistical Applications in Genetics and Molecular Biology 13*, 67–82.

Pritchard, J. K., M. T. Seielstad, A. Perez-Lezaun, and M. W. Feldman (1999). Population growth of human Y chromosomes: A study of Y chromosome microsatellites. *Molecular Biology and Evolution 16*(12), 1791–1798.

Pudlo, P., J. M. Marin, A. Estoup, J. M. Cornuet, M. Gautier, and C. P. Robert (2015). Reliable ABC model choice via random forests. *Bioinformatics 32*(6), 859–866.

Ratmann, O., C. Andrieu, C. Wiuf, and S. Richardson (2009). Model criticism based on likelihood-free inference, with an application to protein network evolution. *Proceedings of the National Academy of Sciences 106*(26), 10576–10581.

Ratmann, O., A. Camacho, A. Meijer, and G. Donker (2013). Statistical modelling of summary values leads to accurate approximate Bayesian computations. *arXiv preprint arXiv:1305.4283*.

Ratmann, O., O. Jorgensen, T. Hinkley, M. Stumpf, S. Richardson, and C. Wiuf (2007). Using likelihood-free inference to compare evolutionary dynamics of the protein networks of *H. pylori* and *P. falciparum*. *PLoS Computational Biology 3*, 2266–2278.

Robert, C. P. (2012). Contribution to the discussion of Fearnhead and Prangle (2012). *Journal of the Royal Statistical Society: Series B 74*, 447–448.

Robert, C. P., J. M. Cornuet, J.-M. Marin, and N. Pillai (2011). Lack of confidence in approximate Bayesian computation model choice. *Proceedings of the National Academy of Sciences 108*(37), 15112–15117.

Ruli, E., N. Sartori, and L. Ventura (2014). Approximate Bayesian computation with proper scoring rules. In *Proceedings of the 47th Scientific Meeting of the Italian Scientific Society*, Cagliari, Italy.

Ruli, E., N. Sartori, and L. Ventura (2016). Approximate Bayesian computation with composite score functions. *Statistics and Computing (online preview) 26*(3), 679–692.

Sedki, M. A. and P. Pudlo (2012). Contribution to the discussion of Fearnhead and Prangle (2012). *Journal of the Royal Statistical Society: Series B 74*, 466–467.

Singh, H., N. Misra, V. Hnizdo, A. Fedorowicz, and E. Demchuk (2003). Nearest neighbor estimates of entropy. *American Journal of Mathematical and Management Sciences 23*, 301–321.

Stocks, M., M. Siol, M. Lascoux, and S. D. Mita (2014). Amount of information needed for model choice in approximate Bayesian computation. *PLoS One 9*(6), e99581.

Stoehr, J., P. Pudlo, and L. Cucala (2014). Adaptive ABC model choice and geometric summary statistics for hidden Gibbs random fields. *Statistics and Computing 25*(1), 129–141.

Varin, C., N. M. Reid, and D. Firth (2011). An overview of composite likelihood methods. *Statistica Sinica 21*(1), 5–42.

Wegmann, D., C. Leuenberger, and L. Excoffier (2009). Efficient approximate Bayesian computation coupled with Markov chain Monte Carlo without likelihood. *Genetics 182*(4), 1207–1218.

Wilson, D. J., E. Gabriel, A. J. H. Leatherbarrow, J. Cheesbrough, S. Gee, E. Bolton, A. Fox, C. A. Hart, P. J. Diggle, and P. Fearnhead (2009). Rapid evolution and the importance of recombination to the gastroenteric pathogen *Campylobacter jejuni. Molecular Biology and Evolution 26*(2), 385–397.

Wood, S. N. (2010). Statistical inference for noisy nonlinear ecological dynamic systems. *Nature 466*(7310), 1102–1104.

Yıldırım, S., S. Singh, T. Dean, and A. Jasra (2015). Parameter estimation in hidden Markov models with intractable likelihoods using sequential Monte Carlo. *Journal of Computational and Graphical Statistics* (*online preview*) *24*(3), 846–865.

6

Likelihood-Free Model Choice

Jean-Michel Marin, Pierre Pudlo, Arnaud Estoup, and Christian Robert

CONTENTS

6.1 Introduction 153
6.2 Simulate Only Simulate 155
6.3 The Curse of Insufficiency 157
6.3.1 Some counter-examples 159
6.3.2 Still some theoretical guarantees 160
6.4 Selecting the Maximum a Posteriori Model via Machine Learning 162
6.4.1 Reconsidering the posterior probability estimation 163
6.4.2 Random forests construction 164
6.4.3 Approximating the posterior probability of the maximum a posteriori 165
6.5 A First Toy Example 166
6.6 Human Population Genetics Example 170
6.7 Conclusion 173
References 174

6.1 Introduction

As it is now hopefully clear from earlier chapters in this book, there exists several ways to set ABC methods firmly within the Bayesian framework. The method has now gone a very long way from the 'trick' of the mid-1990s [1,2], where the tolerance acceptance condition:

$$d(\boldsymbol{y}, \boldsymbol{y}^{\text{obs}}) \leq \epsilon,$$

was a crude practical answer to the impossibility to wait for the event $d(\boldsymbol{y}, \boldsymbol{y}^{\text{obs}}) = 0$ associated with exact simulations from the posterior distribution [3]. Not only do we now enjoy theoretical convergence guarantees [4–6] as the computing power grows to infinity, but we also benefit from new results that set actual ABC implementations, with their finite computing power and strictly positive tolerances, within the range of other types of inference [7–9]. ABC now stands as an inference method that is justifiable on its own ground. This approach may be the only solution available in complex settings, such as those originally tackled in population genetics [1,2], unless one engages into more perilous approximations. The conclusion of this evolution towards mainstream Bayesian inference is quite comforting about the role ABC can play in future computational developments, but this trend is far from delivering the method a blank confidence check, in that some implementations of it will alas fail to achieve consistent inference.

Model choice is actually a fundamental illustration of how much ABC can err away from providing a proper inference when sufficient care is not properly taken. This issue is even more relevant when one considers that ABC is used a lot – at least in population genetics – for the comparison and, hence, the validation of scenarios that are constructed based on scientific hypotheses. The more obvious difficulty in ABC model choice is indeed conceptual rather than computational, in that the choice of an inadequate vector of summary statistics may produce an inconsistent inference [10] about the model behind the data. Such an inconsistency cannot be overcome with more powerful computing tools. Existing solutions avoiding the selection process within a pool of summary statistics are limited to specific problems and difficult to calibrate.

Past criticisms of ABC from the outside have been most virulent about this aspect, even though not always pertinent (see, e.g. [11,12] for an extreme example). It is therefore paramount that the inference produced by an ABC model choice procedure be validated on the most general possible basis for the method to become universally accepted. As we discuss in this chapter, reflecting our evolving perspective on the matter, there are two issues with the validation of ABC model choice: (1) is it not easy to select a good set of summary statistics and (2) even selecting a collection of summary statistics that lead to a convergent Bayes factor may produce a poor approximation at the practical level.

As a warning, we note here that this chapter does not provide a comprehensive survey of the literature on ABC model choice, neither about the foundations (see [13,14]) and more recent proposals (see [15–17]), nor on the wide range of applications of the ABC model choice methodology to specific problems as in, for example, [18,19].

After introducing standard ABC model choice techniques, we discuss the curse of insufficiency. Then, we present the ABC random forest strategy for model choice and consider first a toy example and, at the end, a human population genetics example.

6.2 Simulate Only Simulate

The implementation of ABC model choice should not deviate from the original principle at the core of ABC, in that it proceeds by treating the unknown model index $\mathfrak{M}$ as an extra parameter with an associated prior, in accordance with standard Bayesian analysis. An algorithmic representation associated with the choice of a summary statistic $S(\cdot)$ is thus as follows:

Algorithm 6.1: Standard ABC Model Choice

for $i = 1$ to N **do**
  Generate $\mathfrak{M}$ from the prior $\pi(\mathfrak{M})$
  Generate θ from the prior $\pi_{\mathfrak{M}}(\theta)$
  Generate $\boldsymbol{y}$ from the model $f_{\mathfrak{M}}(\boldsymbol{y}|\theta)$
  Set $\mathfrak{M}^{(i)} = \mathfrak{M}$, $\theta^{(i)} = \theta$ and $s^{(i)} = S(\boldsymbol{y})$
end for
return the values $\mathfrak{M}^{(i)}$ associated with the k smallest distances $d\big(s^{(i)}, S(\boldsymbol{y}^{\text{obs}})\big)$.

In this presentation of the algorithm, the calibration of the tolerance ε for ABC model choice is expressed as a k-nearest neighbours (k-nn) step, following the validation of ABC in this format by [6], and the observation that the tolerance level is chosen this way in practice. Indeed, this standard strategy ensures a given number of accepted simulations is produced. While the k-nn method can be used towards classification and, hence, model choice, we will take advantage of different machine learning tools in Section 6.4. In general, the accuracy of a k-nn method heavily depends on the value of k, which must be calibrated, as illustrated in [20]. Indeed, while the primary justification of ABC methods is based on the ideal case when $\epsilon \approx 0$, hence, k should be taken 'as small as possible', more advanced theoretical analyses of its non-parametric convergence properties led to conclude that ϵ had to be chosen away from zero for a given sample size [4–6]. Rather than resorting to non-parametric approaches to the choice of k, which are based on asymptotic arguments, [20] rely on an empirical calibration of k using the whole simulated sample known as the reference table to derive the error rate as a function of k.

Algorithm 6.1 thus returns a sample of model indices that serves as an approximate sample from the posterior distribution $\pi(\mathfrak{M}|\boldsymbol{y}^{\text{obs}})$ and provides an estimated version via the observed frequencies. In fact, the posterior probabilities can be written as the following conditional expectations:

$$\mathbb{P}(\mathfrak{M} = m \big| S(\mathbf{Y}) = s) = \mathbb{E}\big(\mathbf{1}_{\{\mathfrak{M}=m\}} \big| S(\mathbf{Y}) = s\big).$$

Computing these conditional expectations based on independent and identically distributed (iid) draws from the distribution of $(\mathfrak{M}, S(\mathbf{Y}))$ can be interpreted as a regression problem in which the response is the indicator of whether or not the simulation comes from model m and the covariates are the summary statistics. The iid draws constitute the reference table, which also is the training database for machine learning methods. The process used in the earlier ABC Algorithm 6.1 is a k-nn method, if one approximates the posterior by the frequency of m among the k nearest simulations to s. The proposals of [13] and [21] for ABC model choice are exactly in that vein.

Other methods can be implemented to better estimate $\mathbb{P}(\mathfrak{M} = m | S(\mathbf{Y}) = s)$ from the reference table, the training database of the regression method. For instance, Nadaraya-Watson estimators are weighted averages of the responses, where weights are non-negative decreasing functions (or kernels) of the distance $d(s^{(i)}, s)$. The regression method commonly used (instead of k-nn) is a local regression method, with a multi-nomial link, as proposed by [22] or by [19]: local regression procedures fit a linear model on simulated pairs $(\mathfrak{M}^{(i)}, s^{(i)})$ in a neighbourhood of s. The multi-nomial link ensures that the vector of probabilities has entries between 0 and 1 and sums to 1. However, local regression can prove computationally expensive, if not intractable, when the dimension of the covariate increases. Therefore, [23] proposed a dimension reduction technique based on linear discriminant analysis (an exploratory data analysis technique that projects the observation cloud along axes that maximise the discrepancies between groups, see [24]), which produces to a summary statistic of dimension $M - 1$.

Algorithm 6.2: Local Logistic Regression ABC Model Choice

Generate N samples $(\mathfrak{M}^{(i)}, s^{(i)})$ as in Algorithm 6.1
Compute weights $\omega_i = K_{\mathfrak{h}}(s^{(i)} - S(y^{\text{obs}}))$, where K is a kernel density and $\mathfrak{h}$ is its bandwidth estimated from the sample $(s^{(i)})$
Estimate the probabilities $\mathbb{P}(\mathfrak{M} = m | s)$ by a logistic link based on the covariate s from the weighted data $(\mathfrak{M}^{(i)}, s^{(i)}, \omega_i)$

Unfortunately, all regression procedures given so far suffer from a curse of dimensionality: they are sensitive to the number of covariates, for example, the dimension of the vector of summary statistics. Moreover, as detailed in the following sections, any improvements in the regression method do not change the fact that all these methods aim at approximating $\mathbb{P}(\mathfrak{M} = m | S(\mathbf{Y}) = s)$ as a function of s and use this function at $s = s^{\text{obs}}$, while caution and cross-checking might be necessary to validate $\mathbb{P}(\mathfrak{M} = m | S(\mathbf{Y}) = s^{\text{obs}})$ as an approximation of $\mathbb{P}(\mathfrak{M} = m | \mathbf{Y} = \boldsymbol{y}^{\text{obs}})$.

A related approach worth mentioning here is the expectation propagation ABC (EP-ABC) algorithm of [17], which also produces an approximation of the evidence associated with each model under comparison. Without getting into details, the expectation-propagation approach of [25,26] approximates the posterior distribution by a member of an exponential family, using an iterative and fast moment-matching process that takes only a component of the likelihood product at a time. When the likelihood function is unavailable, [17] propose to instead rely on empirical moments based on simulations of those fractions of the data. The algorithm includes, as a side product, an estimate of the evidence associated with the model and the data, hence, can be exploited for model selection and posterior probability approximation. On the positive side, the EP-ABC is much faster than a standard ABC scheme, does not always resort to summary statistics, or at least to global statistics, and is appropriate for 'big data' settings where the whole data cannot be explored at once. On the negative side, this approach has the same degree of validation as variational Bayes methods [27], which means converging to a proxy posterior that is at best optimally close to the genuine posterior within a certain class, requires a meaningful decomposition of the likelihood into blocks which can be simulated, calls for the determination of several tolerance levels, is critically dependent on calibration choices, has no self-control safety mechanism, and requires identifiability of the models' underlying parameters. Hence, while EP-ABC can be considered for conducting model selection, there is no theoretical guarantee that it provides a converging approximation of the evidence, while the implementation on realistic models in population genetics seems out of reach.

6.3 The Curse of Insufficiency

The paper [10] issued a warning that ABC approximations to posterior probabilities cannot always be trusted in the double sense that (1) they stand away from the genuine posterior probabilities (imprecision), and (2) they may even fail to converge to a Dirac distribution on the true model as the size of the observed dataset grows to infinity (inconsistency). Approximating posterior probabilities via an ABC algorithm means using the frequencies of acceptances of simulations from each of those models. We assumed in Algorithm 6.1 the use of a common summary statistic (vector) to define the distance to the observations, as otherwise the comparison between models would not make sense. This point may sound anti-climactic since the same feature occurs for point estimation, where the ABC estimator is an estimate of $\mathbb{E}[\theta|S(\boldsymbol{y}^{\text{obs}})]$. Indeed, all ABC approximations rely on the posterior distributions knowing those summary statistics, rather than knowing the whole dataset. When conducting point estimation with insufficient statistics, the information content is necessarily degraded. The posterior distribution is then different from the true posterior, but, at least, gathering more observations brings more information about the parameter (and convergence when the number of observations goes to infinity), unless one uses only ancillary statistics. However, while this information impoverishment only has consequences in terms of the precision of the inference for most inferential purposes, it induces a dramatic arbitrariness in the construction of the Bayes factor. To illustrate this arbitrariness, consider the case of starting from a statistic $S(x)$ sufficient for both models. Then, by the factorisation theorem, the true likelihoods factorise as:

$$f_1(x|\theta) = g_1(x)\pi_1(S(x)|\theta) \text{ and } f_2(x|\theta) = g_2(x)\pi_2(S(x)|\theta),$$

resulting in a true Bayes factor equal to:

$$B_{12}(x) = \frac{g_1(x)}{g_2(x)} B^S_{12}(x), \tag{6.1}$$

where the last term, indexed by the summary statistic S, is the limiting (or Monte Carlo error-free) version of the ABC Bayes factor. In the more usual case where the user cannot resort to a sufficient statistic, the ABC Bayes factor may diverge one way or another as the number of observations increases. A notable exception is the case of Gibbs random fields where [13] have shown how to derive inter-model sufficient statistics, beyond the raw sample. This is related to the less pessimistic paper of [28], also concerned with the limiting behaviour for the ratio (6.1). Indeed, these authors reach the opposite conclusion from ours, namely, that the problem can be solved by a sufficiency argument. Their point is that, when comparing models within exponential families (which is the natural realm for sufficient statistics), it is always possible to build an encompassing model with a sufficient statistic that remains sufficient across models.

However, apart from examples where a tractable sufficient summary statistic is identified, one cannot easily compute a sufficient summary statistic for model choice, and this results in a loss of information, when compared with the exact inferential approach, hence, a wider discrepancy between the exact Bayes factor and the quantity produced by an ABC approximation. When realising this conceptual difficulty, [10] felt it was their duty to warn the community about the dangers of this approximation, especially when considering the rapidly increasing number of applications using ABC for conducting model choice or hypothesis testing. Another argument in favour of this warning is that it is often difficult in practice to design a summary statistic that is informative about the model.

Let us signal here that a summary selection approach purposely geared towards model selection can be found in [15]. Let us stress in, and for, this section that the said method similarly suffers from the earlier curse of dimensionality. Indeed, the approach therein is based on an estimate of Fisher's information contained in the summary statistics about the pair $(\mathfrak{M}, \theta)$ and the correlated search for a subset of those summary statistics that is (nearly) sufficient. As explained in the paper, this approach implies that the resulting summary statistics are also sufficient for parameter estimation within each model, which obviously induces a dimension inflation in the dimension of the resulting statistic, in opposition to approaches focussing solely on the selection of summary statistics for model choice, like [16] and [29].

We must also stress that, from a model choice perspective, the vector made of the (exact!) posterior probabilities of the different models obviously constitutes a Bayesian sufficient statistics of dimension $M-1$, but this vector is intractable precisely in cases where the user has to resort to ABC approximations. Nevertheless, this remark is exploited in [16] in a two-stage ABC algorithm. The second stage of the algorithm is ABC model choice with summary statistics equal to approximation of the posterior probabilities. Those approximations are computed as ABC solutions at the first stage of the algorithm. Despite the conceptual attractiveness of this approach, which relies on a genuine sufficiency result, the approximation of the posterior probabilities given by the first stage of the algorithm directly rely on the choice of a particular set of summary statistics, which brings us back to the original issue of trusting an ABC approximation of a posterior probability.

There therefore is a strict loss of information in using ABC model choice, due to the call both to insufficient statistics and to non-zero tolerances (or a imperfect recovery of the posterior probabilities with a regression procedure).

6.3.1 Some counter-examples

Besides a toy example opposing Poisson and geometric distributions to point out the potential irrelevance of the Bayes factor based on poor statistics, [10] goes over a realistic population genetic illustration, where two evolution scenarios involving three populations are compared, two of those populations having diverged 100 generations ago and the third one resulting from a recent admixture between the first two populations (scenario 1) or simply diverging from population 1 (scenario 2) at the same date of five generations in the past. In scenario 1, the admixture rate is 0.7 from population 1. Simulated datasets (100) of the same size as in experiment 1 (15 diploid individuals per population, 5 independent micro-satellite loci) were generated assuming an effective population size of 1000 and a mutation rate of 0.0005. In this experiment, there are six parameters (provided with the corresponding priors): the admixture rate ($\mathcal{U}[0.1, 0.9]$), three effective population sizes ($\mathcal{U}[200, 2000]$), the time of admixture/second divergence ($\mathcal{U}[1, 10]$), and the date of the first divergence ($\mathcal{U}[50, 500]$). While costly in computing time, the posterior probability of a scenario can be estimated by importance sampling, based on 1000 parameter values and 1000 trees per parameter value, thanks to the modules of [30]. The ABC approximation is produced by Do It Yourself Approximate Bayesian Computation (DIYABC) [31], based on a reference sample of two million parameters and 24 summary statistics. The result of this experiment is shown on Figure 6.1, with a clear divergence in the numerical values despite stability in both approximations. Taking the importance sampling approximation as the reference value, the error rates in using the ABC approximation to choose between scenarios 1 and 2 are 14.5% and 12.5% (under scenarios 1 and 2), respectively. Although a simpler experiment with a single parameter and the same 24 summary statistics shows a reasonable agreement between both

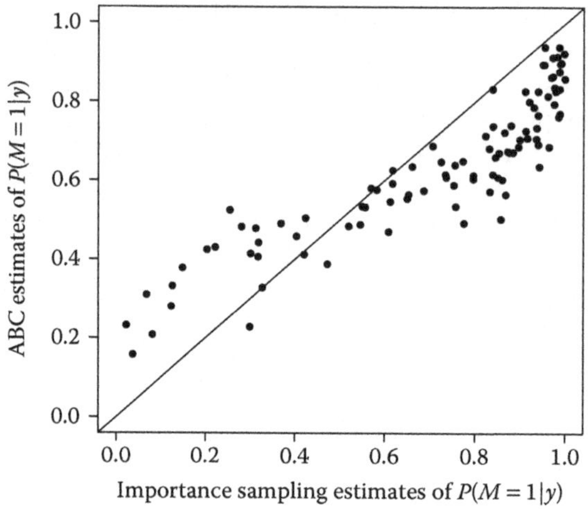

FIGURE 6.1
Comparison of importance sampling (*first axis*) and ABC (*second axis*) estimates of the posterior probability of scenario 1 in the first population genetic experiment, using 24 summary statistics. (From Robert, C.P. et al., *Proc. Natl Acad. Sci.*, 108, 15112–15117, 2011.)

approximations, this result comes as an additional support to our warning about a blind use of ABC for model selection. The corresponding simulation experiment was quite intense, as, with 50 markers and 100 individuals, the product likelihood suffers from an enormous variability that 100,000 particles and 100 trees per locus have trouble addressing despite a huge computing cost.

An example is provided in the introduction of the paper [32], sequel to [10]. The setting is one of a comparison between a normal $\boldsymbol{y} \sim \mathcal{N}(\theta_1, 1)$ model and a double exponential $\boldsymbol{y} \sim \mathcal{L}(\theta_2, 1/\sqrt{2})$ model.[1] The summary statistics used in the corresponding ABC algorithm are the sample mean, the sample median, and the sample variance. Figure 6.2 exhibits the absence of discrimination between both models, since the posterior probability of the normal model converges to a central value around 0.5–0.6 when the sample size grows, irrelevant of the true model behind the simulated datasets.

6.3.2 Still some theoretical guarantees

Our answer to the (well-received) aforementioned warning is provided in [32], which deals with the evaluation of summary statistics for Bayesian model choice. The main result states that, under some Bayesian asymptotic

[1]The double exponential distribution is also called the Laplace distribution, hence, the notation $\mathcal{L}(\theta_2, 1/\sqrt{2})$, with mean θ_2 and variance one.

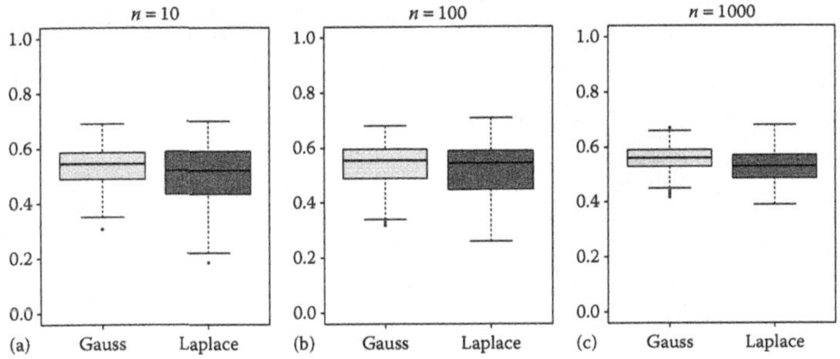

FIGURE 6.2
Comparison of the range of the ABC posterior probability that data are from a normal model with unknown mean θ when the data are made of $n = 10, 100$, and 1000 observations (*a–c respectively*) either from a Gaussian (*lighter*) or Laplace distribution (*darker*) and when the ABC summary statistic is made of the empirical mean, median, and variance. The ABC algorithm generates 10^4 simulations (5,000 for each model) from the prior $\theta \sim \mathcal{N}(0, 4)$ and selects the tolerance ϵ as the 1% distance quantile over those simulations. (From Marin, J.-M. et al., *J. R. Stat. Soc. B*, 76, 833–859, 2014.)

assumptions, ABC model selection only depends on the behaviour of the mean of the summary statistic under both models. The paper establishes a theoretical framework that leads to demonstrate consistency of the ABC Bayes factor under the constraint that the ranges of the expected value of the summary statistic under both models do not intersect. A negative result is also given in [32], which mainly states that, whatever the observed dataset, the ABC Bayes factor selects the model having the smallest effective dimension when the assumptions do not hold.

The simulations associated with the paper were straightforward in that (1) the setup compares normal and Laplace distributions with different summary statistics (including the median absolute deviation), (2) the theoretical results told what to look for, and (3) they did very clearly exhibit the consistency and inconsistency of the Bayes factor/posterior probability predicted by the theory. Both boxplots shown here on Figures 6.2 and 6.3 show this agreement: when using (empirical) mean, median, and variance to compare normal and Laplace models, the posterior probabilities do not select the true model, but instead aggregate near a fixed value. When using instead the median absolute deviation as summary statistic, the posterior probabilities concentrate near one or zero depending on whether or not the normal model is the true model.

It may be objected to such necessary and sufficient conditions that Bayes factors simply are inappropriate for conducting model choice, thus making the whole derivation irrelevant. This foundational perspective is an arguable viewpoint [33]. However, it can be countered within the Bayesian paradygm by the fact that Bayes factors and posterior probabilities are consistent quantities

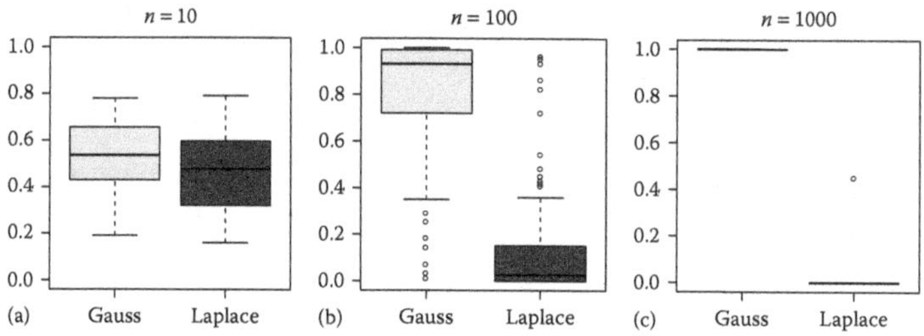

FIGURE 6.3
Same representation as Figure 6.2 when using the median absolute deviation of the sample as its sole summary statistic. (From Marin, J.-M. et al., *J. R. Stat. Soc. B*, 76, 833–859, 2014.)

that are used in conjunction with ABC in dozens of genetic papers. Further arguments are provided in the various replies to both of Templeton's radical criticisms [11,12]. That more empirical and model-based assessments also are available is quite correct, as demonstrated in the multi-criterion approach of [34]. This is simply another approach, not followed by most geneticists so far.

A concluding remark about [32] is that, while the main bulk of the paper is theoretical, it does bring an answer that the mean ranges of the summary statistic under each model must not intersect if they are to be used for ABC model choice. In addition, while the theoretical assumptions therein are not of the utmost relevance for statistical practice, the paper includes recommendations on how to conduct a χ^2 test on the difference of the means of a given summary statistics under both models towards assessing whether or not this summary is acceptable.

6.4 Selecting the Maximum a Posteriori Model via Machine Learning

The earlier sections provide enough arguments to feel less than confident in the outcome of a standard ABC model choice Algorithm 6.1, at least in the numerical approximation of the probabilities $\mathbb{P}(\mathfrak{M} = m | S(\mathbf{Y}) = s^{\text{obs}})$ and in their connection with the genuine posterior probabilities $\mathbb{P}(\mathfrak{M} = m | \mathbf{Y} = \boldsymbol{y}^{\text{obs}})$. There are indeed three levels of approximation errors in such quantities, one due to the Monte Carlo variability, one due to the non-zero ABC tolerance or, more generally, to the error committed by the regression procedure when estimating the conditional expected value, and one due to the curse of insufficiency. While the derivation of a satisfying approximation of the genuine $\mathbb{P}(\mathfrak{M} = m | \mathbf{Y} = \boldsymbol{y}^{\text{obs}})$ seems beyond our reach, we present in the following a novel approach to both construct the most likely model and approximate $\mathbb{P}(\mathfrak{M} = m|S(\mathbf{Y}) = s^{\text{obs}})$ for the most likely model, based on the machine learning tool of random forests.

6.4.1 Reconsidering the posterior probability estimation

Somewhat paradoxically, since the ABC approximation to posterior probabilities of a collection of models is delicate, [20] support inverting the order of selection of the a posteriori most probable model and of approximation of its posterior probability, using the alternative tool of random forests for both goals. The reason for this shift in order is that the rate of convergence of local regression procedure, such as k-nn or the local regression with multi-nomial link, heavily depends on the dimension of the covariates (here the dimension of the summary statistic). Thus, since the primary goal of ABC model choice is to select the most appropriate model, both [20] and [35] argue that one does not need to correctly approximate the probability:

$$\mathbb{P}(\mathfrak{M} = m|S(\mathbf{Y}) \approx s^{\text{obs}}),$$

when looking for the most probable model in the sense of:

$$\mathbb{P}(\mathfrak{M} = m|\mathbf{Y} = \boldsymbol{y}^{\text{obs}}),$$

probability. Stoehr et al. [35] stresses that selecting the most adequate model for the data at hand as the maximum a posteriori (MAP) model index is a classification issue, which proves to be a significantly easier inference problem than estimating a regression function [24,36]. This is the reason why [35] adapt the earlier Algorithm 6.1 by resorting to a k-nn classification procedure, which sums up as returning the most frequent (or majority rule) model index among the k simulations nearest to the observed dataset, nearest in the subspace of the summary statistics. Indeed, generic classification aims at forecasting a variable $\mathfrak{M}$ taking a finite number of values, $\{1, \ldots, M\}$, based on a vector of covariates $S = (S_1, \ldots, S_d)$. The Bayesian approach to classification stands in using a training database (m^i, s^i) made of independent replicates of the pair $(\mathfrak{M}, S(\mathbf{Y}))$ that are simulated from the prior predictive distribution. The connection with ABC model choice is that the latter predicts a model index, $\mathfrak{M}$, from the summary statistic $S(\mathbf{Y})$. Simulations in the ABC reference table can thus be envisioned as creating a learning database from the prior predictive that trains the classifier.

Pudlo et al. [20] widen the paradigm shift undertaken in [35], as they use a machine learning approach to the selection of the most adequate model for the data at hand and exploit this tool to derive an approximation of the posterior probability of the selected model. The classification procedure chosen by [20] is the technique of random forests (RFs) [37], which constitute a trustworthy and seasoned machine learning tool, well adapted to complex settings like those found in ABC settings. The approach further requires no primary selection of a small subset of summary statistics, which allows for an automatic input of summaries from various sources, including softwares like DIYABC [29]. At a first stage, an RF is constructed from the reference table to predict the model index and applied to the data at hand to return a MAP estimate. At a second stage, an additional RF is constructed for explaining the selection error of the MAP estimate, based on the same reference table. When applied to the observed data, this secondary random forest produces an estimate of the posterior probability of the model selected by the primary RF, as detailed in the following [20].

6.4.2 Random forests construction

An RF aggregates a large number of classification trees by adding for each tree a randomisation step to the classification and regression trees (CARTs) algorithm [38]. Let us recall that this algorithm produces a binary classification tree that partitions the covariate space towards a prediction of the model index. In this tree, each binary node is partitioning the observations via a rule of the form $S_j < t_j$, where S_j is one of the summary statistics and t_j is chosen towards the minimisation of an heterogeneity index. For instance, [20] use the Gini criterion [29]. A CART is built based on a learning table, and it is then applied to the observed summary statistic s^{obs}, predicting the model index by following a path that applies these binary rules starting from the tree root and returning the label of the tip at the end of the path.

The randomisation part in RF produces a large number of distinct CARTs by (1) using for each tree a bootstrapped version of the learning table on a bootstrap sub-sample of size N_{boot} and (2) selecting the summary statistics at each node from a random subset of the available summaries. The calibration of a RF thus involves three quantities:

- B, the number of trees in the forest.
- n_{try}, the number of covariates randomly sampled at each node by the randomised CART.
- N_{boot}, the size of the bootstraped sub-sample.

The so-called out-of-bag error associated with an RF is the average number of times a point from the learning table is wrongly allocated, when averaged over trees that exclude this point from the bootstrap sample.

The way [20] build a random forest classifier given a collection of statistical models is to start from an ABC reference table including a set of simulation records made of model indices, parameter values, and summary statistics for the associated simulated data. This table then serves as a training database for a random forest that forecasts a model index based on the summary statistics.

Algorithm 6.3: Random Forest ABC Model Choice

Generate N samples $(\mathfrak{M}^{(i)}, s^{(i)})$ as in Algorithm 6.1 (the reference table)
Construct N_{tree} randomised CART which predicts the model indices using the summary statistics
for $b = 1$ **to** N_{tree} **do**
  draw a bootstrap sub-sample of size N_{boot} from the reference table
  grow a randomised CART T_b
end for
Determine the predicted indices for s^{obs} and the trees $\{T_b; b = 1, \ldots, N_{\text{tree}}\}$
Assign s^{obs} to an indice (a model) according to a majority vote among the predicted indices.

6.4.3 Approximating the posterior probability of the maximum a posteriori

The posterior probability of a model is the natural Bayesian uncertainty quantification [39], since it is the complement of the posterior loss associated with a 0–1 loss $\mathbf{1}_{\mathfrak{M} \neq \hat{\mathfrak{M}}(s^{\text{obs}})}$, where $\hat{\mathfrak{M}}(s^{\text{obs}})$ is the model selection procedure, for example, the RF outcome described in the previous section. However, for reasons described earlier, we are unwilling to trust the standard ABC approximation to the posterior probability as reported in Algorithm 6.1. An initial proposal in [35] is to instead rely on the conditional error rate induced by the k-nn classifier knowing $S(\mathbf{Y}) = s^{\text{obs}}$, namely:

$$\mathbb{P}\left(\mathfrak{M} \neq \widehat{\mathfrak{M}}(s^{\text{obs}}) \middle| s^{\text{obs}}\right),$$

where $\widehat{\mathfrak{M}}$ denotes the k-nn classifier trained on ABC simulations. The aforementioned conditional expected value of $\mathbf{1}_{\{\mathfrak{M} \neq \widehat{\mathfrak{M}}(s^{\text{obs}})\}}$ is approximated in [35], with a Nadaraya-Watson estimator on a new set of simulations, where the authors compare the model index $m^{(i)}$, which calibrates the simulation of the pseudo-data $y^{(i)}$, and the model index $\widehat{\mathfrak{M}}(s^{(i)})$, predicted by the k-nn approach trained on a first database of simulations. However, this first proposal has the major drawback of relying on non-parametric regression, which deteriorates when the dimension of the summary statistic increases. This local error also allows for the selection of summary statistics adapted to s^{obs}, but the procedure of [35] remains constrained by the dimension of the summary statistic, which typically have to be less than ten.

Furthermore, relying on a large dimensional summary statistic – to bypass, at least partially, the curse of insufficiency – was the main reason for adopting a classifier such as RFs in [20]. Hence, the authors proposed to estimate the posterior expectation of $\mathbf{1}_{\mathfrak{M} \neq \hat{\mathfrak{M}}(s^{\text{obs}})}$ as a function of the summary statistics, via another RF construction. Indeed:

$$\begin{aligned}\mathbb{E}[\mathbf{1}_{\mathfrak{M}\neq\hat{\mathfrak{M}}(s^{\text{obs}})}|s^{\text{obs}}] &= \sum_{i=1}^{k}\mathbb{P}[\mathfrak{M}=i|s^{\text{obs}}]\times\mathbf{1}_{\hat{\mathfrak{M}}(s^{\text{obs}})\neq i}\\ &= \mathbb{P}[\mathfrak{M}\neq\hat{\mathfrak{M}}(s^{\text{obs}})|s^{\text{obs}}]\\ &= 1-\mathbb{P}[\mathfrak{M}=\hat{\mathfrak{M}}(s^{\text{obs}})|s^{\text{obs}}]\,.\end{aligned}$$

The estimation of $\mathbb{E}[\mathbf{1}_{\mathfrak{M}\neq\hat{\mathfrak{M}}(s)}|s]$ proceeds as follows:

- Compute the values of $\mathbf{1}_{\mathfrak{M}\neq\hat{\mathfrak{M}}(s)}$ for the trained random forest and all terms in the reference table.
- Train a second RF regressing $\mathbf{1}_{\mathfrak{M}\neq\hat{\mathfrak{M}}(s)}$ on the same set of summary statistics and the same reference table, producing a function $\varrho(s)$ that returns a machine learning estimate of $\mathbb{P}[\mathfrak{M}\neq\hat{\mathfrak{M}}(s)|s]$.
- Apply this function to the actual observations to produce $1-\varrho(s^{\text{obs}})$ as an estimate of $\mathbb{P}[\mathfrak{M}=\hat{\mathfrak{M}}(s^{\text{obs}})|s^{\text{obs}}]$.

6.5 A First Toy Example

We consider in this section a simple uni-dimensional setting with three models where the marginal likelihoods can be computed in closed form.

Under model 1, our dataset is a n-sample from an exponential distribution with parameter θ (with expectation $1/\theta$), and the corresponding prior distribution on θ is an exponential distribution with parameter 1. In this model, given the sample $y=(y_1,\ldots,y_n)$ with $y_i>0$, the marginal likelihood is given by:

$$m_1(\boldsymbol{y})=\Gamma(n+1)\left(1+\sum_{i=1}^{n}y_i\right)^{-n-1}.$$

Under model 2, our dataset is a n-sample from a log-normal distribution with location parameter θ and dispersion parameter equal to 1 [which implies an expectation equal to $\exp(\theta+0.5)$]. The prior distribution on θ is a standard Gaussian distribution. For this model, given the sample $y=(y_1,\ldots,y_n)$ with $y_i>0$, the marginal likelihood is given by:

$$\begin{aligned}m_2(\boldsymbol{y})=\exp&\left[-\left(\sum_{i=1}^{n}\log(y_i)\right)^2/(2n(n+1))-\left(\sum_{i=1}^{n}\log^2(y_i)\right)^2/2\right.\\ &\left.+\left(\sum_{i=1}^{n}\log(y_i)\right)^2/(2n)-\sum_{i=1}^{n}\log(y_i)\right]\times(2\pi)^{-n/2}\times(n+1)^{-1/2}.\end{aligned}$$

Under model 3, our dataset is a n-sample from a gamma distribution with parameter $(2, \theta)$ (with expectation $2/\theta$), and the prior distribution on θ is an exponential distribution with parameter 1. For this model, given the sample $y = (y_1, \ldots, y_n)$with $y_i > 0$, the marginal likelihood is given by:

$$m_3(\boldsymbol{y}) = \exp\left[\sum_{i=1}^{n} \log(y_i)\right] \frac{\Gamma(2n+1)}{\Gamma(2)^n} \left(1 + \sum_{i=1}^{n} y_i\right)^{-2n-1}.$$

We consider three summary statistics:

$$\left(\sum_{i=1}^{n} y_i, \sum_{i=1}^{n} \log(y_i), \sum_{i=1}^{n} \log^2(y_i)\right).$$

These summary statistics are sufficient not only within each model, but also for the model choice problem [28], and the purpose of this example is not to evaluate the impact of a loss of sufficiency.

When running ABC, we set $n = 20$ for the sample size and generated a reference table containing 29,000 simulations (9,676 simulations from model 1, 9,650 from model 2, and 9,674 from model 3). We further generated an independent test dataset of size 1,000. Then, to calibrate the optimal number of neighbours in the standard ABC procedure [13,21], we exploited 1,000 independent simulations.

For each element of the test dataset, as obvious from the above $m_i(\boldsymbol{y})$'s, we can evaluate the exact model posterior probabilities. Figure 6.4 represents the posterior probability of model 3 for every simulation, ranked by model index. In addition, Figure 6.5 gives a plot of the first two LDA projections of the test dataset. Both figures explain why the model choice problem is not easy in this setting. Indeed, based on the exact posterior probabilities, selecting the model associated with the highest posterior probability achieves the smallest prior error rate. Based on the test dataset, we estimate this lower bound as being around 0.245, for example, close to 25%.

Based on a calibration set of 1,000 simulations, and the earlier reference table of size 29,000, the optimal number of neighbours that should be used by the standard ABC model choice procedure, for example, the one that minimises the prior error rate, is equal to 20. In this case, the resulting prior error rate for the test dataset is equal to 0.277.

By comparison, the RF ABC model choice technique of [20] based on 500 trees achieves an error rate of 0.276 on the test dataset. For this example, adding the two LDA components to the summary statistics does not make a difference. This alternative procedure achieves similarly good results in terms of prior error rate, since 0.276 is relatively closed to the absolute lower bound of 0.245. However, as explained in previous sections and illustrated on Figure 6.6, the RF estimates of the posterior probabilities are not to be trusted. In short, a classification tool is not necessarily appropriate for regression goals.

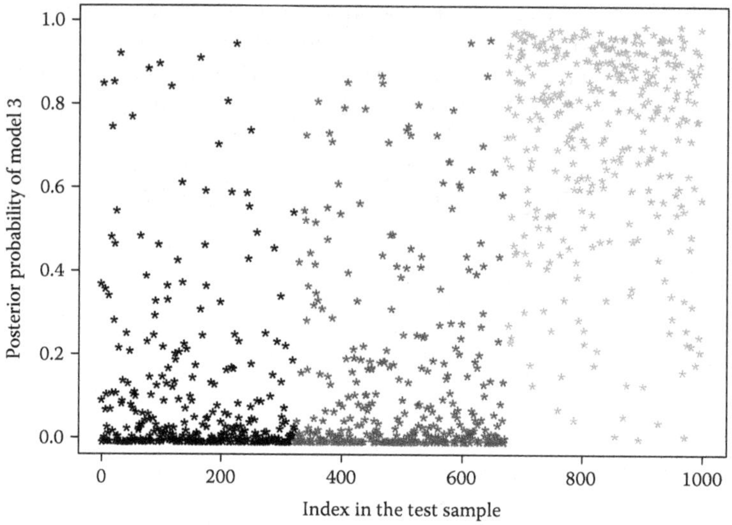

FIGURE 6.4
True posterior probability of model 3 for each term from the test sample. Colour corresponds to the true model index: black for model 1, dark grey for model 2, and light grey for model 3. The terms in the test sample have been ordered by model index to improve the representation.

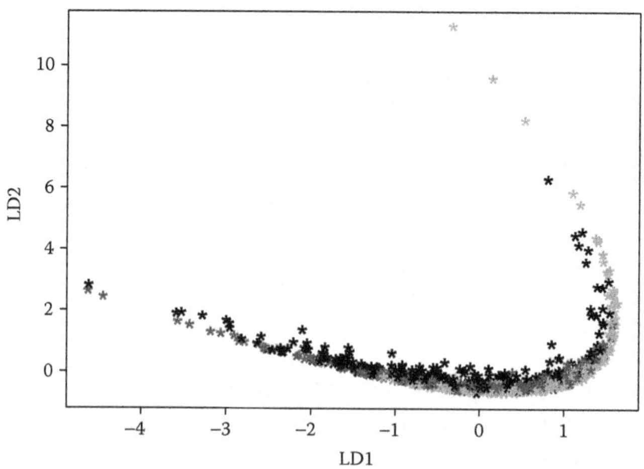

FIGURE 6.5
LDA projection along the first two axes of the test dataset, with the same colour code as in Figure 6.4.

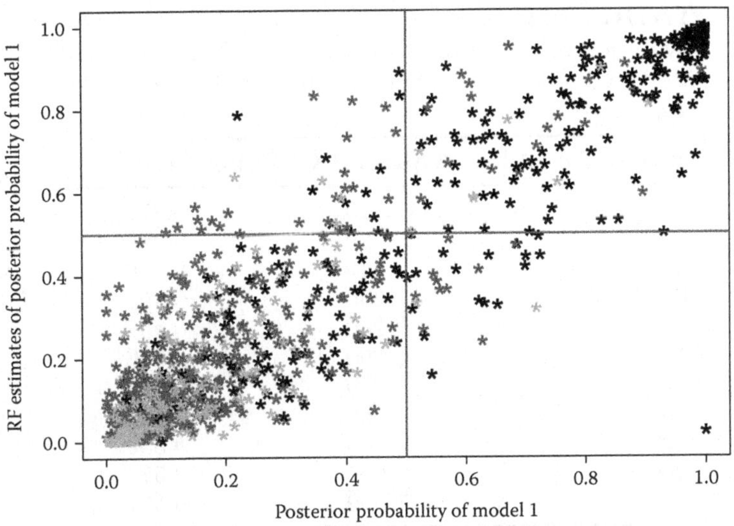

FIGURE 6.6
True posterior probabilities of model 1 against their Random Forest estimates for the test sample, with the same colour code as in Figure 6.4.

A noteworthy feature of the RF technique is its ability to be robust against non-discriminant variates. This obviously is of considerable appeal in ABC model choice since the selection of summary statistics is an unsolved challenge. To illustrate this point, we added to the original set of three summary statistics variables that are pure noise, being produced by independent simulations from standard Gaussian distributions. Table 6.1 shows that the additional error

TABLE 6.1
Evolution of the Prior Error Rate for the RF ABC Model Choice Procedure as a Function of the Number of White Noise Variates

Extra Variables	Prior Error Rate
0	0.276
2	0.283
4	0.288
6	0.272
8	0.280
10	0.286
20	0.318
50	0.355
100	0.391
200	0.419
1,000	0.456

TABLE 6.2
Evolution of the Prior Error Rate for a Standard ABC Model Choice as a Function of the Number of White Noise Variates

Extra Variables	Optimal k	Prior Error Rate
0	20	0.277
2	20	0.368
4	140	0.468
6	200	0.491
8	260	0.492
10	260	0.526
20	260	0.542
50	260	0.548
100	500	0.559
200	500	0.572
1,000	1,000	0.594

due to those irrelevant variates grows much more slowly than for the standard ABC model choice technique, as shown in Table 6.2. In the latter case, a few extraneous variates suffice to propel the error rate above 50%.

6.6 Human Population Genetics Example

We consider here the massive single nucleotide polymorphism (SNP) dataset already studied in [20], associated with a Most Recent Common Ancestor (MRCA) population genetic model corresponding to Kingman's coalescent that has been at the core of ABC implementations from their beginning [1]. The dataset corresponds to individuals originating from four human populations, with 30 individuals per population. The freely accessible public 1000 Genome databases +http://www.1000genomes.org/data has been used to produce this dataset. As detailed in [20], one of the appeals of using SNP data from the 1000 Genomes Project [40] is that such data does not suffer from any ascertainment bias.

The four human populations in this study included the Yoruba population (Nigeria) as representative of Africa, the Han Chinese population (China) as representative of East Asia (encoded CHB), the British population (England and Scotland) as representative of Europe (encoded GBR), and the population of Americans of African ancestry in SW USA (encoded ASW). After applying some selection criteria described in [20], the dataset includes 51,250

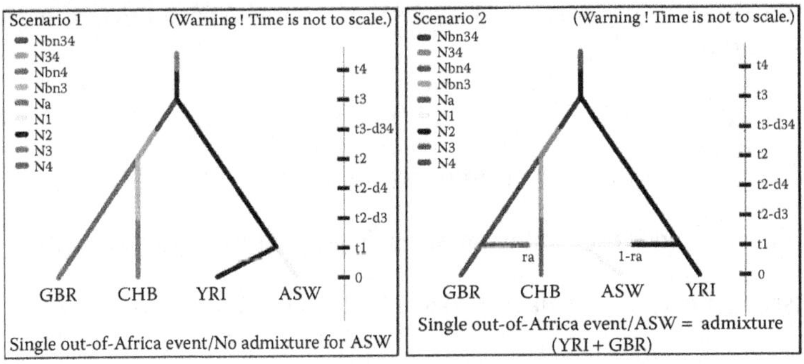

FIGURE 6.7
Two scenarios of evolution of four human populations genotyped at 50,000 SNPs. The genotyped populations are YRI = Yoruba (Nigeria, Africa), CHB = Han (China, East Asia), GBR = British (England and Scotland, Europe), and ASW = Americans of African ancestry (SW USA).

SNP loci scattered over the 22 autosomes with a median distance between two consecutive SNPs equal to 7 kb. Among those, 50,000 were randomly chosen for evaluating the proposed RF ABC model choice method.

In the novel study described here, we only consider two scenarios of evolution. These two models differ by the possibility or impossibility of a recent genetic admixture of Americans of African ancestry in SW USA between their African forebears and individuals of European origins, as described in Figure 6.7. Model 2 thus includes a single out-of-Africa colonisation event giving an ancestral out-of-Africa population with a secondarily split into one European and one East Asian population lineage and a recent genetic admixture of Americans of African origin with their African ancestors and European individuals. RF ABC model choice is used to discriminate among both models and returns error rates. The vector of summary statistics is the entire collection provided by the DIYABC software for SNP markers [29], made of 112 summary statistics described in the manual of DIYABC.

Model 1 involves 16 parameters, while model 2 has an extra parameter, the admixture rate r_a. All times and durations in the model are expressed in number of generations. The stable effective populations sizes are expressed in number of diploid individuals. The prior distributions on the parameters appearing in one of the two models and used to generate SNP datasets are as follows:

1. Split or admixture time t_1, $\mathcal{U}[1, 30]$
2. Split times (t_2, t_3, t_4), uniform on their support $\{(t_2, t_3, t_4) \in [100, 10{,}000]^{\otimes 3} | t_2 < t_3 < t_4\}$

3. Admixture rate (proportion of genes with a non-African origin in model 2) $r_a \sim\sim \mathcal{U}[0.05, 0.95]$

4. Effective population sizes N_1, N_2, N_3, N_3, and N_{34}, $\mathcal{U}$ [1,000, 100,000]

5. Bottleneck durations d_3, d_4, and d_{34}, $\mathcal{U}[5, 500]$

6. Bottleneck effective population sizes Nbn_3, Nbn_4, and Nbn_{34}, $\mathcal{U}[5, 500]$

7. Ancestral effective population size N_a, $\mathcal{U}$[100, 10,000]

For the analyses, we use a reference table containing 19,995 simulations: 10,032 from model 1 and 9,963 from model 2. Figure 6.8 shows the distributions of the first LDA projection for both models, as a byproduct of the simulated reference table. Unsurprisingly, this LDA component has a massive impact on the RF ABC model choice procedure. When including the LDA statistic, most trees (473 out of 500) allocate the observed dataset to model 2. The second random forest to evaluate the local selection error leads a high confidence level: the estimated posterior probability of model 2 is greater than 0.999. Figure 6.9 shows contributions for the most relevant statistics in the forest, stressing once again the primary role of the first LDA axis. Note that using solely this first LDA axis increases considerably the prior error rate.

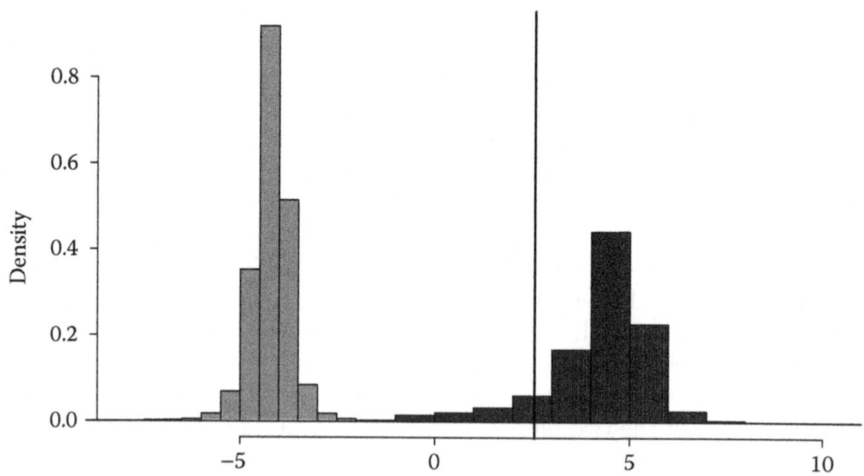

FIGURE 6.8
Distribution of the first LDA axis derived from the reference table, in light grey for model 1 and in dark grey for model 2. The observed dataset is indicated by a black vertical line.

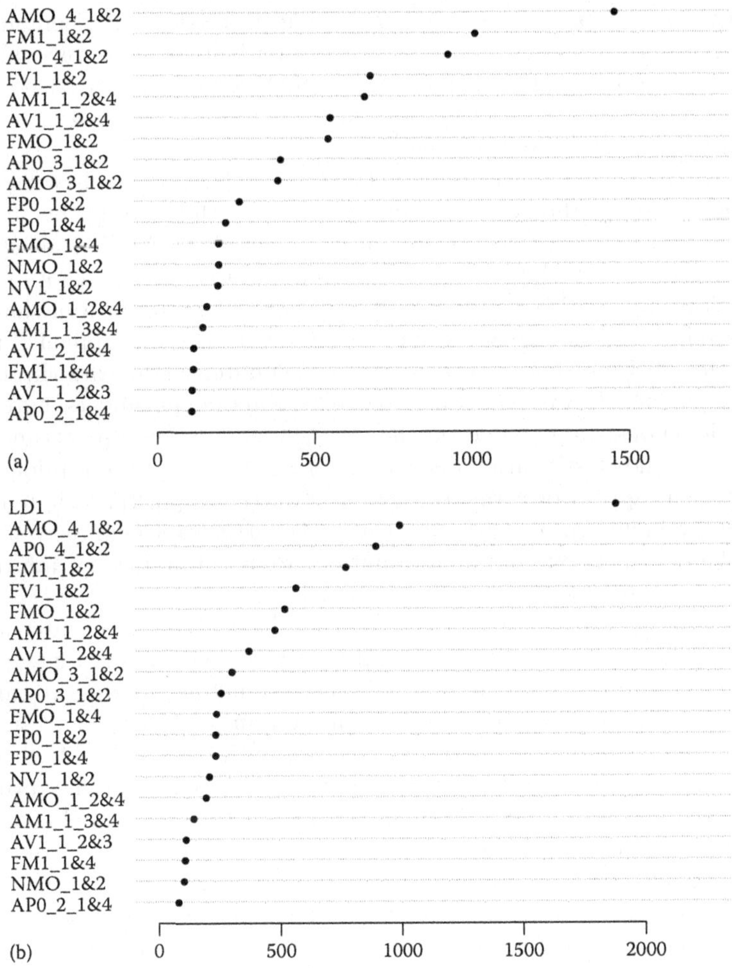

FIGURE 6.9
Contributions of the most frequent statistics in the RF. The contribution of a summary statistic is evaluated as the average decrease in node impurity at all nodes where it is selected, over the trees of the RF when using the 112 summary statistics (a), and when further adding the first LDA axis (b).

6.7 Conclusion

This chapter has presented a solution for conducting ABC model choice and testing that differs from the usual practice in applied fields like population genetics, where the use of Algorithm 6.1 remains the norm. This choice is not due to any desire to promote our own work, but proceeds from a genuine belief that the figures returned by this algorithm cannot be trusted as approximating the actual posterior probabilities of the model. This belief is based on our experience along the years we worked on this problem, as illustrated by the evolution in our papers on the topic.

To move to a machine-learning tool like random forests somehow represents a paradigm shift for the ABC community. For one thing, to gather intuition about the intrinsic nature of this tool and to relate it to ABC schemes is certainly anything but straightforward. For instance, a natural perception of this classification methodology is to take it as a natural selection tool that could lead to a reduced subset of significant statistics, with the side appeal of providing a natural distance between two vectors of summary statistics through the tree discrepancies. However, as we observed through experiments, subsequent ABC model choice steps based on the selected summaries are detrimental to the quality of the classification once a model is selected by the random forest. The statistical appeal of a random forest is on the opposite, that it is quite robust to the inclusion of poorly informative or irrelevant summary statistics and on the opposite able to catch minute amounts of additional information produced by such additions.

While the current state-of-the-art remains silent about acceptable approximations of the true posterior probability of a model, in the sense of being conditional to the raw data, we are nonetheless making progress towards the production of an approximation conditional on an arbitrary set of summary statistics, which should offer strong similarities with the above. That this step can be achieved at no significant extra cost is encouraging for the future.

Another important inferential issue pertaining to ABC model choice is to test a large collection of models. The difficulties to learn how to discriminate between models certainly increase when the number of likelihoods in competition gets larger. Even the most up-to-date machine learning algorithms will loose their efficiency if one keeps constant the number of iid draws from each model, without mentioning that the time complexity will increase linearly with the size of the collection to produce the reference table that trains the classifier. Thus, this problem remains largely open.

References

[1] S. Tavaré, D. Balding, R. Griffith, and P. Donnelly. Inferring coalescence times from DNA sequence data. *Genetics*, 145(26):505–518, 1997.

[2] J. Pritchard, M. Seielstad, A. Perez-Lezaun, and M. Feldman. Population growth of human Y chromosomes: A study of Y chromosome microsatellites. *Molecular Biology and Evolution*, 16(12):1791–1798, 1999.

[3] D. Rubin. Bayesianly justifiable and relevant frequency calculations for the applied statistician. *The Annals of Statistics*, 12(4):1151–1172, 1984.

[4] M. Blum. Approximate Bayesian computation: A non-parametric perspective. *Journal of the American Statistical Association*, 105(491): 1178–1187, 2010.

[5] P. Fearnhead and D. Prangle. Constructing summary statistics for approximate Bayesian computation: Semi-automatic approximate Bayesian computation. *Journal of the Royal Statistical Society: Series B*, 74(3):419–474, 2012.

[6] G. Biau, F. Cérou, and A. Guyader. New insights into approximate Bayesian computation. *Annales de l'Institut Henri Poincaré, Probabilités et Statistiques*, 51(1):376–403, 2015.

[7] S.N. Wood. Statistical inference for noisy nonlinear ecological dynamic systems. *Nature*, 466:1102–1104, 2010.

[8] R. Wilkinson. Approximate Bayesian computation (ABC) gives exact results under the assumption of model error. *Statistical Applications in Genetics and Molecular Biology*, 12(2):129–141, 2013.

[9] R. Wilkinson. Accelerating ABC methods using Gaussian processes. In *Proceedings of the 17th International Conference on Artificial Intelligence and Statistics (AISTATS)*, volume 33 of *AI & Statistics 2014*, pp. 1015–1023. JMLR: Workshop and Conference Proceedings, 2014.

[10] C.P. Robert, J.-M. Cornuet, J.-M. Marin, and N. Pillai. Lack of confidence in ABC model choice. *Proceedings of the National Academy of Sciences*, 108(37):15112–15117, 2011.

[11] A. Templeton. Statistical hypothesis testing in intraspecific phylogeography: Nested clade phylogeographical analysis vs. approximate Bayesian computation. *Molecular Ecology*, 18(2):319–331, 2008.

[12] A. Templeton. Coherent and incoherent inference in phylogeography and human evolution. *Proceedings of the National Academy of Sciences*, 107(14):6376–6381, 2010.

[13] A. Grelaud, J.-M. Marin, C.P. Robert, F. Rodolphe, and F. Tally. Likelihood-free methods for model choice in Gibbs random fields. *Bayesian Analysis*, 3(2):427–442, 2009.

[14] T. Toni and M. Stumpf. Simulation-based model selection for dynamical systems in systems and population biology. *Bioinformatics*, 26(1): 104–110, 2010.

[15] C. Barnes, S. Filippi, M. Stumpf, and T. Thorne. Considerate approaches to constructing summary statistics for ABC model selection. *Statistics and Computing*, 22(6):1181–1197, 2012.

[16] D. Prangle, P. Fearnhead, M. Cox, P. Biggs, and N. French. Semi-automatic selection of summary statistics for ABC model choice. *Statistical Applications in Genetics and Molecular Biology*, 13(1):67–82, 2014.

[17] S. Barthelmé and N. Chopin. Expectation propagation for likelihood-free inference. *Journal of the American Statistical Association*, 109(505):315–333, 2014.

[18] M. Beaumont. Joint determination of topology, divergence time and immigration in population trees. In S. Matsumura, P. Forster, and C. Renfrew (Eds.), *Simulations, Genetics and Human Prehistory*, pp. 134–154. Cambridge, UK: (McDonald Institute Monographs), McDonald Institute for Archaeological Research, 2008.

[19] J.-M. Cornuet, V. Ravigné, and A. Estoup. Inference on population history and model checking using DNA sequence and microsatellite data with the software DIYABC (v1.0). *BMC Bioinformatics*, 11:401, 2010.

[20] P. Pudlo, J.-M. Marin, A. Estoup, J.-M. Cornuet, M. Gautier, and C.P. Robert. Reliable ABC model choice via random forests. *Bioinformatics*, 32(6):859–866, 2016.

[21] T. Toni, D. Welch, N. Strelkowa, A. Ipsen, and M. Stumpf. Approximate Bayesian computation scheme for parameter inference and model selection in dynamical systems. *Journal of the Royal Society Interface*, 6(31):187–202, 2009.

[22] N. Fagundes, N. Ray, M. Beaumont, S. Neuenschwander, F. Salzano, S. Bonatto, and L. Excoffier. Statistical evaluation of alternative models of human evolution. *Proceedings of the National Academy of Sciences*, 104(45):17614–17619, 2007.

[23] A. Estoup, E. Lombaert, J.-M. Marin, C.P. Robert, T. Guillemaud, P. Pudlo, and J.-M. Cornuet. Estimation of demo-genetic model probabilities with approximate Bayesian computation using linear discriminant analysis on summary statistics. *Molecular Ecology Resources*, 12(5):846–855, 2012.

[24] T. Hastie, R. Tibshirani, and J. Friedman. *The Elements of Statistical Learning. Data Mining, Inference, and Prediction.* Springer Series in Statistics. New York: Springer-Verlag, 2 ed., 2009.

[25] T. Minka and J. Lafferty. Expectation-propagation for the generative aspect model. In *Proceedings of the Eighteenth Conference on Uncertainty in Artificial Intelligence*, UAI'02, pp. 352–359. San Francisco, CA: Morgan Kaufmann Publishers Inc., 2002.

[26] M. Seeger. Expectation propagation for exponential families. Technical report. Berkeley, CA: University of California, 2005.

[27] T. Jaakkola and M. Jordan. Bayesian parameter estimation via variational methods. *Statistics and Computing*, 10(1):25–37, 2000.

[28] X. Didelot, R. Everitt, A. Johansen, and D. Lawson. Likelihood-free estimation of model evidence. *Bayesian Analysis*, 6(1):48–76, 2011.

[29] J.-M. Cornuet, P. Pudlo, J. Veyssier, A. Dehne-Garcia, M. Gautier, R. Leblois, J.-M. Marin, and A. Estoup. DIYABC v2.0: A software to make approximate Bayesian computation inferences about population history using single nucleotide polymorphism, DNA sequence and microsatellite data. *Bioinformatics*, 30(8):1187–1189, 2014.

[30] M. Stephens and P. Donnelly. Inference in molecular population genetics. *Journal of the Royal Statistical Society: Series B*, 62(4):605–635, 2000.

[31] J.-M. Cornuet, F. Santos, M. Beaumont, C.P. Robert, J.-M. Marin, D. Balding, T. Guillemaud, and A. Estoup. Inferring population history with DIYABC: A user-friendly approach to approximate Bayesian computation. *Bioinformatics*, 24(23):2713–2719, 2008.

[32] J.-M. Marin, N. Pillai, C.P. Robert, and J. Rousseau. Relevant statistics for Bayesian model choice. *Journal of the Royal Statistical Society: Series B*, 76(5):833–859, 2014.

[33] M. Evans. *Measuring Statistical Evidence Using Relative Belief*. Boca Raton, FL: CRC Press, 2015.

[34] O. Ratmann, C. Andrieu, C. Wiujf, and S. Richardson. Model criticism based on likelihood-free inference, with an application to protein network evolution. *Proceedings of the National Academy of Sciences*, 106(26):10576–10581, 2009.

[35] J. Stoehr, P. Pudlo, and L. Cucala. Adaptive ABC model choice and geometric summary statistics for hidden Gibbs random fields. *Statistics and Computing*, 25(1):129–141, 2015.

[36] L. Devroye, L. Györfi, and G. Lugosi. *A Probabilistic Theory of Pattern Recognition*, volume 31 of *Applications of Mathematics (New York)*. New York: Springer-Verlag, 1996.

[37] L. Breiman. Random forests. *Machine Learning*, 45(1):5–32, 2001.

[38] L. Breiman, J. Friedman, C. Stone, and R. Olshen. *Classification and Regression Trees*. Boca Raton, FL: CRC Press, 1984.

[39] C.P. Robert. *The Bayesian Choice*. New York: Springer-Verlag, 2nd ed., 2001.

[40] 1000 Genomes Project Consortium, G.R. Abecasis, A. Auton, et al. An integrated map of genetic variation from 1,092 human genomes. *Nature*, 491:56–65, 2012.

7

ABC and Indirect Inference

Christopher C. Drovandi

CONTENTS

7.1 Introduction 179
7.2 Indirect Inference 181
7.3 ABC II Methods 185
7.4 Bayesian Indirect Likelihood with a Parametric Auxiliary Model 189
7.5 Further Reading 192
7.6 Examples 196
7.6.1 Infectious disease example 196
7.6.2 Spatial extremes example 199
7.7 Discussion 204
Acknowledgements 205
References 205

7.1 Introduction

Indirect inference (II) is a classical method for estimating the parameter of a complex model when the likelihood is unavailable or too expensive to evaluate. The idea was popularised several years prior to the main developments in approximate Bayesian computation (ABC) by Gourieroux et al. (1993); Smith (1993), where interest was in calibrating complex time series models used in financial applications. The II method became a very popular approach in the econometrics literature [e.g. Smith (1993); Monfardini (1998); Dridi et al. (2007)] in a similar way to the ubiquitous application of ABC to models in population genetics. However, the articles by Jiang and Turnbull (2004) and Heggland and Frigessi (2004) have allowed the II approach to be known and appreciated by the wider statistical community.

In its full generality, the II approach can be viewed as a classical method to estimate the parameter of a statistical model on the basis of a so-called indirect or auxiliary summary of the observed data (Jiang and Turnbull, 2004). A special case of II is the simulated method of moments (McFadden, 1989), where the auxiliary statistic is a set of sample moments. In this spirit, the traditional ABC method may be viewed as a Bayesian version of II, where prior information about the parameter may be incorporated and updated using the information about the parameter contained in the summary statistic. However, much of the II literature has concentrated on developing the summary statistic from an alternative parametric auxiliary model that is analytically and/or computationally more tractable. The major focus of this chapter is on approximate Bayesian methods that harness such an auxiliary model. These are referred to as parametric Bayesian indirect inference (pBII) methods by Drovandi et al. (2016).

One such approach, referred to as ABC II, uses either the parameter estimate or the score of the auxiliary model as a summary statistic for ABC. When the auxiliary parameter is used, the ABC discrepancy function (distance between observed and simulated data) may be based on a direct comparison of the auxiliary parameter estimates [ABC IP where P denotes parameter (Drovandi et al., 2011)] or indirectly via the auxiliary log-likelihood [ABC IL where L denotes likelihood (Gleim and Pigorsch, 2013)]. Alternatively, a discrepancy function can be formulated by comparing auxiliary scores [ABC IS where S denotes score (Gleim and Pigorsch, 2013)]. Another approach, which differs substantially from the ABC II methods in terms of its theoretical underpinnings, uses the likelihood of the auxiliary model as a replacement to the intractable likelihood of the specified (or generative) model provided that a mapping has been estimated between the generative and auxiliary parameters (Reeves and Pettitt, 2005; Gallant and McCulloch, 2009). This method has been referred to as parametric Bayesian indirect likelihood (pBIL) by Drovandi et al. (2016). These methods will be discussed in greater detail in this chapter.

This chapter begins with a tutorial and a summary of the main developments of II in Section 7.2. This is followed by a review of ABC II methods in Section 7.3. Section 7.4 describes the parametric Bayesian indirect likelihood approach to approximate Bayesian inference using ideas from II. Other connections between ABC and II are provided in Section 7.5 for further reading. The methods are illustrated on an infectious disease model and a spatial extremes application in Section 7.6. The chapter is summarised in Section 7.7, which also discusses some possible future directions for utilising or building upon the current Bayesian II literature.

7.2 Indirect Inference

The purpose of this section is to give an overview of the II developments in the classical framework. It is not our intention to provide a comprehensive review of the II literature, but rather to provide a tutorial on the method and to summarise the main contributions.

Assume that there is available a parametric auxiliary model with a corresponding likelihood $p_A(y_{obs}|\phi)$, where ϕ is the parameter of this model. The auxiliary model could be chosen to be a simplified version of the model of interest (the so-called generative model here) or simply a data-analytic model designed to capture the essential features of the data. The majority of the literature on financial time series applications has considered the former; for example, Monfardini (1998) considers autoregressive moving average auxiliary models to estimate the parameters of a stochastic volatility model. In contrast, in an ABC application, Drovandi et al. (2011) use a regression model that has no connection to the assumed mechanistic model, in order to summarise the data.

The main objective of II is to determine the relationship between the generative parameter, θ, and the auxiliary parameter, ϕ. This relationship, denoted here as $\phi(\theta)$, is often referred to as 'the mapping' or binding function in the II literature. If the binding function is known and injective (one-to-one), then the II estimate based on observed data y_{obs} is $\theta_{obs} = \theta(\phi_{obs})$, where $\theta(\cdot)$ is the inverse mapping and ϕ_{obs} is the estimate obtained when fitting the auxiliary model to the data. The II approach essentially answers the question, what is the value of θ that could have produced the auxiliary estimate ϕ_{obs}? In this sense, the II approach acts as a correction method for assuming the wrong model.

Unfortunately the binding function, $\phi(\theta)$, is generally unknown, but it can be estimated via simulation. First, using a similar notation and explanation in Heggland and Frigessi (2004), define an estimating function, $Q(y_{obs}; \phi)$, for the auxiliary model. This could be, for example, the log-likelihood function of the auxiliary model. Before the II process begins, the auxiliary model is fitted to the observed data:

$$\phi_{obs} = \arg\max_{\phi} Q(y_{obs}; \phi).$$

For a particular value of θ, the process involves simulating n independent and identically distributed (iid) datasets from the generative model, $y_{1:n} = (y_1, \ldots, y_n)$. Each replicate dataset y_i, $i = 1, \ldots, n$, has the same dimension as the observed data y_{obs}. Then, the auxiliary model is fitted to this simulated data to recover an estimate of the binding function, $\phi_n(\theta)$:

$$\phi_n(\theta) = \arg\max_{\phi} Q(y_{1:n}; \phi),$$

where $Q(y_{1:n};\phi) = \sum_{i=1}^{n} Q(y_i;\phi)$. The binding function is defined as $\phi(\theta) = \lim_{n\to\infty} \phi_n(\theta)$. There is an alternative representation of $\phi_n(\theta)$. Define the estimated auxiliary parameter based on the ith simulated dataset as:

$$\phi(\theta, y_i) = \arg\max_{\phi} Q(y_i;\phi).$$

Then we obtain $\phi_n(\theta)$ via:

$$\phi_n(\theta) = \frac{1}{n}\sum_{i=1}^{n} \phi(\theta, y_i).$$

When this formulation for $\phi_n(\theta)$ is used, the definition of the binding function can also be represented as $E[\phi(\theta, y)]$ with respect to $f(y|\theta)$. The II procedure then involves solving an optimisation problem to find the θ that generates a $\phi_n(\theta)$ closest to ϕ_{obs}:

$$\theta_{obs,n} = \arg\min_{\theta}\{(\phi_n(\theta) - \phi_{obs})^{\top} W(\phi_n(\theta) - \phi_{obs})\}, \tag{7.1}$$

(Gourieroux et al., 1993), where the superscript $\top$ denotes transpose and $\theta_{obs,n}$ is the II estimator, which will depend on n. Gourieroux et al. (1993) show that the asymptotic properties of this estimator is the same regardless of what form is used for $\phi_n(\theta)$. Note that equation (7.1) assumes that the auxiliary parameter estimates $\phi_n(\theta)$ and ϕ_{obs} are unique. The II estimator should have a lower variance by increasing n, but this will add to the computational cost. Note that the above estimator will depend on the weighting matrix W, which needs to be positive definite. This matrix allows for an efficient comparison when the different components of the auxiliary estimator have different variances and where there is correlation amongst the components. One simple choice for the matrix W is the identity matrix. Another choice is the observed information matrix $J(\phi_{obs})$, which can be used to approximate the inverse of the asymptotic variance of ϕ_{obs}. Discussion on more optimal choices of the weighting matrix (in the sense of minimising the asymptotic variance of the indirect inference estimator) is provided in Gourieroux et al. (1993) and Monfardini (1998), for example.

An alternative approach proposed in Smith (1993) is to set the II estimator as the one that maximises the auxiliary estimating function using the observed data y_{obs} and the estimated mapping:

$$\theta_{obs,n} = \arg\max_{\theta} Q(y_{obs};\phi_n(\theta)). \tag{7.2}$$

This is referred to as the simulated quasi-maximum likelihood (SQML) estimator by Smith (1993), who uses the log-likelihood of the auxiliary model as the estimating function. Gourieroux et al. (1993) show that the estimator in (7.1) is asymptotically more efficient than the one in (7.2), provided that an optimal W is chosen.

An estimator that is quite different to the previous two, suggested by Gallant and Tauchen (1996), involves using the derivative of the estimating function of the auxiliary model. This is defined for some arbitrary dataset y as:

$$S_A(y, \phi) = \left(\frac{\partial Q(y; \phi)}{\partial \phi_1}, \cdots, \frac{\partial Q(y; \phi)}{\partial \phi_{p_\phi}} \right)^\top,$$

where $p_\phi = \dim(\phi)$ and ϕ_i is the ith component of the parameter vector ϕ. The estimator of Gallant and Tauchen (1996) is given by:

$$\theta_{obs,n} = \arg\min_\theta \left\{ \left(\frac{1}{n} \sum_{i=1}^{n} S_A(y_i, \phi_{obs}) \right)^\top \Sigma \left(\frac{1}{n} \sum_{i=1}^{n} S_A(y_i, \phi_{obs}) \right) \right\}. \quad (7.3)$$

The quantity $S_A(y_{obs}, \phi_{obs})$ does not appear in (7.3), as it can be assumed to be 0 by definition. This approach, referred to by Gallant and Tauchen (1996) as the efficient method of moments (EMM) when the estimating function Q is the auxiliary log-likelihood, can be very computationally convenient if there is an analytic expression for S_A, as the method only requires fitting the auxiliary model to data once to determine ϕ_{obs} before the II optimisation procedure in equation (7.3). Alternatively, it would be possible to estimate the necessary derivatives, which still could be faster than continually fitting the auxiliary model to simulated data. The EMM approach is also dependent upon a weighting matrix, Σ. One possible choice for Σ is $J(\phi_{obs})^{-1}$, since $J(\phi_{obs})$ can be used to estimate the variance of the score.

Gourieroux et al. (1993) show that the estimators in (7.1) and (7.3) are asymptotically equivalent for certain choices of the weighting matrices. However, their finite sample performance may differ and thus the optimal choice of estimator may be problem dependent. For example, Monfardini (1998) compares estimation of a stochastic volatility model using II techniques based on auxiliary autoregressive (AR) and autoregressive moving average (ARMA) models. Estimation of the AR model is computationally trivial, so Monfardini (1998) use the estimator in (7.1), whereas the ARMA model is harder to estimate, but has an analytic expression for the score, thus (7.3) is used. A simulation study showed smaller bias for the AR auxiliary model.

As is evident from earlier, a common estimating function is the auxiliary log-likelihood:

$$Q(y_{1:n}; \phi) = \sum_{i=1}^{n} \log p_A(y_i | \phi).$$

However, the user is free to choose the estimating function. Heggland and Frigessi (2004) demonstrate how the estimating function can involve simple summary statistics of the data, for example, the sample moments. This simulated method of moments (McFadden, 1989) involves finding a parameter value that generates simulated sample moments closest to the pre-specified observed sample moments. Thus, the simulated method of moments is a special case of II. Heggland and Frigessi (2004) suggest that the auxiliary statistic should be chosen such that it is sensitive to changes in θ, but is robust to different independent simulated datasets generated based on a fixed θ. Heggland and Frigessi (2004) also show that when the auxiliary statistic is sufficient for θ and has the same dimension as θ, the II estimator has the same asymptotic efficiency as the maximum likelihood estimator (MLE) except for a multiplicative factor of $1+1/n$, which reduces to 1 as $n \to \infty$. Jiang and Turnbull (2004) mention that II estimators could be improved further via a one step Newton–Raphson correction (e.g. Le Cam 1956), but would require some computations involving the complex generative model.

The Bayesian indirect inference procedures summarised in the following are essentially inspired by their classical counterparts earlier. Since a number of methods are surveyed here, all with different acronyms, Table 7.1 defines the acronyms again for convenience together with a description of the methods and key relevant literature.

TABLE 7.1
A List of Acronyms Together with Their Expansions for the Bayesian Likelihood-Free Methods Surveyed in this Chapter. A Description of Each Method is Shown Together with Some Key References

Acronym	Expansion	Description	Key References
ABC II	ABC indirect inference	ABC that uses an auxiliary model to form a summary statistic	Gleim and Pigorsch (2013); Drovandi et al. (2015)
ABC IP	ABC indirect parameter	ABC that uses the parameter estimate of an auxiliary model as a summary statistic	Drovandi et al. (2011); Drovandi et al. (2015)
ABC IL	ABC indirect likelihood	ABC that uses the likelihood of an auxiliary model to form a discrepancy function	Gleim and Pigorsch (2013); Drovandi et al. (2015)
ABC IS	ABC indirect score	ABC that uses the score of an auxiliary model to form a summary statistic	Gleim and Pigorsch (2013); Martin et al. (2014); Drovandi et al. (2015)
BIL	Bayesian indirect likelihood	A general approach that replaces an intractable likelihood with a tractable likelihood within a Bayesian algorithm	Drovandi et al. (2015)

(*Continued*)

TABLE 7.1 (*Continued*)
A List of Acronyms Together with Their Expansions for the Bayesian Likelihood-Free Methods Surveyed in this Chapter. A Description of Each Method is Shown Together with Some Key References

Acronym	Expansion	Description	Key References
pdBIL	Parametric BIL on the full data level	BIL method that uses the likelihood of a parametric auxiliary model on the full data level to replace the intractable likelihood	Reeves and Pettitt (2005); Gallant and McCulloch (2009); Drovandi et al. (2015)
psBIL	Parametric BIL on the summary statistic level	BIL method that uses the likelihood of a parametric auxiliary model on the summary statistic level to replace the intractable likelihood of the summary statistic	Drovandi et al. (2015)
ABC-cp	ABC composite parameter	ABC that uses the parameter of a composite likelihood to form a summary statistic	This chapter, but see Ruli et al. (2016) for a composite score approach
ABC-ec	ABC extremal coefficients	ABC approach of Erhardt and Smith (2012) for spatial extremes models	Erhardt and Smith (2012)

7.3 ABC II Methods

The first of the ABC II methods to appear in the literature uses the parameter of the auxiliary model as a summary statistic. For each dataset that is simulated from the model, $y \sim p(\cdot|\theta)$, the auxiliary model is fitted to this data to produce the simulated summary statistic, $s = \phi_y$. Drovandi et al. (2011) propose to compare this simulated summary statistic to the observed summary statistic, $s_{obs} = \phi_{obs}$, which is obtained by fitting the auxiliary model to the observed data prior to the ABC analysis, using the following discrepancy function:

$$||s - s_{obs}|| = \sqrt{(\phi_y - \phi_{obs})^\top J(\phi_{obs})(\phi_y - \phi_{obs})}, \tag{7.4}$$

where the observed information matrix of the auxiliary model evaluated at the observed summary statistic, $J(\phi_{obs})$, is utilised to provide a natural weighting of the summary statistics that also takes into account any correlations between the components of the auxiliary parameter estimate. This method is referred to by Drovandi et al. (2015) as ABC IP. Instead, if we base the discrepancy on the auxiliary likelihood (Gleim and Pigorsch, 2013), we obtain the ABC IL method:

$$||s - s_{obs}|| = \log p_A(y_{obs}|\phi_{obs}) - \log p_A(y_{obs}|\phi_y).$$

Under some standard regularity conditions, it is interesting to note that the ABC IP discrepancy function appears in the second order term of the Taylor series expansion of the ABC IL discrepancy function. This might suggest that the ABC IL discrepancy function is more efficient than the discrepancy function of ABC IP generally, but this requires further investigation. A common aspect of the ABC IP and ABC IL approaches is that they both use the auxiliary parameter estimate as a summary statistic. As such, these methods involve fitting the auxiliary model to every dataset simulated from the generative model during an ABC algorithm. In one sense, it is desirable to extract as much information out of the auxiliary model as possible. From this point of view, an attractive estimation procedure is maximum likelihood:

$$\phi_{obs} = \arg\max_{\phi \in \Phi} p_A(y_{obs}|\phi),$$

which tends to be more efficient than other simpler estimation approaches, such as the method of moments. However, in most real applications there is not an analytic expression for the auxiliary MLE, and one must then resort to numerical optimisation algorithms [e.g. Drovandi et al. (2011) apply the Nelder-Mead derivative free optimiser]. Having to determine a numerical MLE at every iteration of the ABC algorithm not only slows down the method, but also potentially introduces further issues if the numerical optimiser is prone to getting stuck at local modes of the auxiliary likelihood surface. A pragmatic approach may be to initialise the numerical optimiser at the observed auxiliary parameter estimate, ϕ_{obs}.

In summary, the earlier review reveals that the optimal choice of auxiliary estimator may be a trade-off between the computational cost of obtaining and the statistical efficiency of the chosen auxiliary estimator. One approach to expand on this literature might be to start with a computationally simple, but consistent estimator (e.g. the method of moments) and apply one iteration of a Newton–Raphson method to produce an asymptotically efficient estimator (Le Cam, 1956) in a timely manner. It is important to note that the ABC IP and ABC IL methods are essentially ABC versions of the classical II approaches in Gourieroux et al. (1993) (who compare observed and simulated auxiliary parameters) and Smith (1993) (who maximises the auxiliary log-likelihood), respectively.

A rather different approach to ABC IP and ABC IL considered by Gleim and Pigorsch (2013) uses the score of the auxiliary model as the summary statistic, resulting in the ABC IS procedure. A major advantage of the ABC IS approach is that we always evaluate the score at the observed auxiliary estimate, ϕ_{obs}. Any simulated data, y, obtained during the ABC algorithm can be substituted directly into the auxiliary score to determine the simulated summary statistic, without needing to fit the auxiliary model to the simulated data. In cases where there is an analytic expression for the score, the summary statistic can be very fast to compute (similar to more traditional summary statistics used in ABC), and this leads to substantial computational savings over ABC IP and ABC IL. However, in many applications, the derivatives of the auxiliary likelihood are not available analytically. In such situations, it is necessary to estimate the derivatives numerically [see, e.g. Martin et al. (2014) and Section 7.6.1] using a finite difference strategy, for example. This may contribute another small layer of approximation and add to the computational cost, although the number of likelihood evaluations required to estimate the score is likely to be less than that required to determine the auxiliary MLE.

When the MLE is chosen as the auxiliary estimator, then the observed score (or summary statistic here, s_{obs}) can be assumed to be numerically 0. Therefore, a natural discrepancy function in the context of ABC IS is given by:

$$||s - s_{obs}|| = \sqrt{S_A(y, \phi_{obs})^\top J(\phi_{obs})^{-1} S_A(y, \phi_{obs})}, \tag{7.5}$$

where $s = S_A(y, \phi_{obs})$. We can again utilise the observed auxiliary information matrix to obtain a natural weighting of the summary statistics. The ABC IS approach is effectively an ABC version of the EMM method in Gallant and Tauchen (1996).

The assumptions required for each ABC II approach to behave in a satisfactory manner are provided in Drovandi et al. (2015). In summary, the ABC IP approach requires a unique auxiliary parameter estimator so that each simulated dataset results in a unique value of the ABC discrepancy. ABC IL requires a unique maximum likelihood value and ABC IS requires a unique score (and some other mild conditions) for each simulated dataset generated during the ABC algorithm. Drovandi et al. (2015) consider an example where the auxiliary model is a mixture model, which does not possess a unique estimator due to the well-known label switching issue with mixtures. The ABC IL and ABC IS approaches were more suited to handling the label switching problem.

Martin et al. (2014) contain an important result that shows that the auxiliary score carries the same information as the auxiliary parameter estimate, thus the ABC II approaches will have the same target distribution in the limit as $h \to 0$. Martin et al. (2014) demonstrate that the discrepancy function involving the auxiliary score can be written as a discrepancy function involving the auxiliary parameter estimate, thus ABC will produce the same draws regardless of the choice of summary statistic in the limit as $h \to 0$. Whilst Drovandi et al. (2015) demonstrate empirically that there are differences amongst the ABC II results for $h > 0$, the result of Martin et al. (2014) does provide more motivation for a score approach, which will often be more computationally efficient.

In choosing an auxiliary model in the context of ABC II, it would seem desirable if the auxiliary model gave a good fit to the observed data so that one is reasonably confident that the quantities derived from the auxiliary model capture most of the information in the observed data. In particular, the auxiliary MLE is asymptotically sufficient for the auxiliary model. Thus, assuming some regularity conditions on the auxiliary model, if the generative model is a special case of the auxiliary model then the statistic derived from the auxiliary model will be asymptotically sufficient also for the generative model (Gleim and Pigorsch, 2013). An advantage of the ABC II approach is that the utility of the summary statistic may be assessed prior to the ABC analysis by performing some standard statistical techniques, such as goodness-of-fit and/or residual analyses. For example, Drovandi et al. (2015) consider a chi-square goodness-of-fit test to indicate insufficient evidence against the auxiliary model providing a good description of the data in an application involving a stochastic model of macroparasite population evolution. This is in contrast to more traditional choices of summary statistics (Blum et al., 2013), where it is often necessary to perform an expensive simulation study to select an appropriate summary. The well-known curse of dimensionality issue associated with the choice of summary statistic in ABC (Blum, 2010) can be addressed to an extent in the ABC II approach by choosing a parsimonious auxiliary model which might be achieved by comparing competing auxiliary models through some model selection criterion (e.g. Drovandi et al. (2011) use the Akaike Information Criterion).

Alternatively, it might be more convenient to select the auxiliary model as a simplified version of the generative model so that the auxiliary parameter has the same interpretation as the generative parameter (parameter of the generative model). For example, Section 7.6.1 considers performing inference for a Markov process using the corresponding linear noise approximation as the auxiliary model. This approach has the advantage that the summary statistic will have the same dimension as the parameter of interest [which can be desirable, see e.g. Fearnhead and Prangle (2012), and indeed necessary for some methods (Nott et al., 2014; Li et al., 2015)] and that there will be a strong connection between the generative and auxiliary parameters. The obvious drawback is, assuming the generative model is correct, the simplified version of the model in general will not provide a good fit to the observed data, and ultimately, any asymptotic sufficiency for the auxiliary model is lost for the generative model.

From a classical point of view, it is well known that basing inferences on the wrong model may result in biased parameter estimates. From a Bayesian perspective, the mode and concentration of the posterior distribution may not be estimated correctly when employing an approximate model for inference purposes. As mentioned in Section 7.2, II can be viewed as a method that provides some correction for assuming the wrong/simplified version of the model (Jiang and Turnbull, 2004). It does this by finding the parameter value of the generative model that leads to simulated data where the parameter estimate based on the simplified model applied to the simulated data is closest to that of the observed data. It may also be that applying ABC II in a similar way may lead to posterior modes that are closer to the true posterior mode in general, compared to when inferences are solely based on the misspecified model. Furthermore, ABC has a tendency to provide a less concentrated posterior distribution relative to the true posterior, which depends on how much information has been lost in the data reduction. Thus, using auxiliary parameter estimates of a simplified model as summary statistics in ABC will not lead to over concentrated posteriors, as may be the case if the simplified model was used directly in the Bayesian analysis. Using a summary statistic derived from such an auxiliary model in ABC is yet to be thoroughly explored in the literature [although see Martin et al. (2014) for an example].

Under certain regularity conditions on the auxiliary model that lead to II producing a consistent estimator [e.g. Gourieroux et al. (1993)], ABC II will produce Bayesian consistency in the limit as $h \to 0$ (Martin et al., 2014). Under the regularity conditions, in equation (7.4), $\phi_{obs} \to \phi(\theta_0)$ (where θ_0 is the true value of the parameter), and $\phi_y \to \phi(\theta)$, where $y \sim p(y|\theta)$ as the sample size goes to infinity. Thus, in the limit as $h \to 0$, ABC will only keep $\theta = \theta_0$.

Martin et al. (2014) also take advantage of the strong one-to-one correspondence between the auxiliary and generative parameters in the situation where the auxiliary model is a simplified version of the generative model. Here, Martin et al. (2014) suggest the use of the marginal score for a single auxiliary parameter, which is the score with all other components of the parameter integrated out of the auxiliary likelihood, as a summary statistic to estimate the univariate posterior distribution of the corresponding single generative model parameter.

7.4 Bayesian Indirect Likelihood with a Parametric Auxiliary Model

An alternative to ABC II that has been considered in the literature by Reeves and Pettitt (2005) and Gallant and McCulloch (2009) is to use the likelihood of an auxiliary parametric model as a replacement to that of the intractable generative likelihood, provided a relationship, $\phi(\theta)$, has been estimated between the generative and auxiliary parameters. See also Ryan et al. (2016) for an application of this method to Bayesian experimental design in the presence of an intractable likelihood. This approach is investigated in more theoretical detail in Drovandi et al. (2015) and is referred to as pdBIL (where d stands for data). It is also possible to apply a parametric auxiliary model to the summary statistic likelihood, where some data reduction has been applied. One then obtains psBIL (where s denotes summary statistic), which is discussed briefly later in this section. The pdBIL method could be considered as a Bayesian version of the simulated quasi-maximum likelihood approach in Smith (1993). If the so-called mapping or binding function, $\phi(\theta)$, is known, then the approximate posterior of the pdBIL approach is given by:

$$\pi_A(\theta|y_{obs}) \propto p_A(y_{obs}|\phi(\theta))\pi(\theta),$$

where $\pi_A(\theta|y_{obs})$ denotes the pdBIL approximation to the true posterior and $p_A(y_{obs}|\phi(\theta))$ is the likelihood of the auxiliary model evaluated at the parameter $\phi(\theta)$.

Unfortunately, in practice, the relationship between ϕ and θ will be unknown. More generally, it is possible to estimate $\phi(\theta)$ via simulation of n independent and identically distributed datasets, $y_{1:n} = (y_1, \dots, y_n)$, from the generative model based on a proposed parameter, θ. Then the auxiliary model is fitted to this large dataset to obtain $\phi_n(\theta)$, which could be based on maximum likelihood:

$$\phi_n(\theta) = \arg\max_{\phi} \prod_{i=1}^{n} p_A(y_i|\phi).$$

The target distribution of the resulting method is given by:

$$\pi_{A,n}(\theta|y_{obs}) \propto p_{A,n}(y_{obs}|\theta)\pi(\theta),$$

where

$$p_{A,n}(y_{obs}|\theta) = \int_{y_{1:n}} p_A(y_{obs}|\phi_n(\theta)) \left\{ \prod_{i=1}^{n} p(y_i|\theta) \right\} dy_{1:n},$$

which can be estimated unbiasedly using a single draw of n independent and identically distributed datasets, $y_{1:n} \sim p(\cdot|\theta)$. The introduction of the second subscript n in $p_{A,n}(y_{obs}|\theta)$ highlights that an additional layer of approximation is introduced by selecting a finite value of n to estimate the mapping [see Drovandi et al. (2015) for more details]. The empirical evidence in Drovandi et al. (2015) seems to suggest, in the context of pdBIL, that the approximate posterior becomes less concentrated as the value of n decreases. Therefore, if the auxiliary model chosen is reasonable (discussed later in this section), then better posterior approximations can be anticipated by taking n as large as possible. Initially it would seem apparent that the computational cost of the pdBIL approach would grow as n is increased. However, Drovandi et al. (2015) report an increase in acceptance probability of a Markov chain Monte Carlo (MCMC) algorithm targeting $\pi_{A,n}(\theta|y_{obs})$ as n is increased. Thus, up to a point, n can be raised without increasing the overall computing time since fewer iterations of MCMC will be required to obtain an equivalent effective sample size compared with using smaller values of n. Values of n above a certain limit where the acceptance probability does not increase may reduce the overall efficiency of the approach.

Like ABC IP and ABC IL, pdBIL requires an optimisation step for every simulated dataset, so it can be an expensive algorithm. For a Potts model application with a single parameter, Moores et al. (2015) propose to run pre-simulations across the prior space for a chosen value of n and fit a non-parametric model in order to smooth out the effect of n and recover an estimate of the mapping, denoted $\hat{\phi}(\theta)$. A major computational advantage of this approach is that the estimated mapping can be re-used to analyse multiple observed datasets of the same size. However, devising a useful strategy to extend this idea to higher dimensional problems than that considered by Moores et al. (2015) is an open area of research.

For the pdBIL method to lead to a quality approximation, the approach relies on a quite strong assumption that the auxiliary likelihood acts as a useful replacement likelihood across the parameter space with non-negligible posterior support and that the auxiliary likelihood reduces further in regions of very low posterior support (Drovandi et al., 2015). In contrast, the ABC II methods require that the summary statistic coming from the auxiliary model is informative and an efficient algorithm is available to ensure a close matching between the observed and simulated summary statistics in order to produce a close approximation to the true posterior distribution. Drovandi et al. (2015) demonstrate that under suitable conditions, the pdBIL method will target the true posterior in the limit as $n \to \infty$ if the generative model is nested within the auxiliary model. In this ideal scenario, ABC II methods will not be exact as $h \to 0$ since the quantities drawn from the auxiliary model will not produce a sufficient statistic in general, as the dimension of the statistic will be smaller than the size of the data [however, under suitable regularity conditions, the statistic will be asymptotically sufficient (Gleim and Pigorsch, 2013)]. Of course, it would seem infeasible to find a tractable auxiliary model that incorporates an intractable model as a special case. However, this observation does suggest, in the context of pdBIL method, that a flexible auxiliary model may be useful.

We note that pdBIL is not illustrated empirically in this chapter, but a number of examples are provided in Reeves and Pettitt (2005); Gallant and McCulloch (2009); Drovandi et al. (2015).

We note that a parametric auxiliary model can also be applied at a summary statistic level; that is, when some data reduction technique has been performed. As mentioned earlier, the method is referred to as psBIL in Drovandi et al. (2015). A popular psBIL method in the literature is to assume a multivariate normal auxiliary model. Here, the likelihood of the multivariate normal distribution, with a mean and covariance matrix dependent on θ, is used as a replacement to the intractable summary statistic likelihood. This technique has been referred to as synthetic likelihood (Wood, 2010; Price et al., 2018), and is covered in much greater detail in chapter 12 of Drovandi et al. (2018) (this volume).

7.5 Further Reading

Ruli et al. (2016) propose to use the score of a composite likelihood approximation of the full likelihood as a summary statistic for ABC (referred to as ABC-cs). Methods involving the composite likelihood can be applied when the full data likelihood is intractable, but the likelihood of certain subsets of the data can be evaluated cheaply. Section 7.6.2 considers an example in spatial extremes where composite likelihood methods are applicable. The method of Ruli et al. (2016) has a strong connection with the ABC IS method, but it does not quite fall under the ABC II framework as the composite likelihood is not associated with any parametric auxiliary model, but rather is used as a proxy to the full data likelihood formed by simplifying the dependency structure of the model. Of course, an alternative to Ruli et al. (2016) could use the composite likelihood estimate as a summary statistic and obtain approaches similar to ABC IP and ABC IL (Section 7.6.2 considers a composite likelihood variant on ABC IP). However, as with ABC IS, the approach of Ruli et al. (2016) can be relatively fast if the composite score is easier to obtain than the composite parameter estimator.

Pauli et al. (2011) and Ribatet et al. (2012) consider a Bayesian analysis where they use the composite likelihood directly as a replacement to the true likelihood. A naive application of this can lead to posterior approximations that are incorrectly concentrated, as each data point may appear in more than one composite likelihood component depending on how the data subsets are constructed. However, Pauli et al. (2011) and Ribatet et al. (2012) suggest so-called calibration approaches in an attempt to correct this. The method of Ruli et al. (2016) essentially by-passes this issue by using the composite likelihood to form a summary statistic for ABC. Therefore, Ruli et al. (2016) rely on this summary statistic being approximately sufficient in order to achieve a good approximation to the true posterior distribution. The approach of Ruli et al. (2016) generally falls under the Bayesian indirect inference framework, as it involves simulation from the true model in an attempt to correct an estimator based solely on the composite likelihood. The dimension of the summary statistic will coincide with that of the generative model parameter and there will likely be a strong correspondence between the two.

Forneron and Ng (2016) develop an approach called the reverse sampler (RS) that produces approximate Bayesian inferences that have a strong connection with II. The approach involves solving many II problems with a different random seed each time, and upon re-weighting the resulting solutions,

---

Algorithm 7.1: The reverse sampler of Forneron and Ng (2016)

1: **for** $i = 1$ **to** T where T is the number of samples **do**

2: Solve the II optimisation problem $\theta^{(i)} = \arg\max_{\theta}\{(s(\theta,\xi^{(i)}) - s_{obs})^\top W(s(\theta,\xi^{(i)}) - s_{obs})\}$ where $\xi^{(i)} \sim p(\xi)$. Set $\rho^{(i)} = (s(\theta^{(i)},\xi) - s_{obs})^\top W(s(\theta^{(i)},\xi) - s_{obs})$

3: Set the weight for sample $\theta^{(i)}$ as $w^{(i)} \propto \pi(\theta^{(i)})\mathrm{vol}(s_\theta(\theta^{(i)},\xi^{(i)}))^{-1}$

4: **end for**

---

an independent sample is generated from an approximate posterior. The summary statistic may come from an auxiliary model or could be any summarisation of the full data. The approach is provided in Algorithm 7.1. It is important to note that each II optimisation uses $n = 1$, as in ABC. For this approach, we denote the simulated summary statistic as $s(\theta,\xi)$, where ξ are a set of random numbers generated through the simulation, $\xi \sim p(\xi)$. For each II optimisation procedure, ξ is held fixed. This is equivalent to using the same random seed during each II optimisation.

Denote the sample obtained from solving the ith optimisation problem as $\theta^{(i)}$. After $\theta^{(i)}$ is generated, it must be weighted. One aspect of the weighting is the prior density, $\pi(\theta^{(i)})$. It also involves a Jacobian term and a volume term if the number of summary statistics exceeds the dimension of the parameter. Denote $s_\theta(\theta,\xi)$ as the Jacobian:

$$s_\theta(\theta,\xi) = \frac{\partial s(\theta,\xi)}{\partial \theta},$$

which is a $q \times p$ matrix, where q is the dimension of the summary statistic and p is the dimension of the parameter. That is, the (j,k) element of this matrix is given by $\frac{\partial s_j(\theta,\xi)}{\partial \theta_k}$, where $s_j(\theta,\xi)$ is the function for the jth summary statistic. Then the weight for sample $\theta^{(i)}$ is given by:

$$w^{(i)} \propto \pi(\theta^{(i)})\mathrm{vol}(s_\theta(\theta^{(i)},\xi^{(i)}))^{-1},$$

where

$$\mathrm{vol}(s_\theta(\theta^{(i)},\xi^{(i)})) = \sqrt{\det\left(s_\theta(\theta^{(i)},\xi^{(i)})^\top s_\theta(\theta^{(i)},\xi^{(i)})\right)}.$$

Upon normalisation of the weights, a weighted sample is obtained from an approximate posterior. Forneron and Ng (2016) also include a kernel function to the weights, $K_h(||s_{obs} - s_\theta(\theta^{(i)},\xi^{(i)})||)$ to give higher weight to II optimisation samples that get closer to the observed summary statistic. Note that if $s_\theta(\theta^{(i)},\xi^{(i)})$ is unavailable analytically, it can be estimated via numerical differentiation, for example, finite differencing.

Here we provide an example to obtain some insight into the approximation behaviour of the RS. Consider a dataset $y_1,\ldots,y_N \sim N(\mu,1)$ of length N.

A sufficient statistic is $s = \bar{y} \sim N(\mu, 1/\sqrt{n})$. We may think of the simulated data as a transformation of the parameter and some noise variables, $y_i = \mu + \xi_i$, where $\xi_i \sim N(0, 1)$ and $\xi = (\xi_1, \ldots, \xi_N)$. To obtain the summary statistic, we have $s(\mu, \xi) = \bar{y} = \mu + \bar{\xi}$, where $\bar{\xi}$ is the sample mean of the noise variables. The Jacobian of this transformation is 1. For fixed, $\bar{\xi}^{(i)}$ we obtain $\mu^{(i)} = \bar{y}_{obs} - \bar{\xi}^{(i)}$, in which case $s_{obs} - s(\mu^{(i)}, \xi^{(i)}) = 0$. Effectively the RS algorithm samples from $\mu \sim N(\bar{y}_{obs}, 1/\sqrt{n})$, which is the posterior distribution of μ if the prior is uniform and improper over the real line. If the prior was selected differently then the $\mu^{(i)}$ samples need to be weighted proportional to the prior density $w^{(i)} \propto \pi(\mu^{(i)})$. Assume for the rest of this example that $\pi(\mu) \propto 1$ for $-\infty < \mu < \infty$.

Now consider the two-dimensional summary $(\bar{y}, m)$, where m is the sample median. This remains a sufficient statistic, so that ABC targets the true posterior in the limit as $h \to 0$. Here we have $\bar{y} = \mu + \bar{\xi}$ and $m = \mu + m_\xi$ where $\bar{\xi}$ is the sample mean of the noise variables and m_ξ is the median of the noise variables. We set the discrepancy function as the following:

$$||s_{obs} - s(\mu, \xi)|| = \left\{\bar{y}_{obs} - (\mu + \bar{\xi})\right\}^2 + \left\{m_{obs} - (\mu + m_\xi)\right\}^2.$$

Minimising this with respect to μ for some $\xi^{(i)}$, we obtain $\mu^{(i)} = (\bar{y}_{obs} - \bar{\xi}^{(i)} + m_{obs} - m_\xi^{(i)})/2$. Thus in this case RS does not draw directly from the posterior (despite the fact that the statistic is sufficient). This is confirmed in Figure 7.1(a) where there is departure from the true posterior (although it is better than an RS sampler that just uses the median as a summary). Here, we investigate the impact on the approximate posterior as we discard some of the μ RS samples with the highest discrepancy. Figure 7.1(b) demonstrates improved accuracy by discarding only 50% of the draws. This is equivalent to using a uniform kernel with an appropriate value of h in the RS weighting function of Forneron and Ng (2016).

Also shown in Figure 7.1(a) is the posterior when using five summary statistics (mean, median, min, max, and midrange). The partial derivatives of the summaries (for fixed noise variables) with respect to the parameter are all equal to 1 so that the Jacobian term for all samples is the same so that the weights are the same for each sample. Again using the squared Euclidean distance, it is easy to show what the expression is for μ drawn during the RS sampler. Here, there is quite a large departure from the true posterior, demonstrating that the RS method does suffer from a curse of dimensionality in the summary statistic, as with ABC. Figure 7.1(c) shows that the approximation can be improved by discarding samples with the highest discrepancy values. However, a large proportion (around 95% say) need to be discarded to get close to the true posterior. The results demonstrate that care must be taken when choosing summary statistics for RS, as in ABC.

Figures 7.1(d) and (e) demonstrate that using a local regression adjustment of Beaumont et al. (2002) (as is commonly done in ABC) almost fully corrects the RS approximations. It appears the regression adjustment can be quite useful for RS, as it is for ABC.

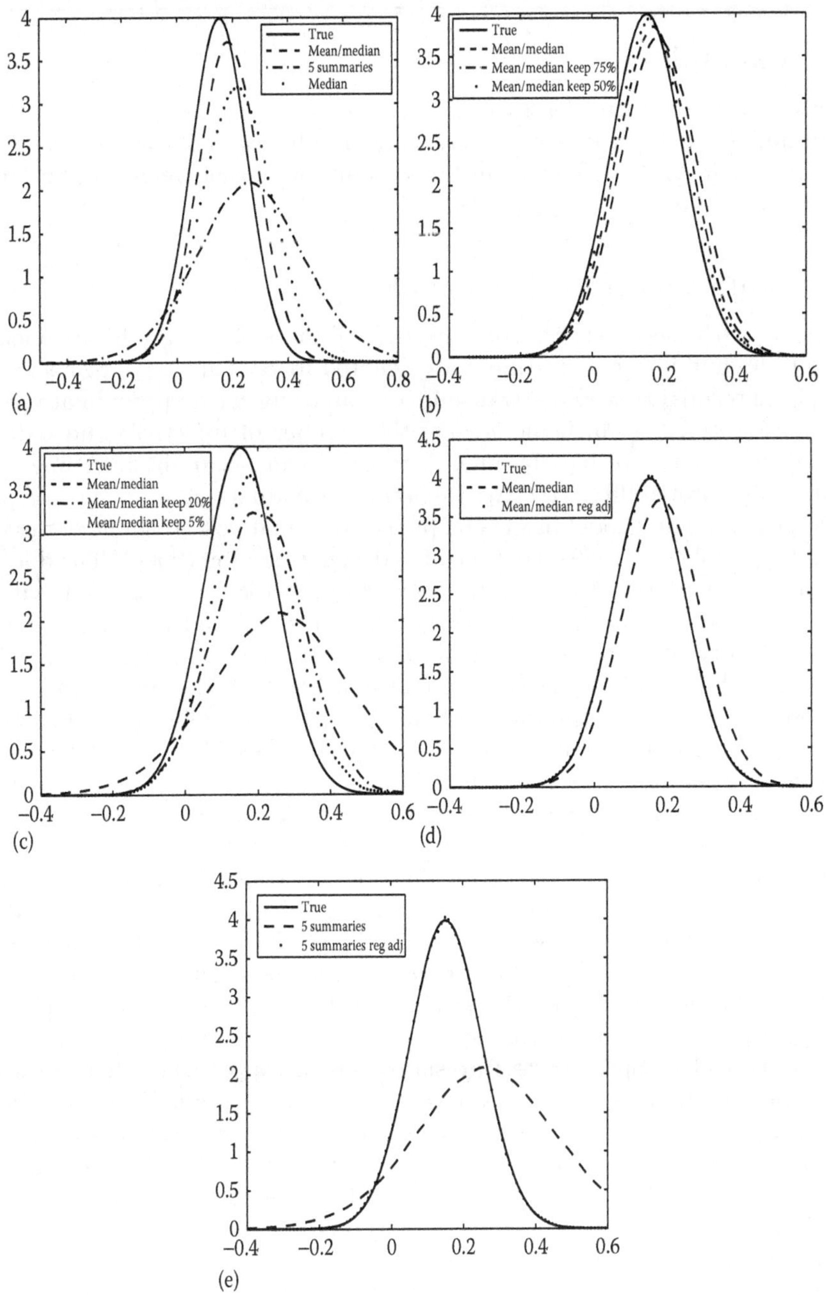

FIGURE 7.1
RS example for the normal location model with different choices of the summary statistics and using rejection and regression adjustment. (a)–(e): see text for description.

7.6 Examples

In this section, we consider two case studies involving real data to demonstrate how ideas from indirect inference can enhance ABC inferences. Note throughout all ABC analyses, we use the uniform kernel weighting function, $K_h(||s - s_{obs}||) = I(||s - s_{obs}|| \leq h)$.

7.6.1 Infectious disease example

Consider a stochastic susceptible-infected (SI) model where the number of susceptibles and infecteds at time t are denoted by $S(t)$ and $I(t)$, respectively. In an infinitesimal time Δ_t, a transmission can occur with approximate probability $\beta S(t)I(t)\Delta_t$, which increments the number of infectives and reduces the number of susceptibles by one. A removal can occur in that time with approximate probability $\gamma I(t)\Delta_t$. Removed individuals play no further part in the process. This model has been applied to a smallpox epidemic dataset in O'Neill and Roberts (1999); Fearnhead and Meligkotsidou (2004). The dataset consists of the removal times in days [see (Becker, 1989, p. 111)]. The village consists of 120 people, one of whom becomes infected and introduces it into the community. We set the first removal at time $t = 0$, where, for simplicity, we assume that there is exactly one infected, i.e. $I(0) = 1$ and $S(0) = 118$. We adopt the same interpretation of the dataset that Golightly et al. (2015) use where the observations refer to daily recordings of $S(t) + I(t)$, which changes only on days where a removal takes place. Thus, we do not assume that the epidemic has necessarily ceased with the last removal recorded in the dataset.

Samples from the true posterior distribution can be obtained using the particle MCMC approach specified in Drovandi et al. (2016) that makes use of the alive particle filter [see Del Moral et al. (2015) for additional information on the alive particle filter]. However, in general, it is very difficult to devise exact and computationally feasible strategies for these Markov process models, especially when the model contains several populations. Therefore, here, we consider approximate Bayesian inference approaches that use a direct approximation to the true stochastic process. A popular and tractable approximation to continuous time Markov chains is the linear noise approximation (LNA). Fearnhead et al. (2014) base their posterior inferences directly on the LNA and thus require that the LNA be a very good approximation to the Markov process, which may not always be the case.

A different approach is to use the auxiliary LNA model to form summary statistics for ABC. In this situation, there are as many summary statistics as generative model parameters and the auxiliary and generative model parameters have the same interpretation. We denote the LNA model parameters as β^A and γ^A, where superscript A denotes the auxiliary model. The auxiliary model likelihood is denoted as $p_A(y|\phi)$ where $\phi = (\beta^A, \gamma^A)$ is the auxiliary parameter. The auxiliary parameter estimate based on the observed data is given by:

$$\phi_{obs} = \arg\max_{\phi} p_A(y_{obs}|\phi).$$

To reduce the computational burden, we consider an ABC IS approach. That is, we consider the score of the LNA model and always use ϕ_{obs} in the score function for the summary statistic:

$$s = S_A(y, \phi_{obs}) = \left(\frac{\partial \log p_A(y|\phi)}{\partial \beta^A}|_{\phi=\phi_{obs}}, \frac{\partial \log p_A(y|\phi)}{\partial \gamma^A}|_{\phi=\phi_{obs}} \right)^\top,$$

where y is a simulated dataset. We assume that ϕ_{obs} has been obtained accurately enough so that we can assume $s_{obs} = S_A(y_{obs}, \phi_{obs}) = (0,0)^\top$. We set the discrepancy function $||\cdot||$ as the L_2 norm of s. However, there is no analytic expression for the score. Therefore, we estimate the score numerically, which requires several evaluations of the LNA likelihood. Thus calculating the summary statistic based on the score is mildly computationally intensive, and given that ABC tends to suffer from poor acceptance rates, we propose a method here to accelerate the algorithm without altering the ABC target distribution.

To improve the computational efficiency of the ABC approach, we propose an implementation of the lazy ABC method of Prangle (2016). Our approach uses a second summary statistic, which we call the lazy summary statistic (denoted by $s_{obs,\text{lazy}}$ and s_{lazy} for the observed and simulated data, respectively), for which computation is trivial once data have been simulated from the generative model, but which is less informative than the originally proposed summary statistic. The discrepancy function for the lazy summary statistic is given by:

$$\rho_{\text{lazy}} = ||s_{\text{lazy}} - s_{obs,\text{lazy}}||.$$

First, a decision on whether or not to compute the actual summary statistic is made on the basis of the distance between the observed and simulated lazy summary statistics. If ρ_{lazy} falls below a threshold, h_{lazy}, then the proposal may have produced simulated data reasonably close to the observed data, and it is then worthwhile to compute the more informative summary statistic. However, if the lazy distance is too high, then the proposed parameter may be rejected and thus calculation of the expensive summary is not required. In order to obtain an algorithm that preserves the original ABC target, it is necessary to include a continuation probability, α, which may require some tuning. Therefore, a proposal that performs poorly in terms of the lazy summary statistic will still make it through to the second stage with probability α. We implement this approach within an MCMC ABC method [Prangle (2016) consider ABC importance sampling], which must include α in the acceptance probability to ensure a theoretically correct method. When $\rho_{\text{lazy}} > h_{\text{lazy}}$, we effectively estimate the ABC likelihood with a random variable that can be two possible values:

$$\begin{matrix} 0 & \text{with probability } 1-\alpha \\ \frac{I(||s-s_{obs}||\leq h)}{\alpha} & \text{with probability } \alpha \end{matrix}.$$

The expected value of this random variable is $I(||s - s_{obs}|| \leq h)$, the desired ABC likelihood. Thus, our implementation of the lazy ABC approach is an instance of the pseudo-marginal method of Andrieu and Roberts (2009). The drawback of this approach is that it inflates the variance of the ABC likelihood. If $\rho_{\text{lazy}} > h_{\text{lazy}}$ and $\rho \leq h$, then the ABC likelihood gets inflated, meaning that the MCMC algorithm may become stuck there for a large number of iterations.

We also incorporate the early rejection strategy of Picchini and Forman (2016). The approach of Picchini and Forman (2016) involves a simple reordering of steps in MCMC ABC where it is possible to reject a proposed parameter prior to simulating data. This approach does not alter the target distribution when a uniform weighting function is applied, as we do here.

For this application, the lazy summary statistic is a scalar, so we choose the absolute value to compute the distance between the observed and simulated lazy summary statistic in line 7 of Algorithm 7.2. For the actual summary statistic, we use the discrepancy function in equation (7.5) at line 18 of Algorithm 7.2, with J set as the identity matrix for simplicity.

The results of using the LNA approximation and the ABC methods are shown in Figure 7.2 (note that the ABC results are based on the lazy ABC implementation). The results are compared with particle MCMC, which has the true posterior as its limiting distribution. The LNA results tend to be overprecise (especially for β), whereas the ABC results tend to be slightly conservative. Note that we also ran ABC using the final observation as the (simple) summary statistic with $h = 0$, which also provides good results. The posterior for β based on the simple summary is similar to that when the LNA summary statistic is used, but is slightly less precise for γ.

In terms of lazy ABC we use the value of $S(t) + I(t)$ at the end of the recording time as the lazy summary statistic. In order to tune h_{lazy}, we run ABC with the summary statistic formed from the LNA approximation only for a small number of iterations and recorded at each iteration the value of the simulated lazy summary statistic and whether or not the proposal was accepted or rejected. From Figure 7.2(c), it is evident that most of the acceptances based on the actual summary statistic occur when the lazy summary statistic is between 70 and 110 (the observed value is 90). Many simulations do not produce a lazy summary statistic within this range, so that early rejection based on this lazy summary statistic seems like a good choice. Therefore, we set the lazy discrepancy to be the absolute value between the last observed and simulated data point and set $h_{\text{lazy}} = 20$. Note that if a proposal does

Algorithm 7.2: MCMC ABC Algorithm Using a Lazy Summary Statistic

Set $C = 1$
Set $\theta^{(0)}$
for $i = 1$ **to** T **do**
    Draw $\theta^* \sim q(\cdot|\theta^{(i-1)})$
    Compute $r = \min\left(1, \frac{\pi(\theta^*)}{C\pi(\theta^{(i-1)})}\right)$
    if $U(0,1) < r$ **then**
        Simulate $y \sim p(\cdot|\theta^*)$
        Compute lazy ABC discrepancy $\rho_{\text{lazy}} = ||s_{\text{lazy}} - s_{obs,\text{lazy}}||$
        if $\rho_{\text{lazy}} > h_{\text{lazy}}$ **then**
            if $U(0,1) < \alpha$ **then**
                Continue and set $C_{\text{prop}} = 1/\alpha$
            else
                Reject early: set $\theta^{(i)} = \theta^{(i-1)}$ and go to the next iteration of the MCMC algorithm
            end if
        else
            Set $C_{\text{prop}} = 1$
        end if
        Compute ABC discrepancy $\rho = ||s - s_{obs}||$
        if $\rho \leq h$ **then**
            $\theta^{(i)} = \theta^*$ and $C = C_{\text{prop}}$
        else
            $\theta^{(i)} = \theta^{(i-1)}$
        end if
    else
        $\theta^{(i)} = \theta^{(i-1)}$
    end if
end for

not satisfy h_{lazy}, then we continue nonetheless with probability $\alpha = 0.1$. This seems like a reasonably conservative choice.

The lazy ABC approach resulted in a very similar acceptance rate to usual ABC, 2.3% and 2.5%, respectively, however, the lazy ABC approach was about 3.5 times faster (roughly 16 hours down to 4.5 hours).

7.6.2 Spatial extremes example

If it exists, the limiting distribution of the maximum of a suitably normalised sequence of independent and identically distributed (multivariate) random variables is in the family of multivariate extreme value distributions (MEVDs). Max-stable processes are the infinite-dimensional generalisation of MEVDs.

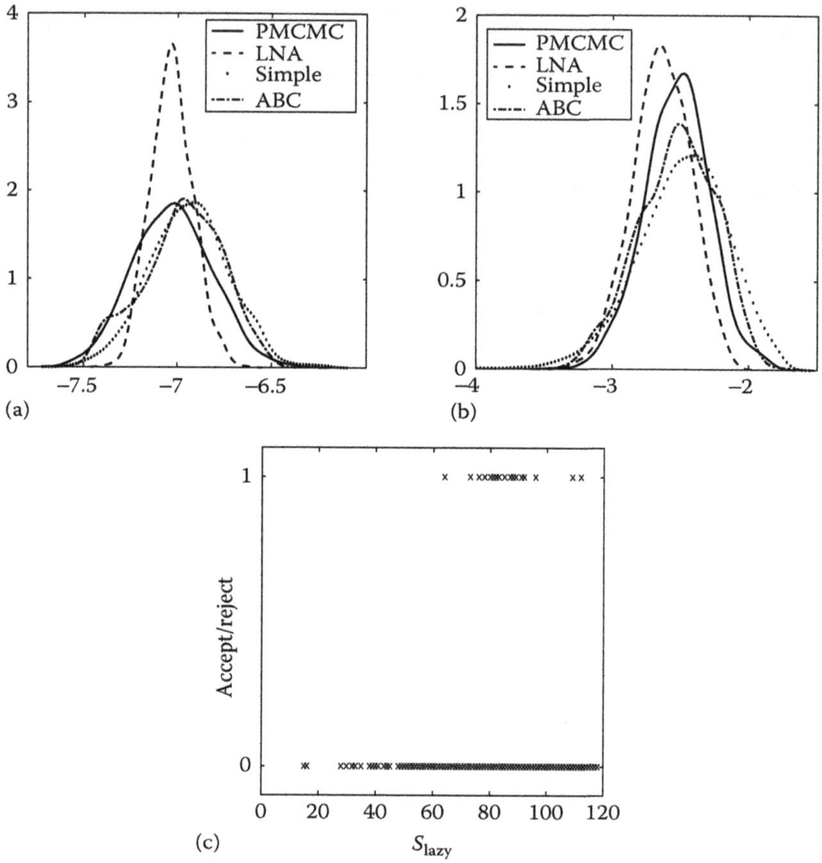

FIGURE 7.2
Results for the epidemic example. Shown are the posterior distributions of (a) $\log \beta$ and (b) $\log \gamma$ based on the particle MCMC approach (solid), LNA approximation (dash), ABC with the final observation as the summary statistic (dot), and ABC with the LNA parameter as a summary statistic (dot-dash). Subfigure (c) shows the results of the pilot run where the x-axis is the value of the lazy summary statistic, whereas the y-axis shows the ABC accept/reject outcome based on the corresponding summary statistic formed by the LNA.

Consider a set of extremal observations $y^t_{obs} = (y^t_{obs}(x_1), \ldots, y^t_{obs}(x_D)) \in \mathbb{R}^D$ collected at spatial locations $x_1, \ldots, x_D \in X \subset \mathbb{R}^p$ at time t. Here, we define an extremal observation as the maximum of a sequence of independent and identically distributed random variables. Assume that T independent and identically distributed extremal observations are taken at each location, indexed by $t = 1, \ldots, T$. We denote the full dataset as $y^{1:T}_{obs}$, whereas all the data at the ith location is $y^{1:T}_{obs}(x_i)$. It is possible to model such data using spatial models called max-stable processes [see, e.g. Schlather (2002) and Davison et al. (2012)]. The max-stable process arises from considering the limiting distribution of the maximum of a sequence of independent and identically distributed random variables. Here, we follow Erhardt and Smith (2014) and focus on a max-stable process where each marginal (i.e. for a particular spatial location) has a unit-Fréchet distribution with cumulative distribution function $G(z) = \exp(-1/z)$. Additional flexibility on each marginal can be introduced via a transformation with location (μ), scale (σ), and shape (ξ) parameters. Assuming that Z has a unit-Fréchet distribution, the random variable:

$$Y = \frac{\sigma}{\xi}(Z^{\xi} - 1) + \mu,$$

has a generalised extreme value distribution. The first step, then, is to estimate the (μ, σ, ξ) parameters for each of the marginals separately based on the T observations $y_{obs}^{1:T}(x_i)$ at each location, $i = 1, \ldots, D$, to transform the data so that they, approximately, follow a unit-Fréchet distribution. The data following this transformation we denote as $z_{obs}^{1:T}(x_i)$.

For simplicity, we consider a realisation of this max-stable process at a particular time point, and thus drop the index t for the moment. Assume that the corresponding random variable for this realisation is denoted $Z = (Z(x_1), \ldots, Z(x_D))$. Unfortunately the joint probability density function of Y is difficult to evaluate for $D > 2$. However, an analytic expression is available for the bivariate cumulative distribution function of any two points, say x_i and x_j (with realisations z_i and z_j), which depends on the distance between the two points, $h = ||x_i - x_j||$:

$$G(z_i, z_j) = \exp\left(-\frac{1}{2}\left[\frac{1}{z_i} + \frac{1}{z_j}\right]\left[1 + \left\{1 - 2(\rho(h)+1)\frac{z_i z_j}{(z_i + z_j)^2}\right\}^{1/2}\right]\right), \tag{7.6}$$

where $\rho(h)$ is the correlation of the underlying process. For simplicity, we consider only the Whittle-Matérn covariance structure:

$$\rho(h) = c_1 \frac{2^{1-\nu}}{\Gamma(\nu)}\left(\frac{h}{c_2}\right)^{\nu} K_{\nu}\left(\frac{h}{c_2}\right),$$

where $\Gamma(\cdot)$ is the gamma function and $K_{\nu}(\cdot)$ is the modified Bessel function of the second kind. Note that there are several other options [see Davison et al. (2012)]. In the above equation, $0 \le c_1 \le 1$ is the sill, $c_2 > 0$ is the range, and $\nu > 0$ is the smooth parameter. The sill parameter is commonly set to $c_1 = 1$, which we adopt here. Therefore, interest is in the parameter $\theta = (c_2, \nu)$. Here, the prior on θ is set as uniform over $(0, 20) \times (0, 20)$.

A composite likelihood can be constructed (Padoan et al., 2010; Ribatet et al., 2012) since there is an analytic expression for the bivariate likelihood (i.e. the joint density of the response at two spatial locations), which can be obtained from the cumulative distribution function in (7.6). The composite likelihood for one realisation of the max-stable process can be derived by considering the product of all possible (unordered) bivariate likelihoods (often referred to as the pairwise likelihood). Then another product can be taken over the T independent realisations of the process. Ribatet et al. (2012) utilise an adjusted composite likelihood directly within a Bayesian algorithm to obtain an approximate posterior distribution for the parameter of the correlation function. We investigate a different approach and use the composite likelihood parameter estimate as a summary statistic for ABC. The composite likelihood can be maximised using the function *fitmaxstab* in the SpatialExtremes package in R (Ribatet et al., 2013). For simplicity, we refer to this approach as ABC-cp [where cp denotes 'composite parameter' to be consistent with Ruli et al. (2016), who refer to their method as ABC-cs ('composite score')].

Our approach is compared with an ABC procedure in Erhardt and Smith (2012), who use a different summary statistic. The method first involves computing the so-called tripletwise extremal coefficients. One extremal coefficient calculation involves three spatial points. Erhardt and Smith (2014) use estimated tripletwise extremal coefficients, which for three spatial locations i, j, k can be obtained using:

$$\frac{1}{\sum_{t=1}^{T} 1/\max(z_{obs}^t(x_i), z_{obs}^t(x_j), z_{obs}^t(x_k))}.$$

The full set of estimated tripletwise coefficients is high-dimensional, precisely $\binom{D}{3}$. The dimension of this summary is reduced by placing the extremal coefficients into K groups, which are selected by grouping similar triangles (formed by the three spatial locations) together. This grouping depends only on the spatial locations and not the observed data. Erhardt and Smith (2012) then use the mean of the extremal coefficients within each group to form a K dimensional summary statistic. For the ABC discrepancy function, Erhardt and Smith (2012) consider the L_1 norm between the observed and simulated K group means. For brevity, we refer to this ABC approach as ABC-ec, where ec stands for 'extremal coefficients'. The reader is referred to Erhardt and Smith (2012) for more details. This method is implemented with the assistance of the ABCExtremes R package (Erhardt, 2013).

There are several issues associated with ABC-ec. First, it can be computationally intensive to determine the K groups. Second, there is no clear way to choose the value of K. There is a trade-off between dimensionality and information loss, which may require investigation for each dataset analysed. Third, only the mean within each group is considered, whereas the variability of the extremal coefficients within each group may be informative too. Finally, there is no obvious ABC discrepancy to apply. In contrast, the ABC-cp offers a low-dimensional summary statistic (same size as the parameter) and a natural way to compare summary statistics through the Mahalanobis distance (using an estimated covariance matrix of what is returned by *fitmaxstab*). However, the tripletwise extremal coefficients consider triples of locations (and so should carry more information compared with the pairwise approach of the composite likelihood), and also we find that computing the summary statistic of ABC-cp using *fitmaxstab* is slower than a C implementation of the tripletwise extremal coefficients calculation (called into R using the Rcpp package (Eddelbuettel et al., 2011)). On the other hand, ABC-cp avoids the expensive clustering of triangles into groups.

For both approaches, an MCMC ABC algorithm was used with proposal distributions carefully chosen to ensure a desired acceptance probability based on the results of some pilot runs. ABC-ec was run for 2,000,000 iterations, and the ABC tolerance chosen resulted in an acceptance rate of roughly 0.8%. Due to the extra computation associated with maximising the composite likelihood at each iteration, ABC-cp was run for 100,000 iterations, with an ABC tolerance that results in an acceptance probability of roughly 8%.

Here, we re-analyse the data considered in Erhardt and Smith (2014), which consists of the maximum annual summer temperature at 39 locations in midwestern United States of America between 1895 and 2009. The data at each spatial location are firstly transformed to approximately unit-Fréchet margins by fitting a generalised extreme value distribution by maximum likelihood, and also taking into account a slight trend in the maximum summer temperatures over time [see Erhardt and Smith (2014) for more details]. The max-stable process is then fitted to this transformed data using the ABC approaches described earlier.

Contour plots of the bivariate posterior distributions for both the ABC-ec and ABC-cp approaches are shown in Figure 7.3. Despite the much higher acceptance rate for ABC-cp, the resulting posterior for ABC-cp is substantially more concentrated compared with the results from ABC-ec. Furthermore, it can be seen that the posterior spatial correlation function is determined much more precisely with ABC-cp.

Despite the encouraging results, a thorough simulation study is required to confirm the ABC-cp approach as a generally useful method for spatial extremes applications. Erhardt and Smith (2012) note that very different parameter configurations can lead to a similar correlation structure, as demonstrated by the 'banana' shape target in Figure 7.3. Therefore it may be inefficient to compare composite parameter summary statistics directly via the Mahalanobis distance. Comparison through the composite likelihood itself may perform better. Alternatively, if an expression for the composite score can be derived, then the ABC-cs method of Ruli et al. (2016) may be a computationally convenient approach for these sorts of models. The ABC-cp method implemented here relies on already available functions in existing R packages. Thus we leave the composite likelihood and score methods for further research.

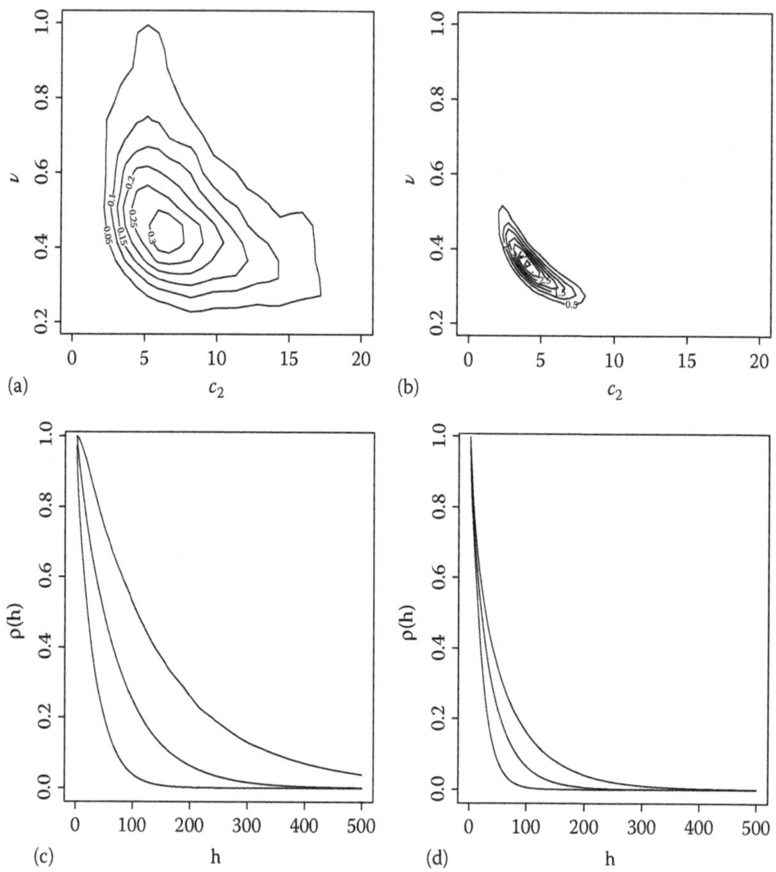

FIGURE 7.3
Posterior results for the spatial extremes example. Shown are the bivariate posterior distributions of (c_2, ν) based on the (a) ABC-ec and (b) ABC-cp approaches. The posterior median and 95% credible interval of the spatial correlation function are shown for the (c) ABC-ec and (d) ABC-cp methods.

7.7 Discussion

In this chapter, we provided a description of the indirect inference method and also detailed links between various likelihood-free Bayesian methods and indirect inference. As highlighted by the examples in this chapter and other applications in articles such as Gallant and McCulloch (2009); Drovandi et al. (2011); Gleim and Pigorsch (2013); Drovandi et al. (2015), it is clear that the tools presented in this chapter can provide useful posterior approximations in complex modelling scenarios.

During this chapter, we have also considered an extension of the reverse sampler of Forneron and Ng (2016) using regression adjustment, an MCMC implementation of the lazy ABC approach of Prangle (2016) and developed an ABC approach for spatial extremes models using the parameter estimate of a composite likelihood as the summary statistic.

A possible avenue for future research involves likelihood-free Bayesian model choice. It is well-known that Bayes factors based on summary statistic likelihoods do not correspond to those based on the full data likelihoods (Robert et al., 2011). It is typically not clear how to choose a summary statistic that is useful for model discrimination [although see, e.g. Didelot et al. (2011); Estoup et al. (2012); Prangle et al. (2014); Martin et al. (2014); Lee et al. (2015) for progress in this area]; a summary statistic that is sufficient for the parameter of each model is still generally not sufficient for the model indicator [see Marin et al. (2013)]. An interesting direction for further research is to explore whether flexible auxiliary models can assist in developing likelihood-free methods that are useful for model selection in terms of Bayes factors and posterior model probabilities.

Acknowledgements

The author is grateful to Kerrie Mengersen for the helpful comments and suggestions on an earlier draft. The author was supported by an Australian Research Council's Discovery Early Career Researcher Award funding scheme (DE160100741). The author is an Associate Investigator of the Australian Centre of Excellence for Mathematical and Statistical Frontiers (ACEMS).

References

Andrieu, C. and G. O. Roberts (2009). The pseudo-marginal approach for efficient Monte Carlo computations. *The Annals of Statistics 37*(2), 697–725.

Beaumont, M. A., W. Zhang, and D. J. Balding (2002). Approximate Bayesian computation in population genetics. *Genetics 162*(4), 2025–2035.

Becker, N. G. (1989). *Analysis of Infectious Disease Data*, Volume 33. New York: Chapman & Hall/CRC Press.

Blum, M. G. B. (2010). Approximate Bayesian computation: A non-parametric perspective. *Journal of the American Statistical Association 105*(491), 1178–1187.

Blum, M. G. B., M. A. Nunes, D. Prangle, and S. A. Sisson (2013). A comparative review of dimension reduction methods in approximate Bayesian computation. *Statistical Science 28*(2), 189–208.

Davison, A. C., S. A. Padoan, and M. Ribatet (2012). Statistical modeling of spatial extremes. *Statistical Science 27*(2), 161–186.

Del Moral, P., A. Jasra, A. Lee, C. Yau, and X. Zhang (2015). The alive particle filter and its use in particle Markov chain Monte Carlo. *Stochastic Analysis and Applications 33*(6), 943–974.

Didelot, X., R. G. Everitt, A. M. Johansen, and D. J. Lawson (2011). Likelihood-free estimation of model evidence. *Bayesian Analysis 6*(1), 49–76.

Dridi, R., A. Guay, and E. Renault (2007). Indirect inference and calibration of dynamic stochastic general equilibrium models. *Journal of Econometrics 136*(2), 397–430.

Drovandi, C. C., A. N. Pettitt, and M. J. Faddy (2011). Approximate Bayesian computation using indirect inference. *Journal of the Royal Statistical Society: Series C (Applied Statistics) 60*(3), 503–524.

Drovandi, C. C., A. N. Pettitt, and A. Lee (2015). Bayesian indirect inference using a parametric auxiliary model. *Statistical Science 30*(1), 72–95.

Drovandi, C. C., A. N. Pettitt, and R. A. McCutchan (2016). Exact and approximate Bayesian inference for low integer-valued time series models with intractable likelihoods. *Bayesian Analysis 11*(2), 325–352.

Eddelbuettel, D., R. François, J. Allaire, J. Chambers, D. Bates, and K. Ushey (2011). Rcpp: Seamless R and C++ integration. *Journal of Statistical Software 40*(8), 1–18.

Erhardt, R. (2013). *ABCExtremes: ABC Extremes.* R package version 1.0.

Erhardt, R. J. and R. L. Smith (2012). Approximate Bayesian computing for spatial extremes. *Computational Statistics & Data Analysis 56*(6), 1468–1481.

Erhardt, R. J. and R. L. Smith (2014). Weather derivative risk measures for extreme events. *North American Actuarial Journal 18*, 379–393.

Estoup, A., E. Lombaert, J.-M. Marin, T. Guillemaud, P. Pudlo, C. P. Robert, and J. Cournuet (2012). Estimation of demo-genetic model probabilities with approximate Bayesian computation using linear discriminant analysis on summary statistics. *Molecular Ecology Resources 12*(5), 846–855.

Fearnhead, P., V. Giagos, and C. Sherlock (2014). Inference for reaction networks using the linear noise approximation. *Biometrics 70*(2), 457–466.

Fearnhead, P. and L. Meligkotsidou (2004). Exact filtering for partially observed continuous time models. *Journal of the Royal Statistical Society: Series B (Statistical Methodology) 66*(3), 771–789.

Fearnhead, P. and D. Prangle (2012). Constructing summary statistics for approximate Bayesian computation: Semi-automatic ABC (with discussion). *Journal of the Royal Statistical Society: Series B (Statistical Methodology) 74*(3), 419–474.

Forneron, J.-J. and S. Ng (2016). A likelihood-free reverse sampler of the posterior distribution. In *Essays in Honor of Aman Ullah*, pp. 389–415. Emerald Group Publishing.

Gallant, A. R. and R. E. McCulloch (2009). On the determination of general scientific models with application to asset pricing. *Journal of the American Statistical Association 104*(485), 117–131.

Gallant, A. R. and G. Tauchen (1996). Which moments to match? *Econometric Theory 12*(4), 657–681.

Gleim, A. and C. Pigorsch (2013). Approximate Bayesian computation with indirect summary statistics. Technical report, University of Bonn.

Golightly, A., D. A. Henderson, and C. Sherlock (2015). Delayed acceptance particle MCMC for exact inference in stochastic kinetic models. *Statistics and Computing 25*(5), 1039–1055.

Gourieroux, C., A. Monfort, and E. Renault (1993). Indirect inference. *Journal of Applied Econometrics 8*(S1), S85–S118.

Heggland, K. and A. Frigessi (2004). Estimating functions in indirect inference. *Journal of the Royal Statistical Society: Series B (Statistical Methodology) 66*(2), 447–462.

Jiang, W. and B. Turnbull (2004). The indirect method: Inference based on intermediate statistics – A synthesis and examples. *Statistical Science 19*(2), 239–263.

Le Cam, L. (1956). On the asymptotic theory of estimation and testing hypotheses. In *Proceedings of the Third Berkeley Symposium on Mathematical Statistics and Probability, Volume 1: Contributions to the Theory of Statistics*, pp. 129–156. Berkeley, CA: The Regents of the University of California.

Lee, X. J., C. C. Drovandi, and A. N. Pettitt (2015). Model choice problems using approximate Bayesian computation with applications to pathogen transmission data sets. *Biometrics 71*(1), 198–207.

Li, J., D. J. Nott, Y. Fan, and S. A. Sisson (2017). Extending approximate Bayesian computation methods to high dimensions via Gaussian copula. *Computational Statistics and Data Analysis, 106*, 77–89.

Marin, J.-M., N. S. Pillai, C. P. Robert, and J. Rousseau (2013). Relevant statistics for Bayesian model choice. *Journal of the Royal Statistical Society: Series B (Statistical Methodology) 76*(5), 833–859.

Martin, G. M., B. P. M. McCabe, W. Maneesoonthorn, and C. P. Robert (2014). Approximate Bayesian computation in state space models. *arXiv:1409.8363*.

McFadden, D. (1989). A method of simulated moments for estimation of discrete response models without numerical integration. *Econometrica: Journal of the Econometric Society 57*(5), 995–1026.

Monfardini, C. (1998). Estimating stochastic volatility models through indirect inference. *The Econometrics Journal 1*(1), C113–C128.

Moores, M. T., C. C. Drovandi, K. L. Mengersen, and C. P. Robert (2015). Pre-processing for approximate Bayesian computation in image analysis. *Statistics and Computing 25*(1), 23–33.

Nott, D. J., Y. Fan, L. Marshall, and S. A. Sisson (2014). Approximate Bayesian computation and Bayes linear analysis: Toward high-dimensional ABC. *Journal of Computational and Graphical Statistics 23*(1), 65–86.

O'Neill, P. D. and G. O. Roberts (1999). Bayesian inference for partially observed stochastic epidemics. *Journal of the Royal Statistical Society: Series A (Statistics in Society) 162*(1), 121–129.

Padoan, S. A., M. Ribatet, and S. A. Sisson (2010). Likelihood-based inference for max-stable processes. *Journal of the American Statistical Association 105*(489), 263–277.

Pauli, F., W. Racugno, and L. Ventura (2011). Bayesian composite marginal likelihoods. *Statistica Sinica 21*, 149–164.

Picchini, U. and J. L. Forman (2016). Accelerating inference for diffusions observed with measurement error and large sample sizes using approximate Bayesian computation. *Journal of Statistical Computation and Simulation 86*(1), 195–213.

Prangle, D. (2016). Lazy ABC. *Statistics and Computing, 26*, 171–185.

Prangle, D., P. Fearnhead, M. P. Cox, P. J. Biggs, and N. P. French (2014). Semi-automatic selection of summary statistics for ABC model choice. *Statistical Applications in Genetics and Molecular Biology 13*(1), 67–82.

Price, L. F., C. C. Drovandi, A. Lee, and D. J. Nott (2018). Bayesian synthetic likelihood. *Journal of Computational and Graphical Statistics 27*(1), 1–11.

Reeves, R. W. and A. N. Pettitt (2005). A theoretical framework for approximate Bayesian computation. In A. R. Francis, K. M. Matawie, A. Oshlack, and G. K. Smyth (Eds.), *Proceedings of the 20th International Workshop on Statistical Modelling*, Sydney, Australia, pp. 393–396.

Ribatet, M., D. Cooley, and A. C. Davison (2012). Bayesian inference from composite likelihoods, with an application to spatial extremes. *Statistica Sinica 22*, 813–845.

Ribatet, M., R. Singleton, and R Core team (2013). *SpatialExtremes: Modelling Spatial Extremes.* R package version 2.0-0.

Robert, C. P., J.-M. Cornuet, J.-M. Marin, and N. S. Pillai (2011). Lack of confidence in approximate Bayesian computation model choice. *Proceedings of the National Academy of Sciences of the United States of America 108*(37), 15112–15117.

Ruli, E., N. Sartori, and L. Ventura (2016). Approximate Bayesian computation using composite score functions. *Statistics and Computing 26*(3), 679–692.

Ryan, C., C. C. Drovandi, and A. N. Pettitt (2016). Optimal Bayesian experimental design for models with intractable likelihoods using indirect inference applied to biological process models. *Bayesian Analysis, 11*(3), 857–883.

Schlather, M. (2002). Models for stationary max-stable random fields. *Extremes 5*(1), 33–44.

Smith, Jr., A. A. (1993). Estimating nonlinear time-series models using simulated vector autoregressions. *Journal of Applied Econometrics 8*(S1), S63–S84.

Wood, S. N. (2010). Statistical inference for noisy nonlinear ecological dynamic systems. *Nature 466*(7310), 1102–1104.

8

High-Dimensional ABC

David J. Nott, Victor M.-H. Ong, Y. Fan, and S. A. Sisson

CONTENTS

8.1 Introduction 211
8.2 Direct ABC Approximation of the Posterior 213
8.2.1 The marginal adjustment strategy 214
8.2.2 A toy example 215
8.2.3 Gaussian copula ABC 219
8.2.4 A multivariate g-and-k model for a foreign currency exchange dataset 222
8.2.5 A non-linear multiple quantile regression analysis 226
8.3 ABC Approximation of the Sampling Distribution of Summary Statistics 229
8.3.1 A flexible regression density estimator 230
8.3.2 Analysis of stereological extremes 232
8.4 Other Approaches to High-Dimensional ABC 235
8.5 Discussion 237
Acknowledgements 237
References 238

8.1 Introduction

Other chapters in this volume have discussed the curse of dimensionality that is inherent to most standard approximate Bayesian computation (ABC) methods. For a p-dimensional parameter of interest $\theta = (\theta_1, \ldots, \theta_p)^\top$, ABC implementations make use of a summary statistic $s = S(y)$ for data $y \in \mathcal{Y}$ of dimension q, where typically $q \geq p$. When either θ or s is high dimensional, standard ABC methods have difficulty in producing simulated summary data that are acceptably close to the observed summary $s_{obs} = S(y_{obs})$, for observed data y_{obs}. This means that standard ABC methods have limited applicability in high-dimensional problems.

More precisely, write $\pi(\theta)$ for the prior, $p(y|\theta)$ for the data model, $p(y_{obs}|\theta)$ for the likelihood function, and $\pi(\theta|y_{obs}) \propto p(y_{obs}|\theta)\pi(\theta)$ for the intractable posterior distribution. Standard ABC methods based on $S(y)$ typically approximate the posterior as $\pi(\theta|y_{obs}) \approx \pi_{ABC,h}(\theta|s_{obs})$, where:

$$\pi_{ABC,h}(\theta|s_{obs}) \propto \int K_h(\|s - s_{obs}\|)p(s|\theta)\pi(\theta)\,ds, \tag{8.1}$$

and where $K_h(\|u\|)$ is a kernel weighting function with bandwidth $h \geq 0$. A Monte Carlo approximation of (8.1) involves a kernel density estimation of the intractable likelihood based on $\|s - s_{obs}\|$, the distance between simulated and observed summary statistics. As a result, the quality of the approximation decreases rapidly as the dimension of the summary statistic q increases, as the distance between s and s_{obs} necessarily increases with their dimension, even setting aside the approximations involved in the choice of an informative $S(y)$.

Several authors (e.g. Blum, 2010; Barber et al., 2015) have given results which illuminate the way that the dimension of the summary statistic q impacts the performance of standard ABC methods. For example, Blum (2010) obtains the result that the minimal mean squared error of certain kernel ABC density estimators is of the order of $N^{-4/(q+5)}$, where N is the number of Monte Carlo samples in the kernel approximation. Barber et al. (2015) consider a simple rejection ABC algorithm where the kernel K_h is uniform and obtain a similar result concerned with optimal estimation of posterior expectations. Biau et al. (2015) extend the analysis of Blum (2010) using a nearest neighbour perspective, which accounts for the common ABC practice of choosing h adaptively based on a large pool of samples (e.g. Blum et al., 2013).

Regression adjustments (e.g. Blum, 2010; Beaumont et al., 2002; Blum and François, 2010; Blum et al., 2013) are extremely valuable in practice for extending the applicability of ABC approximations to higher dimensions, since the regression model has some ability to compensate for the mismatch between the simulated summary statistics s and the observed value s_{obs}. However, except when the true relationship between θ and s is known precisely (allowing for a perfect adjustment), these approaches may only extend ABC applicability to moderately higher dimensions. For example, Nott et al. (2014) demonstrated a rough doubling of the number of acceptably estimated parameters for a fixed computational cost when using regression adjustment compared to just rejection sampling, for a simple toy model. Non-parametric regression approaches are also subject to the curse of dimensionality, and the results of Blum (2010) also apply to certain density estimators which include non-parametric regression adjustments. Nevertheless, it has been observed that these theoretical results may be overly pessimistic in practice for some problems. See Li and Fearnhead (2016) for some recent progress on theoretical aspects of regression adjustment for uncertainty quantification.

This chapter considers the question of whether it may be possible to conduct reliable ABC-based inference for high-dimensional models or when the number of summary statistics $q \geq p$ is large. As a general principle, any methods that improve the efficiency of existing ABC techniques, such as more efficient Monte Carlo sampling algorithms, will as a result help extend ABC methods to higher dimensions, simply because they permit a greater inferential accuracy (measured by an effectively lower kernel bandwidth h) for the same computational overheads. However, there is a limit to the extent to which these improvements can produce substantial high-dimensional gains, as ultimately the bottleneck is determined by the $\|s - s_{obs}\|$ term within the kernel K_h embedded as part of the approximate posterior $\pi_{ABC,h}(\theta|s_{obs})$.

Instead, we examine ways in which the reliance on the q-dimensional comparison $\|s - s_{obs}\|$ can be reduced. One technique for achieving this is by estimating low-dimensional marginal posterior distributions for subsets of θ and then reconstructing an estimate of the joint posterior distribution from these. This approach takes advantage of the fact that the marginal posterior distribution $\pi_{ABC,h}(\theta^{(1)}|s_{obs}) = \int \pi_{ABC,h}(\theta|s_{obs})d\theta^{(2)}$ for some partition of the parameter vector $\theta = (\theta^{(1)^\top}, \theta^{(2)^\top})^\top$ can be much more accurately approximated using ABC directly as $\pi_{ABC,h}(\theta^{(1)}|s^{(1)}_{obs})$, since the corresponding necessary set of summary statistics $s^{(1)} \subset s$ would be a lower-dimensional vector compared with the summary statistics s required to estimate the full joint distribution $\pi_{ABC,h}(\theta|s_{obs})$. The same idea can also be implemented when approximating the likelihood function, where it is the sampling distribution of the summary statistics $p(s|\theta)$ that is approximated based on low-dimensional estimates for subsets of s.

The previous techniques are applicable for general ABC inference problems without any particular exploitable model structure and are the primary focus of this chapter. For models with a known exploitable structure, it may be possible to achieve better results (e.g. Bazin et al., 2010; Barthelmé and Chopin, 2014; White et al., 2015; Ong et al., 2018; Tran et al., 2017), and we also discuss these briefly.

8.2 Direct ABC Approximation of the Posterior

In this section, we consider direct approximation of the posterior distribution $\pi(\theta|s_{obs})$ given the observed summary statistics s_{obs}. We first describe the *marginal adjustment* approach of Nott et al. (2014), in which the standard ABC approximation of the joint posterior distribution is improved by replacing its univariate margins with more precisely estimated marginal approximations. These more precise marginal distributions are obtained by implementing standard ABC methods to construct each univariate marginal posterior separately, for which only low-dimensional summary statistics are required. These univariate marginal posteriors then replace the margins in the original approximate joint posterior sample, via an appropriate replacement of order statistics.

While the marginal adjustment can work well, we show an instructive toy example where this strategy fails to adequately estimate the posterior dependence structure. We subsequently discuss the Gaussian copula ABC approach of (Li et al., 2017), which extends the marginal adjustment to improve estimation of all pairwise dependences of the joint posterior, in combination with the marginal estimates, by use of a meta-Gaussian distribution (Fang et al., 2002). These ideas are illustrated by several examples.

8.2.1 The marginal adjustment strategy

The marginal adjustment method of Nott et al. (2014) is motivated by the following observation. Suppose we wish to estimate accurately the univariate marginal posterior distribution $\pi(\theta_j|s_{obs})$ of the parameter θ_j. If we can find a summary statistic, say $s^{(j)} \subset s$, that is nearly marginally sufficient for θ_j in the data model $p(y|\theta_j)$, then $\pi(\theta_j|s_{obs}) \approx \pi(\theta_j|s_{obs}^{(j)})$, and this summary statistic can be used to obtain marginal ABC posterior inferences about θ_j. Because θ_j is univariate, the summary statistic $s^{(j)}$ can be low dimensional.

Accordingly, the marginal ABC model takes the form:

$$
\begin{aligned}
\pi_{ABC,h}(\theta_j|s_{obs}^{(j)}) &\propto \int K_h(\|s^{(j)} - s_{obs}^{(j)}\|)p(s^{(j)}|\theta_j)\pi(\theta_j)ds^{(j)} \\
&= \int\int K_h(\|s^{(j)} - s_{obs}^{(j)}\|)p(s|\theta)\pi(\theta_{-j}|\theta_j)\pi(\theta_j)d\theta_{-j}ds
\end{aligned}
$$

where θ_{-j} denotes the elements of θ excluding θ_j, and $\pi(\theta_{-j}|\theta_j)$ denotes the conditional prior of θ_{-j} given θ_j.

The idea of Nott et al. (2014) is to exploit the observation that marginal posterior inferences are much easier in the ABC framework, as they only involve a lower-dimensional subset of summary statistics, $s^{(j)} \subset s$. A sample from the joint ABC posterior $\pi_{ABC,h}(\theta|s_{obs})$ is first obtained, and then this joint sample is adjusted so that its marginal distributions match those estimated from the lower-dimensional ABC analyses, $\pi_{ABC,h}(\theta_j|s_{obs}^{(j)})$.

Write $s = (s_1, \dots, s_q)^\top$ for the summary statistics used to approximate the joint posterior $\pi_{ABC,h}(\theta_j|s_{obs})$ and $s^{(j)} = (s_1^{(j)}, \dots, s_{q_j}^{(j)})^\top$ for the summary statistics used to approximate the marginal posterior distribution of θ_j, $\pi_{ABC,h}(\theta_j|s_{obs}^{(j)})$. The marginal adjustment algorithm is then implemented as follows:

1. Using standard ABC methods (including regression adjustments), obtain an approximate sample from the joint posterior distribution $\pi(\theta|s_{obs})$, $\theta^{J1}, \dots, \theta^{Jr}$ say, based on the full summary statistic s.

2. Using standard ABC methods, for each $j = 1, \ldots, p$, obtain an approximate sample from the univariate marginal distribution $\pi(\theta_j | s^{(j)}_{obs})$, $\theta_j^{M1}, \ldots, \theta_j^{Mr'}$ say, based on the lower-dimensional summary statistic $s^{(j)}$.

3. Write $\theta_j^M(k)$ for the k-th order statistic of the (marginally estimated) sample $\theta_j^{M1}, \ldots, \theta_j^{Mr'}$ and $\theta_j^J(k)$ for the k-th order statistic of the (jointly estimated marginal) sample $\theta_j^{J1}, \ldots, \theta_j^{Jr}$. Also write $R(j,k)$ for the rank of θ_j^{Jk} within the sample $\theta_j^{J1}, \ldots, \theta_j^{Jr}$. Define

$$\theta^{Ak} = (\theta_1^M(R(1,k)), \ldots, \theta_p^M(R(p,k)))^\top.$$

Then θ^{Ak}, $k = 1, \ldots, r$, is a marginally adjusted approximate sample from $\pi(\theta | s_{obs})$.

It is worth stating in words what is achieved by the previous step 3. The samples θ^{Ak}, $k = 1, \ldots, r$ are the same as θ^{Jk}, except that componentwise, the order statistics $\theta_j^J(k)$ have been replaced by the corresponding order statistics $\theta_j^M(k)$. If we were to convert the samples θ^{Ak} and θ^{Jk} to ranks componentwise, they would be exactly the same, and so the dependence structure in the original samples θ^{Jk} is preserved in θ^{Ak} in this sense. However, the estimated marginal distribution in θ^{Ak} for θ_j is simply the estimated marginal distribution obtained from the samples $\theta_j^{M1}, \ldots, \theta_j^{Mr'}$, so that the adjusted samples θ^{Ak} give the more precisely estimated marginal distributions from the low-dimensional analyses of step 2, while preserving the dependence structure from the joint samples of step 1.

While it is true that the dependence structure obtained at step 1 may not be well estimated due to standard ABC curse-of-dimensionality arguments, it is also the case that the marginal adjustment improves the estimation of the marginal posterior distributions. These ideas are illustrated in the following example.

8.2.2 A toy example

Following Li et al. (2017), we let the data $y = (y_1, \ldots, y_p)^\top$, $p \geq 2$ follow a $N(\theta, I_p)$ distribution, where $\theta = (\theta_1, \ldots, \theta_p)^\top$ is the parameter of interest and I_p denotes the $p \times p$ identity matrix. The prior $\pi(\theta)$ is specified as the twisted normal form (Haario et al., 1999):

$$\pi(\theta) \propto \exp\left(-\frac{\theta_1^2}{200} - \frac{(\theta_2 - b\theta_1^2 + 100b)^2}{2} - \sum_{j=3}^{p} \theta_j^2 \right)$$

where we set $b = 0.1$, and if $p = 2$ the $\sum_{j=3}^p \theta_j^2$ term is omitted. A contour plot of $\pi(\theta)$ for $p = 2$ is shown in Figure 8.1. This is an interesting example because the likelihood only provides location information about θ.

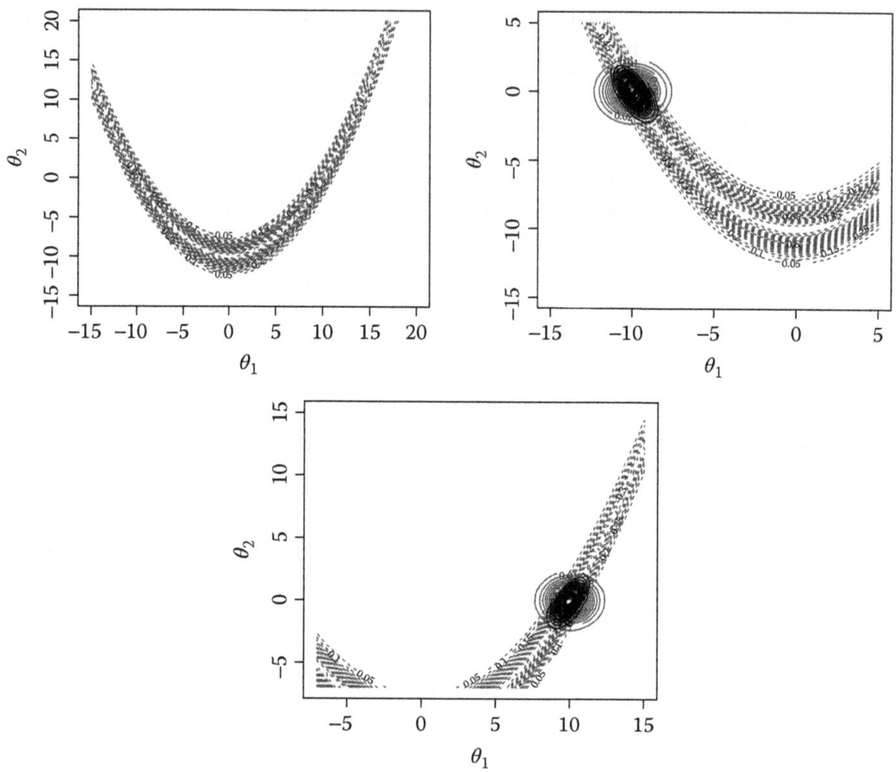

FIGURE 8.1
Contour plots of twisted normal prior distribution $\pi(\theta)$ (grey dashed lines), likelihood (solid grey), and posterior (solid black) for $p = 2$. Middle and right panels illustrate the case when $y_{obs} = (-10, 0)^\top$ and $y_{obs} = (10, 0)^\top$, respectively.

The dependence structure in the posterior comes mostly from the prior, and the association between θ_1 and θ_2 changes direction in the left and right tails of the prior (Figure 8.1). So the posterior dependence changes direction depending on whether the likelihood locates the posterior in the left or right tail of the prior. This feature makes it difficult for standard regression adjustment methods, which merely translate and scale particles [e.g. generated from $(s, \theta) \sim p(s|\theta)\pi(\theta)$], to work in high dimensions.

Figures 8.2 and 8.3 show what happens in an analysis of this example with $p = 5$ and $p = 50$, respectively. Four ABC approximation methods are considered with observed data $y_{obs} = (10, 0, ..., 0)^\top$. The contour plots of the bivariate posterior estimates $\pi(\theta_1, \theta_2|s_{obs})$ are represented by solid lines, while the contour plot of the true bivariate margin is represented by grey dashed lines. For both figures, panel (a) shows the estimates obtained via standard rejection

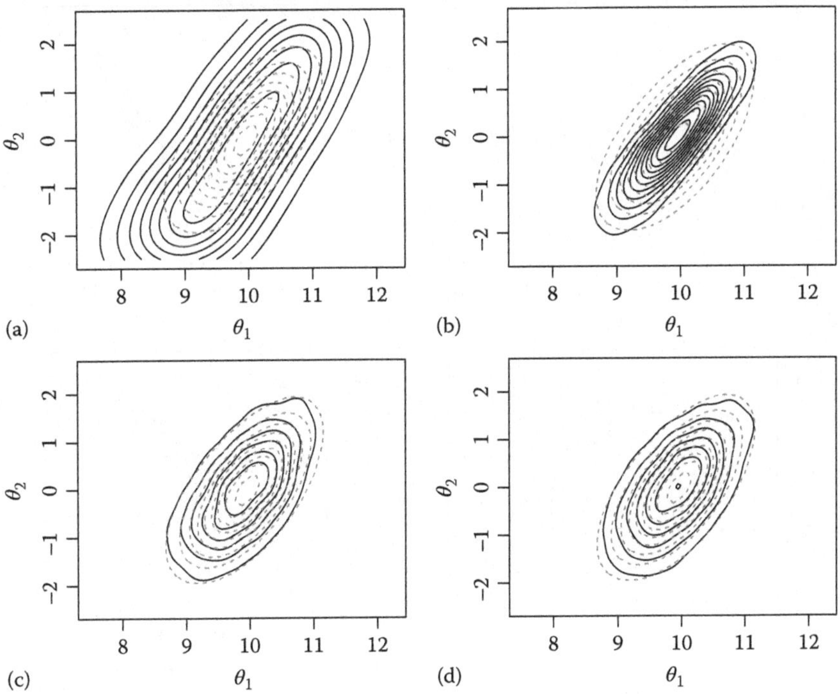

FIGURE 8.2
Contour plots of the (θ_1, θ_2) margin of various ABC posterior approximations for the $p = 5$ dimensional model $\pi(\theta|s_{obs})$ are represented by the black lines. True contours for the bivariate margins are represented by the grey dashed lines. The different ABC approximations approaches are (a) rejection sampling, (b) rejection sampling with marginal adjustment, (c) rejection sampling with regression adjustment, and (d) rejection sampling with regression and marginal adjustment.

ABC, while panels (b), (c), and (d) show estimates obtained after marginal, linear regression and both linear regression and marginal adjustment respectively. Note the regression adjustment step is performed after the rejection sampling stage and before the marginal adjustment.

For the case when $p = 5$ (Figure 8.2), rejection sampling alone captures the correlation between θ_1 and θ_2, but the univariate margins are too dispersed. Performing a marginal adjustment following rejection sampling is not good enough, as it only corrects the margin to the right scale and is not able to recover dependence structure. On the other hand, rejection sampling with linear regression adjustment is able to give a good approximation to the true posterior. Performing a subsequent marginal adjustment (Figure 8.2d) shows no further visual improvement.

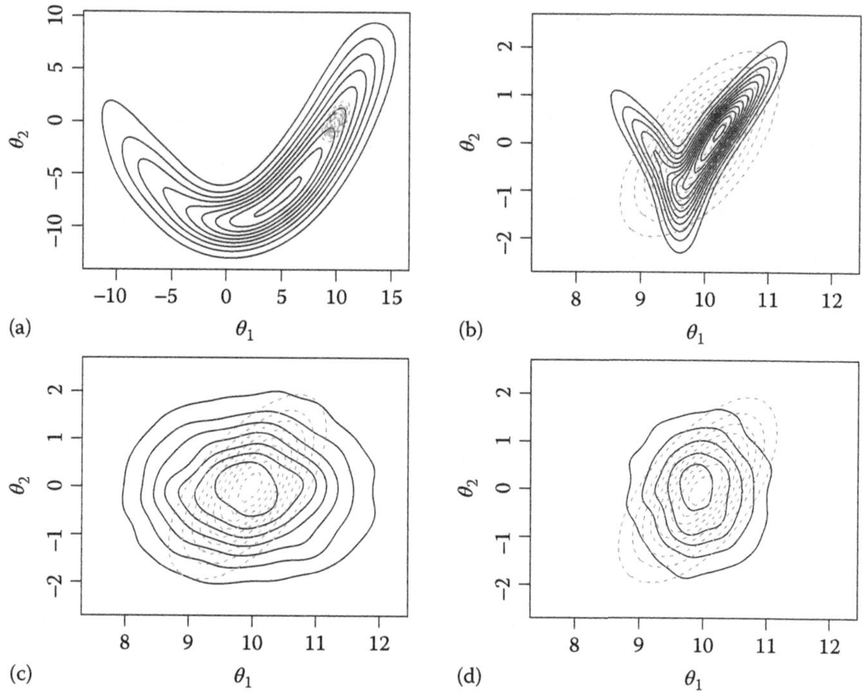

FIGURE 8.3
Contour plots of the (θ_1, θ_2) margin of various ABC posterior approximations for the $p = 50$ dimensional model $\pi(\theta|s_{obs})$ are represented by the black lines. True contours for the bivariate margins are represented by the grey dashed lines. The different ABC approximations approaches are (a) rejection sampling, (b) rejection sampling with marginal adjustment, (c) rejection sampling with regression adjustment, and (d) rejection sampling with regression and marginal adjustment.

The example for $p = 50$ (Figure 8.3) shows both the strengths and limitations of the marginal and regression adjustment strategies. It is very clear that standard rejection ABC estimates do not seem to be using much of the information given by the likelihood, as the posterior estimate follows the shape of the prior distribution $\pi(\theta)$. Performing either regression or marginal adjustment centres the estimates on the right location, but the shape of the contour plots for the adjustments are incorrect. Moreover, applying marginal adjustment after regression adjustment corrects the univariate margins well, but does not recover the dependence structure. In this example, all four approaches are not able to recover the dependence structure of the true bivariate posterior. It is worth noting in this example that for the normal case with $b = 0$, the marginal adjustment approach works very well even in high dimensions.

This example shows some of the limitations of the marginal adjustment strategy. One possible approach to improve the estimation of the dependence structure (not discussed by Nott et al., 2014) is to use the marginal adjustment on a reparameterised parameter vector θ^*, where the margins of θ^* account for the dependence structure in θ, while θ_i^* and θ_j^*, $i \neq j$ remain approximately independent. This approach would require some prior knowledge of the dependence structure.

Since the key idea of the marginal adjustment approach is to build up a more accurate approximation of the joint posterior from estimates of univariate marginal posterior distributions, it is natural to ask if it is possible to consider estimation of marginal posterior distributions of dimension larger than one and to use these to help estimate the joint dependence structure of $\pi(\theta|s_{obs})$ more accurately.

8.2.3 Gaussian copula ABC

One way to implement this idea is the Gaussian copula ABC method of Li et al. (2017). Suppose that $\mathcal{C}(u) = P(U_1 \leq u_1, \ldots, U_p \leq u_p)$ is the distribution function of a random vector $U = (U_1, \ldots, U_p)$, where the marginal distribution of each $U_j \sim U(0,1)$ is uniform. Then $\mathcal{C}(u)$ is called a copula. Multivariate distributions can always be written in terms of a copula, and their marginal distribution functions, which is an implication of Sklar's theorem (Sklar, 1959). This allows for a decoupling of the modelling of marginal distributions and the dependence structure of a multivariate distribution. One useful type of copula derives from a multivariate Gaussian distribution. Suppose that $\eta \sim N(0, C)$ is a p-dimensional multivariate Gaussian random vector, where C is a correlation matrix. The distribution of $U = (\Phi(\eta_1), \ldots, \Phi(\eta_p))^\top$, where $\Phi(\cdot)$ denotes the standard normal distribution function is then a copula. This kind of copula, called a Gaussian copula, characterises the dependence structure of a multivariate Gaussian distribution, and it is parametrised by the correlation matrix C.

Suppose now that we further transform U as $\gamma = (F_1^{-1}(U_1), \ldots, F_p^{-1}(U_p))^\top$, where $F_1(\cdot), \ldots, F_p(\cdot)$ are distribution functions with corresponding density functions $f_1(\cdot), \ldots, f_p(\cdot)$. The components of γ then have the marginal densities $f_1(\cdot), \ldots, f_p(\cdot)$, respectively, and the dependence structure is being described by the Gaussian copula with correlation matrix C. Clearly, if the densities $f_j(\cdot)$, $j = 1, \ldots, p$ are themselves univariate Gaussian, then γ is multivariate Gaussian. A distribution constructed from a Gaussian, copula and given marginal distributions is called *meta-Gaussian* (Fang et al., 2002), and its density function is:

$$h(\gamma) = |C|^{-1/2} \exp\left(\frac{1}{2} z^\top (I - C^{-1}) z\right) \prod_{j=1}^{p} f_j(\gamma_j),$$

where $z = (z_1, \ldots, z_p)^\top$, and $z_j = \Phi^{-1}(F_j(\gamma_j))$.

Li et al. (2017) considered using a meta-Gaussian distribution to approximate the posterior distribution $\pi(\theta|s_{obs})$ in ABC. It is easily seen that a meta-Gaussian distribution is determined by its bivariate marginal distributions, so that if we are prepared to accept a meta-Gaussian approximation to the joint posterior distribution in a Bayesian setting, then it can be constructed based on bivariate posterior marginal estimates. Asymptotically, the posterior will tend to be Gaussian, but a meta-Gaussian approximation may work well even when we are far from this situation since it allows for flexible estimation of the marginal distributions. As with the marginal adjustment, since the bivariate marginal posterior distributions can be estimated using low-dimensional summary statistics, this can help to circumvent the ABC curse of dimensionality in estimation of the joint posterior dependence structure.

As before, write $s^{(j)}$ for the statistics that are informative for ABC estimation of the univariate posterior marginal $\pi(\theta_j|s_{obs})$, and now write $s^{(i,j)}$ for the summary statistics informative for ABC estimation of the bivariate posterior margin $\pi(\theta_i, \theta_j|s_{obs})$, $i \neq j$. Construction of the Gaussian copula ABC approximation to the posterior $\pi(\theta|s_{obs})$ proceeds as follows:

1. Using standard ABC methods (including regression adjustments), for each $j = 1, \ldots, p$, obtain an approximate sample from the univariate marginal distribution $\pi(\theta_j|s_{obs}^{(j)})$, $\theta_j^{U1}, \ldots, \theta_j^{Ur}$ say, based on the lower-dimensional summary statistic $s^{(j)}$. Use kernel density estimation to construct an approximation $\hat{g}_j(\theta_j)$ to $\pi(\theta_j|s_{obs}^{(j)})$.

2. Using standard ABC methods, for $i = 1, \ldots, p-1$ and $j = i+1, \ldots, p$, obtain an approximate sample from the bivariate marginal distribution $\pi(\theta_i, \theta_j|s_{obs}^{(i,j)})$, $(\theta_i^{Bj1}, \theta_j^{Bi1}), \ldots, (\theta_i^{Bjr}, \theta_j^{Bir})$ say, based on the low-dimensional summary statistics $s^{(i,j)}$.

3. Write $R(i,j,k)$ as the rank of θ_i^{Bjk} within the sample $\theta_i^{Bj1}, \ldots, \theta_i^{Bjr}$. With this notation $R(j,i,k)$, $j > i$, is the rank of θ_j^{Bik} within the sample $\theta_j^{Bi1}, \ldots, \theta_j^{Bir}$. Estimate C_{ij} by $\hat{C}_{ij}$, the sample correlation between the vectors:

$$\left(\Phi^{-1}\left(\frac{R(i,j,1)}{r+1}\right), \Phi^{-1}\left(\frac{R(i,j,2)}{r+1}\right), \ldots, \Phi^{-1}\left(\frac{R(i,j,r)}{r+1}\right)\right)^\top$$

   and

$$\left(\Phi^{-1}\left(\frac{R(j,i,1)}{r+1}\right), \Phi^{-1}\left(\frac{R(j,i,2)}{r+1}\right), \ldots, \Phi^{-1}\left(\frac{R(j,i,r)}{r+1}\right)\right)^\top.$$

4. Construct the Gaussian copula ABC approximation of $\pi(\theta|s_{obs})$ as the meta-Gaussian distribution with marginal distributions $\hat{g}_j(\theta_j)$, $j = 1, \ldots, p$ (step 1), and Gaussian copula correlation matrix $\hat{C} = [\hat{C}_{ij}]_{i,j=1,\ldots,p}$, where $\hat{C}_{ij}$, $j > i$, is as in step 2, $\hat{C}_{ji} = \hat{C}_{ij}$ and $\hat{C}_{ii} = 1$.

While the estimated correlation matrix $\hat{C}$ can fail to be positive definite using this procedure (although this did not occur in our analyses), methods to adjust this can be easily implemented (e.g. Løland et al., 2013). Note that by using the approximate posterior sample from $\pi(\theta_i, \theta_j | s_{obs}^{(i,j)})$ from step 2 and the fitted (bivariate) copula model for the pair, it is possible to investigate whether the Gaussian copula dependence structure at least represents the true bivariate posterior dependence structure well (though not the full multivariate dependence structure). This can be supplemented by application specific goodness of fit checking of posterior predictive densities based on the joint copula approximation.

In the twisted normal toy example of Section 8.2.2, the copula strategy can succeed where the marginal adjustment strategy alone fails. Similar to Figure 8.3, Figure 8.4 illustrates both the bivariate estimates of $\pi(\theta_1, \theta_2 | s_{obs})$ based on the Gaussian copula ABC approximation (black solid lines) and the true margins (grey dashed lines) for the $p = 50$ dimensional model. From the contour plots, the ABC copula approximation is able to produce estimates largely similar to the true bivariate margins, in stark contrast to the marginal adjustment alone in Figure 8.3. Thus, in this example where standard ABC sampling with regression and/or marginal adjustment fails, the copula strategy succeeds.

In order to investigate the performance of each ABC posterior estimation method more precisely, we follow Li et al. (2017) and vary the dimension of

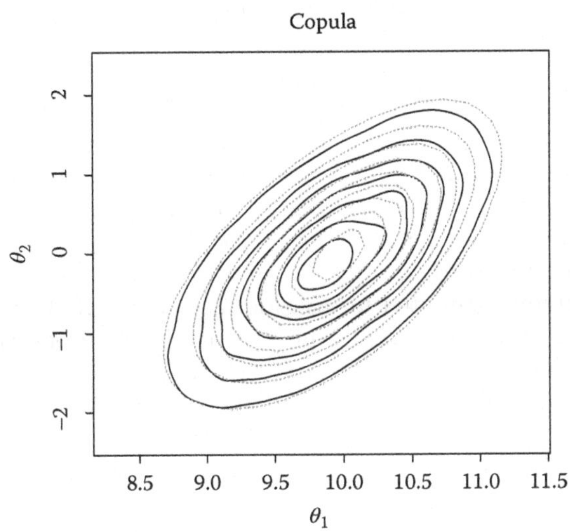

FIGURE 8.4
Contour plots of the (θ_1, θ_2) margin of the Gaussian copula ABC posterior approximation of the $p = 50$ dimensional model $\pi(\theta | s_{obs})$ (black lines). The true contours of $\pi(\theta_1, \theta_2 | s_{obs})$ are represented by grey dashed lines.

the model, p, from 2 to 250. Table 8.1 shows the mean estimated Kullback–Leibler (KL) divergence between the true bivariate margin $\pi(\theta_1, \theta_2|s_{obs})$ and the bivariate margin of the full ABC posterior approximation based on 100 replicate approximations, for all five approaches.

Observe that when the dimension p increases, the performance of the standard rejection ABC approach deteriorates. Adopting any of the adjustment strategies improves the overall performance, but the estimated KL divergences still increase with dimension p up to fixed limits. This suggests that if accurate estimation of the posterior dependence structure is important, then regression and marginal adjustment strategies alone may be limited to low dimensional models. From Table 8.1 it is clear that Gaussian copula ABC outperforms all other methods in terms of KL divergence, and its performance does not deteriorate with increasing dimension, p. This is not surprising, as the Gaussian copula ABC approximation is constructed from bivariate estimates of $\pi(\theta_1, \theta_2|s_{obs})$ and is therefore able to capture the dependence structure of all bivariate pairs of the full posterior distribution $\pi(\theta|s_{obs})$.

In the following sections, we implement Gaussian copula ABC for two real data analyses: an analysis of multivariate currency exchange data, and simultaneous estimation of multiple quantile regressions.

8.2.4 A multivariate *g*-and-*k* model for a foreign currency exchange dataset

The g-and-k distribution (Rayner and MacGillivray, 2002) is a flexible model for univariate data. It is typically specified through its quantile function:

$$Q(p|A, B, g, k) = A + B\left[1 + c\frac{1 - \exp\{-gz(p)\}}{1 + \exp\{-gz(p)\}}\right](1 + z(p)^2)^k z(p), \quad (8.2)$$

where $A, B > 0, g$, and $k > -0.5$ are parameters, respectively, controlling location, scale, skewness, and kurtosis of the distribution. The parameter c is conventionally fixed at 0.8 (resulting in $k > -0.5$), and $z(p)$ denotes the p-quantile of the standard normal distribution. Many distributions can be recovered or well approximated for appropriate values of A, B, g, and k (such as the normal when $g = k = 0$). Despite its attractive properties as a model, inference using the g-and-k distribution is challenging since the density, given by the derivative of the inverse of the quantile function, has no closed form. However, since simulation from the model is trivial by transforming uniform variates on $[0, 1]$ through the quantile function, an ABC implementation is one possible inferential approach. This idea was first explored by Peters and Sisson (2006) and Allingham et al. (2009). Here, we consider a multivariate extension of the model developed by Drovandi and Pettitt (2011). This model has a univariate g-and-k distribution for each margin, and the dependence structure is specified through a Gaussian copula. Note that this use of a Gaussian copula

TABLE 8.1
Estimated Kullback–Leibler Divergence of the (θ_1, θ_2) Margin of Various ABC Posterior Approximation to $\pi(\theta_1, \theta_2 | s_{obs}^{(1,2)})$. Numbers in Parentheses Represent Standard Errors of Mean Divergences over 100 Replications

p	Rejection Only	Marginal	Regression	Regression then Marginal	Copula ABC
2	0.058 (<0.001)	0.040 (<0.001)	0.043 (<0.001)	0.035 (<0.001)	0.039 (<0.001)
5	0.807 (<0.001)	0.053 (0.001)	0.613 (0.002)	0.037 (<0.001)	0.040 (<0.001)
10	1.418 (0.002)	0.100 (0.001)	1.078 (0.002)	0.061 (0.001)	0.040 (<0.001)
15	1.912 (0.002)	0.292 (0.002)	1.229 (0.003)	0.202 (0.001)	0.039 (<0.001)
20	2.288 (0.002)	0.450 (0.001)	1.280 (0.003)	0.292 (0.001)	0.039 (<0.001)
50	3.036 (0.003)	0.520 (0.002)	1.474 (0.009)	0.335 (0.001)	0.040 (<0.001)
100	3.362 (0.002)	0.524 (0.002)	1.619 (0.013)	0.341 (0.001)	0.039 (<0.001)
250	3.663 (0.003)	0.515 (0.002)	1.737 (0.015)	0.344 (0.001)	0.039 (<0.001)

to describe the dependence structure in the data model is distinct from the use of a Gaussian copula to approximate the dependence structure of the posterior distribution.

Suppose that the data are n independent multivariate realisations $y = (y^1, \ldots, y^n)$, where $y^i = (y_1^i, \ldots, y_q^i)^\top$. We assume that marginally, each y_j^i $i = 1, \ldots, n$ follows a g-and-k distribution with parameters (A_j, B_j, g_j, k_j), $j = 1, \ldots, q$. Gaussian copula ABC approximates the joint distribution of y^i by a meta-Gaussian distribution, with Gaussian copula correlation matrix C. For a q-dimensional data model, there are $4q$ marginal parameters and $q(q-1)/2$ distinct parameters in the correlation matrix, giving $p = q(q+7)/2$ parameters in total. We consider an analysis of log daily returns for $q = 16$ currencies (resulting in $p = 184$ parameters) versus the Australian dollar for 1,757 trading days covering the period 1st January 2007 to 31st December 2013 (Reserve Bank of Australia, 2014).

As a prior on C, we adopt the distribution obtained by sampling $V \sim \text{Wishart}(I_q, q)$ and then rescaling V to be a valid correlation matrix with 1's on the diagonal. The priors on A, B, g, and k for each marginal are independent and uniform over the parameter support, although we adopted uniform distributions with ranges of $[-0.1, 0.1]$, $[0, 0.05]$, $[-1, 1]$, and $[-0.2, 0.5]$ for A_j, B_j, g_j, and k_j to produce samples (s, θ) proportional to $p(s|\theta)\pi(\theta)$, but restricted to a region of high posterior density following an initial pilot analysis (see e.g. Fearnhead and Prangle, 2012).

Following the strategy of Li et al. (2017), the following summary statistics were considered informative for each marginal parameter: writing L_{kj}, $k = 1, 2, 3$ for the quantiles and O_{kj}, $k = 1, \ldots, 7$ for the octiles of $y_j^1, \ldots, y_j^n$, the marginally informative summary statistics were chosen as L_{2j} for A_j, $(L_{3j} - L_{1j}, (E_{7j} - E_{5j} + E_{3j} - E_{1j})/(L_{3j} - L_{1j}))^\top$ for B_j, $(L_{3j} + L_{1j} - 2L_{2j})/(L_{3j} - L_{1j})$ for g_j, and $(E_{7j} - E_{5j} + E_{3j} - E_{1j})/(L_{3j} - L_{1j})$ for k_j. These summary statistic choices were guided by similar summary statistics in Drovandi and Pettitt (2011), and preliminary analyses to determine which sets of the distinct summaries were marginally informative for individual parameters. For pairs of parameters, the summary statistics for individual parameters were simply combined. For the correlation parameters in the Gaussian copula, we follow Drovandi and Pettitt (2011) and use the robust normal scores correlation coefficient for the marginal summary statistic.

Contour plots of various estimates of the bivariate (B_1, k_1) posterior marginal distribution using ABC rejection sampling are illustrated in Figure 8.5. The top panels show estimates using the full (p-dimensional) vector of summary statistics with (a) regression adjustment and (b) marginal adjustment, respectively. The performance of each approach individually is poor as the distributions do not exhibit the more accurately estimated dependence structures observed in the remaining panels. These estimates are based on ABC rejection sampling with both marginal and regression adjustments, using (c) the full vector of summary statistics and (d) the marginally

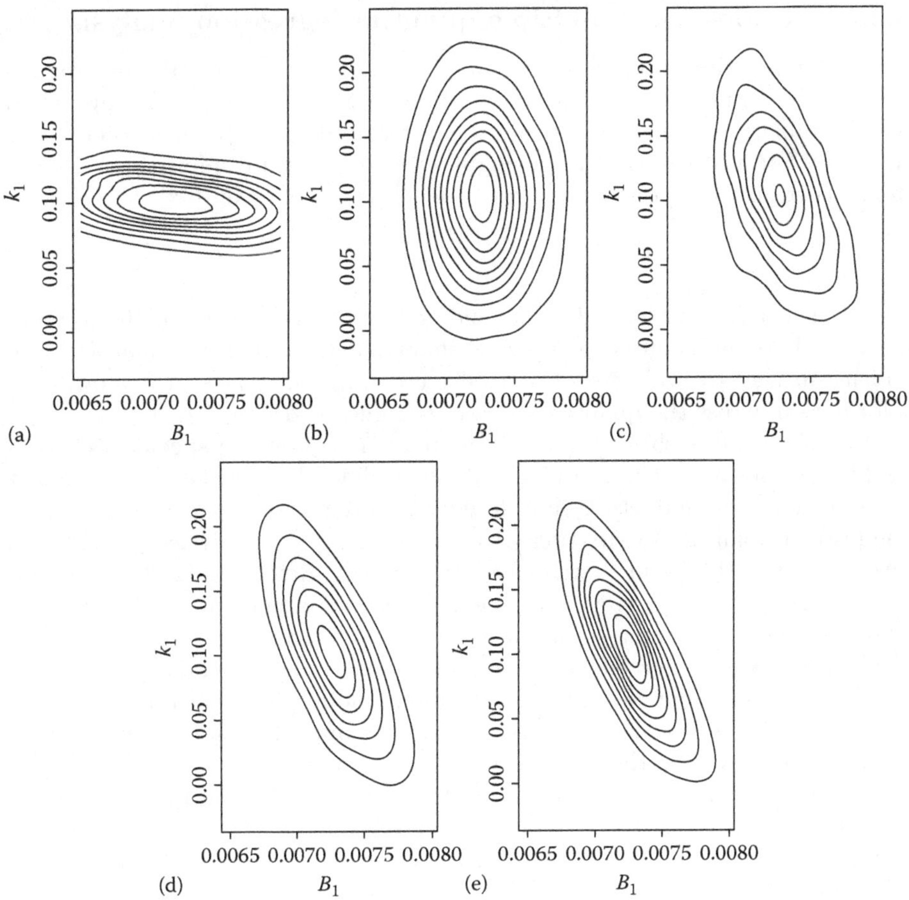

FIGURE 8.5
Contour plots of the (B_1, k_1) margin of rejection sampling-based ABC posterior approximations to the multivariate g-and-k model. Top panels show estimates using (a) regression adjustment, (b) marginal adjustment, and (c) both regression then marginal adjustment, using the full vector of summary statistics. Panel (d) shows the same as (c), but using the lower dimensional vector of summary statistics informative for (B_1, k_1). Panel (e) shows the estimate for the Gaussian copula ABC approximation.

informative summary statistics for (B_1, k_1). The similarity between panels (c) and (d) indicates that the marginally informative summary statistics are indeed highly informative for the parameter pair (B_1, k_1). Finally, panel (e) illustrates the Gaussian copula ABC approximation. The similarity between panels (d) and (e) indicates that the copula model provides an excellent approximation of the bivariate posterior marginal distribution.

8.2.5 A non-linear multiple quantile regression analysis

Quantile regression can provide a robust alternative to standard mean regression. Model estimates obtained at multiple quantile levels can also provide a more complete picture of the conditional distribution between predictor and response. For a regression with a single covariate x, and response y, the linear model corresponding to the τ-th quantile, $Q_y(\tau|x)$, is given by:

$$Q_y(\tau|x) = \alpha_\tau + \beta_\tau x,$$

where the coefficients α_τ and β_τ depend on the quantile level, $\tau \in (0,1)$. Standard methods fit quantile regressions independently for each quantile level, which can lead to problems of quantiles crossing and a lack of borrowing of information across the quantile levels (Rodrigues and Fan, 2017).

Bayesian approaches to quantile regression require the specification of a likelihood. However, exact and tractable likelihood functions are often not available for these models. Quantile regression requires the inversion of many conditional quantile distributions, which are often not analytically available, although numerical grid search can be used (e.g. Reich et al., 2010; Tokdar and Kadane, 2012). However, in the presence of larger datasets, numerical grid searches can become computationally prohibitive, see for example Reich et al. (2010) who suggest using approximations as an alternative.

We consider a dataset for analysing immunodeficiency in infants. In the search for reference ranges to help diagnose infant immunodeficiency, Isaacs et al. (1983) measured the serum concentration of immunoglobulin-G (IgG) in 298 pre-school children. We are interested in estimating the IgG conditional quantiles at the levels $\tau = 0.1, 0.2, 0.3, 0.7, 0.75, 0.8, 0.95$. A quadratic model in age (x) is used to fit the data due to the expected smooth change of IgG with age, so that:

$$Q_y(\tau|x) = \alpha_\tau + \beta_\tau x + \eta_\tau x^2. \tag{8.3}$$

Figure 8.6 illustrates this dataset. The black lines show the separately fitted regression lines for the different quantile levels, based on a frequentist estimator using the `quantreg` package in `R` (Koenker, 2005). Since these curves are fitted separately, no correlation is assumed between the quantile curves, and for close quantile levels τ, the fitted quantile estimates can easily cross each other. In practice, strong correlations can exist between curves close to each other, and the true quantile levels will not cross.

We follow the linearly interpolated likelihood function approach of Feng et al. (2015) as a data model $p(s|\theta)$, while extending their quantile function $Q_y(r|x)$ to contain more than one predictor as in (8.3). For each observed covariate $x_{obs,i}$, $i = 1, \dots, n$, a synthetic data point y_i can be obtained via:

$$y_i = Q_y(\tau_j|x_{obs,i}) + \frac{Q_y(\tau_{j+1}|x_{obs,i}) - Q_y(\tau_j|x_{obs,i})}{\tau_{j+1} - \tau_j}(u_i - \tau_j),$$

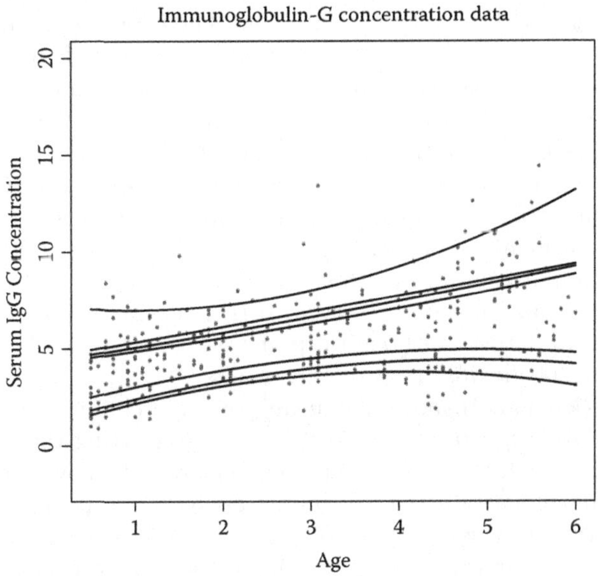

FIGURE 8.6
The immunoglobulin-G dataset. The fitted lines correspond to the classical quantile estimator at the quantile levels $\tau = 0.1, 0.2, 0.3, 0.7, 0.75, 0.8, 0.95$.

where $u_i \sim U(0,1)$, where j is determined so that $\tau_j < u_i < \tau_{j+1}$, and where $Q_\tau(y|x)$ is the model (8.3), which depends on parameters α_τ, β_τ, and η_τ. If $u_i < \tau_1$, y_i is generated from a normal distribution centred on $\bar{y}_{obs}$, with standard deviation three times the sample standard deviation of y_{obs}, and truncated below $Q_y(\tau_1|x)$. Similarly, if $u_i > \tau_m$, we simulate from the same distribution except that it is truncated above $Q_y(\tau_m|x)$. The parameters α_τ, β_τ, and η_τ are sampled from multivariate Gaussian prior distributions $\pi(\theta)$, with mean vector and covariance matrix based on the estimates obtained using `quantreg`. This prior is constrained to satisfy the quantile monotonicity condition so that the fitted quantile regression lines do not cross.

The full vector of summary statistics is constructed as:

$$s = S(y) = (\hat{\alpha}_{\tau_1}, \ldots, \hat{\alpha}_{\tau_m}, \hat{\beta}_{\tau_1}, \ldots, \hat{\beta}_{\tau_m}, \hat{\eta}_{\tau_1}, \ldots, \hat{\eta}_{\tau_m},$$
$$pu_{\tau_1}, \ldots, pu_{\tau_m}, pl_{\tau_1}, \ldots, pl_{\tau_m}, q_1(y), \ldots, q_{100}(y))^\top,$$

where $\hat{\alpha}_\tau$, $\hat{\beta}_\tau$, and $\hat{\eta}_\tau$ are the independent frequentist estimators for α_τ, β_τ, and η_τ at quantile level τ, pu_τ is the proportion of data points above the τth quantile curve, pl_τ is the proportion of data points below the τth quantile curve, and $q_1(y), \ldots, q_{100}(y)$ are the 100 equally spaced quantiles of the data y. The summary statistics for α_{τ_i} are $\hat{\alpha}_{\tau_i}, pu_{\tau_i}, pl_{\tau_i}$, and the closest 20 quantiles $q_1(y), \ldots, q_{100}(y)$ to the level τ_i. Similarly, for β_{τ_j}, the marginally informative summary statistics will be $\hat{\beta}_{\tau_j}, pu_{\tau_j}, pl_{\tau_j}$, and the closest 20 quantiles

$q_1(y), \ldots, q_{100}(y)$ to the level τ_j; and so on. Then for the summaries of the bivariate margin, $(\alpha_{\tau_i}, \beta_{\tau_j})$, we concatenate the two sets of summaries.

The following analysis is based on N = 1,000,000 samples $(s^{(\ell)}, \theta^{(\ell)}) \sim p(s|\theta)\pi(\theta)$, $\ell = 1, \ldots, N$. We specify the smoothing kernel $K_h(\cdot)$ as uniform over the range $(-h, h)$ and determine h as the 0.001 quantile of the Euclidean distances between observed and simulated summary statistics. Our model simultaneously fits the seven quantile levels shown in Figure 8.6, resulting in a $p = 21$ dimensional model with $q = 135$ total summary statistics. Note that with the application of post-hoc adjustments, monotonicity of the conditional quantiles may not be preserved. If this occurs, the offending samples may simply be discarded, although a preferable solution is the development of adjustments that flexibly respect constraints.

Figure 8.6 (left panel) shows the mean predicted conditional quantile estimates for the levels $\tau = 0.1, 0.3, 0.75, 0.95$ based on fitting the seven quantile level model. Although the true quantile curves are not known here, we might expect the independently fitted frequentist estimates to provide a reasonable guide to the truth in this analysis. When the sample size is reasonably large (here $n = 298$), the frequentist approach can produce estimators with good properties (such as a reduced chance for neighbouring quantiles to overlap as n gets large). As a result, in the current example, the frequentist estimates should be expected to produce similar results to the Bayesian approaches, particularly in the non-extreme regions where there is more data. However, the Bayesian analyses naturally enforce non-crossing of quantiles, and so are preferable for this reason, in spite of the approximate posterior. Results from three different ABC variants are shown in Figure 8.7 (left panel). For most quantile levels there are small differences between the marginal univariate quantile estimates, although quantile non-crossing is enforced in each of the Bayesian estimates. For the lower $\tau = 0.1$ quantile, where data are more scarce, increasing the quality of the ABC posterior approximation from standard rejection ABC (dashed line) to regression adjusted ABC (dot-dash line) to regression and marginally adjusted ABC (dotted line), produces a marginal quantile that is increasingly close to the frequentist estimate and which roughly partitions 10% of the data below it. This suggests that there is some ABC approximation error (although this is less obvious in the upper $\tau = 0.95$ quantile), but that this is less apparent the better the ABC approximation becomes.

In the case of these marginal quantile estimates, Gaussian copula ABC produces quantile estimates (not shown) that are highly similar to the regression and marginally adjusted estimates (dotted line). However, the real differences here are in the quality of the dependence structure of the ABC posterior. Figure 8.7 (right panel) shows the correlation in the estimated posterior bivariate margins of (α_i, α_j), (β_i, β_j), and (η_i, η_j) for $i \neq j$ when using Gaussian copula ABC (x-axis) and standard ABC with regression and marginal adjustment using the full vector of summary statistics (y-axis). Here, it is evident that Gaussian copula ABC is able to capture correlations in the bivariate

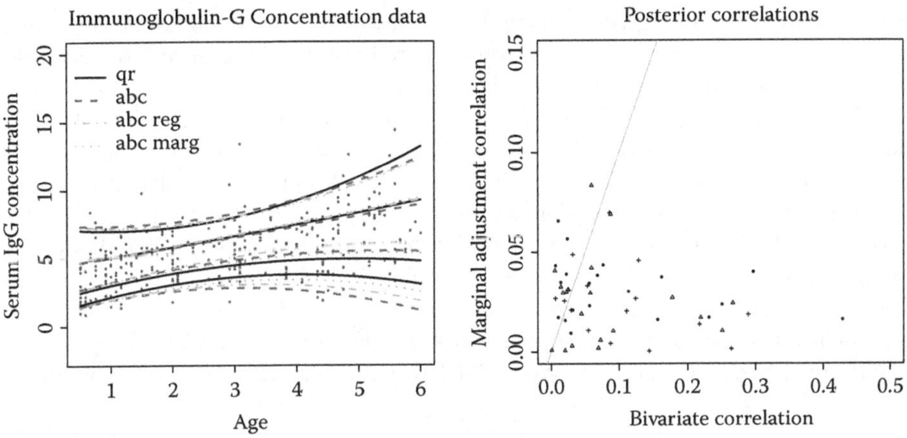

FIGURE 8.7
Left panel: Posterior mean predictive conditional quantile estimates using the full vector of summary statistics based on standard ABC (dashed line), regression adjusted ABC (dot dash line), and both regression and marginally adjusted ABC, for the quantiles $\tau = 0.1, 0.2, 0.3, 0.7, 0.75, 0.8, 0.95$. For clarity, only $\tau = 0.1, 0.3, 0.75, 0.95$ level quantiles are shown. Right panel: Estimated correlation of posterior margins $\pi(\alpha_i, \alpha_j|s_{obs})$ (dot), $\pi(\beta_i, \beta_j|s_{obs})$ (triangle), and $\pi(\eta_i, \eta_j|s_{obs})$ (plus) $i \neq j$, for regression and marginally adjusted ABC with the full vector of summary statistics (y-axis), against that for Gaussian copula ABC (x-axis).

margins that are missed by regular ABC, even when using the univariate marginal adjustment. The quality of the posterior approximation will be vital when considering analyses that critically depend on full, multiple, quantile inference. This lends support to the Gaussian copula approach as a viable ABC model approximation able to capture much of the bivariate dependence structure of $\pi(\theta|s_{obs})$.

8.3 ABC Approximation of the Sampling Distribution of Summary Statistics

An alternative to direct ABC approximation of the posterior distribution $\pi(\theta|s_{obs})$ is to instead approximate the sampling distribution of summary statistics $p(s|\theta)$ (Leuenberger and Wegmann, 2010; Fan et al., 2013), thereby approaching the intractable likelihood problem from the more usual ABC conditional density estimation perspective. The resulting estimated density is then an analytically tractable approximation of the likelihood function for a

Bayesian analysis using conventional Bayesian computational tools. Such approaches may be preferable in problems where inference is required for multiple datasets arising from the same model.

One way to achieve this is to first estimate the joint distribution of (s, θ) flexibly and to then condition on observing $s = s_{obs}$ in the joint model. This approach was considered by Bonassi et al. (2011) using multivariate normal mixture models for the density estimator on (s, θ). Synthetic likelihood (Wood, 2010) is another method that directly approximates the likelihood via an assumed density, such as $p(s|\theta) \approx N_q(\mu(\theta), \Sigma(\theta))$, where the mean $\mu(\theta)$ and covariance matrix $\Sigma(\theta)$ are unknown functions of the parameter θ. Various techniques are then needed to estimate θ. For further details on synthetic likelihoods, See, for example, Wood (2010), Chapters 12 and 20, this volume.

8.3.1 A flexible regression density estimator

We describe the flexible conditional density estimation approach of Fan et al. (2013). As with other ABC density estimators, it is constructed from a sample of N summary statistic and parameter pairs $(s^1, \theta^1), \ldots, (s^N, \theta^N)$ drawn from a distribution $p(s|\theta)h(\theta)$. Note that while the summary statistics are generated given θ from the sampling distribution for the intractable model of interest, the parameters are not necessarily generated from the prior. Instead, $h(\theta)$ is a distribution chosen to reflect the region over which the likelihood should be well approximated. Some rough knowledge of the high likelihood region of the parameter space, perhaps based on an initial pilot analysis, is useful for setting $h(\theta)$. The method of Fan et al. (2013) is based on relating the summary statistics s to θ by regression approximations, and so it is useful if the actual relationships between s and θ are as simple as possible. One convenient procedure to achieve this is the semi-automatic summary statistic approach of Fearnhead and Prangle (2012), which constructs one summary statistic per parameter, where each summary statistic is an estimate of the posterior mean value of the parameter, based on a pilot run. That is, s_k is the univariate summary statistic informative for θ_k, $k = 1, \ldots, p$, with $s^j = (s_1^j, \ldots, s_p^j)$.

The first step is to build marginal regression models for each component of s conditional on θ. The training data $(s_k^1, \theta^1), \ldots, (s_k^N, \theta^N)$ are used to build the marginal model for s_k, resulting in an estimated marginal density $\hat{f}_k(s_k|\theta)$ for s_k. Fan et al. (2013) use a fast variational method for fitting mixture of heteroscedastic regression models (Nott et al., 2012; Tran et al., 2012) for the conditional density estimation.

Then a conditional density estimate for the joint distribution of s given θ is constructed, using a method closely related to that considered in Giordani et al. (2013) for the unconditional case. The data (s^j, θ^j) are transformed to (U^j, θ^j), where $U_k^j = \Phi^{-1}(\hat{F}_k(s_k^j|\theta^j))$, where $\hat{F}_k(s_k|\theta)$ is the distribution function corresponding to the density $\hat{f}_k(s_k|\theta)$. If the marginal densities for each s_k are well estimated, the transformation to U^j makes each component of U^j approximately standard normal regardless of the value of θ. A mixture of normals model is then fitted to the data (U^j, θ^j), $j = 1, \ldots, N$. Write the fitted normal mixture as:

$$\sum_{k=1}^{K} w_k N(\mu_k, \Psi_k),$$

where $N(\mu, \Psi)$ denotes the multivariate normal distribution with mean μ and covariance matrix Ψ, (μ_k, Ψ_k), $k = 1, \ldots, K$ are means and covariances of K normal mixture components, and w_k, $k = 1, \ldots, K$ are mixing weights, $w_k \geq 0$, $\sum_{j=1}^{K} w_j = 1$. The mixture model for the joint distribution of (U, θ) then implies a normal mixture model for the conditional density of $U|\theta$:

$$\sum_{k=1}^{K} w_k^c N(\mu_k^c, \Psi_k^c),$$

where:

$$w_k^c = \frac{w_k \phi(\theta; \mu_k, \Psi_k)}{\sum_{j=1}^{K} w_j \phi(\theta; \mu_j, \Psi_j)},$$

are mixing weights with $\phi(\theta; \mu, \Psi)$ denoting the multivariate normal density function in θ with mean μ and covariance matrix Ψ, and μ_k^c and Ψ_k^c are the conditional mean and covariance of U given θ in the k-th multivariate normal component $N(\mu_k, \Psi_k)$ in the joint mixture model. Write $\hat{g}(U|\theta)$ for the resulting estimated conditional density of U given θ. Inverting the transformation of s to U then produces an estimate of the conditional density of s given θ:

$$\hat{L}(s|\theta) = \hat{g}(U|\theta) \prod_{j=1}^{K} \frac{\hat{f}_j(s_j|\theta)}{\phi(U_j; 0, 1)}. \tag{8.4}$$

An approximation of the observed data likelihood is then given by $\hat{L}(s_{obs}|\theta)$.

The purpose of the transformation from s to U is to simplify the mixture modelling of the joint distribution (U, θ) compared to what would be required to estimate the joint distribution of (s, θ). Note that in $\hat{L}(s|\theta)$, the marginal density of s_k is not exactly $\hat{f}(s_k|\theta)$ due to the fact that the estimated marginal distributions in $\hat{g}(U|\theta)$ are not exactly standard normal. Giordani et al. (2013) suggest replacing the $\phi(U^j; 0, 1)$ in (8.4) by its exact marginal distribution in $\hat{g}(U|\theta)$, but Fan et al. (2013) found that good approximations to $L(s_{obs}|\theta)$ were obtained without this step.

The previous conditional density estimation method seeks to estimate each univariate marginal conditional distribution $s_k|\theta$ arbitrarily well, while approximating the overall joint dependence structure by a mixture of normals model. This approach can work well in relatively high dimensions, in the order of tens to hundreds, provided that the dependence structure is relatively straightforward to capture. This also underlines the importance of techniques that can produce summary statistics with simple relationships to θ, such as the method developed by Fearnhead and Prangle (2012).

8.3.2 Analysis of stereological extremes

To illustrate the regression density estimation approach, we re-analyse a dataset originally analysed using ABC methods by Bortot et al. (2007), and which was previously considered in Chapter 1, this volume. The data comprise information about the intensity and size distribution of inclusions in a three-dimensional block of clean steel, with the recorded observations being the inclusion sizes (above a threshold of $\nu_0 = 5\ \mu$m), and their number, observed in a two-dimensional cross-section.

Bortot et al. (2007) considered models assuming spherical or ellipsoidal inclusion shapes. For the elliptical model, the inclusion size is the length of the major axis of the two-dimensional planar ellipse. In both models, the locations of the inclusions above 5 μm in size follow a Poisson process with intensity λ. Conditional on having an inclusion larger than ν_0, the distribution of the inclusion size is generalised Pareto, with scale parameter $\sigma > 0$ and shape parameter ξ. So in both models there are three parameters, $\theta = (\lambda, \sigma, \xi)^\top$. For the analysis, the priors are $\log \lambda \sim N(0, 100^2)$, $\sigma \sim \text{gamma}(0.01, 0.0001)$, and $\xi \sim N(0, 100^2)$. For the spherical inclusion model, it is possible to directly evaluate the likelihood, but for the ellipsoidal inclusion model, this is not possible and so ABC methods are an attractive option. Here, we focus on the ellipsoidal inclusion model. An analysis of standard rejection ABC with regression adjustment for the spherical model can be found in Erhardt and Sisson (2016).

The high-dimensionality aspect of this analysis comes from the number of summary statistics, rather than the number of parameters. The summary statistics used comprise the logarithm of the number of inclusions observed in the two-dimensional cross-section ($s_1 = 111$), and $s_{j+1} = \log(q_{(j+1)} - q_{(j)})$, $j = 1, \ldots, 111$, where the $q_{(j)}$, $j = 1, \ldots, 112$ are 112 equally spaced quantiles of the observed inclusion sizes. This gives $q = 112$ summary statistics in total and corresponds to conditional density estimation in 112+3=115 dimensions.

The conditional density estimation method requires the choice of $h(\theta)$. This is achieved via a pilot analysis by firstly sampling values (s^i, θ^i), $i = 1, \ldots, n$, where the θ^i are sampled from a uniform distribution over a range wide enough to include the support of the posterior and the s^i are sampled from $p(s|\theta^i)$. The sample mean $\hat{\mu}$ and covariance matrix $\hat{\Sigma}$ are then calculated for those θ values for which $\|s^i - s_{obs}\| \leq 20$. The distribution $h(\theta)$ is then specified as the truncated normal distribution:

$$h(\theta) \propto N(\hat{\mu}, \hat{\Sigma}) I((\theta - \hat{\mu})^\top \hat{\Sigma}^{-1} (\theta - \hat{\mu}) < 9).$$

The conditional density estimation method for estimating $p(s|\theta)$ is then implemented using $N = 5{,}000$ draws $(s^i, \theta^i) \sim p(s|\theta)h(\theta)$, $i = 1, \ldots, N$.

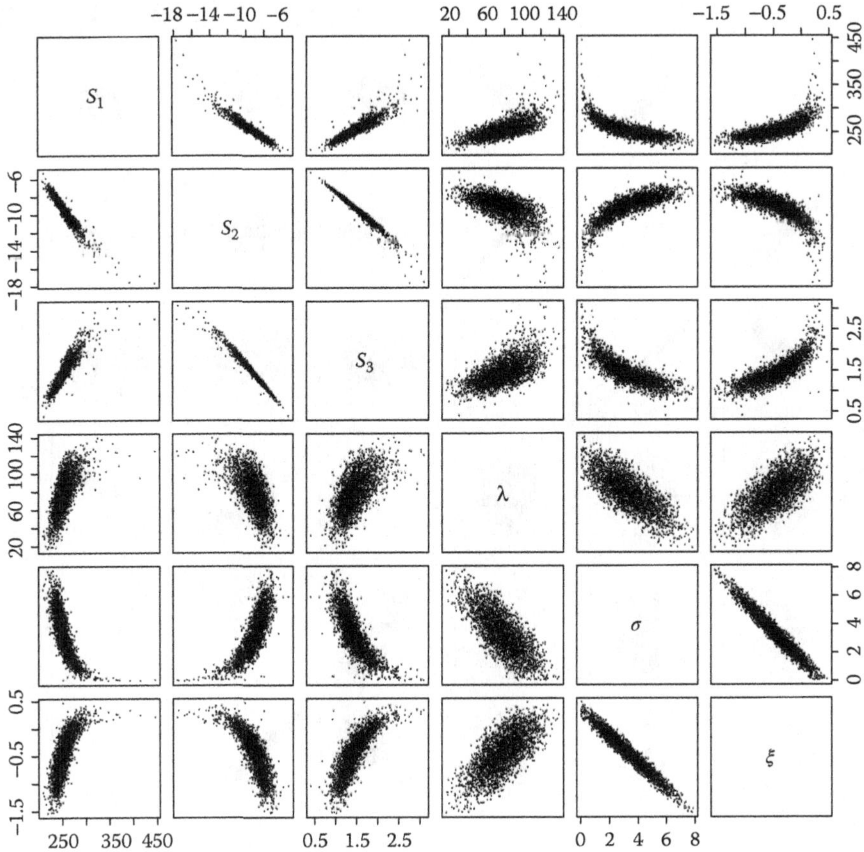

FIGURE 8.8
Pairwise scatterplots between the Fearnhead and Prangle (2012) semi-automatic summary statistics s_1, s_2, and s_3 and the parameters λ, σ, and ξ for the ellipsoidal inclusions model.

For comparison with the 115-dimensional regression density estimation approach, an additional analysis is performed in only six dimensions, using the three-dimensional summary statistics obtained using the semi-automatic method of Fearnhead and Prangle (2012). Figure 8.8 shows pairwise scatterplots of the components of s and θ for the samples generated from $h(\theta)p(s|\theta)$ [plotting the Fearnhead and Prangle (2012) statistics analysis for clarity]. The resulting scatterplots, after fitting the flexible models $f_k(s|\theta)$ to the univariate marginal distributions and transforming to the statistics U, are illustrated in Figure 8.9. Clearly the dependence structure has been greatly simplified, which facilitates the accurate mixture modelling of (U, θ).

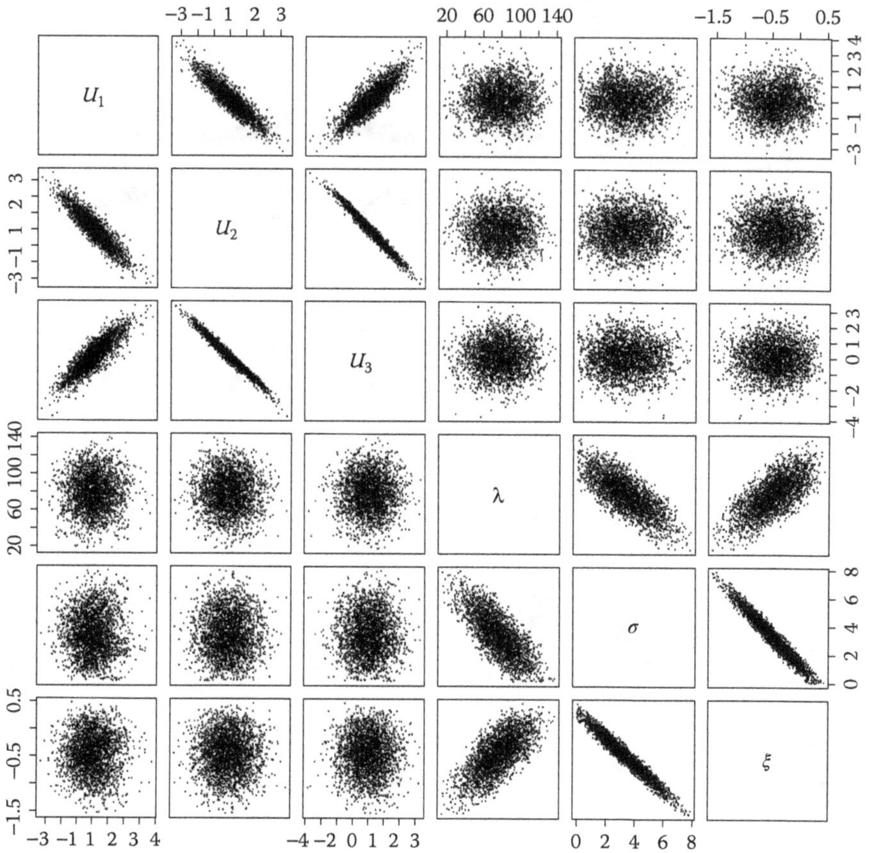

FIGURE 8.9
Pairwise scatterplots between the transformed Fearnhead and Prangle (2012) summary statistics U_1, U_2, and U_3 and the parameters λ, σ, and ξ for the ellipsoidal inclusions model.

The histograms in Figure 8.10 show the regression density estimated marginal posterior distributions obtained by using the original 112 summary statistics (top panels) and the lower dimensional Fearnhead and Prangle (2012) statistics (bottom panels). The solid line illustrates the density estimates obtained by the 'gold standard' ABC-Markov chain Monte Carlo (MCMC) analysis of Bortot et al. (2007) using large computational overheads. It is apparent that even when modelling the original high-dimensional set of summary statistics, reasonable answers are obtained using the regression density approach, although using the same method, but with the Fearnhead and Prangle (2012) summary statistics naturally results in an improved performance.

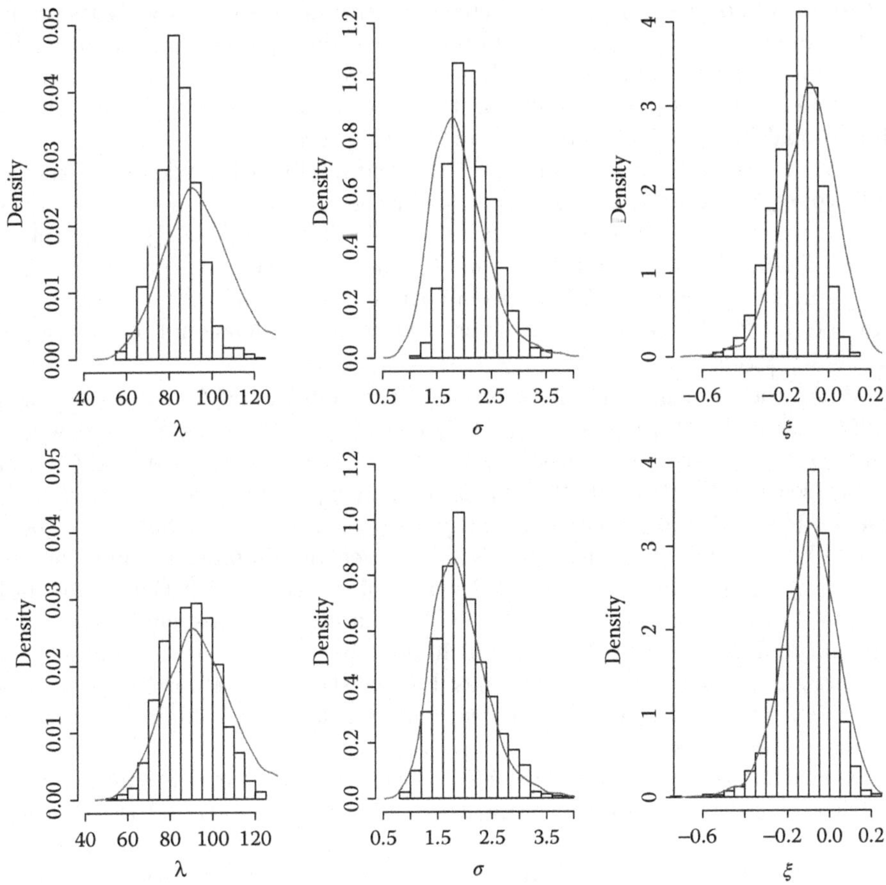

FIGURE 8.10
Histograms illustrating the estimated marginal posterior distributions obtained by regression density estimation for the ellipsoidal inclusions model using 112 summary statistics (top rows) and the three Fearnhead and Prangle (2012) summary statistics (bottom rows). The solid line shows the 'gold standard' marginal densities obtained using the method of Bortot et al. (2007), with a kernel scale parameter of $h = 0.33$.

8.4 Other Approaches to High-Dimensional ABC

Beyond the density estimation techniques described earlier, there are a few alternative approaches for extending ABC analyses to higher dimensions. ABC methods have been previously developed for functional parameters, specifically in the case of non-parametric hierarchical density estimation (Rodrigues et al., 2016). However, while these 'infinite-dimensional' parameters require the development of specialised ABC methods (such as a functional regression adjustment), the dimensionality of these techniques is strictly not high-dimensional in the sense considered in this chapter.

Various possibilities are available when the model of interest has a known and exploitable structure. The simplest of these is where the model factorises into a hierarchical structure $p(s|\theta, \phi) = f(\theta|\phi)\prod_i p_i(s^{(i)}|\theta^{(i)})$ (e.g. Bazin et al., 2010), where $s^{(i)}$ and $\theta^{(i)}$ denote mutually exclusive partitions of s and θ. In this case, the ABC approximation to the joint posterior $\pi(\theta|s_{obs})$ may be naturally constructed using the lower dimensional comparisons $\|s^{(i)} - s^{(i)}_{obs}\|$ only.

When the data model can be written in a conditional factorisation form $p(s|\theta) = p(s_1|\theta)\prod_{i=1}^{q} p(s_i|s_{1:(i-1)}, \theta)$, where $s_{1:k} = (s_1, \ldots, s_k)^\top$, and where conditional simulation from $p(s_i|s_{1:(i-1)}, \theta)$ is possible, Barthelmé and Chopin (2014) (see also White et al., 2015) proposed an expectation-propogation ABC scheme. If s_i is low dimensional, then $p(\theta|s_{i,obs}, s_{1:(i-1),obs})$ (that is, the posterior obtained by matching $\|s_i - s_{i,obs}\|$ based on simulating conditionally on $s_{1:(i-1),obs}$) can be well estimated via regular ABC, which Barthelmé and Chopin (2014) then approximate by a Gaussian density. This leads to a Gaussian approximation of $p(\theta|s_{obs})$, which may be accurate if the number of summary statistics is large. This may be realistic if the summary statistics $S(y) = y$ are the observed data. See Chapter 14, this volume, for further details of this approach.

Kousathanas et al. (2016) consider constructing an MCMC sampler with univariate updates to sample from each univariate conditional distribution $\pi(\theta_i|\theta_{-i}, s_{obs})$. Here, they note that if a low-dimensional summary statistic can be identified that is sufficient for the conditional distribution of $\theta_i|\theta_{-i}$, then an ABC-MCMC sampler can be implemented that compares summary statistics of a much lower dimension than the full vector s at each update step. They demonstrate this approach on a high-dimensional linear model with univariate summary statistics for each parameter update.

Finally, synthetic likelihood methods were discussed in Section 8.3 as a method to approximate the likelihood function using an assumed parametric form, for example, $p(s|\theta) \approx N_q(\mu(\theta), \Sigma(\theta))$ (Wood, 2010). As this technique relies on estimating $\mu(\theta)$ and $\Sigma(\theta)$ for each θ based on a potentially large number of Monte Carlo samples from $p(s|\theta)$, this approach can have high computational overheads. Variational Bayes has only recently been considered as a possible approach for fitting intractable models with synthetic likelihoods, but with greatly reduced computational costs. This then allows higher-dimensional analyses to be implemented. See, for example, Tran et al. (2017) and Ong et al. (2018) for further details on this technique.

8.5 Discussion

Given that a direct ABC approximation of the joint posterior distribution $\pi(\theta|s_{obs})$ involves a kernel density approximation of the likelihood, where the dimensionality involved is the dimension of the summary statistic s, it might initially seem that development of useful, general purpose methods for high-dimensional ABC may not be possible. However, if we are prepared to step away from the limiting comparison of $||s - s_{obs}||$ within the likelihood approximation of standard ABC methods, and build an approximations to $\pi(\theta|s)$ or $p(s|\theta)$ from approximations of lower-dimensional distributions, then it may be possible to develop useful ABC posterior approximations even in high-dimensional settings. The key idea in these approaches is that instead of matching a single vector of summary statistics in high dimensions, $||s - s_{obs}||$, we instead match many different low dimensional summary statistic vectors in constructing our joint posterior approximation. While the methods described here will not always work for posterior distributions with a highly complex dependence structure, or in very high dimensions, further development of related methods using the same 'divide and conquer' strategy may be a promising direction for future research in high-dimensional ABC. This may be particularly true for those methods that are easily parallelisable in their implementation.

Another area that perhaps has good potential for future research involves those techniques related to pseudo-marginal MCMC methods (see Chapter 9, this volume, Andrieu et al., 2019), which is currently seeing a surge of research interest beyond ABC. These methods have opened up ways to perform exact estimation and sampling for models with intractable likelihood functions, the ideas of which can be extended to implement various forms of approximation of posterior distributions. These include synthetic likelihoods (see Chapter 12, this volume) and variational Bayes methods, which can both be fast to implement, and for which the latter tends to underestimate uncertainty. The extension of likelihood-free inference methods to problems of higher dimension is a very active research area and promises to be so for the forseeable future.

Acknowledgements

SAS is supported by the Australian Research Council under the Discovery Project scheme (DP160102544) and the Australian Centre of Excellence in Mathematical and Statistical Frontiers (CE140100049).

References

Allingham, D. R., A. R. King, and K. L. Mengersen (2009). Bayesian estimation of quantile distributions. *Statistics and Computing 19*, 189–201.

Andrieu, C., A. Lee, and M. Vihola (2019). Theoretical and methodological aspects of MCMC computations with noisy likelihoods. In S. A. Sisson, Y. Fan, and M. A. Beaumont (Eds.), *Handbook of Approximate Bayesian Computation*. Chapman & Hall/CRC Press, Boca Raton, FL.

Barber, S., J. Voss, and M. Webster (2015). The rate of convergence for approximate Bayesian computation. *Electronic Journal of Statistics 9*, 80–105.

Barthelmé, S. and N. Chopin (2014). Expectation propagation for likelihood-free inference. *Journal of the American Statistical Association 109*, 315–333.

Barthelmé, S., N. Chopin, and V. Cottet (2019). Divide and conquer in ABC: Expectation-propagation algorithms for likelihood-free inference. In S. A. Sisson, Y. Fan, and M. A. Beaumont (Eds.), *Handbook of Approximate Bayesian Computation*. Chapman & Hall/CRC Press, Boca Raton, FL.

Bazin, E., K. Dawson, and M. A. Beaumont (2010). Likelihood-free inference of population structure and local adaptation in a Bayesian hierarchical model. *Genetics 185*, 587–602.

Beaumont, M. A., W. Zhang, and D. J. Balding (2002). Approximate Bayesian computation in population genetics. *Genetics 162*, 2025–2035.

Biau, G., F. Cérou, and A. Guyader (2015). New insights into approximate Bayesian computation. *Annales de l'Institut Henri Poincaré, Probabilités et Statistiques 51*(1), 376–403.

Blum, M. G. B. (2010). Approximate Bayesian computation: A nonparametric perspective. *Journal of the American Statistical Association 105*, 1178–1187.

Blum, M. G. B. and O. François (2010). Non-linear regression models for approximate Bayesian computation. *Statistics and Computing 20*, 63–75.

Blum, M. G. B., M. A. Nunes, D. Prangle, and S. A. Sisson (2013). A comparative review of dimension reduction methods in approximate Bayesian computation. *Statistical Science 28*, 189–208.

Bonassi, F. V., L. You, and M. West (2011). Bayesian learning from marginal data in bionetwork models. *Statistical Applications in Genetics and Molecular Biology 10*(1).

Bortot, P., S. G. Coles, and S. A. Sisson (2007). Inference for stereological extremes. *Journal of the American Statistical Association 102*, 84–92.

Drovandi, C. C., C. Grazian, K. Mengersen, and C. P. Robert (2019). Approximating the likelihood in approximate Bayesian computation. In S. A. Sisson, Y. Fan, and M. A. Beaumont (Eds.), *Handbook of Approximate Bayesian Computation.* Chapman & Hall/CRC Press, Boca Raton, FL.

Drovandi, C. C. and A. N. Pettitt (2011). Likelihood-free Bayesian estimation of multivariate quantile distributions. *Computational Statistics and Data Analysis 55*, 2541–2556.

Erhardt, R. and S. A. Sisson (2016). Modelling extremes using approximate Bayesian computation. In D. Dey and J. Yan (Eds.), *Extreme Value Modelling and Risk Analysis*, pp. 281–306. Boca Raton, FL: Chapman and Hall/CRC Press.

Fan, Y., D. J. Nott, and S. A. Sisson (2013). Approximate Bayesian computation via regression density estimation. *Stat 2*(1), 34–48.

Fang, H.-B., K.-T. Fang, and S. Kotz (2002). The meta-elliptical distributions with given marginals. *Journal of Multivariate Analysis 82*(1), 1–16.

Fasiolo, M. and S. N. Wood (2019). ABC in ecological modelling. In S. A. Sisson, Y. Fan, and M. A. Beaumont (Eds.), *Handbook of Approximate Bayesian Computation.* Chapman & Hall/CRC Press, Boca Raton, FL.

Fearnhead, P. and D. Prangle (2012). Constructing summary statistics for approximate Bayesian computation: Semi-automatic approximate Bayesian computation. *Journal of the Royal Statistical Society, Series B 74*, 419–474.

Feng, Y., Y. Chen, and X. He (2015). Bayesian quantile regression with approximate likelihood. *Bernoulli 21*(2), 832–580.

Giordani, P., X. Mun, M.-N. Tran, and R. Kohn (2013). Flexible multivariate density estimation with marginal adaptation. *Journal of Computational and Graphical Statistics 22*(4), 814–829.

Haario, H., E. Saksman, and J. Tamminen (1999). Adaptive proposal distribution for random walk Metropolis algorithm. *Computational Statistics 14*, 375–395.

Isaacs, D., D. G. Altman, C. E. Tidmarsh, H. B. Valman, and A. D. B. Webster (1983). Serum immunoglobulin concentration in preschool children measured by laser nephelometry: Reference ranges for IgG, IgA, IgM. *Journal of Clinical Pathology 36*, 1193–1196.

Koenker, R. (2005). *Quantile Regression*, Volume 38 of *Econometric Society Monographs.* Cambridge: Cambridge University Press.

Kousathanas, A., C. Leuenberger, J. Helfer, M. Quinodoz, M. Foll, and D. Wegmann (2016). Likelihood-free inference in high-dimensional models. *Genetics 203*, 893–904.

Leuenberger, C. and D. Wegmann (2010). Bayesian computation and model selection without likelihoods. *Genetics 184*, 243–52.

Li, J., D. J. Nott, Y. Fan, and S. A. Sisson (2017). Extending approximate Bayesian computation methods to high dimensions via Gaussian copula. *Computational Statistics and Data Analysis 106*, 77–89.

Li, W. and P. Fearnhead (2016). Improved convergence of regression adjusted approximate Bayesian computation. *arXiv: 1609.07135*.

Løland, A., R. B. Huseby, N. L. Hjort, and A. Frigessi (2013). Statistical corrections of invalid correlation matrices. *Scandinavian Journal of Statistics 40*(4), 807–824.

Nott, D. J., Y. Fan, L. Marshall, and S. A. Sisson (2014). Approximate Bayesian computation and Bayes linear analysis: Towards high-dimensional ABC. *Journal of Computational and Graphical Statistics 23*(1), 65–86.

Nott, D. J., S. L. Tan, M. Villani, and R. Kohn (2012). Regression density estimation with variational methods and stochastic approximation. *Journal of Computational and Graphical Statistics 21*, 797–820.

Ong, V. M.-H., D. J. Nott, M.-N. Tran, S. A. Sisson, and C. C. Drovandi (2018). Variational Bayes with synthetic likelihood. *Statistics and Computing*, *28*(4), 971–988.

Peters, G. W. and S. A. Sisson (2006). Bayesian inference, Monte Carlo sampling and operational risk. *Journal of Operational Risk 1*, 27–50.

Rayner, G. and H. MacGillivray (2002). Weighted quantile-based estimation for a class of transformation distributions. *Computational Statistics & Data Analysis 39*(4), 401–433.

Reich, B. J., H. D. Bondell, and H. J. Wang (2010). Flexible Bayesian quantile regression for independent and clustered data. *Biostatistics 11*(2), 337–352.

Reserve Bank of Australia (2014). Historical data. http://www.rba.gov.au/statistics/historical-data.html.

Rodrigues, G. S., D. J. Nott, and S. A. Sisson (2016). Functional regression approximate Bayesian computation for Gaussian precess density estimation. *Computational Statistics and Data Analysis 103*, 229–241.

Rodrigues, T. and Y. Fan (2017). Regression adjustment for non crossing Bayesian quantile regression. *Journal of Computational and Graphical Statistics 26*, 275–284.

Sisson, S.A., Y. Fan, and M. A. Beaumont (2019). Overview of approximate Bayesian computation. In S. A. Sisson, Y. Fan, and M. A. Beaumont (Eds.), *Handbook of Approximate Bayesian Computation.* Chapman & Hall/CRC Press, Boca Raton, FL.

Sklar, A. (1959). Fonctions de repartition a n dimensions et leur marges. *Publications de l'Institut de Statistique de l'Université de Paris 8*, 229–231.

Tokdar, S. T. and J. B. Kadane (2012). Simultaneous linear quantile regression: A semiparametric Bayesian approach. *Bayesian Analysis 7*(1), 51–72.

Tran, M.-N., D. J. Nott, and R. Kohn (2012). Simultaneous variable selection and component selection for regression density estimation with mixtures of heteroscedastic experts. *Electronic Journal of Statistics 6*, 1170–1199.

Tran, M.-N., D. J. Nott, and R. Kohn (2017). Variational Bayes with intractable likelihood. *Journal of Computational and Graphical Statistics*, *26*(4), 873–882.

White, S. R., T. Kypraios, and S. P. Preston (2015). Piecewise approximate Bayesian computation: Fast inference for discretely observed Markov models using a factorised posterior distribution. *Statistics and Computing 25*, 289–301.

Wood, S. N. (2010). Statistical inference for noisy nonlinear ecological dynamic systems. *Nature 466*, 1102–1104.

9

Theoretical and Methodological Aspects of Markov Chain Monte Carlo Computations with Noisy Likelihoods

Christophe Andrieu, Anthony Lee, and Matti Vihola

CONTENTS

9.1	The Noisy Likelihood Perspective	244
9.2	Pseudo-Marginal Algorithms	246
9.3	Performance Measures	249
	9.3.1 Approximation of the noiseless algorithm	250
	9.3.2 Rates of convergence	251
	9.3.3 Comparison of algorithms	254
9.4	Strategies to Improve Performance	255
	9.4.1 Averaging estimators	255
	9.4.2 Rejuvenation	257
	9.4.3 Playing with the Us	259
	9.4.3.1 Stratification	260
	9.4.3.2 Introducing dependence between estimators	260
	9.4.4 Locally adaptive ABC-Markov chain Monte Carlo	262
	9.4.5 Inexact algorithms	263
9.5	Remarks	265
References		265

Approximate Bayesian computation (ABC) [1,2] is a popular method for Bayesian inference involving an intractable, or expensive to evaluate, likelihood function, but where simulation from the model is easy. The method consists of defining an alternative likelihood function, which is also in general intractable, but naturally lends itself to pseudo-marginal computations [3], hence, making the approach of practical interest. The aim of this chapter is to show the connections of ABC Markov chain Monte Carlo with pseudo-marginal algorithms, review their existing theoretical results, and discuss how these can inform practice and hopefully lead to fruitful methodological developments.

9.1 The Noisy Likelihood Perspective

Consider some data $y_{\text{obs}} \in \mathcal{Y}$ assumed to arise from a family of probability distributions with densities $\{\ell(\cdot \mid \theta), \theta \in \Theta\}$, with respect to some appropriate reference measure $\lambda(\cdot)$, indexed by some unknown parameter $\theta \in \Theta \subset \mathbb{R}^d$ for some $d \in \mathbb{N}$. In a Bayesian context, θ is ascribed a prior distribution with density $\eta(\cdot)$ (with respect to some appropriate measure) and the posterior distribution has density:

$$\pi(\theta \mid y_{\text{obs}}) \propto \eta(\theta)\ell(y_{\text{obs}} \mid \theta).$$

The intractability of the likelihood function may prevent the implementation of traditional sampling algorithms. To circumvent this problem, it is natural to seek to approximate the desired likelihood function $\ell(y_{\text{obs}} \mid \theta)$, and ABC methods do so by taking advantage of the fact that sampling from the family of distributions $\{\ell(\cdot \mid \theta)\lambda(\cdot), \theta \in \Theta\}$ may be simple. Standard practice consists first of defining a function of the data $s : \mathcal{Y} \to \mathbb{R}^q$ for some $q \in \mathbb{N}_+$, and thereby the 'summary statistics' used to compare datasets. Then a distance $\|\cdot\|$ on $\mathbb{R}^q$ and a 'kernel' $K : \mathbb{R}_+ \to [0, 1]$ are chosen and combined to form $\psi(y^1, y^2) := K(\|s(y^1) - s(y^2)\|)$ for $y^1, y^2 \in \mathcal{Y}$, whose rôle is to measure the strength of the dissimilarity between datasets. One is naturally not constrained to this specific form of ψ, and the only requirement is that $\psi : \mathcal{Y}^2 \to [0, 1]$ and is statistically sensible. Note that there is no loss of generality in choosing the upper bound 1 for ψ since multiplicative constants do not affect Bayes' rule. A standard choice for ψ is:

$$\psi(y^1, y^2) := \mathbb{I}\{\|s(y^1) - s(y^2)\| \le \epsilon\}, \tag{9.1}$$

for some $\epsilon > 0$, although most of this chapter will deal with the general case, rather than this specific choice. Now, given $\psi : \mathcal{Y}^2 \to [0, 1]$, one can the define the 'ABC likelihood' function

$$\ell^{\psi}_{\text{ABC}}(y_{\text{obs}} \mid \theta) := \int \psi(y, y_{\text{obs}})\ell(y \mid \theta)\lambda(\mathrm{d}y). \tag{9.2}$$

One can think of ABC likelihoods arising from the standard choice of ψ as being kernel density estimators of the probability density of the observed summary statistics under the assumed model for the data. The associated posterior distribution has density:

$$\pi(\theta) := \pi_{\text{ABC}}(\theta \mid y_{\text{obs}}) \propto \eta(\theta)\ell^{\psi}_{\text{ABC}}(y_{\text{obs}} \mid \theta),$$

and we will use the simplified notation $\pi(\theta)$ in the remainder of the chapter. It seems at first sight that we have not made any progress since the new likelihood function is now an integral with respect to a distribution whose density is assumed to be intractable. This prevents direct implementation of the workhorse of Markov chain Monte Carlo (MCMC) methodology, the Metropolis–Hastings (MH) algorithm, described in Algorithm 9.1 for a family of proposal distributions $\{q(\theta,\cdot),\theta\in\Theta\}$. For notational simplicity, we adopt the convention that for a random variable $X\sim\varpi(\cdot)$, where $\varpi(\cdot)$ is a probability distribution, $x\sim\varpi(\cdot)$ means that x is a realisation of X, and do not use capital fonts for Greek letters representing random variables.

Algorithm 9.1: Exact ABC-MCMC Update

1. Given θ
2. Sample $\vartheta\sim q(\theta,\cdot)$
3. Return ϑ with probability

$$\min\left\{1,\frac{\eta(\vartheta)\ell^{\psi}_{\mathrm{ABC}}(y_{\mathrm{obs}}\mid\vartheta)q(\vartheta,\theta)}{\eta(\theta)\ell^{\psi}_{\mathrm{ABC}}(y_{\mathrm{obs}}\mid\theta)q(\theta,\vartheta)}\right\}$$

4. Otherwise return θ.

This is where the possibility to sample from $\ell(\cdot\mid\theta)\lambda(\cdot)$ comes into play, in combination with the standard 'auxiliary variable trick'. Define the probability distribution on $\Theta\times\mathcal{Y}$ with the following density

$$\pi(\theta,y)\propto\eta(\theta)\ell(y\mid\theta)\psi(y,y_{\mathrm{obs}}). \tag{9.3}$$

Evidently, this distribution has $\pi(\mathrm{d}\theta)$ as a marginal, and we aim to sample from this joint distribution. The rejection ABC algorithm proceeds, in its simplest form, by sampling $\theta\sim\eta$ and $y\sim\ell(\cdot\mid\theta)\lambda(\cdot)$, and accepting (θ,y) with probability $\psi(y,y_{\mathrm{obs}})$, therefore not requiring the evaluation of the likelihood function. This idea can also be used in the context of the MH algorithm by choosing a family of probability distributions $\{q(\theta,\cdot),\theta\in\Theta\}$ to update the parameter component and $\{\ell(\cdot\mid\theta),\theta\in\Theta\}$ to update the auxiliary dataset component. The resulting update is described in Algorithm 9.2 (we add an index to the auxiliary datasets to indicate the distribution they are sampled from) and first appeared in [4].

We note that in some latent variable models, the ABC posterior π is not approximate. In particular, if $y_{\mathrm{obs}}\sim\psi(y,\cdot)$ and $y\sim\ell(\cdot\mid\theta)\lambda(\cdot)$ for some $\theta\in\Theta$ with s the identity function, then $\ell^{\psi}_{\mathrm{ABC}}(y_{\mathrm{obs}}\mid\theta)$ is the exact, albeit intractable, likelihood function. This has been stressed in [5] and is taken advantage of in, for example, [6] and [7].

Algorithm 9.2: ABC-MCMC

1 Given (θ, y_θ)
2 Sample $\vartheta \sim q(\theta, \cdot)$ and $y_\vartheta \sim \ell(\cdot \mid \vartheta)\lambda(\cdot)$
3 Return (ϑ, y_ϑ) with probability

$$\min\left\{1, \frac{\eta(\vartheta) \times \psi(y_\vartheta, y_{\text{obs}})q(\vartheta, \theta)}{\eta(\theta) \times \psi(y_\theta, y_{\text{obs}})q(\theta, \vartheta)}\right\} \tag{9.4}$$

4 otherwise return (θ, y_θ).

9.2 Pseudo-Marginal Algorithms

We now develop another perspective on Algorithm 9.2, which turns out to be fruitful in many respects and on which the remainder of the chapter is based. As we shall see, this alternative point of view suggests many useful extensions, is both conceptually and notationally much simpler, and in fact covers scenarios of interest beyond ABC. Note, however, that despite the attractive generic nature of this perspective, one should in practice not forget about the initial problem at hand since it may possess additional specific structure one may exploit. The main starting point here is to notice that with $Y \sim \ell(\cdot \mid \theta)\lambda(\cdot)$, then $\psi(Y, y_{\text{obs}})$ is an unbiased estimator of (9.2), and that one can write the joint posterior density (9.3) as follows, in terms of the marginal $\pi(\theta)$:

$$\pi(\theta, y) \propto \eta(\theta)\ell^{\psi}_{\text{ABC}}(y_{\text{obs}} \mid \theta)\frac{\psi(y, y_{\text{obs}})}{\ell^{\psi}_{\text{ABC}}(y_{\text{obs}} \mid \theta)}\ell(y \mid \theta) \propto \pi(\theta)\frac{\psi(y, y_{\text{obs}})}{\ell^{\psi}_{\text{ABC}}(y_{\text{obs}} \mid \theta)}\ell(y \mid \theta),$$

where we have assumed θ such that $\ell^{\psi}_{\text{ABC}}(y_{\text{obs}} \mid \theta) \neq 0$. From this, we conclude that $\pi(\theta)\psi(Y, y_{\text{obs}})/\ell^{\psi}_{\text{ABC}}(y_{\text{obs}} \mid \theta)$ is an unbiased (and non-negative) estimator of $\pi(\theta)$ when $Y \sim \ell(\cdot \mid \theta)\lambda(\cdot)$. Clearly the acceptance probability (9.4) can be equally written as:

$$\min\left\{1, \left[\pi(\vartheta)\frac{\psi(y_\vartheta, y_{\text{obs}})}{\ell^{\psi}_{\text{ABC}}(y_{\text{obs}} \mid \vartheta)}q(\vartheta, \theta)\right] \Big/ \left[\pi(\theta)\frac{\psi(y_\theta, y_{\text{obs}})}{\ell^{\psi}_{\text{ABC}}(y_{\text{obs}} \mid \theta)}q(\theta, \vartheta)\right]\right\},$$

that is, Algorithm 9.2 can be thought of as an approximate implementation of the exact update Algorithm 9.1, where the expression for $\pi(\cdot)$ is replaced with an unbiased, 'noisy' estimator. However, Algorithm 9.2 targets the joint distribution with density $\pi(\theta, y)$ and is therefore exact in the sense that an ergodic Markov chain built on this type of update can produce samples of distribution arbitrarily close to $\pi(\mathrm{d}\theta)$. This remark leads in fact to a far more widely applicable idea [8,3] and the resulting methods are referred to as pseudo-marginal algorithms (see [9,10] for earlier, related, but different ideas). Indeed, assume that for any $\theta \in \Theta$, we can generate 'unbiased measurements' of $\pi(\theta)$ of the form $\pi(\theta) \times W$, where $W \sim Q_\theta$, $Q_\theta(W \geq 0) = 1$, and such that $\mathbb{E}_{Q_\theta}[W] = C$ for some $C > 0$ independent of θ, and consider the probability distribution on $\Theta \times \mathsf{W}$ with density:

$$\tilde{\pi}(\theta, w) = \pi(\theta) \times w \times Q_\theta(w). \tag{9.5}$$

For simplicity, we will hereafter assume that $C = 1$. Note that we do not assume here that $\pi(\theta)$ is tractable, but rather that $\pi(\theta) \times w$ is: W is purely conceptual and implicit in real scenarios. In the ABC scenario described above, W is the positive real valued random variable, such that for any $\theta \in \Theta$ and $A \in \mathcal{B}(\mathbb{R})$, the Borel σ-algebra on $\mathbb{R}$:

$$Q_\theta(W_\theta \in A) = \int \mathbb{I}\{\psi(y, y_{\text{obs}})/\ell^{\psi}_{\text{ABC}}(y_{\text{obs}} \mid \theta) \in A\}\ell(y \mid \theta)\lambda(\mathrm{d}y). \tag{9.6}$$

Now an MH update, with transition probability denoted $\tilde{P}$, targetting this distribution, and with proposal distribution $q(\theta, \vartheta) \times Q_\vartheta(u)$ has acceptance probability:

$$\begin{aligned}\tilde{\alpha}(\theta, w; \vartheta, u) &= \min\left\{1, \frac{\pi(\vartheta) \times u \times Q_\vartheta(u)}{\pi(\theta) \times w \times Q_\theta(w)} \frac{q(\vartheta, \theta) Q_\theta(w)}{q(\theta, \vartheta) Q_\vartheta(u)}\right\} \\ &= \min\left\{1, \frac{\pi(\vartheta) \times u \; q(\vartheta, \theta)}{\pi(\theta) \times w \; q(\theta, \vartheta)}\right\}.\end{aligned}$$

The Markov transition kernel, which we denote by $\tilde{P}$, is described algorithmically in Algorithm 9.3

Algorithm 9.3: Generic Pseudo-Marginal Algorithm

1 Given (θ, w)
2 Sample $\vartheta \sim q(\theta, \cdot)$ and $u \sim Q_\vartheta$
3 Return (ϑ, u) with probability

$$\min\left\{1, \frac{\pi(\vartheta) \times u \; q(\vartheta, \theta)}{\pi(\theta) \times w \; q(\theta, \vartheta)}\right\}$$

4 Otherwise return (θ, w)

Clearly a Markov chain Monte Carlo based on this update is exact in the sense outlined earlier and can be thought of as being an approximation of an exact MH update, which we denote by P, with acceptance probability $\min\{1, r(\theta, \vartheta)\}$, where

$$r(\theta, \vartheta) := \frac{\pi(\vartheta) \; q(\vartheta, \theta)}{\pi(\theta) \; q(\theta, \vartheta)},$$

which would use the exact values of the density $\pi(\theta)$.

What is the interest of these developments? First, $\tilde{\pi}(\theta, w)$ and Algorithm 9.3 are notationally and conceptually simple, equivalent representations of $\pi(\theta, y)$ and Algorithm 9.2, respectively, in that they lead to equivalent algorithms, provided we are only interested in the properties of the chain

$\{\theta_i, i \geq 0\}$. Indeed, let $\{(\check{\theta}_i, Y_i), i \geq 0\}$ and $\{(\theta_i, W_i), i \geq 0\}$ be the Markov chains, such that with μ (resp. ν_θ for any $\theta \in \Theta$) a probability distribution on Θ (resp. on $\mathcal{Y}$) and for any $\theta \in \Theta$ and $A \in \mathcal{B}(\mathbb{R})$:

$$\tilde{\nu}_\theta(\tilde{W}_\theta \in A) := \int \mathbb{I}\{\psi(y, y_{\text{obs}})/\ell_{\text{ABC}}(y_{\text{obs}} \mid \theta) \in A\}\nu_\theta(\text{d}y),$$

$(\check{\theta}_0, Y_0) \sim \mu \times \nu.$, $(\theta_0, W_0) \sim \mu \times \tilde{\nu}.$ and Markov transition probabilities as described in Algorithm 9.2 and Algorithm 9.3, respectively. With $\check{\mathbb{P}}(\cdot)$ and $\mathbb{P}(\cdot)$ the respective probabilities it can be checked easily that for any $m \in \mathbb{N}$, $i_1, i_2, \ldots, i_m \in \mathbb{N}$ and $A_1, A_2, \ldots, A_m \in \mathcal{B}(\Theta)^m$ we have:

$$\check{\mathbb{P}}\left(\check{\theta}_{i_1} \in A_1, \ldots, \check{\theta}_{i_m} \in A_m\right) = \mathbb{P}\left(\theta_{i_1} \in A_1, \ldots, \theta_{i_m} \in A_m\right),$$

which indeed implies that the processes $\{\check{\theta}_i, i \geq 0\}$ and $\{\theta_i, i \geq 0\}$ are probabilistically indistinguishable. In particular, at a conceptual level this tells us on the one hand that the properties of the algorithm (i.e. $\{\check{\theta}_i, i \geq 0\}$) are fully characterised by the properties of the random variables induced by (9.6), but also that the algorithm can be thought of as random perturbation of the exact algorithm with exact acceptance probability $\min\{1, r(\theta, \vartheta)\}$, suggesting links between the exact algorithm and its perturbations.

A second feature of this representation is that it emphasises the fact that the central property exploited in ABC is the possibility to produce unbiased and non-negative estimators of (9.2) cheaply, and such estimators are not restricted to the standard choice $\psi(Y, y_{\text{obs}})$ with $Y \sim \ell(\cdot \mid \theta)\lambda(\cdot)$. For example, one could average N multiple copies of the estimator as proposed in [11] and utilised in, for example, [12]. That is, for any $\theta \in \Theta$ and $Y^1, Y^2, \ldots, Y^N \overset{\text{iid}}{\sim} \ell(\cdot \mid \theta)\lambda(\text{d}\cdot)$ consider the unbiased estimator $T_\theta(N)$ of $\ell_{\text{ABC}}(y_{\text{obs}} \mid \theta)$:

$$T_\theta(N) := \frac{1}{N}\sum_{i=1}^{N} \psi(Y^i, y_{\text{obs}}). \tag{9.7}$$

This leads to the unit expectation, non-negative random variable $W_\theta(N) = T_\theta(N)/\ell^{\psi}_{\text{ABC}}(y_{\text{obs}} \mid \theta)$, and in the light of the earlier, there is no need to check the validity of a noisy MH algorithm that uses this estimator. One can check that the algorithm now targets:

$$\pi(\theta)Q_\theta^N(w^1, w^2, \ldots, w^N)\frac{1}{N}\sum_{i=1}^{N} w^i,$$

but that it is also possible to aggregate, that is, define $w(N) := N^{-1}\sum_{i=1}^{N} w^i$, such that the associated random variable has distribution $\mathcal{Q}_\theta^N\left(W_\theta(N) \in A\right) = \int \mathbb{I}\{w(N) \in A\}Q_\theta^N(\text{d}(w^1, w^2, \ldots, w^N))$. We will return to such aggregation strategies in Section 9.4.1 and discuss alternatives to a product form of the joint distribution of density $Q_\theta^N(w^1, w^2, \ldots, w^N)$, since it is in fact sufficient that its N marginals all be $Q_\theta(\cdot)$ for the above to hold.

9.3 Performance Measures

Before presenting various possible strategies to improve on standard ABC-MCMC algorithms in the next section, we recall here standard performance measures for MCMC algorithms and a summary of some known theoretical results relating (essentially) the properties of $\{W_\theta, \theta \in \Theta\}$ to the performance of pseudo-marginal algorithms. As we shall see, the variability and extreme behaviour of the estimators W_θ play a fundamental part in the (bad) behaviour of ABC-MCMC algorithms. The intuition goes as follows, upon recalling the expression for the acceptance probability of the noisy algorithm (Algorithm 9.3) in terms of the acceptance ratio of the exact algorithm:

$$\min\left\{1, r(\theta,\vartheta)\frac{u}{w}\right\},$$

and that W_θ is a non-negative random variable of expectation one. Realisations of W_θ can take values larger than one and make leaving the state (θ, w) more difficult than for the exact algorithm (since u has to match w), resulting in the familiar 'sticky behaviour' of the ABC-MCMC algorithm (see, e.g. [13]).

For μ, a probability distribution defined on some measurable space $(\mathsf{E}, \mathcal{E})$ and $\Pi : \mathsf{E}\times\mathcal{E} \to [0,1]$ a Markov transition kernel with invariant distribution μ, that is, $\mu\Pi = \mu$, we are interested in this section in two performance measures, that address asymptotic variance and bias:

1. Letting $\Phi_0 \sim \mu$ and $\Phi_n \sim \Pi(\Phi_{n-1}, \cdot)$ for $n \geq 1$ one may be interested, for a function $f : \mathsf{E} \to \mathbb{R}$ in the behaviour of ergodic averages $S_T(\Phi) := T^{-1}\sum_{k=0}^{T-1} f(\Phi_k)$, their asymptotic variance is a natural performance measure, and we will focus on the quantity:

$$\text{var}\,(f, \Pi) := \lim_{T\to\infty} T\text{var}_\mu\big(S_T(\Phi)\big),$$

2. Letting for $x \in \mathsf{E}$, $\mathcal{L}_x(\Phi_n)$ be the law of the n-th state Φ_n of the Markov chain $\{\Phi_i, i \geq 0\}$ where $\Phi_0 = x$, we may be interested in the rate of convergence to equilibrium of the Markov chain. That is, for an appropriate norm $\|\cdot\|_*$ characterise the distance between $\mathcal{L}_x(\Phi_n)$ and μ for all $n \geq 0$ and, for example, establish the existence of $M > 0$, $V : \mathsf{E} \to \mathbb{R}_+$ and $\rho \in [0,1)$ or $\alpha > 0$ or $\{r(n), n \geq 0\}$, such that either of the following inequalities hold for $n \geq 1$:

$$\|\mathcal{L}_x(\Phi_n) - \mu\|_* \leq \begin{cases} M\rho^n & \text{(uniformly ergodic)} \\ MV(x)\rho^n & \text{(geometrically ergodic)} \\ MV(x)n^{-\alpha} & \text{(polynomially ergodic)} \\ V(x)r^{-1}(n), \quad r(n)\to\infty & \text{(ergodic).} \end{cases}$$

The role of V is to take into account the influence of the initialisation on convergence and we leave the norms unspecified, although it may be useful to know that they correspond to the supremum of $|\mathbb{E}(f(\Phi_n)) - \mu(f)|$ for f in certain classes of functions. Such results will therefore provide us with a sense of the speed at which bias vanishes as n increases.

In the sequel, P is the noiseless algorithm, $\tilde{P}$ is the noisy algorithm using the family $\{Q_\theta, \theta \in \Theta\}$, and $\tilde{P}_N$ is the noisy algorithm which averages N samples marginally identically distributed according to $\{Q_\theta, \theta \in \Theta\}$ and joint distributions denoted $\{Q_\theta^N, \theta \in \Theta\}$. We will also consider more general families of noisy algorithms $\{\tilde{P}_\lambda, \lambda \in \mathbb{R}_+\}$ indexed by $\lambda > 0$, that is, using weight distributions $\{Q_\theta^{(\lambda)}, \theta \in \Theta, \lambda \in \Lambda \subset \mathbb{R}_+\}$, with the convention that $\tilde{P}_{\lambda=0} = P$ (i.e. the latter is not defined on the same space as $\tilde{P}_\lambda$, $\lambda \neq 0$).

We start with simple results which confirm that $\tilde{P}$ is a suboptimal approximation of P, but that concentration of W_θ on 1 allows $\tilde{P}$ to approach some performance measures of P arbitrarily closely.

9.3.1 Approximation of the noiseless algorithm

The pseudo-marginal algorithm, Algorithm 9.3, never has a smaller asymptotic variance than its corresponding marginal algorithm [14, Theorem 7], therefore justifying attempts to approximate P in order to improve performance of $\tilde{P}$. A natural class of functions to consider are those with finite second moment under π, that is, functions $\{f : \pi(f^2) < \infty\}$, where $\pi(f^2) := \int f(\theta)^2 \pi(\mathrm{d}\theta)$, or equivalently, the functions f, such that the random variable $f(X)$ has finite variance when $X \sim \pi$, that is, $\mathrm{var}_\pi(f) < \infty$.

Theorem 9.1 (Noiseless is best). *Assume $f : \Theta \to \mathbb{R}$ satisfies $\pi(f^2) < \infty$. The asymptotic variances of f with respect to the pseudo-marginal algorithm $\tilde{P}$ and the marginal algorithm P always satisfy:*

$$\mathrm{var}(f, P) \leq \mathrm{var}(f, \tilde{P}).$$

Under general technical conditions, the asymptotic variance of the pseudo-marginal algorithm converges to the asymptotic variance of the marginal algorithm [14, Theorem 21]. Denote by $\{\tilde{\theta}_k^{(\lambda)}, k \geq 0\}$ the Markov chain with initial distribution $\tilde{\pi}_\lambda$ and kernel $\tilde{P}_\lambda$.

Theorem 9.2. *Assume that $\int |f(\theta)|^{2+\delta}\pi(\theta)\mathrm{d}\theta < \infty$ for some $\delta > 0$, that $\mathrm{var}(f, P) < \infty$ and there exists a constant $\lambda_0 \in \Lambda$ such that:*

$$\lim_{n\to\infty} \sup_{0\leq\lambda\leq\lambda_0} \left| \sum_{k=n}^{\infty} \mathbb{E}[\bar{f}(\tilde{\theta}_0^{(\lambda)})\bar{f}(\tilde{\theta}_k^{(\lambda)})] \right| = 0 \quad \textit{where} \quad \bar{f} = f - \pi(f),$$

and that:

$$\lim_{\lambda\to 0} \int |1 - w| Q_\theta^{(\lambda)}(\mathrm{d}w) = 0 \qquad \textit{for all } \theta \in \Theta.$$

Then:

$$\lim_{\lambda\to 0} \mathrm{var}(f, \tilde{P}_\lambda) = \mathrm{var}(f, P).$$

The first condition simply says that the tails of the integrated autocovariances vanish uniformly for sufficiently good approximations of P, which should be the case, for example, if the approximation does not perturb the ergodicity properties of P significantly. This is further investigated in [14] and related to the results of Section 9.3.2. The second condition is very natural and formalises the idea of concentration of $\tilde{\pi}_\lambda\big(\mathrm{d}(\theta, w)\big)$ on $\pi(\mathrm{d}\theta)\delta_1(\mathrm{d}w)$. The following result formalises the fact that approximations involving unbounded noises are undesirable [3, Theorem 8].

Theorem 9.3. *If the weight distributions are such that:*

$$\pi\big(\{\theta \in \Theta : \textstyle\int_M^\infty Q_\theta(\mathrm{d}w) > 0 \textit{ for all } M < \infty\}\big) > 0,$$

then the pseudo-marginal algorithm cannot be geometrically ergodic.

Corollary 9.1. *Even when P is geometrically ergodic, as soon as:*

$$\left\{\theta \in \Theta : \int_M^\infty Q_\theta(\mathrm{d}w) > 0 \textit{ for all } M < \infty\right\},$$

has a positive π-probability, then for any $N \in \mathbb{N}_+$:

$$\left\{\theta \in \Theta : \int_M^\infty Q_\theta^N(\mathrm{d}w) > 0 \textit{ for all } M < \infty\right\},$$

has a positive π-mass, and $\tilde{P}_N$ cannot be geometrically ergodic for any $N \in \mathbb{N}_+$.

Broadly speaking, the result simply says that boundedness of the weights is required to ensure geometric ergodicity, and the corollary, that while averaging may ensure convergence of the integrated autocovariance, it will not always be the case that one can approach the rate of convergence of P: in fact the result says that one cannot even be geometric. In other words, despite the fact that 'bad events' (e.g. the N weights we have drawn are all large simultaneously) have a vanishing probability of occurrence as N increases, their impact on the long term properties of the algorithm may still be felt. Such bad behaviour will however vanish, for example, for a fixed simulation length T as N increases, provided naturally that the algorithm is not initialised at points corresponding to such bad events.

9.3.2 Rates of convergence

The first result of this section holds under a condition (9.8) stronger than uniform ergodicity for P, but often used in practice to establish this property whenever the space Θ is compact, and the assumption that the noise is uniformly bounded [the condition in [3, Theorem 8] is slightly more general]. In words the result says that uniform ergodicity of the noiseless algorithm is inherited by the noisy algorithm in this scenario.

Theorem 9.4. *Suppose there exists $\epsilon > 0$, a probability measure ν on the measurable space $(\Theta, \mathcal{T})$, such that for any $A \in \mathcal{T}$:*

$$\int_A q(\theta, \mathrm{d}\vartheta) \min\{1, r(\theta, \vartheta)\} \geq \epsilon\nu(A) \qquad \text{for all } \theta \in \Theta, \tag{9.8}$$

and $M < \infty$, such that for all $\theta \in \Theta$, $Q_\theta(W_\theta \leq M) = 1$, then $\tilde{P}$ is also uniformly ergodic.

It should be noted that even in this favourable scenario, $\tilde{P}_N$ may not achieve the rate of convergence of P for any $N \geq 1$ [3, Remark 1]. The interest of the next result is that it establishes in a simple, yet representative, scenario a direct link between the existence of general moments of W_θ and the rate of convergence to equilibrium one may expect from the algorithm.

Theorem 9.5. *Suppose there exists $\epsilon > 0$, a probability measure ν on the measurable space $(\Theta, \mathcal{T})$, such that for any $A \in \mathcal{T}$:*

$$\int_A q(\theta, \mathrm{d}\vartheta) \min\{1, r(\theta, \vartheta)\} \geq \epsilon\nu(A) \qquad \text{for all } \theta \in \Theta,$$

and a non-decreasing convex function $\phi : [0, \infty) \to [1, \infty)$ satisfying:

$$\liminf_{t\to\infty} \frac{\phi(t)}{t} = \infty \qquad \text{and} \qquad M_W := \sup_{\theta\in\Theta} \int \phi(w) Q_\theta(\mathrm{d}w) < \infty. \tag{9.9}$$

Then $\tilde{P}$ is sub-geometrically ergodic with a rate of convergence characterised by ϕ.

Corollary 9.2. *Consider, for example, the case $\phi(w) = w^\beta + 1$ for some $\beta > 1$. Then, there exists $C > 0$, such that for any function $f : \Theta \to \mathbb{R}$ and $n \in \mathbb{N}_+$:*

$$\left|\mathbb{E}(f(\theta_n)) - \pi(f)\right| \leq C\,|f|_\infty\, n^{-(\beta-1)},$$

where $|f|_\infty := \sup_{\theta\in\Theta} |f(\theta)|$. For readers familiar with the total variation distance, this implies that for any $\theta \in \Theta$:

$$\|\mathcal{L}_\theta(\theta_n) - \pi(\cdot)\|_{TV} \leq Cn^{-(\beta-1)}.$$

Other rates of convergence may be obtained for other functions ϕ [15].

These last results will typically hold only in situations where Θ is bounded: extensions to more general scenarios have been considered in [14], but require one to be more specific about the type of MH updates considered and involve

TABLE 9.1
Convergence Inheritance. The Constants are such that $\epsilon, \delta > 0$ and $c \in \mathbb{R}$, While $c(\cdot) : \Theta \to \mathbb{R}_+$ on the Last Line Should Satisfy Some Growth Conditions [14, Theorem 38]. IMH and RWM Stand for Independent MH and Random Walk Metropolis, Respectively

Marginal P	W_θ	Pseudo-marginal $\tilde{P}$
Uniform	$W_\theta \le c$ a.s.	Uniform
Geometric	$W_\theta \le c$ a.s.	Geometric if $\tilde{P}$ positive (conjecture in general [14, Section 3])
Any	W_θ unbounded	Not geometric
Uniform	$\mathbb{E}_{Q_\theta}\left[W_\theta^{1+\epsilon}\right] \le c$	Polynomial
Uniform	Uniform integrability (9.9)	Sub-geometric
IMH	–	IMH
Geometric RWM	$\mathbb{E}_{Q_\theta}\left[W_\theta^{1+\epsilon} + W_\theta^{-\delta}\right] \le c(\theta)$	Polynomial [14, Theorem 38]

substantial additional technicalities. A rough summary of known results is presented in Table 9.1, we refer the reader to [14] for precise statements.

One clear distinction in Table 9.1 is the inheritance of geometric ergodicity of P by (at least) positive $\tilde{P}$ when W_θ is almost surely uniformly bounded, and its failure to do so when W_θ is unbounded, for all θ in some set of positive π-probability. In the case where W_θ is almost surely bounded but not uniformly so, characterisation is not straightforward. Indeed, there are cases where $\tilde{P}$ does inherit geometric ergodicity and cases where it does not, see [14, Remark 14] and [16, Remark 2]. In [16] and its supplement, it is shown that failure to inherit geometric ergodicity in statistical applications is not uncommon when using 'local' proposals such as a random walk, see Theorem 9.9.

One attractive property of uniformly and geometrically ergodic $\tilde{P}$ is that $\mathrm{var}(f, \tilde{P}) < \infty$ for all f with $\mathrm{var}_\pi(f) < \infty$. Conversely, when $\tilde{P}$ is not geometrically ergodic, and the chain is not almost periodic in a particular technical sense, then there do exist f with $\mathrm{var}_\pi(f) < \infty$, such that $\mathrm{var}(f, \tilde{P})$ is not finite (see [17] for more details). It is, however, not straightforward to identify which functions have finite asymptotic variance in the sub-geometric regime. It has recently been shown that uniformly bounded second moments of W_θ and geometric ergodicity of P are sufficient to ensure that all functions f of θ only with $\mathrm{var}_\pi(f) < \infty$ have finite asymptotic variance [18].

Theorem 9.6. *Let P be geometrically ergodic and* $\sup_\theta \mathbb{E}_{Q_\theta}\left[W_\theta^2\right] < \infty$. *Then for any $f : \Theta \to \mathbb{R}$ with* $\mathrm{var}_\pi(f) < \infty$, $\mathrm{var}(f, \tilde{P}) < \infty$.

We note that this result holds even in the case where W_θ is unbounded, in which case $\tilde{P}$ is not geometrically ergodic: the functions f with $\mathrm{var}_\pi(f) < \infty$ that have infinite asymptotic variance in this case are not functions of θ alone and must depend on w.

9.3.3 Comparison of algorithms

The results of the previous section are mostly concerned with comparisons of the noisy algorithm with its noiseless version. We consider now comparing different variations of the noisy algorithm. Intuitively one would, at comparable cost, prefer to use an algorithm which uses the estimators with the lowest variability. It turns out that the relevant notion of variability is the convex order [19].

Definition 9.1. The random variables $W_1 \sim F_1$ and $W_2 \sim F_2$ are *convex ordered*, denoted $W_1 \leq_{cx} W_2$ or $F_1 \leq_{cx} F_2$ hereafter, if for any convex function $\phi : \mathbb{R} \to \mathbb{R}$,

$$\mathbb{E}[\phi(W_1)] \leq \mathbb{E}[\phi(W_2)],$$

whenever the expectations are well defined.

We note that the convex order $W^{(1)} \leq_{cx} W^{(2)}$ of square-integrable random variables automatically implies $\text{var}(W^{(1)}) \leq \text{var}(W^{(2)})$, but that the reverse is not true in general. As shown in [20, Example 13], while the convex order allows one to order performance of competing algorithms, the variance is not an appropriate measure of dispersion.

Let $\bar{P}_1$ and $\bar{P}_2$ be the corresponding competing pseudo-marginal implementations of the MH algorithm targeting $\pi(\cdot)$ marginally sharing the same family of proposal distributions $\{q(\theta, \cdot), \theta \in \Theta\}$, but using two families of weight distributions $\{Q_\theta^{(1)}, \theta \in \Theta\}$ and $\{Q_\theta^{(2)}, \theta \in \Theta\}$. Hereafter the property $Q_\theta^{(1)} \leq_{cx} Q_\theta^{(2)}$ for all $\theta \in \Theta$ is denoted $\{Q_\theta^{(1)}, \theta \in \Theta\} \leq_{cx} \{Q_\theta^{(2)}, \theta \in \Theta\}$. We introduce the notion of the right spectral gap, $\text{Gap}_R(\Pi)$, of a μ-reversible Markov chain evolving on E, noting that the MH update is reversible with respect to its invariant distribution. This can be intuitively understood as follows in the situation where E is finite, in which case the transition matrix Π can be diagonalised (in a certain sense) and its eigenvalues shown to be contained in $[-1, 1]$. In this scenario, $\text{Gap}_R(\Pi) = 1 - \lambda_2$, where λ_2 is the second largest eigenvalue of Π. These ideas can be generalised to more general spaces. The practical interest of the right spectral gap is that it is required to be positive for geometric ergodicity to hold and provides information about the geometric rate of convergence when, for example, all the eigenvalues (in the finite scenario) are non-negative.

Theorem 9.7. *Let $\pi(\cdot)$ be a probability distribution on some measurable space $(\Theta, \mathcal{T})$, and let $\bar{P}_1$ and $\bar{P}_2$ be two pseudo-marginal approximations of P aiming to sample from $\pi(\cdot)$, sharing a common family of marginal proposal probability distributions $\{q(\theta, \cdot), \theta \in \Theta\}$, but with distinct weight distributions satisfying $\{Q_\theta^{(1)}, \theta \in \Theta\} \leq_{cx} \{Q_\theta^{(2)}, \theta \in \Theta\}$. Then,*

1. *for any $\theta, \vartheta \in \Theta$, the conditional acceptance rates satisfy $\alpha_{\theta\vartheta}(\bar{P}_1) \geq \alpha_{\theta\vartheta}(\bar{P}_2)$,*

2. *for any* $f : \Theta \to \mathbb{R}$ *with* $\text{var}_\pi(f) < \infty$, *the asymptotic variances satisfy* $\text{var}(f, \bar{P}_1) \leq \text{var}(f, \bar{P}_2)$,

3. *the spectral gaps satisfy* $\text{Gap}_R(\bar{P}_1) \geq \min\{\text{Gap}_R(\bar{P}_2), 1 - \tilde{\rho}_2^*\}$, *where* $\tilde{\rho}_2^* := \tilde{\pi}_2\text{-ess sup}_{(\theta,w)} \tilde{\rho}_2(\theta, w)$, *the essential supremum of the rejection probability corresponding to* $\bar{P}_2$,

4. *if* $\pi(\theta)$ *is not concentrated on points, that is* $\pi(\{\theta\}) = 0$ *for all* $\theta \in \Theta$, *then* $\text{Gap}_R(\bar{P}_1) \geq \text{Gap}_R(\bar{P}_2)$.

Various applications of this result are presented in [20], including the characterisation of extremal bounds of performance measures for $\{Q_\theta, \theta \in \Theta\}$ belonging to classes of probability distributions (i.e. with given variance). As we shall see in the next section, this result is also useful to establish that averaging estimators always improves the performance of algorithms, that introducing dependence between such copies may be useful, or that stratification may be provably helpful in some situations.

9.4 Strategies to Improve Performance

9.4.1 Averaging estimators

Both intuition and theoretical results indicate that reducing variability of the estimates of $\pi(\cdot)$ in terms of the convex order ensures improved performance. We briefly discuss here some natural strategies. The first one consists of averaging estimators of the density, that is considered for any $\theta \in \Theta$ estimators of $\pi(\theta)$ of the form:

$$\frac{1}{N}\sum_{i=1}^{N} \pi(\theta) W^i = \pi(\theta)\frac{1}{N}\sum_{i=1}^{N} W^i,$$

for $(w^1, w^2, \ldots, w^N) \sim Q_\theta^N(\,\cdot\,)$ for a probability distribution $Q_\theta^N(\,\cdot\,)$ on $\mathbb{R}_+^N$ such that for $i = 1, \ldots, N$ [3]:

$$\int w^i Q_\theta^N(w^1, w^2, \ldots, w^N)\text{d}(w^1, w^2, \ldots, w^N) = 1,$$

and chosen in such a way that it reduces the variability of the estimator. A possibly useful application of this idea in an ABC framework could consist of using a stationary Markov chain with a Q_θ-invariant transition kernel to sample $W^1, W^2, \ldots$, such as a Gibbs sampler, which may not require evaluation of the probability density of the observations.

Increasing N is, broadly speaking, always a good idea, at least provided one can perform computations in parallel at no extra cost.

Proposition 9.1 (see e.g. [19, Corollary 1.5.24]). *For exchangeable random variables* $(W^1, W^2, \ldots)$, *for any* $N \geq 1$*:*

$$\frac{1}{N+1} \sum_{i=1}^{N+1} W^i \leq_{cx} \frac{1}{N} \sum_{i=1}^{N} W^i.$$

Letting $\tilde{P}_N$ denote the noisy transition kernel using N exchangeable random variables $(W^1, W^2, \ldots)$, this leads to the following result [20] by a straightforward application of Theorem 9.7.

Theorem 9.8. *Let* $(W^1, W^2, \ldots)$ *be exchangeable and* f *satisfy* $\pi(f^2) < \infty$. *Then for* $N \geq 2$, $N \mapsto \text{var}(f, \tilde{P}_N)$ *is non-increasing.*

It can also be shown, with an additional technical condition, that $N \mapsto \text{Gap}_R(\tilde{P}_N)$ is non-decreasing, suggesting improved convergence to equilibrium for positive algorithms.

A simple question arising from Theorem 9.8, is whether $\text{var}(f, \tilde{P}_N)$ approaches $\text{var}(f, P)$ as $N \to \infty$ for all f, such that $\pi(f^2) < \infty$ under weaker conditions than in Theorem 9.2. In the ABC setting, however, it can be shown under fairly weak assumptions when Θ is not compact that this is not the case [16, Theorem 2].

Theorem 9.9. *Assume that* ψ *satisfies (9.1),* $\pi(\theta) > 0$ *for all* $\theta \in \Theta$, $\ell_{\text{ABC}}(y_{\text{obs}} \mid \theta) \to 0$ *as* $|\theta| \to \infty$, *and:*

$$\lim_{r \to \infty} \sup_{\theta \in \Theta} q(\theta, B_r^{\complement}(\theta)) = 0,$$

where $B_r^{\complement}(\theta)$ *is the complement of the* $|\cdot|$ *ball of radius* r *around* θ. *Then* $\tilde{P}_N$ *cannot be geometrically ergodic for any* N, *and there exist functions* f *with* $\pi(f^2) < \infty$, *such that* $\text{var}(f, \tilde{P}_N)$ *is not finite.*

The assumptions do not preclude the possibility that P is geometrically ergodic and hence $\text{var}(f, P)$ being finite for all f with $\pi(f^2) < \infty$, and so this result represents a failure to inherit geometric ergodicity in such cases. This result can be generalised to other choices of ψ under additional assumptions, see the supplement to [16].

The inability of $\tilde{P}_N$ more generally to escape the fate of $\tilde{P}_1$ is perhaps not surprising: from Table 9.1, we can see that apart from the case where P is a geometric RWM, the conditions on W_θ are unaffected by simple averaging. Further results in this direction are provided by quantitative bounds on asymptotic variances established by [21, Proposition 4] when ψ satisfies (9.1) and by [22] for general pseudo-marginal algorithms; more detailed results than below can be found in the latter.

Theorem 9.10. *Assume that* $\tilde{P}_N$ *is positive. Then for* $f : \Theta \to \mathbb{R}$ *with* $\text{var}_\pi(f) < \infty$,

$$\text{var}(f, \tilde{P}_1) \leq (2N - 1)\text{var}(f, \tilde{P}_N).$$

This shows that the computational cost of simulating the Markov chain with transition kernel $\tilde{P}_N$ is proportional to N, then there is little to no gain in using $N \geq 1$. This is, however, not always the case: there may be some significant overhead associated with generating the first sample, or parallel implementation may make using some $N > 1$ beneficial [22]. It also implies that $\text{var}(f, \tilde{P}_N) < \infty$ implies $\text{var}(f, \tilde{P}_1) < \infty$ so the class of π-finite variance functions with finite $\text{var}(f, \tilde{P}_N)$ does not depend on N. The selection of parameters governing the concentration of W_θ around 1 in order to maximise computational efficiency has been considered more generally in [23] and [24], although the assumptions in these analyses are less specific to the ABC setting.

We now consider dependent random variables $W^1, \ldots, W^N$. It is natural to ask whether for a fixed $N \geq 1$, introducing dependence can be either beneficial or detrimental. We naturally expect that introducing some form of negative dependence between estimates could be helpful. For probability distributions $F_1, F_2, \ldots, F_N$ defined on some measurable space $(\mathsf{E}, \mathcal{E})$, the associated Fréchet class $\mathcal{F}(F_1, F_2, \ldots, F_N)$ is the set of probability distributions on E^N with $F_1, F_2, \ldots, F_N$ as marginals. There are various ways one can compare the dependence structure of elements of $\mathcal{F}(F_1, F_2, \ldots, F_N)$ and one of them is the supermodular order, denoted:

$$\left(W^1, W^2, \ldots, W^N\right) \leq_{sm} \left(\tilde{W}^1, \tilde{W}^2, \ldots, \tilde{W}^N\right). \tag{9.10}$$

hereafter, (see [19] for a definition). In the case $N = 2$, this can be shown to be equivalent to:

$$\mathbb{E}(f(W^1)g(W^2)) \leq \mathbb{E}(f(\tilde{W}^1)g(\tilde{W}^2)),$$

for any non-decreasing functions $f, g : \mathsf{E} \to \mathbb{R}$ for which the expectations exist. Interestingly, (9.10) implies the following convex order:

$$\sum_{i=1}^{N} W^i \leq_{cx} \sum_{i=1}^{N} \tilde{W}^i,$$

see, for example, the results in [25, Section 9.A]. An immediate application of Theorem 9.7 allows us then to order corresponding noisy algorithms in terms of the dependence structure of $\left(W^1, W^2, \ldots, W^N\right)$ and $\left(\tilde{W}^1, \tilde{W}^2, \ldots, \tilde{W}^N\right)$. We note, however, that it may be difficult in practical situations to check that the supermodular order holds between two sampling schemes.

9.4.2 Rejuvenation

We have seen that introducing multiple copies and averaging improves performance of the algorithm, at the expense of additional computation, which may offset the benefits if parallel architectures cannot be used. In this section, we show that the introduction of $N \geq 2$ copies also allows for the development of other algorithms, which may address the sticky behaviour of some ABC algorithms. We observe that averaging also induces a discrete mixture structure of the distribution targetted:

$$\begin{aligned}\pi(\theta)Q_\theta^N(w^1,w^2,\ldots,w^N)\frac{1}{N}\sum_{i=1}^N w_i &= \sum_{i=1}^N \frac{1}{N}\pi(\theta)Q_\theta^N(w^1,w^2,\ldots,w^N)w^i \\ &= \sum_{k=1}^N \tilde{\pi}(k,\theta,w^1,w^2,\ldots,w^N)\end{aligned}$$

with

$$\tilde{\pi}(k,\theta,w^1,w^2,\ldots,w^N) := \frac{1}{N}\pi(\theta)Q_\theta^N(w^1,w^2,\ldots,w^N)w^k.$$

This is one of the other (hidden) ideas of [26], which can also be implicitly found in [27,28], where such a distribution is identified as target distribution of the algorithm. This means that the mechanism described in Algorithm 9.3 is not the sole possibility in order to define MCMC updates targetting $\pi(\theta)$ marginally. In particular, notice the form of the following two conditional distributions:

$$\tilde{\pi}(k \mid \theta,w^1,w^2,\ldots,w^N) = \frac{w^k}{\sum_{i=1}^N w^i}, \tag{9.11}$$

and with $w^{-k} := \left(w^1,w^2,\ldots,w^k,w^{k+1},\ldots,w^N\right)$:

$$\tilde{\pi}(w^{-k} \mid k,\theta,w^k) = Q_\theta^N(w^{-k} \mid w^k), \tag{9.12}$$

which can be used as Gibbs type MCMC updates leaving $\tilde{\pi}$ invariant. In addition, we have the standard decomposition:

$$\begin{aligned}\tilde{\pi}(k,\theta,w^1,w^2,\ldots,w^N) &= \tilde{\pi}(k \mid \theta,w^1,w^2,\ldots,w^N) \times \tilde{\pi}(\theta,w^1,w^2,\ldots,w^N) \\ &= \tilde{\pi}(k \mid \theta,w^1,w^2,\ldots,w^N) \\ &\quad \times \pi(\theta)Q_\theta^N(w^1,w^2,\ldots,w^N)\frac{1}{N}\sum_{i=1}^N w^i,\end{aligned}$$

which tells us that at equilibrium, k can always be recovered by sampling from (9.11). Then (9.12) suggests that one can rejuvenate $N-1$ of the aggregated pseudo-marginal estimators $w^1,w^2,\ldots,w^N$, provided sampling from $Q_\theta(w_{-k} \mid w_k)$ is simple: this is always the case when independence is assumed. From above, the following MCMC update leaves $\tilde{\pi}(\theta,w^1,w^2,\ldots,w^N)$ invariant:

$$R(\theta,w;\mathrm{d}(\vartheta,u)) := \sum_{k=1}^N \tilde{\pi}(k \mid \theta,w)Q_\theta^N(u^{-k} \mid w^k)\delta_{\theta,w_k}(\mathrm{d}\vartheta \times \mathrm{d}u^k),$$

and is described algorithmically in Algorithm 9.4 [where $\mathcal{P}\left(w^1,w^2,\ldots,w^N\right)$] is the probability distribution of a discrete valued random variable, such that $\mathbb{P}\left(K=k\right) \propto w_k$).

Algorithm 9.4: iSIR Algorithm

Given $\theta, w^1, w^2, \dots, w^N$
Sample $k \sim \mathcal{P}(w^1, w^2, \dots, w^N)$
Sample $u^{-k} \sim Q_\theta^N(\cdot \mid w^k)$ and set $u^k = w^k$
Return $\theta, u^1, \dots, u^N$

Such algorithms are of general interest, and are analysed in [29]. This update can, however, also be intertwined with the standard ABC update. In the context of ABC, this gives Algorithm 9.5.

Algorithm 9.5: iSIR with ABC

Given $\theta, y^1, \dots, y^N$
Sample $k \sim \mathcal{P}(\psi(y^1, y_{\text{obs}}), \psi(y^2, y_{\text{obs}}), \dots, \psi(y^N, y_{\text{obs}}))$
Sample $\tilde{y}^i \sim \ell(\cdot \mid \theta)\lambda(\mathrm{d}\cdot)$, $i \in \{1, \dots, N\} \setminus \{k\}$ and set $\tilde{y}^k = y^k$
Return $\theta, \tilde{y}^1, \dots, \tilde{y}^N$

and intertwining Algorithm 9.5 with Algorithm 9.3 one obtains Algorithm 9.6.

In fact, a recent result [30, Theorem 17] ensures that the resulting algorithm has a better asymptotic variance, the key observation being that one can compare two different inhomogeneous Markov chains with alternating transition kernels $\tilde{P}$ and R. For the standard pseudo-marginal algorithm, R is the identity, whereas for ABC with rejuvenation, R is the iSIR.

9.4.3 Playing with the Us

In practice, simulation of the random variables Y on a computer often involves using d (pseudo-)random numbers uniformly distributed on the unit interval $[0,1]$, which are then mapped to form one Y^i. That is, there is a mapping from the unit cube $[0,1]^d$ to $\mathcal{Y}$, and with an inconsequential abuse of notation, if $U^i \sim \mathcal{U}([0,1]^d)$ then $Y(U^i) \sim \ell(y \mid \theta)\lambda(\mathrm{d}y)$ and

$$T_\theta(N) := \frac{1}{N} \sum_{i=1}^{N} \psi(Y(U^i), y_{\text{obs}}),$$

is an unbiased estimator of (9.2), equivalent in fact to (9.7). This representation is discussed in [31] as a general reparametrisation strategy to circumvent intractability, and we show here how this can be also exploited in order to improve the performance of ABC-MCMC. An extremely simple illustration of this is the situation where $d = 1$ and an inverse cdf method is used, that is $Y(U) = F^{-1}(U)$, where F is the cumulative distribution function (cdf) of Y. This is the case, for example, for the g-and-k model, whose inverse cdf is given by [6].

Algorithm 9.6: ABC-MCMC with Rejuvenation

1 Given $\theta, y^1, \dots, y^N$
2 Sample $k \sim \mathcal{P}(\psi(y^1, y_{\text{obs}}), \psi(y^2, y_{\text{obs}}), \dots, \psi(y^N, y_{\text{obs}}))$
3 Sample $\tilde{y}^i \sim \ell(\cdot \mid \theta)\lambda(\mathrm{d}\cdot)$, $i \in \{1, \dots, N\} \setminus \{k\}$ and set $\tilde{y}^k = y^k$
4 Sample $\vartheta \sim q(\theta, \cdot)$
5 Sample $\check{y}^i \sim \ell(\cdot \mid \vartheta)\lambda(\mathrm{d}\cdot)$, $i \in \{1, \dots, N\}$
6 Return $(\vartheta, \check{y}^1, \dots, \check{y}^N)$ with probability

$$\min\left\{1, \frac{\eta(\vartheta) \times \frac{1}{N}\sum_{i=1}^{N} \psi(\check{y}^i, y_{\text{obs}}) q(\vartheta, \theta)}{\eta(\theta) \times \frac{1}{N}\sum_{i=1}^{N} \psi(\tilde{y}^i, y_{\text{obs}}) q(\theta, \vartheta)}\right\},$$

otherwise return $(\theta, \tilde{y}^1, \dots, \tilde{y}^N)$.

9.4.3.1 Stratification

Stratification is a classical variance reduction strategy with applications to Monte Carlo methods. It proceeds as follows in the present context. Let $\mathcal{A} := \{A_1, \dots, A_N\}$ be a partition of the unit cube $[0, 1]^d$, such that $\mathbb{P}(U^1 \in A_i) = 1/N$ and such that it is possible to sample uniformly from each A_i. Perhaps the simplest example of this is when $\mathcal{A}$ corresponds to the dyadic sub-cubes of $[0, 1]^d$. Let $V^i \sim \mathcal{U}(A_i)$ for $i = 1, \dots, N$ be independent. We may now replace the estimator in (9.7) with:

$$T_\theta^{\text{strat}}(N) := \frac{1}{N}\sum_{i=1}^{N} \psi(Y(V^i), y_{\text{obs}}).$$

It is straightforward to check that this is a non-negative unbiased estimator of $\ell_{\text{ABC}}^{\psi}(y_{\text{obs}} \mid \theta)$, which means that $W_\theta^{\text{strat}}(N) := T_\theta^{\text{strat}}(N)/\ell_{\text{ABC}}^{\psi}(y_{\text{obs}} \mid \theta)$ has unit expectation as required. It has been shown in [20, Section 6.2] that when ψ satisfies (9.1), Theorem 9.7 applies directly in this scenario and that $\tilde{P}^{\text{strat}}$, the approximation of P corresponding to using $\{W_\theta^{\text{strat}}(N), \theta \in \Theta\}$ instead of $\{W_\theta(N), \theta \in \Theta\}$ always dominates $\tilde{P}$. These results extend to more general choices of $\psi(\cdot, \cdot)$, but require a much better understanding of this function.

9.4.3.2 Introducing dependence between estimators

Ideally for the acceptance probability of $\tilde{P}$,

$$\min\left\{1, r(\theta, \vartheta)\frac{w_\vartheta}{w_\theta}\right\},$$

to be reasonably large, we would like w_ϑ large when w_θ is large. That is, we would like large and nefarious realisations of w_θ to be compensated by larger values of w_ϑ. A natural idea is therefore to attempt to introduce 'positive dependence' between the estimators. [32] discuss the introduction of dependence in the discussion of [26], and a more sophisticated methodology that seeks to correlate w_θ and w_ϑ has recently been proposed in [33].

Viewing the introduction of positive dependence very generally, for $\theta, \vartheta \in \Theta$ let $Q_{\theta\vartheta}(dw \times du)$ have marginals $Q_\theta(dw)$ and $Q_\vartheta(dw)$. It is shown in [20] that if in addition:

$$Q_{\theta\vartheta}(A \times B) = Q_{\vartheta\theta}(B \times A), \quad \theta, \vartheta \in \Theta, \quad A, B \in \mathcal{B}(\mathbb{R}_+),$$

then the algorithm with the earlier acceptance ratio and proposal $Q_{\theta\vartheta}(A \times B)/Q_\theta(A)$ remains exact, that is, reversible with respect to (9.5). A natural question is how one may implement the abstract condition earlier. Let $\mathcal{U}(S)$ denote the uniform distribution over the set S. In the context of ABC applications where the observations are functions of $U \sim \mathcal{U}([0,1]^d)$, that is $Y = Y(U)$, it is possible to introduce dependence on the uniforms involved. Assume for now that $d = 1$ and let $C(\cdot,\cdot)$ be a copula, that is a probability distribution on $([0,1]^2, \mathcal{B}([0,1]^2))$ with the uniform distribution on $[0,1]$ as marginals [34]. Some copulas induce positive or negative dependence between the pair of uniforms involved. It is possible to define a partial order among copulas (and in fact more generally for distributions with fixed marginals), which ranks copulas in terms of the 'strength' of the dependence. The concordance order $(W_\theta^{(2)}, W_\vartheta^{(2)}) \leq_c (W_\theta^{(1)}, W_\vartheta^{(1)})$ holds if and only if for any non-decreasing functions for which the expectations exist:

$$\mathbb{E}(f(W_\theta^{(2)})g(W_\vartheta^{(2)})) \leq \mathbb{E}(f(W_\theta^{(1)})g(W_\vartheta^{(1)})).$$

Now let for any $\theta \in \Theta$, $W_\theta = w_\theta(U_1) = \psi \circ y_\theta(U_1)$ and $(U_1, U_2) \sim C(\cdot,\cdot)$. If $C(\cdot,\cdot)$ is symmetric, then for any $\theta, \vartheta \in \Theta$, with $A_\theta^{-1} := \{u \in [0,1] : w_\theta(u) \in A\}$

$$\begin{aligned} Q_{\theta\vartheta}(W_\theta \in A_\theta, W_\vartheta \in A_\vartheta) &= \mathbb{P}(U_1 \in A_\theta^{-1}, U_2 \in A_\vartheta^{-1}) \\ &= C(A_\theta^{-1}, A_\vartheta^{-1}) \\ &= C(A_\vartheta^{-1}, A_\theta^{-1}) \\ &= Q_{\vartheta\theta}(W_\vartheta \in A_\vartheta, W_\theta \in A_\theta), \end{aligned}$$

that is the required condition is satisfied for the algorithm to remain exact. Now, for example, if for any $\theta \in \Theta$ $w_\theta : [0,1] \to \mathbb{R}_+$ is monotone, one should choose a copula which induces positive dependence. Indeed, as stated in [20], $(W_\theta^{(2)}, W_\vartheta^{(2)}) \leq_c (W_\theta^{(1)}, W_\vartheta^{(1)})$ implies that the expected acceptance ratio is larger for the choice $(W_\theta^{(1)}, W_\vartheta^{(1)})$ than the $(W_\theta^{(2)}, W_\vartheta^{(2)})$. However, increasing the expected acceptance probability does not guarantee improved performance of the algorithm: for example, in the limiting case, the Fréchet–Hoeffding bounds will lead to reducible Markov chains in many scenarios. More specifically, consider the copula defined as a mixture of the independent copula and the 'copy' copula with weights $(\lambda, 1-\lambda)$:

$$C_\lambda(A, B) = \lambda\mathcal{U}(A \cap B) + (1-\lambda)\mathcal{U}(A)\mathcal{U}(B).$$

This copula is symmetric for all $\lambda \in [0,1]$ and is such that sampling from its conditionals is simple: copy with probability λ or draw afresh from a uniform. Note that $\lambda = 0$ corresponds to the standard pseudo marginal. It is not difficult to show that if $0 \leq \lambda \leq \lambda' \leq 1$, then $(W_\theta^{(\lambda)}, W_\vartheta^{(\lambda)}) \leq_c (W_\theta^{(\lambda')}, W_\vartheta^{(\lambda')})$, meaning that the conditional acceptance ratio is a non-decreasing function of λ in terms of the concordance order. However, the choice $\lambda = 1$, which corresponds to the upper Fréchet–Hoeffding bound will obviously lead to a reducible Markov chain. This is therefore a scenario where λ needs to be optimised and where adaptive MCMC may be used [35]. Such schemes can naturally be extended to the multivariate scenario, but require the tuning of more parameters and an understanding of the variations of $w_\theta(\cdot)$ along each of its coordinates.

9.4.4 Locally adaptive ABC-Markov chain Monte Carlo

Given the inability of $\tilde{P}_N$ to inherit geometric ergodicity from P in fairly simple ABC settings, one may wonder if an alternative Markov kernel that uses a locally adaptive number of pseudo-samples can. A variety of such kernels are presented in [36], but we restrict our interest here to the '1-hit' ABC-MCMC method proposed in [37] for the specific setting where ψ satisfies (9.1):

Algorithm 9.7: 1-hit ABC-MCMC

1 Given θ
2 Sample $\vartheta \sim q(\theta, \cdot)$
3 With probability

$$1 - \min\left\{1, \frac{\eta(\vartheta)q(\vartheta,\theta)}{\eta(\theta)q(\theta,\vartheta)}\right\},$$

stop and output θ.
4 Sample $Y_\theta \sim \ell(\cdot \mid \theta)\lambda(\mathrm{d}\cdot)$ and $Y_\vartheta \sim \ell(\cdot \mid \vartheta)\lambda(\mathrm{d}\cdot)$ independently until

$$\mathbb{I}\{\|s(Y_\theta) - s(y_{\mathrm{obs}})\| \leq \epsilon\} + \mathbb{I}\{\|s(Y_\vartheta) - s(y_{\mathrm{obs}})\| \leq \epsilon\} \geq 1,$$

and output ϑ if $\mathbb{I}\{\|s(Y_\vartheta) - s(y_{\mathrm{obs}})\| \leq \epsilon\} = 1$. Otherwise, output θ

This algorithm defines a Markov chain evolving on Θ that is not a Metropolis–Hastings Markov chain. Indeed, the algorithm defines a transition kernel $\check{P}$ in which the proposal is q as in Algorithm 9.1, but the acceptance probability is:

$$\min\left\{1, \frac{\eta(\vartheta)q(\vartheta,\theta)}{\eta(\theta)q(\theta,\vartheta)}\right\} \times \frac{\ell^\psi_{\mathrm{ABC}}(y_{\mathrm{obs}} \mid \vartheta)}{\ell^\psi_{\mathrm{ABC}}(y_{\mathrm{obs}} \mid \theta) + \ell^\psi_{\mathrm{ABC}}(y_{\mathrm{obs}} \mid \vartheta) - \ell^\psi_{\mathrm{ABC}}(y_{\mathrm{obs}} \mid \theta)\ell^\psi_{\mathrm{ABC}}(y_{\mathrm{obs}} \mid \vartheta)}.$$

From this, it can be verified directly that the Markov chain is reversible with respect to π.

An interesting feature of this algorithm is that a random number of pseudo-observations from the distributions associated with both θ and ϑ are sampled in what can intuitively be viewed as a race between the two parameter values. In fact, the number of paired samples required for the race to terminate, N, is a geometric random variable depending on both $\ell^{\psi}_{\text{ABC}}(y_{\text{obs}} \mid \theta)$ and $\ell^{\psi}_{\text{ABC}}(y_{\text{obs}} \mid \vartheta)$ in such a way that N is typically larger when these quantities are both small and smaller when either of these are large. This can be interpreted broadly as a local adaptation of the computational effort expended in simulating the Markov chain, in contrast to the fixed N strategy outlined in Section 9.4.1. In [16, Proposition 3], it is shown that the expected computational cost of simulating each iteration of the resulting Markov chain is bounded whenever η defines a proper prior. In [16, Theorem 4], it is shown that $\check{P}$ can inherit (under additional assumptions) geometric ergodicity from P even when Theorem 9.9 holds, that is, $\tilde{P}_N$ is not geometrically ergodic for any N. Consequently, $\check{P}$ can provide superior estimates of expectations, in comparison to $\tilde{P}_N$, for some functions f, such that $\pi(f^2) < \infty$.

9.4.5 Inexact algorithms

A natural question in the context of ABC-MCMC methods is how pseudo-marginal methods compare to inexact variants. Of course, there are a large number of inexact methods, and we consider here only the simple case arising from a small modification of the pseudo-marginal algorithm known as Monte Carlo within Metropolis (MCWM) [38]. In this algorithm, a Markov chain is defined by the transition kernel $\hat{P}_\lambda$, whose algorithmic description is given in Algorithm 9.8. Here, as in the pseudo-marginal approach, one has for any $\lambda > 0$:

$$\mathbb{E}_{Q_\theta^{(\lambda)}}[W_\theta] = 1, \quad \theta \in \Theta,$$

and one defines a Markov chain evolving on Θ as follows:

Algorithm 9.8: MCWM ABC

1. Given θ
2. Sample $\vartheta \sim q(\theta, \cdot)$
3. Sample $W_\theta \sim Q_\theta^{(\lambda)}$
4. Sample $W_\vartheta \sim Q_\vartheta^{(\lambda)}$
5. Return the realisation ϑ with probability

$$\min\left\{1, \frac{\pi(\vartheta) \times W_\vartheta q(\vartheta, \theta)}{\pi(\theta) \times W_\theta q(\theta, \vartheta)}\right\},$$

otherwise return θ.

This Markov chain has been studied in [3], in the case where P is uniformly ergodic, see also [39]. Extensions to the case where P is geometrically ergodic are considered in [40]. One such result is [40, Theorem 4.1]:

Theorem 9.11. *Assume P is geometrically ergodic, that*

$$\lim_{\lambda \to 0} \sup_{\theta \in \Theta} Q_\theta^{(\lambda)}(|W_\theta - 1| > \delta) = 0, \quad \forall \delta > 0,$$

and

$$\lim_{\lambda \to 0} \sup_{\theta \in \Theta} \mathbb{E}_{Q_\theta^{(\lambda)}} \left[W_\theta^{-1} \right] = 1.$$

Then $\hat{P}_\lambda$ is geometrically ergodic for all sufficiently small λ. In addition, under a very mild technical assumption on P:

$$\lim_{\lambda \to 0} \|\pi - \hat{\pi}_\lambda\|_{\mathrm{TV}} = 0,$$

where $\hat{\pi}_\lambda$ is the unique invariant distribution associated with $\hat{P}_\lambda$ for sufficiently small λ.

Results of this kind invite comparison with corresponding results for $\tilde{P}$, the pseudo-marginal Markov transition kernel. While Theorem 9.11 is reassuring, the earlier assumptions correspond to 'uniform in θ' assumptions on $Q_\theta^{(\lambda)}$, similar to assumptions for analysing $\tilde{P}$, that may not hold in practical applications. Nevertheless, there are differences, and $\check{P}$ can be geometrically ergodic when $\tilde{P}$ is not when the same distributions $\{Q_\theta, \theta \in \Theta\}$ are employed for both. On the other hand, examples in which $\tilde{P}$ is geometrically ergodic, but $\hat{P}$ is transient can also be constructed, so a degree of caution is warranted.

This result can be interpreted in the specific case where N is a number of i.i.d. pseudo-samples. When $W_\theta(N)$ is a simple average of N i.i.d. random variables W_θ, one can see that the assumptions of Theorem 9.11 are weaker than corresponding assumptions for $\tilde{P}_N$. In fact, the conditions are satisfied when the weights W_θ are uniformly integrable (rather than being uniformly bounded) and there exist constants $M > 0$, $\beta > 0$ and $\gamma \in (0, 1)$ such that:

$$\sup_{\theta \in \Theta} Q_\theta(W_\theta \leq w) \leq M w^\beta, \qquad w \in (0, \gamma).$$

The latter condition imposes, for example, the requirement that $Q_\theta(W_\theta = 0) = 0$ for all $\theta \in \Theta$, and indeed the MCWM algorithm is not defined without this assumption. When the weights W_θ additionally have a uniformly bounded $1 + k$ moment, then one can obtain bounds on the rate of convergence of $\hat{\pi}_N$ to π in total variation [40, Proposition 4.1].

9.5 Remarks

In this chapter, we have presented a relevant subset of theory and directions in methodological research pertaining to ABC-MCMC algorithms. Given the recent prevalence of applications involving ABC, this survey has not been exhaustive and has focused on specific aspects in which the theory is particularly clear. For example, we have not treated the question of which kernels are most appropriate in various applications, and indeed methodological innovations such as the incorporation of a tolerance parameter ϵ [such as that found in (9.1), but which may parameterise alternative kernels as well] as a state variable of the ABC Markov chain (see, e.g. [41,42]).

Theoretical and methodological advances continue to be made in this area. For example, the spectral gap of the Markov kernel $\tilde{P}_N$ discussed in Section 9.4.1 does not increase as a function of N in general, and it is of interest to consider whether alternative Markov kernels can overcome this issue. In [43], it is shown that in many cases the spectral gap of the ABC-MCMC kernel with rejuvenation in Section 9.4.2 does improve as N grows, and alternative methodology similar to that in Section 9.4.4 is both introduced and analysed. Finally, we note that when the model admits specific structure, alternatives to the simple ABC method presented here may be more computationally efficient. Indeed, in the case where the observations are temporally ordered, then it is natural and typically much more efficient to consider sequential Monte Carlo methods for generating W_θ (see, e.g. [44]).

References

[1] M. A. Beaumont, W. Zhang, and D. J. Balding. Approximate Bayesian computation in population genetics. *Genetics*, 162(4):2025–2035, 2002.

[2] S. Tavare, D. J. Balding, R. Griffiths, and P. Donnelly. Inferring coalescence times from DNA sequence data. *Genetics*, 145(2):505–518, 1997.

[3] C. Andrieu and G. O. Roberts. The pseudo-marginal approach for efficient Monte Carlo computations. *Ann. Statist.*, 37(2):697–725, 2009.

[4] P. Marjoram, J. Molitor, V. Plagnol, and S. Tavaré. Markov chain Monte Carlo without likelihoods. *Proc. Natl. Acad. Sci. USA*, 100(26): 15324–15328, 2003.

[5] R. D. Wilkinson. Approximate Bayesian computation gives exact results under the assumption of model error. *Stat. Appl. Genet. Mol. Biol.*, 12(2):129–141, 2013.

[6] P. Fearnhead and D. Prangle. Constructing summary statistics for approximate Bayesian computation: Semi-automatic approximate Bayesian computation. *J. R. Stat. Soc. Ser. B Stat. Methodol.*, 74(3):419–474, 2012.

[7] T. A. Dean, S. S. Singh, A. Jasra, and G. W. Peters. Parameter estimation for hidden Markov models with intractable likelihoods. *Scand. J. Statist.*, 41(4):970–987, 2014.

[8] M. A. Beaumont. Estimation of population growth or decline in genetically monitored populations. *Genetics*, 164:1139–1160, 2003.

[9] A. Kennedy and J. Kuti. Noise without noise: A new Monte Carlo method. *Physical Review Letters*, 54(23):2473, 1985.

[10] L. Lin and J. Sloan. A stochastic Monte Carlo algorithm. *Phys. Rev. D*, 61(hep-lat/9905033):074505, 1999.

[11] C. Becquet and M. Przeworski. A new approach to estimate parameters of speciation models with application to apes. *Genome research*, 17(10):1505–1519, 2007.

[12] P. Del Moral, A. Doucet, and A. Jasra. An adaptive sequential Monte Carlo method for approximate Bayesian computation. *Stat. Comput.*, 22(5):1009–1020, 2012.

[13] S. A. Sisson, Y. Fan, and M. M. Tanaka. Sequential Monte Carlo without likelihoods. *Proc. Natl. Acad. Sci. USA*, 104(6):1760–1765, 2007.

[14] C. Andrieu and M. Vihola. Convergence properties of pseudo-marginal Markov chain Monte Carlo algorithms. *Ann. Appl. Probab.*, 25(2): 1030–1077, 2015.

[15] R. Douc, G. Fort, E. Moulines, and P. Soulier. Practical drift conditions for subgeometric rates of convergence. *Ann. Appl. Probab.*, 14(3): 1353–1377, 2004.

[16] A. Lee and K. Łatuszyński. Variance bounding and geometric ergodicity of Markov chain Monte Carlo kernels for approximate Bayesian computation. *Biometrika*, 101(3):655–671, 2014.

[17] G. O. Roberts and J. S. Rosenthal. Variance bounding Markov chains. *Ann. Appl. Probab.*, 18(3):1201–1214, 2008.

[18] G. Deligiannidis and A. Lee. Which ergodic averages have finite asymptotic variance? *arXiv preprint arXiv:1606.08373*, 2016.

[19] A. Müller and D. Stoyan. *Comparison Methods for Stochastic Models and Risks.* Hoboken, NJ: Wiley, 2002.

[20] C. Andrieu and M. Vihola. Establishing some order amongst exact approximations of MCMCs. *Ann. Appl. Probab.*, 26(5):2661–2696, 2016.

[21] L. Bornn, N. S. Pillai, A. Smith, and D. Woodard. The use of a single pseudo-sample in approximate Bayesian computation. *Stat. Comput.*, 27:583–590, 2017.

[22] C. Sherlock, A. Thiery, and A. Lee. Pseudo-marginal Metropolis–Hastings using averages of unbiased estimators. *Biometrika*, 104(3):727–734, 2017.

[23] C. Sherlock, A. H. Thiery, G. O. Roberts, and J. S. Rosenthal. On the efficiency of pseudo-marginal random walk Metropolis algorithms. *Ann. Statist.*, 43(1):238–275, 2015.

[24] A. Doucet, M. K. Pitt, G. Deligiannidis, and R. Kohn. Efficient implementation of Markov chain Monte Carlo when using an unbiased likelihood estimator. *Biometrika*, 102(2):295–313, 2015.

[25] M. Shaked and J. G. Shanthikumar. *Stochastic Orders.* New York: Springer, 2007.

[26] C. Andrieu, A. Doucet, and R. Holenstein. Particle Markov chain Monte Carlo methods. *J. R. Stat. Soc. Ser. B Stat. Methodol.*, 72(3):269–342, 2010.

[27] H. Tjelmeland. Using all Metropolis–Hastings proposals to estimate mean values. Technical Report 4, Norwegian University of Science and Technology, Trondhem, Norway, 2004.

[28] H. M. Austad. Parallel multiple proposal MCMC algorithms. Master's thesis, Norwegian University of Science and Technology, 2007.

[29] C. Andrieu, A. Lee, and M. Vihola. Uniform ergodicity of the iterated conditional SMC and geometric ergodicity of particle Gibbs samplers. *Bernoulli*, 24(2):842–872, 2018.

[30] F. Maire, R. Douc, and J. Olsson. Comparison of asymptotic variances of inhomogeneous Markov chains with application to Markov chain Monte Carlo methods. *Ann. Statist.*, 42(4):1483–1510, 2014.

[31] C. Andrieu, A. Doucet, and A. Lee. Discussion of 'Constructing summary statistics for approximate Bayesian computation: semi-automatic approximate Bayesian computation' by Fearnhead and Prangle. *J. R. Stat. Soc. Ser. B Stat. Methodol.*, 74(3):451–452, 2012.

[32] A. Lee and C. Holmes. Discussion of 'particle Markov chain Monte Carlo' by Andrieu, Doucet & Holenstein. *J. R. Stat. Soc. Ser. B Stat. Methodol.*, 72(3):327–329, 2010.

[33] G. Deligiannidis, A. Doucet, and M. K. Pitt. The correlated pseudo-marginal method. *arXiv preprint arXiv:1511.04992*, 2015.

[34] R. B. Nelsen. *An Introduction to Copulas*, volume 139. New York: Springer, 1999.

[35] C. Andrieu and J. Thoms. A tutorial on adaptive MCMC. *Stat. Comput.*, 18(4):343–373, 2008.

[36] A. Lee. On the choice of MCMC kernels for approximate Bayesian computation with SMC samplers. In *Proceedings of the Winter Simulation Conference*, 2012.

[37] A. Lee, C. Andrieu, and A. Doucet. Discussion of 'Constructing summary statistics for approximate Bayesian computation: Semi-automatic approximate Bayesian computation' by Fearnhead and Prangle. *J. R. Stat. Soc. Ser. B Stat. Methodol.*, 74(3):449–450, 2012.

[38] P. D. O'Neill, D. J. Balding, N. G. Becker, M. Eerola, and D. Mollison. Analyses of infectious disease data from household outbreaks by Markov chain Monte Carlo methods. *J. R. Stat. Soc. Ser. C Applied Statistics*, 49(4):517–542, 2000.

[39] P. Alquier, N. Friel, R. Everitt, and A. Boland. Noisy Monte Carlo: Convergence of Markov chains with approximate transition kernels. *Stat. Comput.*, 16(1):29–47, 2016.

[40] F. J. Medina-Aguayo, A. Lee, and G. O. Roberts. Stability of noisy Metropolis–Hastings. *Stat. Comput.*, 26(6):1187–1211, 2016.

[41] P. Bortot, S. G. Coles, and S. A. Sisson. Inference for stereological extremes. *J. Am. Stat. Assoc.*, 102(477):84–92, 2007.

[42] O. Ratmann, C. Andrieu, C. Wiuf, and S. Richardson. Model criticism based on likelihood-free inference, with an application to protein network evolution. *Proc. Natl. Acad. Sci. USA*, 106(26):10576–10581, 2009.

[43] A. Lee, C. Andrieu, and A. Doucet. An active particle perspective of MCMC and its application to locally adaptive MCMC algorithms. In preparation.

[44] J. S. Martin, A. Jasra, S. S. Singh, N. Whiteley, P. Del Moral, and E. McCoy. Approximate Bayesian computation for smoothing. *Stoch. Anal. Appl.*, 32(3):397–420, 2014.

10

Asymptotics of ABC

Paul Fearnhead

CONTENTS

10.1 Introduction 269
10.2 Posterior Concentration 272
10.2.1 Posterior concentration of ABC 272
10.2.2 Rate of concentration 274
10.2.3 Effect of binding function 275
10.2.4 Model error 276
10.3 ABC Posterior and Posterior Mean 277
10.3.1 ABC posterior 277
10.3.2 ABC posterior mean 279
10.4 Monte Carlo Error 281
10.5 The Benefits of Regression Adjustment 283
10.6 Discussion 285
Acknowledgements 286
References 286

10.1 Introduction

This chapter aims to give an overview of recent work on the asymptotics of approximate Bayesian computation (ABC). By asymptotics here we mean how does the ABC posterior, or point estimates obtained by ABC, behave in the limit as we have more data? The chapter summarises results from three papers, Li and Fearnhead (2018a), Frazier et al. (2016) and Li and Fearnhead (2018b). The presentation in this chapter is deliberately informal, with the hope of conveying both the intuition behind the theoretical results from these papers and the practical consequences of this theory. As such, we will not present all the technical conditions for the results we give: The Interested reader should consult the relevant papers for these, and the results we state should be interpreted as holding under appropriate regularity conditions.

We will focus on ABC for a p-dimensional parameter, $\boldsymbol{\theta}$, from a prior $p(\boldsymbol{\theta})$ (we use the common convention of denoting vectors in bold, and we will assume these are column vectors). We assume we have data of size n that is summarised through a d-dimensional summary statistic. The asymptotic results we review consider the limit $n \to \infty$, but assume that the summary statistic is of fixed dimension. Furthermore, all results assume that the dimension of the summary statistic is at least as large as the dimension of the parameters, $d \geq p$ – this is implicit in the identifiability conditions that we will introduce later. Examples of such a setting are where the summaries are sample means of functions of individual data points, quantiles of the data, or, for time-series data, are empirical auto-correlations of the data. It also includes summaries based on fixed-dimensional auxillary models (Drovandi et al., 2015) or on composite likelihood score functions (Ruli et al., 2016).

To distinguish the summary statistic for the observed data from the summary statistic of data simulated within ABC, we will denote the former by $\boldsymbol{s}_{obs}$ and the latter by $\boldsymbol{s}$. Our model for the data will define a probability model for the summary. We assume that this in turn specifies a probability density function, or likelihood, for the summary, $f_n(\boldsymbol{s};\boldsymbol{\theta})$, which depends on the parameter. In some situations, we will want to refer to the random variable for the summary statistic, and this will be $\boldsymbol{S}_{n,\boldsymbol{\theta}}$. As is standard with ABC, we assume that we can simulate from the model, but cannot calculate $f_n(\boldsymbol{s};\boldsymbol{\theta})$.

The most basic ABC algorithm is a rejection sampler (Pritchard et al., 1999), which iterates the following three steps:

(RS1) Simulate a parameter from the prior: $\boldsymbol{\theta}_i \sim p(\boldsymbol{\theta})$.

(RS2) Simulate a summary statistic from the model given $\boldsymbol{\theta}_i$: $\boldsymbol{s}_i \sim f_n(\boldsymbol{s}|\boldsymbol{\theta}_i)$.

(RS3) Accept $\boldsymbol{\theta}_i$ if $||\boldsymbol{s}_{obs} - \boldsymbol{s}_i|| < \epsilon$.

Here, $||\boldsymbol{s}_{obs} - \boldsymbol{s}_i||$ is a suitably chosen distance between the observed and simulated summary statistics, and ϵ is a suitably chosen bandwidth. In the following, we will assume that $||\boldsymbol{x}||$ is either Euclidean distance, $||\boldsymbol{x}||^2 = \boldsymbol{x}^T\boldsymbol{x}$, or a Mahalanobis distance, $||\boldsymbol{x}||^2 = \boldsymbol{x}^T\Gamma\boldsymbol{x}$ for some chosen positive-definite $d \times d$ matrix Γ.

If we define a (uniform) kernel function, $K(\boldsymbol{x})$, to be 1 if $||\boldsymbol{x}|| < 1$ and 0 otherwise, then this rejection sampler is drawing from the following distribution:

$$\pi_{ABC}(\boldsymbol{\theta}) \propto p(\boldsymbol{\theta}) \int f_n(\boldsymbol{s}|\boldsymbol{\theta}) K\left(\frac{\boldsymbol{s}_{obs} - \boldsymbol{s}}{\epsilon}\right) \mathrm{d}\boldsymbol{s}.$$

We call this the ABC posterior. If we are interested in estimating a function of the parameter $h(\boldsymbol{\theta})$, we can use the ABC posterior mean:

$$h_{ABC} = \int h(\boldsymbol{\theta})\pi_{ABC}(\boldsymbol{\theta})\mathrm{d}\boldsymbol{\theta}.$$

In practice, we cannot calculate this posterior mean analytically, but would have to estimate it based on the sample mean of $h(\boldsymbol{\theta}_i)$ for parameter values $\boldsymbol{\theta}_i$ simulated using the previous rejection sampler.

In this chapter, we review results on the behaviour of the ABC posterior, the ABC posterior mean, and Monte Carlo estimates of this mean as $n \to \infty$. In particular, we consider whether the ABC posterior concentrates around the true parameter value in Section 10.2. We then consider the limiting form of the ABC posterior and the frequentist asymptotic distribution of the ABC posterior mean in Section 10.3. For the latter two results, we compare these asymptotic distributions with those of the true posterior given the summary – which is the best we can hope for once we have chosen our summary statistics.

The results in these two sections ignore any Monte Carlo error. The impact of Monte Carlo error on the asymptotic variance of our ABC posterior mean estimate is the focus of Section 10.4. This impact depends on the choice of algorithm we use to sample from the ABC posterior (whereas the choice of algorithm has no effect on the actual ABC posterior or posterior mean that are analysed in the earlier sections). The earlier rejection sampling algorithm is inefficient in the limit as $n \to \infty$, and thus we consider more efficient importance sampling and Markov chain Monte Carlo (MCMC) generalisations in this section.

We then review results that show how post-processing the output of ABC can lead to substantially stronger asymptotic results. The chapter finishes with a discussion that aims to draw out the key practical insights from the theory.

Before we review these results, it is worth mentioning that we can generalise the definition of the ABC posterior, and the associate posterior mean, given earlier. Namely, we can use a more general form of kernel than the uniform kernel. Most of the results we review apply if we replace the uniform kernel by a different kernel, $K(\boldsymbol{x})$, that is monotonically decreasing in $\|\boldsymbol{x}\|$. Furthermore, the specific form of the kernel has little affect on the asymptotic results – what matters most is how we choose the bandwidth and, in some cases, the choice of distance. The fact that most of the theoretical results do not depend on the choice of kernel means that, for concreteness, we will primarily assume a uniform kernel in our following presentation. The exceptions being in Section 10.3 where it is easier to get an intuition for the results if we use a Gaussian kernel. By focussing on these two choices, we do not mean to suggest that they are necessarily better than other choices, it is just that they simplify the exposition. We will return to the choice of kernel in the Discussion.

10.2 Posterior Concentration

The results we present in this section are from Frazier et al. (2016) (though see also Martin et al., 2016) and consider the question of whether the ABC posterior will place increasing probability mass around the true parameter value as $n \to \infty$. It is the most basic convergence result we would wish for, requires weaker conditions than results we give in Section 10.3, and is thus easier to apply to other ABC settings (see, e.g. Marin et al., 2014; Bernton et al., 2017).

We will denote the true parameter value by $\boldsymbol{\theta}_0$. If we define:

$$\Pr_{ABC}(\|\boldsymbol{\theta} - \boldsymbol{\theta}_0\| < \delta) = \int_{\boldsymbol{\theta}:\|\boldsymbol{\theta}-\boldsymbol{\theta}_0\|<\delta} \pi_{ABC}(\boldsymbol{\theta})\mathrm{d}\boldsymbol{\theta},$$

the ABC posterior probability that $\boldsymbol{\theta}$ is within some distance δ of the true parameter value, then for posterior concentration, we want that for any $\delta > 0$:

$$\Pr_{ABC}(\|\boldsymbol{\theta} - \boldsymbol{\theta}_0\| < \delta) \to 1,$$

as $n \to \infty$. That is, for any strictly positive choice of distance, δ, regardless of how small it is, as $n \to \infty$, we need the ABC, posterior to place all its probability on the event that $\boldsymbol{\theta}$ is within δ of the true parameter value.

To obtain posterior concentration for ABC, we will need to let the bandwidth depend on n, and henceforth we denote the bandwidth by ϵ_n.

10.2.1 Posterior concentration of ABC

The posterior concentration result of Frazier et al. (2016) is based upon assuming a law of large numbers for the summary statistics. Specifically, we need the existence of a binding function, $\boldsymbol{b}(\boldsymbol{\theta})$, such that for any $\boldsymbol{\theta}$:

$$\boldsymbol{S}_{n,\boldsymbol{\theta}} \to \boldsymbol{b}(\boldsymbol{\theta}),$$

in probability as $n \to \infty$. If this holds, and the binding function satisfies an identifiability condition: that $\boldsymbol{b}(\boldsymbol{\theta}) = \boldsymbol{b}(\boldsymbol{\theta}_0)$ implies $\boldsymbol{\theta} = \boldsymbol{\theta}_0$, then we have posterior concentration providing the bandwidth tends to zero, $\epsilon_n \to 0$.

To gain some insight into this result and the assumptions behind it, we present an example. To be able to visuallise what is happening, we will assume that the parameter and summary statistic are both one-dimensional. Figure 10.1 shows an example binding function, a value of θ_0 and s_{obs}, and output from the ABC rejection sampler.

As n increases, we can see the plotted points, that show proposed parameter and summary statistic values, converge towards the line that shows the

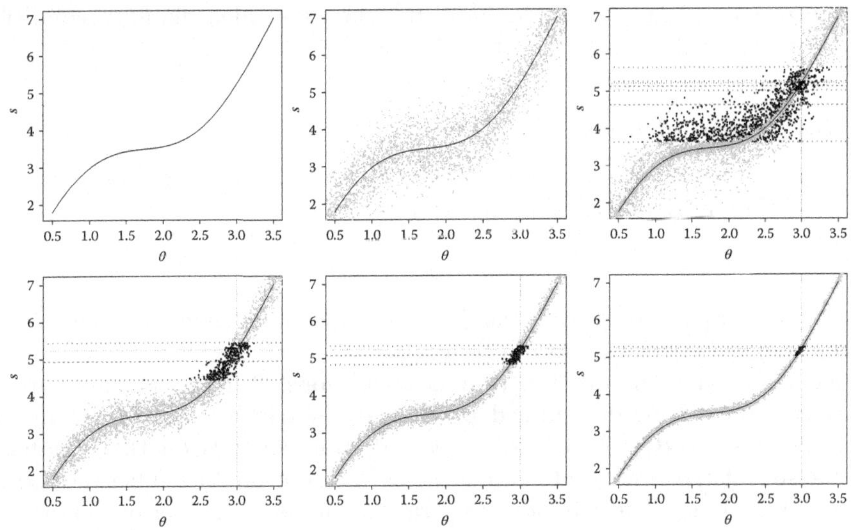

FIGURE 10.1
Example binding function, $b(\theta)$ (top-left plot). Pairs of parameter and summary statistic values proposed by a rejection sampler (top-middle). Output of rejection sampler (top-right): θ_0 and $b(\theta_0)$ (dotted vertical and horizontal lines, respectively); s_{obs} (bold circle and dashed horizontal line) and acceptance region for proposed summaries (bold dashed horizonal lines); and pairs of parameter and summary statistic values accepted (bold) and rejected (grey) by the rejection sampler. Bottom-row plots are the same as top-right plot, but for increasing n and decreasing ϵ_n. Here, and for all plots, our results are for a simple scenario where data are independent and identically distributed (iid) Gaussian with a mean that is a function of the parameter, and the summary statistic is the sample mean. (In this case the binding function is, by definition, equal to the mean function.)

binding function. This stems from our assumption of a law of large numbers for the summaries, so that for each θ value, the summaries should tend to $b(\boldsymbol{\theta})$ as n increases.

We also have that the observed summary statistic, s_{obs}, converges towards $b(\theta_0)$. Furthermore, we are decreasing the bandwidth as we increase n, which corresponds to narrower acceptance regions for the summaries, which means that the accepted summary statistics converge towards $b(\theta_0)$. Asymptotically, only parameter values close to θ_0, which have values $b(\theta)$ which are close to $b(\theta_0)$, will simulate summaries close to $b(\theta_0)$. Hence, the only accepted parameter values will be close to, and asymptotically will concentrate on, θ_0. This can be seen in practice from the plots in the bottom row of Figure 10.1.

The identifiability condition on the binding function is used to ensure that concentration of accepted summaries around $\boldsymbol{b}(\boldsymbol{\theta}_0)$ results in ABC posterior concentration around $\boldsymbol{\theta}_0$. What happens when this identifiability condition does not hold is discussed in Section 10.2.3.

10.2.2 Rate of concentration

We can obtain stronger results by looking at the rate at which concentration occurs. Informally, we can think of this as the supremum of rates, $\lambda_n \to 0$, such that:

$$\Pr_{ABC}(\|\boldsymbol{\theta} - \boldsymbol{\theta}_0\| < \lambda_n) \to 1,$$

as $n \to \infty$. For parametric Bayesian inference with independent and identically distributed data, this rate would be $1/\sqrt{n}$.

Assuming the binding function is continuous at $\boldsymbol{\theta}_0$, then the rate of concentration will be determined by the rate at which accepted summaries concentrate on $\boldsymbol{b}(\boldsymbol{\theta}_0)$. As described earlier, this depends on the variability (or 'noise') of the simulated summaries around the binding function and on the bandwidth, ϵ_n. The rate of concentration will be the slower of the rates at which the noise in the summary statistics and the rate at which ϵ_n tend to 0.

We can see this from the example in Figure 10.2, where we show output from the ABC rejection sampler for different values of n, but with ϵ_n tending

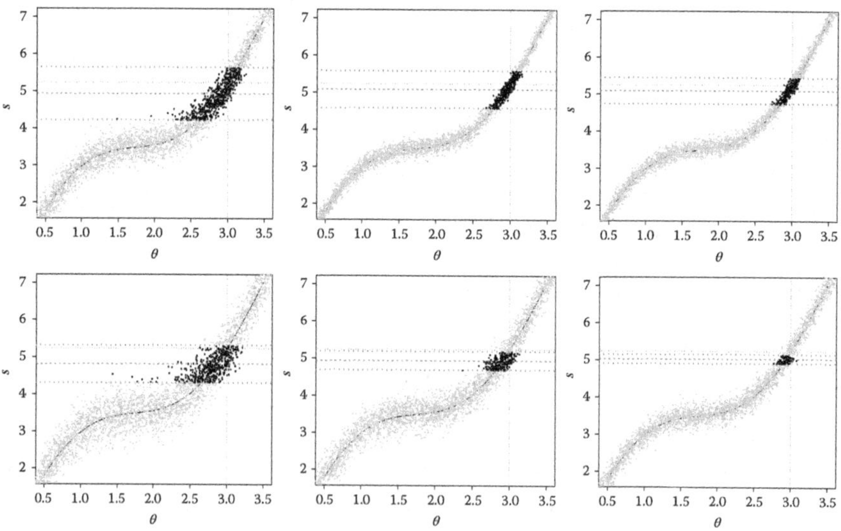

FIGURE 10.2
Example of ABC concentration for differing rates of the noise in the summary statistics and rates of ϵ_n. Plots are as in Figure 10.1. Top-row: noise in summary statistics halving, or equivalently sample size increasing by a factor of 4, and ϵ_n decreasing by $1/\sqrt{2}$ as we move from left to right. Bottom-row: noise in summary statistics decreasing by $1/\sqrt{2}$, or equivalently sample size doubling, and ϵ_n halving as we move from left to right.

to 0 at either a faster or slower rate than that of the noise in the summaries. For each regime, the rate of concentration of both the accepted summaries and of the accepted parameter values is determined by the slower of the two rates.

10.2.3 Effect of binding function

The shape of the binding function for values of $\boldsymbol{\theta}$ for which $\boldsymbol{b}(\boldsymbol{\theta})$ is close to $\boldsymbol{b}(\boldsymbol{\theta}_0)$ affects the ABC posterior, as it affects the range of $\boldsymbol{\theta}$ values that will have a reasonable chance of producing summary statistic values that would be accepted by the ABC rejection sampler.

If the identifiability condition holds and the binding function is differentiable at $\boldsymbol{\theta}_0$, then the value of this gradient will directly impact the ABC posterior variance. This is shown in the top row of Figure 10.3. If this gradient is large (top-left plot) then even quite large differences in summary statistics would correspond to small differences in the parameter, and, hence a small ABC posterior variance. By comparison, if the gradient is small (top-right plot), then large differences in parameters may mean only small differences in summary statistics. In this case, we expect a much larger ABC posterior variance for the same width of the region in which the summary statistics are accepted.

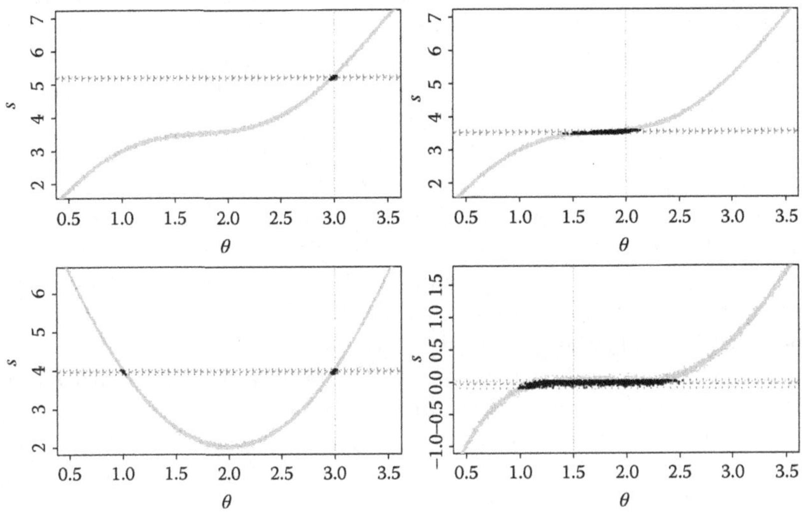

FIGURE 10.3
Example of the effect of the shape of binding function on the ABC posterior (plots are as in Figure 10.1). Top row: gradient of binding function at $b(\theta_0)$ affects the ABC posterior variance, with larger gradient (left-hand plot) resulting in lower ABC posterior variance than smaller gradient (right-hand plot). Bottom row: effect of non-identifiability on the ABC posterior.

The bottom row of Figure 10.3 shows what can happen if the identifiability condition does not hold. The bottom-left plot gives an example where there are two distinct parameter values for which the binding function is equal to $b(\theta_0)$. In this case, we have a bi-modal ABC posterior that concentrates on these two values. The bottom-right plot shows an example where there is a range of parameter values whose binding function value is equal to $b(\theta_0)$, and in this case, the ABC posterior will concentrate on this range of parameter values.

It can be difficult in practice to know whether the identifiability condition holds. In large data settings, observing a multi-modal posterior as in the bottom-left plot of Figure 10.3 would suggest that it does not hold. In such cases, it may be possible to obtain identifiability by adding extra summaries. The wish to ensure identifiability is one reason for choosing a higher-dimensional summary than parameter. However, this does not come without potential cost, as we show in Section 10.3.

10.2.4 Model error

One of the implicit assumptions behind the result on posterior concentration is that our model is correct. This manifests itself within the assumption that as we get more data, the observed summary statistic will converge to the value $\boldsymbol{b}(\boldsymbol{\theta}_0)$. If the model we assume in ABC is incorrect, then this may not be the case (see Frazier et al., 2017, for a fuller discussion of the impact of model error). There are then two possibilities, the first is that the observed summary statistic will converge to a value $\boldsymbol{b}(\tilde{\boldsymbol{\theta}})$ for some parameter value $\tilde{\boldsymbol{\theta}} \neq \boldsymbol{\theta}_0$. In this case, by the earlier arguments, we can still expect posterior concentration but to $\tilde{\boldsymbol{\theta}}$ and not $\boldsymbol{\theta}_0$.

The other possibility is that the observed summary statistic converges to a value that is not equal to $\boldsymbol{b}(\boldsymbol{\theta})$ for any $\boldsymbol{\theta}$. This is most likely to occur when the dimension of the summary statistic is greater than the dimension of the parameter. To give some insight into this scenario, we give an example in Figure 10.4, where we have independent identically distributed data from a Gaussian distribution with mean θ and variance θ^2+2, but our model assumes the mean and variance are θ and θ^2+1, respectively. This corresponds to a wrong assumption about the variance. We then apply ABC with summary statistics that are the sample mean and variance.

As shown in the figure, we still can get posterior concentration in this setting. If we denote the limiting value of the binding function for the true model as b_0, then the posterior concentrates on parameter value, or values, whose binding function value is closest, according to the distance we use for deciding whether to accept simulated summaries, to b_0.

In this second scenario, it may be possible to detect the model error by monitoring the closeness of the accepted summaries to the observed summaries. If the model is correct, then the distance between accepted and observed summaries tends to 0 with increasing n. Whereas in this model error scenario, these distances will tend towards some non-zero constant.

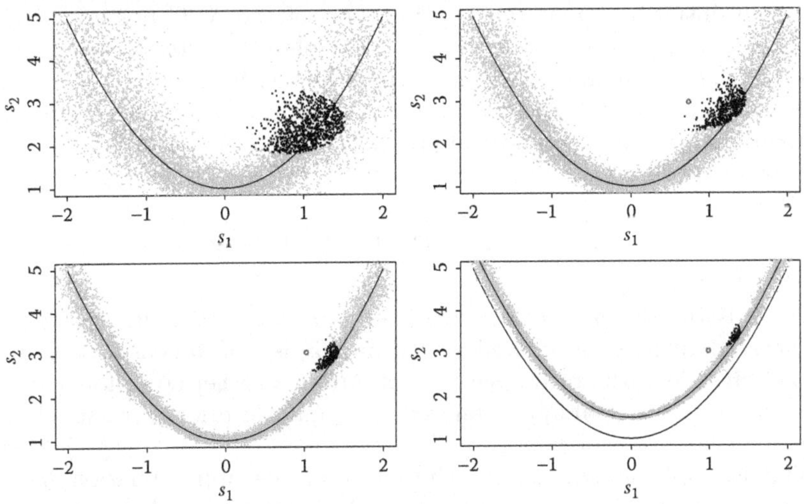

FIGURE 10.4
Example of the effect of model error in ABC for the Gaussian model with incorrect variance described in the text. The plots, from left to right and top to bottom, correspond to increasing sample size. Each plot shows the two-dimensional binding function as we vary θ (line); the observed summary statistic (grey circle) and accepted (black dots) and rejected (grey dots) summary statistic values. (For this model the parameter value used to simulate the summary statistics will be close to the first summary statistic, s_1.)

10.3 ABC Posterior and Posterior Mean

We now consider stronger asymptotic results for ABC. To obtain these results, we need extra assumptions in addition to those required for posterior concentration (see Frazier et al., 2016; Li and Fearnhead, 2018a, for full details). The most important of these is that the summary statistics obey a central limit theorem:

$$\sqrt{n}\left\{\boldsymbol{S}_{n,\boldsymbol{\theta}} - \boldsymbol{b}(\boldsymbol{\theta})\right\} \to \mathrm{N}\left\{0, A(\boldsymbol{\theta})\right\},$$

for some $d \times d$ positive definite matrix $A(\boldsymbol{\theta})$. In the aforementioned central limit theorem, we have assumed a $1/\sqrt{n}$ rate of convergence, but it is trivial to generalise this (Li and Fearnhead, 2018a).

10.3.1 ABC posterior

Under this central limit assumption, we first consider convergence of the ABC posterior. Formal results can be found in Frazier et al. (2016) (but see also Li and Fearnhead, 2018b). Here, we give an informal presentation of these results.

To gain intuition about the limiting form of the ABC posterior, we can use the fact from the previous section that there is posterior concentration around $\boldsymbol{\theta}_0$. Thus, asymptotically, we need only consider the behaviour of the model for $\boldsymbol{\theta}$ close to $\boldsymbol{\theta}_0$. Also, asymptotically, the noise in the summaries is Gaussian. So if we make a linear approximation to $\boldsymbol{b}(\boldsymbol{\theta})$ for $\boldsymbol{\theta}$ close to $\boldsymbol{\theta}_0$, our model will be well approximated by:

$$\boldsymbol{S}_{n,\boldsymbol{\theta}} = \boldsymbol{b}(\boldsymbol{\theta}_0) + D_0(\boldsymbol{\theta} - \boldsymbol{\theta}_0) + \frac{1}{\sqrt{n}}\boldsymbol{Z},$$

where D_0 is the $d \times p$ matrix of first derivatives of $\boldsymbol{b}(\boldsymbol{\theta})$ with respect to $\boldsymbol{\theta}$, with these derivatives evaluated at $\boldsymbol{\theta}_0$; and $\boldsymbol{Z}$ is a d-dimensional Gaussian random variable with covariance matrix $A(\boldsymbol{\theta}_0)$. Furthermore, for $\boldsymbol{\theta}$ close to $\boldsymbol{\theta}_0$, the prior will be well approximated by a uniform prior. For the following, we assume that D_0 is of rank p.

Wilkinson (2013) shows that the effect of the approximation in ABC, whereby we accept simulated summaries which are similar, but not identical, to the observed summary, is equivalent to performing exact Bayesian inference under a different model. This different model has additional additive noise, where the distribution of the noise is given by the kernel, $K(\cdot)$, we use in ABC. So if $\boldsymbol{V}$ is a d-dimensional random variable with density $K(\cdot)$, independent of $\boldsymbol{Z}$, then our ABC posterior will behave like the true posterior for the model:

$$\boldsymbol{S}_{n,\boldsymbol{\theta}} = \boldsymbol{b}(\boldsymbol{\theta}_0) + D_0(\boldsymbol{\theta} - \boldsymbol{\theta}_0) + \frac{1}{\sqrt{n}}\boldsymbol{Z} + \epsilon_n \boldsymbol{V}. \tag{10.1}$$

From Section 10.2.2, we know that the rate of concentration is the slower of the rate of the noise in the summaries, $1/\sqrt{n}$ under our central limit theorem, and the bandwidth ϵ_n. This means that we get different limiting results depending on whether $\epsilon_n = O(1/\sqrt{n})$ or not. This can be seen from (10.1), as whether $\epsilon_n = O(1/\sqrt{n})$ or not will affect whether the $\epsilon_n \boldsymbol{V}$ noise term dominates or not.

If $\sqrt{n}\epsilon_n \to \infty$, so ϵ_n is the slower rate, then to get convergence of the ABC posterior we need to consider the re-scaled variable $\boldsymbol{t} = (\boldsymbol{\theta} - \boldsymbol{\theta}_0)/\epsilon_n$. If we further define $\tilde{\boldsymbol{S}}_{n,\boldsymbol{\theta}} = \{\boldsymbol{S}_{n,\boldsymbol{\theta}} - \boldsymbol{b}(\boldsymbol{\theta}_0)\}/\epsilon_n$, then we can re-write (10.1) as:

$$\tilde{\boldsymbol{S}}_{n,\boldsymbol{\theta}} = D_0\boldsymbol{t} + \boldsymbol{V} + \frac{1}{\epsilon_n\sqrt{n}}\boldsymbol{Z} \to D_0\boldsymbol{t} + \boldsymbol{V}.$$

Thus, the limiting form of the ABC posterior is equivalent to the true posterior for this model, given observation $\tilde{\boldsymbol{s}}_{obs} = \{\boldsymbol{s}_{obs} - \boldsymbol{b}(\boldsymbol{\theta}_0)\}/\epsilon_n$, with a uniform prior for $\boldsymbol{t}$. The shape of this posterior will be determined by the ABC kernel. If we use the standard uniform kernel, then the ABC posterior will asymptotically be uniform. By converting from $\boldsymbol{t}$ to $\boldsymbol{\theta}$, we see that the asymptotic variance for $\boldsymbol{\theta}$ is $O(1/\epsilon_n^2)$ in this case.

The other case is that $\sqrt{n}\epsilon_n \to c$ for some positive, finite constant c. In this case, we consider the re-scaled variable $\boldsymbol{t} = \sqrt{n}(\boldsymbol{\theta} - \boldsymbol{\theta}_0)$ and re-scaled observation $\tilde{\boldsymbol{S}}_{n,\boldsymbol{\theta}} = \sqrt{n}\{\boldsymbol{S}_{n,\boldsymbol{\theta}} - \boldsymbol{b}(\boldsymbol{\theta}_0)\}$. The ABC posterior will asymptotically be equivalent to the true posterior for $\boldsymbol{t}$ under a uniform prior, for a model:

$$\tilde{\boldsymbol{S}}_{n,\boldsymbol{\theta}} = D_0\boldsymbol{t} + \boldsymbol{Z} + \epsilon_n\sqrt{n}\boldsymbol{V} \to D_0\boldsymbol{t} + \boldsymbol{Z} + c\boldsymbol{V},$$

and given an observation $\tilde{\boldsymbol{s}}_{obs} = \sqrt{n}\{\boldsymbol{s}_{obs} - \boldsymbol{b}(\boldsymbol{\theta}_0)\}$.

We make three observations from this. First, if $\epsilon_n = o(1/\sqrt{n})$, so $c = 0$, then using standard results for the posterior distribution of a linear model, the ABC posterior for $\boldsymbol{t}$ will converge to a Gaussian with mean:

$$\{D_0^T A(\theta_0)^{-1} D_0\}^{-1} D_0^T A(\theta_0)^{-1} \tilde{\boldsymbol{s}}_{obs}, \tag{10.2}$$

and variance I^{-1}, where $I = D_0^T A(\boldsymbol{\theta}_0)^{-1} D_0$. This is the same limiting form as the true posterior given the summaries. The matrix I can be viewed as an information matrix, and note that this is larger if the derivatives of the binding function, D_0, are larger; in line with the intuition we presented in Section 10.2.3.

Second, if $c \neq 0$, the ABC posterior will have a larger variance than the posterior given summaries. This inflation of the ABC posterior variance will increase as c increases. In general, it is hard to say the form of the posterior, as it will depend on the distribution of noise in our limiting model, $\boldsymbol{Z} + c\boldsymbol{V}$, which is a convolution of the limiting Gaussian noise of the summaries and a random variable drawn from the ABC kernel.

Finally, we can get some insight into the behaviour of the ABC posterior when $c \neq 0$ if we assume a Gaussian kernel, as again the limiting ABC posterior will be the true posterior for a linear model with Gaussian noise. If the Gaussian kernel has variance Σ, which corresponds to measuring distances between summary statistics using the scaled distance $\|\boldsymbol{x}\| = \boldsymbol{x}^T\Sigma^{-1}\boldsymbol{x}$, then the ABC posterior for $\boldsymbol{t}$ will converge to a Gaussian with mean:

$$\left\{D_0^T(A(\theta_0) + c^2\Sigma)^{-1} D_0\right\}^{-1} D_0^T\{A(\theta_0) + c^2\Sigma\}^{-1}\tilde{\boldsymbol{s}}_{obs}, \tag{10.3}$$

and variance, $\tilde{I}^{-1}$, where:

$$\tilde{I} = D_0^T\{A(\theta_0) + c^2\Sigma\}^{-1} D_0.$$

10.3.2 ABC posterior mean

We now consider the asymptotic distribution of the ABC posterior mean. By this we mean the frequentist distribution, whereby we view the posterior mean as a function of the data and look at the distribution of this under repeated sampling of the data. Formal results appear in Li and Fearnhead (2018a), but we will give informal results, building on the results we gave for the ABC posterior. We will focus on the case where $\epsilon_n = O(1/\sqrt{n})$, but note that results hold for the situation where ϵ_n decays more slowly; in fact, Li and Fearnhead (2018a) show that if $\epsilon_n = o(n^{-3/10})$, then the ABC posterior mean will have the same asymptotic distribution as for the case we consider, where $\epsilon_n = O(1/\sqrt{n})$.

The results we stated for the ABC posterior in Section 10.3.1 for the case $\epsilon_n = O(1/\sqrt{n})$ included expressions for the posterior mean; see (10.2) and (10.3). The latter expression was under the assumption of a Gaussian kernel in ABC, but most of the exposition we give below holds for a general kernel (see Li and Fearnhead, 2018a, for more details).

The first of these, (10.2), is the true posterior mean given the summaries. Asymptotically our re-scaled observation $\tilde{\boldsymbol{s}}_{obs}$ has a Gaussian distribution with mean 0 and variance $A(\boldsymbol{\theta}_0)$ due to the central limit theorem assumption, and the posterior mean for $\boldsymbol{t}$ is a linear transformation of $\tilde{\boldsymbol{s}}_{obs}$. This immediately gives that the asymptotic distribution of the ABC posterior mean of $\boldsymbol{t}$ is Gaussian with mean 0 and variance I^{-1}. Equivalently, for large n, the ABC posterior mean for $\boldsymbol{\theta}$ will be approximately normally distributed with mean $\boldsymbol{\theta}_0$ and variance I^{-1}/n.

The case where $\sqrt{n}\epsilon_n \rightarrow c$ for some $c > 0$ is more interesting. If we have $d = p$, so we have the same number of summaries as we have parameters, then D_0 is a square matrix. Assuming this matrix is invertible, we see that the ABC posterior mean simplifies to $D_0^{-1}\tilde{\boldsymbol{s}}_{obs}$. Alternatively, if $d > p$, but $\Sigma = \gamma A(\boldsymbol{\theta}_0)$ for some scalar $\gamma > 0$, so that the variance of our ABC kernel is proportional to the asymptotic variance of the noise in our summary statistics, then the ABC posterior mean again simplifies; this time to:

$$\left(D_0^T A(\theta_0)^{-1} D_0\right)^{-1} D_0^T A(\theta_0)^{-1} \tilde{\boldsymbol{s}}_{obs}.$$

In both cases, the expressions for the ABC posterior mean are the same as for the $c = 0$ case and are identical to the true posterior mean given the summaries. Thus, the ABC posterior mean has the same limiting Gaussian distribution as the true posterior mean in these cases.

More generally for the $c > 0$ case, the ABC posterior mean will be different from the true posterior mean given the summaries. In particular, the asymptotic variance of the ABC posterior mean can be greater than the asymptotic variance of the true posterior mean given the summaries. Li and Fearnhead (2018a) show that it is always possible to project a $d > p$-dimensional summary to a p-dimensional summary, such that the asymptotic variance of the true posterior mean is not changed. This suggests using such a p-dimensional summary statistic for ABC (see Fearnhead and Prangle, 2012, for a different argument for choosing $d = p$). An alternative conclusion from these results is that one should scale the distance used to be proportional to an estimate of the variance of the noise in the summaries.

It is interesting to compare the asymptotic variance of the ABC posterior mean to the limiting value of the ABC posterior variance. Ideally, these would be the same, as that implies that the ABC posterior is correctly quantifying uncertainty. We do get equality when $\epsilon_n = o(1/\sqrt{n})$; but in other cases we can see that the ABC posterior variance is larger than the asymptotic variance of the ABC posterior mean, and thus ABC over-estimates uncertainty. We will return to this in Section 10.5.

10.4 Monte Carlo Error

The previous section included results on the asymptotic variance of the ABC posterior mean – which gives a measure of accuracy of using the ABC posterior mean as a point estimate for the parameter. In practice, we cannot calculate the ABC posterior mean analytically, and we need to use output from a Monte Carlo algorithm, such as the rejection sampler described in the Introduction. A natural question is what effect does the resulting Monte Carlo error have? And can we implement ABC in such a way that, for a fixed Monte Carlo sample size, the Monte Carlo estimate of the ABC posterior mean is an accurate point estimate? Or do we necessarily require the Monte Carlo sample size to increase as n increases?

Li and Fearnhead (2018a) explore these questions. To do so, they consider an importance sampling version of the rejection sampling algorithm we previously introduced. This algorithm requires the specification of a proposal distribution for the parameter, $q(\boldsymbol{\theta})$, and involves iterating the following N times:

(IS1) Simulate a parameter from the proposal distribution: $\boldsymbol{\theta}_i \sim q(\boldsymbol{\theta})$.

(IS2) Simulate a summary statistic from the model given $\boldsymbol{\theta}_i$: $\boldsymbol{s}_i \sim f_n(\boldsymbol{s}|\boldsymbol{\theta}_i)$.

(IS3) If $\|\boldsymbol{s}_{obs} - \boldsymbol{s}_i\| < \epsilon_n$, accept $\boldsymbol{\theta}_i$, and assign it a weight proportional to $\pi(\boldsymbol{\theta}_i)/q(\boldsymbol{\theta}_i)$.

The output is a set of, N_{acc} say, weighted parameter values which can be used to estimate, for example, posterior means. With a slight abuse of notation, if the accepted parameter values are denoted $\boldsymbol{\theta}^k$ and their weights w_k for $k = 1, \ldots, N_{acc}$, then we would estimate the posterior mean of $\boldsymbol{\theta}$ by:

$$\hat{\boldsymbol{\theta}}_N = \frac{1}{\sum_{k=1}^{N_{acc}} w_k} \sum_{k=1}^{N_{acc}} w_k \boldsymbol{\theta}^k.$$

The use of this Monte Carlo estimator will inflate the error in our point estimate of the parameter by $\text{Var}(\hat{\boldsymbol{\theta}}_N)$, where we calculate variance with respect to randomness of the Monte Carlo algorithm.

If the asymptotic variance of the ABC posterior mean is $O(1/n)$, we would want the Monte Carlo variance to be $O(1/(nN))$. This would mean that the overall impact of the Monte Carlo error is to inflate the mean square error of our estimator of the parameter by a factor $1 + O(1/N)$ (similar to other likelihood-free methods; e.g. Gourieroux et al., 1993; Heggland and Frigessi, 2004).

Now the best we can hope for with a rejection or importance sampler would be equally weighted, independent samples from the ABC posterior. The Monte Carlo variance of such an algorithm would be proportional to the ABC posterior variance. Thus, if we want the Monte Carlo variance to be $O(1/n)$, then we need $\epsilon_n = O(1/\sqrt{n})$, as for slower rates the ABC posterior variance will decay more slowly than $O(1/n)$.

Thus, we will focus on $\epsilon_n = O(1/\sqrt{n})$. The key limiting factor in terms of the Monte Carlo error of our rejection or importance sampler is the acceptance probability. To have a Monte Carlo variance that is $O(1/n)$, we will need an implementation whereby the acceptance probability is bounded away from 0 as n increases. To see whether and how this is possible, we can examine the acceptance criteria in step, (RS3) or (IS3):

$$\|\boldsymbol{s}_{obs} - \boldsymbol{s}_i\| = \|\{\boldsymbol{s}_{obs} - \boldsymbol{b}(\boldsymbol{\theta}_0)\} + \{\boldsymbol{b}(\boldsymbol{\theta}_0) - \boldsymbol{b}(\boldsymbol{\theta}_i)\} + \{\boldsymbol{b}(\boldsymbol{\theta}_i) - \boldsymbol{s}_i\}\|.$$

We need this distance to have a non-negligible probability of being less than ϵ_n. Now, the first and third bracketed terms on the right-hand side will be $O_p(1/\sqrt{n})$ under our assumption for the central limit theorem for the summaries. Thus, this distance is at best $O_p(1/\sqrt{n})$, and if $\epsilon_n = o(1/\sqrt{n})$, the probability of the distance being less than ϵ_n should tend to 0 as n increases.

This suggests we need $\sqrt{n}\epsilon_n \to c$ for some $c > 0$. For this choice, if we have a proposal which has a reasonable probability of simulating $\boldsymbol{\theta}$ values within $O(1/\sqrt{n})$ of $\boldsymbol{\theta}_0$, then we could expect the distance to have a non-zero probability of being less than ϵ_n as n increases. This rules out the rejection sampler or any importance sampler with a pre-chosen proposal distribution. But an adaptive importance sampler that learns a good proposal distribution (e.g. Sisson et al., 2007; Beaumont et al., 2009; Peters et al., 2012) can have this property.

Note that such an importance sampler would need a proposal distribution for which the importance sampling weights are also well behaved. Li and Fearnhead (2018a) give a family of proposal distributions that have both an acceptance probability that is non-zero as $n \to \infty$ and have well-behaved importance sampling weights.

Whilst Li and Fearnhead (2018a) did not consider MCMC based implementations of ABC (Marjoram et al., 2003; Bortot et al., 2007), the intuition behind the results for the importance sampler suggest that we can implement such algorithms in a way that the Monte Carlo variance will be $O(1/(nN))$. For example, if we use a random walk proposal distribution with a variance that is $O(1/n)$, then after convergence the proposed $\boldsymbol{\theta}$ values will be a distance $O_p(1/\sqrt{n})$ away from $\boldsymbol{\theta}_0$ as required. Thus, the acceptance probability should be bounded away from 0 as n increases. Furthermore, such a scaling is appropriate for a random walk proposal to efficiently explore a target whose variance is $O(1/n)$ (Roberts et al., 2001). Note that care would be needed whilst the MCMC algorithm is converging to stationarity, as the proposed parameter values at this stage will be far away from $\boldsymbol{\theta}_0$.

10.5 The Benefits of Regression Adjustment

We finish this chapter by briefly reviewing asymptotic results for a popular version of ABC which post-processes the output of ABC using regression adjustment. This idea was first proposed by Beaumont et al. (2002) (see Nott et al., 2014, for links to Bayes linear methods). We will start with a brief description, then show how using regression adjustment can enable the adjusted ABC posterior to have the same asymptotic properties as the true posterior given the summaries, even if ϵ_n decays slightly slower than $1/\sqrt{n}$.

Figure 10.5 provides an example of the ABC adjustment. The idea is to run an ABC algorithm that accepts pairs of parameters and summaries. Denote these by $(\boldsymbol{\theta}^k, \boldsymbol{s}^k)$ for $k = 1\ldots, N_{acc}$. These are shown in the top-left plot of Figure 10.5. We then fit p linear models that, in turn, aim to predict each component of the parameter vector from the summaries. The output of this fitting procedure is a p-dimensional vector $\hat{\boldsymbol{\alpha}}$, the intercepts in the p linear

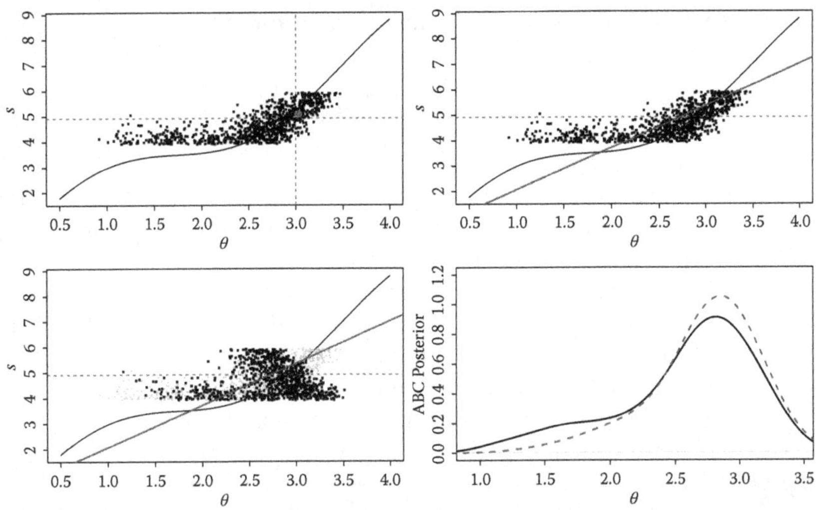

FIGURE 10.5
Example of the regression correction procedure of Beaumont et al. (2002) for a single parameter, single summary statistic. Output of an ABC algorithm (top-left) showing accepted pairs of parameter and summary values (dots), the binding function for this model (solid line), and θ_0 and s_{obs} (circle and also vertical and horizontal lines, respectively). Top-right: the fit from a linear model predicting the parameter value from the summary (solid line). Bottom-left: the adjusted output (dots; with original output in grey); we plot both old and adjusted parameter values against original summary statistic values. Bottom-right: the ABC posterior based on the original accepted parameter values (solid line) and the adjusted values (dashed line).

models, and a $p \times d$ matrix $\hat{B}$, whose ijth entry is the coefficient of the j summary statistic in the linear model for estimating the ith component of $\boldsymbol{\theta}$.

An example of such fit is shown in the top-left hand plot of Figure 10.5. This fit is indicative of biases in our accepted $\boldsymbol{\theta}$, which correspond to different values of the summaries. In our example, the fit suggests that $\boldsymbol{\theta}$ values accepted for smaller, or larger, values of the summary statistic will, on average, be less then, or greater than, the true parameter value. We can then use the fit to correct for this bias. In particular, we can adjust each of the accepted parameter values to $\tilde{\boldsymbol{\theta}}^k$ for $k = 1, \ldots, N_{acc}$ where:

$$\tilde{\boldsymbol{\theta}}^k = \boldsymbol{\theta}^k - \hat{B}(\boldsymbol{s}^k - \boldsymbol{s}_{obs}).$$

The adjusted parameter values are shown in the bottom-left plot of Figure 10.5, and a comparison of the ABC posteriors before and after adjustment are shown in the bottom-right plot. From the latter, we see the adjusted ABC posterior has a smaller variance and has more posterior mass close to the true parameter value.

The vector $\hat{\boldsymbol{\alpha}}$ and the matrix $\hat{B}$ can be viewed as estimates of the vector $\boldsymbol{\alpha}$ and the matrix B that minimises the expectation of:

$$\sum_{i=1}^{p} \left(\boldsymbol{\theta}_i - \boldsymbol{\alpha}_i - \sum_{j=1}^{d} B_{ij} \boldsymbol{S}_j \right)^2,$$

where expectation is with respect to parameter, summary statistic pairs drawn from our ABC algorithm. Li and Fearnhead (2018b) show that if we adjust our ABC output using this optimal B, then, for any $\epsilon_n = o(n^{-3/10})$, the adjusted ABC posterior has the same asymptotic limit as the true posterior given the summaries. Obviously, the asymptotic distribution of the mean of this adjusted posterior will also have the same asymptotic distribution as the mean of the true posterior given the summaries.

The intuition behind this result is that, asymptotically, if we choose $\epsilon_n = o(n^{-3/10})$, then our accepted samples will concentrate around the true parameter value. As we focus on an increasingly small ball around the true parameter value, the binding function will be well approximated by the linear regression model we are fitting. Thus, the regression correction step is able to correct for the biases we obtain from accepting summaries that are slightly different from the observed summary statistics. From this intuition, we see that a key requirement of our model, implicit within the assumptions needed for the theoretical result, is that the binding function is differentiable at the true parameter value: as such, a differentiability condition is needed for the linear regression model to be accurate.

In practice we use an estimate $\hat{B}$, and this will inflate the asymptotic variance of the adjusted posterior mean by a factor that is $1+O(1/N_{acc})$, a similar effect to that of using Monte Carlo draws to estimate the mean. Importantly, we get these strong asymptotic results even when ϵ_n decays more slowly than

$1/\sqrt{n}$. For such a choice, for example, $\epsilon_n = O(n^{-1/3})$, and with a good importance sampling or MCMC implementation, the asymptotic acceptance rate of the algorithm will tend to 1 as n increases.

10.6 Discussion

The theoretical results we have reviewed are positive for ABC. If initially we ignore using regression adjustment, then the results suggest that ABC with $\epsilon_n = O(1/\sqrt{n})$ and with an efficient adaptive importance sampling or MCMC algorithm will have performance that is close to that of using the true posterior given the summaries. Ignoring Monte Carlo error, the accuracy of using the ABC posterior mean will be the same as that of using the true posterior mean if either we have the same number of summaries as parameters, or we choose an appropriate Mahalanobis distance for measuring the discrepancy in summary statistics. However, for this scenario, the ABC posterior will over-estimate the uncertainty in our point estimate. The impact of Monte Carlo error will only be to inflate the asymptotic variance of our estimator by a factor $1+O(1/N)$, where N is the Monte Carlo sample size.

We suggest that this scaling of the bandwidth, $\epsilon_n = O(1/\sqrt{n})$, is optimal if we do not use regression adjustment. Choosing either a faster or slower rate will result in Monte Carlo error that will dominate. One way of achieving this scaling is by using an adaptive importance sampling algorithm and fixing the proportion of samples to accept. Thus, the theory supports the common practice of choosing the bandwidth indirectly in this manner.

Also, based on these results, we suggest choosing the number of summary statistics to be close to, or equal to, the number of parameters and choosing a distance for measuring the discrepancy in summary statistics that is based on the variance of the summary statistics. In situations where there are many potentially informative summary statistics, a dimension reduction approach, that tries to construct low-dimensional summaries that retain the information about the parameters, should be used (e.g. Wegmann et al., 2009; Fearnhead and Prangle, 2012; Blum et al., 2013; Prangle et al., 2014).

The results for ABC with regression adjustment are stronger still. These show that the ABC posterior and its mean can have the same asymptotics as the true ABC posterior and mean given the summaries. Furthermore, this is possible with ϵ_n decreasing more slowly than $1/\sqrt{n}$, in which case the acceptance rate of a good ABC algorithm will increase as n increases. These strong results suggest that regression adjustment should be routinely applied. One word of caution is that the regression adjustment involves fitting a number of linear models to predict the parameters from the summaries. If a large number of summaries are used, then the errors in fitting these models can be large (Fearnhead and Prangle, 2012) and lead to under-estimation of uncertainty in the adjusted posterior (Marin et al., 2016). This again suggests using a small number of summary statistics, close or equal to the number of parameters.

Whilst the choice of bandwidth is crucial to the performance of ABC, and the choice of distance can also have an important impact on the asymptotic accuracy, the actual choice of kernel asymptotically has little impact. It affects the form of the ABC posterior, but does not affect the asymptotic variance of the ABC posterior mean (at least under relatively mild conditions).

These asymptotic results ignore any 'higher-order' effects of the kernel that become negligible as n gets large; so there may be some small advantages of one kernel over another for finite n, but these are hard to quantify. Intuitively, the uniform kernel seems the most sensible choice–as for a fixed acceptance proportion, it accepts the summaries closest to the observed. Furthermore, in situations where there is model error, it is natural to conjecture that a kernel with bounded support, such as the uniform kernel, will be optimal. For such a case, we want to only accept summaries that are $d_0 + O(1/\sqrt{n})$, for some constant distance $d_0 > 0$, away from the observed summary (see Figure 10.4). This is only possible for a kernel with bounded support.

Acknowledgements

This work was supported by Engineering and Physical Sciences Research Council (EPSRC) through the i-like programme grant. It also benefitted from discussions during the Banff International Research Station (BIRS) workshop on Validating and Expanding ABC Methods in February 2017.

References

Beaumont, M. A., Cornuet, J.-M., Marin, J.-M. and Robert, C. P. (2009). Adaptive approximate Bayesian computation. *Biometrika* **96**(4), 983–990.

Beaumont, M. A., Zhang, W. and Balding, D. J. (2002). Approximate Bayesian computation in population genetics. *Genetics* **162**, 2025–2035.

Bernton, E., Jacob, P. E., Gerber, M. and Robert, C. P. (2017). Inference in generative models using the Wasserstein distance. *arXiv:1701.05146*.

Blum, M. G., Nunes, M. A., Prangle, D., Sisson, S. A. et al. (2013). A comparative review of dimension reduction methods in approximate Bayesian computation. *Statistical Science* **28**(2), 189–208.

Bortot, P., Coles, S. G. and Sisson, S. A. (2007). Inference for stereological extremes. *Journal of the American Statistical Association* **102**(477), 84–92.

Drovandi, C. C., Pettitt, A. N., Lee, A. et al. (2015). Bayesian indirect inference using a parametric auxiliary model. *Statistical Science* **30**(1), 72–95.

Fearnhead, P. and Prangle, D. (2012). Constructing summary statistics for approximate Bayesian computation: Semi-automatic approximate Bayesian computation. *Journal of the Royal Statistical Society: Series B (Statistical Methodology)* **74**(3), 419–474.

Frazier, D. T., Martin, G. M., Robert, C. P. and Rousseau, J. (2016). Asymptotic properties of approximate Bayesian computation. *arXiv.1607.06903.*

Frazier, D. T., Robert, C. P. and Rousseau, J. (2017) Model misspecification in ABC: Consequences and diagnostics. *arXiv.1708.01974.*

Gourieroux, C., Monfort, A. and Renault, E. (1993). Indirect inference. *Journal of Applied Econometrics* **8**(S1), S85–S118.

Heggland, K. and Frigessi, A. (2004). Estimating functions in indirect inference. *Journal of the Royal Statistical Society: Series B* **66**, 447–462.

Li, W. and Fearnhead, P. (2018a). On the asymptotic efficiency of approximate Bayesian computation estimators. *Biometrika* **105**(2), 285–299. https://doi.org/10.1093/biomet/asx078.

Li, W. and Fearnhead, P. (2018b). Convergence of regression-adjusted approximate Bayesian computation. *Biometrika*, **105**(2), 301–318. https://doi.org/10.1093/biomet/asx081.

Marin, J.-M., Pillai, N. S., Robert, C. P. and Rousseau, J. (2014). Relevant statistics for Bayesian model choice. *Journal of the Royal Statistical Society: Series B (Statistical Methodology)* **76**(5), 833–859.

Marin, J.-M., Raynal, L., Pudlo, P., Ribatet, M. and Robert, C. P. (2016). ABC random forests for Bayesian parameter inference. *arXiv.1605.05537.*

Marjoram, P., Molitor, J., Plagnol, V. and Tavare, S. (2003). Markov chain Monte Carlo without likelihoods. *Proceedings of the National Academy of Sciences* **100**, 15324–15328.

Martin, G. M., McCabe, B. P., Maneesoonthorn, W. and Robert, C. P. (2016). Approximate Bayesian computation in state space models. *arXiv:1409.8363.*

Nott, D. J., Fan, Y., Marshall, L. and Sisson, S. (2014). Approximate Bayesian computation and Bayes' linear analysis: Toward high-dimensional ABC. *Journal of Computational and Graphical Statistics* **23**(1), 65–86.

Peters, G. W., Fan, Y. and Sisson, S. A. (2012). On sequential Monte Carlo, partial rejection control and approximate Bayesian computation. *Statistics and Computing* **22**(6), 1209–1222.

Prangle, D., Fearnhead, P., Cox, M. P., Biggs, P. J. and French, N. P. (2014). Semi-automatic selection of summary statistics for ABC model choice. *Statistical Applications in Genetics and Molecular Biology* **13**(1), 67–82.

Pritchard, J. K., Seielstad, M. T., Perez-Lezaun, A. and Feldman, M. W. (1999). Population growth of human Y chromosomes: A study of Y chromosome microsatellites. *Molecular Biology and Evolution* **16**, 1791–1798.

Roberts, G. O., Rosenthal, J. S. et al. (2001). Optimal scaling for various Metropolis-Hastings algorithms. *Statistical Science* **16**(4), 351–367.

Ruli, E., Sartori, N. and Ventura, L. (2016). Approximate Bayesian computation with composite score functions. *Statistics and Computing* **26**(3), 679–692.

Sisson, S. A., Fan, Y. and Tanaka, M. M. (2007). Sequential Monte Carlo without likelihoods. *Proceedings of the National Academy of Sciences* **104**(6), 1760–1765.

Wegmann, D., Leuenberger, C. and Excoffier, L. (2009). Efficient approximate Bayesian computation coupled with Markov chain Monte Carlo without likelihood. *Genetics* **182**(4), 1207–1218.

Wilkinson, R. D. (2013). Approximate Bayesian computation (ABC) gives exact results under the assumption of model error. *Statistical Applications in Genetics and Molecular Biology* **12**(2), 129–141.

11

Informed Choices: How to Calibrate ABC with Hypothesis Testing

Oliver Ratmann, Anton Camacho, Sen Hu, and Caroline Colijn

CONTENTS

11.1 Introduction 289
11.2 Equivalence Hypothesis Tests 292
11.3 Equivalence Hypothesis Tests within ABC 294
11.4 A User Guide to the `abc.star` R Package 300
 11.4.1 Testing if locations are similar 300
 11.4.2 Testing if dispersions are similar 303
 11.4.3 Testing if rates are similar 305
11.5 A Worked Example 306
11.6 Discussion and Further Reading 314
Acknowledgements 315
References 315

11.1 Introduction

Approximate Bayesian computations (ABC) proceed by summarising the data, simulating from the model, comparing simulated summaries to observed summaries with an ABC distance function, and accepting the simulated summaries if they do not differ from the observed summaries by more than a user-defined ABC tolerance parameter. These steps are repeated in many Monte Carlo iterations to obtain an approximation to the true posterior density of the model parameters. The process by which precise ABC tolerances and ABC distance functions can be obtained is often referred to as 'ABC calibrations'. The calibrations, that we describe here, apply to the binary ABC accept/reject kernel that evaluates to zero or one. They are based on decision theoretic arguments to construct the ABC accept/reject step, so that the ABC approximation to the posterior density enjoys certain desirable properties. This book chapter aims to give an introduction to ABC calibrations based on hypothesis testing and the `abc.star` **R** package that implements these for the most commonly occurring scenarios.

In the past 15 years, ABC has become a well known inference tool that is used across many applied sciences (Pritchard et al., 1999; Beaumont et al., 2002; Fagundes et al., 2007; Bortot et al., 2007; Ratmann et al., 2007, Cornuet et al., 2008; Luciani et al., 2009; Csilléry et al., 2010; Silk et al., 2011). However, ABC has also been criticised: the choice of summaries remains often arbitrary, summaries can be compared in many possible ways, and the ABC tolerances are vaguely defined as 'small enough' (Barnes et al., 2012; Blum, 2010; Fearnhead and Prangle, 2012; Silk et al., 2013; Dean et al., 2014).

Prangle (2019) discussed in this book approaches for constructing suitable, multi-dimensional summary statistics S to calculate real-valued, observed, and simulated summaries $s_{\text{obs}} \in \mathbb{R}^q$ and $s \in \mathbb{R}^q$. This chapter focuses primarily on the choice of the ABC tolerances $h > 0$ that are used in the ABC kernel to penalise the dissimilarity between the simulated and observed summary statistics. We will see that the approach, that we describe here to specify the tolerances, will also provide unambiguous specifications for comparing simulated to observed summaries.

Let us, for simplicity, consider observed data $\boldsymbol{x}_{1:n} = (x_1, \dots, x_n)$ that consist of n real-valued points x_i and which have an intractable likelihood with parameters $\theta \in \mathbb{R}^p$. ABC approximates the posterior distribution of θ:

$$\pi(\,\theta \mid \boldsymbol{x}_{1:n}\,) = \frac{p(\,\boldsymbol{x}_{1:n} \mid \theta\,)\,\pi(\theta)}{m(\boldsymbol{x}_{1:n})}, \tag{11.1}$$

where $\pi(\theta)$ denotes the prior distribution of θ and $m(\boldsymbol{x}_{1:n}) = \int p(\boldsymbol{x}_{1:n} \mid \theta)\,\pi(\theta)d\theta$. To circumvent likelihood calculations, the vanilla ABC accept/reject sampler simulates data $\boldsymbol{y}_{1:m} = (y_1, \dots, y_m)$ of m real-valued data points and proceeds as follows with m set equal to n:

Set the tolerance $h > 0$;
and calculate the observed summaries $s_{\text{obs}} = S(\boldsymbol{x}_{1:n})$.
for $i = 1$ to N **do**
    repeat
        Sample $\theta^* \sim \pi(\theta)$;
        Simulate $\boldsymbol{y}_{1:m} \sim p(\,\cdot\,|\theta^*\,)$;
        Calculate the simulated summaries $s = S(\boldsymbol{y}_{1:m})$;
        Choose a user-specified, real-valued discrepancy function $d(s, s_{\text{obs}})$;
    until $|d(s, s_{\text{obs}})| < h$
    Set $\theta_{(i)} = \theta^*$.
end for
return $\theta_{(1)}, \dots, \theta_{(N)}$

The tolerances h and discrepancy functions d are specified by the user. In the limit $h \to 0$, the target density of the vanilla ABC sampler approximates the partial posterior distribution:

$$\pi(\,\theta \mid s_{\text{obs}}\,) = \frac{p(\,s_{\text{obs}} \mid \theta\,)\,\pi(\theta)}{m(s_{\text{obs}})}, \tag{11.2}$$

where $m(s_{\text{obs}}) = \int p(\,s_{\text{obs}} \mid \theta\,)\,\pi(\theta)\,d\theta$ (Doksum and Lo, 1990; Wood, 2010). If the summaries are sufficient, the target density of the vanilla ABC sampler approximates in the limit $h \to 0$ the posterior distribution (11.1). In most applications of ABC, h cannot be set to zero because the event that the simulated summaries match the observed summaries is extremely rare. For $h > 0$, ABC can be biased (Blum, 2010; Fearnhead and Prangle, 2012), and in addition, the ABC posterior density is typically broader than (11.1) (Beaumont et al., 2002).

The computational idea behind the ABC calibrations that we will describe in this chapter is as follows. First, we will introduce lower and upper tolerances h^-, h^+ and aim to specify non-symmetric values that offset any bias in the vanilla ABC sampler. Second, we will make the acceptance region $[h^-, h^+]$ wide enough to obtain a pre-specified degree of computational efficiency. Third, we will choose m larger than n in order to offset the broadening effect of the tolerances on the ABC approximation to the posterior density (11.1). We will describe in this chapter how these properties can be obtained if either the data or the summaries have a particularly simple structure. If this is the case, these properties are obtained by (a) choosing specific d and S in line with statistical testing procedures and (b) by setting h^-, h^+, m to specific values h^-_{cali}, h^+_{cali}, m_{cali}. We refer to this process as ABC calibration, call h^-_{cali}, h^+_{cali} the calibrated tolerances, and call m_{cali} the calibrated number of simulations.

ABC calibrations involve a bit of mathematics, but we will proceed gently here. In the first sections, we will assume that the data and simulations are just real values, $\boldsymbol{x}_{1:n} = (x_1, \ldots, x_n)$ and $\boldsymbol{y}_{1:m} = (y_1, \ldots, y_m)$, and that their distribution is particularly simple, such as $x_i \sim \mathcal{N}(\mu, \sigma^2)$. The basic mathematical idea is to interpret the ABC accept/reject step as the outcome of a particular statistical decision procedure. In the case of step 9 earlier, the acceptance region is the interval defined by the tolerances, $[h^-, h^+]$. This corresponds to the structure of classical hypothesis tests. This chapter elaborates on this particular correspondence and is therefore limited to ABC algorithms with the binary accept/reject step. We denote the ABC acceptance event by:

$$T(\boldsymbol{y}_{1:m}, \boldsymbol{x}_{1:n}) \in [\,h^-, h^+\,], \tag{11.3}$$

where the data $\boldsymbol{x}_{1:n}$ are fixed throughout and $\boldsymbol{y}_{1:m}$ are simulated as in step 6. Similarly, we denote the accept/reject step in the aforementioned vanilla ABC sampler that evaluates to either zero or one by:

$$\mathbb{1}\Big\{\,T(\boldsymbol{y}_{1:m}, \boldsymbol{x}_{1:n}) \in [\,h^-, h^+\,]\,\Big\}. \tag{11.4}$$

Here, T is a real-valued hypothesis test statistic that comprises S and d in step 9 earlier. The lower and upper ABC tolerances h^-, h^+ define the rejection region of the hypothesis test. These tolerances are always such that $h^- < h^+$, which corresponds to the case $h > 0$ in step 1 earlier. In Section 11.2, we review so-called equivalence hypothesis tests. Section 11.3 explains how equivalence hypothesis tests fit into ABC. Section 11.4 introduces the `abc.star` **R** package, which provides calibration routines for the most commonly used equivalence hypothesis tests.

In the final Section 11.5, we move from very simple data to a time series example, illustrating how the calibrations could be applied in real-world applications. We relax the assumption that the data and simulations are just real-values following a simple probability law. Instead, we consider arbitrarily complex data and assume there are q observed and simulated summaries that are just real values, for example $\boldsymbol{s}^{\text{obs}}_{i,1:n} = (s^{\text{obs}}_{i,1}, \ldots, s^{\text{obs}}_{i,n})$ and $\boldsymbol{s}^{\text{sim}}_{i,1:m} = (s^{\text{sim}}_{i,1}, \ldots, s^{\text{sim}}_{i,m})$ for $i = 1, \ldots, q$ and $m \geq n$. Typically, we want at least as many summaries as model parameters, $q \geq q$. Next, although the summaries are statistics of the observed and simulated data, we model their sampling distribution directly and assume that the sampling distribution is particularly simple; for example, $s^{\text{sim}}_{i,j} \sim \mathcal{N}(\mu_i, \sigma_i^2)$ for $j = 1, \ldots, m$ and $i = 1, \ldots, q$. The ABC accept-reject step is then a combination of q hypothesis tests, leading to the multivariate accept/reject step:

$$\mathbb{1}\left\{ \bigcap_{i=1}^{q} T_i\left(\boldsymbol{s}^{\text{sim}}_{i,1:m_i}, \boldsymbol{s}^{\text{obs}}_{i,1:n} \right) \in \left[h_i^-, h_i^+ \right] \right\}. \tag{11.5}$$

For each test, the free parameters h_i^-, h_i^+, m_i will be calibrated to specific values $h^-_{i,\text{cali}}, h^+_{i,\text{cali}}, m_{i,\text{cali}}$. We close with a discussion and further reading in Section 11.6.

11.2 Equivalence Hypothesis Tests

Equivalence hypothesis tests are less well known than standard hypothesis tests, such as the T-test or the Mann-Whitney test (Lehmann and Romano, 2005). We are interested in these because the behaviour of ABC algorithms can be quantified in terms of the type-I error probability, the power function, and other properties of equivalence hypothesis tests.

To remind us of the basic terminology (Lehmann and Romano, 2005), consider the following simple scenario. Suppose we have two independent and identically distributed samples $\boldsymbol{x}_{1:n} = (x_1, \ldots, x_n)$, $x_i \sim \mathcal{N}(\theta_x, \sigma^2)$, $\boldsymbol{y}_{1:m} = (y_1, \ldots, y_m)$, $y_i \sim \mathcal{N}(\theta_y, \sigma^2)$, with $n > 1$ and $m > 1$. Here, the two samples are considered random and the model parameters are fixed, with θ_x, θ_y unknown, and σ^2 known. As an example, we consider testing if the population means θ_x, θ_y are similar. To derive a two-sample equivalence test in this setting, the *null and alternative hypotheses* are:

$$\begin{aligned} H_0:&\quad \rho \notin [\tau^-, \tau^+], \quad \rho = \theta_y - \theta_x \\ H_1:&\quad \rho \in [\tau^-, \tau^+], \end{aligned} \tag{11.6}$$

where $\tau^- < 0$ and $\tau^+ > 0$. The interval $[\tau^-, \tau^+]$ is the *equivalence region* and ρ is the *discrepancy parameter*, reflecting the discrepancy between θ_y and θ_x. The *point of equality* is the discrepancy $\rho^\star$ for which θ_y equals θ_x. For (11.6), $\rho^\star = 0$. An equivalence test rejects the null hypothesis if the *test statistic* T falls into the *critical region* $[h^-, h^+]$. We will more typically refer to the critical region as the ABC acceptance region. In the context of ABC, h^-, h^+, τ^-, τ^+, and m are free parameters, which we seek to set to particularly desirable values.

The parameters τ^-, τ^+ are new to ABC. They will be instrumental to calibrating the tolerances h^-, h^+. An equivalence test is said to be *level-α* if, for $\alpha > 0$ and given equivalence region $[\tau^-, \tau^+]$, there are h^-_{cali}, h^+_{cali}, such that $P(\,T \in [h^-_{\text{cali}}, h^+_{\text{cali}}] \mid \rho\,) \leq \alpha$ for all $\rho \in H_0$. It is *size-α* if, for given $\alpha > 0$ and equivalence region $[\tau^-, \tau^+]$, there are h^-_{cali}, h^+_{cali}, such that $\sup_{\rho \in H_0} P(\,T \in [h^-_{\text{cali}}, h^+_{\text{cali}}] \mid \rho\,) = \alpha$. The *power* of the size-α test at a given discrepancy is:

$$\text{pw}(\rho) = P(\,T \in [h^-_{\text{cali}}, h^+_{\text{cali}}] \mid \rho\,). \tag{11.7}$$

For the equivalence problem (11.6), a 'two one-sided' Z-test can be used, similar to the two one-sided T-test first described by Schuirmann (1981) and Kirkwood and Westlake (1981) for the more general case where σ^2 is also not known. The bivariate two-sample test statistic is $T^-(\boldsymbol{y}_{1:m}, \boldsymbol{x}_{1:n}) = \sqrt{nm/(n+m)}(\bar{y} - \bar{x} - \tau^-)/\sigma$ and $T^+(\boldsymbol{y}_{1:m}, \boldsymbol{x}_{1:n}) = \sqrt{nm/(n+m)}(\bar{y} - \bar{x} - \tau^+)/\sigma$. The test rejects the null hypothesis if:

$$\{\, u_{1-\alpha} \leq T^-(\boldsymbol{y}_{1:m}, \boldsymbol{x}_{1:n}) \quad \text{and} \quad T^+(\boldsymbol{y}_{1:m}, \boldsymbol{x}_{1:n}) \leq u_\alpha \,\}, \tag{11.8}$$

where u_α is the lower α-quantile of the normal distribution. As will be apparent later, we are only interested in cases where the power is maximised at the point of equality $\rho^* = 0$. For the two one-sided Z-test, this is the case whenever τ^- is set to $-\tau^+$, for $\tau^+ > 0$. In this case, (11.8) simplifies to:

$$T(\boldsymbol{y}_{1:m}, \boldsymbol{x}_{1:n}) \in [h^-_{\text{cali}}, h^+_{\text{cali}}],$$

where

$$T(\boldsymbol{y}_{1:m}, \boldsymbol{x}_{1:n}) = \bar{y} - \bar{x}, \quad h^+_{\text{cali}} = u_\alpha \frac{\sigma}{\sqrt{nm/(n+m)}} + \tau^+, \quad h^-_{\text{cali}} = -h^+_{\text{cali}}. \tag{11.9}$$

The formula for the ABC acceptance region $[h^-_{\text{cali}}, h^+_{\text{cali}}]$ is exactly such that the Z-test is size-α. The power of the Z-test in (11.9) is:

$$P(\,T \in [h^-_{\text{cali}}, h^+_{\text{cali}}] \mid \rho\,) = F_{\mathcal{N}(0,1)}\left(\,u_\alpha + \frac{\tau^+ - \rho}{\sigma}\sqrt{nm/(n+m)}\,\right) - F_{\mathcal{N}(0,1)}\left(\,u_{1-\alpha} + \frac{\tau^- - \rho}{\sigma}\sqrt{nm/(n+m)}\,\right), \tag{11.10}$$

where $F_{\mathcal{N}(0,1)}$ denotes the cumulative density of the normal distribution.

The Z-test illustrates common features of all equivalence tests that we will consider for ABC inference. The null hypothesis is rejected if T falls *into* the ABC acceptance region, which is the opposite when compared to a test of equality, such as a standard T-test. The tolerances are chosen such that the probability, that T is inside the ABC acceptance region, is small when ρ is large. This is interesting in ABC, because we do not want to accept an ABC iteration frequently, when in fact the discrepancy between the proposed and true (unknown) model parameters is large. The power is a unimodal function that is maximised somewhere in $[\tau^-, \tau^+]$, which is also the opposite compared to a test of equality, where power increases with ρ. This is interesting, because we want the ABC acceptance probability (i.e. the probability to reject the equivalence null hypothesis) to be largest when ρ is small. For these two reasons, we describe the ABC accept/reject step as an equivalence test rather than the standard point null hypothesis tests. We will expand on the advantages of interpreting the ABC accept/reject step as the outcome of an equivalence test in the next section. This will involve calibrating the free parameters such that the power function satisfies particular properties.

Some features of the Z-test are different in other hypothesis testing scenarios. For example, we will see in Sections 11.4.2–11.4.3 that the discrepancy ρ is not necessarily defined as the difference, and that the point of equality $\rho^\star$ is not necessarily equal to zero. Further, the ABC acceptance regions that we consider will not necessarily be symmetric around the point of equality.

11.3 Equivalence Hypothesis Tests within ABC

In this section, we will describe how equivalence tests can be embedded into ABC, explain how the free parameters τ^-, τ^+ and m are calibrated, and outline the resulting ABC approximation to the posterior density. For simplicity, we will focus in this section on the case where the data and simulations are just real values from a normal distribution, and we will focus on the earlier Z-test. In this case, ABC is not needed and estimates of interest, such as posterior densities and maximum likelihood parameter estimates are readily available. Even so, it is still of interest to obtain ABC approximations that match the known posterior density closely in this setting, because otherwise it is difficult to trust ABC in more complex examples.

There is one subtle difference between most equivalence hypothesis tests that can be found in the literature (Wellek, 2003), and those required for ABC. In the literature, test statistics are usually derived for the case where $\boldsymbol{x}_{1:n}$ and $\boldsymbol{y}_{1:m}$ are two random samples. In ABC, the observed data are fixed. This means that we are interested in one-sample test statistics for equivalence hypotheses. Let us consider again scenario (11.6). Instead of θ_x, we consider a pre-specified, fixed reference value ν and test the similarity of θ_y with ν. The reference value ν could be chosen in many ways. Since we want the ABC approximation of the likelihood to match the likelihood, we choose the maximum likelihood estimate of θ_x. Here, the maximum likelihood estimate is the sample mean, $\bar{x} = 1/n \sum_{i=1}^{n} x_i$, and we set $\nu = \bar{x}$. In the more general case, when working with summaries, we do not need to know the maximum likelihood value of the parameter of interest. The reference value will be the maximum likelihood estimate of a parameter from the auxiliary sampling distribution of the summaries, such as the sample mean of summaries. Let us also write θ instead of θ_y. Within ABC, (11.6) changes to:

$$\begin{aligned} H_0: &\quad \rho \notin [\tau^-, \tau^+], \quad \rho = \theta - \bar{x} \\ H_1: &\quad \rho \in [\tau^-, \tau^+], \end{aligned} \tag{11.11}$$

where as before $\tau^+ > 0$ and $\tau^- = -\tau^+$. For $\boldsymbol{x}_{1:n}$ fixed, the one-sample Z-test rejects the null hypothesis:

$$T(\boldsymbol{y}_{1:m}, \boldsymbol{x}_{1:n}) \in [h^-_{\text{cali}}, h^+_{\text{cali}}],$$

where

$$T(\boldsymbol{y}_{1:m}, \boldsymbol{x}_{1:n}) = \bar{y} - \bar{x}, \quad h^+_{\text{cali}} = u_\alpha \frac{\sigma}{\sqrt{m}} + \tau^+, \quad h^-_{\text{cali}} = -h^+_{\text{cali}}. \tag{11.12}$$

The formula for the ABC acceptance region is exactly such that the one-sample Z-test is size-α. We can set the ABC tolerances to the calibrated values h^-_{cali}, h^+_{cali} in (11.12). This way, the ABC tolerances are always chosen in such a way to provide an upper bound α on the misclassification probability that an ABC step is accepted, even though the model parameter used to generate the simulation is very different from the model parameter that underlies the data. This shows that, being able to express the ABC accept/reject step in terms of several free parameters, we can begin to control the behaviour of ABC. The size-α property determines two of the five free ABC parameters.

Furthermore, these ABC tolerances are desirable for several reasons that are related to the power function of the test. The power of the size-α one-sample Z-test in (11.12) is:

$$\begin{aligned} \mathrm{pw}(\rho) &= P(\, T \in [h^-_{\text{cali}}, h^+_{\text{cali}}] \mid \rho\,) \\ &= F_{\mathcal{N}(0,1)}\left(u_\alpha + \frac{\tau^+ - \rho}{\sigma}\sqrt{m}\right) - F_{\mathcal{N}(0,1)}\left(u_{1-\alpha} + \frac{\tau^- - \rho}{\sigma}\sqrt{m}\right). \end{aligned} \tag{11.13}$$

To obtain (11.13), we express the calibrated tolerances h^-_{cali}, h^+_{cali} directly in terms of the free parameters τ^-, τ^+, m.

Importantly, the power function is the ABC approximation to the likelihood in terms of the auxiliary discrepancy parameter ρ. To see this, note that the ABC approximation to the likelihood of the vanilla ABC sampler is, in our notation:

$$\begin{aligned} &P(T \in [h^-_{\text{cali}}, h^+_{\text{cali}}] \mid \theta) \\ &\quad = \int \mathbb{1}\{\, T(\boldsymbol{y}_{1:m}, \boldsymbol{x}_{1:n}) \in [h^-_{\text{cali}}, h^+_{\text{cali}}]\}\, p(\, \boldsymbol{y}_{1:m} \mid \theta\,)\, d\boldsymbol{y}_{1:m}. \end{aligned} \tag{11.14}$$

Accepting an ABC step for θ is equivalent to rejecting the one-sample test for $\rho = \theta - \bar{x}$, and so the ABC approximation to the likelihood is the probability to reject the equivalence null hypothesis, for example, the power. Power functions have been studied extensively in the context of hypothesis testing theory, and we can re-use these results in the context of ABC. One such finding is that the power function, or ABC approximation to the likelihood, is not necessarily maximised at ρ^*, that is when θ equals the maximum likelihood estimate $\bar{x}$. The mode of $\mathrm{pw}(\rho)$ shifts with the choice of τ^- in relation to τ^+. It is also known that for (11.12), $\mathrm{pw}(\rho)$ maximises at ρ^* when τ^- equals $-\tau^+$.

For many other tests, the choice of τ^- is less intuitive. Fortunately, power functions, such as (11.13), are often continuous and monotonic with respect to the free parameters, and so we can use simple Newton-Raphson-type algorithms to calibrate the free parameter τ^- to a value τ^-_{cali}, such that $P(\, T \in [h^-_{\text{cali}}, h^+_{\text{cali}}] \mid \rho\,)$ is largest at ρ^*. We use Brent's method when the calibration values are not analytically known (Press et al., 2007). Thus, the location property determines the third out of the five free parameters.

Next, the power function is also the acceptance probability of the ABC accept/reject step, again, in terms of ρ. The maximum value of the power function, or acceptance probability, depends on the width of the equivalence interval $[\tau^-, \tau^+]$. Let us suppose the free parameter τ^- is already calibrated to a value τ^-_{cali}, such that $\mathrm{pw}(\rho)$ maximises at ρ^* (for the Z-test, it is simply $\tau^-_{\text{cali}} = -\tau^+$). The Z-test is not an unbiased test, and so it can happen for some m, σ, and narrow equivalence intervals that the maximum acceptance probability is zero. We calibrate the free parameter τ^+ to a value τ^+_{cali}, such that the ABC acceptance probability is large at the point of equality, see Figure 11.1a. But note that it is not a good idea to calibrate τ^+, such that $P(T \in [h^-_{\text{cali}}, h^+_{\text{cali}}] \mid \rho^*)$ is very close to one, because in this case $\mathrm{pw}(\rho)$ plateaus around ρ^*, see again Figure 11.1a. This would mean that many values of θ close to $\bar{x}$ are almost equally likely to get accepted with high probability,

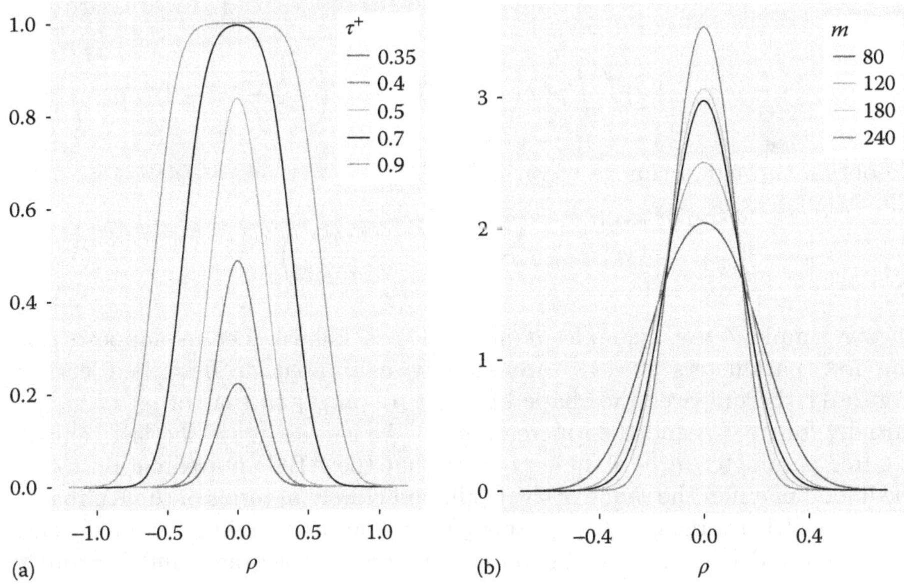

FIGURE 11.1
Power function of the `Z-test`. The power of the equivalence Z-test is unimodal and monotonically decreasing away from the point of equality ρ^* whenever τ^- is set to $-\tau^+$. (a) Power of the Z-test as the equivalence region $[\tau^-, \tau^+]$ widens. In this panel, $\sigma = 1.2$, $m = n = 80$, and τ^- fixed to $-\tau^+$. Power increases with the width of the equivalence region. (b) g_{ABC} density of the Z-test as m increases. In this panel, $\sigma = 1.2$, $n = 80$, $\tau^- = -\tau^+$, and τ^+, such that the maximum acceptance probability is 0.9. The g_{ABC} density is shown for increasing m (shades of grey) and compared against the g density (black).

and we would not be able to recover the actual maximum likelihood estimate with ABC. By default, we suggest calibrating τ^+ to a value τ^+_{cali}, such that $P(T \in [h^-_{\text{cali}}, h^+_{\text{cali}}] \mid \rho^*) = 0.9$. In this case, the power drops relatively sharply away from ρ^*, which makes the maximum likelihood estimate more easily identifiable from ABC output. The value of τ^+ for which the power at ρ^* is 0.9 is not known analytically, so we use Brent's method to determine it numerically. This efficiency property determines the fourth free parameter.

Finally, the ABC approximation to the likelihood (11.13) is usually broader than the likelihood. To see this, consider the power as a density in ρ:

$$g_{\text{ABC}}(\,\rho \mid \boldsymbol{x}_{1:n}\,) = P\left(T \in [h^-_{\text{cali}}, h^+_{\text{cali}}] \mid \rho\right) \Big/ \int P\left(T \in [h^-_{\text{cali}}, h^+_{\text{cali}}] \mid \rho\right)\, d\rho. \tag{11.15}$$

We aim to match it against the likelihood of the data, when considered as a density in terms of ρ. This is straightforward for the example considered here.

From:

$$g(\theta \mid \boldsymbol{x}_{1:n}) = p(\boldsymbol{x}_{1:n} \mid \theta) \Big/ \int p(\boldsymbol{x}_{1:n} \mid \theta) d\theta \propto \exp\left(-\frac{1}{2}\frac{n}{\sigma^2}(\bar{x}-\theta)^2\right), \tag{11.16}$$

we obtain through a change of variables:

$$g(\,\rho \mid \boldsymbol{x}_{1:n}\,) \;\propto\; \exp\left(\,-\frac{1}{2}\frac{\rho^2}{\sigma^2/n}\,\right). \tag{11.17}$$

In our simple Z-test example, σ in (11.17) is known. Let us suppose that the free parameters τ^-, τ^+ are already calibrated as described earlier. Figure 11.1b compares the shape of $g_{\text{ABC}}(\,\rho \mid \boldsymbol{x}_{1:n}\,)$ to that of $g(\,\theta \mid \boldsymbol{x}_{1:n}\,)$, initially for $m = n$ and then increasing m. We can calibrate the last remaining free parameter m to a value m_{cali} so that the ABC approximation to the likelihood matches the shape of the likelihood closely in terms of the Kullback–Leibler (KL) divergence. We do this calibration step again numerically. Let us make the dependence of g_{ABC} on m through the tolerances and T explicit, and denote the minimum KL divergence after calibration by:

$$\varepsilon_{\text{cali}} = \min_m \text{KL}\left(\,g(\,\rho \mid \boldsymbol{x}_{1:n}\,)\,\Big\|\,g_{\text{ABC},m}(\,\rho \mid \boldsymbol{x}_{1:n}\,)\,\right). \tag{11.18}$$

The calibration error $\varepsilon_{\text{cali}}$ is calculated as a by-product of the calibrations. In the more general case, when working with summaries, we do not need to know the likelihood with respect to the data as in (11.16). We will use the likelihood of the summaries under their modelled sampling distribution instead. This shape property determines the last free ABC parameter m.

We are now ready to plug in the calibrated ABC approximation to the likelihood into the ABC posterior density $\pi_{\text{ABC}}(\,\theta \mid \boldsymbol{x}_{1:n}\,)$. After the calibrations, we must transform the auxiliary parameter ρ back to the original parameter of interest, θ. In our example, this transformation is simply the difference $\rho = \theta - \bar{x}$, which is one-to-one. The ABC posterior density, that results from using the calibrated Z-test and substituting ρ for θ, is:

$$\begin{aligned}\pi_{\text{ABC}}(\,\theta \mid \boldsymbol{x}_{1:n}\,) \propto \Bigg[\, & F_{\mathcal{N}(0,1)}\left(u_\alpha + \frac{\tau^+_{\text{cali}} - \theta + \bar{x}}{\sigma}\sqrt{m_{\text{cali}}}\right) \\ & - F_{\mathcal{N}(0,1)}\left(u_{1-\alpha} + \frac{\tau^-_{\text{cali}} - \theta + \bar{x}}{\sigma}\sqrt{m_{\text{cali}}}\right)\Bigg]\;\pi(\theta),\end{aligned} \tag{11.19}$$

which approximates the posterior density:

$$\pi(\,\theta \mid \boldsymbol{x}_{1:n}\,) \propto f_{\mathcal{N}(0,1)}\left(\frac{\bar{x}-\theta}{\sigma}\sqrt{n}\right)\;\pi(\theta). \tag{11.20}$$

Let us first consider the overall difference between $\pi(\theta \mid \boldsymbol{x}_{1:n})$ and $\pi_{\text{ABC}}(\theta \mid \boldsymbol{x}_{1:n})$ in terms of the KL divergence. The calibrations focus only on the ABC approximation to the likelihood, which does not involve the prior density. The prior will usually not influence the calibrations substantially, because the calibration error (11.18) is usually very small (typically well below 0.01), and the same prior term appears in (11.19) and (11.20). We approximate the KL divergence between $\pi(\theta \mid \boldsymbol{x}_{1:n})$ and $\pi_{\text{ABC}}(\theta \mid \boldsymbol{x}_{1:n})$ with that of $g(\theta \mid \boldsymbol{x}_{1:n})$ and $g_{\text{ABC}}(\theta \mid \boldsymbol{x}_{1:n})$, which equals that of $g(\rho \mid \boldsymbol{x}_{1:n})$ and $g_{\text{ABC}}(\rho \mid \boldsymbol{x}_{1:n})$. Hence, to a good approximation:

$$\text{KL}\left(\pi(\theta \mid \boldsymbol{x}_{1:n}) \,\middle\|\, \pi_{\text{ABC}}(\theta \mid \boldsymbol{x}_{1:n})\right) \approx \varepsilon_{\text{cali}}. \tag{11.21}$$

This calibration was subject to the constraint that, in terms of ρ, the ABC maximum likelihood estimate is ρ^*. But then, by the way we chose the reference value ν in (11.11), it follows that the ABC maximum likelihood estimate equals the maximum likelihood estimate:

$$\underset{\theta}{\text{argmax}}\, P(T \in [h^-_{\text{cali}}, h^+_{\text{cali}}] \mid \theta) = \underset{\theta}{\text{argmax}}\, p(\boldsymbol{x}_{1:n} \mid \theta). \tag{11.22}$$

Finally, calibrating such that the tolerances correspond to size-α tests produces a lower bound on the true positive rate of the ABC algorithm in the sense that:

$$P\left(\rho \in [\tau^-_{\text{cali}}, \tau^+_{\text{cali}}] \,\middle|\, T \in [h^-_{\text{cali}}, h^+_{\text{cali}}]\right) \geq 1 - \alpha \big/ P\left(T \in [h^-_{\text{cali}}, h^+_{\text{cali}}]\right), \tag{11.23}$$

where $P(T \in [h^-_{\text{cali}}, h^+_{\text{cali}}])$ is the acceptance probability of the ABC algorithm. If the overall ABC acceptance rate is 10% and $\alpha = 0.05$, then (11.23) implies that the ABC true positive rate is at least 50%. On the other hand, if $\alpha = 0.01$, then (11.23) implies that the ABC true positive rate is at least 90%. For this reason, we use by default $\alpha = 0.01$ to calibrate the tolerances.

These calibration results for the Z-test (11.12) extend to several other testing scenarios, as we describe in the following. One pre-requisite of the numerical calibrations is that the power function must be available analytically or to a good approximation, which is the case for the most commonly occurring testing scenarios (Lehmann and Romano, 2005; Wellek, 2003). A more important pre-requisite is that the data or summaries follow indeed the sampling distribution that is assumed by the underlying testing procedure. For example, to apply the Z-test, the data or summaries must be i.i.d. normal. We will re-visit these issues and how equivalence tests can be combined in Section 11.5 of this chapter.

11.4 A User Guide to the abc.star R Package

This section introduces the numerical calibration routines that are available in the `abc.star` **R** package (Ratmann et al., 2016). The latest version is available from github and can be installed through:

```
> library(devtools)
> install_github("olli0601/abc.star")
```

11.4.1 Testing if locations are similar

Suppose that the data or summaries are normally distributed with known variance. In this case, the one-sample Z-test from Section 11.3 can be used in ABC to test for location equivalence. The corresponding calibration routine is `ztest.calibrate`.

In this section, we focus on the related case that the data or summaries are normally distributed with unknown variances. For simplicity, we write $x_i \sim \mathcal{N}(\theta_0, \sigma_0^2)$, $y_i \sim \mathcal{N}(\theta, \sigma^2)$ and avoid the more cumbersome notation for summaries. The null and alternative hypotheses are (11.11), and the one-sample test statistic is $T(\boldsymbol{y}_{1:m}, \boldsymbol{x}_{1:n}) = \bar{y} - \bar{x}$, where again the fixed reference value is the observed sample mean $\bar{x} = \frac{1}{n}\sum_{i=1}^{n} x_i$, and $\bar{y} = \frac{1}{m}\sum_{i=1}^{m} y_i$. The ABC acceptance region is $[h^-_{\text{cali}}, h^+_{\text{cali}}]$ where:

$$h^+_{\text{cali}} = t_{\alpha,m-1}\frac{\hat{\sigma}}{\sqrt{m}} + \tau^+, \quad h^-_{\text{cali}} = -h^+_{\text{cali}}, \quad \tau^- = -\tau^+, \tag{11.24}$$

and $t_{\alpha,m-1}$ is the lower α-quantile of the t-distribution with $m-1$ degrees of freedom, and $\hat{\sigma}$ is an estimate of the sample standard deviation σ^2 of $\boldsymbol{y}_{1:m}$. The power function is:

$$\begin{aligned} P(T \in [h^-_{\text{cali}}, h^+_{\text{cali}}] \mid \rho) = {} & F_{t_{m-1,\sqrt{m}\rho/\sigma}}\left(t_{\alpha,m-1} + \frac{\tau^+ - \rho}{\hat{\sigma}}\sqrt{m}\right) \\ & - F_{t_{m-1,\sqrt{m}\rho/\sigma}}\left(t_{1-\alpha,m-1} + \frac{\tau^- - \rho}{\hat{\sigma}}\sqrt{m}\right), \end{aligned} \tag{11.25}$$

where $F_{t_{m-1,\sqrt{m}\rho/\sigma}}$ is the cumulative distribution of a non-central t-distribution with $m-1$ degrees of freedom and non-centrality parameter $\sqrt{m}\rho/\sigma$. The non-centrality parameter can be approximated with $\sqrt{m}\rho/\hat{\sigma}$ (Owen, 1965). To derive the $g(\,\rho \mid x_{1:n}\,)$ density, we here need to integrate the additional parameter σ_0^2 out. For the reference prior $\pi(\theta, \sigma_0^2) \propto 1/\sigma_0^2$ which is flat with respect to θ as in (11.16), we obtain in analogy to (11.16–11.17):

$$g(\,\rho \mid \boldsymbol{x}_{1:n}\,) \propto \left(\,S^2(\boldsymbol{x}_{1:n}) + n\rho^2\,\right)^{-\frac{(n-1)+1}{2}},$$

where $S^2(\boldsymbol{x}_{1:n}) = \sum_{i=1}^{n}(x_i - \bar{x})^2$. This means that $\rho/S^2(\boldsymbol{x}_{1:n})\sqrt{n}$ follows a t-distribution with $n-1$ degrees of freedom. To illustrate the available **R** functions, let:

```
> require(abc.star)
#observed and simulated data
> xn  <- 60; xmean<- 1; xsigma<- 1
> yn  <- 60; ymean<- 1.85; ysigma<- 1.2
> obs <- rnorm(xn, xmean, xsigma)
> sim <- rnorm(yn, ymean, ysigma)
#to make this example reproducible
> obs <- (obs - mean(obs))/sd(obs) * xsigma + xmean
> sim <- (sim - mean(sim))/sd(sim) * ysigma + ymean.
```

We will now describe all calibration steps one by one and will end with a function call that calibrates all free parameters in one go. First, we describe how to obtain the ABC tolerances in (11.24) for the default choice on the type-I error bound, $\alpha = 0.01$, and, for example, $\tau^+ = 0.8$:

```
> mutost.calibrate(n.of.y=length(sim), s.of.y=sd(sim),
    what='CR', tau.u=0.8, alpha=0.01)
       c.l        c.u
-0.4295524  0.4295524.
```

Here, the CR option stands for 'critical region' and instructs the calibration routine to compute the ABC acceptance region. The output fields `c.l` and `c.u` return the calibrated lower and upper endpoints of the ABC acceptance region, h^-_{cali} and h^+_{cali}. To obtain α for user-specified ABC tolerances, say $h^- = -0.6$ and $h^+ = 0.6$, we can use:

```
> mutost.calibrate(n.of.y=length(sim), s.of.y=sd(sim), c.u=0.6,
    tau.u=0.8, what='ALPHA')
    alpha
0.1008706.
```

There is a 10% chance that the ABC accept/reject step is accepted when the underlying population mean of the simulated data differs from the observed sample mean by more than `tau.u=0.8`. The power function can be calculated either from h^+ assuming that $h^- = -h^+$ or, alternatively, from α and τ^+ assuming that h^-_{cali} and h^-_{cali} are set as in (11.24):

```
> rho      <- seq(-0.7,0.7,0.01)
> s.of.T   <- sd(sim)/sqrt(length(sim))
> accprob <- mutost.pow(rho, df=length(sim), s.of.T=s.of.T,
    c.u=0.4295524)
# or
> accprob <- mutost.pow(rho, df=length(sim), s.of.T=s.of.T,
    tau.u=0.8, alpha=0.01).
```

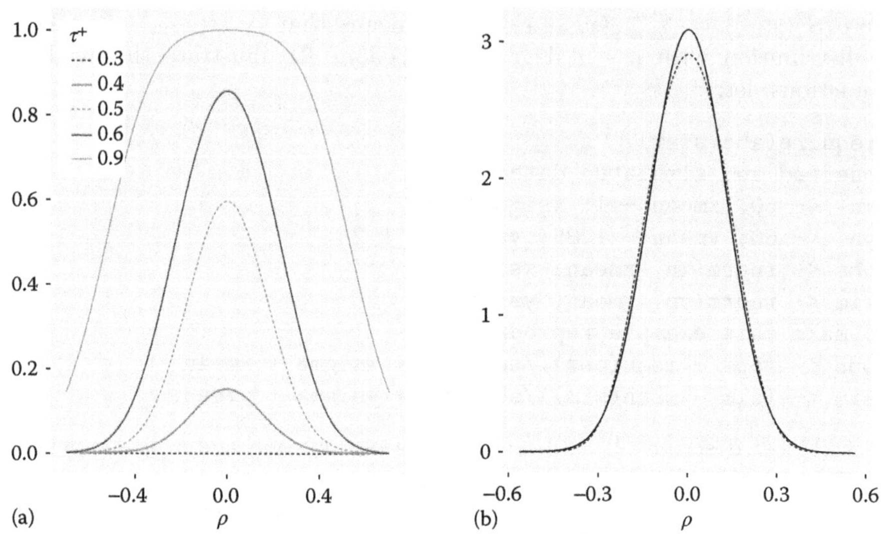

FIGURE 11.2
Calibrated two one-sided T-test. If the data or summaries are normally distributed with unknown variance, the one-sample version of the two one-sided T-test can be used in ABC to test for location equivalence. (a) Corresponding power function for $\alpha = 0.01$ and increasing equivalence regions for the simulated dataset in the main text. The lower endpoint of the equivalence region τ^- is set to $-\tau^+$. Each line in the plot shows the acceptance probability of the corresponding ABC accept/reject step as a function of $\rho = \theta - \bar{x}$. (b) Calibrated g_{ABC} density (grey dashed line) for $\tau^-_{\text{cali}} = -0.375$, $\tau^+_{\text{cali}} = 0.375$, $m_{\text{cali}} = 164$ versus the g density (black).

Figure 11.2a shows the power function for several values of τ^+ and the default $\alpha = 0.01$. The plot shows the acceptance probability of the accept/reject step as ρ changes, each line corresponding to different equivalence regions $[-\tau^+, \tau^+]$. Intuitively, the acceptance probability increases with the width of the equivalence region. If $[\tau^-, \tau^+] = [-0.4, 0.4]$, the calibrated ABC acceptance region is $[h^-_{\text{cali}}, h^+_{\text{cali}}] = [-0.029, 0.029]$, and the acceptance probability is only 15% even when $\rho = 0$. On the other hand, if $[\tau^-, \tau^+] = [-0.9, 0.9]$, the calibrated ABC acceptance region is $[h^-_{\text{cali}}, h^+_{\text{cali}}] = [-0.53, 0.53]$, and the acceptance probability is essentially flat for a large set of θ close to $\bar{x}$. This means that this ABC accept/reject step will not be able to distinguish between population means θ within a relatively large area around $\bar{x}$. Hence, we should aim for smaller ABC tolerances. Finally, if $[\tau^-, \tau^+] = [-0.6, 0.6]$, the acceptance region is $[h^-_{\text{cali}}, h^+_{\text{cali}}] = [-0.23, 0.23]$. The acceptance probability is 85% when $\rho = 0$ and declines markedly as θ departs from $\bar{x}$. This means that there is a high chance to accept when θ is very close to $\bar{x}$, and that this ABC accept/reject step will likely be able to differentiate between population means that are close to $\bar{x}$. This calibration process is automated through:

```
> mutost.calibrate(n.of.y=length(sim),s.of.y=sd(sim),alpha=0.01,
    mx.pw=0.9,what='MXPW')
          c.l              c.u            tau.l             tau.u
-2.588859e-01  2.588859e-01 -6.293335e-01   6.293335e-01.
```

The new output fields `tau.l` and `tau.u` return the calibrated equivalence region $[\tau^-_{\text{cali}}, \tau^+_{\text{cali}}]$. We can also calibrate the length of the simulations m through the `KL` option:

```
> mutost.calibrate(n.of.x=length(obs), s.of.x=sd(obs),
    s.of.y=sd(sim), what='KL')
          c.l              c.u            tau.u                KL  n.of.y
 -0.155010790    0.155010790   0.375163551    0.002321863     164.
```

The new output fields `n.of.y` and `KL` return the calibrated number of simulations m_{cali} and the KL divergence $\varepsilon_{\text{cali}}$ between g and g_{ABC}. Figure 11.2b compares the calibrated ABC approximation of the likelihood when considered as a density in ρ (the g_{ABC} density) to the marginal posterior distribution of ρ under a reference prior (the g density). The KL divergence is small, $\varepsilon_{\text{cali}} = 0.002$. In practice, this is the **R** function call to be used in ABC. The other functions illustrate the intermediate calibration steps.

Note that the two one-sided T-test depends on the standard deviation of the simulated data, $\hat{\sigma}$. As $\hat{\sigma}$ typically changes at every ABC step, the ABC tolerances need to be re-calibrated at every ABC iteration. This is usually not problematic in real-world applications, because 10,000 calls to the `mutost.calibrate` routine take about 1.2 seconds on a standard laptop. We simulate 3–4 times as many data points as in the observed data, calculate $\hat{\sigma}$, calibrate h^-_{cali}, h^+_{cali}, m_{cali}, and then use the first m_{cali} simulated data points in step 9 of the ABC sampler shown at the beginning of this chapter.

11.4.2 Testing if dispersions are similar

Suppose again that the data or summaries are normally distributed with unknown mean and variance, and that we now want to test for dispersion equivalence. We write $y_i \sim \mathcal{N}(\mu, \theta)$, $x_i \sim \mathcal{N}(\mu_0, \theta_0)$. The setup of the test is as follows. Let $\rho = \theta/\hat{\theta}_0$, where $\hat{\theta}_0$ is the maximum likelihood estimate $\frac{1}{n}\sum_{i=1}^{n}(x - \bar{x})^2$. The point of equality is now $\rho^* = 1$. The null hypothesis is $H_0\colon \rho \notin [\tau^-, \tau^+]$ versus $H_1\colon \rho \in [\tau^-, \tau^+]$, where $\tau^- < \rho^* < \tau^+$. The test statistic is:

$$T(\boldsymbol{y}_{1:m}, \boldsymbol{x}_{1:n}) = \frac{S^2(\boldsymbol{y}_{1:m})}{S^2(\boldsymbol{x}_{1:n})} = \frac{\theta}{\hat{\theta}_0}\,\frac{1}{n}\sum_{i=1}^{m}\left(\frac{y_i - \bar{y}}{\theta}\right)^2, \tag{11.26}$$

and the ABC acceptance region is $[h^-_{\text{cali}}, h^+_{\text{cali}}]$, where h^-_{cali} and h^+_{cali} need to be found numerically. The power function of the corresponding size-α test is:

$$P(T \in [h^-_{\text{cali}}, h^+_{\text{cali}}] \mid \rho) = F_{\chi^2_{m-1}}\left(nh^+_{\text{cali}}/\rho\right) - F_{\chi^2_{m-1}}\left(nh^-_{\text{cali}}/\rho\right), \tag{11.27}$$

where $F_{\chi^2_{m-1}}$ is the cumulative density of a χ^2 distribution with $m-1$ degrees of freedom. Following (11.16–11.17), we find:

$$g(\,\rho \mid \boldsymbol{x}_{1:n}\,) \propto \rho^{-(n/2-1)-1} \exp\left(\frac{n}{2}\,/\,\rho\right), \tag{11.28}$$

so ρ is inverse gamma distributed with shape $n/2-1$ and scale $n/2$. In analogy to the previous example, the power can be calculated with `vartest.pow`, and the various calibration routines are available with `vartest.calibrate`. It seems intuitive to use ABC acceptance regions of the form $[1/h^+, h^+]$, $h^+ > 1$. It can be shown that such ABC acceptance regions correspond to equivalence regions $[1/\tau^+, \tau^+]$, $\tau^+ > 1$ (Wellek, 2003). However, Figure 11.3a

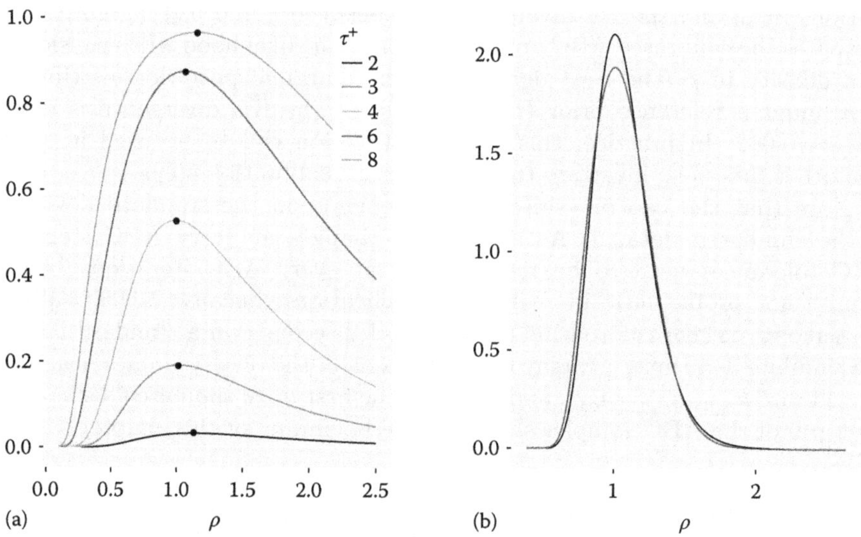

FIGURE 11.3
Calibrated variance test. If the data or summaries are normally distributed with unknown mean and variance, the one sample χ^2-test can be used to test for dispersion equivalence. (a) Power function of the χ^2-test for $\alpha = 0.01$ and increasing equivalence regions of the form $[1/\tau^+, \tau^+]$. Each line in the plot shows the acceptance probability of the corresponding ABC accept/reject step as a function of ρ for increasingly wider equivalence regions. Note that for the choice $\tau^- = 1/\tau^+$, the power is not maximised at the point of equality $\rho^* = 1$ (black squares). (b) Calibrated g_{ABC} density (grey) for $n = 60$, $\hat{\theta}_0 = 1.42^2$, $\tau^-_{\text{cali}} = 0.57$, $\tau^+_{\text{cali}} = 1.75$, $m_{\text{cali}} = 104$ versus the g density (black).

reveals that for this choice, the maximum of the power function is not exactly located at the point of equality (namely, $\rho^* = 1$). This means that such an ABC accept/reject step will most often accept values of θ that do not coincide with the maximum likelihood estimate of the observed data. An analytical solution to τ^- such that the power is maximised at $\rho^* = 1$ is not available, but we can resort to numerical optimisation, again using Brent's method. Therefore, the calibration procedures for the **vartest** involve one additional step, when compared to the calibration procedures for the **mutost**. This additional numerical calibration step is hidden from the user:

```
> n.of.x <- 60
> s.of.x <- 1.42
> vartest.calibrate(n.of.x=n.of.x, s.of.x=s.of.x, what='KL')
         c.l          c.u        tau.l        tau.u       n.of.y           KL
1.396134e+00 2.196636e+00 5.866913e-01 1.755158e+00 1.050000e+02 8.497451e-03.
```

Figure 11.3b compares the calibrated ABC approximation to the likelihood when considered as a density in ρ (the g_{ABC} density) to the likelihood when considered as a density in ρ (the g density). The KL divergence is small, 0.008, indicating a very good approximation. The resulting ABC acceptance region $[h^-_{\text{cali}}, h^+_{\text{cali}}] \approx [1.396, 2.196]$ is far from intuitive, indicating that it would be difficult to achieve the same degree of accuracy through manual selection of the tolerances.

In contrast to the **mutost**, the **vartest** calibration routine only depends on the observed data through n and $S^2(\boldsymbol{x}_{1:n})$. This is known before ABC is started, so that the ABC tolerances have to be calibrated only once before ABC is run, before step 3 in the sampler at the beginning of this chapter.

11.4.3 Testing if rates are similar

We now suppose that the data or summaries are exponentially distributed with unknown rates, and that we want to test for rate equivalence. We write $y_i \sim \text{Exp}(\theta)$, $x_i \sim \text{Exp}(\theta_0)$, where θ is the scale parameter (reciprocal of the rate parameter). The setup of the test is as follows. Let $\rho = \theta/\hat{\theta}_0$, where $\hat{\theta}_0$ is the maximum likelihood estimate $\bar{x} = 1/n \sum_{i=1}^{n} x_i$. The point of equality is $\rho^* = 1$. The null hypothesis is $H_0\colon \rho \notin [\tau^-, \tau^+]$ versus $H_1\colon \rho \in [\tau^-, \tau^+]$, where $\tau^- < \rho^* < \tau^+$. The test statistic is:

$$T(\boldsymbol{y}_{1:m}, \boldsymbol{x}_{1:n}) = \bar{y} \,/\, \bar{x}, \tag{11.29}$$

and the ABC acceptance region is $[h^-_{\text{cali}}, h^+_{\text{cali}}]$, where h^-_{cali} and h^+_{cali} need to be found numerically as well. The power function of the corresponding size-α test is:

$$P(\,T \in [h^-_{\text{cali}}, h^+_{\text{cali}}] \mid \rho\,) = F_{\Gamma_{m,1}}\Big(\, m\, h^+_{\text{cali}} \,/\, \rho \,\Big) - F_{\Gamma_{m,1}}\Big(\, m\, h^-_{\text{cali}} \,/\, \rho \,\Big), \tag{11.30}$$

where $F_{\Gamma_{m,1}}$ is the cumulative density of a gamma distribution with shape m and scale 1. Following (11.16–11.17), we find:

$$g(\rho \mid \boldsymbol{x}_{1:n}) \propto \rho^{-(n-1)-1} \exp(-n/\rho), \tag{11.31}$$

so ρ is inverse gamma distributed with shape $n-1$ and scale n. Comparing (11.26) with (11.29), we see that the rate test is very similar to the dispersion test. The ABC calibrations can be performed through the function `ratetest.calibrate`.

11.5 A Worked Example

In this section, we illustrate the application of ABC calibrations in a real-world example. We consider weekly counts of influenza-like-illness (ILI) in the Netherlands between 1994 and 2009. Figure 11.4 shows weekly counts that could previously be attributed to influenza A subtype H3N2 (in short, H3N2), the most prevalent influenza subtype over the past 40 years. We investigated

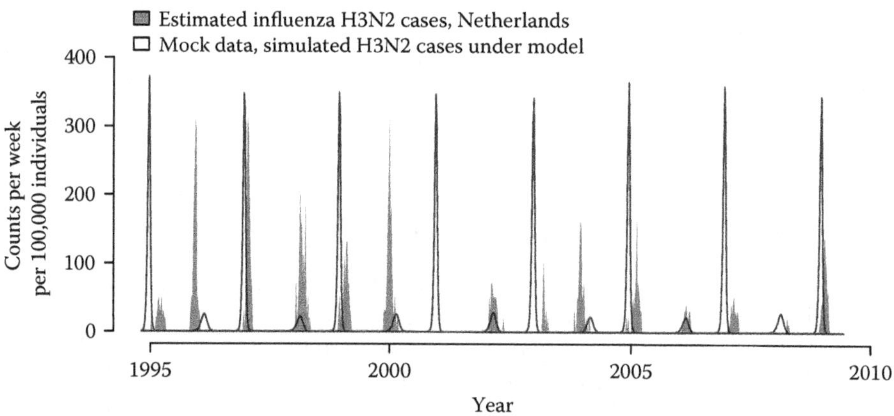

FIGURE 11.4
Empirical and simulated influenza-like-illness time series data from the Netherlands. In the Netherlands, influenza-like-illness counts are obtained through sentinel surveillance of general practices, and extrapolated to the national population. In grey, weekly counts that could be attributed to influenza A subtype H3N2 are shown, as reproduced from (Ratmann et al., 2012). To illustrate the application of ABC calibrations in a real-world example, we considered mock data from a simple mathematical model and aim to re-estimate the known model parameters. As a black line, weekly H3N2 incidence counts that are part of the model output are shown. The simulated data have overall similar magnitude, but show a much more pronounced biennial pattern.

several stochastic simulation models of the H3N2 dynamics previously (Ratmann et al., 2012), and here we focus on the simplest of these. We created mock data from this simple model and compare in this section a previously developed standard ABC inference approach to the calibrated inference approach in their ability to recover the known, true model parameters.

The model describes H3N2's disease dynamics in a population stratified into susceptible (S), infected, but not yet infectious (E), infectious (I_1 and I_2), and immune (R) individuals across two large spatial areas. The stochastic model is easily simulated from Markov transition probabilities derived from the deterministic ordinary differential equations:

$$\begin{aligned}
\frac{dS^{\downarrow}}{dt} &= \mu^{\downarrow}(N^{\downarrow} - S^{\downarrow}) - \lambda_t^{\downarrow}\frac{S^{\downarrow}}{N^{\downarrow}}(I_1^{\downarrow} + I_2^{\downarrow} + M^{\downarrow}) + \gamma R^{\downarrow} \\
\frac{dE^{\downarrow}}{dt} &= \lambda_t^{\downarrow}\frac{S^{\downarrow}}{N^{\downarrow}}(I_1^{\downarrow} + I_2^{\downarrow} + M^{\downarrow}) - (\mu^{\downarrow} + \phi)E^{\downarrow} \\
\frac{dI_1^{\downarrow}}{dt} &= \phi E^{\downarrow} - (\mu^{\downarrow} + 2\nu)I_1^{\downarrow} \\
\frac{dI_2^{\downarrow}}{dt} &= 2\nu I_1^{\downarrow} - (\mu^{\downarrow} + 2\nu)I_2^{\downarrow} \\
\frac{dR^{\downarrow}}{dt} &= 2\nu I_2^{\downarrow} - (\mu^{\downarrow} + \gamma)R^{\downarrow}.
\end{aligned} \tag{11.32}$$

The sink population (indicated by $^{\downarrow}$) represents the Netherlands, and the source population (indicated by $^{\circlearrowleft}$) represents a large viral reservoir in which the virus persists and evolves. Only the rate equations corresponding to the sink population are shown. The transition probabilities in the stochastic model correspond to exponentially distributed event probabilities in a small interval Δt, with mean equal to the inverse of the rates in Greek letters in (11.32) (Gillespie, 2007). Two infectiousness compartments, I_1 and I_2, are used to obtain gamma-distributed infectious periods for the sake of biological realism. The transmission rate $\lambda_t^{\downarrow} = \lambda(1 + \varphi^{\downarrow}\sin(2\pi(t - t^{\downarrow})))$ is seasonally forced with peaks in the winter months in the sink population, so that only the baseline transmission rate λ is a free parameter. By contrast, the transmission rate is constant and equal to λ in the source population. The average incubation period $1/\phi$ and average infectiousness period $1/\nu$ are informed by sharp priors from existing data, while the average duration of immunity $1/\gamma$ is the second free parameter. After the summer months, new influenza seasons are re-seeded from the viral reservoir by the expected number of infected travellers from the source population, $M^{\downarrow} = m^{\downarrow}(I_1^{\circlearrowleft} + I_2^{\circlearrowleft})/N^{\circlearrowleft}$, where $m^{\downarrow}$ is the number of travellers. The birth/death rate in the Netherlands $\mu^{\downarrow}$ and $m^{\downarrow}$ are set to estimates from the literature. To fit (11.32) to the ILI time series, we used a Poisson observation model of new weekly ILI cases with mean $\omega I_1^{\downarrow +}$, where $I_1^{\downarrow +}$ is the number of newly infected individuals in the sink population per week and ω is the reporting rate. The third free parameter is ω. To ease intuition, λ is re-parameterised as the basic re-productive number R_0. Thus, the model parameters of interest to us are $\theta = (R_0, 1/\gamma, \omega)$; i.e. $p = 3$. Pseudo data x were generated for $\theta_0 = (3.5, 10 \text{ years}, 0.08)$, as shown in Figure 11.4.

To re-estimate θ_0, we considered in the standard ABC and calibrated ABC approach six summary statistics. H3N2 annual attack rates are the cumulative incidence per winter season divided by population size. The magnitude, variation, and inter-seasonal correlation in H3N2 annual attack rates are characteristic features of seasonal long-term H3N2 dynamics (Ratmann et al., 2012). To capture these aspects of the data, we considered annual attack rates of the reported ILI time series data (aILI), their first-order differences (fdILI), and estimates of annual population-level attack rates in H3N2 seasons (aINC). On real data, estimates of aINC are obtained from seroprevalence surveys. On the mock data considered here, aINC were calculated directly. Figure 11.5 shows that the aILI, fdILI, and aINC of the simulated data are biennial. For this reason, we formed six summaries corresponding to odd and even values of the aILI, fdILI, and aINC; i.e. $q = 6$. Figure 11.5 shows that, once the time series were split, the sampling distribution of the odd and even values can be modelled with a normal density.

In the standard ABC approach, we compared the means of these six time series summaries. These summaries were combined with the intersection approach, in which an iteration is only accepted when the error between all simulated and observed summaries is sufficiently small. Specifically, let us denote the six observed and simulated summaries by $\boldsymbol{s}^{\text{obs}}_{i,1:n_i} = (s^{\text{obs}}_{i,1}, \ldots, s^{\text{obs}}_{i,n_i})$ and $\boldsymbol{s}^{\text{sim}}_{i,1:n_i} = (s^{\text{sim}}_{i,1}, \ldots, s^{\text{sim}}_{i,n_i})$, where the index runs over even years for three summaries and odd years for the other three summaries. We then used the ABC accept/reject step:

$$\mathbb{1}\left\{ \bigcap_{i=1}^{q} \left\{ h_i^- \leq \left(\bar{s}_i^{\text{sim}} - \bar{s}_i^{\text{obs}}\right) \leq h_i^+ \right\} \right\}, \tag{11.33}$$

where h_i^-, h_i^+ are the user-defined tolerances, and the sample means of the simulated and observed summaries, $\bar{s}_i^{\text{sim}} = \frac{1}{n_i}\sum_{j=1}^{n_i} s^{\text{sim}}_{i,j}$ and $\bar{s}_i^{\text{obs}} = \frac{1}{n_i}\sum_{j=1}^{n_i} s^{\text{obs}}_{i,j}$, are compared to each other.

We used broad prior densities $R_0 \sim \mathcal{U}(1, 8)$, $1/\gamma \sim \mathcal{U}(2, 30)$, $\omega \sim \mathcal{U}(0, 1)$ and implemented an ABC Markov chain Monte Carlo sampler with normal proposal kernel and annealing on the variance of the proposal kernel and the tolerances (Gilks et al., 1996). The ABC tolerances were set such that our ABC MCMC sampler produced an acceptance probability of 5%. In ABC, acceptance probabilities are usually far lower than the recommended range (Gilks et al., 1996). On the mock datasets, the estimated 95% credibility intervals contained θ_0. However, the mean squared error of the posterior mean to θ_0 was large (Figure 11.6a).

In the calibrated approach, we used for each of the six summaries the two one-sided T-test, as described in Section 11.4.1. To verify that this equivalence T-test can be used, we first considered the quantile-quantile plots shown in Figure 11.5b, and then tested at every ABC iteration independence and

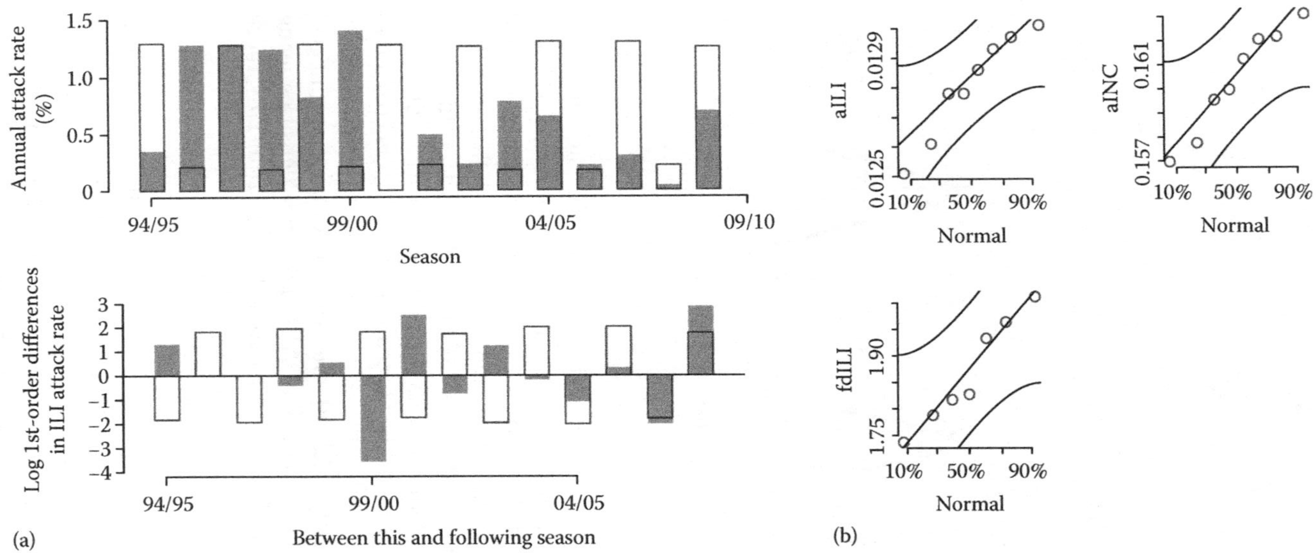

FIGURE 11.5
Features of long-term H3N2 dynamics. The cumulative incidence of influenza H3N2 infections per winter season divided by population size, the so-called annual attack rate, is a characteristic feature of long-term H3N2 disease dynamics. (a) The top panel shows reported annual attack rates (aILI) by winter season for the estimated H3N2 cases in the Netherlands (grey) and the mock data (white bars). The bottom panel shows the first-order differences in reported annual attack rates (fdILI) as a measure of their inter-seasonal variability. These features show strong biennial patterns in the mock data and weakly biennial patterns in the real data. To obtain summary statistics with a unimodal sampling distribution, we considered odd and even seasons separately. (b) Quantile-quantile plots illustrate departures from normality for the three odd summary statistics calculated on the mock data. Bottom and top curves indicate lower 2.5% and upper 97.5% confidence intervals. Normality cannot be rejected for any of the odd and even summary statistics.

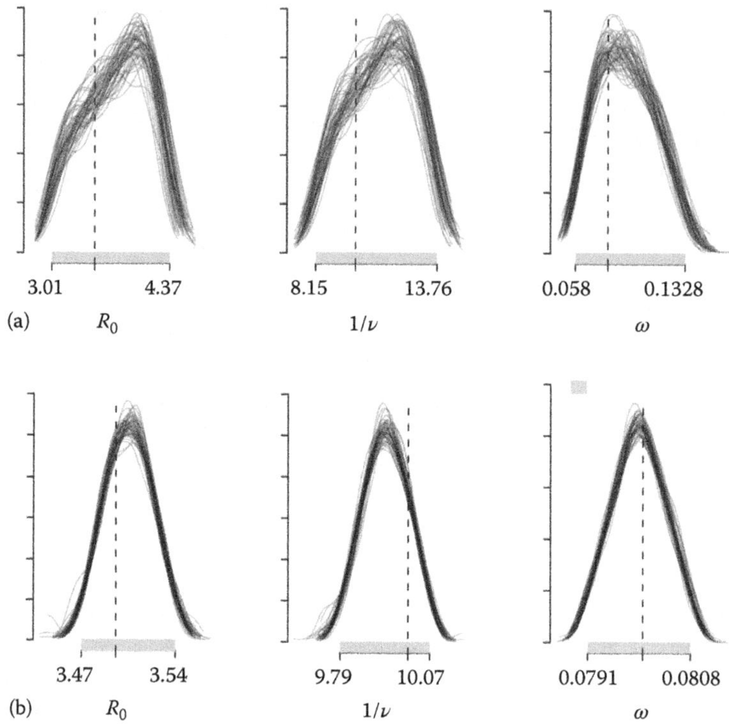

FIGURE 11.6
Comparison of posterior densities obtained with standard ABC and calibrated ABC. Mock data were generated under the spatial SIR-type model (11.32), and then a previously published ABC-MCMC algorithm with user-defined tolerances was used to re-estimate the true model parameters, shown in dashed vertical lines. We then ran a similar ABC-MCMC algorithm with calibrations as described in this chapter. (a) Marginal ABC posterior densities with user-defined tolerances are shown for each of the three model parameters (columns) on 50 replicate mock data sets that were generated from the same model parameter. 95% credibility intervals (grey) and their endpoints (numbers) are highlighted. (b) For comparison, marginal ABC posterior densities using calibrated `mutost` test statistics. The calibrations led to substantially more accurate parameter estimates, see further the main text.

normality of the summary vectors $s^{\text{sim}}_{i,1:m}$, $i = 1, \ldots, 6$ in a pilot run. The p-values of these tests were approximately uniformly distributed, indicating that the two one-sided T-tests are suitable for this application.

From these six size-α hypothesis tests, it is easy to construct a multivariate size-α hypothesis test (Ratmann et al., 2014). Let us denote, for the ith simulated summaries, the population mean of the corresponding, unknown normal sampling distribution by β_i. For the ith observed summaries, the corresponding maximum likelihood estimate is the sample mean $\bar{s}_i^{\text{obs}}$, which is straightforward to calculate. In analogy to the discrepancy parameter ρ defined in (11.11), the unknown, multivariate discrepancy parameter is now:

$$\rho = (\rho_1, \ldots, \rho_q) \in \mathbb{R}^q, \qquad \rho_i = \beta_i - \bar{s}_i^{\text{obs}}, \tag{11.34}$$

and the univariate null and alternative hypotheses are:

$$H_{0i} = \{\, \rho \in \mathbb{R}^q \mid \rho_i \notin [\tau^-, \tau^+] \,\}, \quad H_{1i} = \{\, \rho \in \mathbb{R}^q \mid \rho_i \in [\tau^-, \tau^+] \,\}. \tag{11.35}$$

Now consider the multivariate intersection-union hypotheses:

$$H_0 = \bigcup_{i=1}^{q} H_{0i}, \quad H_1 = \bigcap_{i=1}^{q} H_{1i}. \tag{11.36}$$

The multivariate statistic $T = (T_1, \ldots, T_q)$, which is the vector of the six two one-sided T-tests, with ABC acceptance region $[h_1^-, h_1^+] \times \ldots \times [h_q^-, h_q^+]$ is also level-α and, under further conditions, size-α (Berger, 1982). We suppose further that the sampling distributions of the test statistics are orthogonal to each other, so that:

$$\begin{aligned} p(\boldsymbol{s}_{\text{obs}} \mid \rho\,) &= \prod_{i=1}^{q} p(\boldsymbol{s}_{i,1:n}^{\text{obs}} \mid \rho_i\,), \\ P\Big(T \in [h_1^-, h_1^+] \times \ldots \times [h_q^-, h_q^+] \,\Big|\rho\Big) &= \prod_{i=1}^{q} P\Big(T_i \in [h_i^-, h_i^+] \,\Big|\rho_i\Big). \end{aligned} \tag{11.37}$$

This means that the multivariate ABC acceptance region is calibrated when all univariate acceptance regions are calibrated independently. These calibrations are easy to do: at every ABC iteration, we simulated a time series three times as long as the mock data, calibrated $h_{i,\text{cali}}^-$, $h_{i,\text{cali}}^+$, $\tau_{i,\text{cali}}^-$, $\tau_{i,\text{cali}}^+$, $m_{i,\text{cali}}$ for each summary with the `mutost.calibrate` function, and discarded the excess simulations beyond m_{cali}. It is important to discard excess simulations because otherwise the power function is too tight when compared to the likelihood, as illustrated in Figure 11.1b. The resulting calibrated ABC accept/reject step is:

$$\mathbb{1}\left\{ \bigcap_{i=1}^{q} \left\{ h_{i,\text{cali}}^- \leq \left(\bar{s}_{i,\text{cali}}^{\text{sim}} - \bar{s}_i^{\text{obs}} \right) \leq h_{i,\text{cali}}^+ \right\} \right\}, \tag{11.38}$$

where $\bar{s}_{i,\text{cali}}^{\text{sim}}$ is the sample mean over the $m_{i,\text{cali}}$ simulated summaries. Comparing (11.33) to (11.38), we see that the intersection accept/reject step can be interpreted as the outcome of a multivariate equivalence hypothesis test, with all associated benefits. It also shows that the calibrations can be easily integrated into more complex Markov chain Monte Carlo or sequential Monte Carlo ABC algorithms.

The marginal ABC posterior densities obtained under the calibration approach are reported in Figure 11.6b, and are considerably tighter than those obtained with the previously used standard ABC approach. The mean squared error of the ABC posterior mean of θ_0 over the 50 mock datasets was approximately 20-fold smaller than previously. The reason for this substantial improvement in accuracy is that the calibrated tolerances were substantially smaller than the user-defined tolerances in the standard ABC approach. In effect, the calibration approach separates statistical considerations from computational considerations in the ABC approximation. Previously, the tolerances were set to obtain a pre-specified acceptance probability. With the automatically determined, substantially smaller tolerances $h^-_{i,\text{cali}}$, $h^+_{i,\text{cali}}$, we had no choice but to tune the ABC-MCMC sampler further in order to obtain a similar acceptance probability as before. Note that the power function was calibrated so that the acceptance probability is high for proposed θ that are close to the unknown maximum likelihood estimate; but of course, we still have to propose such θ in the first place. We settled on an adaptive block proposal kernel to obtain an acceptance probability of 5%. With this extra effort into improved mixing of the MCMC sampler, the calibrated ABC posterior distribution was substantially more accurate than the standard ABC approach that we used before.

It is important to note that the calibrations only improve the ABC approximation to the partial posterior density based on the summaries (11.2) and not the posterior density based on the data (11.1). When the summaries are not sufficient, it is common practice to include as many summaries as possible or to construct summaries that are optimal in some sense as discussed by Prangle (2019) in this book. In our example, we opted for the first approach. We used six summaries to estimate three model parameters.

There is one particular recommendation for choosing summaries that emerges from describing the ABC accept/reject step as the outcome of hypothesis tests. The hypothesis tests describe a statistical decision problem on the unknown value of discrepancy parameters ρ in (11.34). Of course, the value of the discrepancies is related to the value of the model parameters θ, but it is not known how. This function is commonly referred to as the 'binding function' in indirect inference approaches (See Drovandi [2019], Chapter 7, this volume). The binding function:

$$\mathfrak{b}\colon \mathbb{R}^p \to \mathbb{R}^q, \quad \mathfrak{b}(\theta) = \rho, \qquad p \leq q, \tag{11.39}$$

must be injective in order to extend (11.21) and (11.22) to the more general case considered here. Intuitively, this property means that we have sufficiently many summaries to detect changes in model parameters θ through the discrepancies ρ. Thus, the number of summaries (or hypothesis tests) should at least equal the number of model parameters, for example, $p \leq q$. The binding function typically only maps to a subset of $\mathbb{R}^q$, and on that subset the binding function is bijective, if it is injective. This means that the inverse of the binding function is well defined, which we need for ABC inference.

It is not so simple to check if high-dimensional binding functions are injective. We can check this condition where it is most important, around the point of equality $\rho^\star$, through the relation:

$$b^{-1}(\rho^*) = \bigcap_{i=1}^{q} b_i^{-1}(\rho_i^*), \tag{11.40}$$

on the inverse of the binding function. Indeed, the one-dimensional level-sets:

$$b_i^{-1}(\rho_i^*) = \{\theta \in \mathbb{R}^p \mid i\text{th element of } b(\theta) \text{ equal to } \rho_i^*\},$$

can be easily estimated with local polynomial regression techniques after ABC has been run (Loader, 1999), and from this, we can calculate their intersection. If the binding function is injective around $\rho^\star$, this intersection is just a single point. Figure 11.7 shows the estimated level sets and their intersections

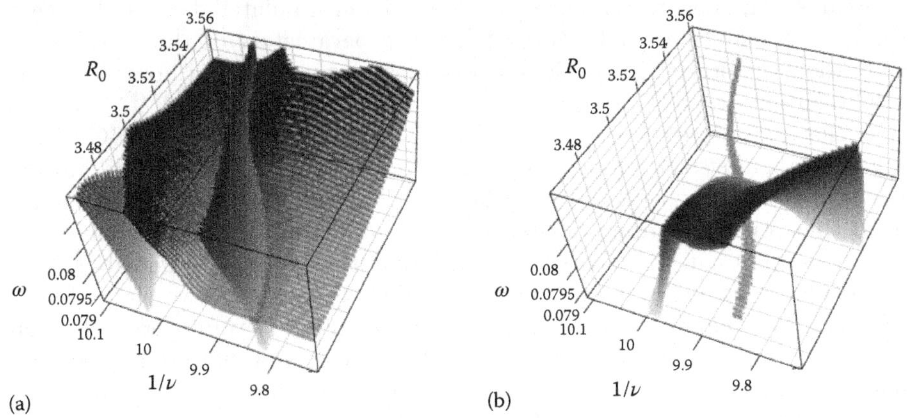

FIGURE 11.7
Checking if sufficiently many summaries are used in ABC. The hypothesis testing approach provides a particular opportunity for checking if sufficiently many summaries are used in ABC. Increasing the number of hypothesis tests, that are used in ABC, changes the binding function from θ to the discrepancy parameters ρ. This function must be injective, which can be checked around $\rho^\star$ through the relation (11.40). We reconstructed the level sets $b_i^{-1}(\rho_i^*)$ for the time series example, using local polynomial regression. (a) Level sets for the odd aILI (black) and the odd fdILI (light grey). The estimated level sets illustrate which θ map to very small errors ρ_i around ρ_i^*. Here, the level sets are highly non-linear and nearly orthogonal to each other. According to the relation (11.40), we calculate their intersection, which results in multiple paths through θ-space. (b) Intersection of level sets from (a) and level set for the odd aINC. According to the relation (11.40), we calculate again their intersection, which results in a very small ball. This indicates that the pre-image $b^{-1}(\rho^*)$ is indeed a single value and that, the binding function is injective around ρ^* for the particular choice of summaries that we used.

for three of the six odd and even summaries that we used for ABC inference. The remaining level sets did not change the intersection shown in Figure 11.7b in dark black. The final intersection maps to a small ball, which indicates that the binding function is indeed injective around ρ^*. If this had not been the case, we would have concluded that more summaries need to be included into ABC inference.

11.6 Discussion and Further Reading

In this chapter, we discussed how one-sample equivalence hypothesis tests can be used within ABC. Their main utility is to inform some of the choices that ABC users need to make when they apply ABC in practice. Statistical decision tests can specify how to compare the observed and simulated data, and further calibrations can be used to specify the free parameters of the test. These calibrations determine in particular how long to simulate and how to set the ABC tolerances.

If the data are of a particularly simple form, the calibrated ABC posterior density enjoys the properties (11.21–11.23). The calibrations can also be extended to the more general case when the data are arbitrarily complex, provided that multiple summaries (such as the annual attack rates in the time series example) can be constructed, and that their sampling distribution is of a simple form. This is not always possible. Where it is, we showed that the quality of the ABC approximation can be sufficiently improved through the calibrations. Particular consideration has to be given that appropriate hypothesis tests are used (i.e. that the sampling distribution of the summaries warrants the use of particular hypothesis tests) and that sufficiently many summaries are used (i.e. so that the underlying binding function is injective). We also required that the test statistics are independent of each other.

For end-users, the `abc.star` package makes available the most commonly used hypothesis tests and their ABC calibration routines. This makes it straightforward to use the calibrations in practice, but some work is needed to identify summaries to which the calibrations can be applied. Several other tests could be implemented and made available for ABC calibrations (Wellek, 2003). For developers, the open source code and package documentation will provide further details on how to extend the existing calibration functions. More technical details are also found in Ratmann et al. (2014).

From a more theoretical perspective, the calibrations fit well into the growing literature that links ABC to indirect inference (See Drovandi [2019] and Mengersen [2019], Chapters 7 and 12, this volume). In its most basic form, indirect inference proceeds by defining an auxiliary probability model for which maximum-likelihood estimates can be obtained and then transports these estimates back to the parameter space of interest (Gouriéroux et al., 1993). In ABC, the sampling distribution of the summaries also defines an auxiliary probability model, and the binding function relates the auxiliary parameters to the parameters of interest. Many aspects of ABC indirect inference have recently received substantial attention in the community (Drovandi et al., 2011, 2015; Gleim and Pigorsch, 2013; Cabras et al., 2014, 2015; Moores et al., 2015), and the calibrations of the free ABC parameters can be embedded into this framework. We only focussed on the requirement that the binding function must be injective around $\rho^{\star}$. Other properties likely influence computational and statistical aspects of ABC and could be more fully characterised along the lines proposed in maximum-likelihood indirect inference (Gallant and Tauchen, 1996; Jiang and Turnbull, 2004; Heggland and Frigessi, 2004; Fermanian and Salanie, 2004).

Further work should explore how our calibration assumptions could be relaxed. For example, multivariate equivalence hypothesis tests have been formulated (Wellek, 2003) and could obviate the requirement that the test statistics are independent of each other. Non-parametric tests also exist (Wellek, 1996; Munk and Czado, 1998), potentially opening up the possibility of calibrating ABC with summaries whose sampling distribution is not necessarily of the simple form that we assumed here.

Perhaps the most fundamental conclusion from the ABC calibrations described in this chapter is that understanding and modelling the distribution of the summaries can provide essential insights into ABC inference. If their distribution is understood, then appropriate statistical decision tests can be used to improve the accuracy of ABC algorithms, and to improve our understanding of the approximations in ABC.

Acknowledgements

OR is supported by the Wellcome Trust (fellowship WR092311MF), AC by the Medical Research Council (fellowship MR/J01432X/1), and CC by the Engineering and Physical Sciences Research Council (EP/K026003/1). We thank the Imperial High Performance Computing Centre, and especially Simon Burbidge for his support.

References

Barnes, C. P., S. Filippi, M. P. Stumpf, and T. Thorne (2012). Considerate approaches to constructing summary statistics for ABC model selection. *Statistics and Computing 22*(6), 1181–1197.

Beaumont, M., W. Zhang, and D. Balding (2002). Approximate Bayesian computation in population genetics. *Genetics 162*, 2025–2035.

Berger, R. (1982). Multiparameter hypothesis testing and acceptance sampling. *Technometrics 24*(4), 295–300.

Blum, M. G. (2010). Approximate Bayesian computation: A nonparametric perspective. *Journal of the American Statistical Association 105*(491), 1178–1187.

Bortot, P., S. Coles, and S. Sisson (2007). Inference for stereological extremes. *Journal of the American Statistical Association 102*(9), 84–92.

Cabras, S., M. E. Castellanos, and E. Ruli (2014). A quasi likelihood approximation of posterior distributions for likelihood-intractable complex models. *METRON 72*, 153–167.

Cabras, S., M. E. C. Nueda, and E. Ruli (2015). Approximate Bayesian computation by modelling summary statistics in a quasi-likelihood framework. *Bayesian Analysis 10*(2), 411–439.

Cornuet, J.-M., F. Santos, M. A. Beaumont, C. P. Robert, J.-M. Marin, D. J. Balding, T. Guillemaud, and A. Estoup (2008). Inferring population history with DIYABC: A user-friendly approach to approximate Bayesian computation. *Bioinformatics 24*(23), 2713–2719.

Csilléry, K., M. G. Blum, O. E. Gaggiotti, and O. François (2010). Approximate Bayesian computation (ABC) in practice. *Trends in Ecology & Evolution 25*(7), 410–418.

Dean, T. A., S. S. Singh, A. Jasra, and G. W. Peters (2014). Parameter estimation for hidden Markov models with intractable likelihoods. *Scandinavian Journal of Statistics 41*(4), 970–987.

Doksum, K. A. and A. Y. Lo (1990). Consistent and robust Bayes procedures for location based on partial information. *The Annals of Statistics 18*(1), 443–453.

Drovandi, C. (2019). ABC and indirect inference. In S. A. Sisson, Y. Fan, and M. A. Beumont (Eds.), *Handbook of Approximate Bayesian Computation.* Chapman & Hall/CRC Press, Boca Raton, FL.

Drovandi, C. C., A. N. Pettitt, and M. J. Faddy (2011). Approximate Bayesian computation using indirect inference. *Journal of the Royal Statistical Society: Series C (Applied Statistics) 60*(3), 317–337.

Drovandi, C. C., A. N. Pettitt, and A. Lee (2015). Bayesian indirect inference using a parametric auxiliary model. *Statistical Science, 30*(1), 72–95.

Fagundes, N. J. R., N. Ray, M. Beaumont, S. Neuenschwander, F. M. Salzano, S. L. Bonatto, and L. Excoffier (2007). Statistical evaluation of alternative models of human evolution. *Proceedings of the National Academy of Sciences USA 104*(45), 17614–17619.

Fearnhead, P. and D. Prangle (2012). Constructing summary statistics for approximate Bayesian computation: Semi-automatic approximate Bayesian computation. *Journal of the Royal Statistical Society: Series B (Statistical Methodology) 74*(3), 419–474.

Fermanian, J.-D. and B. Salanie (2004). A nonparametric simulated maximum likelihood estimation method. *Econometric Theory 20*(4), 701–734.

Gallant, A. R. and G. Tauchen (1996). Which moments to match? *Econometric Theory 12*(4), 657–681.

Gilks, W. R., S. Richardson, and D. J. Spiegelhalter (1996). *Markov Chain Monte Carlo in Practice.* Chapman & Hall: Boca Raton, FL.

Gillespie, D. T. (2007). Stochastic simulation of chemical kinetics. *Annual Review of Physical Chemistry 58*, 35–55.

Gleim, A. and C. Pigorsch (2013). Approximate Bayesian computation with indirect summary statistics. Technical report, University of Bonn, Germany.

Gouriéroux, C., A. Monfort, and E. Renault (1993). Indirect inference. *Journal of Applied Econometrics 8*, S85–118.

Heggland, K. and A. Frigessi (2004). Estimating functions in indirect inference. *Journal of the Royal Statistical Society: Series B (Statistical Methodology) 66*, 447–462.

Jiang, W. and B. Turnbull (2004). The indirect method: Inference based on intermediate statistics: A synthesis and examples. *Statistical Science 19*(2), 239–263.

Kirkwood, T. B. and W. Westlake (1981). Bioequivalence testing – a need to rethink. *Biometrics 37*(3), 589–594.

Lehmann, E. and J. Romano (2005). *Testing Statistical Hypotheses.* Springer: New York.

Loader, C. (1999). *Local Regression and Likelihood.* Springer: New York.

Luciani, F., S. A. Sisson, H. Jiang, A. R. Francis, and M. M. Tanaka (2009). The epidemiological fitness cost of drug resistance in Mycobacterium tuberculosis. *Proceedings of the National Academy of Sciences USA 106*(34), 14711–14715.

Mengersen, K. (2019). Synthetic and empirical likelihoods. In S. A. Sisson, Y. Fan, and M. A. Beumont (Eds.), *Handbook of Approximate Bayesian Computation.* Chapman & Hall/CRC Press, Boca Raton, FL.

Moores, M. T., K. Mengersen, and C. P. Robert (2015). Pre-processing for approximate Bayesian computation in image analysis. *Statistics and Computing 25*(1), 23–33.

Munk, A. and C. Czado (1998). Nonparametric validation of similar distributions and assessment of goodness of fit. *Journal of the Royal Statistical Society: Series B (Statistical Methodology) 60*(1), 223–241.

Owen, D. (1965). A special case of a bivariate non-central t-distribution. *Biometrika 52*(3/4), 437–446.

Prangle, D. (2019). ABC Summary Statistics. In S. A. Sisson, Y. Fan, and M. A. Beumont (Eds.), *Handbook of Approximate Bayesian Computation.* Chapman & Hall/CRC Press, Boca Raton, FL.

Press, W. H., S. A. Teukolsky, W. T. Vetterling, and B. P. Flannery (2007). *Numerical Recipes: The Art of Scientific Computing* (3rd ed.). Cambridge University Press: New York.

Pritchard, J., M. Seielstad, A. Perez-Lezaun, and M. Feldman (1999). Population growth of human Y chromosomes: A study of Y chromosome microsatellites. *Journal of Molecular Biology and Evolution 16*, 1791–1798.

Ratmann, O., A. Camacho, A. Meijer, and G. Donker (2014). Statistical modelling of summary values leads to accurate approximate Bayesian computations. *arXiv.org*, 1305.4283.

Ratmann, O., A. Camacho, H. Sen, and C. Colijn (2016). *abc.star: Calibration procedures for accurate ABC.* https://github.com/olli0601/abc.star.

Ratmann, O., G. Donker, A. Meijer, C. Fraser, and K. Koelle (2012). Phylodynamic inference and model assessment with approximate Bayesian computation: Influenza as a case study. *PLoS Computational Biology 8*(12), e1002835.

Ratmann, O., O. Jørgensen, T. Hinkley, M. P. Stumpf, S. Richardson, and C. Wiuf (2007). Using likelihood-free inference to compare evolutionary dynamics of the protein networks of *H. pylori* and *P. falciparum. PLoS Computational Biology 3*(11), e230.

Schuirmann, D. (1981). On hypothesis testing to determine if the mean of a normal distribution is contained in a known interval. *Biometrics 37*(617), 137.

Silk, D., S. Filippi, and M. P. Stumpf (2013). Optimizing threshold-schedules for approximate Bayesian computation sequential Monte Carlo samplers: Applications to molecular systems. *Statistical Applications in Genetics and Molecular Biology 12*(5), 603–618.

Silk, D., P. D. Kirk, C. P. Barnes, T. Toni, A. Rose, S. Moon, M. J. Dallman, and M. P. Stumpf (2011). Designing attractive models via automated identification of chaotic and oscillatory dynamical regimes. *Nature Communications 2*, 489.

Wellek, S. (1996). A new approach to equivalence assessment in standard comparative bioavailability trials by means of the Mann-Whitney statistic. *Biometrical Journal 38*(6), 695–710.

Wellek, S. (2003). *Testing Statistical Hypotheses of Equivalence.* CRC Press: Boca Raton, FL.

Wood, S. (2010). Statistical inference for noisy nonlinear ecological dynamic systems. *Nature 466*(7310), 1102–1104.

12

Approximating the Likelihood in ABC

Christopher C. Drovandi, Clara Grazian, Kerrie Mengersen, and Christian Robert

CONTENTS

12.1 Introduction 322
12.2 Synthetic Likelihood 322
12.2.1 Classical synthetic likelihood 323
12.2.2 Bayesian synthetic likelihood 325
12.2.3 Accelerating synthetic likelihood 327
12.2.4 Example 328
12.3 Further Reading 338
12.4 Bayesian Empirical Likelihood 338
12.4.1 Empirical likelihood 339
12.4.2 Features of empirical likelihood 341
12.4.3 Estimation 343
12.4.4 Empirical likelihood in practice 345
12.4.5 The BC_{el} algorithm 347
12.4.5.1 Example 1 348
12.4.5.2 Example 2 350
12.4.5.3 Example 3 352
12.4.6 Extensions of the BC_{el} algorithm 352
Acknowledgements 358
Appendix 358
References 361

12.1 Introduction

Approximate Bayesian computation (ABC) is now a mature algorithm for likelihood-free estimation. It has been successfully applied to a wide range of real-world problems for which more standard analytic tools were unsuitable due to the absence or complexity of the associated likelihood. It has also paved the way for a range of algorithmic extensions that take advantage of appealing ideas embedded in other approaches. Despite the usefulness of ABC, the method does have a number of drawbacks. The approach is simulation intensive, requires tuning of the tolerance threshold, discrepancy function and weighting function, and suffers from a curse of dimensionality of the summary statistic. The latter issue stems from the fact that ABC uses a non-parametric estimate of the likelihood function of a summary statistic (Blum, 2010).

In this chapter, we review two alternative methods of approximating the intractable likelihood function for the model of interest, both of which aim to improve computational efficiency relative to ABC. The first is the synthetic likelihood [SL, originally developed by Wood (2010)], which uses a multivariate normal approximation to the summary statistic likelihood. This auxiliary likelihood can be maximised directly or incorporated in a Bayesian framework, which we refer to as BSL. The BSL approach requires substantially less tuning than ABC. Further, BSL scales more efficiently with an increase in the dimension of the summary statistic due to the parametric approximation of the summary statistic likelihood. However, the BSL approach remains simulation intensive. In chapter 20, Fasiolo et al. (2019) apply BSL to dynamic ecological models and compare it with an alternative Bayesian method for state space models. In this chapter, we provide a more thorough review of SL both in the classical and Bayesian frameworks.

The second approach we consider uses an empirical likelihood (EL) within a Bayesian framework [BC_{el}, see Mengersen et al. (2013)]. This approach can in some cases avoid the need for model simulation completely and inherits the established theoretical and practical advantages of synthetic likelihood. This improvement in computational efficiency is at the expense of specification of constraints and making equivalence statements about parameters under the different models. Of note is that the latter enables, for the first time, model comparison using Bayes factors even if the priors are improper. In summary, in the Bayesian context, both of these approaches replace intractable likelihoods with alternative likelihoods that are more manageable computationally.

12.2 Synthetic Likelihood

The first approach to approximating the likelihood that is considered in this chapter is the use of a synthetic likelihood (SL), which was first introduced by Wood (2010). The key idea behind the SL is the assumption that the summary statistic conditional on a parameter value has a multivariate normal distribution with mean vector $\mu(\theta)$ and covariance matrix $\Sigma(\theta)$. That is, we assume that:

$$p(s|\theta) = \mathcal{N}(s; \mu(\theta), \Sigma(\theta)),$$

where $\mathcal{N}$ denotes the density of the multivariate normal distribution. Of course, in general for models with intractable likelihoods, the distribution of the summary statistic is unknown and thus $\mu(\theta)$ and $\Sigma(\theta)$ are generally unavailable. However, it is possible to estimate these quantities empirically via simulation. Consider generating n independent and identically distributed (iid) summary statistic values, $s^{1:n} = (s^1, \ldots, s^n)$, from the model based on a particular value of θ, $s^{1:n} \overset{\text{iid}}{\sim} p(s|\theta)$. Then the mean and covariance matrix can be estimated via:

$$\begin{aligned} \mu(\theta) &\approx \mu_n(\theta) = \frac{1}{n}\sum_{i=1}^{n} s^i, \\ \Sigma(\theta) &\approx \Sigma_n(\theta) = \frac{1}{n-1}\sum_{i=1}^{n}(s^i - \mu_n(\theta))(s^i - \mu_n(\theta))^\top, \end{aligned} \tag{12.1}$$

where the superscript $\top$ denotes transpose. The likelihood of the observed summary statistic, s_{obs}, is estimated via $p_n(s_{obs}|\theta) = \mathcal{N}(s_{obs}; \mu_n(\theta), \Sigma_n(\theta))$. We use the subscript n on $p_n(s_{obs}|\theta)$ to denote the fact that the approximate likelihood will depend on the choice of n. The larger the value of n, the better the mean and covariance parameters of the multivariate normal distribution can be approximated. However, larger values of n need more computation to estimate the likelihood. It is likely that a suitable value of n will be problem dependent, in particular, it may depend on the actual distribution of the summary statistic and also the dimension of the summary statistic. The value of n must be large enough so that the empirical covariance matrix is positive definite.

Note that Wood (2010) described some extensions, such as using robust covariance matrix estimation to handle some non-normality in the summary statistics and robustifying the SL when the observed summary statistic falls in the tails of the summary statistic distribution (i.e. when a poor parameter value is proposed or when the model is mis-specified).

The SL may be incorporated into a classical or Bayesian framework, which are both described in the following. Then, attempts in the literature to accelerate the SL method are described. We finish the section with a real data example in cell biology.

12.2.1 Classical synthetic likelihood

The approach adopted in Wood (2010) is to consider the following estimator:

$$\hat{\theta}_n = \arg\max_{\theta} \mathcal{N}(s_{obs}; \mu_n(\theta), \Sigma_n(\theta)), \tag{12.2}$$

which is the maximum SL estimator. We use the subscript n to denote that the estimator will depend on the value of n, with higher accuracy likely to be obtained for larger values of n. We note that also because the likelihood is stochastic, a different answer will be obtained for fixed n if a different random seed is applied. Since the optimisation in (12.2) is stochastic, Wood (2010) applied a Markov chain Monte Carlo (MCMC) algorithm to explore the space of θ and select the value of θ that produced the highest value of the SL. Some recent applications of the SL method have appeared in Hartig et al. (2014), who used the FORMIND model for explaining complicated biological processes that occur in natural forests, and Brown et al. (2014), who considered models for the transmission dynamics of avian influenza viruses in different bird types.

The synthetic likelihood approach has a strong connection with indirect inference, which is a classical method for obtaining point estimates of parameters of models with intractable likelihoods. In the simulated quasi-maximum likelihood (SQML) approach of Smith (1993), an auxiliary model with a tractable likelihood function, $p_A(y|\phi)$, where ϕ is the parameter of that model, is used. Define the function $\phi(\theta)$ as the relationship between the parameter of the model of interest and the auxiliary model. This is often referred to as the binding function in the indirect inference literature. The SQML method aims to maximise the auxiliary likelihood rather than the intractable likelihood of the model of interest:

$$\hat{\theta} = \max_{\theta} p_A(y_{obs}|\phi(\theta)).$$

Unfortunately, the binding function is typically unavailable. However, it can be estimated by generating n iid datasets, $y_1, \ldots, y_n$, from the model of interest (the generative model) conditional on a value of θ. Define the auxiliary parameter estimate based on the ith simulated dataset as:

$$\phi_{y_i} = \arg\max_{\phi} p_A(y_i|\phi).$$

Then we have:

$$\phi(\theta) \approx \phi_n(\theta) = \frac{1}{n}\sum_{i=1}^{n} \phi_{y_i}.$$

The binding function is defined as $\phi_n(\theta) \to \phi(\theta)$ as $n \to \infty$. The SQML estimator then becomes:

$$\hat{\theta}_n = \max_{\theta} p_A(y_{obs}|\phi_n(\theta)).$$

The synthetic likelihood falls within the SQML framework, but where y_{obs} has been reduced to s_{obs}, and the density of the multivariate normal distribution is used for p_A.

12.2.2 Bayesian synthetic likelihood

An intuitive approach to incorporating SL into a Bayesian framework involves combining the prior $\pi(\theta)$ with the synthetic likelihood, which induces the following approximate posterior:

$$\pi_n(\theta|s_{obs}) \propto \mathcal{N}(s_{obs}; \mu_n(\theta), \Sigma_n(\theta))\pi(\theta),$$

where the subscript n denotes that the approximate posterior depends on the choice of n. Drovandi et al. (2015) consider a general framework called parametric Bayesian indirect likelihood, where the likelihood of some auxiliary model with parameter ϕ, $p_A(y_{obs}|\phi(\theta))$, is used to replace the intractable likelihood of the actual or generative model, $p(y_{obs}|\theta)$. Since the binding function is generally not available in closed form, it can be estimated by simulation via drawing n iid datasets from the generative model and fitting the auxiliary model to this simulated data (as in the SQML method mentioned previously), producing $\phi_n(\theta)$. Drovandi et al. (2015) demonstrate that the resulting approximate posterior depends on n, since in general, $p_A(y_{obs}|\phi_n(\theta))$ is not an unbiased estimate of $p_A(y_{obs}|\phi(\theta))$ even when $\phi_n(\theta)$ is an unbiased estimate of $\phi(\theta)$. We note that when non-negative and unbiased likelihood estimates are used within Monte Carlo methods, such as MCMC (Andrieu and Roberts, 2009) and sequential Monte Carlo (Chopin et al. (2013)) algorithms, the resulting target distribution is the posterior based on the originally intended likelihood function. Such approaches are referred to as pseudo-marginal or exact-approximate methods in the literature (see Chapter 9, this volume for a review of such methods). BSL fits within the pBIL framework, but where the auxiliary model is applied at a summary statistic level rather than the full data level and that the auxiliary model is the multivariate normal distribution, so that the auxiliary parameter estimates have an analytic expression as shown in (12.1). Despite the fact that we use unbiased estimators for $\mu(\theta)$ and $\Sigma(\theta)$ (under the normality assumption), it is clear that $\mathcal{N}(s_{obs}; \mu_n(\theta), \Sigma_n(\theta))$ is not an unbiased estimate of $\mathcal{N}(s_{obs}; \mu(\theta), \Sigma(\theta))$. Therefore, the BSL posterior is inherently dependent on n. However, under the assumption that the model is able to recover the observed statistic, Price et al. (2018) present extensive empirical evidence that the BSL posterior is remarkably insensitive to n. Further, some empirical evidence demonstrates that BSL shows some robustness to the lack of multivariate normality.

Price et al. (2018) developed a new BSL method that uses an exactly unbiased estimator of the normal likelihood, which is developed by Ghurye and Olkin (1969). Using the notation of Ghurye and Olkin (1969), let:

$$c(k, v) = \frac{2^{-kv/2}\pi^{-k(k-1)/4}}{\prod_{i=1}^{k} \Gamma\left(\frac{1}{2}(v - i + 1)\right)},$$

and for a square matrix A, write $\psi(A) = |A|$ if $A > 0$ and $\psi(A) = 0$ otherwise, where $|A|$ is the determinant of A and $A > 0$ means that A is positive definite. The result of Ghurye and Olkin (1969) shows that an exactly unbiased estimator of $\mathcal{N}(s_{obs}; \mu(\theta), \Sigma(\theta))$ is (in the case where the summary statistics are normal and $n > d+3$):

$$\hat{p}_A(s_{obs}|\phi(\theta)) = (2\pi)^{-d/2} \frac{c(d, n-2)}{c(d, n-1)(1-1/n)^{d/2}} |M_n(\theta)|^{-(n-d-2)/2}$$
$$\psi\left(M_n(\theta) - (s_y - \mu_n(\theta))(s_y - \mu_n(\theta))^\top/(1-1/n)\right)^{(n-d-3)/2},$$

where $M_n(\theta) = (n-1)\Sigma_n(\theta)$. It is interesting to note that this estimator is a mixture of a discrete and a continuous random variable (a realisation of the estimator can be identically 0 with positive probability). Thus, if this estimator is used within a Monte Carlo method, the target distribution is proportional to $\mathcal{N}(s_{obs}; \mu(\theta), \Sigma(\theta))\pi(\theta)$ regardless of the value of n (under the multivariate normality assumption). Price et al. (2018) referred to this method as uBSL, where 'u' denotes unbiased.

To sample from the BSL posteriors, an MCMC algorithm can be used, for example. We refer to this as MCMC BSL, which is shown in Algorithm 12.1. Given the insensitivity of the BSL posteriors to the value of n, it is of interest to maximise the computational efficiency of the MCMC method. For large n, the SL is estimated with high precision, but the cost per iteration is high. Conversely, for small n, the cost per iteration is low, but the SL is estimated less precisely, which reduces the MCMC acceptance rate. Price et al. (2018) found empirically that the value of n that leads to an estimated log SL (at a θ

Algorithm 12.1: MCMC BSL algorithm. The inputs required are the summary statistic of the data, s_{obs}, the prior distribution, $p(\theta)$, the proposal distribution q, the number of iterations, T, and the initial value of the chain θ^0. The output is an MCMC sample $(\theta^0, \theta^1, \ldots, \theta^T)$ from the BSL posterior. Some samples can be discarded as burn-in if required

Simulate $s_{1:n} \overset{\text{iid}}{\sim} p(\cdot|\theta^0)$
Compute $\phi^0 = (\mu_n(\theta^0), \Sigma_n(\theta^0))$
for $i = 1$ to T **do**
    Draw $\theta^* \sim q(\cdot|\theta^{i-1})$
    Simulate $s^*_{1:n} \overset{\text{iid}}{\sim} p(\cdot|\theta^*)$
    Compute $\phi^* = (\mu_n(\theta^*), \Sigma_n(\theta^*))$
    Compute $r = \min\left(1, \frac{\mathcal{N}(s_{obs};\mu_n(\theta^*),\Sigma_n(\theta^*))\pi(\theta^*)q(\theta^{i-1}|\theta^*)}{\mathcal{N}(s_{obs};\mu_n(\theta^{i-1}),\Sigma_n(\theta^{i-1}))\pi(\theta^{i-1})q(\theta^*|\theta^{i-1})}\right)$
    if $\mathcal{U}(0,1) < r$ **then**
        Set $\theta^i = \theta^*$ and $\phi^i = \phi^*$
    else
        Set $\theta^i = \theta^{i-1}$ and $\phi^i = \phi^{i-1}$
    end if
end for

with high BSL posterior support) with a standard deviation(SD) of roughly 2 produces efficient results. However, Price et al. (2018) also found that there is a wide variety of n values that lead to similar efficiency. When the unbiased SL is used in place of the SL shown in Algorithm 12.1, the MCMC uBSL algorithm is obtained. In the examples of Price et al. (2018), MCMC BSL and MCMC uBSL have a similar efficiency. We also note that the MCMC BSL posteriors appear to exhibit very slow convergence when starting at a point with negligible posterior support. The reason for this is that the SL is estimated with a large variance when the observed statistic s_{obs} lies in the tail of the actual SL. Thus, additional research is required on more sophisticated methods for sampling from the BSL posteriors.

The BSL method has been applied in the literature. Fasiolo et al. (2019) used BSL for posterior inference for state space models in ecology and epidemiology based on data reduction and compared it with particle Markov chain Monte Carlo (Andrieu et al., 2010). Hartig et al. (2014) implemented BSL for a forest simulation model.

BSL could be seen as a direct competitor with ABC, as they are both simulation-based methods and differ only in the way the intractable summary statistic likelihood is approximated. Importantly, BSL does not require the user to select a discrepancy function, as one is naturally induced via the multivariate normal approximation. The simulated summary statistics in BSL are automatically scaled, whereas an appropriate weighting matrix to compare summary statistics in ABC must be done manually. As noted in Blum (2010) and Drovandi et al. (2015), ABC uses a non-parametric approximation of the summary statistic likelihood based on similar procedures used in kernel density estimation. From this point of view, the ABC approach may be more accurate when the summary statistic s_{obs} is low dimensional, however, the accuracy/efficiency trade-off is less clear when the summary statistic s_{obs} is high dimensional. Price et al. (2018) demonstrated on a toy example that BSL becomes increasingly more computationally efficient than ABC as the dimension of the summary statistic grows beyond 2. Furthermore, Price et al. (2018) demonstrated that BSL outperformed ABC in a cell biology application with 145 summary statistics.

12.2.3 Accelerating synthetic likelihood

As with ABC, the SL method is very simulation intensive. There have been several attempts in the literature to accelerate the SL method by reducing the number of model simulations required. Meeds and Welling (2014) assumed that the summary statistics are independent and during their MCMC BSL algorithm fit a Gaussian process (GP) to each summary statistic output as a function of the model parameter θ. The GP is then used to predict the model output at proposed values of θ, provided that the prediction is accurate enough. If the GP prediction cannot be performed with sufficient accuracy, more model simulations are taken at that proposed θ, and the GP is re-fit for each summary statistic. The independence assumption of the summary statistics is questionable and may overstate the information contained in s_{obs}.

In contrast, Wilkinson (2014) used a GP to model the SL as a function of θ directly and used the GP to predict the SL at new values of θ. The GP is fit using a history matching approach (Craig et al., 1997). Once the final GP fit is obtained, an MCMC algorithm is used with the GP prediction used in place of the SL.

Moores et al. (2015) considered accelerating Bayesian inference for the Potts model, which is a complex single parameter spatial model. Simulations are performed across a pre-defined grid with the mean and standard deviation of the summary statistic (which turns out to be sufficient in the case of the Potts model, as it belongs to the exponential family) estimated from these simulations. Non-parametric regressions are then fitted individually to the mean and standard deviation estimates in order to produce an estimate of the mappings $\mu(\theta)$ and $\sigma(\theta)$ across the space of θ, where σ is the standard deviation of the summary statistic. The regressions are then able to predict the mean and standard deviation of the summary statistic at θ values not contained in the grid. Further, the regression also smooths out the mappings, which are estimated using a finite value of n. The estimated mapping is then used in a sequential Monte Carlo Bayesian algorithm.

12.2.4 Example

Cell motility, cell proliferation, and cell-to-cell adhesion play an important role in collective cell spreading, which is critical to many key biological processes, including skin cancer growth and wound healing (e.g. Cai et al. (2007); Treloar et al. (2013)). The main function of many medical treatments is to influence the biology underpinning these processes (Decaestecker et al., 2007). In order to measure the efficacy of such treatments, it is important that estimates of the parameters governing these cell spreading processes can be obtained along with some characterisation of their uncertainty. Agent-based computational models are frequently used to interpret these cell biology processes since they can produce discrete image-based and movie-based information which is ideally suited to collaborative investigations involving applied mathematicians and experimental cell biologists. Unfortunately, the likelihood functions for these models are computationally intractable, so standard statistical inferential methods for these models are not applicable.

To deal with the intractable likelihood, several papers have adopted an ABC approach to estimate the parameters (Johnston et al., 2014, Vo et al., 2015b). One difficulty with these cell biology applications is that the observed data are typically available as sequences of images and therefore it is not trivial to reduce the dimension of the summary statistic to a suitable level for ABC, while simultaneously retaining relevant information contained in the images. For example, Johnston et al. (2014) considered data collected every 5 minutes for 12 hours, but only analysed images at three time points. Vo et al. (2015a) reduced images initially down to a 15-dimensional summary statistic, but perform a further dimension reduction based on the approach of Fearnhead and Prangle (2012) to ensure there is one summary statistic per parameter.

Here, we will re-analyse the data considered in Treloar et al. (2013) and Vo et al. (2015a). The data consist of images of spatially expanding human malignant melanoma cell populations. To initiate each experiment, either 20,000 or 30,000 cells are approximately evenly distributed within a circular barrier, located at the centre of the well. Subsequently, the barriers are lifted and population-scale images are recorded at either 24 hours or 48 hours, independently. Furthermore, there are two types of experiments conducted. The first uses a treatment in order to inhibit cells giving birth (cell proliferation), while the second does not use the treatment. Each combination of initial cell density, experimental elapsed time, and treatment is repeated three times, for a total of 24 images. The reader is referred to Treloar et al. (2013) for more details on the experimental protocol. For simplicity, we consider here the three images related to using 20,000 initial cells, 24 hours elapsed experimental time, and no cell proliferation inhibitor.

In order to summarise an image, Vo et al. (2015a) considered six sub-regions along a transect of each image. The position of the cells in these regions is mapped to a square lattice. The number of cells in each sub-region is counted, together with the number of isolated cells. A cell is identified as isolated if all of its nearest neighbours (north, south, east, and west) are unoccupied. For each region, these summary statistics are then averaged over the three independent replicates. We refer to these 12 summary statistics as $\{c_i\}_{i=1}^6$ and $\{p_i\}_{i=1}^6$, where c_i and p_i are the number of cells and the percentage of isolated cells (averaged over the three images) for region i, respectively. Vo et al. (2015a) also estimated the radius of the entire cell colony using image analysis. Thus, Vo et al. (2015a) included three additional summary statistics, ($r_{(1)}, r_{(2)}$, and $r_{(3)}$), which are the estimated and ordered radii for the three images. For more details on how these summary statistics are obtained, the reader is referred to Vo et al. (2015a). This creates a total of 15 summary statistics, which is computationally challenging to deal with in ABC. As mentioned earlier, Vo et al. (2015a) found it beneficial to apply the technique of Fearnhead and Prangle (2012), which uses a regression to estimate the posterior means of the model parameters from the initial summaries, which are then used as summary statistics in a subsequent ABC analysis. Here, we attempt to see whether or not BSL is able to accommodate the 15 summary statistics and compare the results with the ABC approach of Vo et al. (2015a).

Treloar et al. (2013) and Vo et al. (2015a) considered a discretised time and space (two-dimensional lattice) stochastic model to explain the cell spreading process of melanoma cells. For more details on this model, the reader is referred to Treloar et al. (2013) and Vo et al. (2015a). The model contains three parameters: P_m (probability that an isolated agent can move to a neighbouring lattice site in one time step), P_p (probability that an agent will attempt to proliferate and deposit a daughter at a neighbouring lattice site within a time step), and q (the strength of cell-to-cell adhesion, that is, cells sticking together). These model parameters can then be related to biologically relevant parameters such as cell diffusivity and the cell proliferation rate. Here, we will report inferences in terms of the parameter $\theta = (P_m, P_p, q)$.

Here, we consider a simulated dataset with $P_m = 0.1$, $P_p = 0.0012$ and $q = 0.2$ [same simulated data as analysed in Vo et al. (2015a)] and the real data. We ran BSL using Algorithm 12.1 with independent $\mathcal{U}(0,1)$ prior distributions on each parameter. We used a starting value and proposal distribution for the MCMC based on the results provided in Vo et al. (2015a), so we do not apply any burn-in. We also applied the uBSL algorithm. The BSL approaches were run with $n = 32, 48, 80$, and 112 (the independent simulations were farmed out across 16 processors of a computer node). To compare the efficiency of the different choices of n, we considered the effective sample size (ESS) for each parameter divided by the total number of model simulations performed multiplied by a large constant scalar to increase the magnitude of the numbers (we refer to this as the 'normalised ESS').

Marginal posterior distributions for the parameters obtained from BSL and uBSL for different values of n are shown in Figures 12.1 and 12.2, respectively. It is evident that the posteriors are largely insensitive to n, which is consistent with the empirical results obtained in Price et al. (2018). The normalised ESS values and MCMC acceptance rates for the BSL approaches are shown in Table 12.1 for different values of n. The efficiency of BSL and uBSL appears to be similar. The optimal value of n out of the trialled values appears to be 32 or 48. However, even $n = 80$ is relatively efficient. For $n = 112$, the increase in acceptance rate is relatively small given the extra amount of computation required per iteration.

We also applied the BSL approaches with similar settings to the real data. The posterior results are presented in Figures 12.3 and 12.4. Again, we found the results are relatively insensitive to n. Table 12.2 suggests that $n = 48$ or $n = 80$ are the most efficient choices for n out of those trialled. However, it is again apparent that there are a wide variety of n values that lead to similar efficiency.

The results, in comparison to those obtained in Vo et al. (2015a), are shown in Figure 12.5 for the simulated data and Figure 12.6 for the real data. From Figure 12.5, it can be seen that the BSL approaches produced results similar to that of ABC for the simulated data. It appears that BSL is able to accommodate the 15 summary statistics directly without further dimension reduction. However, it is clear that the dimension reduction procedure of Vo et al. (2015a) performs well. From Figure 12.6 (real data), it is evident that

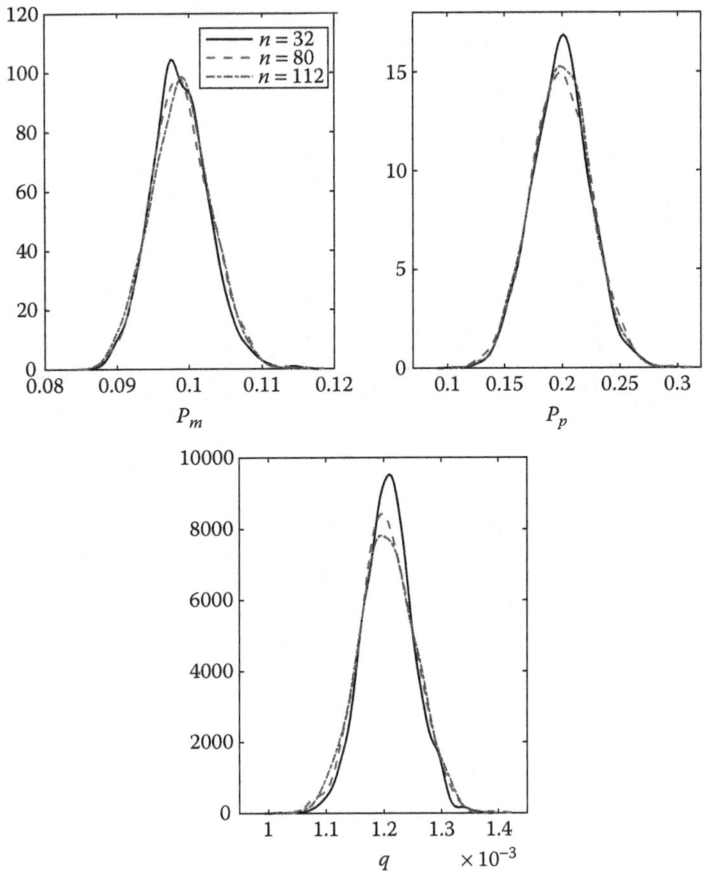

FIGURE 12.1
Posterior estimates for P_m, P_p, and q based on the simulated data for the melanoma cell biology application using MCMC BSL for different values of n.

ABC and the BSL approaches produce similar posterior distributions for P_p and q. For P_m, there is a difference of roughly 0.01 between the posterior means of the BSL and ABC approaches and an increase in precision for BSL. This discrepancy for the real data not apparent in the simulated data requires further investigation. One potential source of error for BSL is the multivariate normal assumption. The estimated marginal distributions of the summary statistics (using $n = 200$) when the parameter is $\theta = (0.1, 0.0015, 0.25)$ is shown in Figure 12.7. All distributions seem quite stable, but there is an indication of non-normality for some of the summary statistics. Given the results in Figures 12.5 and 12.6, it appears that BSL is showing at least some robustness to this lack of normality.

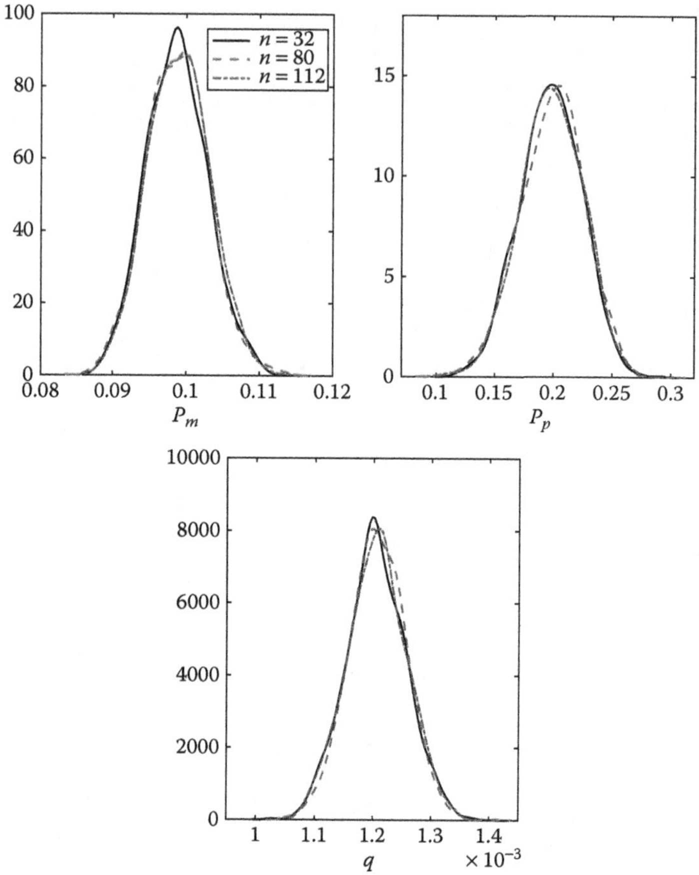

FIGURE 12.2
Posterior estimates for P_m, P_p, and q based on the simulated data for the melanoma cell biology application using MCMC uBSL for different values of n.

TABLE 12.1
Sensitivity of BSL/uBSL to n for the Simulated Data of the Cell Biology Example with Regards to MCMC Acceptance Rate, Normalised ESS for Each Parameter

n	Acceptance Rate (%)	ESS P_m	ESS P_p	ESS q
32	17/17	96/114	86/113	115/126
48	27/32	95/103	93/92	110/115
80	35/37	82/76	74/67	106/89
112	38/40	61/65	61/58	68/70

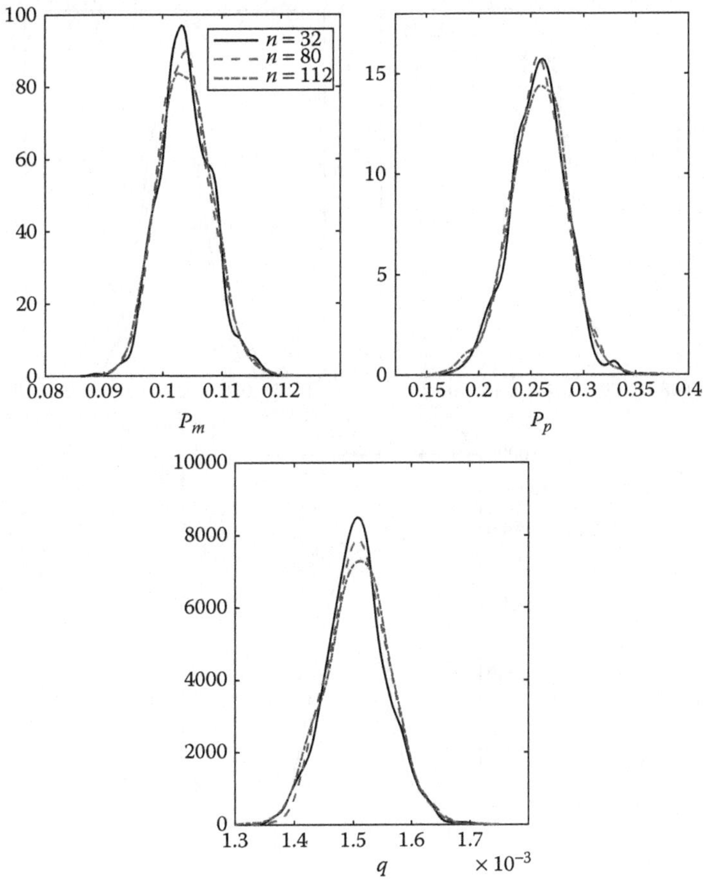

FIGURE 12.3
Posterior estimates for P_m, P_p, and q based on the real data for the melanoma cell biology application using MCMC BSL for different values of n.

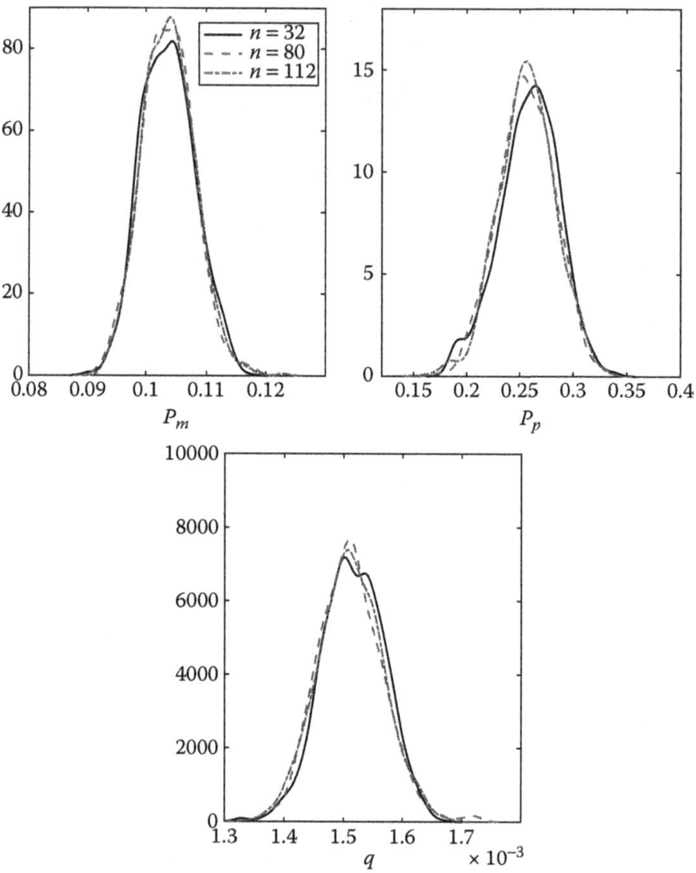

FIGURE 12.4
Posterior estimates for P_m, P_p, and q based on the simulated data for the melanoma cell biology application using MCMC uBSL for different values of n.

TABLE 12.2
Sensitivity of BSL/uBSL to n for the Real Data of the Cell Biology Example with Regards to MCMC Acceptance Rate, Normalised ESS for Each Parameter

n	Acceptance Rate (%)	ESS P_m	ESS P_p	ESS q
32	8/9	46/51	38/45	41/43
48	17/18	76/71	56/63	70/54
80	27/28	66/64	66/60	68/58
112	32/33	58/60	51/54	54/43

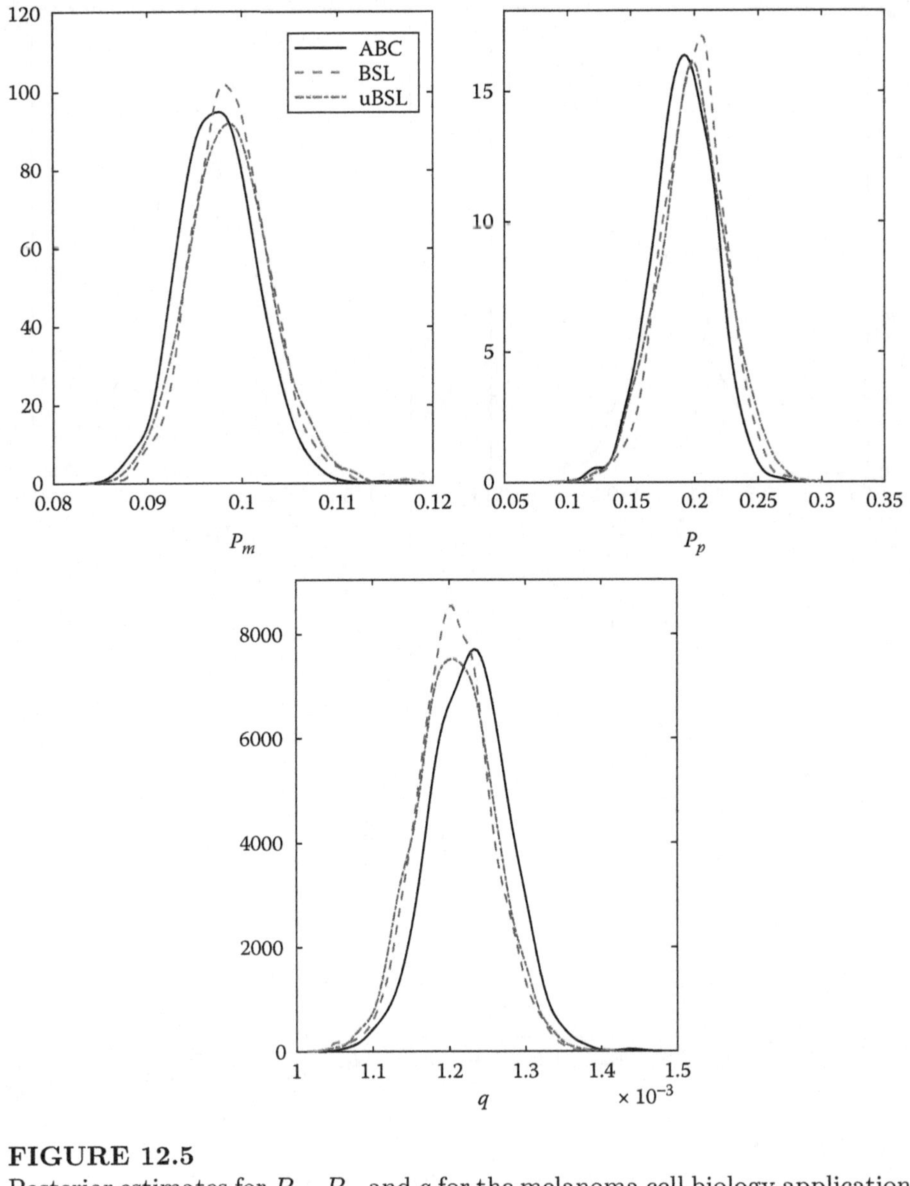

FIGURE 12.5
Posterior estimates for P_m, P_p, and q for the melanoma cell biology application using the ABC approach of Vo et al. (2015a) (solid), BSL (dash), and uBSL (dot-dash) based on simulated data with $P_m = 0.1$, $P_p = 0.0012$, and $q = 0.2$. The BSL results are based on $n = 48$.

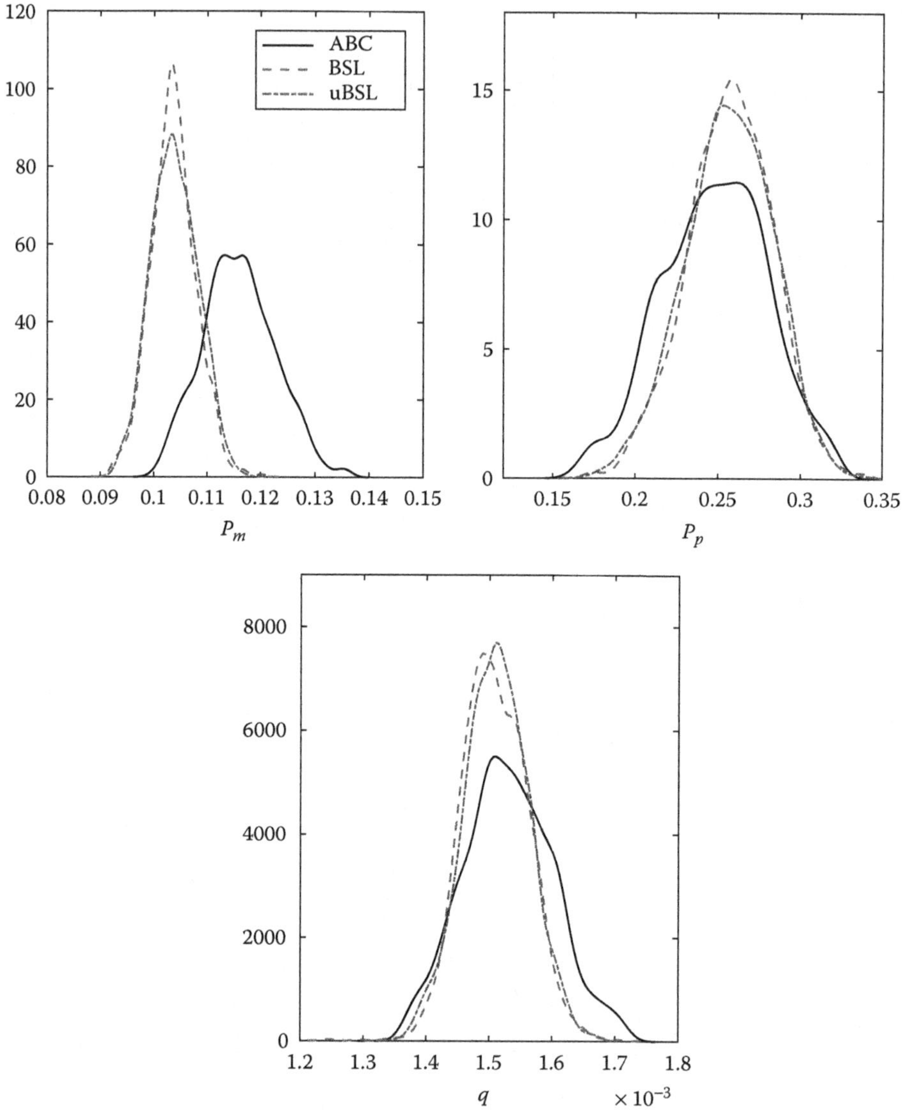

FIGURE 12.6
Posterior estimates for P_m, P_p, and q for the melanoma cell biology application using the ABC approach of Vo et al. (2015a) (solid), BSL (dash), and uBSL (dot-dash) based on real data. The BSL results are based on $n = 48$.

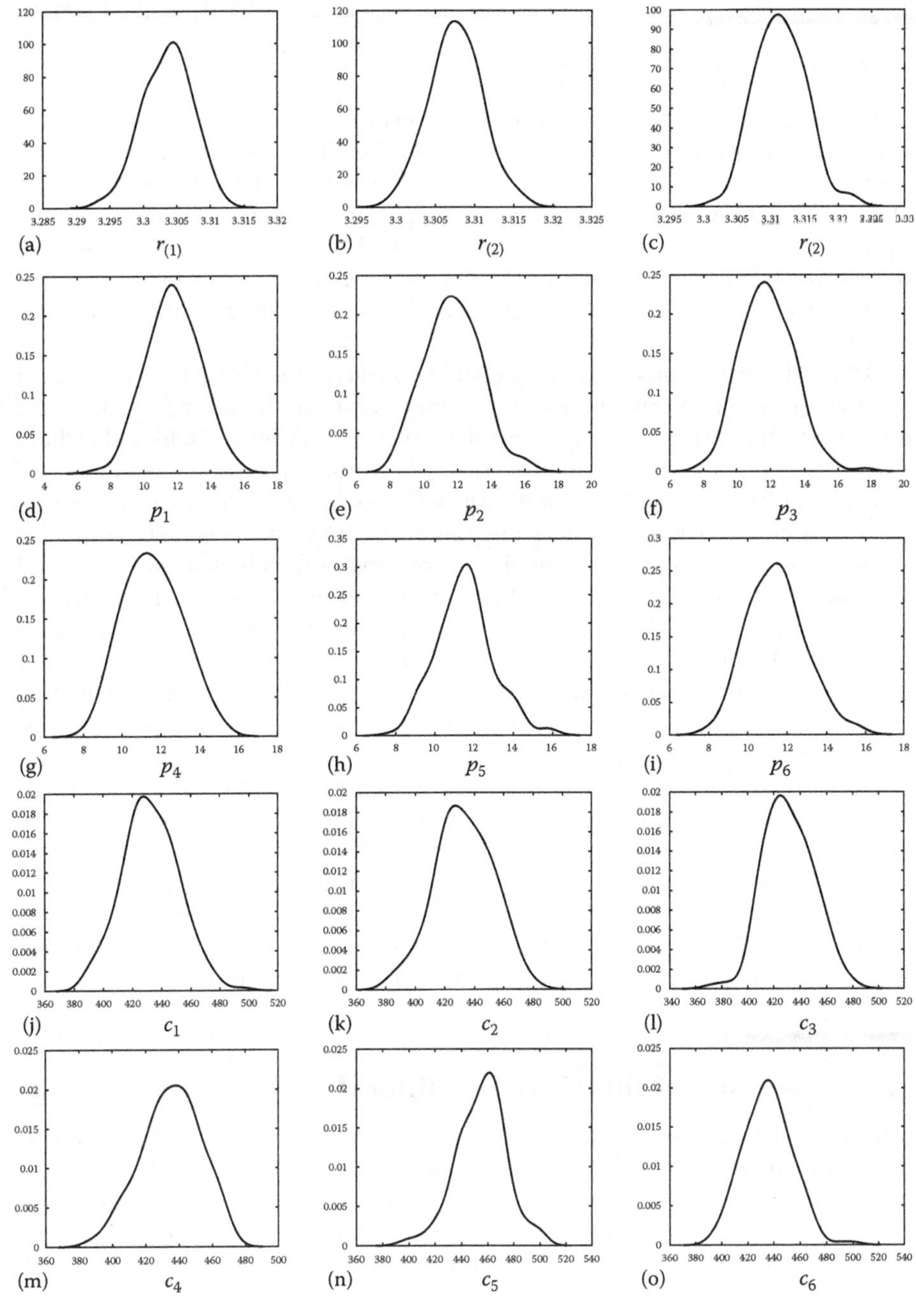

FIGURE 12.7
Estimated marginal distributions of the 15 summary statistics (Panels a-o, with each summary statistic in the x-axis label) using $n = 200$ for the melanoma cell biology applications, when $Pm = 0.1$, $Pp = 0.0015$, and $q=0.25$.

12.3 Further Reading

Ong et al. (2018) developed a stochastic optimisation algorithm to obtain a variational approximation of the BSL posterior. The authors utilise an unbiased estimator of the log of the multivariate normal density due to Ripley (1996, p. 56). Ong et al. (2018) demonstrated that significant computational savings can be achieved relative to MCMC BSL, at the expense of resorting to a parametric approximation of the posterior. This work has been extended by Ong et al. (2017) to higher dimensional summary statistic and parameter spaces.

An et al. (2016) and Ong et al. (2017) considered shrinkage estimators of the covariance matrix of the model summary statistic in order to reduce the number of simulations required to estimate the synthetic likelihood. Pham et al. (2014) replaced the ratio of intractable summary statistic likelihoods of the Metropolis–Hastings ratio in an MCMC algorithm with the outcome of a classification algorithm. Datasets are simulated under the current parameter and proposed parameter with the former observations labelled as class 1 and the latter labelled as class 2. A classifier, Pham et al. (2014) used random forests, is then applied. From the fitted classifier, the odds for the value of the observed summary statistic s_{obs} is computed and used as a replacement to the ratio of intractable likelihoods. Pham et al. (2014) noted that BSL is a special case of this approach when classical quadratic discriminant analysis is adopted as the classifier.

Everitt et al. (2015) suggested that the SL can be used to perform Bayesian model selection in doubly intractable models, which contain an intractable normalising constant that is a function of θ. Such models can occur in complex exponential family models, such as exponential random graph models for social networks and the Potts model for image analysis. Everitt et al. (2015) developed computational algorithms in order to produce an SL approximation to the evidence $p(s_{obs}) = \int_\theta p(s_{obs}|\theta)\pi(\theta)d\theta$ for each model.

12.4 Bayesian Empirical Likelihood

ABC is a popular computational method of choice not only when there is no likelihood, but also when the likelihood is available, but difficult or impossible to evaluate. Another popular idea is to replace the likelihood itself with an empirical alternative. This so-called empirical likelihood (EL) can be embedded within an ABC algorithm or provide an alternative to ABC. The approach is appealing even for less complex models if there is a concern that the model is poorly specified. For instance, if the likelihood is a mixture, but is mis-specified as a single normal distribution, the corresponding parameter estimates, intervals, and inferences may exhibit unacceptably poor behaviour (Chen and Quin, 2003). In this case, normal approximation confidence intervals perform poorly in the area of interest, for example, the lower tail, but intervals based on an EL are shown to perform as well as intervals based on a correctly specified mixture model.

EL has been shown to have good small sample performance compared with methods that rely on asymptotic normality. Moreover, it enables distribution-free tests without simulation and provides confidence intervals and regions that have appealing theoretical and computational properties (Owen, 1988, 2001). Some of these features are discussed in more detail in the following.

Close parallels with the EL approach have been drawn with estimating equations (Qin and Lawless, 2001; Grendar and Judge, 2007), kernel smoothing in regression (Chen and Van Keilegom 2009a, 2009b, Haardle, 1990; Fan and Gijbels, 1996), maximum entropy (Rochet, 2012), and functional data analysis (Lian, 2012). We do not elaborate on these associations in this chapter, but refer the interested reader to the cited references.

EL approaches have been developed in both frequentist and Bayesian contexts. This section provides a brief overview of the method under both paradigms. We then focus on a particular algorithm, BC_{el}, proposed by Mengersen et al. (2013), which was first conceived as part of an ABC algorithm, but was then developed independently of the ABC architecture.

12.4.1 Empirical likelihood

EL has been a topic of active research and investigation for over a quarter of a century. Although similar ideas were established earlier [see, e.g. the proposal of a 'scale-load' method for survey sampling by Hartley and Rao (1968)], EL was formally introduced by Owen (1988) as a form of robust likelihood ratio test.

Following Owen (1988) and Owen (2001), assume that we have iid data $Y_i, i = 1, ..., n$ from a distribution F. An EL denoted $L(F)$ is given by:

$$L(F) = \prod_{i=1}^{n} F(\{y_i\}).$$

The likelihood ratio and corresponding confidence region are given by, respectively:

$$R(F) = L(F)/L(\hat{F}) \quad \text{and}$$
$$\{T(F)|R(F) \geq r\},$$

where $\hat{F}$ is the empirical distribution function and for some appropriate value of r.

Given parameters of interest θ and an appropriate sufficient statistic $T(F)$ for it, a profile likelihood and corresponding confidence region become, respectively;

$$\mathcal{R}(\theta) = \sup\{R(F)|T(F) = \theta\} \quad \text{and}$$
$$\{\theta|\mathcal{R}(F) \geq r\}.$$

If there are no ties, we let $p_i = F(\{y_i\})$, $p_i > 0$, $\sum_{i=1}^n p_i = 1$ and find that

$$L(F) = \prod_{i=1}^{n} p_i \quad ; \quad L(\hat{F}) = \prod_{i=1}^{n} 1/n$$

$$R(F) = \prod_{i=1}^{n} np_i \quad ; \quad \mathcal{R}(\theta) = \sup\left\{\prod_{i=1}^{n} np_i | T(F) = \theta\right\}.$$

Obvious adjustments are made to $L(F)$ and $L(\hat{F})$ if there are ties.

A fundamental result obtained by Owen (1988) is that if the mean θ_0 of the distribution F is finite, and its covariance matrix is finite with rank $q > 0$, then as $n \to \infty$:

$$-2 \log R(\theta_0) \to \chi_q^2.$$

This is the same as that obtained by Wilks' theorem for the parametric setup. Thus, for a $100(1-\alpha)\%$ confidence region, $r = r_0 = X_{q,\alpha}^2/2$.

As a concrete example of EL, suppose that interest is in estimation of the mean, for example, $\theta = E[Y]$. Then, $T(\hat{F}) = n^{-1}\sum_{i=1}^n y_i$, with confidence region and profile likelihood given by, respectively:

$$\left\{\sum_{i=1}^{n} p_i y_i | p_i \geq 0, \sum_{i=1}^{n} p_i = 1, \prod_{i=1}^{n} np_i > r\right\} \quad \text{and}$$

$$R(\theta) = \sup\left\{\prod_{i=1}^{n} np_i | p_i > 0, \sum_{i=1}^{n} p_i = 1, \sum_{i=1}^{n} p_i y_i = \theta\right\}.$$

Thus, under certain conditions, a $(1-\alpha)$-level EL confidence interval for $\theta_0 = \bar{Y}$ is given by:

$$\{\theta | r(\theta) \leq \chi_1^2(\alpha)\},$$

where $r(\theta) = -2\sum \log(n\hat{p}_i)$ is the log EL function and $\chi_1^2(\alpha)$ is the upper α quantile of the χ^2 distribution with one degree of freedom.

The earlier set-up can also be seen as an estimating equation problem, where the true value θ_0 satisfies the estimating equation:

$$E_F[m(Y;\theta)] = 0,$$

with $m(Y;\theta)$ denoting a vector-valued (estimating) function. Hence, we can take $m(Y;\theta) = Y - \theta$ to indicate a vector mean, $m(Y;\theta) = I_{Y \in A} - \theta$ for $Pr(Y \in A)$, $m(Y;\theta) = I_{Y<\theta} - \alpha$ for the αth quantile of Y if Y is continuous, $m(Y;\theta) = I_{Y \leq \theta} - 0.5$ for the median, and so on.

More generally, we have one or more constraints of the form $E_F[h(Y,\theta)] = 0$, where the dimension of h sets the number of constraints in unequivocally defining the parameters of interest θ. Then the EL is defined as:

$$L_{el}(\theta|y) = \max_{p} \prod_{i=1}^{n} p_i,$$

for $p \in [0,1]^n$, with constraints:

$$\sum_{i=1}^{n} p_i = 1; \sum_{i=1}^{n} p_i h(y_i, \theta) = 0.$$

Perhaps surprisingly, there are relatively few Bayesian formulations of EL in the published literature. An earlier Bayesian ABC approach using an approximation of the EL based on the pairwise score equation was proposed by Pauli and Adimara (2010). The authors focused on establishing the validity of the procedure, arguing that its asymptotic properties were preferred over the approximations employed by Pauli et al. (2011). See also Ruli et al. (2016). Owen (2001) (Chapter 9) noted some parallels between EL and the Bayesian bootstrap (Rubin, 1981), and Rochet (2012) has suggested a Bayesian approach to generalised empirical likelihood, and generalised method of moments, via a form of maximum entropy. Chaudhuri and Ghosh (2011) describe Bayesian EL approaches in a spatial modelling context, as discussed in more detail in the following.

More direct research into Bayesian EL comprise a Monte Carlo study (Lazar, 2003) and two probabilistic studies (Schennach, 2005; Grendar and Judge, 2007). In contrast to the reported Bayesian bootstrap-type approaches of Owen (2001), Schennach (2005), and Ragusa (2006), Grendar and Judge (2007, 2009) proposed a Bayesian large deviations (law of large numbers) probabilistic interpretation and justification of EL. They showed that, in a parametric estimating equations setting, the EL method is an asymptotic instance of the Bayesian non-parametric maximum a posteriori approach.

12.4.2 Features of empirical likelihood

Since Owen's paper in 1988, the properties of EL have been comprehensively investigated and reviewed (Hall and La Scala, 1990; Owen, 2001).

EL methods have been favourably contrasted with common alternatives for estimation of complex models. For example, a natural competitor is calibration, which proceeds by choosing, by some method, parameter values that match selected features of the observed data. This can be difficult for richly parametrised models with strong correlation structure. EL can be perceived as a more statistically formal method of calibration in that it uses moments for matching. Another common competitor, maximum likelihood, requires the definition, estimation, and maximisation of a likelihood and can be both analytically and computationally demanding for complex models. In contrast, EL requires only summary (moment) statistics and can perform inference on an approximate likelihood, but inherits the properties of standard likelihood (Owen, 2001). These properties of standard likelihood are principally obtained by appeal to Wilks' theorem (Qin and Lawless, 2001).

As described earlier, likelihood ratio confidence regions can be constructed by EL that often do not require estimation of the variance (Chen and Van Keilegom, 2009a, 2009b) and have the same order of magnitude error as their parametric counterparts. This also applies for more general regression contexts (Chen, 1993, 1994; Chen and Cui, 2006; Chen and Gao, 2007). The confidence regions constructed in this manner respect the boundaries of the support of the target parameter and are more natural in shape and orientation of the data since they contour a log-likelihood ratio. In particular, they are often superior to confidence regions based on asymptotic normality when the sample size is small. The confidence regions can be further improved by applying Bartlett's correction, $(1_a/n)\chi^2_{q,\alpha}$, where a involves higher order moments of Y (DiCiccio et al., 1991).

A key assumption underlying standard EL is that the random variables are independent with a common distribution. An analogue, the weighted EL, or exponentially tilted distribution accommodates data that are independent, but not necessarily identically distributed. This approach was introduced by Schennach (2005) and has been taken up by a large number of authors (Owen, 2001; Kitamura, 2006; Glenn and Zhao, 2007). Chaudhuri and Ghosh (2011) contrast the two approaches as follows. They frame the EL as:

$$l(\theta) = \prod_{i=1}^{m} \hat{w}_i(\theta),$$

where

$$\hat{w}(\theta) = \arg\max_{w \in \mathcal{W}_0} \sum_{i=1}^{m} f\{w_i(\theta)\},$$

for some specified function f. They then consider standard EL as a form of constrained maximum of a non-parametric likelihood since for a given θ, $l(\theta)$ equals the EL when $f(w_i) = \log(w_i)$ and the exponentially tilted likelihood as a form of maximum entropy, such that $f(w_i) = -w_i \log(w_i)$. As these authors discussed, the exponentially tilted likelihood can also be seen as a profile likelihood for θ.

Moreover, Schennach (2005) shows that this re-formulation of the maximisation problem of the EL allows for a probabilistic interpretation which justifies its use in a Bayesian setting. More precisely, the posterior distribution for a parameter of interest θ may be seen as:

$$\pi(\theta|y) = \pi(\theta) \int_{\Psi} L(\theta, \psi|y)\pi(\psi|\theta)d\psi,$$

where ψ represents a (potentially infinite-dimensional) nuisance parameter, which absorbs all those aspects of the model not described by the parameter of interest θ. The information contained in the nuisance parameter may be discretised by a vector of parameter $\xi = (\xi_1, \ldots, \xi_N)$ with $N \to \infty$. The nuisance parmater ξ may then be given a prior which shares the Dirichlet prior's property of providing posteriors which assign probability one to distributions supported by the sample. Schennach (2005) shows then this reformulation has a computationally convenient representation, for which the posterior of the parameter of interest θ may be obtained through:

$$\pi(\theta|y) = \pi(\theta) \prod_{i=1}^{n} p_i^\star,$$

where $p^\star = (p_1^\star, \cdots, p_n^\star)$ are the weights obtained as solution of the maximisation problem:

$$L_{\text{BETEL}}(\theta) = \max_{p^\star} \sum_{i=1}^{n} p_i^\star \log p_i^\star,$$

under constraints $p^\star \in [0,1]^n, \sum_{i=1}^{n} p_i^\star = 1, \sum_{i=1}^{n} p_i^\star h(y_i, \theta) = 0$, where 'BETEL' stands for 'Bayesian exponentially tilted likelihood'. This method may be called 'Bayesian exponentially tilted EL', because it uses the exponential tilting proposed in Efron (1981) and has a Bayesian interpretation. This version of the EL will be used in Section 12.4.5.2.

Glenn and Zhao (2007) examined the robustness properties of the estimates arising from the tilted distribution. For example, whereas the root mean squared error of the EL estimator for the mean increases as the non-iidness of the sample increases, the root mean squared error of the weighted EL estimator remains closer to its theoretical value. Other extensions to standard EL, such as the continuous updating estimator, have also been proposed (Hansen et al., 1996).

In a Bayesian framework, the standard and exponentially tilted likelihoods have been shown to be appropriate for Bayesian inference for a range of set-ups and under certain conditions on the prior, particularly for a prior with sufficiently large variance (Monahan and Boos, 1992; Lazar, 2003; Chaudhuri and Ghosh, 2011).

Notwithstanding these attractions, there are some drawbacks in applying EL. One substantive issue is the formulation of the estimating equations. The number of equations is one issue: there should be at least as many as the dimension of the parameter space, but any more than this (which may be available and desirable from the perspective of model description) has been argued to adversely affect inference (Qin and Lawless, 2001). However, it is suggested by Mengersen et al. (2013) that this concern may not apply in all circumstances, in a Bayesian set-up; this is illustrated in the g-and-k example given in the following.

12.4.3 Estimation

The most common approach to estimation of the EL is through the method of Lagrange multipliers. In general terms, this method aims to maximise $f(x)$ subject to a (multivariate) constraint $g(x) = 0$. This is achieved by finding $x^* = x^*(\lambda)$ maximising $f(x)$-$\lambda' g(x)$, such that $g(x') = 0$. Then x^* solves the constrained problem. Considering again the example of estimating $\theta = E[Y]$, the aim is to maximise:

$$\log R(p_1, .., p_n) = \sum_{i=1}^{n} \log(np_i),$$

under the constraints:

$$n \sum_{i=1}^{n} p_i(Y_i - \theta) = 0, \quad 1 - \sum_{i=1}^{n} p_i = 0.$$

We write:

$$G = \sum_{i=1}^{n} \log(np_i) - n\lambda \sum_{i=1}^{n} (Yi - \mu) - \gamma(1 - \sum_{i=1}^{n} p_i),$$

where λ and γ are the Lagrange multipliers. This can be solved to find a unique solution for $\lambda = \lambda(\theta)$.

There is a range of software for computing the EL, particularly targeted towards specific applications. A helpful repository and description of available code is on Art Owen's website. A powerful library available in the R software is the package 'emplik' (Zhou and Yang, 2014). The underlying computational method is based on the Newton–Lagrange algorithm, whereby the Lagrangian function described earlier is solved by an application of Newton's method, which iteratively uses a second order Taylor approximation of $f(x)$ to find an optimal value x^* satisfying $f'(x^*) = 0$.

For example, the package `el.test` in the `emplik` library conducts a simple EL ratio test that returns -2 log-likelihood ratio (`-2LLR`, which has an approximate chi-squared distribution under the null hypothesis), the associated p-value, the final value of the Lagrange multiplier (`lambda`), the gradient at the maximum (`grad`), the hessian matrix (`hess`), weights on the observations (`wts`), and the number of iterations performed (`nits`).

The following code, provided in the emplik documentation, illustrates two tests on a two-dimensional set of data: (i) $H_0 : \mu_1 = \mu_2$ and (ii) $H_0 : 2\mu_1 - \mu_2 = 0$.

```
# generate data
 x <- matrix(c(rnorm(50,mean=1), rnorm(50,mean=2)), ncol=2,nrow=50)
 y <- 2*x[,1]-x[,2]
# test hypothesis (i)
 el.test(x, mu=c(1,2))
# test hypothesis (ii)
 el.test(y, mu=0)
```

In one realisation of this code, the results of the first test were returned after four iterations, with weights ranging between 0.75 and 1.51, and with $-2\text{LL} = 1.50$ and a p-value of 0.47 under the assumption that -2LL is approximately chi-square under H_0. The second null hypothesis returned a p-value of 0.22.

Examples of the use of the `emplik` library for survival analysis are given by Zhou (2015). Whereas `el.test` requires uncensored data, the packages developed by Zhou and embedded in the `emplik` library enable estimation of hazard functions, cumulative distribution functions, and confidence bands for various types of censored data under a range of survival models.

As an example, the package `em.cen.EM` can be used to test the hypothesis $H_0 : \int g(t)dF(t) = \mu$ versus $H_a : \int g(t)dF(t) \neq \mu$, where $g(t)$ is a user supplied function. For instance, H_0 can be the test about the Kaplan–Meier mean and $g(t) = t$. The myeloma code in the Appendix illustrates this by testing $H_0 : F(10) = 0.2$ in the myeloma dataset incorporated in the `emplik` library. The code also finds the upper and lower confidence limit of a Wilks confidence interval. The output of this analysis provides a value –2LL and a corresponding p-value.

Bayesian EL methods are typically analysed by solving the EL using a Lagrange or similar method, then generating observations from the posterior distribution of the parameters of interest by an MCMC method. A more detailed description of this approach is given in the context of spatial modelling in the next section. An alternative approach, BC_{el}, which employs the `emplik` library to obtain the required likelihood values, is also detailed in a subsequent section.

12.4.4 Empirical likelihood in practice

The EL approach has been shown to be applicable in a broad range of contexts (Qin and Lawless, 2001). For example, following its formulation for estimation in linear regression (Owen, 1991), it has been extended to non-linear, generalised, parameric, non-parametric and semi-parametric models with and without missing data and censoring, time series models, and varying-coefficient models; see the review of Chen and Van Keilegom (2009b) and the references therein. The approach has also been proposed for testing; see again Chen and Van Keilegom (2009b). Einmahl and McKeague (2003) have proposed omnibus tests based on EL for a wide range of hypothesis tests, including symmetry, exponentiality, independence, and change of direction. Tests for stochastic ordering using EL have been proposed by El Barmi and McKeague (2013) and El Barmi and McKeague (2015).

Chaudhuri and Ghosh (2011) have proposed an EL approach for small area estimation and have suggested that the approach is also applicable to general random and mixed-effects models. As the authors argue, EL overcomes the distributional assumptions of the more dominant parametric models as well as the linearity assumptions of the non-parametric models that have been proposed for this problem. In addition, EL avoids the need for resampling methods like jacknife and bootstrap to obtain mean squared error estimation. The authors' methodology is developed using a multivariate-t prior for the parameter vector θ and both the regular and exponentially tilted formulations for the EL.

A Bayesian EL approach for constructing intervals for the analysis of survey data has been explored by Rao and Wu (2010). This work builds on the EL approaches for complex survey analysis proposed by Chen and Sitter (1999) and Wu and Rao (2006). Rao and Wu (2010) provide a clear exposition of EL methods for sample surveys. The authors set up the problem as one in which N_t denotes the number of units $U = \{1, 2, ..., N_t\}$, in a finite population of size $N = \sum_{t=1}^{T} N_t$, that have the value y_t^*, and n_t denotes the number of units in the sample having this value $y_t^*, t = 1, ..., T$. The sample data are then reduced to a set of so-called scale-loads $(n_1, n_2, ..., n_T)', n_t \geq 0, n = \sum_{t=1}^{T} n_t$. Assuming a negligible sampling fraction, the likelihood can be approximated by using the multinomial distribution with a log likelihood given by:

$$l(p) = \sum_{t=1}^{T} n_t \log(p_t),$$

with $p_t = N_t/N$, and the MLE of:

$$\bar{Y} = \sum_{t=1}^{T} p_t y^n,$$

is the sample mean:

$$\bar{y} = \sum_{t=1}^{T} \hat{p}_t y^{n^*}, \quad \hat{p}_t = n_t/n.$$

The authors make the connection with the work of Chen and Sitter (1999) and argue that this 'scale-load' approach is 'in the same spirit' as EL as described by Owen (1988).

As described earlier, survival analysis is another area that lends itself naturally to EL. The popular Kaplan–Meier curve is a non-parametric estimator of the survival function $S(t) = P(T \geq t)$, where T denotes the time to an event. It is conceptually straightforward to see that S can be estimated as a maximum EL estimator. This field has been developed by a number of authors: see, for instance, Wang and Jing (2001) for a general exposition of the survival model, Murphy and van der Vaart (1997) for doubly censored data, Qin and Jing (1994) for Cox modelling using EL, and McKeague and Zhao (2002) for an EL approach to relative survival. The recent text by Zhou (2015) provides an excellent overview of the field, as well as new models and computational algorithms, with associated R code to facilitate implementation. A simple illustration of an EL approach to survival analysis is provided in the next section.

Recent years have also seen an increase in popularity of EL for spatial modelling. Chaudhuri and Ghosh (2011) pioneered a Bayesian EL approach for small area estimation. Their model can accommodate continuous and discrete and area- and unit-level data, random and mixed effects, and the original and exponentially tilted empirical likelihoods.

A similar approach has also been proposed recently by Porter et al. (2015a) for this purpose. The so-called semi-parametric hierarchical EL (SHEL) model can be applied to irregular lattices and irregularly spaced point-referenced data and was shown to have improved mean squared prediction error compared with standard parametric analyses in a simulation study, a large community survey, and a bird survey. In the SHEL model, EL is employed in an empirical data model, which is combined with a parametric process model that accounts for the spatial dependence through a rank-reduced intrinsic conditional autoregressive prior and, finally, with a model at the highest level of the hierarchy describing the parameter.

A companion paper by the same authors (Porter et al., 2015b) extends this work to a multivariate context, with focus on the Fay–Herriot (FH) model, which is a mainstay in small area estimation. The argument is made that this approach encompasses spatial correlation (via the Fay–Herriot model), but avoids the usual restrictive Gaussian distributional assumptions (via EL).

One of the fields in which EL has been prominent is economics and related fields. For example, Riscado (2012) promotes the use of EL as a natural framework for estimation of dynamic stochastic general equilibrium models for macroeconomic analysis, since these models represent complex economic systems as a constrained optimisation problem and can be described as a set of moment conditions. The authors favourably compare EL with calibration and ML approaches, since the model parameters have complex correlation structures that hinder calibration and are typically characterised by non-linear systems of difference equations that have no closed form and hence hinder ML. The likelihood is thus often approximated and then estimated (and maximised) using methods such as the Kalman filter and sequential Monte Carlo. The authors interpret the EL approach as mapping from the set of moment conditions to the stochastic processes of the economic variables and then performing estimation by inverting that mapping. As discussed earlier, the importance and very often the difficulty of defining a set of 'good' moment conditions, or constraints, is highlighted in this setting.

12.4.5 The BC_{el} algorithm

A Bayesian EL algorithm was proposed by Mengersen et al. (2013). It was originally designed in the spirit of ABC, in that it avoids computation of the likelihood, but during its development, the authors realised that simulation from the likelihood could also be avoided and replaced with importance sampling. Thus, the so-called BC_{el} algorithm generates values $\theta_i, i = 1, \ldots, M$ from the prior distribution for θ and uses the values $w_i = L_{el}(\theta_i|y)$ as (unnormalised) weights in an importance sampling (IS) framework. The basic BC_{el} sampler is given in the following. Of course, this importance sampling algorithm will not be efficient if the posterior is very different to the prior. Later, we describe a more sophisticated algorithm based on adaptive multiple importance sampling [AMIS, Cornuet et al. (2012)].

Algorithm 12.2: BC_{el}, Mengersen et al. (2013)

for $i = 1$ to M **do**
  Generate θ_i from the prior distribution $\pi(\cdot)$
  Set the (unnormalised) weight $\omega_i = L_{el}(\theta_i|y)$
end for

12.4.5.1 Example 1

As a concrete example, consider estimation of the population mean θ based on a sample of observations $y_i, i = 1, ..., n$. In this case, two main decisions are required: the prior on θ and the constraints. The computation of the EL $L_{el}(\theta_i|y)$ can be performed using the `el.test` package in the `emplik` library in `R`, as described earlier in this chapter. In this case, the unnormalised weight ω_i is taken to be equal to the value of the empirical likelihood, which is calculated from the value of -2LLR obtained from the `el.test` function.

Suppose that a sample of size 100 is drawn from an (unknown) distribution $N(10, 1)$, and the aim is to estimate the population mean θ. A $N(-10, 30)$ prior is imposed on θ, and a first-moment constraint is chosen, for example, that the sample mean should equal the population mean. For the analysis, it is decided to run $M = 5000$ iterations, noting that in practice a smaller value of M can be used, but care must be taken to check that the weight has not concentrated too strongly on a small number of sampled values of θ. A resampling step can be included to mitigate this, although at a cost of introducing additional variance. In this case, the algorithm becomes:

Algorithm 12.3: BC_{el} Algorithm for Example 1

for $i = 1$ to M **do**
  Generate $\theta_i \sim N(-10, 30)$
  Obtain $-2LL$ from `el.test(y,mu=0)`
  Let $\omega_i = \exp(-0.5 \times (-2LL))$
end for
Resample θ with probability ω
Calculate summary statistics of the resampled values of θ

Example R code for this algorithm is given in the following.

```
data = rnorm(100,10,1)
M = 5000; theta.propose=w=rep(0,M)
for (i in 1:M){
  theta.propose[i] = runif(1,-10,30)
  el = el.test(data,mu=theta.propose[i])
  w[i] = exp(-0.5*(el$'-2LLR'))
}
```

```
theta=sample(theta.propose,M,prob=w,replace=TRUE)
mean(theta); sd(theta); quantile(theta,probs=c(0.1,0.9))
hist(theta, main="",xlab="theta").
```

As noted earlier, the resampling step could be replaced with a weighted mean, standard deviation, and quantiles. One realisation of this code provided the following estimates: $\hat{\theta} = 10.08$; s.d.$(\theta) = 0.12$; 95% CrI$=(9.88, 10.21)$. A histogram of the obtained sample of θ is given in Figure 12.8.

Mengersen et al. (2013) comment on the performance of this algorithm with different constraints, namely, one, two, and three central moments. They observe that one and two constraints work well, but three constraints performed more poorly. This was seen to support the general suggestion by Owen (2001) that the number of constraints should be equal to the number of parameters. Interestingly, this was not seen to be the case for the g-and-k distributions, as described in the next example.

A possible measure of the efficiency of the algorithm is the effective sample size (ESS). The is reportedly a measure of the 'equivalent number of independent observations' in a sample, that is, the value that equates the obtained variance of the estimator of interest with the equivalent variance

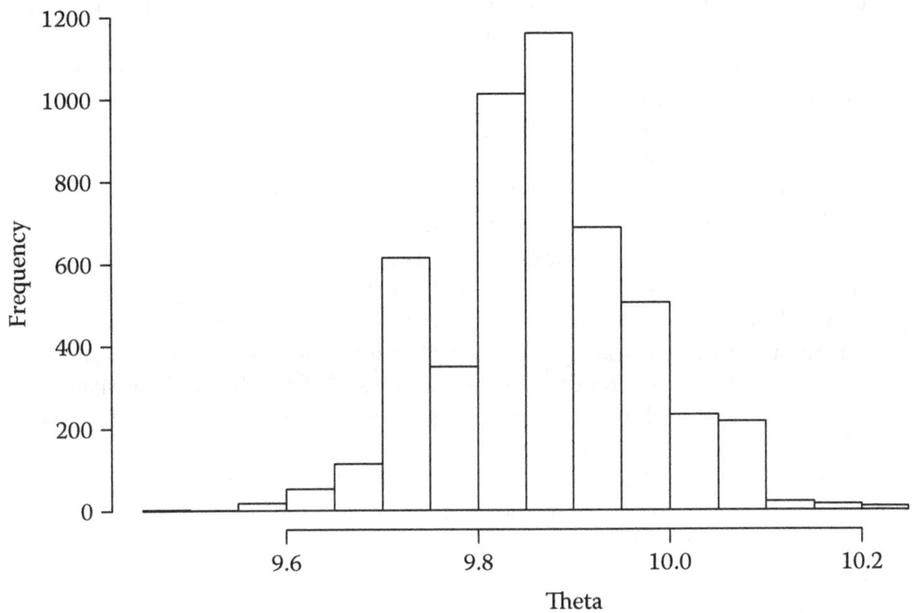

FIGURE 12.8
Histogram of draws from the BC_{el} posterior distribution of θ based on data generated from a $\mathrm{N}(\theta = 10, 1)$ distribution.

assuming an independent sample. For weighted samples as in EL, the ESS can be estimated as:

$$\text{ESS} = 1/\sum_{i=1}^{M}\{w_i/\sum_{j=1}^{M} w_j\}^2.$$

Kish (1965) argues that this substitution (of the EL for the exact likelihood) can be employed in any algorithm that samples from a posterior distribution. For example, it can be employed in composite likelihoods, which are commonly used in areas such as population genetics, where the likelihoods are known, but complex, and, hence, computationally difficult. The 'traditional' composite likelihood approach decomposes the target distribution $\pi(\theta)L(\theta|y)$ into several multivariate Student t distributions. In the BC_{el} approach, the EL is used instead. The computation is achieved using AMIS, which can be parallelised on a multi-core computer. The method can also be tailored for some non-iid problems such as dynamic models with AR structure, although the challenge here is in selecting appropriate constraints; see Mengersen et al. (2013) for details.

12.4.5.2 Example 2

We illustrate the use of BC_{el} by expanding on the discussion by Mengersen et al. (2013) of quantile distributions. These distributions are appealing for ABC in general, and BC_{el} in particular: there is typically no closed form expression for the likelihood, so regular algorithms such as MCMC are not immediately applicable; and it is fast and easy to obtain simulations from a quantile function via an inversion algorithm.

There is a body of literature on using ABC for estimation of quantile distributions. Allingham et al. (2009) proposed an ABC-MCMC algorithm in which draws of the parameters of the quantile distribution are based on a Metropolis algorithm with a Gaussian proposal distribution and are accepted based on the rule $||D - D'|| < h$, where D is the entire set of order statistics, $||\cdot||$ is the Euclidean norm, and h is heuristically chosen after inspection of a histogram of $||D-D'||$ obtained from a preliminary run using a very large value of h. Peters and Sisson (2006) also developed an ABC-MCMC algorithm for complex quantile functions. A range of improvements in the MCMC algorithm, selecting low-dimensional summary statistics, and methods of choosing h have since been suggested (Prangle, 2011; McVinish, 2012). Sequential Monte Carlo approaches for multivariate extensions of quantile distributions have also been proposed (Drovandi and Pettitt, 2011).

The g-and-k distribution is a popular example of a quantile distribution. This is a transformation of the standard normal distribution function, as follows:

$$Q(z(p);\theta) = a + b\left(1 + c\frac{1-\exp(-gz(p))}{1+\exp(-gz(p))}\right)(1+z(p)^2)^k z(p),$$

where $\theta = (a, b, g, k)$ and c is commonly set fixed at 0.8; see Rayner and MacGillivray (2002). Here, p denotes the pth quantile from the g-and-k distribution, and $z(p)$ is the corresponding quantile of the standard normal distribution. Thus, simulation from the g-and-k distribution requires only the generation of uniform$(0, 1)$ variates.

Figure 12.9 shows the estimated cumulative distribution function (cdf) of a standard normal distribution based on a g-and-k approximation, using the basic BC_{el} procedure described in Algorithm 12.2. The parameters of the g-and-k distribution corresponding to a $N(0, 1)$ distribution are $\theta = (0, 1, 0, 0)$. The analysis was based on 1000 observations, 100,000 iterations, and 10,000 resampled parameter values. The percentiles $(0.1, 0.25, 0.5, 0.75, 0.9)$ were chosen arbitrarily to form the constraints for EL, and all parameters were generated from a $U[0, 5]$ prior distribution.

The Bayes factor R code available in the Appendix illustrates the ease with which Bayes Factors (BF) can be computed for g-and-k distributions using EL. The example assumes a true model (Model 1) with $(A, B, g, k) = (0, 1, 1, 0)$ versus two alternatives, $(0, 1, 0.5, 0)$ (Model 2) and $(0, 1, 0, 0)$ (Model 3). Here, all models have zero mean ($A = 0$) and unit variance ($B = 1$), but differ in the degree of skewness, with Model 3 having no skewness ($g = 0$) and, hence, representing a standard normal distribution. The cumulative distributions functions for these three models are depicted in Figure 12.9. Two sample sizes of 100 and 500 and five constraints $(0.1, 0.25, 0.5, 0.75, 0.9)$ are considered. The resultant boxplots shown in Figure 12.10 confirm that Model 1 is

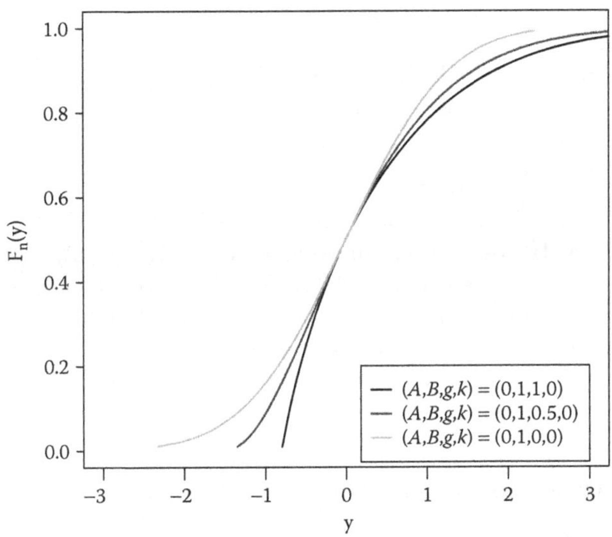

FIGURE 12.9
Cumulative distribution functions for three g-and-k distributions.

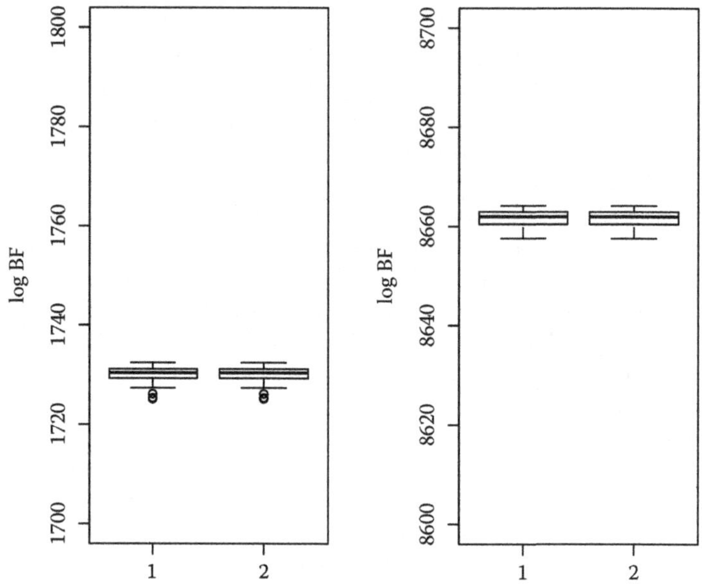

FIGURE 12.10
Boxplots of Bayes factors comparing three g-and-k distributions; data were generated from Model 1. On the x-axis: (1) Model 1 versus Model 2 with sample size $n = 100$, (2) Model 1 versus Model 2 with sample size $n = 500$ on the left and (1) Model 1 versus Model 3 with sample size $n = 100$, (2) Model 1 versus Model 3 with sample size $n = 500$ on the right.

preferred over both of the alternative models, with a stronger log BF obtained for the larger sample size as anticipated.

12.4.5.3 Example 3

Mengersen et al. (2013) also describe a variation on the basic BC_{el} algorithm, which employs AMIS in order to improve computational efficiency over plain importance sampling. The so-called BC_{el}-AMIS sampler employs multivariate Student $t_3(\cdot|m, \Sigma)$ distributions (three degrees of freedom, mean m, covariance matrix Σ) as importance sampling distributions, as described in the following Algorithm 12.4. The output of this algorithm is a weighted sample $\theta_{t,i}$ of size MT_M.

12.4.6 Extensions of the BC_{el} algorithm

Since its introduction, the BC_{el} approach has been applied to a range of problems. For example, Cheng et al. (2014) cite the approach as the foundation for their proposed method for estimating the parameters of the extreme value model of Heffernan and Tawn (2004). Through a large simulation study, the method was found to provide good coverage of credible intervals (CrI), although one of the parameters needed more informative priors under some more challenging setups.

In a second example, Grazian and Liseo (2017) discuss the use of BC_{el} for copula estimation, whereby the marginal likelihood of the quantity of interest is approximated by the EL.

Copula models are an important tool in multivariate analysis: while a huge literature exists about methods of estimating univariate marginal distributions, the problem of estimating the dependence structure of a multivariate distribution is more complex. Copula models allow for separately working with the univariate marginals and the joint distribution. They are widely used in many applications, including actuarial sciences (Embrechts et al., 2002), epidemiology (Clayton, 1978), finance (Cherubini et al., 2004), hydrology (Salvadori and De Michele, 2007), among others.

Algorithm 12.4: BC_{el}-AMIS

for $i = 1$ to M **do**
  Generate $\theta_{1,i}$ from the prior distribution $\pi(\cdot)$.
  Set the weight $\omega_{1,i} = L_{el}(\theta_{1,i}|y)$.
end for
for $t = 2$ to T_M **do**
  Compute the weighted mean m_t and weighted variance matrix Σ_t of the $\theta_{s,i}(1 \leq s \leq t-1, 1 \leq i \leq M)$.
  Denote by $q_t(\cdot)$ the density of $t_3(\cdot|m_t, \Sigma_t)$.
  for $(i = 1$ to $M)$ **do**
    Generate $\theta_{t,i}$ from $q_t(\cdot)$.
    Set $\omega_{t,i} = \pi(\theta_{t,i})L_{el}(\theta_{t,i}|y)/\Sigma_{s=1}^{t-1} q_s(\theta_{t,i})$.
  end for
  for $(r = 1$ to $t-1)$ **do**
    for $(i = 1$ to $M)$ **do**
      Update the weight of $\theta_{r,i}$ as $\omega_{r,i} = \pi(\theta_{r,i})L_{el}(\theta_{r,i}|y)/\Sigma_{s=1}^{t-1} q_s(\theta_{r,i})$.
    end for
  end for
end for.

A copula model is a way of representing the joint distribution of a random vector $X = (X_1, \ldots, X_d)$. Given a d-variate cumulative distribution function F which depends on some parameter ψ, it is possible to show (Sklar, 2010) that there always exists a d-variate function $C_\psi : [0,1]^d \to [0,1]$, such that:

$$F(x_1, \ldots, x_d; \lambda_1, \ldots, \lambda_d, \psi) = C_\psi(F_1(x_1; \lambda_1), \ldots, F_d(x_d; \lambda_d)),$$

where F_j is the marginal distribution of X_j, indexed by a parameter λ_j, and ψ is a parameter characterising the joint distribution.

In other terms, the copula C is a distribution function with uniform margins on $[0, 1]$, which takes value from the univariate $F_1, F_2, \ldots, F_d$ (which may be of the same form or may differ in terms of the parameters or of the forms) in order to produce the d-variate distribution F. The resulting model is very flexible, because it may utilise different types of marginal distributions and dependence structures.

Many different types of copula functions have been proposed in the literature; see Joe (2015) for a review. An example is the Clayton copula, defined in the general d-dimension case as:

$$C(\mathbf{u}) = (u_1^{-\psi} + u_2^{-\psi} + \cdots + u_d^{-\psi} - d + 1)^{-\frac{1}{\psi}},$$

where $\psi \in [-1, \infty) \setminus \{0\}$ is a one-dimensional parameter. The Clayton copula is characterised by lower-tail dependence (that approaches 1 as $\psi \to \infty$) and no upper-tail dependence. A representation of the Clayton copula (obtained through simulation) is available in Figure 12.11.

The frequentist standard method of estimating copula models is the 'inference from the margins' (IFM) approach (Joe, 2015), for example a two-step procedure, where first the marginal distribution functions are separately estimated, either in a parametric or in a non-parametric way (depending on the information available on the marginals), and then the copula function is estimated. Bayesian alternatives have been explored, nevertheless, they are still limited. The reader may refer to Smith (2013) for a review.

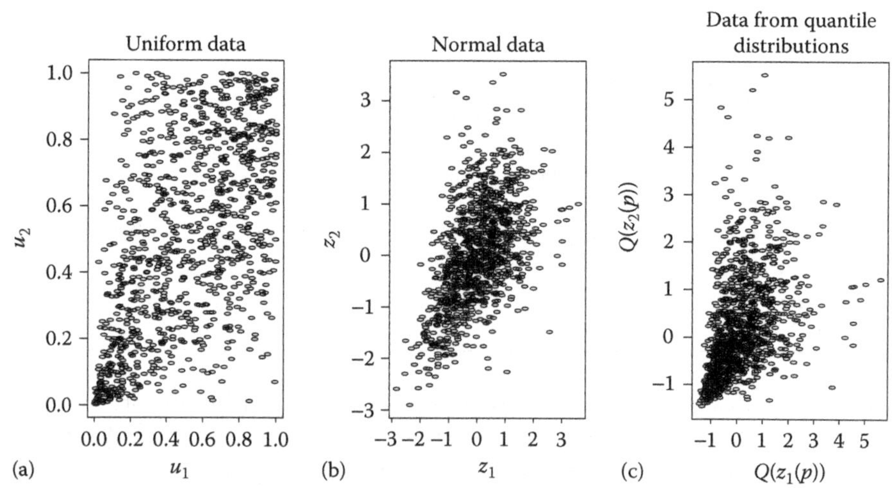

FIGURE 12.11
Scatterplots of the first two variables in the generation procedure: data from a Clayton copula with $\rho = 0.5$, (a) transformed to normal data, (b) and then, to data from a g-and-k distribution with $a = 0$, $b = 1$, $g = 0.5$, and $k = 0$ (c).

In some cases, the interest of the analysis is in a function of interest θ of the copula and not in the complete dependence structure; this may be due to weak information about the type of structure or to the need of a low-dimensional quantification of the dependence. Some typical quantities of interest are, for example, tail dependence indices, Spearman's ρ, or Kendall's τ. While tail dependence indices represent, in the bivariate case, the probability that a random variable exceeds a certain threshold given that another random variable has already exceeded that threshold (Großmaß, 2007), Spearman's ρ and Kendall's τ are measures of rank correlation, which are both expressible in terms of the copula C. For example, the Spearman's ρ in the bivariate case is defined as:

$$\rho = 12 \int_{[0,1]^2} C(u,v)\, du\, dv - 3 = 12 \int_{[0,1]^2} uv\, dC(u,v) - 3. \tag{12.3}$$

In this case, Grazian and Liseo (2017), in the same spirit of the IFM method, propose to first estimate the marginal distributions and then study the interest measure of multivariate dependence with an approximate Bayesian approach based on an estimation of the likelihood of θ via EL (the authors use its Bayesian modification described in Schennach (2005) and in Section 12.4.2). In this way, it is possible to avoid the complete definition of the dependence structure (usually difficult to be determined) and elicite the prior distribution only for the quantity of interest, in order to reduce the bias derived from wrong distributional assumptions. Moreover their Bayesian approach avoids the loss of information of the IFM method and may be proved to be consistent.

The *BCOP* ('Bayesian computation for copulas') algorithm follows and its final output will then be a posterior sample drawn from an approximation of the posterior distribution of the quantity of interest θ (see Algorithm 12.5).

This approach presents several advantages with respect to classical approaches to copula estimation. First, it may be applied to a generic dimension d, while in the literature there is a huge difference in terms of consistency results on the proposed estimators between the bivariate and the multivariate case. The authors have applied the *BCOP* algorithm to a maximum dimension equal to 50 with no loss of precision and with a reasonable computational expense (it has to be noted that the algorithm may be easily parallelised in the first step of estimation of the marginals). Second, the method provides a quantification of the error of estimation, not easily available in the classical approach [see Schmid and Schmidt (2007) for the Spearman's ρ and Schmidt and Stadtmüller (2006) for the tail dependence indices]. Third, it avoids the specification of the particular copula function which describes the dependence structure; this is particularly important in absence of information on it, since methods to select the copula function are not yet fully developed.

Since the interest is in small dimension parameter (often only one measure of dependence), the choice of the constraints should be easy; unfortunately, in practical applications, these conditions might hold only asymptotically. This is the case, for example, of the Spearman's ρ: its sample counterpart ρ_n is only

Algorithm 12.5: *BCOP* algorithm, Grazian and Liseo (2017)

1 Given a $n \times d$ dataset $x = \{x_1, \ldots, x_n\}'$ and marginal posterior samples $\{\lambda_1^{(s)}, \ldots, \lambda_d^{(s)}\}$ for $s = 1, \cdots, S$

for $s = 1, \ldots, S$ **do**

Use the s-th row of the posterior simulation $\{\lambda_1^{(s)}, \lambda_2^{(s)}, \ldots, \lambda_d^{(s)}\}$ to create a matrix of uniformly distributed data $u_{ij}^{(s)} = F_j(x_{ij}; \lambda_j^{(s)})$

$$u^{(s)} = \begin{pmatrix} u_{11}^{(s)} & u_{12}^{(s)} & \ldots & u_{1d}^{(s)} \\ u_{21}^{(s)} & u_{22}^{(s)} & \ldots & u_{2d}^{(s)} \\ \ldots & \ldots & u_{ij}^{(s)} & \ldots \\ u_{n1}^{(s)} & u_{n2}^{(s)} & \ldots & u_{nd}^{(s)} \end{pmatrix}.$$

end for

Given a prior distribution $\pi(\theta)$ for the quantity of interest ϕ,

for $b = 1, \ldots, B$ **do**

Draw $\theta^{(b)} \sim \pi(\theta)$

for $s = 1, \ldots, S$ **do**

Compute $L_{BEL}(\theta^{(b)}; u^{(s)}) = \omega_{bs}$

Take the average weight $\omega_b = S^{-1} \sum_{s=1}^{S} \omega_{bs}$

end for

end for

Output $(\theta^{(b)}, \omega_b), b = 1, \ldots, B$.

an asymptotically unbiased estimator of ρ, so the moment condition is strictly valid only for large samples.

Grazian and Liseo (2017) also apply the method to a real dataset based on the study of the dependence among five Italian financial institutes, where the returns are supposed to marginally follow a *generalized autoregressive conditional heteroskedasticity (GARCH)*(1,1) model with Student's t innovations. They show how it is possible to obtain an approximated posterior distribution of the Spearman's ρ of the financial asset returns of these institutes with Algorithm 12.5.

As an application, consider the setting of Section 12.4.5.2, where five sets of observations are simulated from g-and-k identical, but not independent quantile distributions with $a = 0$, $b = 1$, $g = 0.5$ and $k = 0$. The dependence structure is described by a multivariate Clayton copula (McNeil and Nešlehová, 2009) with true unknown multivariate ρ equal to 0.5. There are many ways to extend the bivariate Spearman's ρ defined in (12.3) to the multivariate case and they are not in general equivalent; nevertheless it is often of interest in many fields of application to describe the dependence with a low-dimensional quantity, for example, in the multivariate analysis of financial asset returns, where there is the need to express the amount of dependence in a portfolio by a single number. Here, the following is considered:

$$\rho = \frac{\int_{[0,1]^d} (C(u) - \Pi(u))\, du}{\int_{[0,1]^d} (M(u) - \Pi(u))\, du} = h(d) \left\{ 2^d \int_{[0,1]^d} C(u) du - 1 \right\}, \qquad (12.4)$$

where $M(u) = \min(u_1, u_2, \ldots, u_d)$ is the upper Fréchet–Hoeffding bound, and $h(d) = (d+1)/\{2^d - (d+1)\}$. For a review of the definitions of the Spearman's ρ in the literature, one may refer to Schmid and Schmidt (2007).

Uniform data have been generated from a multivariate Clayton copula with $\rho = 0.5$ and then inverted in order to obtain data from the corresponding quantile distributions. Figure 12.11 shows the correlation between the first two sets of observations generated with this procedure.

Figure 12.12 described the approximation to the posterior distribution of ρ, as defined in (12.4), obtained via Algorithm 12.5: it is possible to see that the posterior distribution is concentrated around the true value from which the data have been generated.

The R code used is available in the Appendix ('Copula code').

As noted earlier, one of the key considerations in developing and implementing BC_{el} is the choice of constraints for the EL. This consideration is not particular to BC_{el}, but applies to all EL methods. However, the difference here is that the selected constraints must be also applicable to the ABC context.

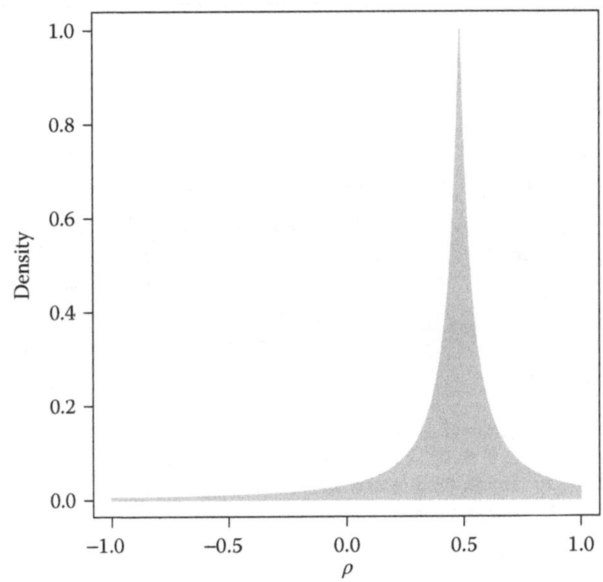

FIGURE 12.12
Approximation of the posterior distribution of the Spearman's ρ as defined in (12.3) for the data described in Figure 12.11.

With this goal in mind, Ruli et al. (2016) advocate the use of scaled composite likelihood score functions as summary statistics in ABC. The scaling takes into account a measure of the relative amount of information provided by the different parameters. They argue that the corresponding ABC procedure is therefore invariant to re-parametrisation and accommodates automatically the curvature of the posterior distribution. This approach is argued to be an improvement over that proposed by Pauli et al. (2011) and more 'fully ABC' than the BC_{el} approach.

Acknowledgements

Christopher Drovandi was supported by an Australian Research Council's Discovery Early Career Researcher Award funding scheme (DE160100741). Kerrie Mengersen gratefully acknowledges support from the Australian Research Council. Christopher Drovandi is an Associate Investigator and Kerrie Mengersen is a Chief Investigator of the Australian Centre of Excellence for Mathematical and Statistical Frontiers (ACEMS).

Appendix

Myeloma code

```
data(myeloma)
survtimes <- myeloma[,1]     # survival times
censtatus <- myeloma[,2]     # vital status (0=alive, 1=dead)
myfun1 <- function(t){ as.numeric(t <= 10) }
el.cen.EM(fun=myfun1, x=survtimes, d=censtatus, mu=0.2)
```

Bayes factor code

```
# test Model 1 (A,B,g,k)=(0,1,1,0) [skew]
# versus Model 2 (0,1,0.5,0) [less skew] and
# Model 3 (0,1,0,0) [standard normal]
# Compare B12=el_1/el_2 and B13=el1/el3
# Two sample sizes: n=100, 1000
library(emplik)
# set qc; traditionally set at 0.8
  qc=0.8
# specify the models of interest; qp1 is the 'true' model
  qp1=c(0,1,1,0) ;  qp2=c(0,1,0.5,0) ; qp3=c(0,1,0,0)
# specify the quantiles for each model
```

```
  refp=c(0.1,0.25,0.5,0.75,0.9)
  refq=qnorm(refp)
  simq1=qp1[1]+qp1[2]*(1+qc*((1-exp(-qp1[3]*refq))/(1+exp(-qp1[3]*refq))))
              *((1+refq^2)^qp1[4])*refq
  simq2=qp2[1]+qp2[2]*(1+qc*((1-exp(-qp2[3]*refq))/(1+exp(-qp2[3]*refq))))
              *((1+refq^2)^qp2[4])*refq
  simq3=qp3[1]+qp3[2]*(1+qc*((1-exp(-qp3[3]*refq))/(1+exp(-qp3[3]*refq))))
              *((1+refq^2)^qp3[4])*refq
# set sample size
  nob=c(100, 500)  # no. observations
  lennob=length(nob)
  nrep=100         # replicates of BF12
# set up matrices and vectors
  BF12=logBF12=BF13=logBF13=matrix(0,nrep,lennob)
  th1=th2=th3=rep(0,nrep)
# compute BF using el.test based on true parameters for M1 vs M2, M3
  for (nk in 1:lennob){
    dth1=dth2=dth3=matrix(0,nrow=nob[nk],ncol=length(refp))
    for (repk in 1:nrep){
       # generate reference data
       zp=qnorm(runif(nob[nk]))
       dob=qp1[1]+qp1[2]*(1+qc*((1-exp(-qp1[3]*zp))/
             (1+exp(-qp1[3]*zp))))*((1+zp^2)^qp1[4])*zp
       for (k in 1:nob[nk]){
         for (j in 1:length(refp)){
           dth1[k,j] = (dob[k]<simq1[j])*1
           dth2[k,j] = (dob[k]<simq2[j])*1
           dth3[k,j] = (dob[k]<simq3[j])*1
         }}
       th1=el.test(dth1,mu=refp)
       th2=el.test(dth2,mu=refp)
       th3=el.test(dth3,mu=refp)
       thll1=th1$'-2LLR' ;  thll2=th2$'-2LLR' ;  thll3=th3$'-2LLR'
       logBF12[repk,nk] = -0.5*(thll1 - thll2)
       logBF13[repk,nk] = -0.5*(thll1 - thll3)
       BF12[repk,nk] = exp(logBF12[repk,nk])
       BF13[repk,nk] = exp(logBF13[repk,nk])
       }} #end of repk, nk
par(mfrow=c(2,2))
boxplot(logBF123[,c(1,3)],ylim=c(1700,1800),
     #xlab="1=M1 v M2,n=100; 2=M1 v M2,n=500; 3=M1 v M3, n=100; 4=M1 v M3,
     n=500", ylab="log BF")
boxplot(logBF123[,c(2,4)], ylim=c(8600,8700),
     #xlab="1=M1 v M2,n=100; 2=M1 v M2,n=500; 3=M1 v M3, n=100; 4=M1 v M3,
     n=500", ylab="log BF")
```

Copula code

```
### Function to generate from a quantile function

quantile.fun=function(z,A,B,g,k,c=0.8)
{
val = A + B * ( 1 + c * (1-exp(-g*z)) / (1+exp(-g*z)) ) *
( 1+z^2 )^k * z
```

```
return(val)
}

### Simulations from the copula with a fixed Spearman's rho

# Generation from the copula
library(copula)
cc=claytonCopula(d=5,param=1.076)
uu=rCopula(1000,cc)

# Generation from the normal
z=matrix(NA,nrow=1000,ncol=5)
for(i in 1:5)
{
z[,i]=qnorm(uu[,i])
}

# Generation from the quantile distribution
quant.sim=matrix(NA,nrow=1000,ncol=5)
for(i in 1:5)
{
quant.sim[,i]=quantile.fun(z[,i],A=0,B=1,g=0.5,k=0)
}

#### (Nonparametric) estimation of the marginals

n=1000
F.hat=matrix(NA,nrow=1000,ncol=5)
for(i in 1:5)
{
for(j in 1:1000)
{
F.hat[j,i]=sum(quant.sim[,i]<quant.sim[j,i])/n
}
}

#### BCOP for the Spearman's rho

n=dim(F.hat)[1]
d=dim(F.hat)[2]
S=10^5

# Ranks
U.hat=matrix(NA,ncol=d,nrow=n)
for(i in 1:d)
{
U.hat[,i]=rank(F.hat[,i])/n
}
VV1=apply(1-U.hat,1,prod)
VV2=apply(U.hat,1,prod)

# Frequentist estimate
const=(d+1)/(2^d-(d+1))
estim1=const*(2^d/n*sum(VV1)-1)
```

```
# BCOP

rho=runif(S, -1,1)
omega=rep(0,S)

for (s in 1:S)
{
est=estim1 - rho[s]
omega1[s]<-exp(-EL(est)$elr)
}

rho.sim=cbind(rho, omega)

plot(rho.sim[,1],rho.sim[,2],type="h",
xlab=expression(rho),ylab="Density",main="",col="grey")

par(mfrow=c(1,3))
plot(uu[,1],uu[,2],xlab=expression(u[1]),ylab=expression(u[2]),
main="Uniform data")
plot(z[,1],z[,2],xlab=expression(z[1]),ylab=expression(z[2]),
main="Normal data")
plot(quant.sim[,1],quant.sim[,2],xlab=expression(Q(z[1](p))),
ylab=expression(Q(z[2](p))),main="Data from quantile distributions")
```

References

Allingham, D., R. A. R. King, and K. L. Mengersen (2009). Bayesian estimation of quantile distributionss. *Statistics and Computing 19*, 189–201.

An, Z., L. F. South, D. J. Nott, and C. C. Drovandi (2016). Accelerating Bayesian synthetic likelihood with the graphical lasso. https://eprints.qut.edu.au/102263/.

Andrieu, C., A. Doucet, and R. Holenstein (2010). Particle Markov chain Monte Carlo methods (with discussion). *Journal of the Royal Statistical Society: Series B (Statistical Methodology) 72*(3), 269–342.

Andrieu, C. and G. O. Roberts (2009). The pseudo-marginal approach for efficient Monte Carlo computations. *The Annals of Statistics 37*(2), 697–725.

Blum, M. G. B. (2010). Approximate Bayesian computation: A nonparametric perspective. *Journal of the American Statistical Association 105*(491), 1178–1187.

Brown, V. L., J. M. Drake, H. D. Barton, D. E. Stallknecht, J. D. Brown, and P. Rohani (2014). Neutrality, cross-immunity and subtype dominance in avian influenza viruses. *PLoS One 9*(2), 1–10.

Cai, A. Q., K. A. Landman, and B. D. Hughes (2007). Multi-scale modeling of a wound-healing cell migration assay. *Journal of Theoretical Biology 245*(3), 576–594.

Chaudhuri, S. and M. Ghosh (2011). Empirical likelihood for small area estimation. *Biometrika 98*, 473–480.

Chen, J. and R. R. Sitter (1999). A pseudo empirical likelihood approach to the effective use of auxiliary information in complex surveys. *Statistica Sinica 9*, 385–406.

Chen, S. (1993). On the accuracy of empirical likelihood confidence regions for linear regression model. *Annals of the Institute for Statistical Mathematics 45*, 621–637.

Chen, S. (1994). Empirical likelihood confidence intervals for linear regression coefficients. *Journal of Multivariate Analysis 49*, 24–40.

Chen, S. and H. Cui (2006). On Bartlett correction of empirical likelihood in the presence of nuisance parameters. *Biometrika 93*, 215–220.

Chen, S. and J. Gao (2007). An adaptive empirical likelihood test for parametric time series regression models. *Journal of Econometrics 141*, 950–972.

Chen, S. and J. Quin (2003). Empirical likelihood-based confidence intervals for data with possible zero observations. *Statistics & Probability Letters 65*, 29–37.

Chen, S. and I. Van Keilegom (2009a). A goodness-of-fit test for parametric and semi-parametric models in multiresponse regression. *Bernoulli 15*, 955–976.

Chen, S. and I. Van Keilegom (2009b). A review on empirical likelihood methods for regression. *Test 18*, 415–447.

Cheng, L., E. Gilleland, M. Heaton, and A. AghaKouchak (2014). Empirical Bayes estimation for the conditional extreme value model. *Stat 3*, 391–406.

Cherubini, U., E. Luciano, and W. Vecchiato (2004). *Copula Methods in Finance.* Hoboken, NJ: John Wiley & Sons.

Chopin, N., P. E. Jacob, and O. Papaspiliopoulos (2013). SMC^2: An efficient algorithm for sequential analysis of state space models. *Journal of the Royal Statistical Society: Series B (Statistical Methodology) 75*(3), 397–426.

Clayton, D. (1978). A model for association in bivariate life tables and its application in epidemiological studies of familial tendency in chronic disease incidence. *Biometrika 65*(1), 141–151.

Cornuet, J., J.-M. MARIN, A. Mira, and C. P. Robert (2012). Adaptive multiple importance sampling. *Scandinavian Journal of Statistics 39*(4), 798–812.

Craig, P. S., M. Goldstein, A. H. Seheult, and J. A. Smith (1997). Pressure matching for hydrocarbon reservoirs: A case study in the use of Bayes linear strategies for large computer experiments. In *Case studies in Bayesian Statistics*, pp. 37–93. New York: Springer-Verlag.

Decaestecker, C., O. Debeir, P. Van Ham, and R. Kiss (2007). Can anti-migratory drugs be screened in vitro? A review of 2D and 3D assays for the quantitative analysis of cell migration. *Medicinal Research Reviews 27*(2), 149–176.

DiCiccio, T., P. Hall, and J. Romano (1991). Empirical likelihood is Bartlett-correctable. *Annals of Statistics 19*, 1053–1061.

Drovandi, C. C. and A. N. Pettitt (2011). Estimation of parameters for macroparasite population evolution using approximate Bayesian computation. *Biometrics 67*(1), 225–233.

Drovandi, C. C., A. N. Pettitt, and A. Lee (2015). Bayesian indirect inference using a parametric auxiliary model. *Statistical Science 30*(1), 72–95.

Efron, B. (1981). Nonparametric standard errors and confidence intervals. *Canadian Journal of Statistics 9*, 139–172.

Einmahl, J. and I. McKeague (2003). Empirical likelihood based hypothesis testing. *Bernoulli 9*(2), 267–290.

El Barmi, H. and I. McKeague (2013). Empirical likelihood-based tests for stochastic ordering. *Bernoulli 19*(1), 295–307.

El Barmi, H. and I. McKeague (2015). Testing for uniform stochastic ordering via empirical likelihood. *Annals of the Institute of Statistical Mathematics* 68, 955–976.

Embrechts, P., A. Mcneil, and D. Straumann (2002). Correlation and dependence properties in risk management: Properties and pitfalls. M. A. H. Dempster (Ed.), *Risk Management: Value at Risk and Beyond*, pp. 176–223. Cambridge, UK: Cambridge University Press.

Everitt, R. G., A. M. Johansen, R. E., and M. Evdemon-Hogan (2015). Bayesian model comparison with intractable likelihoods. http://arxiv.org/abs/1504.00298.

Fan, J. and I. Gijbels (1996). *Local Polynomial Modelling and its Applications.* London, UK: Chapman & Hall.

Fasiolo, M., N. Pya, and S. Wood (2019). Statistical inference for highly nonlinear dynamical models in ecology and epidemiology. *Statistical Science 31*(1), 96–118.

Fearnhead, P. and D. Prangle (2012). Constructing summary statistics for approximate Bayesian computation: Semi-automatic ABC (with discussion). *Journal of the Royal Statistical Society: Series B (Statistical Methodology) 74*(3), 419–474.

Ghurye, S. G. and I. Olkin (1969). Unbiased estimation of some multivariate probability densities and related functions. *The Annals of Mathematical Statistics 40*(4), 1261–1271.

Glenn, N. and Y. Zhao (2007). Weighted empirical likelihood estimates and their robustness properties. *Computational Statistics and Data Analysis 51*, 5130–5141.

Grazian, C. and B. Liseo (2017). Approximate Bayesian computation for copula estimation. *Statistica 75*(1), 111–127.

Grendar, M. and G. Judge (2007). A Bayesian large deviations probabilistic interpretation and justification of empirical likelihood. Technical report, Department of Agricultural and Resource Economics, University of California Berkeley, CA.

Grendar, M. and G. Judge (2009). Asymptotic equivalence of empirical likelihood and Bayesian MAP. *The Annals of Statistics 37*, 2445–2457.

Großmaß, T. (2007). Copulae and tail dependence. PhD thesis, Humboldt University of Berlin, Germany.

Haardle, W. (1990). *Applied Nonparametric Regression.* Cambridge, UK: Cambridge University Press.

Hall, P. and B. La Scala (1990). Methodology and algorithms of empirical likelihood. *International Statistical Review 58*, 109–127.

Hansen, L., J. Heaton, and A. Yaron (1996). Finite-sample properties of some alternative GMM estimators. *Journal of Business and Economic Statistics 14*, 262–280.

Hartig, F., C. Dislich, T. Wiegand, and A. Huth (2014). Technical note: Approximate Bayesian parametrization of a process-based tropical forest model. *Biogeosciences 11*, 1261–1272.

Hartley, H. and J. Rao (1968). A new estimation theory for sample surveys. *Biometrika 55*, 547–557.

Heffernan, J. E. and J. A. Tawn (2004). A conditional approach for multivariate extreme values (with discussion). *Journal of the Royal Statistical Society: Series B (Statistical Methodology) 66*(3), 497–546.

Joe, H. (2015). *Dependence Modeling with Copulas*, Volume 134. Boca Raton, FL: CRC Press.

Johnston, S. T., M. J. Simpson, D. L. S. McElwain, B. J. Binder, and J. V. Ross (2014). Interpreting scratch assays using pair density dynamics and approximate Bayesian computation. *Open Biology 4*(9), 140097.

Kish, L. (1965). *Survey Sampling.* New York: John Wiley & Sons.

Kitamura, Y. (2006). Empirical likelihood methods in econometrics: Theory and practice. Technical report, Cowles Foundation for Research in Economics, Yale University, New Haven, CT.

Lazar, N. (2003). Bayesian empirical likelihood. *Biometrika 90*, 319–326.

Lian, H. (2012). Empirical likelihood confidence intervals for nonparametric functional data analysis. *Journal of Statistical Planning and Inference 142*, 1669–1677.

McKeague, I. and Y. Zhao (2002). Simultaneous confidence bands for ratios of survival functions via empirical likelihood. *Statistics and Probability Letters 60*, 405–415.

McNeil, A. J. and J. Nešlehová (2009). Multivariate archimedean copulas, d-monotone functions and l_1-norm symmetric distributions. *The Annals of Statistics 37*, 3059–3097.

McVinish, R. (2012). Improving ABC for quantile distributions. *Statistics and Computing 22*(6), 1199–1207.

Meeds, E. and M. Welling (2014). GPS-ABC: Gaussian process surrogate approximate Bayesian computation. In N. L. Zhang and J. Tian (Eds.), *Uncertainty in Artificial Intelligence Proceedings of the Thirtieth Conference*, pp. 593–602. Arlington, VA: AUAI Press.

Mengersen, K., P. Pudlo, and C. Robert (2013). Approximate Bayesian computation via empirical likelihood. *Proceedings of the National Academy of Science 110*, 1321–1326.

Monahan, J. and D. Boos (1992). Proper likelihoods for Bayesian analysis. *Biometrika 79*(2), 271–278.

Moores, M. T., C. C. Drovandi, K. L. Mengersen, and C. P. Robert (2015). Pre-processing for approximate Bayesian computation in image analysis. *Statistics and Computing 25*(1), 23–33.

Murphy, S. and W. van der Vaart (1997). Semiparametric likelihood ratio inference. *Annals of Statistics 25*(4), 1471–1509.

Ong, V. M.-H., M.-N. Tran, S. A. Sisson, and C. C. Drovandi (2018). Variational Bayes with synthetic likelihood. *Statistics and Computing 28*(4), 971–988.

Ong, V. M. H., D. J. Nott, M.-N. Tran, S. A. Sisson, and C. C. Drovandi (2017). Likelihood-free inference in high dimensions with synthetic likelihood. https://eprints.qut.edu.au/112213/.

Owen, A. (1988). Empirical likelihood ratio confidence intervals for a single functional. *Biometrika 75*, 237–249.

Owen, A. (1991). Empirical likelihood for linear models. *Annals of Statistics 19*, 1725–1747.

Owen, A. (2001). *Empirical Likelihood.* New York: Chapman & Hall/CRC Press.

Pauli, F. and G. Adimara (2010). Bayesian inference with a pairwise likelihood: An approach based on empirical likelihood. *Proceedings of the 45th Scientific Meeting of the Italian Statistical Society, 53*, Padova, Italy.

Pauli, F., W. Racugno, and L. Ventura (2011). Bayesian composite marginal likelihoods. *Statistica Sinica 21*, 149–164.

Peters, G. W. and S. A. Sisson (2006). Bayesian inference, Monte Carlo sampling and operational risk. *Journal of Operational Risk 1*(3), 27–50.

Pham, K. C., D. J. Nott, and S. Chaudhuri (2014). A note on approximating ABC-MCMC using flexible classifiers. *Stat 3*(1), 218–227.

Porter, A., S. Holan, and C. Wikle (2015a). Bayesian semiparametric hierarchical empirical likelihood spatial models. *Journal of Statistical Planning and Inference 165*, 78–90.

Porter, A., S. Holan, and C. Wikle (2015b). Multivariate spatial hierarchical Bayesian empirical likelihood methods for small area estimation. *STAT 4*(1), 108–116.

Prangle, D. (2011). Summary statistics and sequential methods for approximate Bayesian computation. Technical report, Lancaster University, UK.

Price, L. F., C. C. Drovandi, A. Lee, and D. J. Nott (2018). Bayesian synthetic likelihood. *Journal of Computational and Graphical Statistics 27*(1), 1–11.

Qin, J. and B. Jing (1994). Empirical likelihood for cox regression model under random censorship. *Annals of Statistics 22*, 300–325.

Qin, J. and J. Lawless (2001). Empirical likelihood and general estimating equations. *Communications in Statistics – Simulation and Computation 30*, 79–90.

Rao, J. and C. Wu (2010). Bayesian pseudo-empirical-likelihood intervals for complex surveys. *Journal of the Royal Statistical Society, Series B 72*, 533–544.

Ragusa, G. (2006). Bayesian likelihoods for moment condition models. Technical report, University of California, Irvine, CA.

Rayner, G. D. and H. L. MacGillivray (2002). Numerical maximum likelihood estimation for the g-and-k and generalized g-and-h distributions. *Statistics and Computing 12*(1), 57–75.

Ripley, B. D. (1996). *Pattern Recognition and Neural Networks.* Cambridge UK: Cambridge University Press.

Riscado, S. (2012). *DSGE Models in Macroeconomics: Estimation, Evaluation, and New Developments* (*Advances in Econometrics, Volume 28*), Chapter On the Estimation of Dynamic Stochastic General Equilibrium Models: An Empirical Likelihood Approach. Bingley, UK: Emerald Group Publishing.

Rochet, P. (2012). Bayesian interpretation of generalized empirical likelihood by maximum entropy. Technical report.

Rubin, D. (1981). Bayesian bootstrap. *Annals of Statistics 9*, 130–134.

Ruli, E., N. Sartori, and L. Ventura (2016). Approximate Bayesian computation using composite score functions. *Statistics and Computing To appear. 26*(3), 679–692.

Salvadori, G. and C. De Michele (2007). On the use of copulas in hydrology: Theory and practice. *Journal of Hydrologic Engineering 12*(4), 369–380.

Schennach, S. (2005). Bayesian exponentially tilted empirical likelihood. *Biometrika 92*, 31–46.

Schmid, F. and R. Schmidt (2007). Multivariate extensions of Spearman's rho and related statistics. *Statistics and Probability Letters 77*(4), 407–416.

Schmidt, R. and U. Stadtmüller (2006). Non-parametric estimation of tail dependence. *Scandinavian Journal of Statistics 33*(2), 307–335.

Sklar, M. (2010). Fonctions de répartition a n dimensions et leurs marges. *Publications de l'Institut de statistique de l'Université de Paris 54*(1–2): 3–6. With an introduction by Denis.

Smith, Jr., A. A. (1993). Estimating nonlinear time-series models using simulated vector autoregressions. *Journal of Applied Econometrics 8*(S1), S63–S84.

Smith, M. S. (2013). Bayesian approaches to copula modelling. In P. Damien, P. Dellaportas, N. G. Polson, and D. A. Stephens (Eds.), *Bayesian Theory and Applications*, pp. 336–358. Oxford, UK: Oxford University Press.

Treloar, K. K., M. J. Simpson, P. Haridas, K. J. Manton, D. I. Leavesley, D. L. S. McElwain, and R. E. Baker (2013). Multiple types of data are required to identify the mechanisms influencing the spatial expansion of melanoma cell colonies. *BMC Systems Biology 7*(1), 137.

Vo, B. N., C. C. Drovandi, A. N. Pettitt, and G. J. Pettet (2015a). Melanoma cell colony expansion parameters revealed by approximate Bayesian computation. *PLoS Computational Biology 11*(12), e1004635.

Vo, B. N., C. C. Drovandi, A. N. Pettitt, and M. J. Simpson (2015b). Quantifying uncertainty in parameter estimates for stochastic models of collective cell spreading using approximate Bayesian computation. *Mathematical Biosciences 263*, 133–142.

Wang, Q. and B. Jing (2001). Empirical likelihood for a class of functionals of survival distribution with censored data. *Annals of the Institute of Statistical Mathematics 53*, 517–527.

Wilkinson, R. (2014). Accelerating ABC methods using Gaussian processes. *Journal of Machine Learning Research 33*, 1015–1023.

Wood, S. N. (2010). Statistical inference for noisy nonlinear ecological dynamic systems. *Nature 466*, 1102–1107.

Wu, C. and J. N. K. Rao (2006). Pseudo-empirical likelihood ratio confidence intervals for complex surveys. *Canadian Journal of Statistics 34*, 359–375.

Zhou, M. (2015). *Empirical Likelihood Method in Survival Analysis.* London, UK: Chapman & Hall/CRC Press.

Zhou, M. and Y. Yang (2014). *emplik: Empirical likelihood ratio for censored/truncated data.* R package version 0.9-9-6.

13

A Guide to General-Purpose ABC Software

Athanasios Kousathanas, Pablo Duchen, and Daniel Wegmann

CONTENTS

13.1 Introduction 370
13.2 Toy Models 371
    13.2.1 Observed data and summary statistics 371
    13.2.2 Generating simulations 372
13.3 Parameter Inference 373
    13.3.1 Rejection algorithm 374
    13.3.2 Post-sampling adjustments 378
    13.3.3 Multi-dimensional posteriors 380
    13.3.4 Validation of parameter estimation 382
        13.3.4.1 Using a wrong model 382
        13.3.4.2 Cross-validation/accuracy of point estimates 384
        13.3.4.3 Checking for biased posteriors 386
13.4 Model Choice 387
    13.4.1 Inferring Bayes factors 388
    13.4.2 Model choice validation 390
    13.4.3 Choosing summary statistics 392
        13.4.3.1 Statistics for parameter inference 392
        13.4.3.2 Statistics for model choice 394
13.5 Generating Simulations 395
    13.5.1 Generating simulations for rejection 396
    13.5.2 Performing Markov chain Monte Carlo 399
    13.5.3 A population genetics example 402
Appendix 407
References 409

13.1 Introduction

There are currently many programs available to conduct ABC analyses (Table 13.1). However, most programs are specific to a particular problem, and the large majority for questions that typically arise in a population-genetics setting. These include programs to infer demographic histories (e.g. *ONeSAMP*, *PopABC*, *msABC*, or *DIYABC*), to infer F-statistics (*ABC4F*), or to infer parental contributions in an admixture event (*2BAD*). However, there exists also programs for phylogeographic inference (*msBayes*), Systems Biology (*ABC-SysBio*), or the inference under stochastic differential equations (the MATLAB® package *abc-sde*).

A major benefit of such specific programs is that a model parameterisation suitable for simulation-based inference has already been worked out, and a set of summary statistics informative for the problem identified. However, a particularly powerful aspect of ABC is its application to almost any inference problem and model, and we will thus focus here on general purpose ABC software designed to be helpful in a large array of ABC applications, either to conduct the whole analysis pipeline or at least specific parts of it.

Specifically, we will discuss two similar ABC pipelines, one provided through the command line program *ABCtoolbox* (version 2.0, Wegmann et al., 2010), and the other by means of combining the two R packages *abc* (version 2.0, Csilléry et al., 2012) and *EasyABC* (version 1.4, Jabot et al., 2013). Both pipelines offer very similar features and build around a similar logic: (1) they

TABLE 13.1
General and Specific Purpose ABC Software

Software	Purpose	Reference
ABCtoolbox	General	Wegmann et al. (2010)
abc	General	Csilléry et al. (2012)
ABC-EP	General	Barthelmé and Chopin (2011)
ABC_distrib	General	Beaumont et al. (2002)
ABCreg	General	Thornton (2009)
EasyABC	General	Jabot et al. (2013)
DIYABC	Population genetics	Cornuet et al. (2008)
msABC	Population genetics	Pavlidis et al. (2010)
ONeSAMP	Population genetics	Tallmon et al. (2008)
PopABC	Population genetics	Lopes et al. (2009)
2BAD	Population genetics	Bray et al. (2010)
ABC4F	Population genetics	Foll et al. (2008)
msBayes	Phylogeography	Hickerson et al. (2007)
ABC-SysBio	Systems Biology	Liepe et al. (2010)
abc-sde	Stochastic differential equations	Picchini (2014)

both provide utilities to generate a large set of simulations using external simulation software that employs a variety of sampling algorithms, (2) they both offer algorithms to infer parameters from such a set of simulations, (3) they provide tools to conduct model choice from such sets of simulations performed under different models, and finally, (4) they both offer a series of functions to validate estimations as well as model choice. However, these pipelines differ in specific implementations of some algorithms, which we will outline in the following, as well as the general way users interact with them. The packages *EasyABC* and *abc* are used within the statistical software environment R and are, hence, well suited for people writing their simulation programs in R or for those familiar with the handling of datasets in this environment. In contrast, *ABCtoolbox* is written purely in `C++` and run from the command line, resulting in generally faster execution and making it particularly well suited when command-line programs are used to conduct simulations or calculate summary statistics.

In the remainder of this chapter, we will walk the reader through the general usage of these two ABC pipelines. In order to give the reader a chance to replicate our analysis easily, we will use rather simple models and always give a detailed description of all settings used, along with all the code required to replicate the described analyses. We will begin with parameter inference, as we believe this is the step for which the programs discussed here will be used most often. However, we will also discuss how to use these pipelines to perform model choice, and how to conduct simulations using existing software.

13.2 Toy Models

We will introduce the usage of the program *ABCtoolbox* and the R packages *EasyAbc* and *abc* through their application to the problem of inferring the mean (μ) and variance (σ^2) of a normal (model A) and a uniform (model B) distribution from a random sample. We will then further show how to use these ABC pipelines to distinguish between these models using model choice.

Using such simple models has two major benefits. First, it will allow us to compare the ABC estimates with those obtained from full-likelihood solutions. Second, it is straightforward and quick to generate data under these models using just a few lines of code, for instance, when using the free statistical programming language R, which we will do here. However, note that we will also discuss how to generate simulations with existing programs in Section 13.5.

13.2.1 Observed data and summary statistics

We will begin by generating a dataset of 100 random samples under model A (the normal distribution) for which we would like to infer parameters. This is readily done in R as follows:

TABLE 13.2
Observed Statistics of Test Data Set

Mean	Var	Median	Min	Max	Range	Q1	Q3
0.102	1.14	0.0788	−2.02	3.16	5.18	−0.598	0.799

```
sampleSize <- 100;
data.obs <- rnorm(sampleSize, mean=0, sd=1);
```

Note that for this test example, we do know the true parameters $\theta = (\mu, \sigma^2)$ that we want to estimate. This will allow us to test the accuracy of the estimation.

Next, we will define a function to calculate summary statistics both on the observed, as well as simulated datasets. As summary statistics, we will use here the sample mean, variance, and median, along with the smallest (min), largest (max) value, the range (max–min), and the first and third quartile (Q1 and Q3). A vector containing these summary statistics is readily generated in R using the following function:

```
calc.stats <- function (x){
  S <- c(mean(x), var(x), median(x), range(x), max(x)-min(x),
  quantile(x, probs=c(0.25, 0.75)));
  names(S) <- c("mean", "var", "median", "min", "max", "range", "Q1", "Q3")
  return(S);
}
```

This will now allow us to calculate these summary statistics on the observed data set (object data.obs), save them in the variable S_{obs} (object S.obs), and to also save them in a file called `normal.obs`.:

```
S.obs <- calc.stats(data.obs);
write.table(t(S.obs), file="normal.obs", quote=F, row.names=F);
```

While each run will produce different values, we provide the values we obtained and will use in the following in Table 13.2 to allow for the replication of all our results.

13.2.2 Generating simulations

We will next generate a large number (10,000) of simulations with parameter values drawn from prior distributions and calculate the associated summary statistics for each simulation. In order to allow for a direct comparison between the models, we will assume uniform prior distributions for the mean $\mu \sim U[-1, 1]$ and variance $\sigma^2 \sim U[0.1, 4]$ for both models. We also set the internal random number seed generator equal to one so that the reader can reproduce exactly our results. Simulations for the normal model (model A) are then generated as follows:

```
set.seed(1)
nsim <- 10000;
P.normal <- data.frame(mu=runif(nsim, min=-1, max=1), sigma2=runif(nsim,
    min=0.1 , max=4));
S.normal <- data.frame(matrix(data=0, ncol=length(S.obs), nrow=nsim));
names(S.normal) <- names(S.obs);
for ( i in 1:nsim ) {
  y <- rnorm(sampleSize, mean = P.normal$mu[i], sd=sqrt(P.normal$sigma2[i])
      );
  S.normal[i,] <- calc.stats(y);
}
write.table(cbind(P.normal, S.normal), file="simNorm.txt", quote=F, row.
    names=F);
```

Again, we saved the simulations also in a text file (`simNorm.txt`) in order to use them with *ABCtoolbox*. Note that *ABCtoolbox* requires the parameters and statistics in the same file, which is achieved by binding the data frames containing the parameters (`P.normal`) and statistics (`S.normal`) together using `cbind()`.

Generating simulations under the uniform model (model B) is achieved similarly. However, since the R function `runif()` requires the two limits a, b of the uniform distribution, rather than the mean and variance, we need to calculate them from the parameters μ and σ^2 after each draw as $a = \mu - \sqrt{3\sigma^2}$ and $b = \mu + \sqrt{3\sigma^2}$, respectively.

```
nsim <- 10000;
P.unif <- data.frame(mu=runif(nsim, min=-1, max=1), sigma2=runif(nsim, min
    =0 , max=4));
S.unif <- data.frame(matrix(data=0, ncol=length(S.obs), nrow=nsim));
names(S.unif)<-names(S.obs);
for ( i in 1: nsim ) {
  y <- runif(sampleSize, min=P.unif$mu[i]-sqrt(3*P.unif$sigma2[i]), max=P.
      unif$mu[i]+sqrt(3*P.unif$sigma2[i]));
  S.unif[i,] <- calc.stats(y);
}
write.table(cbind(P.unif, S.unif), file="simUnif.txt", quote=F, row.names=F
    );
```

13.3 Parameter Inference

The estimation of posterior distributions is straightforward with both *ABCtoolbox* and the R package *abc*. As a common feature, they both implement the basic rejection algorithm originally introduced by Tavaré et al. (1997) and Pritchard et al. (1999), but they differ in the post-sampling adjustment algorithms they offer. Specifically, the package *abc* implements the original post-sampling adjustment based on a local linear regression initially introduced by Beaumont et al. (2002), as well as an extension to non-linear models with heteroscedastic variance (Blum and François, 2010). In contrast, *ABCtoolbox* offers an implementation of the general linear model adjustment algorithm (ABC-GLM) introduced by Leuenberger and Wegmann (2010). In this section, we will discuss differences between these algorithms, as well as how to use them in the present example.

13.3.1 Rejection algorithm

abc

To start using *abc*, the package has first to be installed and loaded in R, which is done by the following two commands:

```
install.packages("abc");
library(abc);
```

To now conduct an ABC rejection on the data simulated under the normal model (model A), simply use the function `abc()` with the argument `method="rejection"` and by specifying the tolerance to be applied.

```
rejection <- abc(S.obs,P.normal,S.normal,tol=0.01,method="rejection");
```

Here, `S.obs`, `P.normal`, and `S.normal` refer to the vector of observed summary statistics S_{obs} and the data frames containing the simulated parameters and summary statistics, respectively, as generated under Section 13.2. The additional argument `tol` specifies the fraction of simulations to be retained based on their distance to the observed summary statistics. A tolerance of 0.01, for example, indicates that the posterior density will be estimated from the parameter values of the 1% of all simulations that produced summary statistics closest to the observed summary statistics based on an euclidean distance metric.

The package *abc* offers an internal plotting function `hist.abc()` to display posterior distributions. Since this function overloads the basic `hist()` function of R, it can be called on an `abc` object by simply typing:

```
hist(rejection);
```

Alternatively, it is also possible to use general R functions such as `hist()` or `density()` to plot posterior distributions. The results we obtained for the observed data shown earlier is plotted using `density` in Figure 13.1.

ABCtoolbox

In contrast to the R packages discussed here, the program *ABCtoolbox* is a program to be used from the command line, preferentially in a UNIX/Linux environment. While it can be run from the Windows command prompt, it is recommend to use the *cygwin* Unix-like interface on a Windows computer to benefit from all features of *ABCtoolbox*.

ABCtoolbox accepts input settings both directly from the command line, as well as through an input file. A list of all arguments relevant for estimation

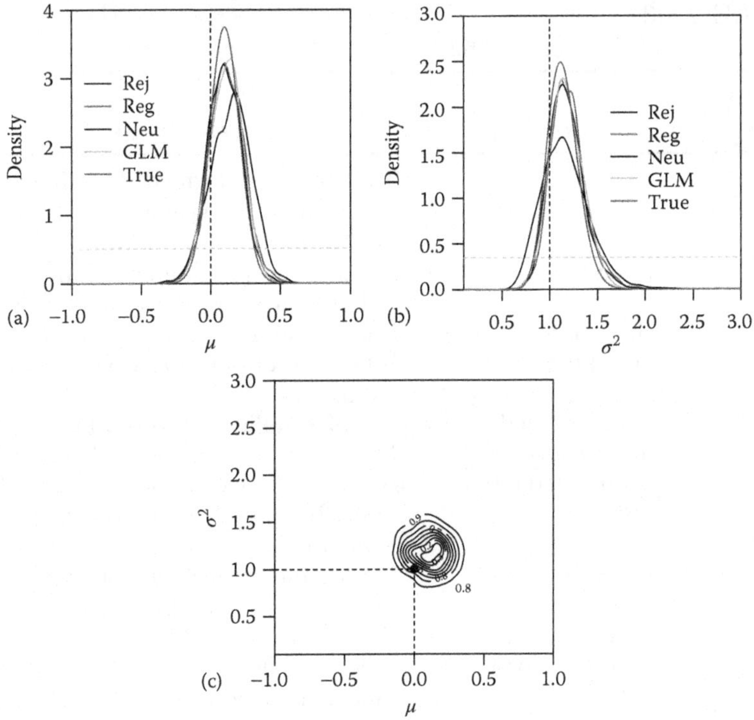

FIGURE 13.1
Posterior densities for the mean (a) and variance (b) parameters of the normal distribution model produced by a variety of methods. Dashed grey lines indicate prior densities. (c) Joint posterior density for μ and σ^2 produced with *ABCtoolbox*. Dashed black lines indicate true parameter values for all panels.

and discussed in this chapter is provided in Table 13.3. To perform parameter estimation for the normal distribution example, for instance, the following input file may be used:

```
task estimate
simName simNorm.txt
obsName normal.obs
params 1-2
maxReadSims 10000
numRetained 100
writeRetained
maxCor 1.0
```

Here, the argument `task` is set to `estimate` in order to run *ABCtoolbox* in estimation mode. Then, the argument `simName` specifies the name of the file containing the performed simulations. This file is requested to contain the used model parameter values together with the associated statistics, the names of

TABLE 13.3
ABCtoolbox Settings for Estimation

Setting Type	Setting	Description
Basic	task	Task to be performed. Possible options are: simulate, estimate, findStatsModelChoice.
	params	Specify the parameter columns in the file that contains the simulations.
	simName	File containing simulations.
	obsName	File containing observed summary statistics.
	numRetained	No. of simulations to retain.
	maxReadSims	Maximum number of read simulations.
	pruneCorrelated Stats	Remove statistics that are correlated, possible options 0 (retain), or 1 (remove).
	maxCor	Maximum acceptaple correlation coefficient between statistics.
	outputPrefix	Prefix for output files.
	writeRetained	Indicate whether to write retained simulations which can be used to obtain ABC rejection posteriors.
	standardiseStats	Standardise statistics.
Validation	obsPValue	The number of retained datasets for testing how well the inferred GLM model fits the observed data in multidimensional space (Section 13.3.4.1).
	tukeyPValue	The number of retained datasets for performing the Tukey test (Section 13.3.4.1).
	modelChoice Validation	The number of cross-validation replicates for validating model choice (Section 13.4.2).
	randomValidation	The number of cross-validation replicates for random parameter validation (Section 13.3.4.3).
	retainedValidation	The number of cross-validation replicates for retained parameter validation (Section 13.3.4.3).

(*Continued*)

TABLE 13.3 (*Continued*)
ABCtoolbox Settings for Estimation

Setting Type	Setting	Description
Posterior density	posteriorDensity Points	Number of points to estimate posterior density.
	diracPeakWidth	Smoothing parameter for posterior densities.
	jointPosteriors	Comma separated list of parameters for which the joint posterior is to be evaluated (Section 13.3.3).
	jointPosterior Density Points	No. of points to evaluate joint posterior (Section 13.3.3).

which are provided in the first line. Similarly, the file specified with the argument `obsName` should contain the summary statistics S_{obs} calculated from the observed data, again with the first line of the file containing the names of the statistics and the second line the associated values. Using the argument `params`, *ABCtoolbox* is further told which columns of the simulation file contain the model parameters to be estimated. Note that the simulation file may contain an arbitrary number of additional columns that will be ignored if they are neither specified to be model parameters with the argument `params` nor summary statistics also present in the file with the observed summary statistics.

The additional required arguments `maxReadSims` and `numRetained` specify the maximum number of lines (simulations) that will be read from the simulation file and the number of simulations to be retained in the rejection step, respectively. Finally, the argument `maxCor` specifies the maximal allowed correlation between summary statistics. If we also add the argument `pruneCorrelatedStats`, then the analysis will be performed by using statistics that their pairwise correlation do not exceed the `maxCor` threshold. We will discuss the issues with the following correlated statistics Section 13.4.3, but set this option here to 1 in order to include all statistics in the calculations and to avoid *ABCtoolbox* complaining about the presence of highly correlated statistics in our toy models.

To run *ABCtoolbox* with this input file (assuming it was saved under the name `estimate.input`), simply run in the command:

```
./ABCtoolbox estimate.input
```

Alternatively, the example can be run without using an input file by specifying the commands in the command line as:

```
./ABCtoolbox task=estimate simName=simNorm.txt obsName=normal.obs
params=1-2 maxReadSims=10000 numRetained=100 writeRetained maxCor=1.0
```

TABLE 13.4
ABCtoolbox Estimation Output Files

File Type	File Tag	Content
Basic	MarginalPosterior Characteristics	Characteristics of marginal posterior distributions (e.g. mode, mean, quantiles).
	BestSimsParam Stats	Retained simulations from ABC-rejection.
	MarginalPosterior Densities	GLM-adjusted marginal posterior densities.
	jointPosterior	GLM-adjusted joint posterior estimates.
	modelFit	Model choice results including Bayes factors and posterior support for compared models.
Parameter validation	RandomValidation	Results from random validation.
	RetainedValidation	Results from retained validation.
Model choice validation	modelChoice Validation	Results for model choice validation.

The output of such a run is found in a series of files, the names of which begin with a prefix that can be set with the argument `outputPrefix`, a tag referring to its content, and a number referring to the dataset and model for which the estimation has been conducted. While an exhaustive list of all output filename tags discussed in this chapter is given in Table 13.4, we will focus here on the file with tag `BestSimsParamStats`, which contains the retained simulations and is used to plot the rejection posterior.

The posterior estimates for μ and σ^2 obtained from the ABC-rejection algorithm are readily plotted in R using the `density()` function.

```
par(mfrow=c(1,2))
ABCrej<-read.delim("ABC_GLM_model0_BestSimsParamStats_Obs0.txt",sep="\t");
plot(density(ABCrej$mu,from=-1,to=1),main=expression(mu));
plot(density(ABCrej$sigma2,from=0,to=4),main=expression(sigma^2));
```

The results we obtained for the observed data shown earlier are plotted using `density` in Figure 13.1.

13.3.2 Post-sampling adjustments

Posterior distributions estimated with the rejection algorithm tend to be much broader than the true posterior distributions. This is shown for the normal model in Figure 13.1, but has been observed generally and is due to the often relatively large distance thresholds leading to parameter values resulting in summary statistics rather distant from S_{obs} to be accepted. Obviously, this loss of precision can be reduced by being more restrictive in accepting simulations, but this may require unrealistically computational efforts, particularly in more complex models.

An alternative is to correct for the effect of using large thresholds by exploiting the often simpler relationship between model parameters and summary statistics locally around the observed summary statistics. In a landmark paper, Beaumont et al. (2002) assume a linear relationship between model parameters and summary statistics locally among the retained simulations and proposed to use this relationship to project the parameter values of all retained parameter values to S_{obs}. More recently, Blum and François (2010) introduced an extension of this approach by fitting a non-linear, heteroscedastic model using neural networks. Both of these algorithms are implemented in the R package *abc*.

In contrast, *ABCtoolbox* offers an implementation of the ABC-General Linear Model (GLM) algorithm introduced by Leuenberger and Wegmann (2010) that estimates a local likelihood function instead of directly targeting the posterior distribution. While potentially slightly slower, this formulation is flexible in the choice of prior distributions and allows for model choice based on the marginal density. In practice, however, all mentioned post-sampling adjustment algorithms tend to give very similar results and the reader is advised to validate any estimation carefully in which these algorithms produce diverging estimates.

abc

The two post-sampling adjustments implemented in the R package *abc* are used by simply choosing the appropriate `method` when calling the `abc()` function. There are three different methods available: `loclinear`, `ridge`, and `neuralnet`, which correspond, respectively, to the classic regression adjustment introduced by Beaumont et al. (2002), a version of this algorithm using a ridge regression to deal with extensive collinearity among statistics and the non-linear, heteroscedastic regression proposed by Blum and François (2010). When using the loclinear method, if a warning appears regarding the collinearity of the design matrix, then we recommend to use the ridge method instead.

The following commands will perform posterior estimation using these algorithms on the toy model introduced earlier.

```
regression <- abc(S.obs,P.normal,S.normal,tol=0.01,method="loclinear");
neural <- abc(S.obs,P.normal,S.normal,tol=0.01,method="neuralnet");
```

The built-in function `hist()` can then again be used to plot the estimated posterior distributions.

```
hist(regression);
hist(neural);
```

Another function provided by the package *abc* is `plot.abc`, which can be used to plot the densities of the estimated posterior distributions together with additional, informative plots such as the prior distribution, the distribution of euclidean distances, and the residuals of the regression. Since this function overloads the standard R function `plot()`, it is simply used as follows:

```
plot(regression,param=P.normal);
plot(neural,param=P.normal);
```

Alternatively, the estimated posterior distributions can also be plotted using the R function **density**. For that purpose, one has to access specific elements of the object returned by the **abc()** function, namely the projected model parameter values as **adj.values**, as well as their weights **weights**. The following R commands, for instance, plot the posterior densities obtained via the regression and neural network adjustment for μ:

```
plot(density(regression$adj.values[,1], weights=regression$weights/sum(
     regression$weights)),main=expression(mu));
plot(density(neural$adj.values[,2], weights=neural$weights/sum(neural$
     weights)),main=expression(sigma^2));
```

Posterior distributions plotted using these functions are compared to those obtained through other methods in Figure 13.1. Note that the object returned also contains the retained model parameter values in the element **unadj.values** that can be used to plot the rejection posterior distribution.

```
plot(density(regression$unadj.values[,1]),main=expression(mu));
plot(density(regression$unadj.values[,2]),main=expression(sigma^2));
```

ABCtoolbox

When running *ABCtoolbox* in **estimation** mode, the ABC-GLM adjustment introduced by Leuenberger and Wegmann (2010) is performed automatically and the results available in the output file with tag **MarginalPosteriorDensities**. To plot the posterior estimates in R, simply load that file and use the function **density**.

```
par(mfrow=c(1,2))
ABCglm <- read.delim("ABC_GLM_model0_MarginalPosteriorDensities_Obs0.txt");
plot(ABCglm$mu,ABCglm$mu.density,type="l");
plot(ABCglm$sigma2,ABCglm$sigma2.density,type="l");
```

13.3.3 Multi-dimensional posteriors

abc

Apart from marginal posterior distributions, both the R package *abc* as well as *ABCtoolbox* are capable of estimating multi-dimensional posterior distributions. In the case of *abc*, however, the posterior densities have to be estimated using standard functions of R, such as **kde2d()** for two-dimensional posterior distributions. In our toy model, for instance, using:

```
posterior2d <- kde2d(regression$adj.values[,1],regression$adj.values[,2],n
     =100);
contour(posterior2d,xlab=expression(mu),ylab=expression(sigma^2));
```

where, `regression$adj.values[,1]` represents the projected model parameter values for μ, `regression$adj.values[,2]` represents the projected model parameter values for σ^2, and `n=100` specifies the number of marginal grid points to be used for the density estimation. These R commands will produce a plot similar to the one in Figure 13.1c.

ABCtoolbox

To generate multi-dimensional posterior densities on a grid, simply add the argument `jointPosteriors`, followed by the names of the model parameters for which the multi-dimensional posterior distribution is to be estimated. In addition, the number of marginal grid points has to be specified using `jointPosteriorDensityPoints`. The joint posterior estimates for the parameters μ and σ^2 of the normal distribution model, for instance, are thus estimated by simply running *ABCtoolbox* with a modified `estimate.input` input file that contains these two additional lines:

```
jointPosteriors mu,sigma2
jointPosteriorDensityPoints 100
```

Note that the total number of grid points grows exponentially with the dimensionality. In this previous example, the density will be evaluated at $100 \times 100 = 10^4$ positions. Running such a command for a four-dimensional posterior will already result in 10^8 positions at which the density has to be estimated.

When running *ABCtoolbox* with the additional arguments `jointPosteriors` and `jointPosterior DensityPoints`, the posterior density at each grid point will be written to an output file with tag `jointPosterior`. The resulting joint posterior of μ and σ^2 can then be plotted in R using the function `contour()`.

```
plot2D <- read.delim("ABC_GLM_model0_jointPosterior_1_2_Obs0.txt");
x <- unique(plot2D$mu);S.unif$var
y <- unique(plot2D$sigma2);
z_density <- matrix(data=plot2D$density,nrow=length(x),ncol=length(y),byrow
     =F);
contour(x,y,z_density,xlab=expression(mu),ylab=expression(sigma^2));
```

Since densities may be hard to interpret, *ABCtoolbox* also calculates and prints the smallest high posterior density interval (HDI) including each grid point to the same output file. The HDI corresponds to a posterior credible interval in the multi-dimensional parameter space and, hence, allows the generation of contour plots where contour lines indicate posterior credible intervals as follows:

```
z_HPD <- matrix(data=plot2D$HDI,nrow=length(x),ncol=length(y),byrow=F);
contour(x,y,z_HPD,xlab=expression(mu),ylab=expression(sigma^2));
```

The two-dimensional posterior distribution we obtained this way for the normal distribution example is given in Figure 13.1c.

Note that using a grid evaluation is not suitable to estimate multi-dimensional densities in high dimensions, as the total number of grid points grows exponentially with the dimensionality. In this earlier example, the density will be evaluated at $100 \times 100 = 10^4$ positions. Running such a command for a four-dimensional posterior will already result in 10^8 positions at which the density has to be estimated. An alternative is to generate samples from the joint posterior distribution from which densities are estimated using kernel estimators. To generate samples from high-dimensional posterior distributions after post-sampling adjustment, *ABCtoolbox* also implements an MCMC algorithm. While we will not discuss that algorithm here, we refer the user to the manual of *ABCtoolbox* for more details.

13.3.4 Validation of parameter estimation

13.3.4.1 Using a wrong model

An essential first validation step is to check whether the observed statistics can be reproduced by the examined model. A failure of the model to reproduce some of the statistics may indicate that a model is either not reflecting reality close enough or that inappropriate prior distributions have been used (e.g. too narrow distributions). More importantly, all post-sampling adjustments assume that the model fitted to the model parameters and summary statistics can be used to either accurately project retained simulations to S_{obs} (the methods implemented in *abc*), or is an accurate description of the likelihood of S_{obs} with the parameter range of the retained simulations (the method implemented in *ABCtoolbox*). A violation of these assumptions leads to an extrapolation to an area of the summary statistics space for which no samples have been obtained, and, hence, is prone to biased inference.

Checking if the observed summary statistics S_{obs} are within the range of summary statistics generated by the model is, however, a bit tricky in higher dimensions. For instance, consider the marginal summary statistics distributions shown in Figure 13.2 for the normal and uniform toy models, respectively. These distributions were plotted in R using the `density()` function directly from the simulated data.

```
plot(density(S.normal$var), col='black');
lines(density(S.unif$var), col='grey');
abline(v=S.obs[2])
```

These plots suggest that both summary statistics are readily generated by both models. However, a mismatch might manifest when looking at combinations of summary statistics. To check this we can plot two-dimensional distributions of pairs of summary statistics using the R functions `kde2d` and `contour`.

```
d <- kde2d(S.unif$var, S.unif$range, n=100);
contour(d$x, d$y, d$z);
```

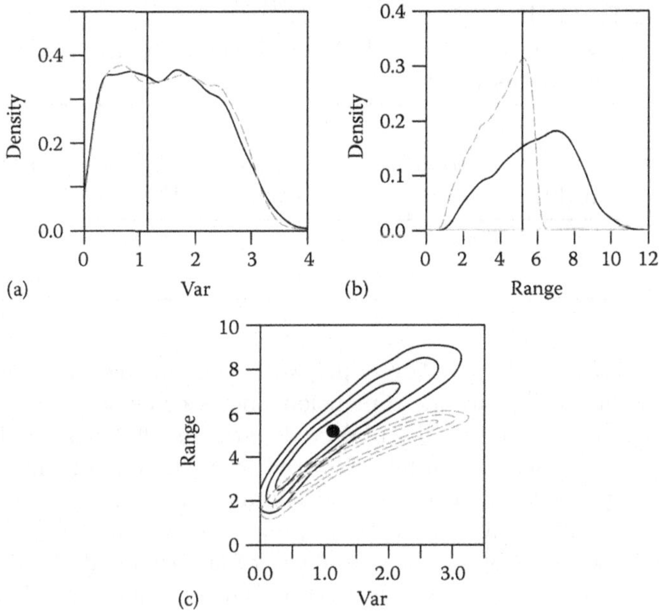

FIGURE 13.2
The distributions of simulated versus observed statistics `variance` (a) and `range` (b) and their joint distribution (c) for the normal (black) and uniform (grey) distribution models. Observed data are shown with a black vertical line in panels a and b, and with a black dot in panel c.

As is shown in the rightmost panel in Figure 13.2, the combination of the two observed summary statistics `variance` and `range` can indeed not be reproduced by the uniform model. However, this fact is not visible when looking at marginal densities only.

ABCtoolbox

Since visual inspection is only fruitful for a limited number of dimensions, *ABCtoolbox* offers two statistical tests for assessing whether a given model can reproduce the observed data S_{obs} in the multi-dimensional statistics space. The first test compares the marginal density (also called 'marginal likelihood') of the observed data to the marginal density of the retained simulations. The fraction of retained simulations with smaller or equal marginal density than the observed data is then provided as *the marginal density P-value*, where small values indicate a poor fit of the model to the observed data.

The second test evaluates how central the observed data lay within the multi-dimensional cloud of retained simulations by reporting the fraction of retained simulations with smaller or equal Tukey depth than the observed data as the *Tukey P-value* (Cuesta-Albertos and Nieto-Reyes (2008);

TABLE 13.5
Observed P-Value and Tukey P-Value Results for Normal Distribution Example

Model	Marginal Density	Marginal Density P-Value	Tukey Depth	Tukey P-Value
1	1158.75	0.098	0.13	0.96
2	4.16×10^{-12}	0	0	0

Adrion et al. (2014)). The Tukey depth is a common measure of centrality analogous to the median in one dimension and is calculated by *ABCtoolbox* for a retained simulation (or the observed data) as the smallest fraction of retained simulations which can be separated from the rest of the simulations using a hyper plane through the chosen simulation (or the observed data). Again, a low Tukey P-value indicates a poor fit of the model, since the observed data appears to be at the periphery of the retained cloud. However, note that the opposite is not necessarily true. Indeed, even a poor model (e.g. a model producing summary statistics at random) may be capable of generating a cloud of summary statistics surrounding S_{obs} and will thus pass both tests.

To perform these tests, simply call *ABCtoolbox* with the arguments `marDensPValue` and `tukeyPValue`, where each of them indicates the number of retained simulations to be used when calculating the respective P-value. When adding the following two lines to the input file `estimate.input`, for instance, *ABCtoolbox* will use 1000 retained simulations to evaluate the P-values.

```
marDensPValue 1000
tukeyPValue 1000
```

The results we obtained for these tests for our toy models are shown in Table 13.5. As expected from the visual inspection in Figure 13.2, the uniform model is not capable of reproducing the observed data and, hence, fails both tests.

13.3.4.2 Cross-validation/accuracy of point estimates

The accuracy of posterior point estimates is generally assessed by estimating the parameters for datasets for which the true parameter values are known. This is readily done in an ABC setting as a leave-one-out test in which one of the provided simulations is randomly chosen and all other are used to infer the parameter estimates for this data (often called 'pseudo-observed' data). The inferred posterior point estimates, such as the maximum a posteriori (MAP or posterior mode), the posterior mean or the posterior median are then plotted against the parameter values used to generate the data (referred to as the 'true parameters'). This process (also called 'cross-validation') is then repeated for many 'pseudo-observed' datasets to obtain a general measure of accuracy. The procedure may also be repeated to test specific ABC settings such as the effect of the choice of tolerance or the number of available simulations.

abc

To use this cross-validation algorithm with the R package *abc*, simply call the function `cv4abc()` with the arguments matching those of the estimation plus the additional argument `nval`, which specifies how many pseudo-observed datasets are to be used. However, note that the observed data does not have to be provided, as they are not used in cross-validation. The following code, for instance, will conduct cross-validation on the normal distribution example for the neural network estimation algorithm based on 100 individual pseudo-observed datasets.

```
cv.neural <- cv4abc(P.normal, S.normal, tols=0.1, statistic="mode",method="
    neuralnet", nval=100);
```

Such a call will return an object containing both the true parameter values (element `true`), as well as the estimated parameter values (element `estim`), which can be used to estimate correlations among them and plot for visual inspection. A plot such as the one shown in Figure 13.3 is generated by:

```
plot(cv.neural);
```

ABCtoolbox

ABCtoolbox offers two flavours of this cross-validation algorithm: either by picking simulations randomly or by picking simulations only among those

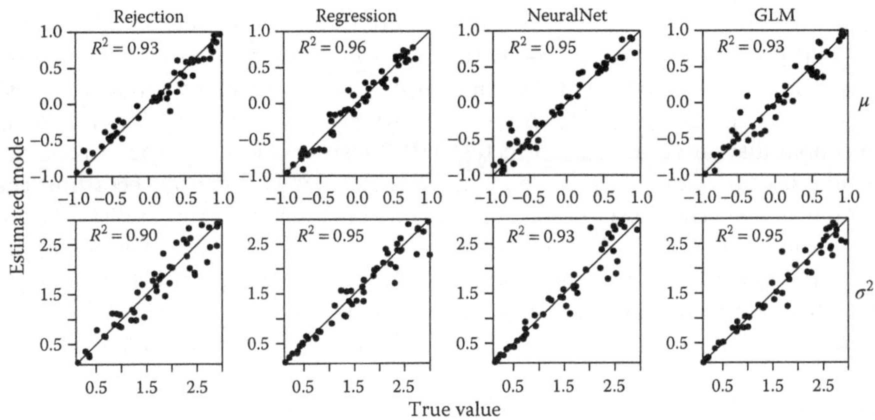

FIGURE 13.3
Parameter validation using different methods. The estimated posterior mode for mean and variance is plotted against the true values.

that were retained. The former, which is invoked through the argument **randomValidation**, corresponds to picking parameter values from the prior distribution and is thus informative above the overall accuracy of the ABC estimation under the chosen model. The latter, which is invoked using the argument **retainedValidation**, is informative about the accuracy of the estimation for the parameter space leading to similar data as the one observed.

```
task estimate
simName simNorm.txt
obsName normal.obs
params 1-2
maxReadSims 10000
numRetained 1000
maxCor 1.0
randomValidation 1000
retainedValidation 1000
```

When running *ABCtoolbox* with one or both of those arguments, an additional output file with tag **randomValidation** or **retainedValidation** is generated that contains the true parameter values along with a series of posterior point estimates for each parameter, namely the MAP (or mode), as well as the posterior mean and median. This file can then be loaded into R to generate plots such as those shown in Figure 13.3 as follows:

```
Random_validation <- read.delim("ABC_GLM_model0_RandomValidation.txt");
Retained_validation <- read.delim("ABC_GLM_model0_RetainedValidation_Obs0.
    txt");
plot(Random_validation$mu, Random_validation$mu_mode);
plot(Retained_validation$mu, Retained_validation$mu_mode);
```

13.3.4.3 Checking for biased posteriors

Pseudo-observed datasets can be used equally to detect potential biases in the marginal posterior distributions. If the posterior distributions of a parameter were unbiased, the position of the true parameters across many replicates must be given by the posterior densities. We proposed to test this directly using the probability integral transform test (PIT histogram or coverage property) Wegmann et al. (2009); Prangle et al. (2014). This is done by recording the position of the true parameter value in the cumulative posterior distribution (the posterior quantile) for each pseudo-observed dataset. In case posteriors were unbiased, these posterior quantiles must be uniformly distributed between 0 and 1. Similarly, the smallest high posterior density intervals (HDI) containing the true parameter value must be distributed uniformly.

ABCtoolbox

The procedure for this is the same as the one described in Section 13.3.4.2. The same output file will contain information that allows us to check for biased posteriors. For instance, the output from these analyses can be used to

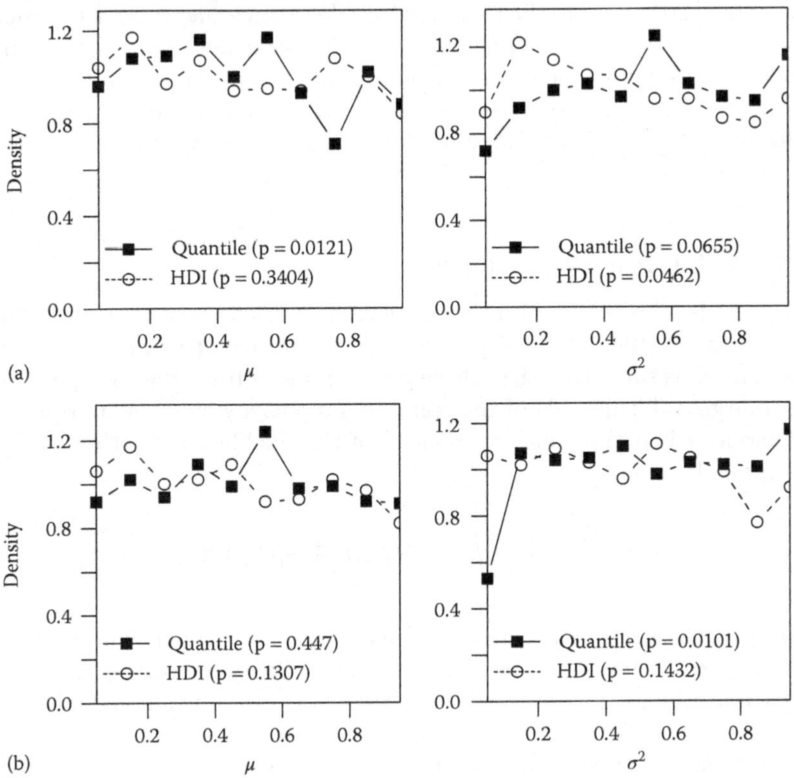

FIGURE 13.4
Parameter validation testing uniformity of the distribution of posterior quantiles and HDI when using random (a) and retained (b) simulations.

determine whether the posterior quantiles and HDI are uniformly distributed either by visual inspection or by performing a statistical test such as the Kolmogorov–Smirnov test:

```
hist(Random_validation$mu_HDI)
ks.test(Random_validation$mu_HDI,"punif")
```

The result of the random and retained validation analyses for the normal distribution example can be seen in Figure 13.4a and b, respectively.

13.4 Model Choice

While model choice is commonly used in Bayesian statistics, it is contentious in an ABC setting due to the problem that even summary statistics sufficient for both models may lead to biased inferences (Didelot et al., 2011;

Robert et al., 2011). Nonetheless, ABC model choice has been used successfully in practice and both ABC pipelines discussed here offer algorithms to conduct model choice when simulations from multiple models are available. However, the user is advised to validate any ABC model choice carefully, and we will discuss here tools provided by the R package *abc*, as well as *ABCtoolbox* to aid in that crucial step.

13.4.1 Inferring Bayes factors

Bayesian model choice relies on the estimation of model posterior probabilities or Bayes factors (the ratios of posterior probabilities of competing models). In standard Bayesian statistics, these are estimated from the marginal densities (or marginal likelihood) of the compared models, where the marginal density of model i is defined as the integral of the likelihood function, weighted by the prior distribution:

$$\mathbb{P}(M_i|D) = \int \mathbb{P}(D|\theta_i, M_i)\mathbb{P}(\theta_i|M_i)d\theta_i.$$

Importantly, the marginal density is thus affected by the choice of prior distribution, and, hence, what is evaluated in a Bayesian setting is thus the combination of a stochastic model and the prior distribution specified on its parameters.

In case the likelihood function is not available for analytical evaluation, both posterior probabilities and Bayes factors can be estimated using ABC. The algorithm implemented in *ABCtoolbox*, for instance, fits a likelihood model to the retained data, and then uses this model to analytically calculate the marginal density for each model. In contrast, the algorithms available in the R package *abc* use the fact that the marginal density is proportional to the fraction of simulations that resulted in simulations close to the observed data S_{obs} when the simulations are generated according to the prior distribution. The posterior probabilities of the different models are thus estimated from the relative number of simulations being close to S_{obs} either from direct counting or through a regression adjustment similar to parameter inference. Note that it is crucial for both algorithms that the exact same summary statistics have been calculated under both models.

To illustrate the model choice algorithms implemented, we will attempt to estimate which of the two toy models introduced earlier (the normal and uniform model) was used to generate the observed data. We refer the reader to Section 13.2 for more details on those models.

abc

In order to conduct model choice with *abc*, the summary statistics of all models have to be concatenated into a single data frame or matrix. In addition, a vector indicating for each simulation the model under which it has been generated has to be created. For our toy models, this is simply achieved as follows:

```
allSimulations <- rbind(S.normal,S.unif);
index <- c(rep("norm",dim(S.normal)[1]),rep("unif",dim(S.unif)[1]));
```

The actual model choice is then conducted using the function `postpr`, which takes as arguments the observed summary statistics S_{obs}, the object containing the summary statistics for all models, and the index vector. In addition, the tolerance for the rejection step, as well as the method for estimating Bayes factors has to be provided. In total, *abc* offers three such methods. The simplest is method `rejection`, which estimates posterior probabilities of the different models directly from the relative proportions of accepted simulations. The two other methods, `mnlogistic` and `neuralnet` attempt to correct for the often large tolerance values by estimating the relative densities of retained simulations at S_{obs} using either a multi-nomial logistic regression Beaumont (2008); Fagundes et al. (2007) or neural networks François and Guillaume (2011), respectively.

The following command will run model choice using the `neuralnet` method on our toy models and, using the function `summary()`, print the results to screen in a nice format.

```
model.choice <- postpr(S.obs, index=index, sumstat=allSimulations, tol
     =0.1, method="neuralnet");
summary(model.choice);
```

The results for our toy models are shown in Table 13.6. As is expected from the observation that the uniform model fails to reproduce the observed summary statistics, the preferred model for this data is the normal model. However, note that the results for the rejection method, which is also run by default when performing a `mnlogistic` or `neuralnet` estimation, is much less clear due to the relatively large tolerance applied here.

ABCtoolbox

To perform model choice with *ABCtoolbox*, simply provide the arguments `simName` and `params` for multiple models using semicolons. To run model

TABLE 13.6
Results of Model Choice by Toy Models

	Posterior Probability		Bayes Factor
Model	**Normal**	**Uniform**	**Normal versus Uniform**
abc (rejection)	0.65	0.395	1.53
abc (neuralnet)	1.00	0.00	1.1×10^6
ABCtoolbox (GLM)	1.00	9.79×10^{-13}	9.79×10^{13}

choice on our toy models, for instance, the input file `estimate.input` provided abode is modified as follows:

```
task estimate
simName simNorm.txt;simUnif.txt
obsName normal.obs
params 1-2;1-2
maxReadSims 10000
numRetained 1000
maxCor 1.0
```

When running *ABCtoolbox* with such an input file, an additional file with tag `modelFit` is generated. This file contains the marginal densities, Bayes factors, and posterior probabilities for each model. The results from this file obtained for our toy models are shown in Table 13.6, clearly indicating that the normal model is a much better fit.

13.4.2 Model choice validation

As was shown recently, model choice conducted with ABC may lead to biased or even wrong posterior probabilities, even if the summary statistics are sufficient for all models compared (Robert et al. (2011)). Consider two models $\mathcal{M}_1$ and $\mathcal{M}_2$ of shared parameters θ. If a set of summary statistics S was sufficient for both models, then the likelihood of the summary statistics and the likelihood of the full data are proportional for both models:

$$\mathbb{P}(D|\theta, \mathcal{M}_1) = c_1 \mathbb{P}(S|\theta, \mathcal{M}_1)$$
$$\mathbb{P}(D|\theta, \mathcal{M}_2) = c_2 \mathbb{P}(S|\theta, \mathcal{M}_2).$$

However, there is no guarantee that the two proportionality constants c_1 and c_2 are identical, which leads to the Bayes factors that are off by c_1/c_2. Therefore, careful validation is a key and compulsory step of any ABC model choice analysis. An initial first test may be to evaluate the power of choosing the correct model by means of pseudo-observed datasets. Such a cross-validation, which is offered by both *abc*, as well as *ABCtoolbox*, simply picks random simulations among those provided from both models, conducts model choice, and records how frequently the correct model was preferred. In addition, *ABCtoolbox* provides means to test for biases in the obtained posterior probabilities by comparing the ABC posterior probability (termed 'p_{ABC}') against those empirically expected ($p_{empirical}$) (Peter et al., 2010).

abc

The R package *abc* contains the function `cv4postpr` to conduct cross-validation for model choice. This function randomly picks one simulation from the file containing all simulations, performs model choice using the chosen simulation as pseudo-observed data, and records which model obtained the highest posterior probability. This is then repeated many times to determine the confusion matrix. As an example, consider the following call to `cv4postpr` to conduct 100 such replicates on our toy models using the index vector created earlier. To then print the confusion matrix, one may use the function `summary()` and to obtain a graphical representation the built in `plot()` function as follows.

```
cv.model.choice <- cv4postpr(index,allSimulations, nval=100, tols=0.1,
     method="neuralnet")
summary(cv.model.choice);
plot(cv.model.choice);
```

ABCtoolbox

To perform model choice validation with ABCtoolbox, simply add the argument `modelChoiceValidation` followed by the number of pseudo-observed datasets to be used to the estimation file. To use 1000 pseudo-observed datasets, for instance, you may add the following line to the input file:

```
modelChoiceValidation 1000
```

ABCtoolbox will then perform cross-validation and write the results to two different files. The first has the tag `confusionMatrix` containing the confusion matrix (fraction of correctly and incorrectly inferred models), as well as statistics calculated from it. For the toy model, for instance, we learn from this file that the normal model is correctly identified from data generated under that model in >99% of the cases.

The second file with tag `modelChoiceValidation` contains the raw results from the model choice validation and can be used for a more detailed validation analyses. For instance, we have recently proposed to compare the estimated model posterior probabilities with the empirical ones, an analysis that can reveal biases in ABC model choice (Peter et al., 2010; Chu et al., 2013). The basic logic of this analysis is that among all pseudo-observed datasets that resulted in an ABC posterior probability $p_{ABC} = x$ in favor of, say, model 1, a fraction x should have been generated under model 1. To test for this, data sets are binned according to their ABC posterior probabilities p_{ABC}, and the empirical posterior probabilities $p_{empirical}$ are then estimated as the fraction of simulations within each bin that were indeed generated with model 1.

The *ABCtoolbox* package includes the Rscript `Make_Model_choice_power_plot.r` to conduct this analysis and to produce the plot shown in Figure 13.5. For our toy models, it appears that there is a slight bias towards the normal model. This is evident from the fact that when $p_{ABC} = 0.5$, the datasets were actually generated from the uniform model in about 70% of the cases. However, among the datasets that resulted in very high p_{ABC} ($\geq$0.99), the vast majority was generated under that model. Therefore, we have high confidence in the model choice results of our observed data, which produced $\geq$0.99 posterior probability support for the normal distribution model (Table 13.6).

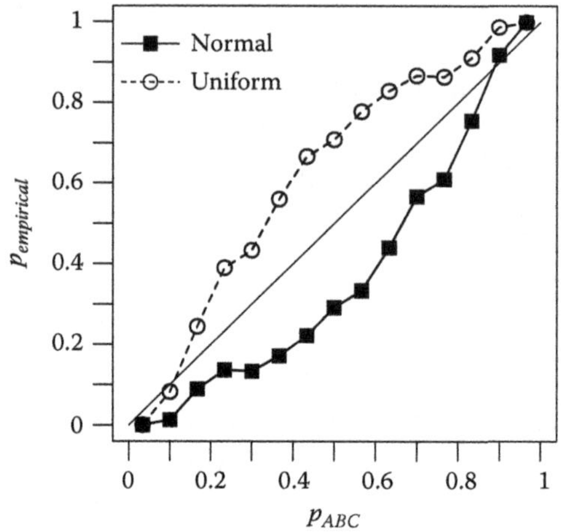

FIGURE 13.5
Posterior probability for normal and uniform distribution models estimated by *ABCtoolbox* (p_{ABC}) versus an empirical estimate of the same probability through simulation ($p_{empirical}$).

13.4.3 Choosing summary statistics

13.4.3.1 Statistics for parameter inference

The choice of summary statistics is crucial in any ABC inference, in that too few summary statistics are likely to miss out on important information and too many introduce harmful noise to the estimation (Wegmann et al., 2009; Beaumont, 2008; Blum et al., 2013). To date, many methods have been proposed to choose informative summary statistics from a larger set (see Blum et al. (2013) for a review), but we will focus here on those available through the *ABCtoolbox* package, in particular the use of linear combinations of summary statistics.

The use of such linear combinations was first introduced by Wegmann et al. (2009), who proposed to find them by means of partial least squares (PLS) regression. Broadly speaking, PLS is similar in spirit to a principal component analysis, but instead of finding linear combinations that maximise the variance explained in the summary statistics space, PLS components are chosen such that they maximise the product of the variance among summary statistics and the covariance between parameters and statistics (Tenenhaus et al., 1995). Recently, alternative means of finding linear combinations of summary statistics have been proposed, such as through boosting (Aeschbacher et al., 2012) or by regressing summary statistics on to posterior means inferred from an initial set of simulations (Fearnhead and Prangle, 2012).

While all these methods are readily used with *ABCtoolbox* once the linear combinations have been found, we will illustrate the usage of this functionality

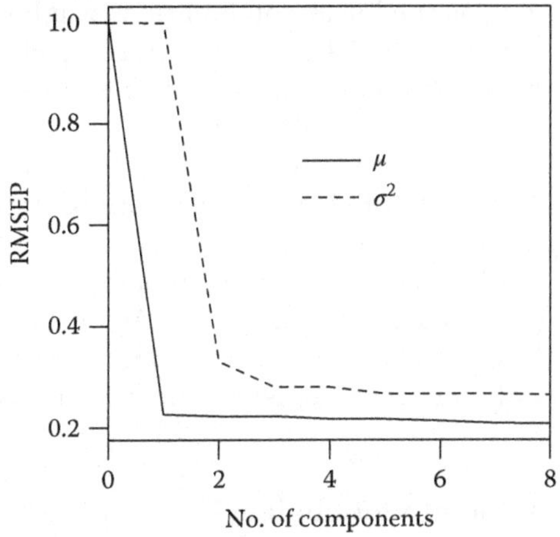

FIGURE 13.6
Number of PLS components versus root mean square error of prediction (RMSEP) for the normal distribution model.

based on the PLS approach, which is easy to implemevR-package 'pls'. In fact, the *ABCtoolbox* package provides an R-script to perform this analysis taking as input the simulations file (`simNorm.txt`). Performing a PLS analysis on the simulations from the normal distribution example reveals that two PLS components are sufficient for explaining the variance of the parameters of the normal distribution (Figure 13.6). This result is expected since the mean of a sample and the mean and variance of a sample are sufficient statistics for estimating, respectively, the mean and variance parameters of a normal distribution.

Any definition of linear combinations resulting from such a PLS or any other approach can then be used to transform the statistics of a set of simulations and the observed data using *ABCtoolbox* and then used in parameter inference. The PLS R-script mentioned earlier, for instance, writes the resulting PLS components to the file `PLSdef.txt`, which is then provided to *ABCtoolbox* and run in the `transform` mode as follows:

```
./ABCtoolbox task=transform linearComb=PLSdef.txt input=simNorm.txt output=
    simNorm.pls numLinearComb=2
./ABCtoolbox task=transform linearComb=PLSdef.txt input=normal.obs output=
    normal.obs.pls numLinearComb=2
```

Note that while we provided all arguments to *ABCtoolbox* on the command line, they may equally well be given in an input file. Using the transformed summary statistics in the estimation step is then straightforward: simply provide the transformed files using the arguments `simName` and `obsName`. As mentioned earlier, using alternative ways to find linear combinations is compatible with *ABCtoolbox*, as long as the linear combinations can be written in a definition file as the one created by the PLS script. Alternatively, the statistics may also be transformed with different software and then provided to *ABCtoolbox* (or the R package *abc*) for the estimation step.

One issue with using linear combinations of summary statistics is that information that arises from non-linear combinations of statistics are not taken into account. In such situations, it may be beneficial to increase the summary statistics space through combinations of summary statistics before finding linear combinations (Aeschbacher et al., 2012). As outlined with an example in Section 13.5.3, *ABCtoolbox* has an option (`doBoosting`) to also generate all pairwise products of summary statistics when generating simulations for this purpose.

13.4.3.2 Statistics for model choice

Finding appropriate combinations of statistics for model choice is particularly tricky. Just as for parameter inference, too few statistics may fail to capture important information, while too many are likely adding non-informative noise leading to large estimation variance and potentially a bias. Unfortunately, the methods introduced previously for finding good summary statistics for parameter inference are not readily extended to the problem of model choice. If one aims for using linear combinations, the most obvious choice is linear discriminant analysis (LDA), as was recently proposed by Estoup et al. (2012). To use LDA for model choice with the programs discussed here is similar to the use of linear combinations for parameter inference, in that the summary statistics of the observed and simulated data have to be transformed as explained in Section 13.4.3.1 before running either *ABCtoolbox* or the R package *abc* to perform model choice.

As an alternative to LDA, *ABCtoolbox* offers a greedy search algorithm to identify the combination of statistics having the largest power to discriminate between models. This search is done iteratively, by firstly evaluating the power of each single statistic and then adding additional statistics until no increase in power is observed. To perform this type of analysis, *ABCtoolbox* has to be run in the `findStatsModelChoice` mode and by providing the simulation files for at least two models, as well as the parameters required to perform the estimation. In addition, and using the argument `modelChoiceValidation`, one also needs to specify the number of simulations to be used as pseudo-observed data in each iteration to evaluate the power. As an example, consider the following input file for performing this type of analysis to find summary statistics appropriate for contrasting the normal versus uniform distribution toy models. We have added argument `maxCorSSFinder`, and set it equal to one in order to include combinations of statistics that are highly correlated in the greedy search. A lower threshold might be appropriate if many summary statistics are used in order to speed up the search.

TABLE 13.7
Combinations of Statistics Sorted by Estimated Power to Distinguish Models

Rank	Power	Largest Pairwise Correlation	No. Statistics	Statistics
1	1	0.966	3	mean,var,range
6	0.999	0.966	2	var,range
82	0.755	0	1	range

```
task findStatsModelChoice
simName simNorm.txt;simUnif.txt
obsName simple.obs
maxCor 1.0
maxCorSSFinder 1.0
params 1-2;1-2
numRetained 1000
maxReadSims 10000
outputPrefix ABC_searchStats
modelChoiceValidation 1000
```

The results of this analysis are written to a file with tag `searchStat sgreedySearch`. Part of this file is shown in Table 13.7. As shown there, the power to distinguish between these models is very high for multiple sets of summary statistics. Generally, it is recommended to choose the smallest among all sets with highest power, which would be the set consisting of the statistics `var` and `range`. As is shown in Figure 13.2, the two-dimensional distribution of these two statistics is indeed rather different between the models.

13.5 Generating Simulations

For simple models, such as the normal distribution example that we examined in the previous section, it is relatively easy to perform the simulations using custom scripts written in scripting languages such as R. However, for realistically complex models, we often rely on specialised programs for performing simulations. Moreover, for certain ABC variants such as ABC-MCMC, the simulation and estimation procedures are inherently linked, thus requiring running the program that performs simulations jointly with the program that performs ABC. When choosing the appropriate program to do simulations, we should keep in mind that interpreted languages, such as R, are generally inefficient. In most cases, compiled languages, such as C, should be preferred. In this section, we will illustrate how to use the program *ABCtoolbox*, as well as the R package *EasyABC* to automate and streamline the simulation process and to perform more sophisticated ABC techniques, such as ABC-MCMC.

13.5.1 Generating simulations for rejection

We will first focus on how to use these two ABC pipelines to generate simulations from parameter values drawn from prior distributions. The such generated simulations are then ready to be used with all the estimation techniques introduced earlier.

EasyABC

This R package allows the user to launch simulations from an external program and to retrieve the output of these simulations in a format ready for post-processing or to dynamically perform ABC-MCMC. To achieve this, the user has to provide both a list containing the definitions of the prior distributions, as well as a model definition. The list containing prior distributions simply contains the names of the desired distributions, along with their arguments. For instance, a list defined as:

```
prior <- list(c("unif",-1,1),c("unif",0.1,4));
```

will imply that there are two model parameters with uniform priors bounded at -1 and 1, and 0.1 and 4, respectively.

The model may be either an R function taking the parameters as arguments and returning a vector of summary statistics or the name of an executable that will be used to generate the simulation. In case an executable is given, it is assumed that this executable will read the model parameters to be used from a file called `input` and write the resulting summary statistics to a file called `output`. These files are read and written dynamically as *EasyABC* concatenates into a list the parameters sampled from the prior and the simulated summary statistics. As an example, consider an executable R-script named `generate_norm_EasyABC.R` that wraps a program to run the simulation of a normal distribution which is a model with two parameters:

```
#!/usr/bin/Rscript
param<-scan("input")
sampleSize <- 100;
data <- rnorm(sampleSize, mean=param[1], sd=param[2]);

calc.stats <- function (x){
  S <- c(mean(x), var(x), median(x), range(x), max(x)-min(x),
  quantile(x, probs=c(0.25, 0.75)));
  names(S) <- c("mean", "var", "median", "min", "max", "range", "Q1", "Q3")
  return(S);
}

sim <- calc.stats(data);
write.table(t(sim), file="output", quote=F, row.names=F,col.names=F);
```

The user should make the script executable like this:

```
chmod +x generate_norm_EasyABC.R
```

EasyABC is used to generate simulations (in this case 10^3) with the priors defined earlier as follows:

```
ABC_sim <- ABC_rejection(model=binary_model('./generate_norm_EasyABC.R'),
    prior=prior,nb_simul=1000)
```

which should take approximately 2 minutes to finish. Note that using the internal function of R to generate 10^3 deviates from a normal distribution would take less than 1 second to complete. Therefore, using an external program would be advised only if an R function for performing the simulations cannot be devised (e.g. for complex population genetics simulations, see the following).

The earlier command would only generate the simulations. To perform rejection and obtain the posterior distribution of parameters, we need to specify the observed summary statistics with argument `summary_stat_target` and the tolerance value with argument `tol` as follows:

```
ABC_sim <- ABC_rejection(model=binary_model('./generate_norm.R'),prior=
    prior,nb_simul=1000,summary_stat_target=sum_stat_obs,tol=0.1);
```

ABCtoolbox

An even more advanced and feature-rich way of using existing programs to generate simulations is offered by *ABCtoolbox*. To do so, *ABCtoolbox* has to be run in `simulate` mode, specified with the argument `task`. Similarly to *EasyABC*, the user then needs to specify both the model parameters and their prior distributions, as well as how to use existing programs to generate simulations using values drawn from the prior.

The model parameters and their priors have to be provided through an external file referred to as the `est` file, the name of which is provided with the argument `estName`. This file is structured in three distinct sections called `[PARAMETERS]`, `[RULES]`, and `[COMPLEX PARAMETERS]`. Only the first of those is mandatory and contains the definitions of the model parameters for which estimations are to be carried out. These model parameters and their prior distributions are declared using multiple columns as explained in Table 13.8. In brief, the first column indicates whether or not a model parameter is to be truncated to an integer value, the second column lists the name of the parameter, and the third column the prior distribution function. The remaining columns contain the parameters for this distribution, for instance the lower and upper bound, as well as the mean and standard deviation for a normal prior. The last column specifies whether or not the parameter values are to be printed to the output file.

TABLE 13.8
Declaration of Parameters in the `estName` File

Column	Content
1	Indicator 1/0 for being integer or rational number.
2	Name of the parameter.
3	Type of prior (see *ABCtoolbox* Manual for the types of supported priors).
4 -	Parameters for prior (for example min,max for uniform prior).
Last	Indicator output/hide for whether to print the parameter in the output file.

As an example, consider the following `est` file.

```
[PARAMETERS]
0 PARAM_A unif -1 1 output
0 PARAM_B norm -10 10 1 2 output
[RULES]
PARAM_A > PARAM_B
[COMPLEX PARAMETERS]
0 PARAM_B_SCALED = exp(PARAM_B) / PARAM_A
```

Here, we made use of the optional `[RULES]` section to limit the simulations to cases where `PARAM_A` is larger than `PARAM_B`. In addition, we benefited from `[COMPLEX PARAMETERS]` section to define a new variable `PARAM_B_SCALED`, which will always be set to the exponential of `PARAM_B`, divided by the value of `PARAM_A`. *ABCtoolbox* will understand most mathematical symbols and offers a wide variety of functions in this section, which allows for the definition of prior distributions and model parameterisation in a different way than required by the simulation software.

To demonstrate the use of *ABCtoolbox* to perform simulations, we can use a slightly modified simulation script named `generate_norm_ABCtoolbox.R` to generate deviates from a normal distribution similarly to the procedure described earlier for *EasyABC*:

```
#!/usr/bin/Rscript
args = commandArgs(trailingOnly=TRUE)
param1=as.numeric(args[1])
param2=as.numeric(args[2])
sampleSize <- 100;
data <- rnorm(sampleSize, mean=param1, sd=param2);

calc.stats <- function (x){
  S <- c(mean(x), var(x), median(x), range(x), max(x)-min(x),
  quantile(x, probs=c(0.25, 0.75)));
  names(S) <- c("mean", "var", "median", "min", "max", "range", "Q1", "Q3")
  return(S);
}

sim <- calc.stats(data);
write.table(t(sim), file="summary_stats-temp.txt", quote=F, row.names=F);
```

To specify how *ABCtoolbox* is to interact with the external simulation software, the arguments `simProgram` and `simArgs` are used, where the former defines the name of the executable to be used and the latter the arguments to be passed to that executable. These arguments may contain tags referring to model parameters listed in the `est` file, as well as any other string. The appropriate input file for *ABCtoolbox* may thus look as follows:

```
task simulate
obsName normal.obs
estName Rules.est
numSims 1000
simProgram generate_norm_ABCtoolbox.R
simArgs PARAM_A PARAM_B
```

The `Rules.est` file that contains the definitions of parameters and their priors for should be specified like this:

```
[PARAMETERS]
0 PARAM_A unif -1 1 output
0 PARAM_B unif 0.1 4 output
```

In this example, the parameters are read by the simulation program directly from the command line. In case the simulation program reads the parameters from a specific input file, *ABCtoolbox* can be set up to scan such a file and to replace all occurrences of model parameter tags defined in the `est` file with their current values, and to save the result to a new file, which is then passed to the simulation program. To make use of this feature, the name of the input file has to be specified with the argument `simInputName`, and the tag `SIMINPUTNAME` may then be used to refer to the newly created input file among the arguments passed to the simulation program.

Moreover, the output of the simulation program is stored in a file named 'summary_stats-temp.txt' which is read by *ABCtoolbox* by default, but a different name could be specified with argument `sumStatName`. In case the simulation program is generating data instead of directly summary statistics itself, *ABCtoolbox* can run additional programs to do extra operations on the output of the simulation program, such as the calculation of summary statistics. Such a program can be defined with the argument `sumStatProgram`, and the command-line arguments for the program are set with `sumStatArgs`. Note that `sumStatProgram` will always run after `simProgram`. A list of commonly used arguments when running *ABCtoolbox* in `simulate` mode are listed in Table 13.9.

13.5.2 Performing Markov chain Monte Carlo

Several other likelihood-free algorithms have been proposed that overcome the inherently low acceptance rates of rejection algorithms, among them a Markov chain Monte Carlo sampler (ABC-MCMC) first introduced by Marjoram et al. (2003), a Gibbs sampler using parameter-specific statistics (ABC-PaSS;

TABLE 13.9
ABCtoolbox Settings for Simulation

Setting Type	Setting	Description
Basic	task	Possible options simulate and estimate.
	samplerType	Possible sampler types are standard, MCMC, PaSS, and PMC.
	numSims	No. of simulations to perform.
	outName	Prefix for output files.
	estName	Filename containing definitions of priors for parameters and rules.
	simProgram	Program to perform simulations.
	simArgs	Arguments for simulation program.
	obsName	File containing observed summary statistics.
	sumStatProgram	Program to be run after simProgram. For example, a script calculating summary statistics.
	sumStatArgs	Arguments for sumStatProgram.
	sumStatName	File containing simulated summary statistics.
	doBoxCox	Do boxcox transformation.
	linearCombName	File containing linear combinations for transformation of statistics. (e.g. PLS components).
	doBoosting	Use all product combinations of statistics as additional statistics.
MCMC	numCaliSims	No. of calibration simulations.
	thresholdProp	Tolerance proportion of calibration simulations.
	rangeProp	Range of proposals.
	startingPoint	Starting location set from a random simulation (random) or the simulation with the minimum distance to the observed data (best).
	mcmcSampling	Interval between iterations that are printed in the output file.

Kousathanas et al. (2016)) and sequential Monte Carlo or particle samplers (ABC-PMC; (Sisson et al., 2007; Beaumont et al., 2009)). While both the R package *EasyABC*, as well as *ABCtoolbox* offer several types of algorithms, we will focus here on the use of ABC-MCMC with these tools.

The basic idea of ABC-MCMC is to replace the likelihood ratio in the Hastings ratio of a classic MCMC by an acceptance-rejection step using some tolerance ϵ. Such an ABC-MCMC chain is then generating samples directly from $\mathbb{P}(\|S - S_{obs})\| < \epsilon|\theta)$, where θ is the vector of model parameters, S and S_{obs} the simulated and observed vectors of summary statistics, respectively, and $\|\cdot\|$ some distance measure in the summary statistics space. Such an algorithm was shown to require much less simulations than standard ABC methods to obtain equally good posterior estimates (Marjoram et al., 2003). However, it turned out to be relatively tricky to tune this algorithm to perform properly since the acceptance rate of such an algorithm is directly given by the absolute likelihood, rather than the relative likelihood as in standard MCMC. A result of this is that ABC-MCMC chains may easily get stuck in regions of low likelihood, requiring a careful choice of both the tolerance ϵ as well as the initial starting positions. To improve the performance of this algorithm, we have proposed to tune the ABC-MCMC algorithm by means of an initial training set of simulations (Wegmann et al., 2009), which has been adopted by both *EasyABC* as well as *ABCtoolbox*. Specifically, the idea of such a calibration step is to choose a tolerance value ϵ that will result in sufficiently high acceptance rates and to find starting values in high likelihood regions. As with the classic rejection algorithm, it may be useful to transform summary statistics when calculating distances (Wegmann et al., 2009), and, hence, both *EasyABC* as well as *ABCtoolbox* offer to specify such transformations to be used during an ABC-MCMC chain.

While generally faster, an important issue with ABC-MCMC as well as Sequential Monte Carlo algorithms is that their output can not be directly used for validation. Instead, validation has to be done by repeating the whole process using pseudo-observed datasets, which may easily eat away the computational benefit of using these methods.

EasyABC

In *EasyABC*, the ABC-MCMC algorithm is offered through the function `ABC_mcmc()`, which takes similar arguments as the function to perform the rejection algorithm, namely a list with prior definitions as well as a model, but also requires the vector containing the observed summary statistics to be specified using the argument `summary_stat_target`. In addition, several arguments for tuning the actual MCMC run are required. As an example, consider the following R code to generate posterior samples using ABC-MCMC for our normal toy model, using the function `calc.stats()`, and the vector of observed summary statistics `S.obs` introduced previously:

```
#define model
toy_model <- function(x){
  data <- rnorm(100, x[1], sqrt(x[2]));
  return(calc.stats(data));
}
toy_prior <- list(c("unif",-1,1),c("unif",0.1,4));

#run ABC-MCMC
ABC_posterior <- ABC_mcmc(method="Wegmann", model=toy_model, prior=toy_
```

```
    prior, n_between_sampling=1,n_rec=10000, summary_stat_target=S.obs, n_
    calibration=10000, tolerance_quantile=0.1, numcomp=2);
```

Here, the argument `n_rec` specifies that 10,000 samples are to be generated. Further, the arguments `n_calibration` and `tolerance_quantile` specify that the ABC-MCMC chain will be calibrated from 10,000 simulations conducted under the prior, of which a fraction of 0.1 will be retained to calibrate the MCMC chain. Finally, the argument `numcomp` specifies that the total set of summary statistics is to be transformed into two PLS components .

Since an ABC-MCMC run is generating posterior samples, the output can be used directly to plot posterior distributions.

```
par(pty="s",mfrow=c(1,2));
plot(density(ABC_posterior$param[,1],from=-1,to=1,adjust=3),main="",xlab=
    expression(mu));
plot(density(ABC_posterior$param[,2],from=0.1,to=4,adjust=3),main="",xlab=
    expression(sigma^2));
```

ABCtoolbox

To perform the ABC-MCMC algorithm with *ABCtoolbox*, a few arguments have to be added to the input file shown earlier for standard sampling. First, the argument `samplerType` has to be set to `MCMC`. Then, the arguments `numCaliSims`, `thresholdProp`, and `rangeProp` are used to specify the number of simulations to be used for calibration, the fraction of those simulations to be used to calibrate the threshold, and the fraction of the standard deviation of parameter values among these retained simulations to be used to propose new values during the MCMC chain, respectively. To transform the summary statistics during the MCMC chain, a file with the definition of linear combinations can be provided with the argument `linearCombName`. To use PLS transformations, for instance, an initial set of calibration simulations can be used to find appropriate PLS components as discussed earlier, and the resulting PLS definition file is then provided using this argument. For an example of an input file, we refer the reader to the population genetics example discussed in the following.

13.5.3 A population genetics example

Here, we will illustrate how to implement techniques described in the previous sections to estimate important aspects of the recent human demographic history from an allele frequency data set made publicly available by Boyko et al. (2008). Specifically, we will use the site-frequency spectrum (SFS) for synonymous sites obtained for a sample of 24 African Americans (from Table S2 in Boyko et al. (2008)) to infer the parameters of a simple population genetic model. The SFS is an information rich summary of allele frequency data and synonymous sites in a gene are sites where any point mutation would lead to the same amino acid, thus likely to evolve neutrally, which is an assumption we have to make for demographic inference. Our model assumes an ancestral African population of size N_{ANC}, which experienced an instantaneous change in size t generations ago to N_{CUR}. We note that there are multiple full-likelihood solutions available to infer the parameters of this simple model from SFS data that might outperform ABC (e.g. Excoffier et al., 2013; Gutenkunst et al. 2009). However, the goal here is to provide a detailed step-by-step, guide to using *ABCtoolbox* for demographic inference, for which we prefer a simple model that is fast to run. The benefit of ABC over the full-likelihood approaches lies in its flexibility, and working through this rather simple example will illustrate all aspects necessary to build even more complex models that may violate the assumptions of available full-likelihood solutions. In Table 13.10, we provide a look-up table of all the files that will be described and used in this example.

Following Boyko et al. (2008), we parameterised the time of the size change in units of the current population size ($\tau = t/(2 \times N_{CUR})$), and to allow the simulations to be performed in a time reasonable for an illustrative example, we downsampled the original data from the original 5 million sites to the SFS of only 10,000 sites shown in Table 13.11. From this data, we then calculated the set of classic population genetic summary statistics shown in Table 13.12.

TABLE 13.10
Files Required to Run the Full Population Genetics Example

Filename	**Description**	**Source**
popgen.obs	Observed summary statistics	Section 13.5.3
popgen.est	Rules file for ABCtoolbox containing definition of priors	Section 13.5.3
popgen.input	ABCtoolbox input file	Section 13.5.3
fsc25221	fastsimcoal2 executable	http://cmpg.unibe.ch/software/fastsimcoal2/
popgen.par	fastsimcoal2 input file	Chapter appendix
calcPopstats.pl	Perl script to calculate summary statistics from fastsimcoal2 output	Chapter appendix
findPLS.r	R-script to find PLS components	https://bitbucket.org/phaentu/abctoolbox-public/
PLSdef_popgen.txt	file containing PLS definitions generated with findPLS.r	Chapter appendix

TABLE 13.11
Downsampled Synonymous SFS

Site class	0	1	2	3	4	5	6	7	8	9	10	11	12	13	14	15	16	17	18	19	20	21	22	23	24
Site count	9906	7	5	2	0	1	1	0	0	1	0	0	0	0	0	0	0	0	0	0	0	0	0	0	77

TABLE 13.12
Summary Statistics for Synonymous Sites

Statistic	Header Tag	Value
No. of singletons	sfs1	7
No. of segregating sites (S)	S	17
Average pairwise diversity (π)	pi	3.06
Waterson's theta	theta	4.55
Tajima's D	taj_D	−1.17

These summary statistics should then be stored in the file `popgen.obs` for later usage as follows:

Observed file popgen.obs.

```
sfs1 S pi thita taj_D
7 17 3.06 4.55 -1.17
```

The first step always consists of defining the model parameters in the `est` file. For the model concerned here, we will use the file `popgen.est` provided in the following.

Rules file popgen.est.

```
[PARAMETERS]
0 LOG10_N_CUR unif 2 6 output
0 LOG10_OMEGA unif -3 3 output
0 TAU unif 0 1 output
0 MUTRATE fixed 2.5e-8 hide
[COMPLEX PARAMETERS]
1 N_CUR = pow10(LOG10_N_CUR) hide
1 T1 = TAU * 2 * N_CUR hide
0 OMEGA = pow10(LOG10_OMEGA) hide
```

As can be seen from this file, we decided to put uniform priors on the logarithm of the current population size N_{CUR} and the relative size of the ancestral population $\omega = \frac{N_{ANC}}{N_{CUR}}$, but a uniform prior on the relative age of the population size change τ.

To generate simulations under this model, we will make use of the program *fastsimcoal2* (*v*.2.5.2.21 downloaded from http://cmpg.unibe.ch/software/fastsimcoal2/) that allows to simulate SFSs under various demographic scenarios (Excoffier et al., 2013) . However, *fastsimcoal2* requires the parameters to be specified differently from our priors, and we thus make use of the `[COMPLEX PARAMETERS]` section to transform our model parameters appropriately. Specifically, we have to provide N_{CUR} and the population size change on the natural scale and further the age of the population size changes in generations. We then prepare the input file `popgen.par` for *fastsimcoal2* that specifies this model, using the parameter tags defined in the input file. While explaining the details of how to use *fastsimcoal2* for such a model is beyond the scope of this chapter, we provide the corresponding input file in the Appendix and refer the reader to the manual of *fastsimcoal2* for more details. To calculate summary statistics from the simulated data, we will use the custom perl script `calcPopstats.pl` also provided in the Appendix to this chapter.

In order to use a PLS transformation during the MCMC chain, we first generated an initial set of 1000 simulations using *ABCtoolbox* using the following input file:

ABCtoolbox input file to perform simulations with fastsimcoal2.

```
task simulate
obsName popgen.obs
estName popgen.est
numSims 1000
outName popgen_PLS
simInputName popgen.par
simProgram ./fsc25221
simArgs -i popgen-temp.par -s 0 -d -n 1 -q -x
sumStatProgram calcPopstats.pl
sumStatArgs popgen-temp/popgen-temp_DAFpop0.obs
doBoosting
```

We specify how *ABCtoolbox* is interacting with *fastsimcoal2* (executable `fsc25221`) with arguments `simProgram` and `simArgs`. We further specify our custom perl script *calcPopstats.pl* with argument `sumStatProgram`, which calculates summary statistics from the output of *fastsimcoal2*. While we refer the reader to the manual of *fastsimcoal2* for the details on the command line used, we note that the output written by *fastsimcoal2* will be located in a sub-directory (`popgen-temp`) and have a specific name (`popgen-temp_DAFpop0.obs`). We thus provide the path to this file to our perl script `calcPopstats.pl` via command line arguments (using `sumStatArgs`). In contrast to previously discussed input files, we also added the additional argument `doBoosting`, which will tell *ABCtoolbox* to also add all squares and pair-wise products of calculated summary statistics as additional summary statistics. This often proves helpful in exploiting non-linear relationships between parameters and statistics when finding linear combinations.

PLS components are then readily identified by following the steps discussed in Section 13.4.3.1. By looking at the Root-Mean-Squared Error (RMSE) plot (Figure 13.7a), we found that 4 PLS components are sufficient to capture the information contained in the total of 20 summary statistics (including the boosted ones). Having the appropriate PLS definition file `PLSdef_popgen.txt`

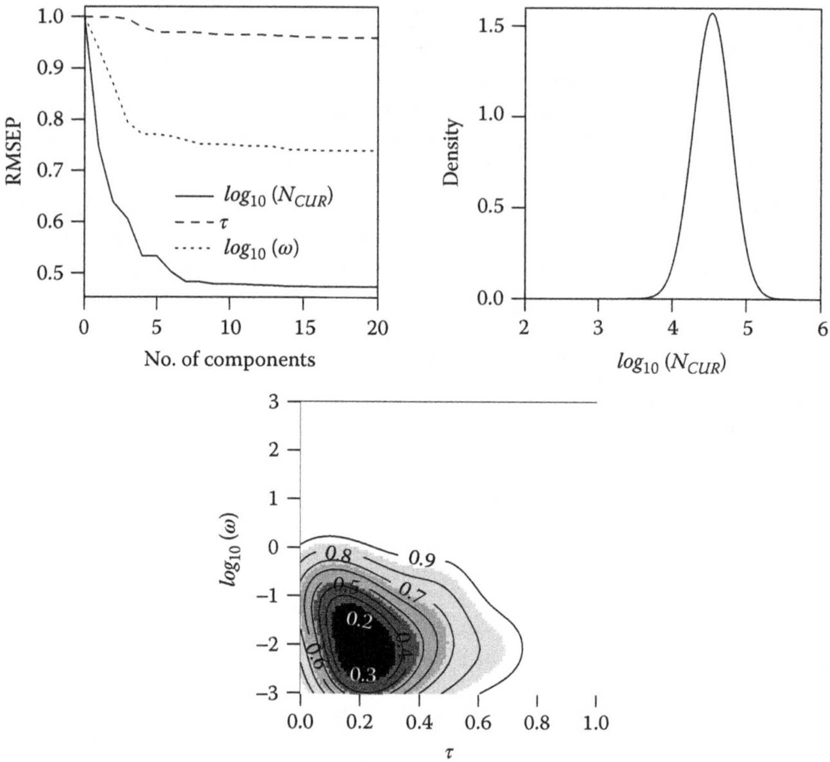

FIGURE 13.7
Root mean squared error of prediction (RMSEP) as a function of the number of components used for each parameter (top left), marginal posterior estimate for parameter $log_{10}(N_{CUR})$ (top right) and joint posterior for τ and $log_{10}(\omega)$ (bottom). Solid contour lines specify highest posterior density intervals.

at hand, we can then set up *ABCtoolbox* to run an ABC-MCMC chain using the following input file:

ABCtoolbox input file to perform ABC-MCMC with fastsimcoal2 and PLS-transformed statistics.

```
task simulate
samplerType MCMC
obsName popgen.obs
estName popgen.est
numSims 10000
outName popgen_MCMC
simInputName popgen.par
simProgram ./fsc25221
simArgs -i popgen-temp.par -s 0 -d -n 1 -q -x
sumStatProgram calcPopstats.pl
```

```
sumStatArgs popgen-temp/popgen-temp_DAFpop0.obs
doBoosting
numCaliSims 1000
thresholdProp 0.1
rangeProp 1
linearCombName PLSdef_popgen.txt
doBoxCox
```

This input file differs from the previous one in that the argument **samplerType** was added to instruct *ABCtoolbox* to run an ABC-MCMC chain, and in that the arguments required for the calibration step (**numCaliSims**, **thresholdProp** and **rangeProp**), and those to use linear combinations of summary statistics (**linearCombName** and **doBoxCox**) were added. Note that the R-script *find-PLS.r* to find linear combinations provided by *ABCtoolbox* performs a Box-Cox transformation on the summary statistics, and, hence, in order to use the generated PLS definition file, *ABCtoolbox* needs to perform a similar transformation first, which is requested with the argument **doBoxCox**. If the user does not wish to perform a PLS transformation and simply use the raw statistics for inference then they can omit arguments **linearCombName** and **doBoxCox** from the input file.

While the output of the ABC-MCMC run (**popgen_MCMC_sampling1.txt**) already corresponds to samples taken from the posterior distribution $\mathbb{P}(\|S - S_{obs}\| < \epsilon | N_{CUR}, \omega, \tau)$, an additional improvement may be achieved by conducting an ABC-GLM estimation with an additional rejection step that will further reduce the threshold ϵ and, hence, the accuracy of the posterior. Such an analysis can be conducted as described in Section 13.3.2, and the resulting posteriors may then be plotted in R.

The posterior estimates we obtained for parameters N_{CUR}, ω, and τ (Figure 13.7b and c) indicated a large population expansion that happened ~13,000 generations ago to a present effective population size of ~32,000. The credible intervals of the posterior estimates for ω and especially τ are large (Figure 13.7c) due to the small size of the downsampled dataset. Additionally, the statistics used here seem not to be informative for parameter τ as indicated by the PLS analysis in Figure 13.7a. However, the more precise estimates for N_{CUR} and ω are in good agreement with the findings of Boyko et al. (2008), who used a maximum likelihood method on the full data.

Appendix

Here, we provide additional files required to replicate our population genetics example. First, we provide the input file **popgen.par** for *fastsimcoal2* specifying the population genetics model used. Note that we decided to simulate the SFS using ten independent loci with 1000 sites each.

fastsimcoal input file.

```
//Number of population samples (demes)
1
//Population effective sizes (number of genes)
N_CUR
//Sample sizes
24
//Growth rates negative growth implies population expansion
0
//Number of migration matrices : 0 implies no migration between demes
0
//historical event: time, source, sink, migrants, new size, new growth rate
    ,migr.matrix
1 historical events
T1 0 0 1 OMEGA 0 0
//Number of independent loci [chromosome]
10 0
//Per chromosome: Number of linkage blocks
1
//per Block: data type, num loci, rec. rate and mut rate
DNA 1000 0.00000 MUTRATE 0.33
```

Further, we provide the custom perl script `calcPopstats.pl` used to calculate summary statistics from site frequency spectra simulated with the program *fastsimcoal2*.

Perl script to calculate statistics from SFS.

```
#!/usr/bin/perl -w
use strict;
#read fastsimcoal output SFS file
my $sfsfile=$ARGV[0];
open(FILE,"<",$sfsfile) or die "can't open SFS file";
open (OUT, ">","summary_stats-temp.txt") or die "can't open sum-stats file"
    ;
my ($firstline,$header,$sfsline)=(<FILE>,<FILE>,<FILE>);
#split sfsline into sfs
my @SFS=split /\t/,$sfsline;
my @stats;
#calculate stats
my ($sum,$S,$a1,$a2,$taj_D)=(0,0,0,0,0);
my $n=@SFS-2;
my ($b1,$b2)=(($n+1)/(3*($n-1)),2*($n**2+$n+3)/(9*$n*($n-1)));
#No. Segregating. sites S
for (my $i=1;$i<$n;$i++){
$sum=$sum+$i*($n-$i)*$SFS[$i];
$S=$S+$SFS[$i];
($a1,$a2)=($a1+1/$i,$a2+1/$i**2);
}
#Thita and pi
my ($thita,$pi)=($S/$a1,2*$sum/($n*($n-1)));
#Tajima's D
my ($c1,$c2)=($b1-1/$a1,$b2-($n+2)/($a1*$n)+$a2/($a1**2));
my ($e1,$e2)=($c1/$a1,$c2/($a1**2+$a2));
```

```
if($S>0) {$taj_D=($pi-$S/$a1)/sqrt($e1*$S+$e2*$S*($S-1));}
#print out stats
@stats=($SFS[1],$S,$pi,$thita,$taj_D);
print OUT join("\t","sfs1","S","pi","thita","taj_D"),"\n",join("\t",@stats,
     "\n");
close(FILE);close(OUT);system("rm $sfsfile");
```

Finally, we provide the file **PLSdef_popgen.txt** specifying the PLS transformation of the 20 statistics (7 polymorphism statistics + their products) to four components. The first six columns in the file specify the boxcox transformation of the statistic, and the remaining 4 columns specify the PLS components.

The file PLSdef_popgen.txt containing the definitions of the PLS transformations for ABCtoolbox.

```
sfs1 1140 0 -17.58 1.06 0.06 0.07 0.22 0.27 -0.16 0.23
S 4478 0 -10.3 1.12 0.12 0.13 0.25 0.1 -0.3 0.11
pi 1730.18 0 -11.52 1.1 0.09 0.11 0.26 0.05 -0.27 0.07
thita 1199.16 0 -10.3 1.12 0.12 0.13 0.25 0.1 -0.3 0.11
taj_D 2.87 -2.19 0.61 1.48 0.51 0.19 0.11 -0.38 -0.53 -0.5
sfs1_X_sfs1 1299600 0 -20 1.02 0.01 0.02 0.21 0.29 0.13 -0.14
sfs1_X_S 5104920 0 -20 1.03 0.01 0.03 0.25 0.16 0.18 -0.12
sfs1_X_pi 1426520 0 -20 1.03 0.02 0.03 0.25 0.12 0.19 -0.11
sfs1_X_thita 1367040 0 -20 1.03 0.01 0.03 0.25 0.16 0.18 -0.12
sfs1_X_taj_D 634.38 -559.99 -0.61 1.48 0.66 0.09 0.15 -0.36 0.45 -0.01
S_X_S 20052500 0 -18.79 1.06 0.04 0.06 0.26 0.03 0.06 -0.08
S_X_pi 7455340 0 -20 1.06 0.03 0.05 0.26 -0.02 0.08 -0.07
S_X_thita 5369820 0 -18.79 1.06 0.04 0.06 0.26 0.03 0.06 -0.08
S_X_taj_D 8737.33 -1796.63 -9.09 1.21 0.61 0.06 0.18 -0.4 0.16 0.14
pi_X_pi 2993510 0 -20 1.05 0.02 0.04 0.26 -0.07 0.12 -0.05
pi_X_thita 1996450 0 -20 1.06 0.03 0.05 0.26 -0.02 0.08 -0.07
pi_X_taj_D 3508.27 -349.33 -13.94 1.14 0.35 0.05 0.18 -0.39 0.07 0.16
thita_X_thita 1437980 0 -18.79 1.06 0.04 0.06 0.26 0.03 0.06 -0.08
thita_X_taj_D 2339.76 -481.12 -9.09 1.21 0.61 0.06 0.18 -0.4 0.16 0.14
taj_D_X_taj_D 8.21 0 -6.67 1.12 0.13 0.12 0.06 -0.25 -0.3 0.78
```

References

Adrion, J. R., A. Kousathanas, M. Pascual, H. J. Burrack, N. M. Haddad, A. O. Bergland, H. Machado et al. Drosophila suzukii: The genetic footprint of a recent, worldwide invasion. *Molecular Biology and Evolution*, 31(12):3148–3163, 2014. doi:10.1093/molbev/msu246.

Aeschbacher, S., M. A. Beaumont, and A. Futschik. A novel approach for choosing summary statistics in approximate Bayesian computation. *Genetics*, 192:1027–1047, 2012. doi:10.1534/genetics.112.143164.

Barthelmé, S. and N. Chopin. ABC-EP: Expectation propagation for likelihood-free Bayesian computation. In *Proceedings of the 28th International Conference on Machine Learning*, Bellevue, WA, 2011.

Beaumont, M. A. Joint determination of topology, divergence time, and immigration in population trees. In S. Matsumura, P. Forster, and C. Renfrew (Eds.) *Simulation, Genetics and Human Prehistory*. McDonald Institute for Archaeological Research, Cambridge, UK, 2008.

Beaumont, M. A., J.-M. Cornuet, J.-M. Marin, and C. P. Robert. Adaptive approximate Bayesian computation. *Biometrika*, 96(4):983–990, 2009. doi:10.1093/biomet/asp052.

Beaumont, M. A., W. Zhang, and D. J. Balding. Approximate Bayesian computation in population genetics. *Genetics*, 162:2025–2035, 2002.

Blum, M. G. B. and O. François. Non-linear regression models for approximate Bayesian computation. *Statistics and Computing*, 20(1):63–73, 2010. doi:10.1007/s11222-009-9116-0.

Blum, M. G. B., M. A. Nunes, D. Prangle, and S. A. Sisson. A comparative review of dimension reduction methods in approximate Bayesian computation. *Statistical Science*, 28(2):189–208, 2013. doi:10.1214/12-STS406.

Boyko, A. R., S. H. Williamson, A. R. Indap, J. D. Degenhardt, R. D. Hernandez, K. E. Lohmueller, M. D. Adams et al. Assessing the evolutionary impact of amino acid mutations in the human genome. *PLoS Genetics*, 4(5):e1000083, 2008. doi:10.1371/journal.pgen.1000083.

Bray, T. C., V. C. Sousa, B. Parreira, M. W. Bruford, and L. Chikhi. 2BAD: An application to estimate the parental contributions during two independent admixture events. *Molecular Ecology Resources*, 10(3):538–541, 2010.

Chu, J. H., D. Wegmann, C. F. Yeh, R. C. Lin, X. J. Yang, F. M. Lei, C. T. Yao, F. S. Zou, and S. H. Li. Inferring the geographic mode of speciation by contrasting autosomal and sex-linked genetic diversity. *Molecular biology and evolution*, 30(11):2519–2530, 2013. doi:10.1093/molbev/mst140.

Cornuet, J. M., F. Santos, M. A. Beaumont, C. P. Robert, J. M. Marin, D. J. Balding, T. Guillemaud, and A. Estoup. Inferring population history with *DIY ABC*: A user-friendly approach to approximate Bayesian computation. *Bioinformatics*, 24(23):2713–2719, 2008.

Csilléry, K., O. François, and M. G. B. Blum. abc: An R package for approximate Bayesian computation (ABC). *Methods in Ecology and Evolution*, 3:475–479, 2012.

Cuesta-Albertos, J. A. and A. Nieto-Reyes. The random tukey depth. *Computational Statistics & Data Analysis*, 52(11):4979–4988, 2008. doi:10.1016/j.csda.2008.04.021.

Didelot, X., R. G. Everitt, A. M. Johansen, and D. J. Lawson. Likelihood-free estimation of model evidence. *Bayesian Analysis*, 6(1):49–76, 2011. doi:10.1214/11-BA602.

Estoup, A., E. Lombaert, J. M. Marin, T. Guillemaud, P. Pudlo, C. P. Robert, and J. M. Cornuet. Estimation of demo-genetic model probabilities with Approximate Bayesian Computation using linear discriminant analysis on summary statistics. *Molecular Ecology Resources*, 12(5):846–55, 2012. doi:10.1111/j.1755-0998.2012.03153.x.

Excoffier, L., I. Dupanloup, E. Huerta-Sánchez, V. C. Sousa, and M. Foll. Robust demographic inference from genomic and SNP data. *PLoS Genetics*, 9(10):e1003905, 2013. doi:10.1371/journal.pgen.1003905.

Fagundes, N. J. R., N. Ray, M. A. Beaumont, S. Neuenschwander, F. M. Salzano, S. L. Bonatto, and L. Excoffier. Statistical evaluation of alternative models of human evolution. *Proceedings of the National Academy of Sciences of the United States of America*, 104(45):17614–17619, 2007.

Fearnhead, P. and D. Prangle. Constructing summary statistics for approximate Bayesian computation: Semi-automatic approximate Bayesian computation. *Journal of the Royal Statistical Society: Series B (Statistical Methodology)*, 74(3):419–474, 2012. doi:10.1111/j.1467-9868.2011.01010.x.

Foll, M., M. A. Beaumont, and O. Gaggiotti. An approximate Bayesian computation approach to overcome biases that arise when using amplified fragment length polymorphism markers to study population structure. *Genetics*, 179:927–939, 2008.

François, O. and G. Laval. Deviance information criteria for model selection in approximate Bayesian computation. *Statistical Applications in Genetics and Molecular Biology*, 10(1):1–25, 2011.

Gutenkunst, R. N., R. D. Hernandez, S. H. Williamson, and C. D. Bustamante. Inferring the joint demographic history of multiple populations from multidimensional SNP frequency data. *PLoS Genetics*, 5(10):e1000695, 2009. doi:10.1371/journal.pgen.1000695.

Hickerson, M. J., E. Stahl, and N. Takebayashi. msBayes: Pipeline for testing comparative phylogeographic histories using hierarchical approximate Bayesian computation. *BMC Bioinformatics*, 8:268, 2007.

Jabot, F., T. Faure, and N. Dumoulin. EasyABC: Performing efficient approximate Bayesian computation sampling schemes using r. *Methods in Ecology and Evolution*, 4(7):684–687, 2013. doi:10.1111/2041-210X.12050.

Kousathanas, A., C. Leuenberger, J. Helfer, M. Quinodoz, M. Foll, and D. Wegmann. Likelihood-free inference in high-dimensional models. *Genetics*, 203(2):893–904, 2016. doi:10.1534/genetics.116.187567.

Leuenberger, C. and D. Wegmann. Bayesian computation and model selection without likelihoods. *Genetics*, 184(1):243–252, 2010. doi:10.1534/genetics.109.109058.

Liepe, J., C. Barnes, E. Cule, K. Erguler, P. Kirk, T. Toni, and M. P. H. Stumpf. ABC-SysBio – approximate Bayesian computation in Python with GPU support. *Bioinformatics*, 26(14):1797–1799, 2010.

Lopes, J. S., D. Balding, and M. A. Beaumont. PopABC: A program to infer historical demographic parameters. *Bioinformatics*, 25(20):2747–2749, 2009.

Marjoram, P., J. Molitor, V. Plagnol, and S. Tavaré. Markov chain Monte Carlo without likelihoods. *Proceedings of the National Academy of Sciences*, 100(26):15324–15328, 2003. doi:10.1073/pnas.0306899100.

Pavlidis, P., S. Laurent, and W. Stephan. msABC: A modification of Hudson's ms to facilitate multi-locus ABC analysis. *Molecular Ecology Resources*, 10(4):723–727, 2010.

Peter, B. M., D. Wegmann, and L. Excoffier. Distinguishing between population bottleneck and population subdivision by a Bayesian model choice procedure. *Molecular Ecology*, 19(21):4648–60, 2010. doi:10.1111/j.1365-294X.2010.04783.x.

Picchini, U. Inference for SDE models via approximate Bayesian computation. *Journal of Computational and Graphical Statistics*, 23(4):1080–1100, 2014.

Prangle, D., M. G. B. Blum, G. Popovic, and S. A. Sisson. Diagnostic tools for approximate Bayesian computation using the coverage property. *Australian and New Zealand Journal of Statistics*, 56(4):309–329, 2014. doi:10.1111/anzs.12087.

Pritchard, J. K., M. T. Seielstad, A. Perez-Lezaun, and M. W. Feldman. Population growth of human Y chromosome: A study of Y chromosome microsatellites. *Molecular Biology and Evolution*, 16(12):1791–1798, 1999.

Robert, C. P., J. M. Cornuet, J. M. Marin, and N. S. Pillai. Lack of confidence in approximate Bayesian computation model choice. *Proceedings of the National Academy of Sciences*, 108(37):15112–15117, 2011. doi:10.1073/pnas.1102900108.

Sisson, S. A., Y. Fan, and M. M. Tanaka. Sequential Monte Carlo without likelihoods. *Proceedings of the National Academy of Sciences of the United States of America*, 104(6):1760–1765, 2007. doi:10.1073/pnas.0607208104.

Tallmon, D. A., A. Koyuk, G. Luikart, and M. A. Beaumont. ONeSAMP: A program to estimate effective population size using approximate Bayesian computation. *Molecular Ecology Resources*, 8:299–301, 2008.

Tavaré, S., D. J. Balding, R. C. Griffiths, and P. Donnelly. Inferring coalescence times from DNA sequence data. *Genetics*, 145:505–518, 1997.

Tenenhaus, M., J. P. Gauchi, and C. Ménardo. Régression PLS et applications. *Revue de Statistique Appliquée*, 43(1):7–64, 1995.

Thornton, K. R. Automating approximate Bayesian computation by local linear regression. *BMC Genetics*, 10:35, 2009.

Wegmann, D., C. Leuenberger, and L. Excoffier. Efficient approximate Bayesian computation coupled with Markov chain Monte Carlo without likelihood. *Genetics*, 182(4):1207–1218, 2009. doi:10.1534/genetics.109.102509.

Wegmann, D., C. Leuenberger, S. Neuenschwander, and L. Excoffier. ABC-toolbox: A versatile toolkit for approximate Bayesian computations. *BMC Bioinformatics*, 11(1):116, 2010. doi:10.1186/1471-2105-11-116.

14

Divide and Conquer in ABC: Expectation-Propagation Algorithms for Likelihood-Free Inference

Simon Barthelmé, Nicolas Chopin, and Vincent Cottet

CONTENTS

14.1 Introduction 416
14.2 Expectation-Propagation Algorithms 417
  14.2.1 General presentation 417
  14.2.2 Properties of exponential families 418
  14.2.3 Site update 419
  14.2.4 Gaussian sites 420
  14.2.5 Order of site updates: Sequential EP, parallel EP, and block-parallel EP 420
  14.2.6 Other practical considerations 423
  14.2.7 Theoretical properties of expectation propagation 423
14.3 Applying Expectation Propagation in ABC 424
  14.3.1 Principle 424
  14.3.2 Practical considerations 426
  14.3.3 Speeding up parallel expectation propagation-ABC in the iid case 426
14.4 Application to Spatial Extremes 428
  14.4.1 Background 428
  14.4.2 Summary statistics 429
  14.4.3 Numerical results on real data 429
  14.4.4 Expectation-propagation convergence 431
14.5 Conclusion 431
Acknowledgements 432
References 432

14.1 Introduction

A standard ABC algorithm samples in some way from the pseudo-posterior:

$$p_\epsilon^{\text{std}}(\boldsymbol{\theta}|\mathbf{y}^*) \propto p(\boldsymbol{\theta}) \int p(\mathbf{y}|\boldsymbol{\theta})\mathbb{I}_{\{\|s(\mathbf{y})-s(\mathbf{y}^\star)\|\le\epsilon\}}\, d\mathbf{y}, \tag{14.1}$$

where $p(\mathbf{y}|\boldsymbol{\theta})$ denotes the likelihood of data $\boldsymbol{y} \in \mathcal{Y}$ given parameter $\boldsymbol{\theta} \in \Theta$, $\mathbf{y}^*$ is the actual data, s is some function of the data called a 'summary statistic', and $\epsilon > 0$. As discussed elsewhere in this book, there are various ways to sample from (14.1), for example, rejection, Markov Chain Monte Carlo (MCMC) (Marjoram et al., 2003), Sequential Monte Carlo (SMC) (Sisson et al., 2007; Beaumont et al., 2009; Del Moral et al., 2012), and so on, but they all require simulating a large number of complete datasets $\boldsymbol{y}^j$ from the likelihood $p(\mathbf{y}|\boldsymbol{\theta})$, for different values of $\boldsymbol{\theta}$. This is typically the bottleneck of the computation. Another drawback of standard ABC is the dependence on s: as $\epsilon \to 0$, $p_\epsilon^{\text{std}}(\boldsymbol{\theta}|\mathbf{y}^*) \to p(\boldsymbol{\theta}|s(\boldsymbol{y}^\star)) \neq p(\boldsymbol{\theta}|\boldsymbol{y}^\star)$, the true posterior distribution, and there is no easy way to choose s such that $p(\boldsymbol{\theta}|s(\boldsymbol{y}^\star)) \approx p(\boldsymbol{\theta}|\boldsymbol{y}^\star)$.

In this paper, we assume that the data may be decomposed into n 'chunks', $\boldsymbol{y} = (y_1, \ldots, y_n)$, and that the likelihood may be factorised accordingly:

$$p(\mathbf{y}|\boldsymbol{\theta}) = \prod_{i=1}^{n} f_i(y_i|\boldsymbol{\theta}),$$

in such a way that it is possible to sample pseudo-data y_i from each factor $f_i(y_i|\boldsymbol{\theta})$. The objective is to approximate the pseudo-posterior:

$$p_\epsilon(\boldsymbol{\theta}|\boldsymbol{y}^\star) \propto p(\boldsymbol{\theta}) \prod_{i=1}^{n} \left\{ \int f_i(y_i|\boldsymbol{\theta})\mathbb{I}_{\{\|s_i(y_i)-s_i(y_i^\star)\|\le\epsilon\}}\, dy_i \right\},$$

where s_i is a 'local' summary statistic, which depends only on y_i. We expect the bias introduced by the n local summary statistics s_i to be much smaller than the bias introduced by the global summary statistic s. In fact, there are practical cases where we may take $s_i(y_i) = y_i$, removing this bias entirely.

Note that we do not restrict to models such that the chunks y_i are independent. In other words, we allow each factor f_i to implicitly depend on other data-points. For instance, we could have a Markov model, with $f_i(y_i|\boldsymbol{\theta}) = p(y_i|y_{i-1}, \boldsymbol{\theta})$, or even a model with a more complicated dependence structure, say $f_i(y_i|\boldsymbol{\theta}) = p(y_i|y_{1:i-1}, \boldsymbol{\theta})$. The main requirement, however, is that we are able to sample from each factor $f_i(y_i|\boldsymbol{\theta})$. For instance, in the Markov case, this means we are able to sample from the model realisations of variable y_i, conditional on $y_{i-1} = y_{i-1}^\star$ and $\boldsymbol{\theta}$.

Alternatively, in cases where the likelihood does not admit a simple factorisation, one may replace it by some factorisable pseudo-likelihood; for example, a marginal composite likelihood:

$$p^{\text{MCL}}(\boldsymbol{y}|\boldsymbol{\theta}) = \prod_{i=1}^{n} p(y_i|\boldsymbol{\theta}),$$

where $p(y_i|\boldsymbol{\theta})$ is the marginal density of variable y_i. Then one would take $f_i(y_i|\boldsymbol{\theta}) = p(y_i|\boldsymbol{\theta})$ (assuming we are able to simulate from the marginal distribution of y_i). Conditional distributions may be used as well; see Varin et al. (2011) for a review of composite likelihoods. Of course, replacing the likelihood by some factorisable pseudo-likelihood adds an extra level of approximation, and one must determine in practice whether the computational benefits are worth the extra cost. Estimation based on composite likelihoods is generally consistent, but their use in a Bayesian setting results in posterior distributions that are overconfident (the variance is too small, as dependent data are effectively treated as independent observations).

Many authors have taken advantage of factorisations to speed up ABC. ABC strategies for hidden Markov models are discussed in Dean et al. (2014) and Yıldırım et al. (2014); see the review of Jasra (2015). White et al. (2015) describe a method based on averages of pseudo-posteriors, which in the Gaussian case reduces to just doing one pass of parallel EP. Ruli et al. (2016) use composite likelihoods to define low-dimensional summary statistics.

We focus on expectation-propagation (EP, Minka, 2001), a widely successful algorithm for variational inference. In Barthelmé and Chopin (2014), we showed how to adapt EP to a likelihood-free setting. Here, we extend this work with a focus on a parallel variant of EP (Cseke and Heskes, 2011) that enables massive parallelisation of ABC inference. For textbook descriptions of EP, see for example, Section 10.7 of Bishop (2006) or Section 13.8 of Gelman et al. (2014).

The chapter is organised as follows. Section 14.2 gives a general presentation of both sequential and parallel EP algorithms. Section 14.3 explains how to adapt these EP algorithms to ABC contexts. It discusses in particular some ways to speed up EP-ABC. Section 14.4 discusses how to apply EP-ABC to spatial extreme models. Section 14.5 concludes.

We use the following notations throughout: bold symbols refer to vectors or matrices, for example, $\boldsymbol{\theta}$, $\boldsymbol{\lambda}$, $\boldsymbol{\Sigma}$. For data-points, we use (bold) $\boldsymbol{y}$ to denote complete datasets and y_i to denote data 'chunks', although we do not necessarily assume the y_i's to be scalars. The letter p typically refers to probability densities relative to the model: $p(\boldsymbol{\theta})$ is the prior, $p(y_1|\boldsymbol{\theta})$ is the likelihood of the first data chunk, and so on. The transpose of matrix $\mathbf{A}$ is denoted $\mathbf{A}^t$.

14.2 Expectation-Propagation Algorithms

14.2.1 General presentation

Consider a posterior distribution $\pi(\boldsymbol{\theta})$ that may be decomposed into $(n+1)$ factors:

$$\pi(\boldsymbol{\theta}) \propto \prod_{i=0}^{n} l_i(\boldsymbol{\theta}),$$

where, say, $l_0(\boldsymbol{\theta})$ is the prior, and $l_1, \ldots, l_n$ are n contributions to the likelihood. EP (Minka, 2001) approximates π by a similar decomposition:

$$q(\boldsymbol{\theta}) \propto \prod_{i=0}^{n} q_i(\boldsymbol{\theta}),$$

where each 'site' q_i is updated in turn, conditional on the other factors, in a spirit close to a coordinate-descent algorithm.

To simplify this rather general framework, one often assumes that the q_i belong to some exponential family of distributions $\mathcal{Q}$ (Seeger, 2005):

$$q_i(\boldsymbol{\theta}) = \exp\left\{\boldsymbol{\lambda}_i^t \boldsymbol{t}(\boldsymbol{\theta}) - \phi(\boldsymbol{\lambda}_i)\right\},$$

where $\boldsymbol{\lambda}_i \in \mathbb{R}^d$ is the natural parameter, $\boldsymbol{t}(\boldsymbol{\theta})$ is some function $\Theta \to \mathbb{R}^d$, and ϕ is known variously as the *log-partition function* or the *cumulant function*: $\phi(\boldsymbol{\lambda}) = \log\left[\int \exp\left\{\boldsymbol{\lambda}^t \boldsymbol{t}(\boldsymbol{\theta})\right\} d\boldsymbol{\theta}\right]$. Working with exponential families is convenient for a number of reasons. In particular, the global approximation q is automatically in the same family, and with parameter $\boldsymbol{\lambda} = \sum_{i=0}^{n} \boldsymbol{\lambda}_i$:

$$q(\boldsymbol{\theta}) \propto \exp\left\{\left(\sum_{i=0}^{n} \boldsymbol{\lambda}_i\right)^t \boldsymbol{t}(\boldsymbol{\theta})\right\}.$$

The next section gives additional properties of exponential families upon which EP relies. Then Section 14.2.3 explains how to perform a site update, that is, how to update $\boldsymbol{\lambda}_i$, conditional on the $\boldsymbol{\lambda}_j$, $j \neq i$, so as, informally, to make q progressively closer and closer to π.

14.2.2 Properties of exponential families

Let $\mathrm{KL}(\pi||q)$ be the Kullback–Leibler divergence of q from π:

$$\mathrm{KL}(\pi||q) = \int \pi(\boldsymbol{\theta}) \log \frac{\pi(\boldsymbol{\theta})}{q(\boldsymbol{\theta})}\, d\boldsymbol{\theta}.$$

For a generic member $q_{\boldsymbol{\lambda}}(\boldsymbol{\theta}) = \exp\left\{\boldsymbol{\lambda}^t \boldsymbol{t}(\boldsymbol{\theta}) - \phi(\boldsymbol{\lambda})\right\}$ of our exponential family $\mathcal{Q}$, we have:

$$\frac{d}{d\boldsymbol{\lambda}} \mathrm{KL}(\pi||q_\lambda) = \frac{d}{d\boldsymbol{\lambda}} \phi(\boldsymbol{\lambda}) - \int \pi(\boldsymbol{\theta}) \boldsymbol{t}(\boldsymbol{\theta})\, d\boldsymbol{\theta}, \tag{14.2}$$

where the derivative of the partition function may be obtained as:

$$\frac{d}{d\boldsymbol{\lambda}} \phi(\boldsymbol{\lambda}) = \int \boldsymbol{t}(\boldsymbol{\theta}) \exp\left\{\boldsymbol{\lambda}^t \boldsymbol{t}(\boldsymbol{\theta}) - \phi(\boldsymbol{\lambda})\right\} d\boldsymbol{\theta} = \mathbb{E}_{\boldsymbol{\lambda}}\left\{\boldsymbol{t}(\boldsymbol{\theta})\right\}. \tag{14.3}$$

Let $\boldsymbol{\eta} = \boldsymbol{\eta}(\boldsymbol{\lambda}) = \mathbb{E}_{\boldsymbol{\lambda}}\left\{\boldsymbol{t}(\boldsymbol{\theta})\right\}$; $\boldsymbol{\eta}$ is called the moment parameter, and there is a one-to-one correspondence between $\boldsymbol{\lambda}$ and $\boldsymbol{\eta}$; abusing notations, if $\boldsymbol{\eta} = \boldsymbol{\eta}(\boldsymbol{\lambda})$, then $\boldsymbol{\lambda} = \boldsymbol{\lambda}(\boldsymbol{\eta})$. One may interpret (14.2) as follows: finding the $q_{\boldsymbol{\lambda}}$ closest to π (in the Kullback–Leibler sense) amounts to perform *moment matching*, that is, to set $\boldsymbol{\lambda}$ such that the expectation of $\boldsymbol{t}(\boldsymbol{\theta})$ under π and under $q_{\boldsymbol{\lambda}}$ match.

To make this discussion more concrete, consider the Gaussian case:

$$q_{\boldsymbol{\lambda}}(\boldsymbol{\theta}) \propto \exp\left\{-\frac{1}{2}\boldsymbol{\theta}^t\boldsymbol{Q}\boldsymbol{\theta} + \boldsymbol{r}^t\boldsymbol{\theta}\right\}, \quad \boldsymbol{\lambda} = \left(\boldsymbol{r}, -\frac{1}{2}\boldsymbol{Q}\right), \quad t(\boldsymbol{\theta}) = \left(\boldsymbol{\theta}, \boldsymbol{\theta}\boldsymbol{\theta}^t\right),$$

and the moment parameter is $\boldsymbol{\eta} = (\boldsymbol{\mu}, \boldsymbol{\Sigma} + \boldsymbol{\mu}\boldsymbol{\mu}^t)$, with $\boldsymbol{\Sigma} = \boldsymbol{Q}^{-1}$, $\boldsymbol{\mu} = \boldsymbol{Q}^{-1}\boldsymbol{r}$. [More precisely, $\boldsymbol{\theta}^t\boldsymbol{Q}\boldsymbol{\theta} = \text{trace}(\boldsymbol{Q}\boldsymbol{\theta}\boldsymbol{\theta}^t) = \text{vect}(\boldsymbol{Q})^t\text{vect}(\boldsymbol{\theta}\boldsymbol{\theta}^t)$, so the second component of $\boldsymbol{\lambda}$ [respectively, $\boldsymbol{t}(\boldsymbol{\theta})$] should be $-(1/2)\text{vect}(\boldsymbol{Q})$ [respectively $\text{vect}(\boldsymbol{\theta}\boldsymbol{\theta}')$]. But, for notational convenience, our derivations will be in terms of matrices $\boldsymbol{Q}$ and $\boldsymbol{\theta}\boldsymbol{\theta}'$, rather than their vectorised versions.]

In the Gaussian case, minimising $\text{KL}(\pi||q_{\boldsymbol{\lambda}})$ amounts to taking $\boldsymbol{\lambda}$, such that the corresponding moment parameter $(\boldsymbol{\mu}, \boldsymbol{\Sigma} + \boldsymbol{\mu}\boldsymbol{\mu}^t)$ is such that $\boldsymbol{\mu} = \mathbb{E}_\pi[\boldsymbol{\theta}]$, $\boldsymbol{\Sigma} = \text{Var}_\pi[\boldsymbol{\theta}]$. We will focus on the Gaussian case in this paper (i.e. EP computes iteratively a Gaussian approximation of π), but we go on with the more general description of EP in terms of exponential families, as this allows for more compact notations, and also because we believe that other approximations could be useful in the ABC context.

14.2.3 Site update

We now explain how to perform a site update for site i, that is, how to update given $\boldsymbol{\lambda}_i$, assuming $(\boldsymbol{\lambda}_j)_{j\neq i}$ is fixed. Consider the 'hybrid' distribution:

$$\begin{aligned} h(\boldsymbol{\theta}) \propto q(\boldsymbol{\theta})\frac{l_i(\boldsymbol{\theta})}{q_i(\boldsymbol{\theta})} &= l_i(\boldsymbol{\theta})\prod_{j\neq i} q_j(\boldsymbol{\theta}) \\ &= l_i(\boldsymbol{\theta})\exp\left\{\left(\sum_{j\neq i}\boldsymbol{\lambda}_j\right)^t \boldsymbol{t}(\boldsymbol{\theta})\right\}; \end{aligned}$$

that is, h is obtained by replacing site q_i by the true factor l_i in the global approximation q. The hybrid can be viewed as a 'pseudo-posterior' distribution, formed of the product of a 'pseudo-prior' q_i and a single likelihood site l_i. The update of site i is performed by minimising $\text{KL}(h||q)$ with respect to $\boldsymbol{\lambda}_i$ (again, assuming the other $\boldsymbol{\lambda}_j$, $j \neq i$, are fixed). Informally, this may be interpreted as a local projection (in the Kullback–Leibler sense) of π to $\mathcal{Q}$.

Given the properties of exponential families laid out in the previous section, one sees that this site update amounts to setting $\boldsymbol{\lambda}_i$ so that $\boldsymbol{\lambda} = \sum_j \boldsymbol{\lambda}_j$ matches $\mathbb{E}_h[\boldsymbol{t}(\boldsymbol{\theta})]$, the expectation of $t(\boldsymbol{\theta})$ with respect to the hybrid distribution. In addition, one may express the update of $\boldsymbol{\lambda}_i$ as a function of the current values of $\boldsymbol{\lambda}_i$ and $\boldsymbol{\lambda}$, using the fact that $\sum_{j\neq i}\boldsymbol{\lambda}_j = \boldsymbol{\lambda} - \boldsymbol{\lambda}_i$, as done in Algorithm 14.1.

In practice, the feasibility of EP for a given posterior is essentially determined by the difficulty to evaluate, or approximate, the integral (14.4).

Algorithm 14.1: Generic Site Update in EP

Function SiteUpdate($i, l_i, \boldsymbol{\lambda}_i, \boldsymbol{\lambda}$):

1. Compute:

$$\boldsymbol{\lambda}^{\text{new}} := \boldsymbol{\lambda}\left(\mathbb{E}_h[\boldsymbol{t}(\boldsymbol{\theta})]\right), \quad \boldsymbol{\lambda}_i^{\text{new}} := \boldsymbol{\lambda}^{\text{new}} - \boldsymbol{\lambda} + \boldsymbol{\lambda}_i,$$

where $\boldsymbol{\eta} \to \boldsymbol{\lambda}(\boldsymbol{\eta})$ is the function that maps the moment parameters to the natural parameters (for the considered exponential family, see previous section) and:

$$\mathbb{E}_h\left[\boldsymbol{t}(\boldsymbol{\theta})\right] = \frac{\int \boldsymbol{t}(\boldsymbol{\theta}) l_i(\boldsymbol{\theta}) \exp\left\{(\boldsymbol{\lambda} - \boldsymbol{\lambda}_i)^t \boldsymbol{t}(\boldsymbol{\theta})\right\} d\boldsymbol{\theta}}{\int l_i(\boldsymbol{\theta}) \exp\left\{(\boldsymbol{\lambda} - \boldsymbol{\lambda}_i)^t \boldsymbol{t}(\boldsymbol{\theta})\right\} d\boldsymbol{\theta}}. \tag{14.4}$$

2. Return $\boldsymbol{\lambda}_i^{\text{new}}$, and optionally $\boldsymbol{\lambda}^{\text{new}}$ (as determined by syntax, i.e. either $\boldsymbol{\lambda}_i^{\text{new}} \leftarrow$ SiteUpdate($i, l_i, \boldsymbol{\lambda}_i, \boldsymbol{\lambda}$), or ($\boldsymbol{\lambda}_i^{\text{new}}, \boldsymbol{\lambda}^{\text{new}}$) $\leftarrow$ SiteUpdate ($i, l_i, \boldsymbol{\lambda}_i, \boldsymbol{\lambda}$)).

Note the simple interpretation of this quantity: this is the posterior expectation of $\boldsymbol{t}(\boldsymbol{\theta})$, for pseudo-prior q_{-i} and pseudo-likelihood the likelihood factor $l_i(\boldsymbol{\theta})$. (In the EP literature, the pseudo-prior q_{-i} is often called the 'cavity distribution', and the pseudo-posterior $\propto q_{-i}(\boldsymbol{\theta})l_i(\boldsymbol{\theta})$ the 'tilted or hybrid distribution'.)

14.2.4 Gaussian sites

In this paper, we will focus on Gaussian approximations; that is, $\mathcal{Q}$ is the set of Gaussian densities:

$$q_{\boldsymbol{\lambda}}(\boldsymbol{\theta}) \propto \exp\left\{-\frac{1}{2}\boldsymbol{\theta}^t \boldsymbol{Q}\boldsymbol{\theta} + \boldsymbol{r}^t\boldsymbol{\theta}\right\}, \quad \boldsymbol{\lambda} = \left(\boldsymbol{r}, -\frac{1}{2}\boldsymbol{Q}\right),$$

and EP computes iteratively a Gaussian approximation of π, obtained as a product of Gaussian factors. For this particular family, simple calculations show that the site updates take the form given by Algorithm 14.2.

In words, one must compute the expectation and variance of the pseudo-posterior obtained by multiplying the Gaussian pseudo-prior q_{-i} and likelihood l_i.

14.2.5 Order of site updates: Sequential EP, parallel EP, and block-parallel EP

We now discuss in which *order* the site updates may be performed; i.e. should site updates be performed sequentially, or in parallel, or something in between?

Algorithm 14.2: EP Site Update (Gaussian Case)

Function SiteUpdate$(i, l_i, (\boldsymbol{r}_i, \boldsymbol{Q}_i), (\boldsymbol{r}, \boldsymbol{Q}))$:

1. Compute:

$$\begin{aligned} Z_h &= \int q_{-i}(\boldsymbol{\theta}) l_i(\boldsymbol{\theta})\, d\boldsymbol{\theta} \\ \boldsymbol{\mu}_h &= \frac{1}{Z_h} \int \boldsymbol{\theta} q_{-i}(\boldsymbol{\theta}) l_i(\boldsymbol{\theta})\, d\boldsymbol{\theta} \\ \boldsymbol{\Sigma}_h &= \frac{1}{Z_h} \int \boldsymbol{\theta}\boldsymbol{\theta}^t q_{-i}(\boldsymbol{\theta}) l_i(\boldsymbol{\theta})\, d\boldsymbol{\theta} - \boldsymbol{\mu}_h \boldsymbol{\mu}_h^t, \end{aligned}$$

   where $q_{-i}(\boldsymbol{\theta})$ is the Gaussian density:

$$q_{-i}(\boldsymbol{\theta}) \propto \exp\left\{ -\frac{1}{2} \boldsymbol{\theta}^t (\boldsymbol{Q} - \boldsymbol{Q}_i) \boldsymbol{\theta} + (\boldsymbol{r} - \boldsymbol{r}_i)^t \boldsymbol{\theta} \right\}.$$

2. Return $(\boldsymbol{r}_i^{\text{new}}, \boldsymbol{Q}_i^{\text{new}})$, and optionally $(\boldsymbol{r}^{\text{new}}, \boldsymbol{Q}^{\text{new}})$ (according to syntax as in Algorithm 14.1), where:

$$\begin{aligned} (\boldsymbol{Q}^{\text{new}}, \boldsymbol{r}^{\text{new}}) &= \left(\boldsymbol{\Sigma}_h^{-1}, \boldsymbol{\Sigma}_h^{-1} \boldsymbol{\mu}_h \right), \\ (\boldsymbol{Q}_i^{\text{new}}, \boldsymbol{r}_i^{\text{new}}) &= (\boldsymbol{Q}_i + \boldsymbol{Q}^{\text{new}} - \boldsymbol{Q}, \boldsymbol{r}_i + \boldsymbol{r}^{\text{new}} - \boldsymbol{r}). \end{aligned}$$

The initial version of EP, as described in Minka (2001), was purely sequential (and will therefore be referred to as 'sequential EP' from now on): one updates $\boldsymbol{\lambda}_0$ given the current values of $\boldsymbol{\lambda}_1, \ldots, \boldsymbol{\lambda}_n$, then one updates $\boldsymbol{\lambda}_1$ given $\boldsymbol{\lambda}_0$ (as modified in the previous update) and $\boldsymbol{\lambda}_2, \ldots, \boldsymbol{\lambda}_n$, and so on; see Algorithm 14.3. Since the function SiteUpdate $(i, l_i, \boldsymbol{\lambda}_i, \boldsymbol{\lambda})$ computes the updated version of both $\boldsymbol{\lambda}_i$ and $\boldsymbol{\lambda} = \sum_{j=0}^n \boldsymbol{\lambda}_j$, $\boldsymbol{\lambda}$ changes at each call of SiteUpdate.

Algorithm 14.3 is typically run until $\boldsymbol{\lambda} = \sum_{i=0}^n \boldsymbol{\lambda}_i$ stabilises in some sense.

The main drawback of sequential EP is that, given its sequential nature, it is not easily amenable to parallel computation. Cseke and Heskes (2011)

Algorithm 14.3: Sequential EP

Require: initial values for $\boldsymbol{\lambda}_0, \ldots, \boldsymbol{\lambda}_n$
  $\boldsymbol{\lambda} \leftarrow \sum_{i=0}^n \boldsymbol{\lambda}_i$
  repeat
    for $i = 0$ to n **do**
      $(\boldsymbol{\lambda}_i, \boldsymbol{\lambda}) \leftarrow$ SiteUpdate $(i, l_i, \boldsymbol{\lambda}_i, \boldsymbol{\lambda})$
    end for
  until convergence
  return $\boldsymbol{\lambda}$.

proposed a parallel EP algorithm, where all sites are updated in parallel, independently of each other. This is equivalent to update the sum $\boldsymbol{\lambda} = \sum_{i=0}^{n} \boldsymbol{\lambda}_i$ only after all the sites have been updated; see Algorithm 14.4.

Algorithm 14.4: Parallel EP

Require: initial values for $\boldsymbol{\lambda}_0, \ldots, \boldsymbol{\lambda}_n$
  $\boldsymbol{\lambda} \leftarrow \sum_{i=0}^{n} \boldsymbol{\lambda}_i$
  repeat
    for $i = 0$ to n **do** (parallel)
      $\boldsymbol{\lambda}_i \leftarrow \mathrm{SiteUpdate}\,(i, l_i, \boldsymbol{\lambda}_i, \boldsymbol{\lambda})$
    end for
    $\boldsymbol{\lambda} \leftarrow \sum_{i=0}^{n} \boldsymbol{\lambda}_i$
  until convergence
  return $\boldsymbol{\lambda}$.

Parallel EP is 'embarrassingly parallel', since its inner loop performs $(n+1)$ independent operations. A drawback of parallel EP is that its convergence is typically slower (i.e. requires more complete passes over all the sites) than sequential EP. Indeed, during the first pass, all the sites are provided with the same initial global approximation $\boldsymbol{\lambda}$, whereas in sequential EP, the first site updates allow to refine progressively $\boldsymbol{\lambda}$, which makes the following updates easier.

We now propose a simple hybrid of these two EP algorithms, which we call 'block-parallel EP'. We assume we have n_{core} cores (single processing units) at our disposal. For each block of n_{core} successive sites, we update these n_{core} sites in parallel, and then update the global approximation $\boldsymbol{\lambda}$ after these n_{core} updates; see Algorithm 14.5.

Algorithm 14.5: Block-Parallel EP

Require: initial values for $\boldsymbol{\lambda}_0, \ldots, \boldsymbol{\lambda}_n$
  $\boldsymbol{\lambda} \leftarrow \sum_{i=0}^{n} \boldsymbol{\lambda}_i$
  repeat
    for $k = 1$ to $\lceil (n+1)/n_{\text{core}} \rceil$ **do**
      for $i = (k-1)n_{\text{core}}$ to $(kn_{\text{core}} - 1) \wedge n$ **do** (parallel)
        $\boldsymbol{\lambda}_i \leftarrow \mathrm{SiteUpdate}\,(i, l_i, \boldsymbol{\lambda}_i, \boldsymbol{\lambda})$
      end for
      $\boldsymbol{\lambda} \leftarrow \sum_{i=0}^{n} \boldsymbol{\lambda}_i$
    end for
  until convergence
  return $\boldsymbol{\lambda}$.

Quite clearly, block-parallel EP generalises both sequential EP (take $n_{\text{core}} = 1$) and parallel EP (take $n_{\text{core}} = n+1$). This generalisation is useful in any situation where the actual number of cores n_{core} available in a given architecture is such that $n_{\text{core}} \ll (n+1)$. In this way, we achieve essentially the same speed-up as parallel EP in terms of parallelisation (since only n_{core} cores are available anyway), but we also progress faster thanks to the sequential nature of the successive block updates. We shall discuss more specifically in the next section the advantage of block-parallel EP over standard parallel EP in an ABC context.

14.2.6 Other practical considerations

Often, the prior, which was identified with l_0 in our factorisation, already belongs to the approximating parametric family: $p(\boldsymbol{\theta}) = q_{\boldsymbol{\lambda}_0}(\boldsymbol{\theta})$. In that case, one may fix beforehand $q_0(\boldsymbol{\theta}) = l_0(\boldsymbol{\theta}) = p(\boldsymbol{\theta})$, and update only $\boldsymbol{\lambda}_1, \ldots, \boldsymbol{\lambda}_n$ in the course of the algorithm, while keeping $\boldsymbol{\lambda}_0$ fixed to the value given by the prior.

EP also provides at no extra cost an approximation of the normalising constant of π: $Z = \int_{\boldsymbol{\theta}} \prod_{i=0}^{n} l_i(\boldsymbol{\theta})\, d\boldsymbol{\theta}$. When π is a posterior, this can be used to approximate the marginal likelihood (evidence) of the model. See, for example, Barthelmé and Chopin (2014) for more details.

In certain cases, EP updates are 'too fast', in the sense that the update of difficult sites may lead to, for example, degenerate precision matrices (in the Gaussian case). One well known method to slow down EP is to perform fractional updates (Minka, 2004); that is, informally, update only a fraction $\alpha \in (0,1]$ of the site parameters; see Algorithm 14.6.

Algorithm 14.6: Generic Site Update in EP (Fractional Version, Requires $\boldsymbol{\alpha \in (0,1]}$)

Function SiteUpdate$(i, l_i, \boldsymbol{\lambda}_i, \boldsymbol{\lambda})$:

1. Compute:

$$\boldsymbol{\lambda}^{\text{new}} := \alpha \boldsymbol{\lambda}\left(\mathbb{E}_h[\boldsymbol{t}(\boldsymbol{\theta})]\right) + (1-\alpha)\boldsymbol{\lambda}, \quad \boldsymbol{\lambda}_i^{\text{new}} := \boldsymbol{\lambda}_i + \alpha\left\{\boldsymbol{\lambda}\left(\mathbb{E}_h[\boldsymbol{t}(\boldsymbol{\theta})]\right) - \boldsymbol{\lambda}\right\}$$

with $\mathbb{E}_h[\boldsymbol{t}(\boldsymbol{\theta})]$ defined in (14.3), see step 1 of 14.1.

2. As Step 2 of Algorithm 14.1.

In practice, reducing α is often the first thing to try when EP either diverges or fails because of non-invertible matrices (in the Gaussian case). Of course, the price to pay is that with a lower α, EP may require more iterations to converge.

14.2.7 Theoretical properties of expectation propagation

EP is known to work well in practice, sometimes surprisingly so, but it has proved quite resilient to theoretical study. In Barthelmé and Chopin (2014), we could give no guarantees whatsoever, but since then the situation has improved. The most important question concerns the quality of the approximations produced by EP. Under relatively strong conditions Dehaene and Barthelmé (2018) were able to show that Gaussian EP is asymptotically exact in the large-data limit. This means that if the posterior tends to a Gaussian (which usually happens in identifiable models), then EP will recover the exact posterior. Dehaene and Barthelm (2015) show further that EP recovers the mean of the posterior with an error that vanishes in $\mathcal{O}(n^{-2})$, where n is the number of data-points. The error is up to an order of magnitude lower than what one can expect from the canonical Gaussian approximation, which uses the mode of the posterior as an approximation to the mean.

However, in order to have an EP approximation, one needs to find one in the first place. The various flavours of EP (including the ones described here) are all relatively complex fixed-point iterations, and their convergence is hard to study. Dehaene and Barthelmé (2018) show that parallel EP converges in the large-data limit to a Newton iteration and inherits the potential instabilities in Newton's method. Just like Newton's method, non-convergence in EP can be fixed by slowing down the iterations, as described earlier.

The general picture is that EP should work very well if the hybrids are well-behaved (log-concave, roughly). Like any Gaussian approximation, it can be arbitrarily poor when used on multi-modal posterior distributions, unless the modes are all equivalent.

Note finally that the earlier results apply to variants of EP where hybrid distributions are tractable (meaning their moments can be computed exactly). In ABC applications that is not the case, and we will incur additional Monte Carlo error. As we will explain, part of the trick in using EP in ABC settings is finding ways of minimising that additional source of errors.

14.3 Applying Expectation Propagation in ABC

14.3.1 Principle

Recall that our objective is to approximate the ABC posterior:

$$p_\epsilon(\boldsymbol{\theta}|\boldsymbol{y}^\star) \propto p(\boldsymbol{\theta}) \prod_{i=1}^{n} \left\{ \int f_i(y_i|\boldsymbol{\theta}) \mathbb{I}_{\left\{ \left\| s_i(y_i) - s_i(y_i^\star) \right\| \leq \epsilon \right\}} \, dy_i \right\},$$

for a certain factorisation of the likelihood and for a certain collection of local summary statistics s_i. This immediately suggests using EP on the following collection of sites:

$$l_i(\boldsymbol{\theta}) = \int f_i(y_i|\boldsymbol{\theta}) \mathbb{I}_{\left\{ \left\| s_i(y_i) - s_i(y_i^\star) \right\| \leq \epsilon \right\}} \, dy_i,$$

for $i = 1, \ldots, n$. For convenience, we focus on the Gaussian case (i.e. the l_i's will be approximated by Gaussian factors q_i), and assume that the prior $p(\boldsymbol{\theta})$ itself is already Gaussian, and does not need to be approximated.

From Algorithm 14.2, we see that, in this Gaussian case, it is possible to perform a site update provided that we are able to compute the mean and variance of a pseudo-posterior, corresponding to a Gaussian prior q_{-i}, and likelihood l_i.

Algorithm 14.7 describes a simple rejection algorithm that may be used to perform the site update. Using this particular algorithm inside sequential EP leads to the EP-ABC algorithm derived in Barthelmé and Chopin (2014). We stress, however, that one may generally use any ABC approach to perform such a site update. The main point is that this local ABC problem is much simpler than ABC for the complete likelihood for two reasons. First, the pseudo-prior q_{-i} is typically much more informative than the true prior $p(\boldsymbol{\theta})$, because q_{-i} approximates the posterior of all the data minus y_i. Thus, we are much less likely to sample values of $\boldsymbol{\theta}$ with low likelihood. Second, even for a fixed $\boldsymbol{\theta}$, the probability that $\|s_i(y_i) - s_i(y_i^\star)\| \leq \epsilon$ is typically much larger than $\|s(\boldsymbol{y}) - s(\boldsymbol{y}^\star)\| \leq \epsilon$, as s_i is generally of lower dimension than s.

Algorithm 14.7: Local ABC Algorithm to Perform Site Update

Function SiteUpdate$(i, f_i, (\boldsymbol{r}_i, \boldsymbol{Q}_i), (\boldsymbol{r}, \boldsymbol{Q}))$:

1. Simulate $\boldsymbol{\theta}^{(1)}, \ldots, \boldsymbol{\theta}^{(M)} \sim N(\boldsymbol{\mu}_{-i}, \boldsymbol{\Sigma}_{-i})$ where $\boldsymbol{\Sigma}_{-i}^{-1} = \boldsymbol{Q} - \boldsymbol{Q}_i$, $\boldsymbol{\mu}_{-i} = \boldsymbol{\Sigma}_{-i}(\boldsymbol{r} - \boldsymbol{r}_i)$.

2. For each $m = 1, \ldots M$, simulate $y_i^{(m)} \sim f_i(\cdot|\boldsymbol{\theta}^{(m)})$.

3. Compute

$$M_{\text{acc}} = \sum_{m=1}^{M} \mathbb{I}\left\{\left\|s_i(y_i^{(m)}) - s_i(y_i^\star)\right\| \leq \epsilon\right\}$$

$$\hat{\boldsymbol{\mu}}_h = \frac{1}{M_{\text{acc}}} \sum_{m=1}^{M} \boldsymbol{\theta}^{(m)} \mathbb{I}\left\{\left\|s_i(y_i^{(m)}) - s_i(y_i^\star)\right\| \leq \epsilon\right\}$$

$$\hat{\boldsymbol{\Sigma}}_h = \frac{1}{M_{\text{acc}}} \sum_{m=1}^{M} \boldsymbol{\theta}^{(m)} \left[\boldsymbol{\theta}^{(m)}\right]^t \mathbb{I}\left\{\left\|s_i(y_i^{(m)}) - s_i(y_i^\star)\right\| \leq \epsilon\right\} - \hat{\boldsymbol{\mu}}_h \hat{\boldsymbol{\mu}}_h^t.$$

4. Return $(\boldsymbol{r}_i^{\text{new}}, \boldsymbol{Q}_i^{\text{new}})$, and optionally $(\boldsymbol{r}^{\text{new}}, \boldsymbol{Q}^{\text{new}})$ (according to syntax as in Algorithm 14.1), where:

$$(\boldsymbol{Q}^{\text{new}}, \boldsymbol{r}^{\text{new}}) = \left(\hat{\boldsymbol{\Sigma}}_h^{-1}, \hat{\boldsymbol{\Sigma}}_h^{-1} \hat{\boldsymbol{\mu}}_h\right),$$

$$(\boldsymbol{Q}_i^{\text{new}}, \boldsymbol{r}_i^{\text{new}}) = \left(\boldsymbol{Q}_i + \boldsymbol{Q}^{\text{new}} - \boldsymbol{Q}, \boldsymbol{r}_i + \boldsymbol{r}^{\text{new}} - \boldsymbol{r}\right).$$

14.3.2 Practical considerations

We have observed that in many problems the acceptance rate of Algorithm 14.7 may vary significantly across sites, so, instead of fixing M, the number of simulated pairs $(\boldsymbol{\theta}^{(m)}, y_i^{(m)})$, to a given value, we recommend to sample until the number of accepted pairs (i.e. the number of $(\boldsymbol{\theta}^{(m)}, y_i^{(m)})$, such that $\left\| s_i(y_i^{(m)}) - s_i(y_i^{(m)}) \right\| \leq \epsilon$) equals a certain threshold M_0.

Another simple way to improve EP-ABC is to generate the $\boldsymbol{\theta}^{(m)}$ using quasi-Monte Carlo for distribution $N(\boldsymbol{\mu}_{-i}, \boldsymbol{\Sigma}_{-i})$, we take $\boldsymbol{\theta}^m = \boldsymbol{\mu}_{-i} + \boldsymbol{L}\boldsymbol{\Phi}^{-1}(\boldsymbol{u}^m)$, where $\boldsymbol{\Phi}^{-1}$ is the Rosenblatt transformation (multi-variate quantile function) of the unit normal distribution of dimension $\dim(\boldsymbol{\theta})$, $\boldsymbol{L}\boldsymbol{L}^t = \boldsymbol{\Sigma}_{-i}$ is the Cholesky decomposition of $\boldsymbol{\Sigma}_{-i}$, and the $\boldsymbol{u}^{(m)}$ is a low-discrepancy sequence, such as the Halton sequence; see, for example, Chapter 5 in Lemieux (2009) for more background on low-discrepancy sequences and quasi-Monte Carlo.

Regarding ϵ, our practical experience is that finding a reasonable value through trial and error is typically much easier with EP-ABC than with standard ABC. This is because the y_i's are typically of much lower dimension than the complete dataset $\boldsymbol{y}$. However, one more elaborate recipe to calibrate ϵ is to run EP-ABC with a first value of ϵ, then set ϵ to the minimal value such that the proportion of simulated y_i at each site, such that $\| s_i(y_i) - s_i(y_i^\star) \| \leq \epsilon$ is above, say, 5%. Then one may start over with this new value of ϵ.

Another direction suggested by Mark Beaumont in a personal communication is to correct the estimated precision matrices for bias, using formula (4) from Paz and Sánchez (2015).

14.3.3 Speeding up parallel expectation propagation-ABC in the iid case

This section considers the iid case, for example, the model assumes that the y_i are iid (independent and identically distributed), given $\boldsymbol{\theta}$: then

$$p(\boldsymbol{y}|\boldsymbol{\theta}) = \prod_{i=1}^{n} f_1(y_i|\boldsymbol{\theta}),$$

where f_1 denotes the common density of the y_i. In this particular case, each of the n local ABC posteriors, as described by Algorithm 14.7, will use pseudo-data from the *same* distribution (given $\boldsymbol{\theta}$). This suggests recycling these simulations across sites.

Barthelmé and Chopin (2014) proposed a recycling strategy based on sequential importance sampling. Here, we present an even simpler scheme that may be implemented when parallel EP is used. At the start of iteration t of parallel EP, we sample $\boldsymbol{\theta}^{(1)}, \ldots, \boldsymbol{\theta}^{(M)} \sim N(\boldsymbol{\mu}, \boldsymbol{\Sigma})$, the current *global* approximation of the posterior. For each $\boldsymbol{\theta}^m$, we sample $y^{(m)} \sim f_1(y|\boldsymbol{\theta}^m)$. Then, for each site i, we can compute the first two moments of the hybrid distribution by simply doing an importance sampling step, from $N(\boldsymbol{\mu}, \boldsymbol{\Sigma})$ to $N(\boldsymbol{\mu}_{-i}, \boldsymbol{\Sigma}_{-i})$, which is obtained by dividing the density of $N(\boldsymbol{\mu}, \boldsymbol{\Sigma})$ by factor q_i. Specifically, the weight function is:

$$\frac{|\boldsymbol{Q}_{-i}| \exp\left\{-\frac{1}{2}\boldsymbol{\theta}^t\boldsymbol{Q}_{-i}\boldsymbol{\theta} + \boldsymbol{r}_{-i}^t\boldsymbol{\theta}\right\}}{|\boldsymbol{Q}| \exp\left\{-\frac{1}{2}\boldsymbol{\theta}^t\boldsymbol{Q}\boldsymbol{\theta} + \boldsymbol{r}^t\boldsymbol{\theta}\right\}} = \frac{|\boldsymbol{Q} - \boldsymbol{Q}_i|}{|\boldsymbol{Q}|} \exp\left\{\frac{1}{2}\boldsymbol{\theta}^t\boldsymbol{Q}_i\boldsymbol{\theta} - \boldsymbol{r}_i^r\boldsymbol{\theta}\right\},$$

since $\boldsymbol{Q} = \boldsymbol{Q}_i + \boldsymbol{Q}_{-i}$, $\boldsymbol{r} = \boldsymbol{r}_i + \boldsymbol{r}_{-i}$. Note that further savings can be obtained by retaining the samples for several iterations, regenerating only when the global approximation has changed too much relative to the values used for sampling. In our implementation, we monitor the drift by computing the effective sample size of importance sampling from $N(\boldsymbol{\mu}, \boldsymbol{\Sigma})$ (the distribution of the current samples) for the new global approximation $N(\boldsymbol{\mu}', \boldsymbol{\Sigma}')$.

We summarise the so-obtained algorithm as Algorithm 14.8. Clearly, recycling allows us for a massive speed-up when the number n of sites is large, as we re-use the same set of simulated pairs $(\boldsymbol{\theta}^{(m)}, y^{(m)})$ for all the n sites. In turn, this allows us to take a larger value for M, the number of simulations, which leads to more stable results.

Algorithm 14.8: Parallel EP-ABC with Recycling (iid Case)

Require: M (number of samples), initial values for $(\boldsymbol{r}_i, \boldsymbol{Q}_i)_{i=0,\ldots,n}$ (note $(\boldsymbol{r}_0, \boldsymbol{Q}_0)$ stays constant during the course of the algorithm, as we have assumed a Gaussian prior with natural parameter $(\boldsymbol{r}_0, \boldsymbol{Q}_0)$):

repeat

  $\boldsymbol{Q} \leftarrow \sum_{i=0}^n \boldsymbol{Q}_i$, $\boldsymbol{r} \leftarrow \sum_{i=0}^n \boldsymbol{r}_i$, $\boldsymbol{\Sigma} \leftarrow \boldsymbol{Q}^{-1}$, $\boldsymbol{\mu} \leftarrow \boldsymbol{\Sigma}\boldsymbol{r}$

  for $m = 1, \ldots, M$ **do**

    $\boldsymbol{\theta}^{(m)} \sim N(\boldsymbol{\mu}, \boldsymbol{\Sigma})$

    $y^{(m)} \sim f_1(y|\boldsymbol{\theta}^{(m)})$

  end for

  for $i = 1, \ldots, n$ **do**

    for $m = 1, \ldots, M$ **do**

      $w^{(m)} \leftarrow \frac{|\boldsymbol{Q}-\boldsymbol{Q}_i|}{|\boldsymbol{Q}|} \exp\left\{\frac{1}{2}(\boldsymbol{\theta}^{(m)})^t\boldsymbol{Q}_i\boldsymbol{\theta}^{(m)} - \boldsymbol{r}_i^t\boldsymbol{\theta}^{(m)}\right\} \mathbb{I}\left\{\left\|s_i(y_i^{(m)}) - s_i(y_i^\star)\right\| \leq \epsilon\right\}$

    end for

    $\hat{Z} \leftarrow M^{-1} \sum_{m=1}^M w^{(m)}$

    $\hat{\boldsymbol{\mu}} \leftarrow (M\hat{Z})^{-1} \times \sum_{m=1}^M w^{(m)}\boldsymbol{\theta}^{(m)}$

    $\hat{\boldsymbol{\Sigma}} \leftarrow (M\hat{Z})^{-1} \times \sum_{m=1}^M w^{(m)}\boldsymbol{\theta}^{(m)} \left[\boldsymbol{\theta}^{(m)}\right]^t - \hat{\boldsymbol{\mu}}\hat{\boldsymbol{\mu}}^t$

    $\boldsymbol{r}_i \leftarrow \hat{\boldsymbol{\Sigma}}^{-1}\hat{\boldsymbol{\mu}} - \boldsymbol{r}_{-i}$

    $\boldsymbol{Q}_i \leftarrow \hat{\boldsymbol{\Sigma}}^{-1} - \boldsymbol{Q}_{-i}$

  end for

until Stopping rule (e.g. changes in $(\boldsymbol{r}, \boldsymbol{Q})$ have become small).

We have advocated parallel EP in Section 14.2 as a way to parallelise the computations over the n sites. Given the particular structure of Algorithm 14.8, we see that it is also easy to parallelise the simulation of the M pairs $(\boldsymbol{\theta}^{(m)}, y^{(m)})$ that is performed at the start of each EP iteration; this part is usually the bottleneck of the computation. In fact, we also observe that Algorithm 14.8 performs slightly better than the recycling version of EP-ABC (as described in Barthelmé and Chopin 2014) even on a non-parallel architecture.

14.4 Application to Spatial Extremes

We now turn our attention to likelihood-free inference for spatial extremes, following Erhardt and Smith (2012), see also Prangle (2016).

14.4.1 Background

The data $\boldsymbol{y}$ consist of n iid observations y_i, typically observed over time, where $y_i \in \mathbb{R}^d$ represents some maximal measure (e.g. rainfall) collected at d locations x_j (e.g. in $\mathbb{R}^2$). The standard modelling approach for extremes is to assign to y_i a max-stable distribution (i.e. a distribution stable by maximisation, in the same way that Gaussians are stable by addition). In the spatial case, the vector y_i is composed of d observations of a max-stable process $x \to Y(x)$ at the d locations x_j. A general approach to defining max-stable processes is (Schlather, 2002):

$$Y(x) = \max_k \{s_k \max(0, Z_k(x))\}, \tag{14.5}$$

where $(s_k)_{k=1}^{\infty}$ is the realisation of a Poisson process over $\mathbb{R}^+$ with intensity $\Lambda(\mathrm{d}s) = \mu^{-1}s^{-2}\mathrm{d}s$ (if we view the Poisson process as producing a random set of 'spikes' on the positive real line, then s_1 is the location of the first spike, s_2 the second, etc), $(Z_k)_{k=1}^{\infty}$ is a countable collection of iid realisations of a zero-mean, unit-variance stationary Gaussian process, with correlation function $\rho(h) = \mathrm{Corr}(Z_k(x), Z_k(x'))$ for x, x', such that $\|x - x'\| = h$, and $\mu = \mathbb{E}[\max(0, Z_k(x))]$. Note that $Y(x)$ is marginally distributed according to a unit Fréchet distribution, with Cumulative Density Function (CDF) $F(y) = \exp(-1/y)$.

As in Erhardt and Smith (2012), we will consider the following parametric Whittle-Matérn correlation function:

$$\rho_{\boldsymbol{\theta}}(h) = \frac{2^{1-\nu}}{\Gamma(\nu)}\left(\frac{h}{c}\right)^{\nu} K_\nu\left(\frac{h}{c}\right), \quad c, \nu > 0,$$

where K_ν is the modified Bessel function of the third kind. We take $\boldsymbol{\theta} = (\log\nu, \log c)$, so that $\boldsymbol{\Theta} = \mathbb{R}^2$. (We will return to this logarithmic parametrisation later.)

The main issue with spatial extremes is that, unless $d \leq 2$, the likelihood $p(\boldsymbol{y}|\boldsymbol{\theta})$ is intractable. One approach to estimate $\boldsymbol{\theta}$ is pairwise marginal composite likelihood (Padoan et al., 2010). Alternatively, (14.5) suggests a simple way to simulate from $p(\boldsymbol{y}|\boldsymbol{\theta})$, at least approximately (e.g. by truncating the domain of the Poisson process to $[0, S_{\max}]$). This motivates likelihood-free inference (Erhardt and Smith, 2012).

14.4.2 Summary statistics

One issue, however, with likelihood-free inference for this class of models is the choice of summary statistics: Erhardt and Smith (2012) compare several choices and find that the one that performs best is some summary of the clustering of the $d(d-1)(d-2)/6$ triplet-wise coefficients:

$$\frac{n}{\sum_{i=1}^{n} \{\max(y_i(x_j), y_i(x_k), y_i(x_l))\}^{-1}}, \quad 1 \leq j < k < l \leq d.$$

But computing these coefficients require $\mathcal{O}(d^3)$ operations, and may actually be more expensive than simulating the data itself: Prangle (2016) observes in a particular experiment that the cost of computing these coefficients is already more than twice the cost of simulating data for $d = 20$. As a result, the overall approach of Erhardt and Smith (2012) may take several days to run on a single-core computer.

In contrast, EP-ABC allows us to define local summary statistics, $s_i(y_i)$, that depend only on one data-point y_i. We simply take $s_i(y_i)$ to be the (2-dimensional) OLS (ordinary least squares) estimate of regression:

$$\log|F(y_i(x_j)) - F(y_i(x_k))| = a + b\log\|x_j - x_k\| + \epsilon_{jk}, \quad 1 \leq j < k \leq d.$$

where F is the unit Fréchet CDF. The madogram function $h \to \mathbb{E}[|Y(x) - Y(x')|]$, for $\|x - x'\| = h$, or its empirical version, is a common summary of spatial dependencies (for extremes). Here, we take the $F-$madogram, for example, $Y(x)$ is replaced by $F(Y(x)) \sim U[0,1]$, because $Y(x)$ is Fréchet and thus: $\mathbb{E}[|Y(x)|] = +\infty$.

14.4.3 Numerical results on real data

We now apply EP-ABC to the rainfall dataset of the SpatialExtremes R package (available at http://spatialextremes.r-forge.r-project.org/), which records maximum daily rainfall amounts over the years 1962–2008 occurring during June–August at 79 sites in Switzerland. We ran sequential EP with recycling and quasi-Monte Carlo (see discussion in Section 14.3.2). Figure 14.1 plots the EP-ABC posterior for $\epsilon = 0.2$, 0.05, and 0.02. A $N(0,1)$ prior was used for both components of $\boldsymbol{\theta} = (\log\nu, \log c)$.

Each run took about 3 hours on our desktop computer and generated about 10^5 data-points (i.e. realisations $y_i \in \mathbb{R}^d$, where d is the number of

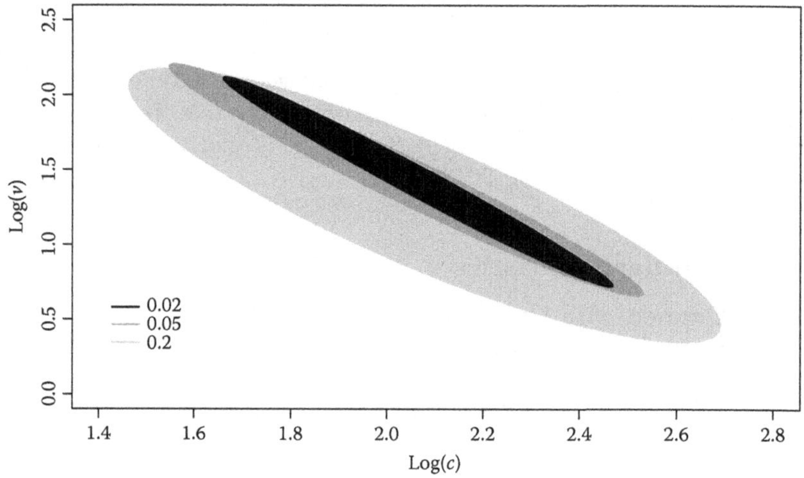

FIGURE 14.1
Fifty percent credible ellipses of the Gaussian approximation of the posterior computed by EP-ABC for different values of ϵ and rainfall dataset.

stations). As a point of comparison, we ran Erhardt and Smith (2012)'s R package for a week on the same computer, which led to the generation of 5×10^4 complete datasets (i.e. $\approx 4 \times 10^6$ data-points). However, the ABC posterior approximation obtained from the 100 generated datasets that were closest to the data, relative to their summary statistics, was not significantly different from the prior.

Finally, we discuss the strong posterior correlations between the two parameters that are apparent in Figure 14.1. Figure 14.2 plots a heat map of functions $(\nu, c) \to \int |\rho_{\nu,c} - \rho_{\nu_0,c_0}|$ and $(\log \nu, \log c) \to \int |\rho_{\nu,c} - \rho_{\nu_0,c_0}|$, for

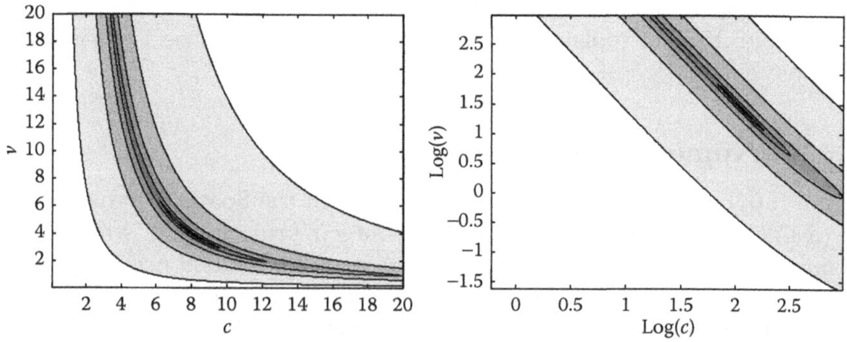

FIGURE 14.2
Heat map of functions $(\nu, c) \to \int |\rho_{\nu,c} - \rho_{\nu_0,c_0}|$ (left panel) and $(\log \nu, \log c) \to \int |\rho_{\nu,c} - \rho_{\nu_0,c_0}|$, for $(\nu_0, c_0) = (8, 4)$ (right panel).

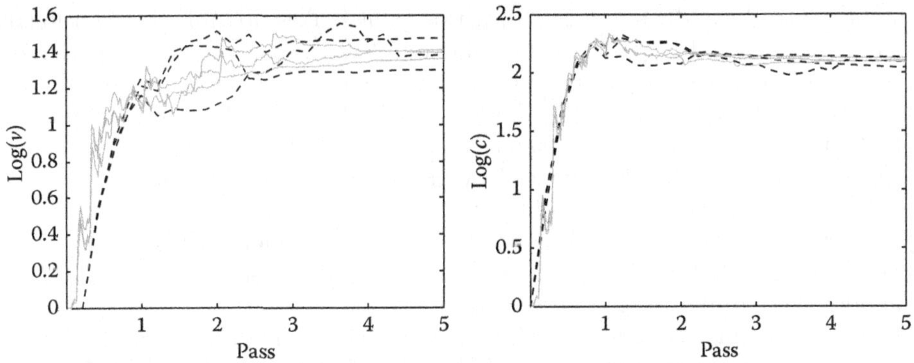

FIGURE 14.3
Posterior mean of $\log \nu$ (left panel) and $\log c$ (right panel) as a function of the number of passes (one pass equals $n = 47$ site updates) for three runs of the sequential version (solid grey line) and block-parallel version ($n_{\text{core}} = 10$, dashed black line) of EP-ABC applied to rainfall dataset ($\epsilon = 0.05$).

$(\nu_0, c_0) = (8, 4)$. The model appears to be nearly non-identifiable, as values of (ν, c) that are far away may produce correlation functions that are nearly indistinguishable. In addition, the parametrisation $\boldsymbol{\theta} = (\log \nu, \log c)$ has the advantage of giving an approximately Gaussian shape to contours, which is clearly helpful in our case given that EP-ABC generates a Gaussian approximation. Still, it is interesting to note that EP-ABC performs well on such a nearly non-identifiable problem.

14.4.4 Expectation-propagation convergence

Finally, we compare the convergence (relative to the number of iterations) of the standard version, and the block-parallel version (described in Section 14.2.5) of EP-ABC, on the rainfall dataset discussed earlier. Figure 14.3 plots the evolution of the posterior mean of both parameters ν (left panel) and c (right panel), relative to the number of site updates, for three runs of both versions, and for $\epsilon = 0.05$.

We took $n_{\text{core}} = 10$ (i.e. blocks of ten sites are updated in parallel), although both algorithms were run on a single core. We see that both algorithms essentially converge at the same rate. Thus, if implemented on a 10-core machine, the block-parallel version should offer essentially a $\times 10$ speed-up.

14.5 Conclusion

Compared to standard ABC, the main drawback of EP-ABC is that it introduces an extra level of approximation, because of its EP component. On the other hand, EP-ABC strongly reduces, or sometimes removes entirely, the bias introduced by summary statistics, as it makes possible to use n local summaries, instead of just one for the complete dataset. In our experience [see e.g. the examples in Barthelmé and Chopin (2014)], this bias reduction more than compensates the bias introduced by EP. But the main advantage of EP-ABC is that it is much faster than standard ABC. Speed-ups of more than 100 are common, as evidenced by our spatial extremes example.

We have developed a MATLAB® package, available at https://sites.google.com/site/simonbarthelme/software, that implements EP-ABC for several models, including spatial extremes. The current version of the package includes the parallel version described in this paper.

An interesting direction for future work is to integrate current developments on model emulators into EP-ABC. Model emulators are Machine Learning (ML) algorithms that seek to learn a tractable approximation of the likelihood surface from samples (Wilkinson, 2014). A variant directly learns an approximation of the posterior distribution, as in Gutmann and Corander (2015). Heess et al. (2013) introduce a more direct way of using emulation in an EP context. Their approach is to consider each site as a mapping between the parameters of the pseudo-prior and the mean and covariance of the hybrid and to learn the parameters of that mapping. In complex, but low-dimensional, models typical of ABC applications, this viewpoint could be very useful and deserves to be further explored.

Acknowledgements

We are very grateful to Mark Beaumont, Dennis Prangle, and Markus Hainy for their careful reading and helpful comments that helped us to improve this chapter.

The second author is partially supported by ANR (Agence Nationale de la Recherche) grant ANR-11-IDEX-0003/Labex Ecodec/ANR-11-LABX-0047 as part of Investissements d'Avenir program.

References

Barthelmé, S. and Chopin, N. (2014). Expectation propagation for likelihood-free inference. *J. Am. Stat. Assoc.*, 109(505):315–333.

Beaumont, M. A., Cornuet, J.-M., Marin, J.-M., and Robert, C. P. (2009). Adaptive approximate Bayesian computation. *Biometrika*, 96(4):983–990.

Bishop, C. M. (2006). *Pattern Recognition and Machine Learning.* Information Science and Statistics. Springer, New York.

Cseke, B. and Heskes, T. (2011). Approximate marginals in latent Gaussian models. *J. Mach. Learn. Res.*, 12:417–454.

Dean, T. A., Singh, S. S., Jasra, A., and Peters, G. W. (2014). Parameter estimation for hidden Markov models with intractable likelihoods. *Scand. J. Stat.*, 41(4):970–987.

Dehaene, G. P. and Barthelm, S. (2015). Bounding errors of expectation-propagation. In Cortes, C., Lawrence, N. D., Lee, D. D., Sugiyama, M., and Garnett, R. (Eds.), *Advances in Neural Information Processing Systems 28: Annual Conference on Neural Information Processing Systems 2015*, December 7–12, Montreal, Quebec, Canada, pp. 244–252. Curran Associates, Red Hook, NY.

Dehaene, G. and Simon, B. (2018). Expectation propagation in the large data limit. *J. R. Stat. Soc.: Series B Stat. Methodol.*, 80(1):199–217.

Del Moral, P., Doucet, A., and Jasra, A. (2012). An adaptive sequential Monte Carlo method for approximate Bayesian computation. *Stat. Comput.*, 22(5):1009–1020.

Erhardt, R. J. and Smith, R. L. (2012). Approximate Bayesian computing for spatial extremes. *Comput. Stat. Data Anal.*, 56(6):1468–1481.

Gelman, A., Carlin, J. B., Stern, H. S., Dunson, D. B., Vehtari, A., and Rubin, D. B. (2014). *Bayesian Data Analysis*, 3rd ed. Texts in Statistical Science Series. CRC Press, Boca Raton, FL.

Gutmann, M. U. and Corander, J. (2015). Bayesian optimization for likelihood-free inference of simulator-based statistical models. *ArXiv preprint 1501.03291.*

Heess, N., Tarlow, D., and Winn, J. (2013). Learning to pass expectation propagation messages. In Burges, C., Bottou, L., Welling, M., Ghahramani, Z., and Weinberger, K. (Eds.), *Advances in Neural Information Processing Systems 26*, pp. 3219–3227. Curran Associates, Red Hook, NY.

Jasra, A. (2015). Approximate Bayesian computation for a class of time series models. *International Statistical Review*, 83:405–435.

Lemieux, C. (2009). *Monte Carlo and Quasi-Monte Carlo Sampling* (*Springer Series in Statistics*). Springer, Berlin, Germany.

Marjoram, P., Molitor, J., Plagnol, V., and Tavaré, S. (2003). Markov chain Monte Carlo without likelihoods. *Proceedings of the National Academy of Sciences*, 100(26):15324–15328.

Minka, T. (2004). Power EP. Technical report, Department of Statistics, Carnegie Mellon University, Pittsburgh, PA.

Minka, T. P. (2001). Expectation propagation for approximate Bayesian inference. *Proceedings of Uncertainty in Artificial Intelligence*, 17:362–369.

Padoan, S. A., Ribatet, M., and Sisson, S. A. (2010). Likelihood-based inference for max-stable processes. *J. Am. Stat. Assoc.*, 105(489):263–277.

Paz, D. J. and Sánchez, A. G. (2015). Improving the precision matrix for precision cosmology. *Monthly Notices of the Royal Astronomical Society*, 454(4):4326–4334.

Prangle, D. (2016). Lazy ABC. *Statistics and Computing*, *26*(1–2), 171–185.

Ruli, E., Sartori, N., and Ventura, L. (2016). Approximate Bayesian computation with composite score functions. *Statistics and Computing*, *26*(3), 679–692.

Schlather, M. (2002). Models for stationary max-stable random fields. *Extremes*, 5(1):33–44.

Seeger, M. (2005). Expectation propagation for exponential families. Technical report, University of California, Berkeley, CA.

Sisson, S. A., Fan, Y., and Tanaka, M. M. (2007). Sequential Monte Carlo without likelihoods. *Proc. Natl. Acad. Sci. USA*, 104(6):1760–1765 (electronic).

Varin, C., Reid, N., and Firth, D. (2011). An overview of composite likelihood methods. *Statist. Sinica*, 21(1):5–42.

White, S. R., Kypraios, T., and Preston, S. P. (2015). Piecewise approximate Bayesian computation: Fast inference for discretely observed Markov models using a factorised posterior distribution. *Stat. Comput.*, 25(2):289–301.

Wilkinson, R. D. (2014). Accelerating ABC methods using Gaussian processes. *ArXiv preprint 1401.1436.*

Yıldırım, S., Singh, S. S., Dean, T., and Jasra, A. (2014). Parameter estimation in hidden Markov models with intractable likelihoods using sequential Monte Carlo. *J. Comput. Graph. Stat.*, 24(3):846–865.

Part II

Applications

15

Sequential Monte Carlo-ABC Methods for Estimation of Stochastic Simulation Models of the Limit Order Book

Gareth W. Peters, Efstathios Panayi, and Francois Septier

CONTENTS

15.1 Introduction to Intractable Likelihood Models for High Frequency Financial Market Dynamics 438
- 15.1.1 Introduction to the limit order book and related multi-queue simulation models 439
- 15.1.2 Features of significance in limit order book stochastic processes: Liquidity 441

15.2 Bayesian Models with Intractable Likelihoods for High Frequency Financial Market Dynamics 442
- 15.2.1 Limit order book agent-based model 443
  - 15.2.1.1 Stochastic agent representation: Liquidity providers and demanders 445
- 15.2.2 Bayesian model formulation of the stochastic agent limit order book model representation 451
  - 15.2.2.1 Posterior models for computationally intractable likelihoods 452
- 15.2.3 Summary statistics in Bayesian limit order book models 454

15.3 Estimation of Bayesian Limit Order Book Stochastic Agent Model via Population-Based Samplers 455
- 15.3.1 Brief overview of sequential Monte Carlo-based samplers 456
- 15.3.2 Sequential Monte Carlo samplers for intractable likelihood Bayesian models 458
  - 15.3.2.1 Adaptive schedules: choice of sequence of ABC distributions via annealed tolerance schedule 460
  - 15.3.2.2 Choice of mutation kernel 461

15.4 Application to Equities Limit Order Book Data: Data Description ... 464
15.5 Results ... 465
15.5.1 Estimation algorithm configuration ... 467
15.5.2 Final particle fitness and distributions of parameters ... 468
15.5.3 Results comparison to multi-objective evolutionary algorithm-II procedure ... 470
15.6 Conclusion ... 474
References ... 476

15.1 Introduction to Intractable Likelihood Models for High Frequency Financial Market Dynamics

In this chapter, we consider classes of models that have been recently developed for quantitative finance that involve modelling a highly complex multi-variate, multi-attribute stochastic process known as the limit order book (LOB). The LOB is the primary data structure recorded each day intra-daily for the majority of assets on electronic exchanges around the world in which trading takes place. As such, it represents one of the most important fundamental structures to study from a stochastic process perspective if one wishes to characterise features of stochastic dynamics for price, volume, liquidity, and other important attributes for a traded asset. In this paper, we aim to adopt the model structure recently proposed by Panayi and Peters (2015), which develops a stochastic model framework for the LOB of a given asset and to explain how to perform calibration of this stochastic model to real observed LOB data for a range of different assets. One can consider this class of problems as truly a setting in which both the likelihood is intractable to evaluate pointwise, but trivial to simulate, and in addition the amount of data is massive. This is a true example of big-data application, as for each day and for each asset one can have anywhere between 100,000–500,000 data vectors for the calibration of the models.

The class of calibration techniques we will consider here involves an approximate Bayesian computation (ABC) re-formulation of the indirect inference framework developed under the multi-objective optimisation formulation proposed recently by Panayi and Peters (2015). To facilitate an equivalent comparison for the two frameworks, we also adopt a re-formulation of the class of genetic stochastic search algorithms utilised by Panayi and Peters (2015), known as NGSA-II (Deb et al., 2002). We adapt this widely utilised stochastic genetic search algorithm from the multi-objective optimisation algorithm literature to allow it to be utilised as a mutation kernel in a class of sequential Monte Carlo samplers (SMC sampler) algorithms in the ABC context. We begin with the problem and model formulation, then we discuss the estimation frameworks and finish with some real data simulation results for equity data from a highly utilised pan-European secondary exchange formerly known as Chi-X, before it was recently aquired by BATS to form BATS Chi-X Europe in 2014.*

15.1.1 Introduction to the limit order book and related multi-queue simulation models

The structure of a financial market dictates the form of interaction between buyers and sellers. Markets for financial securities generally operate either as quote driven markets, in which specialists (dealers) provide 2-way prices, or order driven markets, in which participants can trade directly with each other by expressing their trading interest through a central matching mechanism. The LOB is an example of the latter, and indicatively, the Helsinki, Hong Kong, Shenzhen, Swiss, Tokyo, Toronto, and Vancouver Stock Exchanges, together with Euronext and the Australian Securities Exchange operate as pure LOBs, while the New York Stock Exchange, NASDAQ, and the London Stock Exchange also operate a hybrid LOB system (Gould et al., 2013), with specialists operating for the less liquid securities.

We will consider trading activity in the context of the LOB in this chapter. Market participants are typically allowed to place two types of orders on the venue: limit orders, where they specify a price over which they are unwilling to buy (or a price under which they are unwilling to sell), and market orders, which are executed at the best available price. Market orders are executed immediately, provided there are orders of the same size on the opposite side of the book. Limit orders to buy (sell) are only executed if there is opposite trading interest in the order book at, or below (above), the specified limit price. If there is no such interest, the order is entered into the limit order book, where orders are displayed by price, then time priority.

Figure 15.1 shows an example snapshot of the order book for a particular stock, as traded on the Chi-X exchange, at a particular instance of time. A market order to buy 200 shares would result in three trades: 70 shares at 2,702, another 100 shares at 2,702, and the remaining 30 at 2,704. A limit order to sell 300 shares at 2,705, on the other hand, would not be executed immediately, as the highest order to buy is only at 2,700 cents. It would instead enter the limit order book on the right hand side, second in priority at 2,705 cents after the order for 120 shares, which is already in the book.

LOB simulation models aim to generate the trading interactions observed in such a LOB and allow for a realistic description of the intra-day trading process. In particular, the models simulate the behaviour of individual market participants, usually based on the behaviour of various classes of real traders. The price of the traded financial asset is then determined from the limit and market orders submitted by these traders. Depending on the model,

*https://www.batstrading.co.uk/

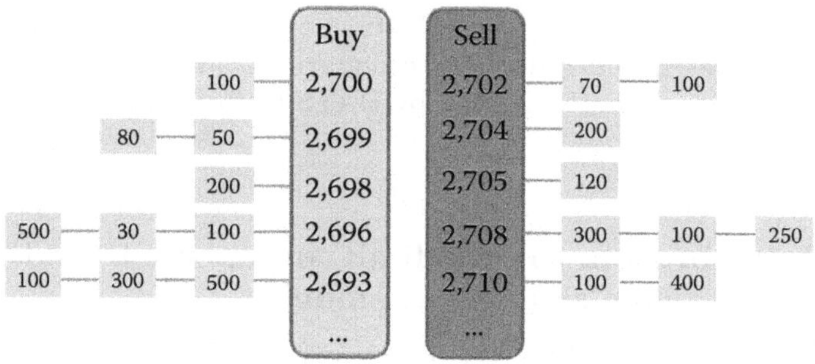

FIGURE 15.1
An example of the state of the Chi-X order book. The left hand side of the book is the buying interest, while the right hand side is the selling interest. The highest bid (order to buy) is for 100 shares at 2,700 cents, while there are two lowest offers (orders to sell) for 70 and 100 shares at 2,702. Orders are prioritised by price, then time.

the instantaneous price is either considered to be the mid-point between the highest bid price and lowest ask price or the last traded price.

Because of the practical interest of modelling the intra-day dynamics of both stock prices and available volumes in the LOB, as well as the difficulty of traditional economic models based on rationality to reproduce these dynamics, there have been a multitude of research approaches attempting to address this gap. On one hand, there have been the zero-intelligence approaches [e.g. Maslov (2000)], which generally consist of a single, unsophisticated type of trading agent who submits orders randomly, possibly subject to (very few) constraints, like budget considerations. This minimum amount of discipline imposed by their actions is therefore often sufficient to reproduce some commonly observed features of financial markets, such as fat tails in the distribution of returns (Maslov, 2000). Later models (LiCalzi and Pellizzari, 2003; Fricke and Lux, 2015) also considered more realistic trading behaviour, where agents act based on their perceived value of the financial asset. However, LiCalzi and Pellizzari (2003) noted that these additional considerations regarding agent behaviour did not necessarily lead to a more realistic stock price output, and that it was likely the imposed market structure that had the largest effect on reproducing the price features observed.

One other prominent strand of research pertains to the modelling of the LOB as a set of queues at each price and on both the buy (bid) and sell (ask) side. In the models of Cont et al. (2010) and Cont and De Larrard (2013), a power law characterising the intensities of the limit order processes is found to be a good emprical fit. However, their assumption of independence between the activity at each level and for each event type is unlikely to hold in modern LOBs, due to the presence of algorithmic trading strategies, which induce different types of dependence structures. It is clear from observed empirical LOB data that non-trivial dependence structures are present, and as such, ignoring these features in the model formulation will result in inadequate representations of the stochastic LOB process being modelled.

In the LOB simulation model introduced in Panayi and Peters (2015), they attempted to provide both a richer description of the LOB market structure and its constituent agents, but also consider the dependence between the trading activity at different levels. The main components of the proposed model are detailed in the following sections, before it is extended into an ABC posterior formulation for estimation and inference purposes.

15.1.2 Features of significance in limit order book stochastic processes: Liquidity

In order to motivate the development of our stochastic representative agent-based model from a liquidity demand and supply perspective, it is important to also comment on how most modern electronic exchanges operate. We argue that it is especially important to understand this from a liquidity perspective and to incorporate such features into a stochastic model of the LOB. To our knowledge, there have been no previous attempts to build-agent based models based around liquidity provision characteristics other than the work of Panayi and Peters (2015), which we extend into a Bayesian formulation with SMC-ABC inference in this chapter.

To put this in context, we note that the notion of liquidity can take many forms in a LOB stochastic process. Indeed, the seminal paper of Kyle (1985) acknowledged the difficulty in capturing the liquidity of a financial market in a single metric, and identified tightness, depth, and resiliency as three main properties that characterise the liquidity of a limit order book, see more detailed discussion on these in Panayi et al. (2015b). Tightness and depth have been mainstays of the financial literature (and indeed, are easily captured through common liquidity measures, such as the spread and depth, respectively) and there have been substantial literature in studying the intra-day variation and commonality in these measures. In addition, the resilience of liquidity in the LOB of a financial asset has recently begun to receive attention, see Panayi et al. (2015b), who provided a review of the state of the art in liquidity resilience and noted that the extant definitions seemed to be divided into two categories: In the first, definitions provided by Kyle (1985) and Obizhaeva and Wang (2013) were related to price evolution and, specifically, to the return of prices to a steady state. The second category of definitions, proposed by Garbade and Garbade (1982) and Harris (2003) was concerned with liquidity replenishment.

To further understand the role of liquidity in a model for LOB, one must also be aware of the fact that financial exchanges have different modes of operation, or market models, for different assets, and Panayi et al. (2015a) provide an extended discussion of the circumstances under which each is appropriate.

The German electronic trading system Xetra supports the following modes, for example*: continuous trading in connection with auctions (e.g. opening and closing auction and, possibly, one or more intra-day auctions); mini-auction in connection with auctions; and one or more auctions at pre-defined points in time.

In brief, this is often determined by an asset's liquidity in the prevailing period, and an exchange will endeavour to choose a market model that facilitates trading in the asset. Xetra, for example, offers continuous trading for the most liquid assets, while the second most liquid category of assets is supplemented with a 'Designated Sponsor', with market-making obligations, typically including maximum spread, minimum quote size, and effective trading times, they receive a discount or re-imbursement of transaction fees in return. Certain securities feature a single market maker, while less liquid assets operate instead in a 'continuous auction' mode, with the assistance of specialist. In order to decide the market model, Xetra classifies assets quarterly by averaging their liquidity over the period.

Liquidity provision in the equities and foreign exchange markets has increasingly become the domain of high-frequency traders. The SEC (Securities et al., 2010) reports that these traders contribute approximately half of the total traded volume of all participants. Because these traders operate in the millisecond domain, we can reasonably assume a high degree of automation in their trading, and we will try to capture the aggregate liquidity providing activity of this group of representative agents in our model. Even though for liquid assets, these firms have no legal obligation to provide uninterrupted liquidity, for the period under consideration there are no major structural breaks, we can assume that this group of liquidity providers will be operating in the market on the LOB for the assets under consideration throughout the majority of the trading day.

In this chapter, we develop a LOB stochastic agent model based on liquidity provision for the class of assets operating under a continuous trading mode, which will be the case for the majority of assets traded on major indices, for example.

15.2 Bayesian Models with Intractable Likelihoods for High Frequency Financial Market Dynamics

In this section, we develop a new class of Bayesian models that can be utilised to study the dynamics of LOB intra-daily. We start by presenting the stochastic multi-variate order flow model, which we develop as a 'stochastic

*Xetra trading models, available at http://www.xetra.com/xetra-en/trading/trading-models

representative agent' model framework. This presentation is largely based on Panayi and Peters (2015), as this is the model we will be working with. The model is then re-formulated as an intractable likelihood stochastic model and developed into an approximate Bayesian computational (posterior model).

15.2.1 Limit order book agent-based model

We consider the intra-day LOB activity in fixed intervals of time $\ldots, [t-1,t), [t,t+1), \ldots$. For every interval $[t,t+1)$, we allow the total number of levels on the bid or ask sides of the LOB to be dynamically adjusted as the simulation evolves. These LOB levels are defined with respect to two reference prices, equal to $p_{t-1}^{b,1}$ and $p_{t-1}^{a,1}$, for example, the price of the highest bid and lowest ask price at the start of the interval. We consider these reference prices to be constant throughout the interval $[t-1,t)$, and, thus, the levels on the bid side of the book are defined at integer number of ticks away from $p_{t-1}^{a,1}$, while the levels on the ask side of the book are defined at integer number of ticks away from $p_{t-1}^{b,1}$.

This does not mean that we expect the best bid and ask prices to remain constant, just that we model the activity (i.e. limit order arrivals, cancellations, and executions) according to the distance in ticks from these reference prices during this period. We note that it is of course possible that the volume at the best bid price is consumed during the interval, and that limit orders to sell are posted at this price, which would be considered at 0 ticks away from the reference price. To allow for this possibility, we actively model the activity at $-l_d+1, \ldots, 0, \ldots, l_p$ ticks away from each reference price. Here, the p subscript will refer to passive orders, for example, orders which would not lead to immediate execution, if the reference prices remained constant and d refers to direct, or aggressive orders (those which would be executed immediately), where it is again understood that they are aggressive with respect to the reference prices at the start of the period. Therefore, we actively model the activity at a total $l_t = l_p + l_d$ levels on the bid and ask.

We assume that activity that occurs further away is uncorrelated with the activity close to the top of the book, and therefore unlikely to have much of an impact on price evolution and the properties of the volume process. Therefore, the volume resting outside the actively modelled LOB levels $(-l_d+1, \ldots, 0, \ldots, l_p)$ on the bid and ask is assumed to remain unchanged until the agent interactions brings those levels inside the band of actively modelled levels, at which time they will again dynamically evolve. Such a set of assumptions is consistent with observed stylised features of all LOB for all modern electronic exchanges.

We distinguish different order types that agents may perform as follows: place the order 'at market' or 'at limit'. Market orders provide instruction to execute, as quickly as possible, a transaction at the present, or market, price.

Conversely, a limit order provides instruction to only execute at or under a purchase price or at or above a sales price. In practice, there are many different variations of such order types, we focus on the most fundamental vanilla passive and aggressive limit orders and standard immediate effect market orders.

To present the details of the simulation framework, including the stochastic model components for each agent, for example, liquidity providers and liquidity demanders, we first define the following notation:

1. $\boldsymbol{V}_t^a = (V_t^{a,-l_d+1}, \ldots, V_t^{a,l_p})$ – the random vector for the number of existing orders resting at each level on the ask side at time t at the actively modelled levels of the LOB at time t.

2. $\boldsymbol{N}_t^{LO,a} = (N_t^{LO,a,-l_d+1}, \ldots, N_t^{LO,a,l_p})$ – the random vector for the number of limit orders entering the limit order book on the ask side at each level in the interval $[t-1, t)$.

3. $\boldsymbol{N}_t^{C,a} = (N_t^{C,a,1}, \ldots, N_t^{C,a,l_t})$ – the random vector for the number of limit orders cancelled on the ask side in the interval $[t-1, t)$.

4. $N_t^{MO,a}$ – the random variable for the number of market orders submitted by liquidity demanders in the interval $[t-1, t)$.

We consider the processes for limit orders and market orders, as well as cancellations to be linked to the behaviour of real market participants in the LOB. In the following, we model the aggregation of the activity of two classes of liquidity motivated agents, namely liquidity providers and liquidity demanders. As we model LOB activity in discrete time intervals, we process the aggregate activity at the end of each time interval in the following order:

1. Limit order arrivals – passive – by the liquidity provider agent.

2. Limit order arrivals – aggressive or direct – by the liquidity provider agent.

3. Cancellations by the liquidity provider agent.

4. Market orders by the liquidity demander agent.

The rationale for this ordering is that the vast majority of limit order submissions and cancellations are typically accounted for by the activity of high-frequency traders, and many resting orders are cancelled before slower traders can execute against them. In addition, such an ordering allows us to condition on the state of the LOB, so that we do not have more cancellations at a particular level than the orders resting at that level. This is generally appropriate, as the time interval we consider can be made as small as desired for a given simulation.

15.2.1.1 Stochastic agent representation: Liquidity providers and demanders

We assume liquidity providers are responsible for all market-making behaviour (i.e. limit order submissions and cancellations on both the bid and ask side of the LOB). After liquidity is posted to the LOB, liquidity seeking market participants, such as mutual funds using some execution algorithm, can take advantage of the resting volume with market orders. For market makers, achieving a balance between volume executed on the bid and the ask side can be profitable. However, there is also the risk of adverse selection, for example, trading against a trader with superior information. This may lead to losses if, for example, a trader posts multiple market orders that consume the volume on several levels of the LOB. The risk of adverse selection as a result of asymmetric information is one of the basic tenets of market microstructure theory (O'hara, 1995). To reduce this risk, market makers cancel and re-submit orders at different prices and/or different sizes.

Definition 15.1 (Limit order submission process for the liquidity provider agent) *Consider the limit order submission process of the liquidity provider agent to include both passive and aggressive limit orders on the bid and ask sides of the book, which are assumed to have the following stochastic model structure:*

1. *Let the multi-variate path-space random matrix* $\boldsymbol{N}_{1:T}^{LO,k} \in \mathbb{N}_{+}^{l_t \times T}$ *be constructed from random vectors for the numbers of limit order placements* $\boldsymbol{N}_{1:T}^{LO,k} = \left(\boldsymbol{N}_1^{LO,k}, \boldsymbol{N}_2^{LO,k}, \ldots, \boldsymbol{N}_T^{LO,k}\right)$. *Furthermore, assume these random vectors for the number of orders at each level at time t are each conditionally dependent on a latent stochastic process for the intensity at which the limit orders arrive, given by the random matrix* $\boldsymbol{\Lambda}_{1:T}^{LO,k} \in \mathbb{R}_{+}^{l_t \times T}$ *and on the path-space by* $\boldsymbol{\Lambda}_{1:T}^{LO,k} = \left(\boldsymbol{\Lambda}_1^{LO,k}, \boldsymbol{\Lambda}_2^{LO,k}, \ldots, \boldsymbol{\Lambda}_T^{LO,k}\right)$. *In the following,* $k \in \{a, b\}$ *indicates the respective process on the ask and bid side.*

2. *Assume the conditional independence property for the random vectors given by:*

$$\left[\boldsymbol{N}_s^{LO,k} | \boldsymbol{\Lambda}_s^{LO,k}\right] \perp\!\!\!\perp \left[\boldsymbol{N}_t^{LO,k} | \boldsymbol{\Lambda}_t^{LO,k}\right], \quad \forall s \neq t, \quad s, t \in \{1, 2, \ldots, T\} \tag{15.1}$$

3. *For each time interval* $[t-1, t)$ *from the start of trading on the day, let the random vector for the number of new limit orders placed in each actively modelled level of the limit order book, for example, the price points corresponding to ticks* $(-l_d + 1, \ldots, 0, 1, \ldots, l_p)$, *be denoted by* $\boldsymbol{N}_t^{LO,k} = (N_t^{LO,k,-l_d+1}, \ldots, N_t^{LO,k,l_p})$, *and assume that these random vectors satisfy the conditional independence property:*

$$\left[N_t^{LO,k,s} | \Lambda_t^{LO,k,s}\right] \perp\!\!\!\perp \left[N_t^{LO,k,q} | \Lambda_t^{LO,k,q}\right], \quad \forall s \neq q, \quad s, q \in \{-l_d + 1, \ldots, 0, 1, \ldots, l_p\} \tag{15.2}$$

4. *Assume the random vector* $\boldsymbol{N}_t^{LO,k} \in \mathbb{N}_+^{l_t}$ *is distributed according to a multi-variate generalised Cox process with conditional distribution* $\boldsymbol{N}_t^{LO,k} \sim \mathcal{GCP}\left(\boldsymbol{\lambda}_t^{LO,k}\right)$ *given by:*

$$\mathbb{P}r\left(N_t^{LO,k,-l_d+1} = n_1, \ldots, N_t^{LO,k,l_p} = n_{l_t} \middle| \boldsymbol{\Lambda}_t^{LO,k} = \boldsymbol{\lambda}_t^{LO,k}\right) = \prod_{s=-l_d+1}^{l_p} \frac{\left(\lambda_t^{LO,k,s}\right)^{n_s}}{n_s!} \exp\left[-\lambda_t^{LO,k,s}\right] \tag{15.3}$$

5. *Assume the independence property for random vectors of latent intensities unconditionally according to:*

$$\boldsymbol{\Lambda}_s^{LO,k} \perp\!\!\!\perp \boldsymbol{\Lambda}_t^{LO,k}, \ \ \forall s \neq t, \ \ s,t \in \{1,2,\ldots,T\} \tag{15.4}$$

6. *Assume that the intensity random vector* $\boldsymbol{\Lambda}_t^{LO,k} \in \mathbb{R}_+^{l_t}$ *is obtained through an element-wise transformation of the random vector* $\boldsymbol{\Gamma}_t^{LO,k} \in \mathbb{R}^{l_t}$*, where for each element we have the mapping:*

$$\Lambda_t^{LO,k,s} = \mu_0^{LO,k,s} F\left(\Gamma_t^{LO,k,s}\right) \tag{15.5}$$

where we have $s \in \{-l_d+1, \ldots, l_p\}$*, baseline intensity parameters* $\left\{\mu_0^{LO,k,s}\right\} \in \mathbb{R}_+$*, and a strictly monotonic mapping* $F : \mathbb{R} \mapsto [0,1]$.

7. *Assume the random vector* $\boldsymbol{\Gamma}_t^{LO,k} \in \mathbb{R}$ *is distributed according to a multi-variate skew-t distribution* $\boldsymbol{\Gamma}_t^{LO,k} \sim MSt(\boldsymbol{m}^k, \boldsymbol{\beta}^k, \nu^k, \Sigma^k)$ *with location parameter vector* $\boldsymbol{m}^k \in \mathbb{R}^{l_t}$*, skewness parameter vector* $\boldsymbol{\beta}^k \in \mathbb{R}^{l_t}$*, degrees of freedom parameter* $\nu^k \in \mathbb{N}_+$*, and* $l_t \times l_t$ *covariance matrix* Σ^k*. Hence,* $\boldsymbol{\Gamma}_t^{LO,k}$ *has density function:*

$$f_{\boldsymbol{\Gamma}_t^{LO,k}}\left(\boldsymbol{\gamma}_t; \boldsymbol{m}^k, \boldsymbol{\beta}^k, \nu^k, \Sigma^k\right) = \frac{cK_{\frac{\nu^k+l_t}{2}}\left(\sqrt{(\nu^k+Q(\boldsymbol{\gamma}_t,\boldsymbol{m}^k))[\boldsymbol{\beta}^k]^T[\Sigma^k]^{-1}\boldsymbol{\beta}^k}\right)\exp\left(\boldsymbol{\gamma}_t-\boldsymbol{m}^k\right)^T\left[\Sigma^k\right]^{-1}\boldsymbol{\beta}^k}{\left(\sqrt{(\nu^k+Q(\boldsymbol{\gamma}_t,\boldsymbol{m}^k))[\boldsymbol{\beta}^k]^T[\Sigma^k]^{-1}\boldsymbol{\beta}^k}\right)^{-\frac{\nu^k+l_t}{2}}\left(1+\frac{Q(\boldsymbol{\gamma}_t,\boldsymbol{m}^k)}{\nu^k}\right)^{\frac{\nu^k+l_t}{2}}} \tag{15.6}$$

where $K_v(z)$ *is a modified Bessel function of the second kind given by:*

$$K_v(z) = \frac{1}{2}\int_0^\infty y^{v-1}e^{-\frac{z}{2}(y+y^{-1})}dy \tag{15.7}$$

and c is a normalisation constant. We also define the function $Q(\cdot,\cdot)$ as follows:

$$Q(\gamma_t, \boldsymbol{m}^k) = (\gamma_t - \boldsymbol{m}^k)^T \left[\Sigma^k\right]^{-1} (\gamma_t - \boldsymbol{m}^k) \tag{15.8}$$

This model admits skew-t marginals and a skew-t copula, see Smith et al. (2012) for details. Importantly, this stochastic model admits the following scale mixture representation:

$$\boldsymbol{\Gamma}_t^{LO,k} \stackrel{d}{=} \boldsymbol{m}^k + \boldsymbol{\beta}^k W + \sqrt{W}\boldsymbol{Z} \tag{15.9}$$

with inverse-gamma random variable $W \sim IGa\left(\frac{\nu^k}{2}, \frac{\nu^k}{2}\right)$ and independent Gaussian random vector $\boldsymbol{Z} \sim N\left(\boldsymbol{0}, \Sigma^k\right)$.

8. *Assume that for every element $N_t^{LO,k,s}$ of order counts from the random vector $\boldsymbol{N}_t^{LO,k}$, there is a corresponding random vector $\boldsymbol{O}_t^{LO,k,s} \in \mathbb{N}_+^{N_t^{LO,k,s}}$ of order sizes. We assume that the element $O_{i,t}^{LO,k,s}, i \in \left\{1,\ldots,N_t^{LO,k,s}\right\}$ is distributed as $O_{i,t}^{LO,k,s} \sim H(\cdot)$. Furthermore, we assume that order sizes are unconditionally independent $O_{i,t}^{LO,k,s} \perp\!\!\!\perp O_{i',t}^{LO,k,s}$ for $i \neq i'$, $s \neq s'$ and $t \neq t'$.*

Remark 15.1 *Under this proposed model for market maker liquidity activity, the number of limit orders placed by the liquidity providers in the market has an appropriate dynamic intensity structure that can evolve intra-daily to reflect the changing nature of liquidity provided by market makers throughout the trading day. In addition, the number of limit orders placed at each level of the bid and ask also allow for the model to capture the observed dependence structures in order placements in each level of the bid and ask regularly seen in empirical analysis of high-frequency LOB data. The dependence structure utilised is based on a skew-t copula which allows non-exchangeability of the stochastic intensity on the bid and ask at each level of the book, as well as asymmetry in the tail dependence features. This means that when large movements by market makers to replenish liquidity after a liquidity drought occurs intra-daily, such as after a large market order execution, the model can produce such replenishment on just the bid or just the ask depending on the situation. Under the skew-t copula structure used in this model, the liquidity provider agent does not automatically replenish both sides of the book equally likely, as would occur under a standard t-copula structure.*

We now define the second component of the liquidity provider agents, namely, the cancellation process. The cancellation process has the same stochastic process model specification as the aforementioned limit order submission process, including a skew-t dependence structure between the stochastic intensities at each LOB level on the bid and ask. We therefore only specify the differences unique to the cancellation process relative to the order placement model definition in the following specification, to avoid repetition.

Definition 15.2 (Limit order cancellation process for liquidity provider agent) *Consider the limit order cancellation process of the liquidity provider agent to have an identically specified stochastic model structure as the limit order submissions. The exception to this pertains to the assumption that the number of cancelled orders in each interval at each level is right-truncated at the total number of orders at that level.*

1. *As for submissions, we assume for cancellations a multi-variate path-space random matrix* $\boldsymbol{N}_{1:T}^{C,k} \in \mathbb{N}_{+}^{l_t \times T}$ *constructed from random vectors for the number of cancelled orders given by* $\boldsymbol{N}_{1:T}^{C,k} = \left(\boldsymbol{N}_1^{C,k}, \boldsymbol{N}_2^{C,k}, \dots, \boldsymbol{N}_T^{C,k}\right)$. *Furthermore, assume for these random vectors for the number of cancelled orders at each of the* l_t *levels, the latent stochastic process for the intensity is given by the random matrix* $\boldsymbol{\Lambda}_{1:T}^{C,k} \in \mathbb{R}_{+}^{l_t \times T}$ *and given on the path-space by* $\boldsymbol{\Lambda}_{1:T}^{C,k} = \left(\boldsymbol{\Lambda}_1^{C,k}, \boldsymbol{\Lambda}_2^{C,k}, \dots, \boldsymbol{\Lambda}_T^{C,k}\right)$.

2. *Assume that for the random vector* $\tilde{\boldsymbol{V}}_t^k$ *for the volume resting in the LOB after the placement of limit orders we have* $\tilde{\boldsymbol{V}}_t^k = \boldsymbol{V}_{t-1}^k + \boldsymbol{N}_t^{LO,k}$, *and that the random vector* $\boldsymbol{N}_t^{C,k} \in \mathbb{N}_{+}^{l_t}$ *is distributed according to a truncated multi-variate generalised Cox process with conditional distribution* $\boldsymbol{N}_t^{C,k} | \tilde{\boldsymbol{V}}_t^k = \underline{v} \sim \mathcal{GCP}\left(\boldsymbol{\lambda}_t^{C,k}\right) \mathbb{I}(\boldsymbol{N}_t^{C,k} < \underline{v})$ *(with* $\underline{v} = (v_{-l_d+1}, \dots, v_{l_p})$*) given by:*

$$\mathbb{P}r\left(N_t^{C,k,-l_d+1} = n_{-l_d+1}, \dots, N_t^{C,k,l_p} = n_{l_p} \,\middle|\, \boldsymbol{\Lambda}_t^{C,k} = \boldsymbol{\lambda}_t^{C,k}, \tilde{\boldsymbol{V}}_t^k = \underline{v}\right)$$
$$= \prod_{s=-l_d+1}^{l_p} \frac{\frac{(\lambda_t^{C,k,s})^{n_s}}{n_s!}}{\sum_{j=0}^{v_s} \frac{(\lambda_t^{C,k,s})^j}{j!}} \tag{15.10}$$

3. *Assume that for the cancellation count* $N_t^{C,k,s}$, *the orders with highest priority are cancelled from level s (which are also the oldest orders in their respective queue). Assume also that cancellations always remove an order in full, for example, there are no partial cancellations.*

Remark 15.2 *Cancellations are a critical part of a market makers ability to modulate and adjust their liquidity activity to avoid large losses in trades that would otherwise be executed under an adverse selection setting. Under this proposed model for market maker liquidity removal activity (cancellations), the number of limit orders cancelled by the liquidity providers in the market has an appropriate dynamic intensity structure that can evolve intra-daily to reflect the changing nature of liquidity demand throughout the trading day. In addition, the number of limit orders cancelled at each level of the bid and ask also allow for the model to capture the observed dependence structures in order* *cancellations at each level of the bid and ask. The dependence structure utilised is based on a skew-t copula which allows non-exchangeability of the stochastic intensity on the bid and ask at each level of the book, as well as asymmetry in the tail dependence features. This means that when large price movements occur in the LOB, market makers need to adjust their LOB volumes and profile by cancelling existing resting orders and creating new orders. This typically occurs many times throughout the trading day, and the ability to do this with an appropriate dependence structure is critical. In addition, the number of cancelled orders needs to preserve the principle of volume preservation, that is the upper bound on the total number of limit orders that may be cancelled at any given time is based on the instantaneous resting volume in the book at the given time instant.*

We complete the specification of the representative agents by considering the specification of the liquidity demander agent. In addition to market makers who are incentivised to place orders intra-daily in the limit order book, by exchanges in which they operate, there are also other market participants who trade for other reasons. These other market participants include hedge funds, pension funds, and other types of large investors, typically we refer to such groups of traders as liquidity demanders. They absorb liquidity throughout the day by purchasing resting orders in the limit order book. These purchases are most often made through market orders or aggressive limit orders. In this chapter, we assume that all such activities can be adequately modelled by a stochastic liquidity demander agent making dynamically evolving decisions to place market orders, as specified in the following.

Definition 15.3 (Market order submission process for liquidity demander agent) *Consider a representative agent for the liquidity providers to be composed of a* **market order** *component, which has the following stochastic structure:*

1. *Assume a path-space random vector $\mathbf{N}_{1:T}^{MO,k} \in \mathbb{N}_{+}^{1\times T}$ for the number of market orders constructed from the random variables for the number of market orders in each interval $\mathbf{N}_{1:T}^{MO,k} = \left(N_1^{MO,k}, N_2^{MO,k}, \ldots, N_T^{MO,k}\right)$. Furthermore, assume that for these random variables, the latent stochastic process for the intensity is given by random variable $\mathbf{\Lambda}_{1:T}^{MO,k} \in \mathbb{R}_{+}^{l_t \times T}$ and given on the path-space by $\mathbf{\Lambda}_{1:T}^{MO,k} = \left(\Lambda_1^{MO,k}, \Lambda_2^{MO,k}, \ldots, \Lambda_T^{MO,k}\right)$.*

2. *Assume the conditional independence property for the random variables:*

$$\left[N_s^{MO,k} | \Lambda_s^{MO,k}\right] \perp\!\!\!\perp \left[N_t^{MO,k} | \Lambda_t^{MO,k}\right], \quad \forall s \neq t, \quad s,t \in \{1,2,\ldots,T\} \tag{15.11}$$

3. *Assume that for the random variable $\tilde{R}_t^k$ for the volume resting on the opposite side of the LOB after the placement of limit orders and cancellations we have $\tilde{R}_t^k = \Sigma_{s=1}^{l_p}\left[\tilde{V}_{t-\Delta t}^{k',s} - N_t^{C,k',s}\right]$, where $k' = a$ if $k = b$, and*

vice-versa, and that the random variable $N_t^{MO,k} \in \mathbb{N}_+$ *is distributed according to a truncated generalised Cox process with conditional distribution* $N_t^{MO,k} | \tilde{R}_t^k = r \sim \mathcal{GCP}\left(\lambda_t^{MO,k}\right) \mathbb{1}(N_t^{MO,k} < r)$ *given by:*

$$\mathbb{P}r\left(N_t^{MO,k} = n \middle| \Lambda_t^{MO,k} = \lambda_t^{MO,k}, \tilde{R}_t^k = r\right) = \frac{\frac{(\lambda_t^{MO,k})^n}{n!}}{\sum_{j=0}^{r} \frac{(\lambda_t^{MO,k})^j}{j!}} \tag{15.12}$$

4. *Assume the independence property for random vectors of latent intensities unconditionally according to:*

$$\Lambda_s^{MO,k} \perp\!\!\!\perp \Lambda_t^{MO,k}, \quad \forall s \neq t, \quad s, t \in \{1, 2, \ldots, T\} \tag{15.13}$$

5. *Assume that for each intensity random variable* $\Lambda_t^{MO,k} \in \mathbb{R}_+$, *there is a corresponding transformed intensity variable* $\Gamma_t^{MO,k} \in \mathbb{R}$, *and the relationship for each element is given by the mapping:*

$$\Lambda_t^{MO,k} = \mu_0^{MO,k} F\left(\Gamma_t^{MO,k}\right) \tag{15.14}$$

for some baseline intensity parameter $\mu_0^{MO,k} \in \mathbb{R}_+$ *and strictly monotonic mapping* $F : \mathbb{R} \mapsto [0, 1]$.

6. *Assume that the random variables* $\Gamma_t^{MO,k} \in \mathbb{R}$, *characterising the intensity before transformation of the generalised Cox-process, are distributed in interval* $[t-1, t)$ *according to a uni-variate skew-t distribution* $\Gamma_t^{MO,k} \sim St(m_t^{MO,k}, \beta^{MO,k}, \nu^{MO,k}, \sigma^{MO,k})$.

7. *Assume that for every element* $N_t^{MO,k}$ *of market order counts, there is a corresponding random vector* $\boldsymbol{O}_t^{MO,k,s} \in \mathbb{N}_+^{N_t^{MO,k}}$ *of order sizes. We assume that the element* $O_{i,t}^{MO,k}, i \in \left\{1, \ldots, N_t^{MO,k}\right\}$ *is distributed according to* $O_{i,t}^{MO,k} \sim H(\cdot)$. *Assume also that market order sizes are unconditionally independent* $O_{i,t}^{MO,k} \perp\!\!\!\perp O_{i',t}^{MO,k}$ *for* $i \neq i'$ *or* $t \neq t'$.

We denote the LOB state for the real dataset at time t on a given day by the random vector $\boldsymbol{L}_t$, and this corresponds to the prices and volumes at each level of the bid and ask. Utilising the stochastic agent-based model specification described earlier, and given a parameter vector $\boldsymbol{\theta}$, which will generically represent all parameters of the liquidity providing and liquidity demanding agent types, one can then also generate simulations of intra-day LOB activity and arrive at the synthetic state $\boldsymbol{L}_t^*(\boldsymbol{\theta})$. The state of the simulated LOB at time t is obtained from the state at time $t-1$ and a set of stochastic components, denoted generically by $\boldsymbol{X}_t$, which are obtained from a single stochastic realisation of the following components of the agent-based models:

- Limit order submission intensities $\boldsymbol{\Lambda}_t^{LO,b}$, $\boldsymbol{\Lambda}_t^{LO,a}$, order numbers $\boldsymbol{N}_t^{LO,b}$, $\boldsymbol{N}_t^{LO,a}$, and order sizes $\boldsymbol{O}_{i,t}^{LO,a,s}, \boldsymbol{O}_{j,t}^{LO,b,s}$, where $s = -l_d + 1 \ldots l_p, i = 1 \ldots N_t^{LO,a,s}, j = 1 \ldots N_t^{LO,b,s}$.

- Limit order cancellation intensities $\boldsymbol{\Lambda}_t^{C,b}$, $\boldsymbol{\Lambda}_t^{C,a}$, and numbers of cancellations $\boldsymbol{N}_t^{C,b}$, $\boldsymbol{N}_t^{C,a}$.

- Market order intensities $\boldsymbol{\Lambda}_t^{MO,b}$, $\boldsymbol{\Lambda}_t^{MO,a}$, numbers of market orders $\boldsymbol{N}_t^{MO,b}$, $\boldsymbol{N}_t^{MO,a}, \boldsymbol{V}_t^{MO,b}, \boldsymbol{V}_t^{MO,a}$, and market order sizes $\boldsymbol{O}_{i,t}^{MO,a}, \boldsymbol{O}_{j,t}^{MO,b}, i = 1 \ldots N_t^{MO,a}, j = 1 \ldots N_t^{MO,b}$.

These stochastic features are combined with the previous state of the LOB, $\boldsymbol{L}_{t-1}^*(\boldsymbol{\theta})$, to produce the new state $L_t^*(\boldsymbol{\theta})$ for a given set of parameters $\boldsymbol{\theta}$, given by:

$$\boldsymbol{L}_t^*(\boldsymbol{\theta}) = G(\boldsymbol{L}_{t-1}^*(\boldsymbol{\theta}), \boldsymbol{X}_t) \tag{15.15}$$

$G(\cdot)$ is a transformation that maps the previous state of the LOB and the activity generated in the current step into a new step, much the same way as the matching engine updates the LOB after every event. As we model the activity in discrete intervals, however, the LOB is only updated at the end of every interval, and the incoming events (limit orders, market orders, and cancellations) are processed in the order specified in Section 15.2.1. Conditional then on a realisation of these parameters $\boldsymbol{\theta}$, the trading activity in the LOB can be simulated for a single trading day, and the complete procedure is described in the algorithm set out in Panayi and Peters (2015).

15.2.2 Bayesian model formulation of the stochastic agent limit order book model representation

In this section, we consider the class of LOB stochastic models developed in the previous section, and we detail a Bayesian model formulation under an ABC framework. Methods for Bayesian modelling in the presence of computationally intractable likelihood functions are of growing interest. These methods may arise either because the likelihood is truly intractable to evaluate pointwise, or, in our case, it may be that the likelihood is so complex in terms of model specification and costly to evaluate pointwise, that one has to resort to alternative methods to perform estimation and inference. Termed *likelihood-free samplers* or ABC methods, simulation algorithms such as Sequential Monte Carlo Samplers have been adapted for this setting, see, for instance, Peters et al. (2012a).

We start by re-calling a few basics. Typically, Bayesian inference proceeds via the posterior distribution, generically denoted by $\pi(\boldsymbol{\theta}|\boldsymbol{y}) \propto f(\boldsymbol{y}|\boldsymbol{\theta})\pi(\boldsymbol{\theta})$, the updating of prior information $\pi(\boldsymbol{\theta})$ for a parameter $\boldsymbol{\theta} \in E$ through the likelihood function $f(\boldsymbol{y}|\boldsymbol{\theta})$ after observing data $\boldsymbol{y} \in \mathcal{Y}$. Numerical algorithms, such as importance sampling, Markov chain Monte Carlo (MCMC) and SMC, are commonly employed to draw samples from the posterior $\pi(\boldsymbol{\theta}|\boldsymbol{y})$.

Remark 15.3 (Note on data vector) *In the context of this chapter the data $\boldsymbol{y}$ are the entire order book structure for a given asset over a day as summarised by:*

- *Limit order submission order numbers $\boldsymbol{N}_t^{LO,b}$, $\boldsymbol{N}_t^{LO,a}$, and order sizes $\boldsymbol{O}_{i,t}^{LO,a,s}, \boldsymbol{O}_{j,t}^{LO,b,s}$, where $s = -l_d + 1 \ldots l_p, i = 1 \ldots N_t^{LO,a,s}, j = 1 \ldots N_t^{LO,b,s}$.*
- *Limit order numbers of cancellations $\boldsymbol{N}_t^{C,b}$, $\boldsymbol{N}_t^{C,a}$.*
- *Numbers of market orders $\boldsymbol{N}_t^{MO,b}$, $\boldsymbol{N}_t^{MO,a}, \boldsymbol{V}_t^{MO,b}, \boldsymbol{V}_t^{MO,a}$, and market order sizes $\boldsymbol{O}_{i,t}^{MO,a}, \boldsymbol{O}_{j,t}^{MO,b}, i = 1 \ldots N_t^{MO,a}, j = 1 \ldots N_t^{MO,b}$.*

The resulting observation vector $\boldsymbol{y}_t$, at time t, is then the concatenation of all these variables. These stochastic features are obtained at sampling rate t within the market hours of the trading day, typically say every 5–30 seconds for the 8.5 hour trading day, producing a total of between 1,000–6,000 vector valued observations per day.

Clearly, even evaluating a likelihood on this many records, even if it could be written down, which in many LOB models built on queues like the one in this chapter will not be the case, would still be a challenging task. Generically, we will denote in the following the collection of all observations $\boldsymbol{y}$ of the LOB for an asset on a given day, and by $\boldsymbol{\theta}$, the set of all parameters that are utilised to parameterise the LOB stochastic model.

There is growing interest in posterior simulation in situations where the likelihood function is computationally intractable, for example, $f(\boldsymbol{y}|\boldsymbol{\theta})$ may not be numerically evaluated pointwise. As a result, sampling algorithms based on repeated likelihood evaluations require modification for this task. They employ generation of auxiliary datasets under the model as a means to circumvent (intractable) likelihood evaluation.

15.2.2.1 Posterior models for computationally intractable likelihoods

In essence, likelihood-free methods first reduce the observed data, $\boldsymbol{y}$, to a low-dimensional vector of summary statistics $t_y = T(\boldsymbol{y}) \in \mathcal{T}$, where $\dim(\boldsymbol{\theta}) \leq \dim(t_y) << \dim(\boldsymbol{y})$. Accordingly, the true posterior $\pi(\boldsymbol{\theta}|\boldsymbol{y})$ is replaced with a new posterior $\pi(\boldsymbol{\theta}|t_y)$. These are equivalent if t_y is sufficient for $\boldsymbol{\theta}$, and $\pi(\boldsymbol{\theta}|t_y) \approx \pi(\boldsymbol{\theta}|\boldsymbol{y})$ is an approximation, if there is some loss of information through t_y. The new target posterior, $\pi(\boldsymbol{\theta}|t_y)$, still assumed to be computationally intractable, is then embedded within an augmented model from which sampling is viable. Specifically, the joint posterior of the model parameters $\boldsymbol{\theta}$ and auxiliary data $t \in \mathcal{T}$ given observed data t_y are:

$$\pi(\boldsymbol{\theta}, t|t_y) \propto K_\epsilon(t_y - t) f(t|\theta)\pi(\theta) \tag{15.16}$$

where $t \sim f(t|\boldsymbol{\theta})$ may be interpreted as the vector of summary statistics $t = T(x)$ computed from a dataset simulated according to the model $\boldsymbol{x} \sim f(\boldsymbol{x}|\boldsymbol{\theta})$. Assuming such simulation is possible, data-generation under the model, $t \sim f(t|\boldsymbol{\theta})$, forms the basis of computation in the likelihood-free setting. The target marginal posterior $\pi_M(\boldsymbol{\theta}|t_y)$ for the parameters $\boldsymbol{\theta}$ is then obtained as:

$$\pi_M(\boldsymbol{\theta}|t_y) = c_M \int_{\mathcal{T}} K_\epsilon(t_y - t) f(t|\boldsymbol{\theta})\pi(\boldsymbol{\theta})dt \tag{15.17}$$

where $(c_M)^{-1} = \int_E \int_{\mathcal{T}} K_\epsilon(t_y - t) f(t|\boldsymbol{\theta})\pi(\boldsymbol{\theta})dtd\boldsymbol{\theta}$ normalises (15.17), such that it is a density in $\boldsymbol{\theta}$ (Reeves and Pettitt, 2005; Sisson et al., 2007; Blum, 2010; **?**; Wilkinson, 2013). The function $K_\epsilon(t_y - t)$ is a standard kernel function, with scale parameter $\epsilon \geq 0$, which weights the intractable posterior with high density in regions $t \approx t_y$, where auxiliary and observed datasets are similar. As such, $\pi_M(\boldsymbol{\theta}|t_y) \approx \pi(\boldsymbol{\theta}|t_y)$ forms an approximation to the intractable posterior via (15.17) through standard smoothing arguments (Blum, 2010). In the case as $\epsilon \to 0$, so that $K_\epsilon(t_y - t)$ becomes a point mass at the origin (i.e. $t_y = t$) and is zero elsewhere, if t_y is sufficient for θ, then the intractable posterior marginal $\pi_M(\boldsymbol{\theta}|t_y) = \pi(\boldsymbol{\theta}|t_y) = \pi(\boldsymbol{\theta}|\boldsymbol{y})$ is recovered exactly (although small h is usually impractical). Various choices of smoothing kernel K have been examined in the literature (Beaumont et al., 2002; Marjoram et al., 2003; Peters et al., 2012a).

For our discussion on likelihood-free or ABC samplers, it is convenient to consider a generalisation of the joint distribution (15.16) incorporating $S \geq 1$ auxiliary summary vectors:

$$\pi_J(\boldsymbol{\theta}, t^{1:S}|t_y) \propto \tilde{K}_\epsilon(t_y, t^{1:S}) f(t^{1:S}|\boldsymbol{\theta})\pi(\boldsymbol{\theta})$$

where $t^{1:S} = (t^1, \ldots, t^S)$ and $t^1, \ldots, t^S \sim f(t|\boldsymbol{\theta})$ are S independent datasets generated from the (intractable) model. As the auxiliary datasets are, by construction, conditionally independent given $\boldsymbol{\theta}$, we have $f(t^{1:S}|\boldsymbol{\theta}) = \prod_{s=1}^S f(t^s|\boldsymbol{\theta})$. We follow Del Moral et al. (2012) and specify the kernel $\tilde{K}$ as $\tilde{K}_\epsilon(t_y, t^{1:S}) = S^{-1}\sum_{s=1}^S K_\epsilon(t_y - t^s)$, which produces the joint posterior:

$$\pi_J(\boldsymbol{\theta}, t^{1:S}|t_y) = c_J \left[\frac{1}{S}\sum_{s=1}^S K_\epsilon(t_y - t^s)\right]\left[\prod_{s=1}^S f(t^s|\boldsymbol{\theta})\right]\pi(\boldsymbol{\theta}) \tag{15.18}$$

with $c_J > 0$ the appropriate normalisation constant, where in (15.18) we extend the uniform kernel choice of $K(t_y - t^s)$ by Del Moral et al. (2012) to the general case. It is easy to see that, by construction, $\int_{\mathcal{T}^S} \pi_J(\boldsymbol{\theta}, t^{1:S}|t_y)dt^{1:S} = \pi_M(\boldsymbol{\theta}|t_y)$ admits the distribution (15.17) as a marginal distribution (cf. Del Moral et al., 2012). The case $S = 1$ with $\pi_J(\theta, t^{1:S}|t_y) = \pi(\theta, t|t_y)$ corresponds to the more usual joint posterior (15.16) in the likelihood-free setting. We note that recent studies, such as that by Bornn et al. (2014), have demonstrated that it may not be beneficial to sample multiple data replications at each instance of time, for example, in the case of importance sampling and MCMC, it has been found to be better to select $S = 1$, even though the individual representation has a higher variance approximation, the overall computational efficiency and accuracy of the ABC method is improved in such a setting. This may not be true for SMC samplers and remains to be studied further.

There are two obvious approaches to posterior simulation from $\pi_M(\boldsymbol{\theta}|t_y) \approx \pi(\boldsymbol{\theta}|t_y)$ as an approximation to $\pi(\boldsymbol{\theta}|\boldsymbol{y})$. The first approach proceeds by sampling directly on the augmented model $\pi_J(\boldsymbol{\theta}, t^{1:S}|t_y)$, realising joint samples $(\boldsymbol{\theta}, t^{1:S}) \in E \times \mathcal{T}^S$ before *a posteriori* marginalisation over $t^{1:S}$ (i.e. by discarding the t^s realisations from the sampler output). In this approach, the summary quantities $t^{1:S}$ are treated as parameters in the augmented model.

The second approach is to sample from $\pi_M(\boldsymbol{\theta}|t_y)$ directly, a lower-dimensional space, by approximating the integral (15.17) via Monte Carlo integration in lieu of each posterior evaluation of $\pi_M(\boldsymbol{\theta}|t_y)$. In this case:

$$\pi_M(\boldsymbol{\theta}|t_y) \propto \pi(\boldsymbol{\theta}) \int_{\mathcal{T}} K_\epsilon(t_y - t) f(t|\boldsymbol{\theta}) dt \approx \frac{\pi(\boldsymbol{\theta})}{S} \sum_{s=1}^{S} K_\epsilon(t_y - t^s) := \hat{\pi}_M(\boldsymbol{\theta}|t_y) \tag{15.19}$$

where $t^1, \ldots, t^S \sim f(t|\boldsymbol{\theta})$. This expression, examined by various authors (Marjoram et al., 2003; Reeves and Pettitt, 2005; Ratmann et al., 2009; Toni et al., 2009; Peters et al., 2012a), requires multiple generated datasets $t^1, \ldots, t^S$ for each evaluation of the marginal posterior distribution $\pi_M(\boldsymbol{\theta}|t_y)$. As with standard Monte Carlo approximations, $\text{Var}[\hat{\pi}_M(\boldsymbol{\theta}|t_y)]$ reduces as S increases, with $\lim_{S\to\infty} \text{Var}[\hat{\pi}_M(\boldsymbol{\theta}|t_y)] = 0$. For the marginal posterior distribution, the quantities $t^{1:S}$ serve only as a means to estimate $\pi_M(\boldsymbol{\theta}|t_y)$ and do not otherwise enter the model explicitly. The number of samples S directly impacts on the variance of the estimation.

15.2.3 Summary statistics in Bayesian limit order book models

In terms of ABC methods, it is important to consider carefully the choice of summary statistics, and how they should be designed to enter into the ABC posterior model. Ideally, we would seek sufficient statistics for the stochastic model, however, in many realistic practical settings where ABC is particularly useful, this may be difficult to achieve, see discussions in Nunes and Balding (2010). Other alternatives are therefore typically adopted, for instance, in some cases the model is sufficiently well specified that specific summary statistics may arise as sensible model-based choices from an interpretation perspective, such as those discussed in financial-and insurance-based examples in Peters et al. (2012b) and Peters et al. (2010). In other cases, one may adopt automatic summary statistics procedures, such as those described in Fearnhead and Prangle (2010) and the literature therein. In Blum and François (2010) and Fan et al. (2013) regression-based approaches to summary statistics are utilised, and, in Blum et al. (2013), discussions on dimension reduction approaches are considered. In this chapter, we have decided to utilise model-based summary statistics which are obtained via an application specific dimension reduction approach. This was decided due to its practicality for interpretation of practitioners and to make results obtained readily comparable to the multi-objective indirect inference procedures studied on these LOB models in Panayi and Peters (2015).

We note that the liquidity provision stochastic representative agent model for the LOB is a complex model structure which will not readily admit sufficient statistics for applications in ABC. It is therefore important to consider carefully what would be appropriate choices for summary statistics in the model. In this chapter, we explore the notion of model-based summary statistics based around first a dimension reduction of the LOB stochastic process to a subset of important features representative of key attributes of the LOB process from the practitioners' and regulators' perspective. In this context, the idea is to take the LOB model stochastic process structure and transform this stochastic process with multiple components to summary processes throughout the trading day. It is suggested that in practice these may correspond to attributes of interest, such as:

- Volume-based processes, total volume on bid, total volume on ask, individual volumes on each level of the book for the ask, and bid etc.
- Fair price, for example, mid price induced by LOB stochastic process, returns process for mid price.
- Round-trip cost-based prices, such as XLM measures of liquidity, as well as liquidity resilience measures, such as those discussed in detail in Panayi et al. (2015b).

Having obtained these dimension reductions of the complex LOB stochast process, one can then construct summary statistics based on a model fit either regression, kernel density, or, in our case, time series-based regression models. We provide detailed analysis of our choices in this regard in the results section.

15.3 Estimation of Bayesian Limit Order Book Stochastic Agent Model via Population-Based Samplers

Population-based likelihood-free samplers were introduced to circumvent poor mixing in MCMC samplers (Sisson et al., 2007; Beaumont et al., 2009; Toni et al., 2009; Del Moral et al., 2012; Peters et al., 2012a). These samplers propagate a population of *particles*, $\boldsymbol{\theta}^{(1)}, \ldots, \boldsymbol{\theta}^{(N)}$, with associated importance weights $W(\boldsymbol{\theta}^{(i)})$, through a sequence of related densities $\phi_1(\boldsymbol{\theta}_1), \ldots, \phi_T(\boldsymbol{\theta}_T)$, which defines a smooth transition from the distribution ϕ_1, from which direct sampling is available, to ϕ_T the target distribution.

For likelihood-free or ABC samplers, ϕ_k is defined by allowing $K_{\epsilon_n}(t_y - t)$ to place greater density on regions for which $t_y \approx t$ as k increases (that is, the bandwidth ϵ_n decreases with n). Hence, we denote $\pi_{J,n}(\boldsymbol{\theta}, t^{1:S}|t_y) \propto \tilde{K}_{\epsilon_n}(t_y, t^{1:S}) f(t^{1:S}|\boldsymbol{\theta})\pi(\boldsymbol{\theta})$ and $\pi_{M,n}(\boldsymbol{\theta}|t_y) \propto \pi(\boldsymbol{\theta}) \int_{\mathcal{T}^S} \tilde{K}_{\epsilon_n}(t_y, t^{1:S}) f(t^{1:S}|\boldsymbol{\theta}) dt^{1:S}$ for $n = 1, \ldots, T$, under the joint and marginal posterior models, respectively. Since it will be assumed that a significant description of such methodologies is provided in other chapters in this book, we only briefly discuss this family of sampling methods in the following, instead focusing on our applications of such methods in this chapter.

15.3.1 Brief overview of sequential Monte Carlo-based samplers

SMC methods emerged out of the fields of engineering, probability, and statistics in recent years. Variants of the methods sometimes appear under the names of particle filtering or interacting particle systems (Doucet et al., 2001; Del Moral, 2004; Ristic et al., 2004), and their theoretical properties have been extensively studied by Crisan and Doucet (2002), Del Moral (2004), Chopin (2004), and Künsch (2005). In the last few years, Chopin (2002), Neal (2001), Del Moral et al. (2006), Peters (2005), Peters et al. (2012a), and Targino et al. (2015), amongst others, have developed generalisations to the SMC algorithm to the case where the target distributions π_n are all defined on the same support E, for example, no longer a product space formulation. This generalisation, termed the 'SMC *sampler*', adapts the SMC algorithm to the more popular setting in which the state space E remains static, typically arising in applications of MCMC algorithms.

Analogously with standard SMC algorithms, the SMC sampler is developed to generate weighted samples (termed *particles*) from a sequence of distributions π_n, for $n = 1, \ldots, T$, where π_T may be of particular interest. We refer to π_T as the target distribution, such as a posterior distribution for model parameters. Hence, given a sequence of distributions $\{\pi_n(d\boldsymbol{\theta})\}_{n=1}^{T}$, the aim is to develop a large collection of N-weighted random samples at each time n denoted by $\left\{W_n^{(i)}, \boldsymbol{\Theta}_n^{(i)}\right\}_{i=1}^{N}$, such that $W_n^{(i)} > 0$ and $\sum_{i=1}^{N} W_n^{(i)} = 1$. These importance weights and samples, denoted by $\left\{W_n^{(i)}, \boldsymbol{\Theta}_n^{(i)}\right\}_{i=1}^{N}$, are known as particles (hence, the name often given to such algorithms as particle filters or interacting particle systems). For such approaches to be sensible, we would require that the empirical distributions constructed through these samples should converge asymptotically ($N \to \infty$) to the target distribution π_n for each time n. This means that for any π_n integrable function, denoted, for example, by $\phi(\boldsymbol{\theta}) : E \to \mathbb{R}'$, one would have the following convergence:

$$\sum_{i=1}^{N} W_n^{(i)} \phi\left(\boldsymbol{\theta}_n^{(i)}\right) \xrightarrow{a.s.} \mathbb{E}_{\pi_n}\left[\phi(\boldsymbol{\Theta})\right] \tag{15.20}$$

In the SMC sampler algorithm is constructed by introducing a sequence of backward kernels L_k, to obtain new distributions:

$$\widetilde{\pi}_n(\boldsymbol{\theta}_1, \ldots, \boldsymbol{\theta}_n) = \pi_n(\boldsymbol{\theta}_n) \prod_{k=1}^{n-1} L_k\left(\boldsymbol{\theta}_{k+1}, \boldsymbol{\theta}_k\right) \tag{15.21}$$

may be defined for the *path* of a particle $(\boldsymbol{\theta}_1, \ldots, \boldsymbol{\theta}_n) \in E^n$ through the sequence $\pi_1, \ldots, \pi_n$. The only restriction on the backward kernels is that the correct marginal distributions $\int \widetilde{\pi}_n(\boldsymbol{\theta}_1, \ldots, \boldsymbol{\theta}_n) d\boldsymbol{\theta}_1, \ldots, d\boldsymbol{\theta}_{n-1} = \pi_n(\boldsymbol{\theta}_n)$ are available. Within this framework, one may then work with the constructed sequence of distributions, $\widetilde{\pi}_n$, under the standard SMC algorithm.

In summary, the SMC Sampler algorithm involves three stages:

1. *Mutation*, whereby the particles are moved from $\boldsymbol{\theta}_{n-1}$ to $\boldsymbol{\theta}_n$ via a mutation kernel $M_n(\boldsymbol{\theta}_{n-1}, \boldsymbol{\theta}_n)$.

2. *Correction*, where the particles are re-weighted with respect to π_n via the incremental importance weight (Equation 15.22).

3. *Selection*, where according to some measure of particle diversity, commonly the effective sample size, the weighted particles may be re-sampled in order to reduce the variability of the importance weights.

In more detail, suppose that at time $n-1$, the distribution $\widetilde{\pi}_{n-1}$ can be approximated empirically by $\widetilde{\pi}_{n-1}^N$ using N-weighted particles. These particles are first propagated to the next distribution $\widetilde{\pi}_n$ using a mutation kernel $M_n(\boldsymbol{\theta}_{n-1}, \boldsymbol{\theta}_n)$, and then assigned new weights $W_n = W_{n-1} w_n\left(\boldsymbol{\theta}_1, \ldots \boldsymbol{\theta}_n\right)$, where W_{n-1} is the weight of a particle at time $n-1$, and w_n is the incremental importance weight given by:

$$\begin{aligned} w_n\left(\boldsymbol{\theta}_1, \ldots, \boldsymbol{\theta}_n\right) &= \frac{\widetilde{\pi}_n\left(\boldsymbol{\theta}_1, \ldots, \boldsymbol{\theta}_n\right)}{\widetilde{\pi}_{n-1}\left(\boldsymbol{\theta}_1, \ldots, \boldsymbol{\theta}_{n-1}\right) M_n\left(\boldsymbol{\theta}_{n-1}, \boldsymbol{\theta}_n\right)} \\ &= \frac{\pi_n\left(\boldsymbol{\theta}_n\right) L_{n-1}\left(\boldsymbol{\theta}_n, \boldsymbol{\theta}_{n-1}\right)}{\pi_{n-1}\left(\boldsymbol{\theta}_{n-1}\right) M_n\left(\boldsymbol{\theta}_{n-1}, \boldsymbol{\theta}_n\right)} \end{aligned} \tag{15.22}$$

The resulting particles are now weighted samples from $\widetilde{\pi}_n$. Consequently, from Equation (15.22), under the SMC sampler framework, one may work directly with the marginal distributions $\pi_n(\boldsymbol{\theta}_n)$, such that $w_n(\boldsymbol{\theta}_1, \ldots, \boldsymbol{\theta}_n) = w_n(\boldsymbol{\theta}_{n-1}, \boldsymbol{\theta}_n)$. While the choice of the backward kernels L_{n-1} is essentially arbitrary, their specification can strongly affect the performance of the algorithm, as will be discussed in the following sub-sections. The basic version of the SMC sampler algorithm therefore proceeds explicitly as given in Algorithm 15.1.

Algorithm 15.1: Sequential Monte Carlo samplers

1. Initialise the particle system;
   (a) Set $n = 1$.
   (b) For $i = 1, \ldots, N$, draw initial particles $\boldsymbol{\Theta}_1^{(i)} \sim p(\boldsymbol{\theta})$.
   (c) Evaluate incremental importance weights $\left\{w_1\left(\boldsymbol{\Theta}_1^{(i)}\right)\right\}$ using Equation (15.22) and normalise the weights to obtain $\left\{W_1^{(i)}\right\}$.

   Iterate the following steps through each distribution in sequence $\{\pi_t\}_{n=2}^{T}$.
2. Re-sampling;
   (a) If the effective sampling size $(ESS) = \frac{1}{\sum_{i=1}^{N}\left(w_n^{(i)}\right)^2} < N_{eff}$ is less than a threshold N_{eff}, then re-sample the particles via the empirical distribution of the weighted sample either by multi-nomial or stratified methods; see discussions on unbiased re-sampling schemes by Künsch (2005) and Del Moral (2004).
3. Mutation and correction;
   (a) Set $n = n + 1$, if $n = T + 1$, then stop.
   (b) For $i = 1, \ldots, N$ draw samples from mutation kernel $\boldsymbol{\Theta}_n^{(i)} \sim M_n\left(\boldsymbol{\Theta}_{n-1}^{(i)}\right)$.
   (c) Evaluate incremental importance weights $\left\{w_n\left(\boldsymbol{\Theta}_n^{(i)}\right)\right\}$ using Equation (15.22) and normalise the weights to obtain $\left\{W_n^{(i)}\right\}$ via

$$W_n^{(i)} = W_{n-1}^{(i)} \frac{w_n^{(i)}\left(\boldsymbol{\Theta}_{n-1}, \boldsymbol{\Theta}_n\right)}{\sum_{j=1}^{N} W_{n-1}^{(i)} w_n^{(i)}\left(\boldsymbol{\Theta}_{n-1}, \boldsymbol{\Theta}_n\right)} \tag{15.23}$$

15.3.2 Sequential Monte Carlo samplers for intractable likelihood Bayesian models

Formalising this in the context of SMC samplers for ABC posterior settings, the particle population $\boldsymbol{\theta}_{n-1}$ drawn from the distribution $\phi_{n-1}(\boldsymbol{\theta}_{n-1})$ at time $n-1$ is mutated to $\phi_n(\boldsymbol{\theta}_n)$ by the kernel $M_n(\boldsymbol{\theta}_{n-1}, \boldsymbol{\theta}_n)$. The weights for the mutated particles θ_n may be obtained as $W_n(\boldsymbol{\theta}_n) = W_{n-1}(\boldsymbol{\theta}_{n-1}) w_n\left(\boldsymbol{\theta}_{n-1}, \boldsymbol{\theta}_n\right)$, where, for the marginal model sequence $\pi_{M,n}(\boldsymbol{\theta}_n | t_y)$, the incremental weight is:

$$w_n\left(\boldsymbol{\theta}_{n-1},\boldsymbol{\theta}_n\right)=\frac{\pi_{M,n}(\boldsymbol{\theta}_n|t_y)L_{n-1}\left(\boldsymbol{\theta}_n,\boldsymbol{\theta}_{n-1}\right)}{\pi_{M,n-1}(\boldsymbol{\theta}_{n-1}|t_y)M_n\left(\boldsymbol{\theta}_{n-1},\boldsymbol{\theta}_n\right)}\approx\frac{\hat{\pi}_{M,n}(\boldsymbol{\theta}_n|t_y)L_{n-1}\left(\boldsymbol{\theta}_n,\boldsymbol{\theta}_{n-1}\right)}{\hat{\pi}_{M,n-1}(\boldsymbol{\theta}_{n-1}|t_y)M_n\left(\boldsymbol{\theta}_{n-1},\boldsymbol{\theta}_n\right)} \tag{15.24}$$

where, following (15.19), and setting the kernel bandwidth to an ABC tolerance level ϵ_n, we obtain

$$\hat{\pi}_{M,n}(\boldsymbol{\theta}_n|t_y):=\frac{\pi(\boldsymbol{\theta})}{S}\sum_{s=1}^{S}K_{\epsilon_n}(t_y-t^s)$$

which is proportional to an (unbiased) estimate of $\pi_{M,n}(\boldsymbol{\theta}_n|t_y)$ based on S Monte Carlo draws $t^1,\ldots,t^S\sim f(t|\boldsymbol{\theta}_n)$. Here, $L_{n-1}\left(\boldsymbol{\theta}_n,\boldsymbol{\theta}_{n-1}\right)$ is a reverse-time kernel describing the mutation of particles from $\phi_n(\boldsymbol{\theta}_n)$ at time n to $\phi_{n-1}(\boldsymbol{\theta}_{n-1})$ at time $n-1$. As with the ABC-MCMC algorithm, the incremental weight (15.24) consists of the 'biased' ratio $\hat{\pi}_{M,n}(\boldsymbol{\theta}_n|t_y)/\hat{\pi}_{n-1}(\boldsymbol{\theta}_{M,n-1}|t_y)$ for finite $S\geq 1$.

If we now consider a sequential Monte Carlo sampler under the joint model $\pi_{J,n}(\boldsymbol{\theta},t^{1:S}|t_y)$, with the natural mutation kernel factorisation:

$$M_n[(\boldsymbol{\theta}_{n-1},t_{n-1}^{1:S}),(\boldsymbol{\theta}_n,t_n^{1:S})]=M_n(\boldsymbol{\theta}_{n-1},\boldsymbol{\theta}_n)\prod_{s=1}^{S}f(t_n^s|t_y)$$

(and similarly for L_{n-1}), following the form of (15.24), the incremental weight is exactly:

$$w_n\left[(\boldsymbol{\theta}_{n-1},t_{n-1}^{1:S}),(\boldsymbol{\theta}_n,t_n^{1:S})\right]=\frac{\frac{1}{S}\sum_s K_{\epsilon_n}(t_y-t_n^s)\pi(\boldsymbol{\theta}_n)L_{n-1}\left(\boldsymbol{\theta}_n,\boldsymbol{\theta}_{n-1}\right)}{\frac{1}{S}\sum_s K_{\epsilon_{n-1}}(t_y-t_{n-1}^s)\pi(\boldsymbol{\theta}_{n-1})M_n\left(\boldsymbol{\theta}_{n-1},\boldsymbol{\theta}_n\right)} \tag{15.25}$$

Hence, as the incremental weights (15.24, 15.25) are equivalent, they induce identical SMC algorithms for both marginal and joint models $\pi_M(\boldsymbol{\theta}|t_y)$ and $\pi_J(\boldsymbol{\theta},t^{1:S}|t_y)$. As a result, while applications of the marginal sampler targeting $\pi_M(\boldsymbol{\theta}|y)$ are theoretically biased for finite $S\geq 1$, as before, they are in practice unbiased through association with the equivalent sampler on joint space targeting $\pi_J(\boldsymbol{\theta},t^{1:S}|t_y)$.

We note that a theoretically unbiased sampler targeting $\pi_M(\boldsymbol{\theta}|t_y)$, for all $S\geq 1$, can be obtained by careful choice of the kernel $L_{n-1}(\boldsymbol{\theta}_n,\boldsymbol{\theta}_{n-1})$. For example, Peters (2005), Peters et al. (2012a), and Targino et al. (2015) all use the sub-optimal approximate optimal kernel given by:

$$L_{n-1}(\boldsymbol{\theta}_n,\boldsymbol{\theta}_{n-1})=\frac{\pi_{M,n-1}(\boldsymbol{\theta}_{n-1}|t_y)M_n(\boldsymbol{\theta}_{n-1},\boldsymbol{\theta}_n)}{\int\pi_{M,n-1}(\boldsymbol{\theta}_{n-1}|t_y)M_n(\boldsymbol{\theta}_{n-1},\boldsymbol{\theta}_n)d\boldsymbol{\theta}_{n-1}} \tag{15.26}$$

from which the incremental weight (15.24) is approximated by:

$$\begin{aligned}w_n(\boldsymbol{\theta}_{n-1},\boldsymbol{\theta}_n) &= \pi_{M,n}(\boldsymbol{\theta}_n|t_y)\Big/\int\pi_{M,n-1}(\boldsymbol{\theta}_{n-1}|t_y)M_n(\boldsymbol{\theta}_{n-1},\boldsymbol{\theta}_n)d\boldsymbol{\theta}_{n-1}\\ &\approx \hat{\pi}_{M,n}(\boldsymbol{\theta}_n|t_y)\Big/\sum_{i=1}^{N}W_{n-1}(\boldsymbol{\theta}_{n-1}^{(i)})M_n(\boldsymbol{\theta}_{n-1}^{(i)},\boldsymbol{\theta}_n) \qquad (15.27)\end{aligned}$$

Under this choice of backward kernel, the weight calculation is now unbiased for all $S \geq 1$, since the approximation $\hat{\pi}_{M,n-1}(\boldsymbol{\theta}|y)$ in the denominator of (15.24) is no longer needed.

15.3.2.1 Adaptive schedules: choice of sequence of ABC distributions via annealed tolerance schedule

In this section, we consider how to take the ABC posterior distribution and construct the sequence of distributions that are required for the SMC sampler. That is, we address the question of how to develop an ABC-specific sequence of target distributions. We have chosen to design this sequence by following what we call 'ABC reverse annealing'. In particular, we construct a sequence of target posterior distributions $\{\phi_n\}_{n\geq 0}$, which are constructed based on strictly decreasing tolerance values, generically denoted by the sequence $\{\epsilon_n\}_{n\geq 0}$, such that $\epsilon_1 > \epsilon_2 > \ldots > \epsilon_n > \ldots > \epsilon_T$. We obtain this sequence of ABC posterior distributions by considering the ϕ_n, which was defined with respect to the ABC likelihood involving a kernel. If we consider the kernel to have a decreasing bandwidth given by $K_{\epsilon_n}(t_y - t)$, then we will progressively place greater emphasis, for example, density on regions for which $t_y \approx t$ as n increases (that is, the bandwidth ϵ_n decreases with n). Hence, we denote the two possible ABC constructions one may consider under the joint and marginal posterior models respectively, in the SMC Samplers procedure as follows:

$$
\begin{aligned}
\pi_{J,n}(\boldsymbol{\theta}, t^{1:S}|t_y) &\propto \tilde{K}_{\epsilon_n}(t_y, t^{1:S}) f(t^{1:S}|\boldsymbol{\theta})\pi(\boldsymbol{\theta}) \\
\pi_{M,n}(\boldsymbol{\theta}|t_y) &\propto \pi(\boldsymbol{\theta}) \int_{\mathcal{T}^S} \tilde{K}_{\epsilon_n}(t_y, t^{1:S}) f(t^{1:S}|\boldsymbol{\theta}) dt^{1:S}
\end{aligned}
\tag{15.28}
$$

Now the aspect of this procedure that makes it adaptive is the selection of the size of the discrepancy between π_n and π_{n+1}, for each $n \in \{1, 2 \ldots, T\}$, as well as the final stopping point. In this paper, we propose to perform adaption of the ABC target distribution sequence at every step. The aim is to progressively select a sequence of distributions online, such that the discrepancy between the next distribution and the previous, as controlled by the tolerance ϵ_n sequence, is controlled by the 'fitness' or efficiency of the particle approximation of the previous target distribution in the sequence. A good approximation would indicate that one may take a larger step, whilst a poorer approximation indicates that smaller steps should be taken.

Formally, we perform the adaption such that a new tolerance ϵ_n, at iteration n, is generated through a particle system-based quantile matching procedure. The procedure adopted considers the new tolerance to be obtained as the solution for ϵ in the following equation:

$$
\widehat{q}_{n-1} = \epsilon\sqrt{2}\text{erf}^{-1}\left[2F\left(q\right) - 1\right] \tag{15.29}
$$

where q is a user specified quantile level, F is the CDF of a normal distribution with mean 0, and standard deviation ϵ and $\widehat{q}_{n-1}$ is the particle population quantile estimate obtained from the ABC posterior approximation after correction stage in the SMC sampler ABC algorithm. In this way, the tolerance schedule is continually adapting to the local particle approximation performance. In practice, it is computationally efficient to employ the following adaptive schedule for the tolerance in the ABC posterior, where we ensure that the sequence of distributions is designed such that the new tolerance is calculated as a strictly decreasing schedule given by:

$$\epsilon_n = \min\left((1-\alpha)\epsilon_{n-1}, \epsilon^*\right) \tag{15.30}$$

where $\alpha \in (0,1)$.

15.3.2.2 Choice of mutation kernel

There are many choices for mutation kernel that could be considered when designing an SMC sampler algorithm, see discussions on such choices in, for instance, references such as Peters (2005), Del Moral et al. (2006), Peters et al. (2012a), and Targino et al. (2015). The choice of kernel is often critical to select in order for the algorithm to perform well. In what follows, we present a particular choice we have developed specifically for the limit order book application in this chapter. This is a specialised choice of mutation kernel we adopted from the genetic search literature (Li and Zhang, 2009) which involves combination of mutation and cross-over operators for the particle mutation in the SMC sampler. In order to utilise this class of mutation operator in SMC sampler settings, we had to formally write down not just the mutation and cross-over operators in a structural form, as typically specified in the NGSAII class of genetic optimisation algorithms, but to also define their distributional form.

15.3.2.2.1 Genetic mutation and cross-over operators

In this class of mutation kernel in the SMC sampler, we consider the class of genetic algorithm type mutations. In particular, we describe the class of MOEA mutation and cross-over operators widely used in stochastic search algorithms, introduced in the method of Deb et al. (2002). This class of mutation kernel is the most widely used operator in multi-objective optimisation and we demonstrate its adaption to the SMC sampler framework.

A disadvantage of the NSGA-II operators is that they are only able to mutate binary, integer, or real encodings of the output parameter vectors, whereas the stochastic process for the limit order submission activity by liquidity providers requires the specification of a positive definite and symmetric covariance matrix for the generation of intensities from a multi-variate skew-t distribution. The positive definiteness and symmetry constraints of the covariance matrix will not be preserved if one simply employs the evolutionary operators above to produce new sets of covariance matrix candidate solutions. For this reason, Panayi and Peters (2015) introduced a new covariance mutation operator, which generates new candidate covariance matrices which remains in the manifold of positive definite matrices.

15.3.2.2.2 Simulated binary crossover

From two previous particles $\boldsymbol{\theta}_{n-1}^{(i)}, \boldsymbol{\theta}_{n-1}^{(j)}$, a new solution $\boldsymbol{\theta}_n^{(i)}$ is formed, where the k-th elements is crossed as follows:

$$\theta_n^{(i,k)} = \frac{1}{2}[(1-\bar{\beta})\theta_{n-1}^{(i,k)} + (1+\bar{\beta})\theta_{n-1}^{(j,k)}] \tag{15.31}$$

Here, $\bar{\beta}$ is a random sample from a distribution with density:

$$\bar{\beta} = \begin{cases} (\alpha u)^{\frac{1}{\eta_c+1}}, & \text{if } u \leq \frac{1}{\alpha} \\ \left(\frac{1}{2-\alpha u}\right)^{\frac{1}{\eta_c+1}}, & \text{otherwise} \end{cases}$$

where $u \sim U(0,1)$ and $\alpha = 2 - \beta^{-(\eta_c+1)}$, with:

$$\beta = 1 + \frac{2}{\theta_{n-1}^{(j,k)} - \theta_{n-1}^{(i,k)}} \min\left[\left(\theta_{n-1}^{(i,k)} - \theta_{k_L}\right), \left(\theta_{k_U} - \theta_{n-1}^{(j,k)}\right)\right] \tag{15.32}$$

This would produce a mutation kernel for this type of move at SMC sampler iteration n for the k-th element of the i-th particle vector, which would be updated according to a density given by:

$$M_n(\theta_n^{(i,k)}|\theta_{n-1}^{(i,k)}, \theta_{n-1}^{(j,k)})$$

$$= \frac{2}{\theta_{n-1}^{(j,k)} - \theta_{n-1}^{(i,k)}} \left[\mathbb{1}_{\theta_n^{(i,k)} \in \left[\theta_{n-1}^{(i,k)}, \frac{\theta_{n-1}^{(i,k)}+\theta_{n-1}^{(j,k)}}{2}\right]} \left(\frac{\eta_c+1}{\alpha} \cdot \frac{\theta_{n-1}^{(i,k)} + \theta_{n-1}^{(j,k)} - \theta_n^{(i,k)}}{\theta_{n-1}^{(j,k)} - \theta_{n-1}^{(i,k)}}\right)^{\eta_c} \right.$$

$$\left. + \mathbb{1}_{\theta_n^{(i,k)} \in \left[\frac{\theta_{n-1}^{(i,k)}+\theta_{n-1}^{(j,k)}}{2} - \frac{1}{2(2-\alpha)^{\frac{1}{\eta_c+1}}}, \theta_{n-1}^{(i,k)}\right]} \left(\frac{\eta_c+1}{\alpha} \cdot \frac{\theta_{n-1}^{(i,k)} + \theta_{n-1}^{(j,k)} - \theta_n^{(i,k)}}{\theta_{n-1}^{(j,k)} - \theta_{n-1}^{(i,k)}}\right)^{\eta_c+2} \right]$$

We use the cross-over operator with probability $p_c = 0.7$ and a distribution index $\eta_c = 5$. Every element k of the i-th particle vector is crossed with probability 0.5.

15.3.2.2.3 Polynomial mutation

The mutation operator perturbs elements of the solution, according to the distance from the boundaries:

$$\theta_n^{(i,k)} = \theta_{n-1}^{(i,k)} + \bar{\delta}(\theta_{k_U} - \theta_{k_L})$$

where we have for $\bar{\delta}$:

$$\bar{\delta} = \begin{cases} \left[2\gamma + (1-2\gamma)(1-\delta)^{\eta_m+1}\right]^{\frac{1}{\eta_m+1}} - 1 & \text{if } \gamma < 0.5 \\ 1 - \left[2(1-\gamma) + 2(\gamma-0.5)(1-\delta)^{\eta_m+1}\right]^{\frac{1}{\eta_m+1}} & \text{if } \gamma \geq 0.5 \end{cases}$$

with

$$\delta = \min\left[\left(\theta_n^{(i,k)} - \theta_{k_L}\right), \left(\theta_{k_U} - \theta_n^{(i,k)}\right)\right]$$

where, $\gamma \sim U(0,1)$.

This would produce a mutation kernel for this type of move at SMC sampler iteration n for the k-th element of the i-th particle vector, which would be updated according to a density given by:

$$M_n(\theta_n^{(i,k)}|\theta_{n-1}^{(i,k)}) = \frac{1}{\theta_{k_U} - \theta_{k_L}}\left[\mathbb{1}_{\theta_n^{(i,k)} \leq \theta_{n-1}^{(i,k)}}\left(\frac{(\eta_m+1)(\bar{\delta}+1)^{\eta_m}}{2(1-(1-\delta)^{\eta_m+1})}\right) + \mathbb{1}_{\theta_n^{(i,k)} > \theta_{n-1}^{(i,k)}}\left(\frac{(\eta_m+1)(1-\bar{\delta})^{\eta_m}}{2(1-(1-\delta)^{\eta_m+1})}\right)\right]$$

The distribution index $\eta_m = 10$. The polynomial mutation operator is used with probability $p_m = 0.2$.

15.3.2.2.4 Covariance mutation operator

In the t-th generation of the MOEA, we generate $\left\{\Sigma_t^{(i)}\right\}, i = 1 \dots N$ from a mixture distribution $M_n(\Sigma_{n,i})$ defined as follows:

$$M_n(\Sigma_t^{(i)}) = (1-w_1)\mathcal{IW}(\Psi_n, p_1) + w_1\mathcal{IW}(\Psi, p_2)$$

where $\mathcal{IW}$ denotes the inverse Wishart distribution, p_1, p_2 are degrees of freedom parameters with $p_2 < p_1$, and, where w_1 is small so that sampling from the second distribution happens infrequently. Here, Ψ denotes an uninformative positive definite matrix, with the effect that sampling from the second distribution leads to moves away from the local region being explored. Ψ_t is also a positive definite matrix, fitted based on moment matching to the sample mean of the successfully proposed candidate solutions in the previous stage of the multi-objective optimisation as follows:

$$\Psi_n = \frac{1}{\sum_{s=1}^{n} w^s}\sum_{s=1}^{n} w^s \frac{1}{\sum_{i=1}^{N}\frac{1}{r_s^{(i)}}}\sum_{i=1}^{N}\frac{1}{r_s^{(i)}}\tilde{\Sigma}_n^{(i)}$$

where $r_s^{(i)}$ is the non-domination rank of the i-th solution in the s-th generation, and w^s with $w < 1$ is an exponential weighting factor.

15.4 Application to Equities Limit Order Book Data: Data Description

The data employed in this study constitutes the intra-day trading activity on the European multi-lateral trading facility Chi-X Europe between January and April 2012. Chi-X Europe operated as an individual entity from 2007, before being purchased by BATS Europe at the end of the trading period under consideration. We note that Chi-X Europe is a secondary exchange, for example, the securities that are traded on the exchange are listed and primarily traded on national/supranational exchanges, including the London Stock Exchange, Euronext, Deutsche Boerse, and the SIX Swiss Exchange, amongst others. However, it maintains a significant proportion of the daily trading activity in each of these markets, between 20% and 35% in most cases.*

The complete dataset covers over 1,300 assets, primarily stocks, but also including exchange-traded funds and American depositary receipts. For the purposes of this study, we select one of the most commonly traded stocks in the French CAC 40 Index, namely BNP Paribas SA. Figure 15.2 shows the evolution of the LOB on a typical day for this asset based on real market observation data from the LOB. We also present a heatmap of the inside spread $S_t = P_t^{a,1} - P_t^{b,1}$ over the 2 month period February to March 2012. The inside spread is the most common measure of 'liquidity', for example, the relative ease with which one can buy or sell a financial asset.

Chi-X Europe operates both a visible and a hidden order book, and traders have the option to route orders to the hidden book, if they meet certain conditions relating to order type and size. The dataset consists of only data in the visible book, after it has been processed by the exchange's matching engine. That is, while the exchange allows for a range of order types with time-in-force modifiers, the processed data consist of the timestamps and order sizes of limit order submissions, executions and cancellations. However, this data are sufficient to construct a much more detailed picture of the state of the LOB than is typically available in previous studies (which only consider aggregate volume in either the first level or the first five levels), as we can disaggregate the volumes per level up to any depth in the LOB.

The raw, unevenly spaced data are thus used to construct the state of the LOB at each event timestamp (these are accurate up to millisecond precision). Because of our interest in fitting the auxiliary models describing price and volume dynamics (these are outlined in Section 15.5), however, we sub-sample the process at regular 10 second intervals, in order to then extract the price and volume variables of interest. Thus, from an irregularly spaced process

*http://www.liquidmetrix.com/liquidmetrix/battlemap

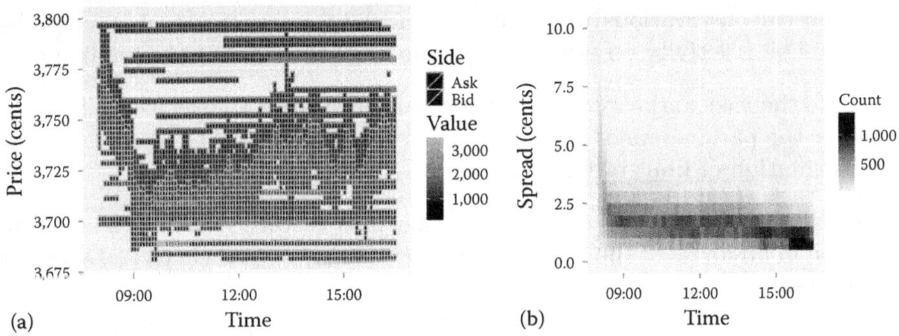

FIGURE 15.2
(a) A representation of real market data intra-day LOB states obtained from the trading activity for asset BNP Paribas SA on the 5th of March 2012. The shading of each box indicates the volume available at that price, which is volume available to buy for light grey-bordered boxes and volume available to sell for dark grey-bordered boxes. (b) A heatmap of the intra-day spread for the period February to March 2012 for asset BNP Paribas SA.

typically containing between 50,000 and 500,000 events every day, we extract a regular timeseries of the auxiliary model variables for the purposes of our estimation.

15.5 Results

The results presented in this section may be compared to those obtained from indirect inference procedures as reported in Panayi and Peters (2015). To achieve comparison we have also provided the results for what they termed the benchmark 'reference' model, which makes a series of assumptions in order to simplify estimation and model structure.

Benchmark Model Assumptions: The basic reference model is constructed from the earlier model under the following assumptions, for more details see Section 15.5.1 of Panayi and Peters (2015).

- We assume that the associated limit order submission distributions for the bid and ask have common parameter value settings. In addition, market order submission distributions for the bid and ask are also assumed to have common parameter value settings. This is reasonably consistent with empirical observations for a large number of assets when observing the submission activity on either side of the LOB throughout the trading day.

- Since the vast majority of orders get cancelled prior to execution, we consider the parameters of the distribution of cancellations to also match the distribution of limit order placements.

- We also set $\boldsymbol{m} = \mathbf{0}$ and consider the skewness vector, $\boldsymbol{\gamma}$, to take a common value in all levels of the bid and ask, such that $\boldsymbol{\gamma} = \gamma\mathbf{1}$, where $\mathbf{1}$ is a vector of ones.

- The monotonic mapping $F(\cdot)$, transforming the random variables $\Gamma^{LO,k,s}$, $\Gamma^{C,k,s}$ and $\Gamma^{MO,k}$ into intensity random variables $\Lambda^{LO,k,s}$, $\Lambda^{C,k,s}$ and $\Lambda^{MO,k}$ is set as the CDF of the standard normal. This transformation is necessary in order to ensure that intensities are positive and to bound the event counts.

- For the baseline intensities of limit order activity at each level, we assume that they will be the same for the passive limit orders on both sides, for example, $\mu_0^{LO,a,1} = \cdots = \mu_0^{LO,a,l_p} = \mu_0^{LO,b,1} = \cdots = \mu_0^{LO,b,l_p} = \mu_0^{LO,p}$, while aggressive limit orders will have a different limit order intensity, for example, $\mu_0^{LO,a,0} = \cdots = \mu_0^{LO,a,-l_d+1} = \mu_0^{LO,b,0} = \cdots = \mu_0^{LO,b,-l_d+1} = \mu_0^{LO,d}$. Market order baseline intensities are also equal on either side, for example, $\mu_0^{MO,a} = \mu_0^{MO,b} = \mu_0^{MO}$. The cancellation baseline activity will be the same as the submission baseline activity.

- Finally, we assume constant order sizes, for example, $O_{i,t}^{LO,k,s} = c = O_{j,t}^{MO,k}$ for all $i \in \left\{1,\ldots,N_t^{LO,k,s}\right\}$, $j \in \left\{1,\ldots,N_t^{MO,k}\right\}$, $k \in \{a,b\}$, $s \in \{-l_d+1,\ldots,l_p\}$ and $t \in \{1,\ldots,T\}$.

This basic reference model has the following parameter vector $\left\{\mu_0^{LO,p}, \mu_0^{LO,d}, \mu_0^{MO}, \gamma_0, \nu, \sigma^{MO}\right\}$, as well as the covariance matrix Σ to be estimated, see details in Panayi and Peters (2015). The results will be presented in terms of the ABC marginal posterior distributions of the individual parameters of the LOB simulation and in terms of the resulting stochastic agent-based LOB model to reasonably produce realistic features of the simulated LOB intra-daily.

In our introduction to likelihood-free methods in Section 15.2.2.1, we discussed the reduction of the observed data $\boldsymbol{y}$ to a low-dimensional vector of summary statistics t_y. We are interested specifically in two of the most commonly studied LOB characteristics, which correspond to the volatility in the log returns obtained from the price process dynamic (as obtained from half the inside spread) and the evolution of the volume resting on the LOB (as measured by the instantaneous aggregate total volume on the bid and ask at levels 1 to 5). The summaries we adopt at this stage are less standard in

ABC applications since they employ a functional (i.e. regression model based) summary of features of observable LOB process. In this case the summary information becomes the model characterisation (dimensional reduction) captured by the estimated model parameters fit to the real LOB data and the simulated LOB data for price or volume dynamic. Specifically we have:
Auxiliary model 1 - price features: If we denote the mid-price as $p_t^{mid} = \frac{p_t^{a,1}+p_t^{b,1}}{2}$, then the log return is defined as

$$r_t = \ln \frac{p_t^{mid}}{p_{t-\Delta_t}^{mid}}$$

where Δ_t is a suitable interval, in our case 1 minute. We fit a GARCH(1,1) model for this aspect of the data parameterised by $\widehat{\boldsymbol{\beta}}_1$.
Auxiliary model 2 - volume features: We fit an MA(1) model to the detrended total volume (i.e. an ARIMA(0,1,1) model) in the first five levels on both the bid and ask side parameterised by $\widehat{\boldsymbol{\beta}}_2$, in order to capture the time series structure of the LOB volumes.

The auxiliary models are fit to both the real y and simulated data y^*, and for the distance, we estimate the Euclidean distances between the auxiliary parameter vectors

$$\mathcal{D}_1 = \mathcal{D}\left(\widehat{\boldsymbol{\beta}}_1\left(\boldsymbol{y}\right), \widehat{\boldsymbol{\beta}}_1(\boldsymbol{y}^*(\boldsymbol{\theta}))\right)$$
$$\mathcal{D}_2 = \mathcal{D}\left(\widehat{\boldsymbol{\beta}}_2\left(\boldsymbol{y}\right), \widehat{\boldsymbol{\beta}}_2(\boldsymbol{y}^*(\boldsymbol{\theta}))\right)$$

15.5.1 Estimation algorithm configuration

To perform the estimation, there are also a number of inputs to the SMC sampler ABC algorithm that we specify, including the number of particles, the tolerance schedule forced decrement amount, and the total number of iterations over which to run the estimation. Specifically, we have for our estimation procedure:

- The procedure was run for 20 iterations.
- The tolerance schedule employed was the forced decrement schedules specified in Section 15.3.2.1, with a decrement parameter $\alpha = 0.1$.
- We obtain results using 50, 100, and 200 particles per iteration.
- We also tested the quality of the results for a series of quantile levels for the tolerance, for example, $q_{0.5}, q_{0.75}$, and $q_{0.9}$.

Carrying out the estimation procedure for each configuration above indicated that the best results (in terms of the lowest values of D_1, D_2) were obtained

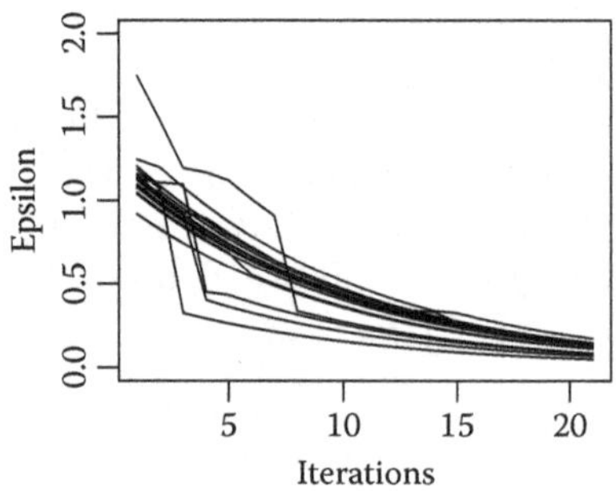

FIGURE 15.3
The adaptively estimated tolerance schedule obtained from multiple SMC sampler-ABC runs on real data for BNP Paribas on 05/03/2012 specified in Section 15.3.2.1.

for a quantile level $q_{0.9}$ for the tolerance and 200 particles. We repeated the estimation procedure 20 times with this configuration and the earlier configuration, and Figure 15.3 shows the evolution of the tolerance in the ABC posterior in the case of the forced tolerance schedule, when the estimation is run for $T = 20$ iterations.

We note that the mutation operator for the covariance matrix, specified in Section 15.3.2.2, which was composed of both an exploration and a mutation component, could lead to particle degeneracy in higher dimensions. Consequently, in practice, it can be computationally more efficient to simplify the mutation kernel for the covariance matrix to a static mutation kernel, which would eliminate the prior weighting in the numerator and denominator of each incremental particle weight. When this was performed, it produced particle systems with less degeneracy issues in higher dimensions.

Secondly, due to the nature of the crossover operator, it is possible for a particle to cross with an identical particle, for example, if the two particles were produced in the re-sampling step of the previous iteration. In our estimation, we explicitly exclude this possibility and, where a particle is chosen to cross with an identical particle, it is instead mutated using the operator specified in Section 15.3.2.2.

15.5.2 Final particle fitness and distributions of parameters

Having run the SMC sampler-ABC algorithm on the BNP Paribas LOB data for 05/03/2012, we obtained estimates of the posterior for the agent-based LOB simulation model. The first set of results shows the accuracy of the LOB model to replicate features of the real LOB stochastic process relating to price

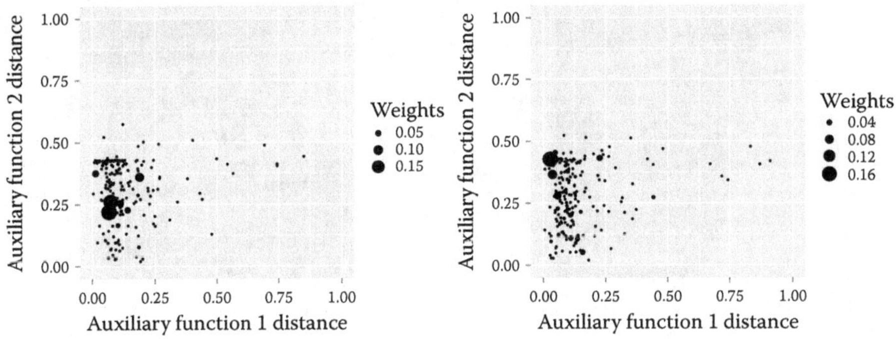

FIGURE 15.4
The realised objective function (distance metrics $\mathcal{D}_1$ and $\mathcal{D}_2$) values from each particle in the SMC Sampler-ABC algorithm at the final iteration for independent trials on the real data for BNP Paribas on 05/03/2012. The x axis is the GARCH(1,1) model parameter distance discrepancies for the intra-day volatility dynamic of the price process. The y axis is the ARIMA(0,1,1) model parameter distance discrepancies for the intra-day volume process dynamics.

and volume dynamics. This is clearly illustrated in Figure 15.4 in terms of the values of the objective functions $\mathcal{D}_1, \mathcal{D}_2$ for each of the particles at the final iteration stage of the SMC sampler-ABC algorithm. This is the standard way in which results for optimisation using MOEAs are presented [see discussion in Panayi and Peters (2015)], in order to show the Pareto optimal front that is obtained in that setting, see the discussion in the Section 15.5.3.

We also present realisations of the LOB intra-day evolution for both the particle with the highest weight and the weighted mean of the particles in Figure 15.5. We note that there are differences in the intra-day dynamics of the simulated financial market resulting from different repetitions of the estimation procedure. However, we note that for a subset of particles, we can recover price and volume dynamics that are similar to those observed in the real market (an example of which we had seen in Figure 15.2). In Figure 15.6 we present examples of the simulated LOB order books from two posterior estimators, the mode Maximum a-posteriori (MAP) estimators and the mimimum mean squared error (MMSE) posterio mean estimators. Furthermore, we also present in Figure 15.7 heat maps of the solutions for the intra day spreads simulated from the LOB solutions.

To complete the analysis, we also illustrate the median of the resulting marginal posterior distributions for the model parameters obtained from 20 independent runs of the SMC sampler-ABC algorithm for the BNP Paribas data on 05/03/2012. These results are presented in Figure 15.8.

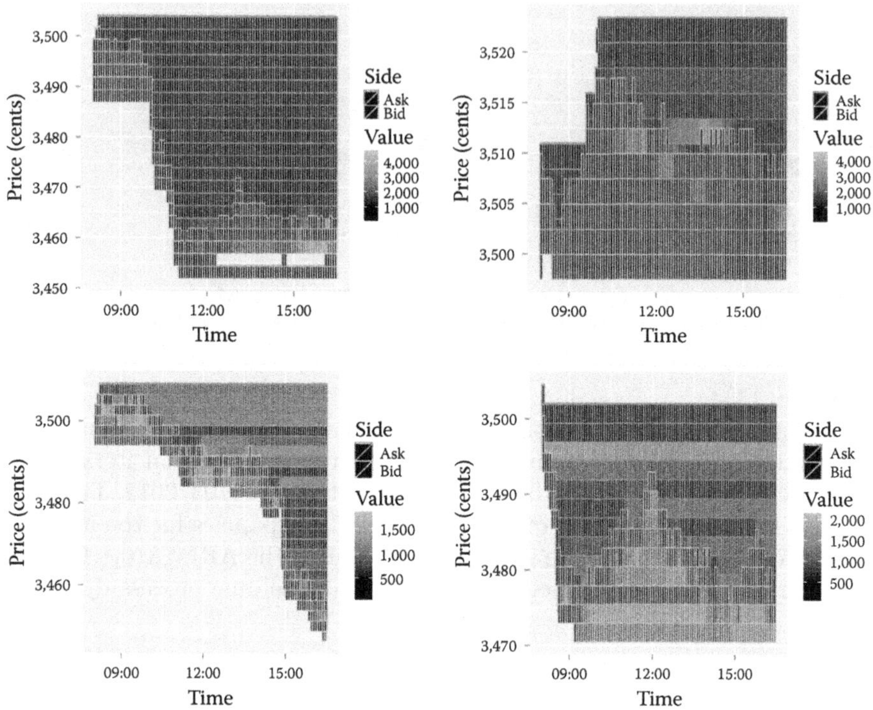

FIGURE 15.5
Representations of simulated intra-day LOB states obtained from using the (Top): Maximum a-posteriori (MAP) particle from a single estimation procedure and (Bottom): Minimum Mean Squared Error (MMSE) particle estimates.

15.5.3 Results comparison to multi-objective evolutionary algorithm-II procedure

The method introduced in Panayi and Peters (2015) is a combination of simulation-based indirect inference (II) and multi-objective optimisation, denoted the 'multi-objective-II estimation framework'. In common with ABC, II is used when one cannot write down the likelihood of the data generating model in closed form, but realisations are easily obtained via simulation given model parameters $\boldsymbol{\theta}$. II introduces a new, 'auxiliary' model (with parameter vector $\boldsymbol{\beta}$), which is fit to a transform of both the real and simulated data [$\boldsymbol{y}$ and $\boldsymbol{y}^*(\boldsymbol{\theta})$, respectively] and the objective is to find the model parameter vector $\hat{\boldsymbol{\theta}}$, which minimises some distance metric $D(\boldsymbol{\beta}(\boldsymbol{y}), \boldsymbol{\beta}(\boldsymbol{y}^*(\boldsymbol{\theta})))$.

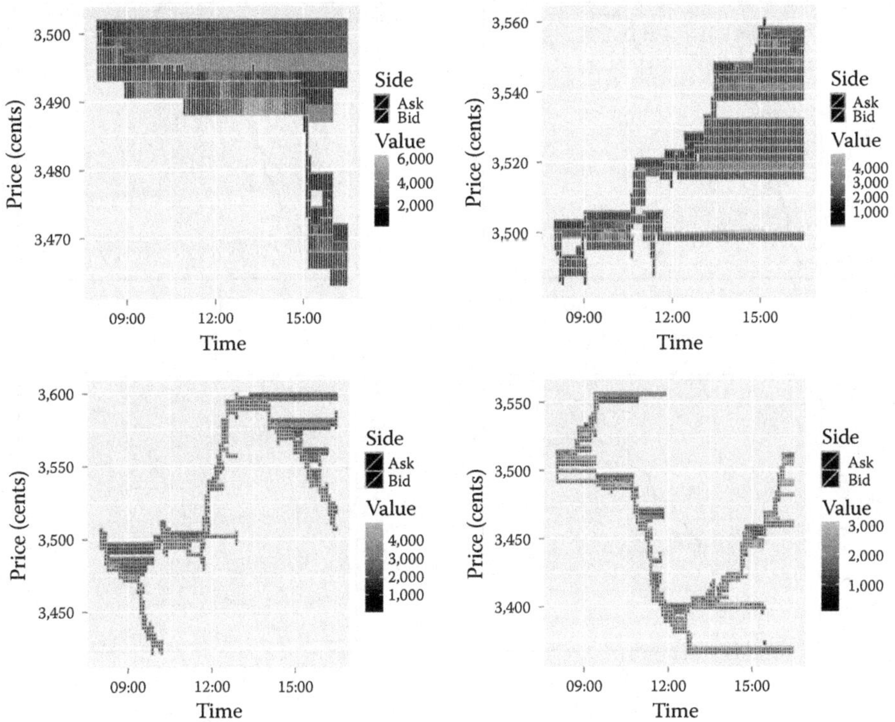

FIGURE 15.6
Representations of simulated intra-day LOB states obtained from using the (Top): MAP particle from a single estimation procedure and (Bottom): MMSE particle estimates.

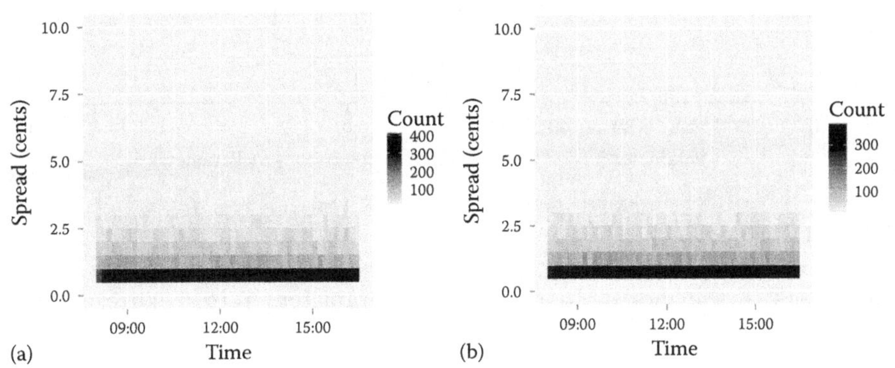

FIGURE 15.7
Heatmaps of the intra-day value of the spread for (a) The MAP particle from the estimation procedure and (b) MMSE particle estimates.

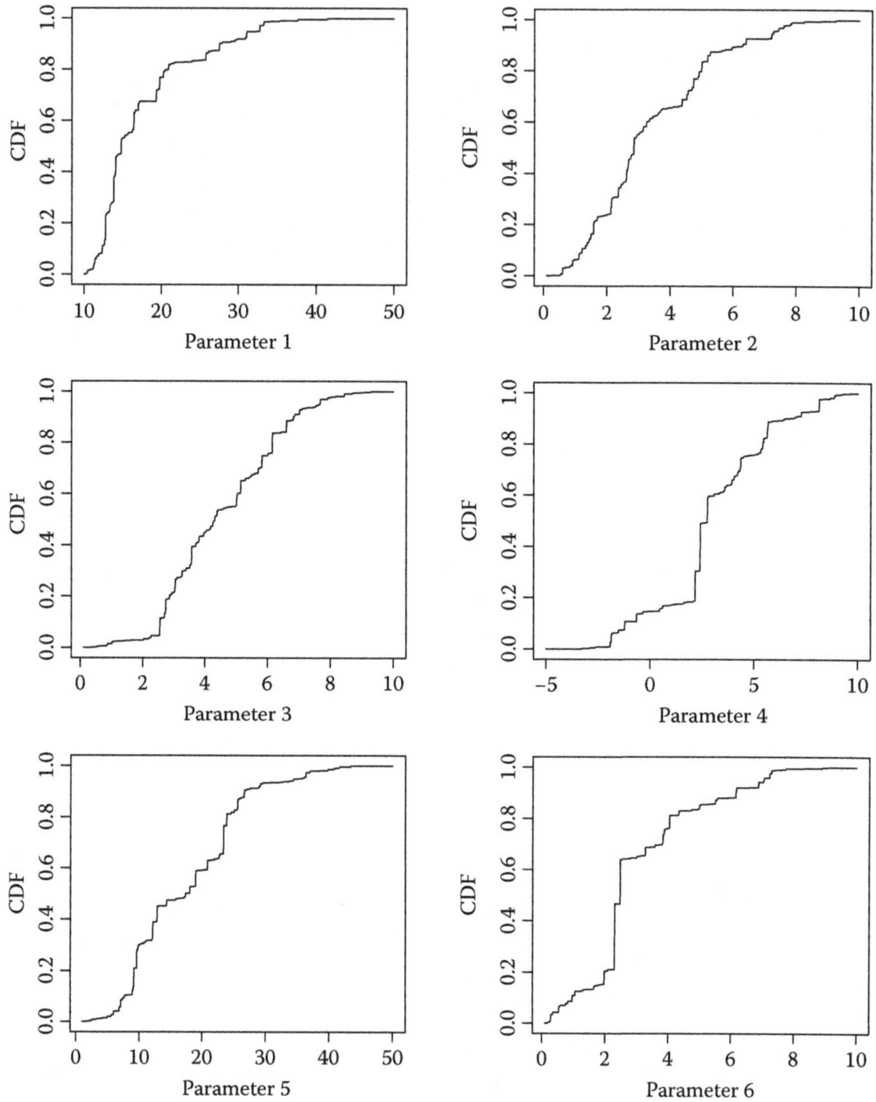

FIGURE 15.8
Median of the CDFs for every iteration of the estimation procedure for the parameters of the model. In the figures, parameters 1 to 6 correspond to $\left\{\mu_0^{LO,p}, \mu_0^{LO,d}, \mu_0^{MO}, \gamma_0, \nu, \sigma^{MO}\right\}$, respectively.

The multi-objective extension to the standard II procedure pertains to the objective function $D(\boldsymbol{\beta}(\boldsymbol{y}), \boldsymbol{\beta}(\boldsymbol{y}^*(\boldsymbol{\theta})))$. Where standard II procedures consider a scalar output of the objective function, the multi-objective-II method considers a vector-valued output, where each element of the vector pertains to a different feature of the LOB stochastic process. In this framework, the search is then for *non-dominated* parameter vectors, for examples, such that there is no parameter vector in the search space that can unilaterally improve a single criterion (objective function element) without worsening another criterion. The procedure uses the same mutation and cross-over kernels outlined in Section 15.3.2.2 and outputs a set of *Pareto optimal* solutions, see details in Panayi and Peters (2015).

Where the SMC-ABC method returns a family of particles and associated weights, the MOEA-II procedure returns a family of particles and their non-domination rank. Our comparison is then between the highest weighted particles returned from the former procedure, and the non-dominated particles returned from the latter. The results have been found to be comparable between the two methods, both in terms of achieving similar objective function values and in terms of producing simulations which resemble real financial markets. That is, while not all highest weight/non-dominated particles will give rise to realistic financial market simulations, there is a subset that do.

While we have tried to provide a fair comparison between the two methods by utilising the same mutation and cross-over operators, we should highlight some differences between the MOEA-II procedure and the SMC-ABC procedure presented in this paper. Firstly, the MOEA-II procedure did not suffer from particle degeneracy when using the adaptive mutation kernel for the covariance mutation operator, and, thus, the operator in Section 15.3.2.2 was utilised as described. Secondly, where the probability of crossover between particles in the MOEA-II procedure was set at the default value of 0.7 in every iteration, this was found to cause additional particle degeneracy issues and, thus, the probability was reduced to 0.05 (with the additional exclusion of crossing with identical particles described at the front of this section).

In addition to these practical implementation considerations, the studies performed also demonstrated several structural features for the Bayesian LOB model when we applied the reference 'Benchmark Model' from Panayi and Peters (2015). For instance, under the benchmark model parameterisation, we found that the Bayesian formulation was able to reproduce LOB daily profiles from MAP or MMSE posterior parameter estimates, which were consistent with those that one observes in real LOB activity in a trading day. This gives us confidence that even with the least flexible family of models in our specified Bayesian LOB model class, we can capture sensible dynamics for the LOB when calibrated to real market data and then used to simulate a LOB. In addition, the particles that were on the leading first and second efficient frontiers and weighted reasonably, when used to simulate the dynamics of the

LOB, were also producing implied mid price dynamics and liquidity profiles based on bid and ask volumes commensurate with those one can observe in practice. This tells us that we can definitely start to consider such models for applications, such as market simulators for testing market trading, execution, and market making strategies.

15.6 Conclusion

This chapter has proposed a stochastic agent-based liquidity supply and demand-based simulation model to characterise the LOB for an asset traded on an electronic exchange. The calibration of this model to real market LOB data has been performed via a posterior inference procedure that adopted an ABC structure due to the complexity of writing down the resulting likelihood for the LOB agent simulation model. The estimation of the posterior distribution was then shown how to be performed via an adaptive SMC sampler-ABC algorithm. The results were tested on real data and compared to an indirect inference procedure with multi-objective optimisation features.

Such a model is important for many applications in high frequency finance which require the ability to calibrate a realistic model to the LOB stochastic process on a daily basis. These types of models will find important applications in better understanding and assessing performance of trade strategy selection and risk assessment, brokerage strategy design, regulation impacts, and so on. To date, no realistic LOB simulation frameworks have been developed that take into account the complete structure of the LOB stochastic process from a constructive and interpretable approach. In this chapter, we have developed such a structure through a representative agent-based model that incorporates modern market participants that one finds in both primary and secondary MTF electronic exchanges, such features include the presence of two representative types of agent, the liquidity providers and the liquidity demanders. Each such representative agent has the ability to place passive and aggressive limit orders, as well as cancellations and market orders with each such order type having an associated bid or ask tick level and volume. The representative agents are modelled via a stochastic process rather than a simple set of heuristic rules, typically utilised in other agent-based models. The outputs of the stochastic models for each agent then operate in a mechanistic manner, as would typically occur in agent models to produce an update to the LOB being simulated.

Since the agent model we adopt is not based on the typical framework of many agents interacting with simple heuristic rules, but instead, we have a small set of representative agent populations, each characterised by a stochastic model structure, this model can be calibrated in a statistical manner.

This is an important advantage that the SMC-ABC methodology allows one to undertake calibration and estimation of the representative agent model for the LOB in a structured and rigorous manner, unlike typical approaches to calibration of agent models in the literature. This is where the application of ABC is critical, typically, the agent models have simple heuristic calibrations because the models are too complex to do formal inference, we believe that the representative agent-based stochastic models we have developed find a compromise between the attributes an agent model provides and the ability to rigorously calibrate such models. It is precisely the ability to readily simulate realisations of the order book structure from our parameterised stochastic agent models, without having to write the likelihood which would be intractable, that makes it so amenable to ABC-based statistical inference.

Overall, the chapter has demonstrated that even with a simplified benchmark version of the Bayesian LOB simulation model proposed, it can be fitted accurately to different observed LOB regimes on a daily basis via the methodology of SMC-ABC. In particular, we have demonstrated that the utilisation of model-based summary profiles of mid price/returns and the volume profile based LOB summaries carry sufficient information to make for accurate practical calibrations of an ABC model on real LOB equity data. In the future, it would be important to further extend these characteristic features/summary models/summary statistics of the observed LOB for use in the SMC-ABC calibration and simulation.

There are numerous future extensions that could be made to this model with regard to the application, the methodology, and the study of the model attributes. We mention a few next. One such class of extensions of this work could seek to combine the attributes of multi-objective optimisation solutions with those of the SMC-ABC-based solution. For instance, it would be practically useful to develop a probabilistic representation of particles on Pareto-dominated efficient frontiers based on each of the objective functions developed for assessing a multi-objective optimisation criterion in the LOB simulation characterisation. Such a combined representation would allow for volume, liquidity, and price-based LOB features to be more readily discerned in their contribution to the particle weights and probabilitistic representations. In addition, we believe it would allow for more appropriate probabilistic choice of points on the frontier, which can then be used in these applications for simulation, scenario analysis, and forecasting. The ability to probabilisitically undertake these tasks is beneficial for many applications in financial mathematics requiring accurate LOB simulation frameworks, these include assessment of trading strategies, assessment and development of market making strategies, exchanges can assess performance statistically of market making behaviours of designated sponsors to decide renumeration and compensation for such risk taking activities, assessment of regulatory impacts under new clearing and exchange regulations from central banks, and of course assessment of optimal execution and brokerage strategies in different market regimes.

Other future extensions to the agent model that would be of relevance to extending this framework would be to include additional order types such as iceberg orders, fill or kill orders, and so on. In addition, features such as tick size change at different price levels of the mid would be important to incorporate for practical extensions. In addition, it would be important in some applications to have a combination of continous trading models, as well as auction mechanisms built into the model. This would align with features observed in real electronic exchanges, such as Xetra, where they offer a range of specialised trading models adapted to the needs of its various trading groups, as well as the different assets classes. The models differ according to: market type (e.g. number of trading parties); the transparency level of available information pre- and post-trade; the criteria of the order prioritisation; price determination rules; and the form of order execution. For equity trading, the following trading models are supported: continuous trading in connection with auctions (e.g. opening and closing auction, and possibly, one or more intra-day auctions); mini-auction in connection with auctions; and one or more auctions at pre-defined points in time. Future extensions will seek to explore aspects of these components in the simulation based LOB models.

References

Beaumont, M. A., J.-M. Cornuet, J.-M. Marin, and C. P. Robert. Adaptive approximate Bayesian computation. *Biometrika*, 96(4):983–990, 2009.

Beaumont, M. A., W. Zhang, and D. J. Balding. Approximate Bayesian computation in population genetics. *Genetics*, 162(4):2025–2035, 2002.

Blum, M. G. B. Approximate Bayesian computation: A nonparametric perspective. *Journal of the American Statistical Association*, 105(491):1178–1187, 2010.

Blum, M. G. B. and O. François. Non-linear regression models for approximate Bayesian computation. *Statistics and Computing*, 20(1):63–73, 2010.

Blum, M. G. B., M. A. Nunes, D. Prangle, and S. A. Sisson. A comparative review of dimension reduction methods in approximate Bayesian computation. *Statistical Science*, 28(2):189–208, 2013.

Bornn, L., N. Pillai, A. Smith, and D. Woodard. The use of a single pseudo-sample in approximate Bayesian computation. *arXiv preprint arXiv:1404.6298*, 2014.

Chopin, N. A sequential particle filter method for static models. *Biometrika*, 89(3):539–552, 2002.

Chopin, N. Central limit theorem for sequential Monte Carlo methods and its application to Bayesian inference. *Annals of statistics*, 32(6):2385–2411, 2004.

Cont, R. and A. De Larrard. Price dynamics in a Markovian limit order market. *SIAM Journal on Financial Mathematics*, 4 (1):1–25, 2013.

Cont, R., S. Stoikov, and R. Talreja. A stochastic model for order book dynamics. *Operations Research*, 58(3):549–563, 2010.

Crisan, D. and A. Doucet. A survey of convergence results on particle filtering methods for practitioners. *IEEE Transactions on Signal Processing*, 50(3):736–746, 2002.

Deb, K., A. Pratap, S. Agarwal, and T. Meyarivan. A fast and elitist multiobjective genetic algorithm: NSGA-II. *IEEE Transactions on Evolutionary Computation*, 6(2):182–197, 2002.

Del Moral, P. *Feynman-Kac Formulae.* Springer, New York, 2004.

Del Moral, P., A. Doucet, and A. Jasra. Sequential Monte Carlo samplers. *Journal of the Royal Statistical Society: Series B (Statistical Methodology)*, 68(3):411–436, 2006.

Del Moral, P., A. Doucet, and A. Jasra. An adaptive sequential Monte Carlo method for approximate Bayesian computation. *Statistics and Computing*, 22(5): 1009–1020, 2012.

Doucet, A., N. De Freitas, and N. Gordon. *An Introduction to Sequential Monte Carlo Methods.* Springer, New York, 2001.

Fan, Y., D. J. Nott, and S. A. Sisson. Approximate Bayesian computation via regression density estimation. *Stat*, 2(1):34–48, 2013.

Fearnhead, P. and D. Prangle. Constructing summary statistics for approximate Bayesian computation: Semi-automatic ABC. *arXiv preprint arXiv:1004.1112*, 2010.

Fearnhead, P. and D. Prangle. Constructing summary statistics for approximate Bayesian computation: Semi-automatic approximate Bayesian computation. *Journal of the Royal Statistical Society: Series B (Statistical Methodology)*, 74(3):419–474, 2012.

Fricke, D. and T. Lux. The effects of a financial transaction tax in an artificial financial market. *Journal of Economic Interaction and Coordination*, 10(1):119–150, 2015.

Garbade, K. D. and K. Garbade. *Securities Markets*. McGraw-Hill, New York, 1982.

Gould, M. D., M. A. Porter, S. Williams, M. McDonald, D. J. Fenn, and S. D. Howison. Limit order books. *Quantitative Finance*, 13(11):1709–1742, 2013.

Harris, L. *Trading and Exchanges: Market Microstructure for Practitioners*. Oxford University Press, Oxford, UK, 2003.

Künsch, H. R. Recursive Monte Carlo filters: Algorithms and theoretical analysis. *Annals of Statistics*, 33(5):1983–2021, 2005.

Kyle, A. S. Continuous auctions and insider trading. *Econometrica: Journal of the Econometric Society*, 53:1315–1335, 1985.

Li, H. and Q. Zhang. Multiobjective optimization problems with complicated Pareto sets, MOEA/D and NSGA-II. *IEEE Transactions on Evolutionary Computation*, 13(2):284–302, 2009.

LiCalzi, M. and P. Pellizzari. Fundamentalists clashing over the book: A study of order-driven stock markets. *Quantitative Finance*, 3(6):470–480, 2003.

Marjoram, P., J. Molitor, V. Plagnol, and S. Tavaré. Markov chain Monte Carlo without likelihoods. *Proceedings of the National Academy of Sciences*, 100(26):15324–15328, 2003.

Maslov, S. Simple model of a limit order-driven market. *Physica A: Statistical Mechanics and its Applications*, 278(3):571–578, 2000.

Neal, R. M. Annealed importance sampling. *Statistics and Computing*, 11(2):125–139, 2001.

Nunes, M. A. and D. J. Balding. On optimal selection of summary statistics for approximate Bayesian computation. *Statistical Applications in Genetics and Molecular Biology*, 9(1):1–16, 2010.

Obizhaeva, A.A. and J. Wang. Optimal trading strategy and supply/demand dynamics. *Journal of Financial Markets*, 16(1):1–32, 2013.

O'hara, M. *Market Microstructure Theory*, Volume 108. Blackwell, Cambridge, MA, 1995.

Panayi, E. and G. W. Peters. Stochastic simulation framework for the Limit Order Book using liquidity motivated agents. *Available at SSRN 2551410*, 2015.

Panayi, E., G. W. Peters, J. Danielsson, and J.-P. Zigrand. Designating market maker behaviour in limit order book markets. *arXiv preprint arXiv:1508.04348*, 2015a.

Panayi, E., G. W. Peters, and I. Kosmidis. Liquidity commonality does not imply liquidity resilience commonality: A functional characterisation for ultra-high frequency cross-sectional lob data. *Quantitative Finance*, 15(10):1737–1758, 2015b.

Peters, G. W. Topics in sequential Monte Carlo samplers. M.Sc, University of Cambridge, Department of Engineering, 2005.

Peters, G. W., Y. Fan, and S. A. Sisson. On sequential Monte Carlo, partial rejection control and approximate Bayesian computation. *Statistics and Computing*, 22(6): 1209–1222, 2012a.

Peters, G. W., S. A. Sisson, and Y. Fan. Likelihood-free Bayesian inference for α-stable models. *Computational Statistics & Data Analysis*, 56(11):3743–3756, 2012b.

Peters, G. W., M. V. Wüthrich, and P. V. Shevchenko. Chain ladder method: Bayesian bootstrap versus classical bootstrap. *Insurance: Mathematics and Economics*, 47(1):36–51, 2010.

Ratmann, O., C. Andrieu, C. Wiuf, and S. Richardson. Model criticism based on likelihood-free inference, with an application to protein network evolution. *Proceedings of the National Academy of Sciences*, 106(26):10576–10581, 2009.

Reeves, R. W. and A. N. Pettitt. A theoretical framework for approximate Bayesian computation. *Proceedings of the 20th International Workshop Statistical Modelling*, Sydney, Australia, pp. 393–396, 2005a.

Ristic, B., S. Arulampalam, and N. Gordon. *Beyond the Kalman Filter: Particle Filters for Tracking Applications*, Volume 685. Artech house, Boston, MA, 2004.

Sisson, S. A., Y. Fan, and M. M. Tanaka. Sequential Monte Carlo without likelihoods. *Proceedings of the National Academy of Sciences*, 104(6):1760–1765, 2007.

Smith, M. S., Q. Gan, and R. J. Kohn. Modelling dependence using skew t copulas: Bayesian inference and applications. *Journal of Applied Econometrics*, 27(3): 500–522, 2012.

Targino, R. S., G. W. Peters, and P. V. Shevchenko. Sequential Monte Carlo Samplers for capital allocation under copula-dependent risk models. *Insurance: Mathematics and Economics*, 61(1):206–226, 2015.

Toni, T., D. Welch, N. Strelkowa, A. Ipsen, and M. P. H. Stumpf. Approximate Bayesian computation scheme for parameter inference and model selection in dynamical systems. *Journal of the Royal Society Interface*, 6(31):187–202, 2009.

US Securities, Exchange Commission, the Commodity Futures Trading Commission et al. Findings regarding the market events of May 6, 2010. *Report of the Staffs of the CFTC and SEC to the Joint Advisory Committee on Emerging Regulatory Issues*, 2010.

Wilkinson, R. D. Approximate Bayesian computation (ABC) gives exact results under the assumption of model error. *Statistical Applications in Genetics and Molecular Biology*, 12(2):129–141, 2013.

16

Inferences on the Acquisition of Multi-Drug Resistance in Mycobacterium Tuberculosis *Using Molecular Epidemiological Data*

Guilherme S. Rodrigues, Andrew R. Francis, S. A. Sisson, and Mark M. Tanaka

CONTENTS

16.1 Introduction 482
16.2 Data 484
16.3 Model 485
16.3.1 New infections 490
16.3.2 Cure, recovery, and death 492
16.3.3 Detection and treatment 492
16.3.4 Acquisition of drug resistance 493
16.3.5 Mutation of the marker 494
16.3.6 Initial conditions of the model 495
16.4 Inference with Approximate Bayesian Computation 496
16.4.1 Summary statistics 498
16.4.2 Parameter specifications and prior distributions 499
16.5 Competing Models of Resistance Acquisition 501
16.5.1 Can resistance to both drugs be acquired simultaneously? 503
16.5.2 Is resistance to both drugs acquired at equal rates? ... 504
16.5.3 The relative contribution of transmission and treatment failure to multi-drug–resistant tuberculosis 505
16.6 Conclusions 506
Acknowledgements 508
References 508

16.1 Introduction

Tuberculosis (TB) is a lung disease caused by the bacterium *Mycobacterium tuberculosis*, which kills around 1.5 million people each year and remains a serious challenge for global public health (WHO, 2015). Antibiotic drugs for treating TB have been available since the mid twentieth century, and currently implemented strategies for TB control rely on the efficacy of these drugs. Treatment of TB involves combination therapy – in which multiple drugs are administered together in part to improve killing efficacy. The 'first-line' drugs used in combination to treat tuberculosis are rifampicin, isoniazid, pyrazinamide, ethambutol, and streptomycin.

As with most other pathogens, resistance to antibiotic drugs has rapidly evolved in *M. tuberculosis*. Streptomycin was the first of the first-line drugs to be developed and deployed in 1943, but resistance was observed before the end of that decade (Mitchison, 1951; Gillespie, 2002). Of particular concern is the rise of bacterial strains resistant to multiple drugs, as cases caused by them are difficult to treat successfully. Multidrug resistance (MDR) is defined as resistance to both rifampicin and isoniazid. These are the two most effective drugs against tuberculosis (when the strain is not resistant). Currently, 3.3% of new TB cases are multi-drug resistant (WHO, 2015). The occurrence of MDR-TB strains that have additional resistance (called extensively drug resistant and totally drug resistant) are particularly problematic and have the potential to cause large outbreaks that are difficult to control (Gandhi et al., 2006). A better understanding of how multiple drug resistance evolves would aid efforts to contain resistance and control tuberculosis.

Genetic studies have established that many independent mutation events have led to resistance (Ramaswamy and Musser, 1998). Although this suggests that mutation of genes is an important source of resistance, model-based analysis of molecular data has revealed that among resistant cases, most are due to the transmission of already-resistant bacteria (Luciani et al., 2009). It is therefore of interest to investigate whether or not this finding also holds for multi-drug resistant tuberculosis.

The rates at which resistance evolves against different drugs vary. For instance, isoniazid resistance is known to be acquired faster than rifampicin resistance (Ford et al., 2013; Gillespie, 2002; Nachega and Chaisson, 2003). The rates of mutation to resistance per cell generation are low in absolute value, for example, for isoniazid, the rate is around 3×10^{-8} and for rifampicin, it is around 2×10^{-10} (David, 1970; Gillespie, 2002), although there is a high degree of variation across different lineages of *M. tuberculosis* (Ford et al., 2013). One might therefore expect that double resistance of these drugs (MDR) evolves at an exceedingly low rate (Nachega and Chaisson, 2003). However, MDR strains often occur at appreciable frequencies (Zhao et al., 2012; Anderson et al., 2014), and a recent study has presented a theoretical model showing how double resistance can evolve rapidly within hosts (Colijn et al., 2011).

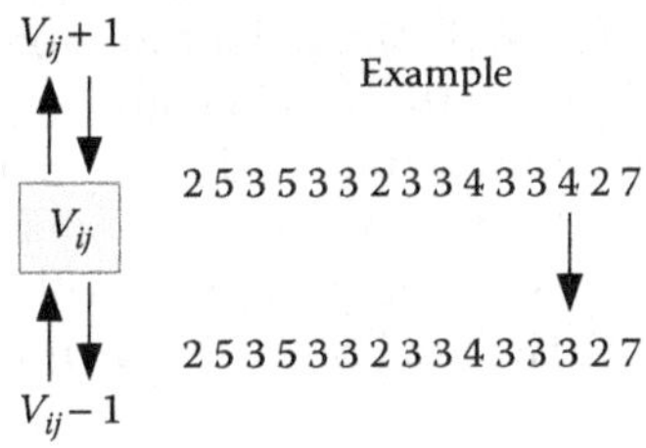

FIGURE 16.1
Variable numbers of tandem repeats (VNTRs) loci mutate in a stepwise manner so that the number of repeat units at a locus increases or decreases. In our analysis, we assume that when mutation occurs at a locus j in genotype i, the repeat number V_{ij} increases or decreases by a single copy. We further assume that a single unit (repeat number of 1) is an absorbing boundary. The hypothetical example shows how mutation at locus number 13 creates a new VNTR genotype.

It would be useful to establish whether such fast direct acquisition of double resistance can be detected in bacterial isolates from epidemiological studies.

To characterise patterns of TB transmission and drug resistance in a given geographic region, bacterial isolates from TB patients are often genotyped using molecular markers known as variable numbers of tandem repeats (VNTRs), which are repeated genetic sequences that exhibit variation across isolates. The source of this variation is mutation at the VNTR genetic loci, which leads to the expansion or contraction of repeat numbers at those loci (Figure 16.1). A scheme for discriminating effectively among a set of isolates involves considering repeat numbers at multiple VNTR sites. This molecular typing scheme is called multi-locus VNTR analysis; in the context of tuberculosis epidemiology, it is often known as mycobacterial interspersed repetitive units-VNTR (Mazars et al., 2001; Supply et al., 2006). Typing techniques such as MLVA have been useful for tracking particular strains and understanding how drug resistance evolves and disseminates at the epidemiological level (Monteserin et al., 2013; Anderson et al., 2014).

Here, we investigate the rates of drug resistance acquisition in a natural population using molecular epidemiological data from Bolivia (Monteserin et al., 2013). First, we study the rate of direct acquisition of double resistance from the double sensitive state within patients and compare it to the rates of evolution to single resistance. In particular, we address whether or not double resistance can evolve directly from a double sensitive state within a given host. Second, we aim to understand whether the differences in mutation rates to rifampicin and isoniazid resistance translate to the epidemiological scale. Third, we estimate the proportion of MDR TB cases that are due to the transmission of MDR strains compared to acquisition of resistance through evolution. To address these problems, we develop a model of TB transmission in which we track the evolution of resistance to two drugs and the evolution of VNTR loci. However, the available data (see Section 16.2) is incomplete, in that it is recorded only for a fraction of the population and at a single point in time. The likelihood function induced by the proposed model is computationally prohibitive to evaluate and accordingly impractical to work with directly. We therefore approach statistical inference using approximate Bayesian computation techniques.

16.2 Data

The dataset we use is taken from a study of tuberculosis in Bolivia (Monteserin et al., 2013). Bolivia has a population of 11 million people and a TB incidence of 120 per 100,000 per year. This rate is comparable to the global incidence of TB (133 per 100,000 per year) and to the rate in Peru, but is 3–6 times the TB incidence in neighbouring countries Brazil, Paraguay, Uruguay, Argentina, and Chile (WHO, 2015). In the molecular epidemiological study, the investigators genotyped 100 isolates collected in 2010, which represented an estimated 1.1% of the cases in Bolivia at the time of the study (Monteserin et al., 2013). Each isolate was tested for drug sensitivity to five drugs. Here, we focus on resistance against the two drugs isoniazid and rifampicin used to define multidrug resistance. Of the 100 isolates, 14 were found to be MDR, that is, resistant to both of these drugs, 78 were sensitive to both drugs and the remaining 8 were resistant to isoniazid, but sensitive to rifampicin. No isolates were resistant to rifampicin while being sensitive to isoniazid.

In addition to these drug resistance profiles, each isolate was genotyped using 15 VNTR loci. For example, an isolate in the dataset, which was resistant to isoniazid, but sensitive to rifampicin, had the following 15 repeat numbers for its 15 VNTR loci: 143533233433527, which together constitute its genotype. Variation in these genotypes occurs through a process of mutation in which repeat numbers increase or decrease (see Figure 16.1).

Let g be the number of distinct genotypes present in a sample, and label the resistance profiles by (0, INH, RIF, MDR), where 0 denotes sensitivity to both drugs, INH denotes resistance to isoniazid and sensitivity to rifampicin, RIF denotes resistance to rifampicin and sensitivity to isoniazid, and MDR denotes resistance to both drugs. The observed data $\mathbf{X}_{obs}$ are then a $g \times 4$ matrix of counts, such that each row gives the distribution of isolates across the four resistance profiles for a given genotype and each column gives the distribution of isolates across genotypes for a given resistance profile. The sum of entries in a particular row is the number of isolates with that genotype, while the sum of entries in a particular column is the number of isolates with that resistance profile. The dataset also includes a $g \times 15$ matrix of repeat numbers from the VNTR genotyping.

The Bolivian dataset is displayed in full in Table 16.1, which shows all $g = 66$ distinct genotypes and classifies all 100 isolates according to genotype and resistance profile. The $\mathbf{X}_{obs}$ matrix is formed by combining the 0, INH, RIF, and MDR columns.

16.3 Model

In this section, we introduce a model that incorporates both VNTR-based genotyping and drug resistance states. The dynamic variables of the model correspond to numbers of cases of untreated and treated tuberculosis, their resistance states, and VNTR genotypes associated with these infections in the population. We will now briefly describe processes involved in the model and provide further details in the following subsections.

An untreated case of TB can become detected and treated, and treatment involves a combination of drugs including the two in question. Drug sensitive strains can acquire resistance under treatment with some probability and thereby change their resistance state. Treated and untreated cases can infect susceptible individuals and convert them to untreated cases. We disregard latent infections for simplicity (although latency is an important feature of the natural history of tuberculosis) and focus on active infections which are the larger source of new infections. Treated and untreated individuals can also recover or die. Treated individuals enjoy an additional probability of recovery that depends on the efficacy of the drugs, which in turn depends on the sensitivity or resistance of the infecting strain. Treated and untreated cases are also associated with a VNTR genotype, and this genotype evolves over time according to a stepwise mutation process for each locus. Figure 16.2 shows the broad structure of the model with respect to treatment and resistance states, while suppressing details of transmission, recovery, death, and mutation of the VNTR loci.

At the end of the period of evolution, a simple random sample of 100 isolates is taken without replacement from the population, which matches the sample size of the Bolivian dataset. The following provides a full description of the generative process for the observable data.

Let G be the number of distinct genotypes in the population (the number of distinct genotypes in the *sample* is g) and L be the number of VNTR loci used in the genotyping scheme. For the Bolivian dataset $L = 15$. In the model, the variable G is unknown and varies dynamically. We maintain three matrices which change through time: a $G \times L$ matrix, $\mathbf{V}$, which describes the VNTR genotypes; a $G \times 4$ matrix, $\mathbf{U}$, which describes the numbers of *untreated* cases of tuberculosis classified according to VNTR genotype and resistance state; and a $G \times 4$ matrix $\mathbf{T}$, which describes the numbers of *treated* cases of tuberculosis, again classified according to VNTR genotype and resistance state. It will be

TABLE 16.1
Molecular Dataset Compiled from Monteserin et al. (2013)

Genotype	0	INH	RIF	MDR	Genotype	0	INH	RIF	MDR
253533233433427	4	0	0	0	243413342212437	1	0	0	0
253533233433327	3	0	0	0	233312442212437	1	0	0	0
253533233433527	11	1	0	0	233313441212437	1	0	0	0
253533233433525	3	0	0	0	233413442212248	1	0	0	0
143533233433527	2	1	0	0	233413442212249	1	0	0	0
253333244232232	1	0	0	0	233213442212349	1	0	0	0
25333324423-232	1	0	0	0	231413542212335	1	0	0	0
254333243232342	0	2	0	2	232433242212436	1	0	0	0
263532232423139	3	0	0	0	234413442212436	1	0	0	0
223413442212437	2	0	0	0	434433452212427	1	0	0	0
233413542212347	0	1	0	2	256432342122237	2	1	0	0
244333244232332	0	0	0	1	256433342123236	2	0	0	0
244333244232322	1	0	0	0	247432342122136	1	0	0	0
245333244242332	1	0	0	0	268432252122227	0	1	0	0
254333244232232	0	0	0	1	268632252122227	1	0	0	0
254333244232332	1	0	0	0	221313352122338	0	0	0	1
254333244242332	1	0	0	0	263532233423148	1	0	0	0
253333244242232	1	0	0	0	360332233423138	1	0	0	0
252333243232232	0	0	0	1	263513233523344	1	0	0	0
252333243232332	0	0	0	1	253523233433527	1	0	0	0
251333243242332	1	0	0	0	253533232433527	1	0	0	1

(*Continued*)

TABLE 16.1 (*Continued*)
Molecular Dataset Compiled from Monteserin et al. (2013)

Genotype	0	INH	RIF	MDR	Genotype	0	INH	RIF	MDR
252333243262222	0	0	0	1	253533232433427	1	0	0	0
244233234222322	1	0	0	0	253523133433527	1	0	0	0
233373242232325	1	0	0	0	253533133433527	1	0	0	0
252343242232524	1	0	0	0	353533233433427	0	0	0	1
25233234423251a	1	0	0	0	253533233433837	1	0	0	0
35234234423251a	1	0	0	0	253533233433237	1	0	0	0
233413442212338	0	1	0	0	254533233433537	0	0	0	1
233413442212335	1	0	0	0	253533233433536	1	0	0	0
233413442212337	1	0	0	0	252533233433428	1	0	0	0
21341344221233a	1	0	0	0	253534233433325	1	0	0	0
213413442212327	0	0	0	1	243533232433737	1	0	0	0
233413442212437	3	0	0	0	242433433433436	1	0	0	0

Note: All isolates were classified according to their genotype and resistance profile. The symbol 'a' represents ten repeat units and '-' represents missing data. The entries in the four columns sum to the total number of isolates, 100.

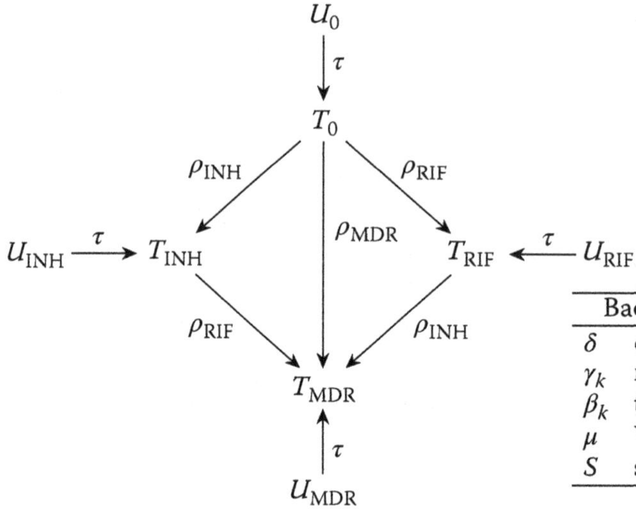

Background parameters	
δ	death rate
γ_k	recovery rate
β_k	transmission rate
μ	VNTR mutation rate
S	susceptible population

FIGURE 16.2
Model structure for numbers of untreated (U_k) and treated (T_k) cases and per capita rates of conversion (within-host substitution) among resistance classes. Rates are ρ_{INH} and ρ_{RIF} for acquisition of resistance to isoniazid and rifampicin, respectively, and ρ_{MDR} for single step acquisition of resistance to both drugs. Detection (and treatment) of cases is shown with arrows labelled with τ. Background parameters are shown in the table to the right, with rates per capita per unit time, and resistance states $k = 0$, INH, RIF, MDR. The mutation process of the VNTR locus is described in Section 16.3.5.

useful to define a $G \times 4$ matrix, $\mathbf{W}$, whose entries are the total numbers of both treated and untreated cases: $\mathbf{W} = \mathbf{U} + \mathbf{T}$.

As it will be helpful to be able to pick out columns of these matrices, we adopt notation for the standard basis vectors of $\mathbb{R}^n$. Let e_i denote the i-th basis (column) vector, so that $e_i = (0, \ldots, 0, 1, 0 \ldots, 0)^\top$, with the 1 in the i-th position. This allows us, for instance, to write the columns of the matrix $\mathbf{T}$ corresponding to each resistance state as $\mathbf{T}_0 = \mathbf{T}\, e_1$, $\mathbf{T}_{\text{INH}} = \mathbf{T}\, e_2$, $\mathbf{T}_{\text{RIF}} = \mathbf{T}\, e_3$ and $\mathbf{T}_{\text{MDR}} = \mathbf{T}\, e_4$, with similar notation for other matrices (note that the dimension of the e_i is left open, but inferred from the matrix multiplication, in this case, they are in $\mathbb{R}^4$).

Further, writing $\mathbf{1}_i$ for the column vector in $\mathbb{R}^i$ whose entries are all 1, then the product $\mathbf{T}\,\mathbf{1}_4$ is a $G \times 1$ column vector whose entries are the numbers of individual cases of each VNTR genotype in the treated population, and the product $\mathbf{1}_G^\top \mathbf{T}\,\mathbf{1}_4$ is the sum of all the entries in $\mathbf{T}$ (the size of the treated population). Thus, we can write the size of the susceptible population, S, as:

$$S = N - \mathbf{1}_G^\top \mathbf{W}\,\mathbf{1}_4,$$

where $\mathbf{1}_G^\top \mathbf{W}\, \mathbf{1}_4$ is the size of the infected population, and where N is the total population size which remains constant. We treat N as modelling the set of all individuals who come in contact with infectious cases, and so we exclude individuals who either do not encounter infectious cases or are otherwise protected from infection. This variable, therefore, may be smaller than the actual population size.

The components of each vector $\mathbf{T}_k$ for k = 0, INH, RIF, or MDR, are integers, representing the number of individual cases for each genotype. In the schematic diagram of the model in Figure 16.2, we use $T_k = \mathbf{1}_G^\top \mathbf{T}_k$ to represent the total population number of treated individuals with resistance state k, with similar notation $U_k = \mathbf{1}_G^\top \mathbf{U}_k$ to represent the untreated populations. The matrix notation is gathered and shown in Table 16.2.

The arrows between populations in Figure 16.2 represent the directional rates of detection and treatment τ and acquisition of resistance to each drug or set of drugs, so that ρ_{INH} and ρ_{RIF} represent rates of acquisition of resistance to isoniazid and rifampicin, respectively, and ρ_{MDR} the rate of double acquisition.

In this model time is discrete, and during each time step the following events take place in sequence.

1. Disease transmission giving rise to new cases;
2. Natural recovery, cure, or death of cases;
3. Detection of cases, which are then treated with drugs;

TABLE 16.2
Summary of Linear Algebra Notation

Symbol	Meaning
$\mathbf{V}$	$G \times L$ matrix describing the VNTR genotypes.
$\mathbf{U}$, $\mathbf{T}$, $\mathbf{W}$	$G \times 4$ matrices of untreated, treated, and total cases, respectively, with columns corresponding to resistance profiles.
$\mathbf{U}_k$, $\mathbf{T}_k$, $\mathbf{W}_k$	$G \times 1$ column vector for resistance profile k of untreated, treated, and total cases, respectively.
U_k, T_k, W_k	Total population sizes of untreated, treated, and total with resistance profile k.
$\mathbf{U}_{i,k}$, $\mathbf{T}_{i,k}$, $\mathbf{W}_{i,k}$	(i, k) entries of the matrices $\mathbf{U}$, $\mathbf{T}$, $\mathbf{W}$: the number of cases in each category with genotype i and resistance profile k.
$\mathbf{1}_i$	$i \times 1$ column vector whose entries are all 1.
e_i	Column vector whose entries are 0 except for 1 in position i. Dimension determined by context.

4. Conversion among resistant profiles in treated cases due to acquisition of resistance; and

5. Mutation of the genetic marker (multiple VNTR loci).

The remainder of this section provides details of how each of these events are modelled. Readers wishing to focus on the statistical aspects of the ABC inference can skip these subsections and go directly to Section 16.4.

We regard the earlier process as a discrete-time stochastic model rather than a discrete-time approximation of a continuous-time stochastic process with rates approximating probabilities, although the latter interpretation becomes more appropriate as the time step length decreases. Here, rates of events will be treated as probabilities, which again is appropriate when time steps are short. The rate parameters are measured in years, but we make time steps 1/12 of a year.

A summary of all model parameters, both fixed and to be estimated, and their meanings, is provided in Table 16.3.

16.3.1 New infections

In our model, new infections occur by mass action. The per capita rate at which a susceptible individual becomes infected by a case with resistance profile k is given by β_k/N times the number of infected cases in state k. The transmission parameters β_k are scaled by $1/N$ for convenience since realistic values of β_k/N are typically very small, and this ensures that the β_k are on the natural 'per person per unit time' scale.

The acquisition of resistance to antibiotics often comes at a cost to the fitness of the bacterium, and we implement this fitness cost by assuming the transmission rate is lower for cases that carry resistance. Specifically, we assume a cost of c 'per drug', so that if β_0 is the transmission rate of sensitive cases, then cases resistant to one drug transmit at rate $\beta_{\text{INH}} = \beta_{\text{RIF}} = (1-c)\beta_0$ and cases resistant to two drugs at $\beta_{\text{MDR}} = (1-c)^2\beta_0$. Here, $c = 0.1$ is considered known and fixed based on previous analyses of molecular epidemiological data (Luciani et al., 2009).

We now construct an expression for the average transmission probability across the infected population. The matrix $\mathbf{W}$ records all infected cases with different resistance states in each column, and the individuals corresponding to these columns have different transmission rates $\beta = (\beta_0, \beta_{\text{INH}}, \beta_{\text{RIF}}, \beta_{\text{MDR}})^\top$. If we write D_β for the diagonal matrix whose entries are from β, then the matrix $\mathbf{W}\, D_\beta$ is the infected population matrix $\mathbf{W}$ whose columns have been scaled by the entries of β (the relevant transmission rates). The expression:

$$p = \frac{1}{N}\mathbf{1}_G^{\text{T}}\,\mathbf{W}\,D_\beta\,\mathbf{1}_4,$$

then gives the average transmission rate per susceptible individual. Since the population size N is usually large and the time steps are short, the value for

TABLE 16.3
Summary of Model Parameters

Symbol	Meaning	Fixed Value
δ	Rate of death and natural recovery	0.52
γ_0	Cure rate for resistance profile 0, when treated	0.5
γ_{INH}	Cure rate for resistance profile INH, when treated	0.25
γ_{RIF}	Cure rate for resistance profile RIF, when treated	0.25
γ_{MDR}	Cure rate for resistance profile MDR, when treated	0.05
N	Total susceptible population size in absence of disease	10^4
τ	Treatment and detection rate	0.5
c	Cost of resistance	0.1
Symbol	**Meaning**	**Prior**
β_0	Transmission rate for resistance profile 0	Gamma*
μ	Mutation rate of VNTR per locus per unit time	$U(0,1)$
ρ_{INH}	Rate of acquisition of resistance to INH	$U(0,1)$
ρ_{RIF}	Rate of acquisition of resistance to RIF	$U(0,1)$
ρ_{MDR}	Rate of acquisition of resistance to INH and RIF	$U(0,1)$

Note: The top set of parameters are given fixed values, whereas the bottom set of parameters are allocated prior distributions and estimated using ABC. Fixed values and priors are justified in Section 16.4.2. Rates are in units of per capita per year, but the time unit is set to 1/12 year in simulations. * Specifically, β_0 is assumed to follow a (shifted) gamma prior defined as $\beta_0 - 0.68 \sim \text{Gamma}(\text{shape} = 2, \text{rate} = 0.73)$. See Section 16.4.2 for further details.

p will nearly always be small. Accordingly, and to ensure that it does not exceed 1, we model the probability of transmission per susceptible individual as $\tilde{p} = \min\{1, p\}$.

At each time step, the number B of new infections is a random variable distributed as:

$$B \sim \text{Binomial}(S, \tilde{p}).$$

These B new infections are then allocated across VNTR genotypes and resistance profiles according to the proportions represented by the matrix $\mathbf{W}\, D_\beta$. That is, a multinomial random sample distributes B according to the existing infected population and their relative transmission rates, so that the resulting allocation is a $G \times 4$ matrix Δ_β. Finally, as new infections are all assumed to be initially undetected, they are allocated to the untreated subpopulation, so that the matrix $\mathbf{U}$ is updated to $\mathbf{U} \to \mathbf{U} + \Delta_\beta$.

16.3.2 Cure, recovery, and death

Infected individuals who are untreated (the population represented by the counts $\mathbf{U}$) recover or die at rate $\delta = \delta_r + \delta_d$ per case per time unit, where $\delta_r > 0$ is the rate of recovery and $\delta_d > 0$ is the rate of death due to any cause. The rate of cure due to successful treatment may vary according to resistance profile, so this rate is given by γ_k for $k = 0$, INH, RIF, MDR. The number of cures, recoveries, and deaths in a time step is given by:

$$R \sim \text{Binomial}(U, \delta),$$

for the untreated population, where $U = \mathbf{1}_G^\top \mathbf{U}\, \mathbf{1}_4$ is the total number of all untreated cases, and:

$$C_k \sim \text{Binomial}(T_k, \delta + \gamma_k),$$

for the treated population, where T_k is the number of treated individuals with resistance profile k (as defined at the start of this section). The R *untreated* recovered individuals are distributed across both VNTR genotypes and resistance profiles with a multinomial distribution according to the counts given in $\mathbf{U}$. These are recorded in the $G \times 4$ update matrix Δ_δ (so that the sum of the entries in Δ_δ is $R = \mathbf{1}_G^\top \Delta_\delta \mathbf{1}_4$). Similarly, the C_k *treated* recovered individuals of resistance profile k are distributed across the VNTR genotypes according to the distribution observed in $\mathbf{T}_k$. These recovered counts for *all* resistance profiles are recorded in the G update matrix $\Delta_{\delta+\gamma}$, which is constructed from the column vectors of recovered treated counts for profile k in the order $k = 0$, INH, RIF, and MDR. The matrices $\mathbf{U}$ and $\mathbf{T}$ are then updated to $\mathbf{U} \to \mathbf{U} - \Delta_\delta$ and $\mathbf{T} \to \mathbf{T} - \Delta_{\delta+\gamma}$ respectively. If the last instance of any genotype is removed by cure, recovery, or death, the matrices $\mathbf{U}, \mathbf{T}$, and $\mathbf{V}$ are adjusted by removing the rows corresponding to those genotypes, and the number of genotypes is updated with $G \to G - 1$.

Similarly to the case of new infections (Section 16.3.1), we assume that the recovery rate due to treatment depends only on the number of drugs the infecting strain is resistant to. Specifically, this implies that $\gamma_{\text{INH}} = \gamma_{\text{RIF}}$.

16.3.3 Detection and treatment

In this model, the detection of cases and the commencement of treatment are combined as a single process. Detected cases are transferred from the untreated class to the treated class. We denote this combined detection and treatment rate, per case, per unit time, as $\tau > 0$. With this rate, we draw D individuals to transfer between untreated and treated populations, where:

$$D \sim \text{Binomial}(U, \tau).$$

These D individuals are then allocated across VNTR genotypes and resistance profiles according to the observed distribution of untreated cases, $\mathbf{U}$. As before, this results in a $G \times 4$ update matrix Δ_τ, which we use to update $\mathbf{U} \to \mathbf{U} - \Delta_\tau$ and $\mathbf{T} \to \mathbf{T} + \Delta_\tau$.

16.3.4 Acquisition of drug resistance

Individual treated cases are able to convert from one resistance profile to another through adaptive evolution. That is, under drug treatment, natural selection acts to favour increasing levels of resistance. As a result of this process, individuals may move from the $k = 0$ resistance profile (sensitive to both drugs) to one of the other three resistance profiles: INH, RIF, or MDR (resistance to one or both drugs). Individuals may also move from resistance to exactly one of the drugs (INH or RIF) to the multiple drug resistance profile MDR. We respectively denote the rate of acquisition of resistance to INH or RIF by ρ_{INH} and ρ_{RIF} and denote the rate of acquisition of resistance from individuals in the sensitive population to both drugs simultaneously by ρ_{MDR}. These conversions and rates are illustrated schematically in Figure 16.2.

To model resistance acquisition, we select individuals to move between resistance profiles in the treated population, for example, between columns in the matrix $\mathbf{T}$. Acquiring resistance to the drug rifampicin will result in individuals moving from the column $\mathbf{T}_0$ to $\mathbf{T}_{\text{RIF}}$ and from $\mathbf{T}_{\text{INH}}$ to $\mathbf{T}_{\text{MDR}}$ at a rate ρ_{RIF}. Similarly, acquiring resistance to the drug isoniazid results in individuals moving from the column $\mathbf{T}_0$ to $\mathbf{T}_{\text{INH}}$ and from $\mathbf{T}_{\text{RIF}}$ to $\mathbf{T}_{\text{MDR}}$ at a rate ρ_{RIF}. Simultaneous acquisition of resistance to both drugs moves individuals from the column $\mathbf{T}_0$ to $\mathbf{T}_{\text{MDR}}$ at the rate ρ_{MDR}. These movements occur between columns, but not across rows (infections do not change VNTR genotypes through this process).

Mechanistically, we can obtain the number of cases of genotype i transitioning from resistance profile k to resistance profile k', denoted $A_{i,k \to k'}$, as:

$$
\begin{aligned}
A_{i,0\to *} &\sim \text{Multinomial}(\mathbf{T}_{i,0}, \rho_{0\to *}) \\
A_{i,\text{INH}\to\text{MDR}} &\sim \text{Binomial}(\mathbf{T}_{i,\text{INH}}, \rho_{\text{RIF}}) \\
A_{i,\text{RIF}\to\text{MDR}} &\sim \text{Binomial}(\mathbf{T}_{i,\text{RIF}}, \rho_{\text{INH}}),
\end{aligned}
$$

where $A_{i,0\to *} = (A_{i,0\to\text{INH}}, A_{i,0\to\text{RIF}}, A_{i,0\to\text{MDR}}, A_{i,0\to 0})^\top$ is the vector of cases transitioning from sensitivity, $\mathbf{T}_{i,k}$ is the entry of the matrix $\mathbf{T}$ corresponding to the genotype i and resistance profile k (Table 16.2), and $\rho_{0\to *} = (\rho_{\text{INH}}, \rho_{\text{RIF}}, \rho_{\text{MDR}}, 1 - \sum_k \rho_k)^\top$ is the vector of probabilities of these events.

If we denote $\Delta_{k\to k'}$ as column vectors of counts of movements from resistance profile k to k' across all G genotypes, we can then construct the overall $G \times 4$ update matrix:

$$
\Delta_\rho = (\Delta_0 \mid \Delta_{\text{INH}} \mid \Delta_{\text{RIF}} \mid \Delta_{\text{MDR}}),
$$

from the column vectors Δ_k, which denote the total population change for resistance profile k, where:

$$\begin{aligned}
\Delta_0 &= -(\Delta_{0\to\text{INH}} + \Delta_{0\to\text{RIF}} + \Delta_{0\to\text{MDR}}) \\
\Delta_{\text{INH}} &= \Delta_{0\to\text{INH}} - \Delta_{\text{RIF}\to\text{MDR}} \\
\Delta_{\text{RIF}} &= \Delta_{0\to\text{RIF}}\Delta_{\text{INH}\to\text{MDR}} \\
\Delta_{\text{MDR}} &= \Delta_{0\to\text{MDR}} + \Delta_{\text{RIF}\to\text{MDR}} + \Delta_{\text{INH}\to\text{MDR}}.
\end{aligned}$$

The population of treated cases is then updated to $\mathbf{T} \to \mathbf{T} + \Delta_\rho$.

16.3.5 Mutation of the marker

The set of $L = 15$ VNTR loci constitute the genetic marker used to genotype bacterial isolates (see Section 16.2). Each genotype is a list of numbers of tandem repeat units at the L loci. The states of all VNTRs in the infected population are given by the $G \times L$ matrix $\mathbf{V}$ with elements V_{ij} describing the repeat number of locus j in genotype i. Each locus mutates through a stepwise mutation process at rate μ per locus per case per unit time. When mutation occurs, the repeat number V_{ij} at a locus j of genotype i changes by $+1$ or -1, each with probability 0.5. A repeat number of 1 is treated as an absorbing boundary (i.e. there is zero probability of the repeat number increasing from 1 to 2) because at state 1 there is no longer a genetic sequence that is tandemly repeated and no mechanism such as replication slippage acts to expand it from 1 to 2.

Mutation of the marker has the effect of moving cases between the rows of the matrix $\mathbf{W}$. We first identify the number of mutation events in the population, M, where $M \sim \text{Binomial}(S, \mu)$ and $S = N - \mathbf{1}_G^{\text{T}}\,\mathbf{W}\,\mathbf{1}_4$ is the size of the susceptible population (see Section 16.3). The M cases are then distributed across the population of VNTR genotypes and resistance profiles, according to the entries of the matrices $\mathbf{T}$ and $\mathbf{U}$. Each individual case undergoing mutation corresponds to a specific entry in either $\mathbf{T}$ or $\mathbf{U}$. This entry is described by its VNTR genotype $\mathbf{V}_i = (V_{i,1}, \ldots, V_{i,L})$, where $L = 15$ for the Bolivian data, and its resistance profile, k = 0, INH, RIF, MDR. The result of the mutation is a change to the VNTR genotype, which is represented by a change in the repeat number at a single locus, V_{ij}, by ± 1. This may or may not result in a VNTR genotype that is already present in the population.

If the new VNTR genotype already appears as a row in the matrix $\mathbf{V}$ as an existing type in the data, then there is no change to $\mathbf{V}$. The matrix $\mathbf{T}$ or $\mathbf{U}$ on the other hand is changed by subtracting 1 from one entry and adding one to another entry in the same column (the resistance profile, k, does not change). In matrix terms, supposing the change is to a treated case, this can be described by updating $\mathbf{T} \to \mathbf{T} - e_{i,j} + e_{i,k}$, where $e_{i,j}$ is the matrix whose entries are zero except for a 1 in the (i, j)-th position and where the VNTR genotype changes from row j to row k.

If the new VNTR genotype does not already appear in the population, then the matrix $\mathbf{V}$ is expanded to include a new row describing the new genotype, so that $\mathbf{V}$ becomes a $(G+1) \times 15$ matrix. The update for $\mathbf{T}$ or $\mathbf{U}$ is the same as described earlier except that now both matrices are $(G+1) \times 4$ dimensional. Subsequent to this update, we increment $G \to G+1$. If mutation of a VNTR genotype removes the last instance of the original genotype from $\mathbf{U}$ and $\mathbf{T}$, the corresponding rows of matrices $\mathbf{V}$, $\mathbf{U}$, and $\mathbf{T}$ are deleted, requiring the update $G \to G-1$.

16.3.6 Initial conditions of the model

The model covers the period from when drugs are introduced at time $t = 0$ to when sampling occurs. Since the main first-line anti-tuberculosis drugs were discovered/developed in the 1940s to early 1960s, we assumed treatment commenced around 1960 and ran the simulation for a period of 50 years. We assumed that both drugs, isoniazid and rifampicin, were introduced at the same time and are administered together in combination therapy. The standard course of treatment includes both drugs along with other first-line drugs (WHO, 2015).

We assume that at the start of the process all cases are sensitive to both drugs, and that the number of cases is at equilibrium in the absence of treatment and resistance. To compute this equilibrium state, we consider the differential equation describing the deterministic version of the model ignoring VNTR genotypes. Namely:

$$\frac{dU}{dt} = (\beta_0/N)SU - \delta U,$$

where $S = N - U$ and t indicates time. Setting dU/dt to zero and solving for the dynamic variables, we obtain equilibrium values of

$$\hat{U} = N\left(1 - \frac{\delta}{\beta_0}\right) \qquad \text{and} \qquad \hat{S} = \frac{\delta N}{\beta_0},$$

for $U > 0$.

The basic reproduction number of a pathogen R_0 is defined to be the average number of new infectious cases caused by a single infection in a completely susceptible population. In our model, before there is any treatment, assuming all cases are doubly susceptible, a single case on average persists for $1/\delta$ years and generates $S\beta_0/N$ new cases per unit time, but since $S = N$ in a wholly susceptible population then $R_0 = \beta_0/\delta$.

All cases are initially untreated and sensitive. From time $t = 0$, treatment in the population commences. To reintroduce into the model genetic variation at the marker loci, the initial distribution of genotype clusters is a random sample drawn from the infinite alleles model from population genetic theory (Ewens, 1972; Hubbell, 2001; Luciani et al., 2008). The infinite alleles model depends on a single parameter, the diversity parameter, which we set to $2\hat{U}\mu L$, where $\hat{U}$ is the number of cases, taken from the equilibrium value described earlier, μ is the mutation rate per VNTR locus, and L is the number of VNTR loci used in genotyping isolates. To initialise the multi-locus VNTR genotypes, each genotype is a sequence of random integers, of length L, with each VNTR number V_{ij} drawn from a discrete uniform distribution over $\{1, \ldots, 10\}$. Although the initial distribution of genotype clusters is set under the infinite alleles model, the mutation process for VNTRs brings the distribution in line with the stepwise model over time.

The initial conditions are a function of the parameters which are set according to the priors specified in Section 16.4.2.

16.4 Inference with Approximate Bayesian Computation

For the model in Section 16.3, when the data are only observed at a single point in time, the cost of evaluating the likelihood function is computationally prohibitive. This results from the 'incomplete' nature of the observed data (see Section 16.2) in the sense that we only have access to a snapshot of the population, via the observed sample, at the time the study was conducted, with no direct measurements of the system as it progressed. Computing the likelihood then requires integrating over all potential trajectories the population could have gone through before reaching its final, observed state.

As such, we adopt approximate Bayesian computation (ABC) methods as a means of performing Bayesian statistical inference for the unknown model parameters $\theta = (\beta_0, \mu, \rho_{\text{INH}}, \rho_{\text{RIF}}, \rho_{\text{MDR}})^\top$. As observed in other chapters in this Handbook, the ABC approximation to the true posterior distribution is given by:

$$\pi_{ABC}(\theta|s_{obs}) \propto \pi(\theta) \int K_h(\|s - s_{obs}\|)p(s|\theta)ds,$$

where $\pi(\theta)$ is the prior distribution, $s = S(\mathbf{X})$ is a vector of summary statistics with $s_{obs} = S(\mathbf{X}_{obs})$, $p(s|\theta)$ is the computationally intractable likelihood function for the summary statistics s, and $K_h(u) = K(u/h)/h$ is a standard smoothing kernel with scale parameter $h > 0$. In the following analyses, we used the uniform kernel on $[-h, h]$ for $K_h(u)$. The quality of the ABC approximation depends on the information loss in the summary statistics s over the full dataset $\mathbf{X}$ and the size of the kernel scale parameter h with smaller h producing greater accuracy and increased computational cost. Choice of both s and h are typically driven by the amount of expert knowledge and computation available for the analysis.

For the present analysis, we implement a version of a simple ABC importance sampling algorithm, as outlined in the box. Given a suitable importance sampling distribution $q(\theta)$, the algorithm produces a set of weighted samples from the ABC approximation to the true posterior

$(\theta^{(1)}, w^{(1)}), \ldots, (\theta^{(\tilde{N})}, w^{(\tilde{N})}) \sim \pi_{ABC}(\theta|s_{obs})$. As with standard importance sampling, suitable choice of $q(\theta)$ is important to avoid high variance in the importance weights and also to avoid needlessly generating datasets $s = S(\mathbf{X}^{(i)})$, $\mathbf{X}^{(i)} \sim p(\mathbf{X}|\theta)$ for which $s^{(i)}$ and s_{obs} will never be close.

Algorithm 16.1: ABC Importance Sampling Algorithm

Inputs:

- A target posterior density $\pi(\theta|\mathbf{X}_{obs}) \propto p(\mathbf{X}_{obs}|\theta)\pi(\theta)$, consisting of a prior distribution $\pi(\theta)$ and a procedure for generating data under the model $p(\mathbf{X}_{obs}|\theta)$.
- A proposal density $q(\theta)$, with $q(\theta) > 0$ if $\pi(\theta|\mathbf{X}_{obs}) > 0$.
- An integer $\tilde{N} > 0$.
- An observed vector of summary statistics $s_{obs} = S(\mathbf{X}_{obs})$.
- A kernel function $K_h(u)$ and scale parameter $h > 0$.

Sampling:
For $i = 1, \ldots, \tilde{N}$:

1. Generate $\theta^{(i)} \sim q(\theta)$ from sampling density q.
2. Generate $\mathbf{X}^{(i)} \sim p(\mathbf{X}|\theta^{(i)})$ from the likelihood.
3. Compute the summary statistics $s^{(i)} = S(\mathbf{X}^{(i)})$.
4. Assign $\theta^{(i)}$ the weight $w^{(i)} \propto K_h(\|s^{(i)} - s_{obs}\|)\pi(\theta^{(i)})/q(\theta^{(i)})$.

Output:
A set of weighted parameter vectors $\{(\theta^{(i)}, w^{(i)})\}_{i=1}^{\tilde{N}} \sim \pi_{ABC}(\theta|s_{obs})$.

To determine a suitable importance sampling distribution $q(\theta)$, we adopt a two-stage procedure, following the approach of Fearnhead and Prangle (2012). In the first stage, we perform a pilot ABC analysis using a sampling distribution that is diffuse enough to easily encompass the ABC posterior approximation obtained for a moderate value of the kernel scale parameter h. We specified $q(\theta) \propto \pi(\theta)I(\theta \in A)$, which is proportional to the prior, but restricted to the hyper-rectangle A. Here, A is constructed as the smallest credible hyper-rectangle that we believe contains the ABC posterior approximation. As such, this $q(\theta)$ will identify the general region in which $\pi_{ABC}(\theta|s_{obs})$ is located. Specifically, for

$\theta = (\beta_0, \mu, \rho_{\text{INH}}, \rho_{\text{RIF}}, \rho_{\text{MDR}})^\top$, we adopt $q(\theta) = \tilde{\pi}_{15}(\beta_0) \times U(0, .005) \times U(0, .01) \times U(0, .005) \times U(0, .001)$, where $\tilde{\pi}_{15}(\beta_0)$ is the prior $\pi(\beta_0)$ for β_0 specified in Section 16.4.2, but truncated to exclude density above the point $\beta_0 = 15$.

For posterior distributions with strong dependence between parameters, defining $q(\theta)$ over such a hyper-rectangle may be inefficient, as it will cover many regions of effectively zero posterior density. Accordingly, we construct the sampling distribution for the second stage, with the lowest value of h, as a kernel density estimate of the previous ABC estimate of the posterior distribution: $q(\theta) = \sum_i w^{(i)} L(\theta|\theta^{(i)})$, where L is a suitable kernel density (not to be confused with the kernel K_h). This approach follows the ideas behind the sequential Monte Carlo-based ABC samplers of Sisson et al. (2007) and others. At each stage, the kernel scale parameter h is decreased, and determined as the value which results in ~2,000 posterior samples with non-zero weights, for the given computational budget.

To ensure greater efficiency at each stage, we also performed a non-linear regression adjustment using a neural network with a single hidden layer (see Blum and François, 2010; Csilléry et al., 2012; Beaumont et al., 2002), as implemented in the *R* package `abc`. The adjustment used logistic transformations for the response.

For samples drawn from the final importance sampling distribution $q(\theta)$, the data generation procedure took on average ~40 seconds in *R*. This is computationally expensive from an ABC context and could be reduced by recoding the simulator in a compiled language such as *C* or by adapting the 'lazy ABC' ideas of Prangle (2016) to terminate early those simulations that are likely to be rejected. In this implementation, we performed importance sampling from each distribution $q(\theta)$ in parallel on multiple nodes of a computational cluster.

16.4.1 Summary statistics

Considering the matrix structure of the observed data $\mathbf{X}_{obs}$ (see Section 16.2), we determine the information content in $\mathbf{X}$ as if it was the design matrix of a regression model and summarise it accordingly. Specifically, we define the summary statistics $s = S(\mathbf{X})$ to be the upper-triangular elements of the matrix:

$$(\mathbf{1}_g|\mathbf{X})^\top(\mathbf{1}_g|\mathbf{X}),$$

where the vertical lines denote the addition of an extra column. The added columns of ones enrich the set of summary statistics by including the row and column totals of $\mathbf{X}$. Alternatively, these summary statistics can be described as:

1. g: the number of distinct genotypes in the sample.

2. n_k: the number of isolates with resistance profile k = 0, INH, RIF, and MDR.

3. $c_{k,k'} = (\mathbf{X}_k)^\top \mathbf{X}_{k'}$: the dot product between the resistance profiles of k and k' within $\mathbf{X}$.

Note that these summary statistics are over-specified in that $n_0 + n_{\text{INH}} + n_{\text{RIF}} + n_{\text{MDR}}$ equals the total number of isolates sampled from the population, which is known and equal to the number of isolates in the observed data sample (100 for the Bolivian data). Accordingly, and without loss of generality, we remove n_{MDR} as a summary statistic to avoid collinearity. In combination, this set of 14 summary statistics efficiently encapsulates the available information about the covariance structure of the original dataset $\mathbf{X}$, the distribution of the isolates among the different resistance profiles and the degree of diversity of isolates within the sample.

For the Bolivian dataset, there are $g = 68$ distinct genotypes, $n_0 = 78$ sensitive isolates, $n_{\text{INH}} = 8$ isolates resistant to isoniazid only, $n_{\text{RIF}} = 0$ isolates resistant to rifampicin only, and $n_{\text{MDR}} = 16$ doubly resistant isolates (Table 16.1). The remaining statistics, $c_{k,k'}$, are computed as:

	0	INH	RIF	MDR
0	232	15	0	1
INH	–	10	0	6
RIF	–	–	0	0
MDR	–	–	–	18

Finally, in order to reduce the impact of summary statistics operating on different scales, we compare simulated and observed summary statistics within the kernel $K_h(\|s - s_{obs}\|)$ via the $L_{\frac{1}{2}}$ norm:

$$\|s - s_{obs}\| = \|S(\mathbf{X}) - S(\mathbf{X}_{obs})\| = \left(\sum_{j=1}^{\dim(s)} [S(\mathbf{X})_j - S(\mathbf{X}_{obs})_j]^{\frac{1}{2}} \right)^2,$$

where $\dim(s) = 14$ is the number of summary statistics. Alternative approaches could rescale the statistics via an appropriate covariance matrix (e.g. Luciani et al., 2009; Erhardt and Sisson, 2016) or use other norms; however the results in the following section proved to be robust to more structured comparisons, so we did not pursue this further. In particular, the following results were robust to these choices because of the use of a good (non-linear) regression adjustment, which greatly improves the ABC posterior approximation and which has a larger impact on this approximation than the choice of metric $\|\cdot\|$.

16.4.2 Parameter specifications and prior distributions

Of the 13 model parameters (Table 16.3), eight of these are known well enough for the purposes of our analysis to fix their values. Namely, the parameters $(\delta, \gamma_0, \gamma_{\text{INH}}, \gamma_{\text{RIF}}, \gamma_{\text{MDR}}, N, \tau, c)^\top$ are set to these fixed values. We justify our choices for these following values. The remaining five parameters $\theta = (\beta_0, \mu, \rho_{\text{INH}}, \rho_{\text{RIF}}, \rho_{\text{MDR}})^\top$ are to be estimated and require a prior distribution specification.

The rate of death or recovery, δ, was fixed and set to be $\delta = 0.52$ per case per year following Dye and Espinal (2001) and Cohen and Murray (2004). Similarly, following Dye and Espinal (2001), untreated individuals are detected and treated at rate $\tau = 0.5$ per case per year. The rates of recovery due to treatment, γ_k, for resistance profiles $k = 0$, INH, RIF, and MDR, can be written in terms of the probability of treatment success:

$$p_k = \frac{\delta_r + \gamma_k}{\delta_d + \delta_r + \gamma_k}.$$

We set the cure rates to be $\gamma_0 = 0.5, \gamma_{\text{INH}} = \gamma_{\text{RIF}} = 0.25$, and $\gamma_{\text{MDR}} = 0.05$, which, by using $\delta_r = 0.2$ (Dye and Espinal, 2001; Cohen and Murray, 2004), corresponds to treatment success probabilities of approximately $p_0 = 0.69$, $p_{\text{INH}} = p_{\text{RIF}} = 0.58$, and $p_{\text{MDR}} = 0.44$. These values are within the supported ranges in the literature, namely, $p_0 = 0.45 - 0.75$, $p_{\text{INH}} = p_{\text{RIF}} = 0.3 - 0.6$, and $p_{\text{MDR}} = 0.05 - 45$ (Blower and Chou, 2004). We chose higher values within these ranges since Blower and Chou (2004) explored a wide range of possibilities in models including epidemiologically pessimistic scenarios.

The fitness cost of drug resistance, c, was fixed and set to be $c = 0.1$ based on estimates by Luciani et al. (2009). To set the total population size N, we first observe that because the sample of 100 isolates represents $\sim$1.1% of the population, this implies that the infected population is 9,091. We expect that the number of susceptible individuals who are exposed to disease is somewhat higher than this. Accordingly, we assumed that the total size of the population susceptible to tuberculosis is $N = 10{,}000$. Larger total population sizes can be used, at the price of greater computational overheads for generating data under the model.

Previous work estimated rates of resistance acquisition by mutation to be around $0.0025 - 0.02$ per case per year (Luciani et al., 2009). The rate of mutation of the VNTR loci in *M. tuberculosis* was estimated to be around 10^{-3} per locus per case per year (Reyes and Tanaka, 2010; Aandahl et al., 2012; Ragheb et al., 2013), but lower estimates have also been found (Wirth et al., 2008; Supply et al., 2011). All of these mutation rates are much lower than 1. We treat these mutation rate parameters as probabilities and conservatively set the standard uniform distribution as a wide prior on each parameter. That is, for the acquisition of resistance to isoniazid or rifampicin (or both), we specify priors for the rates of resistance acquisition as $\rho_{\text{INH}}, \rho_{\text{RIF}}, \rho_{\text{MDR}} \sim U(0,1)$. Similarly, for the mutation rate of the VNTR molecular marker, μ, we use the prior $\mu \sim U(0,1)$.

The transmission parameter for doubly sensitive strains β_0 is given the shifted gamma prior:

$$\beta_0 - 0.68 \sim \text{Gamma}(\text{shape} = 2, \text{rate} = 0.73),$$

where the parameters are chosen such that the resulting prior distribution of the basic reproduction number R_0 closely resembles the distribution obtained in a numerical analysis of tuberculosis dynamics by Blower et al. (1995). Note that the prior on β_0 is shifted in order ensure the realistic condition that $R_0 > 1$. A value of R_0 lower than unity would lead to extinction of *M. tuberculosis*.

We reiterate that we interpret the rate parameters as probabilities per time step and handle the parameters so that their values remain in (0,1). This approximation increases in accuracy as the time unit decreases. Here, we divide the natural time unit of one year into new units of 1/12 year per time step.

16.5 Competing Models of Resistance Acquisition

We estimate the rates of acquisition of drug resistance to rifampicin and isoniazid by fitting the model described in Section 16.3 to the Bolivian data (Monteserin et al., 2013) with the ABC method described in Section 16.4. Additionally, by constraining particular resistance-acquisition parameters ρ_k to produce meaningful submodels of the full model, we are able to examine two specific biological questions. The relationships between the two submodels and the full model are illustrated in Figure 16.3. First, we ask whether it is possible for multi-drug resistance to evolve directly from doubly sensitive bacteria or whether this direct conversion does not occur (i.e. $\rho_{\text{MDR}} = 0$: Submodel 1). Second, we ask whether differences between rates of mutation to rifampicin and isoniazid resistance are apparent at the epidemiological scale (i.e. $\rho_{\text{INH}} = \rho_{\text{RIF}} = \rho_{\text{single}}$: Submodel 2).

Figure 16.4 illustrates the ABC marginal posterior density estimates of each parameter under the three different sets of model assumptions. Under the

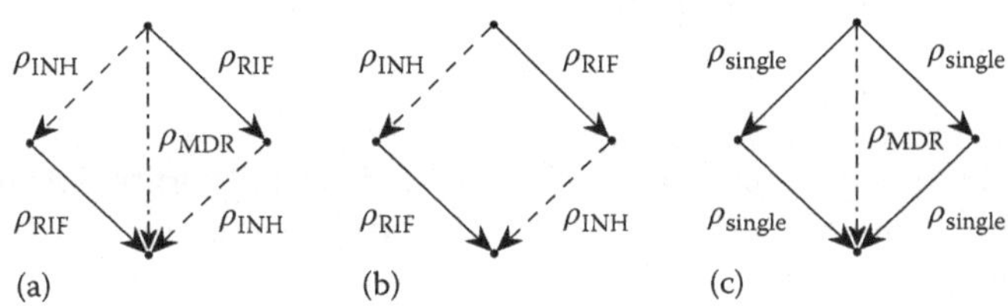

FIGURE 16.3
Three candidate models of acquisition of multiple drug resistance. (a) The full model: two different rates of conversion leading to acquisition of resistance and a rate of conversion from resistance profile 0 to resistance profile MDR. This model is also shown in Figure 16.2. (b) Submodel 1: no direct conversion from resistance profile 0 to resistance profile MDR ($\rho_{\text{MDR}} = 0$). (c) Submodel 2: same rate of conversion for the two drugs ($\rho_{\text{INH}} = \rho_{\text{RIF}} = \rho_{\text{single}}$).

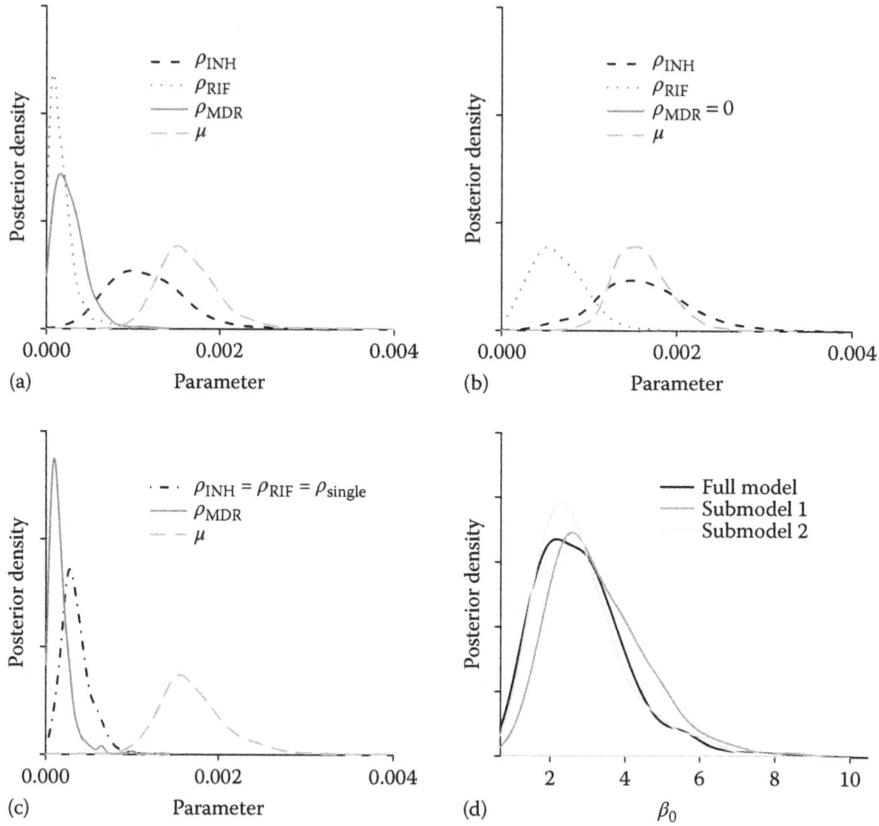

FIGURE 16.4
Estimated ABC marginal posterior densities for each estimated parameter under (a) the full model, (b) Submodel 1 ($\rho_{\text{MDR}} = 0$), and (c) Submodel 2 ($\rho_{\text{INH}} = \rho_{\text{RIF}} = \rho_{\text{single}}$). Panel (d) shows the estimated ABC marginal posterior density of the transmission rate β_0 of the sensitive strain for each model structure.

full model, there is a clear visual difference between the rates of mutation of rifampicin and isoniazid resistance, with the latter occurring at a much higher rate. In contrast, the rate of simultaneous resistance acquisition appears to be higher than that for rifampicin alone. When eliminating the possibility of simultaneous acquisition of multiple drug resistance $\rho_{\text{MDR}} = 0$ (Submodel 1), ρ_{INH} and ρ_{RIF} both increase, relative to the full model, to compensate for the imposed restriction when fitting to the observed data (Figure 16.4b). Similarly, when we fix the identity $\rho_{\text{INH}} = \rho_{\text{RIF}} = \rho_{\text{single}}$ (Submodel 2) to impose a single rate of resistance acquisition, the posterior density of this parameter moves to intermediate values compared to the two distinct rates of acquisition estimated under the full model (Figure 16.4c). The estimated posterior densities for the

TABLE 16.4
ABC Posterior Means with Lower and Upper Limits of the 95% HPD Credible Intervals for Each Parameter of Each Fitted Model

	ρ_{INH}	ρ_{RIF}	ρ_{MDR}	μ	β_0
Full model					
Posterior mean	1.14×10^{-3}	1.67×10^{-4}	2.62×10^{-4}	1.64×10^{-3}	2.85
CI lower limit	3.40×10^{-4}	3.82×10^{-6}	3.93×10^{-6}	1.11×10^{-3}	0.97
CI upper limit	1.94×10^{-3}	4.28×10^{-4}	5.81×10^{-4}	2.40×10^{-3}	5.33
Submodel 1					
Posterior mean	1.60×10^{-3}	6.37×10^{-4}	–	1.59×10^{-3}	3.29
CI lower limit	4.55×10^{-4}	1.27×10^{-4}	–	1.03×10^{-3}	1.20
CI upper limit	2.49×10^{-3}	1.24×10^{-3}	–	2.19×10^{-3}	5.78
Submodel 2					
Posterior mean	3.46×10^{-4}	–	1.56×10^{-4}	1.70×10^{-3}	2.81
CI lower limit	7.26×10^{-5}	–	6.62×10^{-7}	1.10×10^{-3}	0.86
CI upper limit	6.90×10^{-4}	–	3.76×10^{-4}	2.54×10^{-3}	5.20

transmission (β_0) and mutation (μ) parameters are visually similar across all models. ABC marginal posterior means and highest posterior density (HPD) credible intervals for all models are reported in Table 16.4.

16.5.1 Can resistance to both drugs be acquired simultaneously?

To determine whether resistance to both drugs can evolve directly from a double sensitive strain within an infection, we compare Submodel 1 ($\rho_{\text{MDR}} = 0$) against the full model. Formal standard Bayesian model comparison typically occurs through Bayes factors. In the ABC framework, this task is complicated by the need to perform ABC with summary statistics that are informative for the model indicator parameter, in addition to those informative for the model specific parameters. Such summary statistics can not only be difficult to identify, but the resulting composite vector of summary statistics can be high dimensional, which may then produce more inaccurate inference than if each model was analysed independently. See, for example, Robert et al. (2011), Marin et al. (2014) and Marin et al. (2019, Chapter 6, this volume) for a discussion of these issues. A useful alternative is to consider posterior predictive checks or related goodness-of-fit tests (e.g. Thornton and Andolfatto, 2006; Csilléry et al., 2010; Aandahl et al., 2012; Prangle et al., 2014).

Figure 16.5 shows the posterior predictive distribution of the summary statistics ($n_0, n_{\text{INH}} + n_{\text{RIF}}, n_{\text{MDR}}$) described in Section 16.4.1, for the full model [panel (a)] and Submodel 1 [panel (b)], where a darker intensity indicates higher density. This predictive distribution graphically illustrates

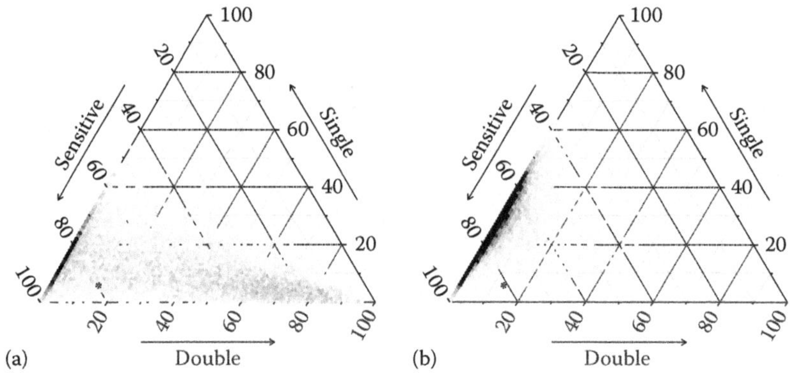

FIGURE 16.5
Posterior predictive distribution of $(n_0, n_{\text{INH}} + n_{\text{RIF}}, n_{\text{MDR}})$ under the full model (a) and Submodel 1 (b). Darker intensity indicates higher posterior density. The asterisk (*) indicates the observed data (78, 8, 16).

each model's ability to generate the observed summary statistics (78, 8, 16), indicated by the asterisks, which represent the number of individuals in the sample sensitive to both drugs (n_0), resistant to a single drug $(n_{\text{INH}} + n_{\text{RIF}})$, and resistant to both drugs (n_{MDR}).

The predictive distributions for each model are diffuse, particularly for the full model. This variability is expected given that the sample size is small (100 isolates) and that the evolution of drug resistance from sensitivity is a relatively rare stochastic event. In the case of Submodel 1 [Figure 16.5 panel (b)] where we impose the condition $\rho_{\text{MDR}} = 0$, the density of samples is shifted away from the bottom-right corner, which represents double resistance. This pattern is due to the lack of the direct route to multi-drug resistance. The observed data (asterisk) is in a region of low posterior predictive density under Submodel 1, and so we conclude that this model is not particularly supported by the data. In contrast, the observed data lie more clearly within a moderately high density region of the posterior predictive under the full model [Figure 16.5 panel (a)]. This analysis therefore suggests that of the two competing hypotheses, it is more likely that resistance to both drugs can be acquired simultaneously $(\rho_{\text{MDR}} > 0)$ than otherwise. Note, however, that this direct route is not the only possible path to double resistance, which can still occur in stages through single resistance.

16.5.2 Is resistance to both drugs acquired at equal rates?

In order to determine whether the rates of acquisition of resistance to the two drugs are equal $(\rho_{\text{INH}} = \rho_{\text{RIF}})$, we compare Submodel 2 against the full model. Figure 16.6 depicts the posterior predictive distribution of $(n_{\text{INH}}, n_{\text{RIF}})$

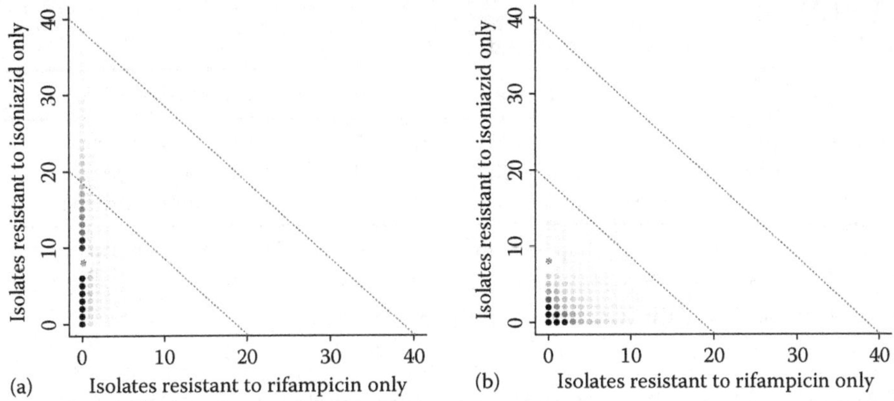

FIGURE 16.6
Posterior predictive distribution of $(n_{\mathrm{INH}}, n_{\mathrm{RIF}})$ under the full model (a) and Submodel 2 (b) Darker intensity indicates higher posterior predictive density. The asterisk (*) indicates the observed data (8, 0).

under each model – the number of cases resistant only to isoniazid (n_{INH}) and the number of cases resistant only to rifampicin (n_{RIF}) in the sample. The observed values of these summary statistics are $n_{\mathrm{INH}} = 8$ for isoniazid and $n_{\mathrm{RIF}} = 0$ for rifampicin, illustrated as the asterisk in Figure 16.6. As Submodel 2 does not favor any drug over the other, the predictive surface is symmetric with respect to the line $n_{\mathrm{INH}} = n_{\mathrm{RIF}}$. The extra flexibility provided by the full model shifts the predictive distribution towards the observed data. While the distribution under the full model comfortably accommodates the empirical point in a high density region, the predictive distribution under Submodel 2 is much more diffuse. This indicates that while the observed data are not unsupported under Submodel 2, it is far more likely to be observed under the full model. As a result, we conclude that the evidence favours the drugs being acquired at different rates, specifically, isoniazid resistance evolves faster than rifampicin resistance.

16.5.3 The relative contribution of transmission and treatment failure to multi-drug–resistant tuberculosis

In addition to estimating the rates of acquisition of drug resistance and assessing whether rates differ, we may also consider where doubly resistant cases come from. That is, estimation of the relative contribution to multi-drug resistant cases of transmission of existing MDR-TB strains compared to treatment failure leading to evolution of multi-drug resistance. The posterior predicted samples generated under the full model provide a clear portrait of the relative

TABLE 16.5
Contributions to MDR-TB from Alternative Sources

Source	Median	Mean	95% Credible Interval
Transmission	0.9975	0.9655	(0.7826, 0.9999)
Conversion in one step	0.0023	0.0284	(0.0000, 0.1667)
Conversion in two steps	0.0000	0.0060	(0.0000, 0.0073)

Note: This table contains the posterior medians and means and lower and upper limits of the 95% HPD credibility intervals for the proportion of double resistance cases originating from each possible source.

contribution of the different paths to achieving double resistance (see e.g. Luciani et al., 2009 for an additional illustration of this procedure).

Table 16.5 shows the means, medians and the 95% HPD credible intervals for the predicted proportion of cases of double resistance from each potential source. These proportions are obtained conditionally on there being at least one case of double resistance in the predictive sample. Simulated samples of this nature account for 99.67% of all predictive samples. The predictive distributions of the proportions are highly asymmetric (not shown), making the median a more reliable point estimate than the mean.

In the overwhelming majority of posterior predictive samples, direct *transmission* was the main source of acquisition of double resistance, followed by *conversion* in a single step directly from a sensitive profile (from profile 0 to MDR), and conversion in two steps via a state of resistance to a single drug (from profile 0 to INH to MDR, or from 0 to RIF to MDR). This analysis corroborates the finding from Section 16.5.1 that ρ_{MDR} is most likely positive, and furthermore, that this path is likely to be of even greater importance than conversion in two steps.

16.6 Conclusions

In this chapter we have estimated epidemiological parameters describing the acquisition of multi-drug resistance in *M. tuberculosis* from molecular epidemiological data (Monteserin et al., 2013) using approximate Bayesian computation. The underlying model is intended to capture essential processes that give rise to the data, namely, transmission of the disease, recovery or death, and within-host evolution giving rise to drug resistance and new genotypes at the molecular marker loci. From this analysis, we may draw three major biological conclusions about the manner in which drug resistance arises.

First, there is an asymmetry in the acquisition of resistance to isoniazid and rifampicin. Specifically, isoniazid resistance occurs approximately an order of magnitude more frequently than resistance against rifampicin (see Table 16.4). This asymmetry in rates is consistent with *in vitro* (i.e. through laboratory experiments) microbiological estimates of mutation rates per cell generation which find around 1 to 2 orders of magnitude difference between the two rates (David, 1970; Ford et al., 2013).

Second, the analysis supports the occurrence of *direct conversion* from doubly drug sensitive to doubly resistant (MDR) infections. This may be initially unintuitive because under mutation alone, if mutation occurs at rate ρ per gene per unit time, the rate of appearance of double mutants is ρ^2, which would be vanishingly small if ρ is low. However, using a mathematical model, Colijn et al. (2011) argued that direct conversion can occur surprisingly fast because resistant cells are sometimes present at low frequencies in a within-host population even before treatment commences. Our analysis of data at the epidemiological level is consistent with that theoretical result. This direct conversion to double resistance is epidemiologically important, as it accelerates the accumulation of resistance, in that resistance evolution does not have to take place sequentially. Once double resistant mutants appear, transmission of these mutants further increases their prevalence in the population.

Third, the overwhelming majority of cases of multi-drug resistant tuberculosis come from transmission of already multi-drug resistant strains (see Table 16.5), a finding that is consistent with those of Luciani et al. (2009). This large contribution of transmission occurs despite the 10% transmission cost of each resistance, which results in a ~20% cost for MDR-TB. This implies that in controlling drug resistance, although there is widespread concern about treatment failure leading to rising resistance, most resistant cases may be due to transmission. Therefore, although it is important to support treatment adherence, public health efforts may benefit from focusing more on preventing disease transmission. That is, control measures that reduce the incidence of new cases are likely to help reduce MDR-TB.

By developing epidemiological models with evolutionary processes, we have been able to estimate parameters describing how drug resistance – particularly multidrug resistance – emerges in *M. tuberculosis*. Although there is existing knowledge of rates of mutation to resistant states *in vitro*, there is a need to assess the extent to which those rates translate to the epidemiological level. Large scale molecular epidemiological models, such as those presented here, are highly complex and multi-dimensional and, as such, likelihood-based analyses are not straightforward mathematically or computationally. In such cases, approximate Bayesian computation methods present a practical and viable approach to making statistical inferences, particularly as continually advancing molecular technologies require dynamical models to be extended and refined.

Acknowledgements

GSR is funded by the Coordination for the Improvement of Higher Education Personnel (CAPES) Foundation via the Science Without Borders program (BEX 0974/13-7). SAS is supported by the Australian Research Council under the Discovery Project scheme (DP160102544), and the Australian Centre of Excellence in Mathematical and Statistical Frontiers (CE140100049). MMT is supported by grant DP170101917 from the Australian Research Council. This research includes computations using the Linux computational cluster Katana supported by the Faculty of Science, UNSW Australia.

References

Aandahl, R. Z., J. F. Reyes, S. A. Sisson, and M. M. Tanaka (2012). A model-based Bayesian estimation of the rate of evolution of VNTR loci in *Mycobacterium tuberculosis*. *PLoS Computational Biology 8*(6), e1002573.

Anderson, L. F., S. Tamne, T. Brown, J. P. Watson, C. Mullarkey, D. Zenner, and I. Abubakar (2014). Transmission of multidrug-resistant tuberculosis in the UK: A cross-sectional molecular and epidemiological study of clustering and contact tracing. *The Lancet Infectious Diseases 14*(5), 406–415.

Beaumont, M. A., W. Zhang, and D. J. Balding (2002). Approximate Bayesian computation in population genetics. *Genetics 162*(4), 2025–2035.

Blower, S. M. and T. Chou (2004). Modeling the emergence of the 'hot zones': Tuberculosis and the amplification dynamics of drug resistance. *Nature Medicine 10*(10), 1111–1116.

Blower, S. M., A. R. McLean, T. C. Porco, P. M. Small, P. C. Hopewell, M. A. Sanchez, and A. R. Moss (1995). The intrinsic transmission dynamics of tuberculosis epidemics. *Nature Medicine 1*(8), 815–821.

Blum, M. G. B. and O. François (2010). Non-linear regression models for approximate Bayesian computation. *Statistics and Computing 20*, 63–73.

Cohen, T. and M. Murray (2004). Modeling epidemics of multidrug-resistant *M. tuberculosis* of heterogeneous fitness. *Nature Medicine 10*(10), 1117–1121.

Colijn, C., T. Cohen, A. Ganesh, and M. Murray (2011). Spontaneous emergence of multiple drug resistance in tuberculosis before and during therapy. *PLoS One 6*(3), e18327.

Csilléry, K., M. G. B. Blum, O. E. Gaggiotti, and O. François (2010). Approximate Bayesian computation in practice. *Trends in Ecology and Evolution 25*, 410–418.

Csilléry, K., O. François, and M. G. B. Blum (2012). abc: An R package for approximate Bayesian computation (ABC). *Methods in Ecology and Evolution 3*(3), 475–479.

David, H. L. (1970). Probability distribution of drug-resistant mutants in unselected populations of *Mycobacterium tuberculosis*. *Applied Microbiology 20*(5), 810–814.

Dye, C. and M. A. Espinal (2001). Will tuberculosis become resistant to all antibiotics? *Proceedings of the Royal Society of London B: Biological Sciences 268*(1462), 45–52.

Erhardt, R. and S. A. Sisson (2016). Modelling extremes using approximate Bayesian computation. In D. Dey and J. Yan (Eds.), *Extreme Value Modelling and Risk Analysis: Methods and Applications*, Volume 281–306. New York: Chapman & Hall/CRC Press.

Ewens, W. J. (1972). The sampling theory of selectively neutral alleles. *Theoretical Population Biology 3*(1), 87–112.

Fearnhead, P. and D. Prangle (2012). Constructing summary statistics for approximate Bayesian computation: Semi-automatic approximate Bayesian computation (with discussion). *Journal of the Royal Statistical Society: Series B 74*, 419–474.

Ford, C. B., R. R. Shah, M. K. Maeda, S. Gagneux, M. B. Murray, T. Cohen, J. C. Johnston, J. Gardy, M. Lipsitch, and S. M. Fortune (2013). *Mycobacterium tuberculosis* mutation rate estimates from different lineages predict substantial differences in the emergence of drug-resistant tuberculosis. *Nature Genetics 45*(7), 784–790.

Gandhi, N. R., A. Moll, A. W. Sturm, R. Pawinski, T. Govender, U. Lalloo, K. Zeller, J. Andrews, and G. Friedland (2006). Extensively drug-resistant tuberculosis as a cause of death in patients co-infected with tuberculosis and HIV in a rural area of South Africa. *The Lancet 368*(9547), 1575–1580.

Gillespie, S. H. (2002). Evolution of drug resistance in *Mycobacterium tuberculosis*: Clinical and molecular perspective. *Antimicrobial Agents and Chemotherapy 46*(2), 267–274.

Hubbell, S. P. (2001). *The Unified Neutral Theory of Biodiversity and Biogeography (MPB-32)*, Volume 32. Princeton, NJ: Princeton University Press.

Luciani, F., A. R. Francis, and M. M. Tanaka (2008). Interpreting genotype cluster sizes of *Mycobacterium tuberculosis* isolates typed with *IS*6110 and spoligotyping. *Infection, Genetics and Evolution 8*(2), 182–190.

Luciani, F., S. A. Sisson, H. Jiang, A. R. Francis, and M. M. Tanaka (2009). The epidemiological fitness cost of drug resistance in *Mycobacterium tuberculosis*. *Proceedings of the National Academy of Sciences United States of America 106*(34), 14711–14715.

Marin, J.-M., N. Pillai, C. P. Robert, and J. Rousseau (2014). Relevant statistics for Bayesian model choice. *Journal of the Royal Statistical Society: Series B 76*, 833–859.

Marin, J.-M., P. Pudlo, and C. P. Robert (2019). Likelihood-free Model Choice. In S. A. Sisson, Y. Fan, and M. A. Beaumont (Eds.), *Handbook of Approximate Bayesian Computation*. Chapman & Hall/CRC Press, Boca Raton, FL.

Mazars, E., S. Lesjean, A. L. Banuls, M. Gilbert, V. Vincent, B. Gicquel, M. Tibayrenc, C. Locht, and P. Supply (2001). High-resolution minisatellite-based typing as a portable approach to global analysis of *Mycobacterium tuberculosis* molecular epidemiology. *Proceedings of the National Academy of Sciences of the United States of America 98*(4), 1901–1906.

Mitchison, D. A. (1951). The segregation of streptomycin-resistant variants of *Mycobacterium tuberculosis* into groups with characteristic levels of resistance. *Microbiology 5*(3), 596–604.

Monteserin, J., M. Camacho, L. Barrera, J. C. Palomino, V. Ritacco, and A. Martin (2013). Genotypes of *Mycobacterium tuberculosis* in patients at risk of drug resistance in Bolivia. *Infection, Genetics and Evolution 17*, 195–201.

Nachega, J. B. and R. E. Chaisson (2003). Tuberculosis drug resistance: A global threat. *Clinical Infectious Diseases 36*(Supplement 1), S24–S30.

Prangle, D. (2016). Lazy ABC. *Statistics and Computing 26*(1–2), 171–185.

Prangle, D., M. G. B. Blum, G. Popovic, and S. A. Sisson (2014). Diagnostic tools for approximate Bayesian computation using the coverage property. *Australia and New Zealand Journal of Statistics 56*, 309–329.

Ragheb, M. N., C. B. Ford, M. R. Chase, P. L. Lin, J. L. Flynn, and S. M. Fortune (2013). The mutation rate of mycobacterial repetitive unit loci in strains of *M. tuberculosis* from cynomolgus macaque infection. *BMC Genomics 14*, 145.

Ramaswamy, S. and J. M. Musser (1998). Molecular genetic basis of antimicrobial agent resistance in *Mycobacterium tuberculosis*: 1998 update. *Tubercle and Lung Disease 79*(1), 3–29.

Reyes, J. F. and M. M. Tanaka (2010). Mutation rates of spoligotypes and variable numbers of tandem repeat loci in *Mycobacterium tuberculosis. Infection, Genetics and Evolution 10*(7), 1046–1051.

Robert, C. P., J.-M. Corunet, J.-M. Marin, and N. Pillai (2011). Lack of confidence in approximate Bayesian computational (ABC) model choice. *Proceedings of the National Academy of Sciences of the United States of America 108*, 15112–15117.

Sisson, S. A., Y. Fan, and M. M. Tanaka (2007). Sequential Monte Carlo without likelihoods. *Proceedings of the National Academy of Sciences of the United States of America 104*(6), 1760–1765. Errata (2009), *106*, 16889.

Supply, P., C. Allix, S. Lesjean, M. Cardoso-Oelemann, S. Rüsch-Gerdes, E. Willery, E. Savine et al. (2006). Proposal for standardization of optimized mycobacterial interspersed repetitive unit-variable-number tandem repeat typing of *Mycobacterium tuberculosis. Journal of Clinical Microbiology 44*(12), 4498–4510.

Supply, P., S. Niemann, and T. Wirth (2011). On the mutation rates of spoligotypes and variable numbers of tandem repeat loci of *Mycobacterium tuberculosis. Infection, Genetics and Evolution 11*(2), 251–252.

Thornton, K. and P. Andolfatto (2006). Approximate Bayesian inference reveals evidence for a recent, severe bottleneck in a Netherlands population of *Drosophila melanogaster. Genetics 172*, 1607–1619.

WHO (2015). Global tuberculosis report 2015. Technical report, World Health Organization.

Wirth, T., F. Hildebrand, C. Allix-Béguec, F. Wölbeling, T. Kubica, K. Kremer, D. van Soolingen et al. (2008). Origin, spread and demography of the *Mycobacterium tuberculosis* complex. *PLoS Pathogens 4*(9), e1000160.

Zhao, Y., S. Xu, L. Wang, D. P. Chin, S. Wang, G. Jiang, H. Xia et al. (2012). National survey of drug-resistant tuberculosis in China. *New England Journal of Medicine 366*(23), 2161–2170.

17

ABC in Systems Biology

Juliane Liepe and Michael P.H. Stumpf

CONTENTS

17.1 Introduction 513
17.2 Parameter Estimation for Dynamical Systems 515
17.2.1 Inference for dynamical systems 515
17.2.2 Posterior analysis for dynamical systems 517
17.3 ABC Model Selection in Systems Biology 519
17.3.1 Dynamical systems and model selection 520
17.3.2 ABC for design 522
17.4 Network Analysis and Inference 522
17.4.1 Models of network evolution 523
17.4.2 Distances and summaries of network data for ABC 523
17.5 Multi-Scale Models 525
17.5.1 Inference for biological multi-scale models 526
17.5.2 ABC models of cell migration 527
17.6 ABC and Beyond 532
References 533

17.1 Introduction

Systems biology aims to capture the fundamental molecular and cellular processes underlying living systems [1]. Mathematical models that describe how cells, or indeed complex multi-cellular organisms, function are increasingly gaining prominence in biological research. While statistics and probability have been pivotal in evolutionary and population biology for well over a century, mathematical models have entered the cellular and molecular life sciences only relatively recently. In order to understand how complex molecular machines [2] such as ribosomes and proteasomes, function, or how essential cellular processes, such as metabolism, replication, cell-cycle, and response to environmental signals and stresses, are controlled, we need mathematical models. They open up the possibility of stating mechanistic hypotheses in concise, clear, and quantitative terms. Looking at the agreement between suitably calibrated mathematical models and experimental data is one way in which we can test our understanding and successively improve our models.

Iteration between experiment and theoretical analysis and prediction has become a hallmark of systems biology studies. In contrast to the physical sciences, however, it is typically not possible to use first principles to construct suitable mathematical models for biological phenomena. Instead, we have to look at interaction and relationships between different *species*. Here *species* can refer to actual biological species in the contexts of population biology; different cell-types in developmental biology and immunology; or different molecules in biochemistry and molecular biology. Typically, we aim to characterise their interactions using ordinary, partial, or stochastic differential equations, but in many other instances all we can currently do – due to our current state of understanding and/or the nature of the available data – is to find adequate statistical descriptions for the dependencies among these species [3,4].

The models that we are dealing with in many biological situations must really be seen as first attempts at modelling these phenomena mathematically [5,6]. Models are thus under continuous revision, in many cases even the appropriate modelling framework – for example, ordinary versus stochastic differential equations; Boolean networks, stochastic Petri-nets of differential equations – is far from clear at the outset. The need to adapt to the challenges posed by real-world biological systems is quite considerable, and it is rare that a single mathematical approach suffices to make progress in the analysis of biological systems.

Cutting edge scientific problems often defy conventional statistical approaches. But, without sound inference procedures and reliable statistical analysis, it is in turn hard to make progress in such areas of research. Here, *approximate Bayesian computation* (ABC) provides an ideal framework to progress with challenging research problems [7]. While there is a lot of interest in ABC as an inferential framework in its own right [8–10], here we stress the utilitarian aspects of ABC: it allows us to solve challenging inference tasks in situations where it is either hard or impossible to evaluate likelihoods. In such situations, it is typically still possible to simulate the mathematical models and use ABC to compare simulated and observed data. With this framework in place, it is then possible to, for example, fit models to data [11–13]; design better, more discriminatory experiments [14]; choose rationally from different mechanistic models [16,15]; and investigate which models (or interventions) are most likely to result in certain types of behaviour [17].

17.2 Parameter Estimation for Dynamical Systems

Most computational analyses in systems biology can be divided into two classes: those aiming to describe and analyse the structure of biological networks; and those that model explicitly the temporal behaviour of a biological system, such as a signalling or gene regulation network. The former is typically done using graphical models – with some exceptions discussed in Section 17.4 – and can therefore be handled with classical or exact Bayesian approaches. Here, we will therefore deal with the latter.

17.2.1 Inference for dynamical systems

We are interested in how the state of some system, $X(t)$, changes with time t. Often the state of the (typically vector-valued random variable) X is not observable directly, and, instead, we observe $Y \sim g(X(t) = x(t); \eta)$. Both deterministic and stochastic dynamical systems are considered in systems biology, while traditionally deterministic approaches, typically implemented via ordinary differential equations (ODE), have predominated, more recently, there has been a noticeable increase in stochastic studies.

For ODEs:

$$\frac{dy(t)}{dt} = f(y, t; \theta), \tag{17.1}$$

a likelihood has to be defined via an error model, and we typically assume that discrete observations $y(t_i)$ for $1 \leq i \leq n$ are subject to some noise, for example:

$$y_{obs}(t_i) = y(t_i) + \epsilon(t_i),$$

where $\epsilon(t_i)$ is drawn from some suitable probability model [18–20]. For pragmatic reasons, $\epsilon \sim \mathcal{N}(0, \sigma^2)$, is the conventional choice, and for such a model, but also more complicated models, it is straightforward to define likelihoods (via the error model) and use conventional Bayesian analyses. ABC methods have, however, been also applied in this context [21], but their real potential lies in inference for stochastic dynamical systems, described here and for multi-scale systems (which are the topic of Section 17.5).

Here, for sake of concreteness, we focus on systems that are described by stochastic differential equations [22, 70], and we use a state-space representation where:

$$dX(t) = f(X(t), t; \theta)dt + dW_t \tag{17.2}$$

$$Y \sim g(X(t) = x(t); \eta). \tag{17.3}$$

W_t denotes the classical Wiener process, and in many applications a simplified form for the observation function $g(X, \eta)$ is assumed, with the assumption of normal noise being a conventional and attractive choice. The error model, however, also introduces a probability model for the observed data and, in applications involving ODEs, can suffice to define a likelihood. For SDEs (or stochastic processes more generally), however, the evaluation of the likelihood is complicated, especially if suitable approximation schemes, such as the linear noise approximation [23] cannot be applied. This is typically the case for highly non-linear dynamics and low molecular abundances.

For data, $y_{obs}(t_i)$, observed at discrete time points, $\tau = \{t_1, \ldots, t_n\}$, we need to evaluate the transition probabilities, assuming that the state of the system evolves as a continuous-time Markov process, as:

$$p(Y(t_{i+1})) = y_{obs}(t_{i+1}|P(Y(t_i)) = y_{obs}(t_i). \tag{17.4}$$

But this is an integral over all possible paths from the state of the system at time t_i to the state at time t_{i+1}. Even in approximation, the evaluation of such *bridging processes* is computationally costly and rarely affordable for realistic systems of biological interest [25,24]. Thus, the likelihood of such a system, given by:

$$p(y_{obs}|\theta) = \prod_{i=1}^{n-1} p(Y(t_{i+1}) = y_{obs}(t_{i+1})|P(Y(t_i)) = y_{obs}(t_i), \tag{17.5}$$

is typically impossible to evaluate, and computation of the posterior:

$$\pi(\theta|y_{obs}) \propto p(y_{obs}|\theta)\pi(\theta), \tag{17.6}$$

becomes computationally cumbersome or impossible.

Approximation schemes, such as the linear noise approximation offer some alternatives, but are typically limited to relatively benign dynamical regimes, and other approximation to chemical master equations (see e.g. [26,27]) are also either computationally demanding, capable of dealing only with simplified dynamics, or their convergence to the true dynamics is hard to control or verify.

By contrast, ABC is capable (subject to the constraints discussed in the following) of coping with simulated data from models with very complicated dynamics [28]. The appropriate ABC posterior, corresponding to the model in (17.5) is given by:

$$\pi_{ABC}(\theta|y_{obs}) \propto K_h(||y - y_{obs}||)p(y|\theta)\pi(\theta). \tag{17.7}$$

This is easy to calculate and allows us to obtain (approximate) posterior estimates. Quite generally, once the technical parameters have been set up, ABC works in an essentially straightforward manner. This should not detract, however, from the fact that getting different algorithms, such as ABC-sequential Monte Carlo (SMC) [29,21], the current work-horse of ABC-based inference in systems biology, to converge while maintaining good coverage of the parameter space can be non-trivial.

17.2.2 Posterior analysis for dynamical systems

Dynamical systems pose considerable inferential challenges, and a rich literature [30–32] has been building up around concepts such as identifiability, inferability, and sloppiness. Common to these three closely related notions – even though the relationship is rarely if ever explored – is that local, point-estimates are: (i) potentially poor representations of the true parameter and (ii) hide the fact that many similar parameters would be capable of describing the data equally well. This should be reason enough to consider interval estimators and, in particular, Bayesian methods from the outset. While notions such as identifiability and sloppiness have been considered in depth – though perhaps not always satisfactorily – in the context of ODE models, many of the same problems will also carry through to stochastic modelling approaches [33].

As a simple example, we consider a hypothetical gene regulatory network (Figure 17.1a). In this model, a protein P_1 is produced from its mRNA M.

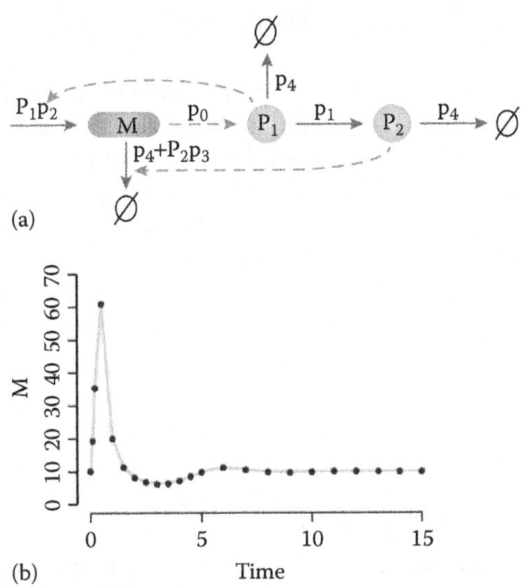

FIGURE 17.1
Gene regulatory model. (a) The cartoon depicts the model described by equations 17.8–17.10. mRNA (M) produces the protein P_1, which can be modified and results in the production of protein P_2. P_1 is required to produce M, while P_2 enhances the degradation of M. P_1 and P_2 are degraded with a constant rate (the $\emptyset$ symbol denotes that a reaction is a death process). (b) Shown are measurements of M over time. The data were generated from the corresponding ODE model with parameters $(p_1, ..., p_5) = (10.0, 0.5, 10.0, 2.0, 1.0)$ and initial conditions $(M, P_1, P_2)=(10, 5, 0)$. This dataset was used to estimate the model parameters.

Once P_1 is produced, it allows the production of its own mRNA. P_1 can also be post-translationally modified and transformed into protein P_2, which in turn degrades M. The production of P_1 is therefore self-regulated via post-translational modifications and a positive feedback loop. Both proteins, as well as the mRNA, are furthermore degraded at a constant rate. The described system contains three species (P_1, P_2, and M) and five parameters ($p_1, \ldots, p_5$). The corresponding stochastic model using SDEs can be written as Wilkinson 2011 [22]):

$$\frac{dm}{dt} = P_1 p_2 - M(p4 + P_2 p_3) + dW_t^{(1)} \tag{17.8}$$

$$\frac{dP1}{dt} = Mp_0 - P_1(p_1 + p_4) + dW_t^{(3)}, \tag{17.9}$$

and

$$\frac{dP2}{dt} = P_1 p_1 - P_2 p_4 + dW_t^{(3)}. \tag{17.10}$$

We can, for the purpose of this example, collect data for mRNA generated by the corresponding ODE model using the parameters $(p_1, \ldots, p_5) =$ (10.0, 0.5, 10.0, 2.0, 1.0) and the initial conditions (M, P_1, P_2) = (10, 5, 0) (Figure 17.1b). We now aim to infer the posterior parameter distribution of this model given the generated data. We use the Python package ABC-SysbBio [34,35], which implements ABC-SMC. The fitted trajectories are shown in Figure 17.2a and the resulting posterior distributions in Figure 17.2b. Simulation of trajectories from the posterior distribution allows us to understand, how much information the data carry about predicting the behaviour of the three species. The resulting confidence intervals for M are narrow, as can be seen in Figure 17.2a. This is expected because the data describe the temporal behaviour of M. On the contrary, we observe larger confidence intervals for the prediction of P_1 and P_2. In order to better predict these two species' behaviours, we would need to use a more informative dataset [14]. The marginal posterior parameter distributions indicate that p_0 and p_2 are not well inferred. However, the pairwise density plot of these two parameters shows strong non-linear correlations, for example, if one of the two parameters were to be known, the remaining parameter could be identified (Figure 17.2b).

This posterior spread seems ubiquitous in parameter inference for dynamical systems [30–32], and while experimental design can increase our ability to measure parameter – such as different initial conditions, knock-out and knock-down mutants, as well as small molecules that interfere with the dynamics of a system – not all informative experiments can be carried out for practical or ethical reasons.

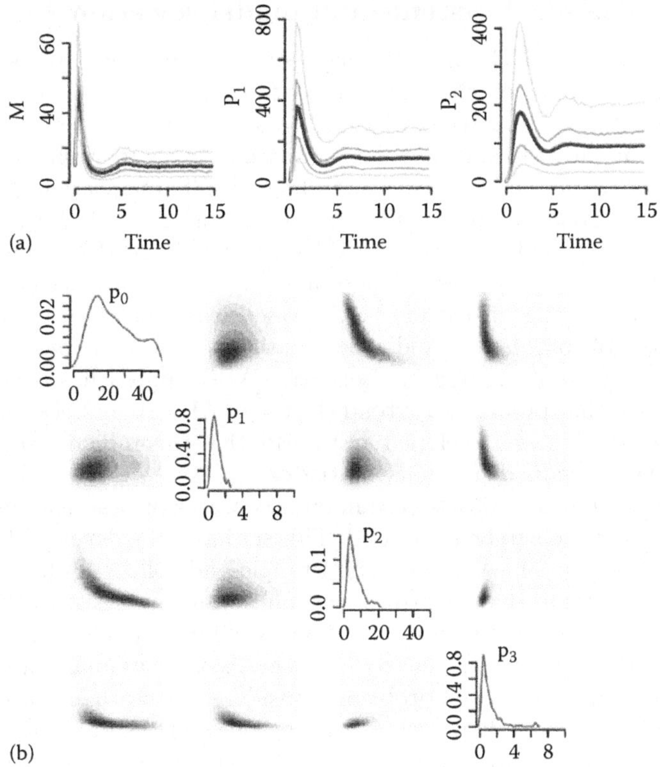

FIGURE 17.2
Analysis of the posterior distributions. (a) Once the posterior parameter distribution is obtained, one can simulate all species of the model based on samples from the posterior distribution. Shown are the mean (dark black), the 25% and 75%-iles (dark grey), and the 5% and 95%-iles (light grey) of 500 simulated trajectories. (b) The marginal posterior distributions (diagonal) and the pairwise density plots (off-diagonal) contain insights about identifiability and correlations. Parameter p_4 was assumed to be known and is therefore not inferred.

17.3 ABC Model Selection in Systems Biology

In any fledgling scientific discipline, quantitative mechanistic models must be subject to revision and cycles of improvement. In systems biology, ABC has therefore been attracting considerable attention as a tool for model selection [36]. Choosing from a set of competing hypotheses (or models) is central to many systems biology applications [37].

17.3.1 Dynamical systems and model selection

Model selection in ABC contexts has been an active field of research and has thrown up some of the most important questions in ABC inference [8–10,16,38]. When summary statistics, $s = s(y)$, are used, then, even for sufficient summary statistics, it may not be possible to compare different models. Different models m_1 and m_2, corresponding to different models $f_1(y|\theta_1)$ and $f_2(y|\theta_2)$, with potentially quite different parameter sets θ_1 and θ_2, will tend to have different sufficient statistics s_1 and s_2, respectively. As a rule, the union of these statistics $s_1 \cup s_2$ does not offer a sufficient statistic across models [9].

This would be the death of ABC model selection save for a relatively limited class of problems (and one where other exact mechanisms are available). A range of pragmatic alternative ABC approaches have, however, been published that promise to extend the use of ABC to model selection problems. Barnes et al. have developed an information-theoretical framework that constructs vector-valued summary statistics $s = (s_1, s_2, \ldots, s_q)$ that are sufficient or nearly sufficient (i.e. to within an acceptable or practical limit) within and across the models to be compared. This set has to be developed in practice in a greedy fashion [39]. Alternatively, Prangle and colleagues have developed an efficient algorithm that identifies suitable summary statistics (for parameter inference and model selection) based on a preliminary set of simulation runs that are used to construct a (nearly) sufficient set of summary statistics [8].

The most important set of problems considered by mathematical models in systems biology relates, however, to the temporal change of molecular abundancies in response to environmental or developmental signals (such as growth factors in tissues that determine proliferation or differentiation of cells). For such dynamical systems, there are no summary statistics that would represent the data in a meaningful or helpful manner better than the data do themselves. We are thus able to compare and contrast mechanistic models in systems biology for many of the problems of interest. Here, the ability to analyse biological data in a meaningful manner has, however, also changed the types of problems considered, and many more experimental studies seek to collect time-resolved data for that purpose.

For given observations, y_{obs}, and m competing models, m_i with $f_i(y|\theta_i)$ (such as different ODE or SDE models) and $i = 1, 2, \ldots, m$, we can therefore evaluate the ABC marginal likelihoods $\pi_{ABC}(m_i|y_{obs})$. There are different ways in which this quantity can be evaluated. Perhaps the easiest way is to augment the parameter vector by a discrete model indicator [40]. Thus, inference in ABC-SMC (or any other computational scheme) can proceed on the joint model-parameter space $\theta^* = (i, \theta)$, where θ^* now contains all parameters considered in all the models [16].

We thus have for N draws from:

$$\pi_{ABC}(m_j|y_{obs}) \propto \sum_{i=1}^{N} K_h(||y - y_{obs}||)p_i(y|(i, \theta_i))\pi_i(((\theta_i))\delta_{i,j}\pi(m_i), \quad (17.11)$$

where $\delta_{i,j}$ is the normal Kronecker delta, and we assume that the parameter and model priors can be written as $\pi(\theta_i)\pi(m_i)$. That is, the fraction of particles that correspond to model j determines the posterior model probability.

On this joint space, model selection is easily incorporated into ABC-SMC or other computational frameworks [16]. In practice, the problems of monitoring convergence can be quite considerable, and, ideally, inference across models should be accompanied by parameter analysis for each model considered [41]. And, of course, ABC model selection shows the samc dependence on the choice of parameter priors for each model as conventional, exact Bayesian model selection.

We illustrate an example of ABC-SMC for model selection [35] by returning to the gene regulation model from Section 17.2.1 and the generated data describing the time course of mRNA. One could ask whether the post-translational modification of P_1 resulting in P_2 is necessary to explain the data. A second, simpler model (Figure 17.3a) is then defined as:

$$\frac{dm}{dt} = P_1 p_2 - M p4 + dW_t$$

$$\frac{dP1}{dt} = M p_0 - P_1 p_4 + dW_t$$

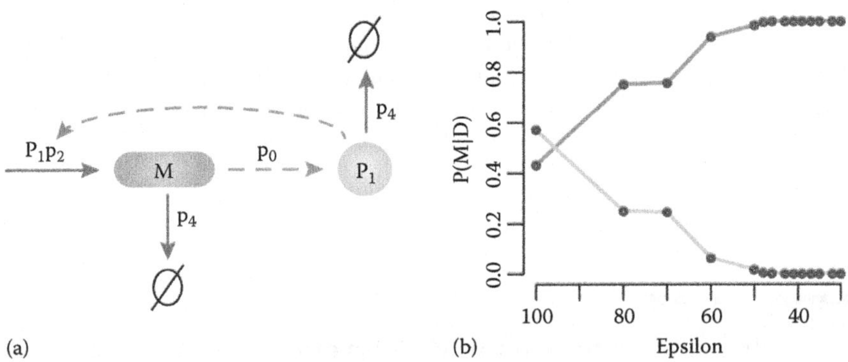

FIGURE 17.3
Model selection in the ABC-SMC framework. (a) A second gene regulation model was constructed and compared to model 1. The second model does not contain the post-translational modification and is therefore a simplification of model 1: mRNA (M) produces the protein P_1, which is required to produce M. Both, M and P_1 are degraded with constant rates. (b) Shown is the evolution of the posterior model probability for model 1 (dark grey) and model 2 (light grey) in the ABC-SMC framework. With decreasing tolerance (epsilon), model 1 is more likely to represent the data.

Using an initial model probability of (M_1, M_2)=(0.5, 0.5), we infer the posterior model distribution (Figure 17.3b). We find that over the ABC-SMC populations the data are more likely produced by M_1. This is not too surprising, because the data were generated by the corresponding deterministic model (ODE) of model 1. However, this also means that model 1 can not be reduced to model 2 in order to describe the dynamical system of interest.

17.3.2 ABC for design

Model selection aims to identify which model among a set of candidates best explains the available data; in Bayesian model selection, we assign a probability to each model, $p(m_i|y_{obs})$. In synthetic biology, the aim is to develop rationally engineered biological systems (or circuits) that are able to carry out specific functions [42]. Biosensors that allow cells to signal the presence of certain environmental conditions are typical examples, as are organisms that produce desirable high-value compounds using biomolecular pathways that do not occur naturally. In such a synthetic biology setting, we may have different competing design ideas, which we can also represent in terms of mathematical models, m_i. But instead of data, y_{obs}, we now have a set of objectives or desired system behaviours [43].

If we can encode such outputs in a convenient way, for example, by providing surrogate data, y, then we can apply ABC to identify that design that is best able to satisfy our specifications. Here, $p_{ABC}(m_i|y)$ is in all respects equivalent to the posterior distribution (17.11); the only difference is that we replace observed data with the *data that we would like to see*. In traditional synthetic biology settings, optimisation is often used to identify models that fulfil given design objectives [44], but the Bayesian perspective also offers an assessment of the robustness of the different design alternatives [43].

17.4 Network Analysis and Inference

Molecular networks have become a popular organisational tool in systems biology [45]. They allow researchers to gain a level of control over the diversity of molecular processes occurring inside cells, and they can form the basis for the mathematical characterisation of biological systems. But because these networks typically consist of thousands of nodes or vertices $v \in \mathcal{V}$ that are connected by edges $e_{ij} = (v_i, v_j) \in \mathcal{E}$, the level of analysis is typically more coarse grained than for the dynamical systems considered earlier. Instead of detailed mechanistic analysis, we are typically confined to calculation of graph theoretical summary statistics and statistical dependencies among nodes. Analyses are now becoming more sophisticated and detailed, especially as the combination of different types of networks (e.g. transcriptional regulation, protein-protein interaction and metabolic networks) opens up new perspectives and throws up new modelling challenges [46].

Here, for concreteness, we will be considering binary networks, where the entries in the adjacency, $A = (a_{ij})$, are either 1, if an edge is present between nodes i and j, or zero otherwise. This is particularly useful for protein-interaction networks, but more generally, we can use this binary representation for any network structure; any quantitative aspects of interaction strengths *etc.* can be specified via a separate matrix, for example, a weight matrix, $W = (w_{ij})$, and the whole quantitative network is then given by the Hadamard product of adjacency and weight matrix, $A \circ W$.

17.4.1 Models of network evolution

Just as ABC methods first emerged in population and evolutionary genetics, so did ABC inference enter the systems biology arena first as a tool to analyse the evolution of biological networks [47]. The way in which networks evolve is typically also modelled as a Markov process, although this generally takes the form of a discrete time Markov process, where each time-step involves a change to the network structure. In particular, the evolution of protein-protein interaction networks has attracted much attention. These networks aim to describe the whole set of potential PPIs in a given organism, data are available for an increasing number of species, but the most complete networks are still in brewer's yeast, *Saccharomyces cerevisiae.*

The most popular models include: (i) attachment of a new node to the existing network, where the node is connected to existing nodes by a fixed or random number of edges; (ii) duplication of an existing node with inheritance of some or all of its edges; or (iii) 'rewiring' of edges [49,48]. Thus, the Markov process operates on the network or a convenient representation, such as the adjacency matrix. Early analyses of networks, including their evolution have focussed on the degree distribution, $p(d)$, the probability for a node to have degree, d, here, the degree of a node is the number of nodes in the network that a node is connected to. This is because it is particularly straightforward to write down a chemical master equation for the degree distribution as new nodes are added to the network:

$$p_{t+1}(d) \propto \sum_{i=0}^{t} p_t(i)q(d,i) - p_t(d)q(i,d),$$

where $q(i,j)$ is the probability that a node with degree j will become a node with degree j given the evolutionary process [50].

17.4.2 Distances and summaries of network data for ABC

The degree distribution can, of course, be used to determine a likelihood and it is indeed possible to use this to define a composite likelihood. But it only captures a single aspect, and arguably not a very important or decisive one. For some models it is possible to write down likelihoods of networks, and importance sampling approaches have been used to obtain some insights into the evolution of networks under a duplication-attachment model [51]. But in addition to the computational cost of the likelihood evaluation, the likelihood is not identified for all network structures, and there remains a core network that can not further be decomposed in a manner compatible with the duplication attachment model.

For these reasons, ABC offers an attractive framework in which we can study the evolution of biological networks (or at least evolutionary models of the evolution of models). The primary problem is related to the choice of summary statistics: summary statistics for networks can be expensive to calculate (many scale $O(N^3)$, where N is the number of nodes in the network), and only capture aspects of the network; there is, crucially, no sufficient summary statistic for a network.

The first attempt at using ABC to determine the parameters of a plausible model of network evolution therefore relied on a collection of summary statistics. Ratman and colleagues [47] therefore used extensive simulations to explore the feasibility of their approach – where they also took into consideration that the available data on protein-protein interactions is incomplete. However, this adds to the computational complexity, which is already quite considerable.

An alternative way of applying ABC to network evolution has considered the *edit distance* between networks [52]. For two networks with adjacency matrices $A^1 = (a^1_{ij})$ and $A^2 = (a^2_{ij})$ this is defined as:

$$d_e = \sum_{i,j=1}^{N} \left(a^1_{ij} - a^2_{ij}\right)^2. \tag{17.12}$$

Equation (17.12) assumes that the nodes are labelled or ordered suitably. While the nodes in the real data will be labelled, for example, by protein ID, sequence accession number, or similar, the nodes in the simulated network are not labelled in a natural or meaningful way. Instead, we would have to identify the permutation of labels of the nodes of the simulated network $\mathcal{G}'$ that minimises the edit distance between observed adjacency matrix, A_{obs} and the adjacency matrix of the simulated network, A_θ. This, however, can be approximated conveniently by the spectral distance:

$$d_\lambda = \sum_{i=1}^{N} \left(\lambda^{(i)}_{\text{obs}} - \lambda^{(i)}_{\theta}\right)^2, \tag{17.13}$$

which provides a lower bound on the edit distance, $d_\lambda \leq d_e$. Thus, with the help of equation (17.13), we have an approximation to the distance between the data [52], which is, of course, very different from a distance of summary statistics of the data.

Using the ABC-SMC estimator proposed by Toni et al. [21] we can thus obtain parameter estimates and, for the distance defined by (17.13), also choose

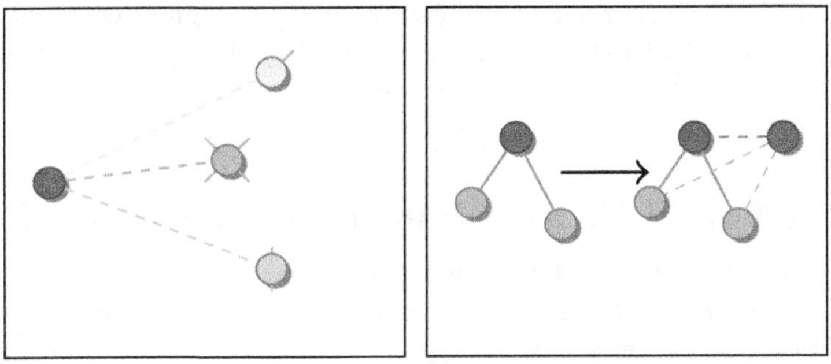

(a) (b)

FIGURE 17.4
Examples of network evolution models. (a) The preferential attachment model has been used to model generic, but non-biological network growth models. Here, nodes are added to the network and connected to existing nodes proportionally to their degrees. (b) In the duplication-attachment model, networks grow by duplication of existing nodes and (partial) inheritance of edges by copies. These are among the simplest models that can and have been used to investigate network evolution.

from different models. For the two simplistic models shown in Figure 17.4, for example, we find that for any real biological network $p_{ABC}(m_{DA}) \approx 1$, if only the preferential attachment model is considered as an alternative. If we consider more complicated models that include duplication, attachment, and divergence (rewiring) of edges, then this situation changes: the DA model retains some posterior probability but is generally [except for yeast, where $p_{ABC}(m_{DA}) \approx 0.4$] loosing out to more complicated models [52]. Here, as elsewhere, it is important to remember that these models are vast oversimplifications of a much more complicated and historically contingent evolutionary process.

17.5 Multi-Scale Models

Observing processes inside living organisms – whether they are bacteria, archaea single celled eukaryotes, or more complex multi-cellular beings – poses a number of challenges. Keeping the organism alive during observations is one of them; while this is not always required, many physiological processes can only be meaningfully studied in living organisms [53]. This is one, but certainly not the only reason, for the rise in so-called multi-scale modelling approaches [54]. Here we adopt a pragmatic definition of what constitutes a multi-scale process: one where events at one spatial (or spatio-temporal) scale affect or need to be observed at a different, typically higher level. An example of this is provided by the movement of cells through their environment: we observe the pattern of migration that is governed by a set of complex molecular processes occurring inside (or on the surface of) the cell.

17.5.1 Inference for biological multi-scale models

Common to multi-scale models is thus a juncture between the processes that we can measure and those that we want to understand [54]. From this, stem a range of practical problems, as well as a formidable inferential problem. The former includes, for example, the potential need to smooth data.

In discrete time a hidden Markov model (HMM) could represent a two-scale process (see Figure 17.5), here, however, X_i can represent a biological network or dynamical system, and Y_i is an observation (potentially vector-valued) at a typically higher level that is shaped by processes affecting X_i. Y_i and X_i are linked by some observation function:

$$Y_i \sim g(X_i; \eta).$$

HMMs are widely used in computational statistics (including in computational biology) and signal processing, and a wealth of statistical techniques have been developed that allow us to infer the parameters (and structure) of HMMs for sufficiently simple Markov processes describing the temporal evolution of X_i. Here, however, we have potentially much more complicated structures that determine how X_i changes.

We are after a posterior $p(x, \theta|y)$, assuming that the observation function $g(x; \eta)$ is sufficiently simple and well characterised; often, in fact, we can treat this as a deterministic function that maps $X_i \in \mathbb{R}^m$ to $Y_i \in \mathbb{R}^n$ (typically with $n \ll m$). Note that Y_i may in many cases be a summary statistic – or should at least be understood as one; for example, there is a potential deterioration in information about X_i and θ. But generally, evaluation of the likelihood for multi-scale systems will be difficult or impossible, which makes ABC approaches an attractive alternative [7].

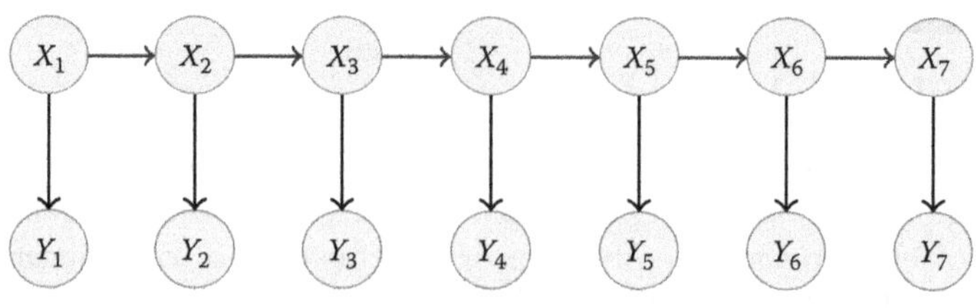

FIGURE 17.5
The structure of a hidden Markov model (HMM). Observations of states Y_i need to be linked to the underlying, but non-directly observable state X_i, which is assumed to be given by a Markov process.

Analogously to the state-space models discussed in Section 17.2.1 we write down an approximate posterior of the form:

$$p_{ABC}(\theta|y_{obs}) \propto K_h(||y - y_{obs}||)g(y|x)p(x|\theta)\pi(\theta), \tag{17.14}$$

which assumes that the parameters regarding the observation process, η, are known – which may be the case, often η can be assumed to be normally distributed if it represents measurement noise. Otherwise we write:

$$p_{ABC}(\theta, \eta|y_{obs}) \propto K_h(||y - y_{obs}||)g(y|x, \eta)p(x|\theta)\pi(\theta)\pi(\eta). \tag{17.15}$$

Here, $g(y|x, \eta)$ typically is a function connecting spaces of different dimensionality. The mutual information between the random variables A and B tells us how much knowing the state of the observation A reduces the uncertainty about B:

$$I(A, B) = H(A) - H(A|B),$$

where $H(A)$ is the entropy of A and $H(A|B)$ is the conditional entropy of A given B. Y will be sufficient about the parameters θ if and only if:

$$I(Y, \theta) = I(X, \theta);$$

but since Y is generated from X in a manner dependent on some additional parameters, η, this will rarely be fulfilled.

This offers, we feel, exciting opportunities for future research in ABC. While this problem is related to the problem of finding sufficient statistics for ABC inference, the nature of Y is restricted by the experimental and biological problem. Whereas, we can typically investigate and combine different summary statistics until sufficiency or near-sufficiency has been achieved [8,39,55–57], here, we have no or only little choice. This clearly poses challenges on how to set up, apply, and interpret ABC inference for most multi-scale problems.

17.5.2 ABC models of cell migration

Perhaps the simplest example of a multi-scale process that is biologically meaningful is given by cellular migration [13,15,58,59]. Many cells, ranging from bacteria and amoeba to immune cells in multi-cellular organisms, move towards (or away) from certain environmental signals. A wound in a multi-cellular organism, for example, results in the release of cytokines that can attract immune cells. Here, the two scales are the organism or tissue and the immune cells, respectively. What makes this problem more tractable than the models alluded to earlier is that the model for the migrating cells can be relatively simple: for example, a gradient in the attractant can determine the direction of the next move.

Migration and models of migratory processes can, if experimentally resolved in sufficient detail, produce large amounts of data, and there are many different summary statistics that can be used to characterise migration [60]. Average displacement, its variance, the distribution over angles between successive moves, directionality coefficients, etc. can all be used to quantify aspects of the movement. Sufficiency of these is, however, only given for some types of models.

We illustrate how ABC aids the analysis of cell migration data using the example of immune cell migration of macrophages in response to acute injury [13]. At the wound, a cytokine is produced and released into the tissue, which guides the macrophages towards the wound. We are interested in learning the diffusive behaviour of that cytokine over time based on macrophage migration data (Figure 17.6). The macrophage is described as a circle centred at (x, y) with radius r. In presence of the cytokine, the macrophage detects the concentration on its cell surface and migrates towards higher cytokine concentration. In our system, we can assume that the cytokine concentration is constant for a given distance, x, from the wound, which means that the macrophage can be simplified to its centre coordinates (x, y), a front $x - r$, and a rear $x + r$. To model the directionality of the macrophage, we now need to detect the cytokine concentration at the front and back of the cell. The detection of the concentration can be modelled assuming simple receptor-ligand binding kinetics, as we assume that this process is much

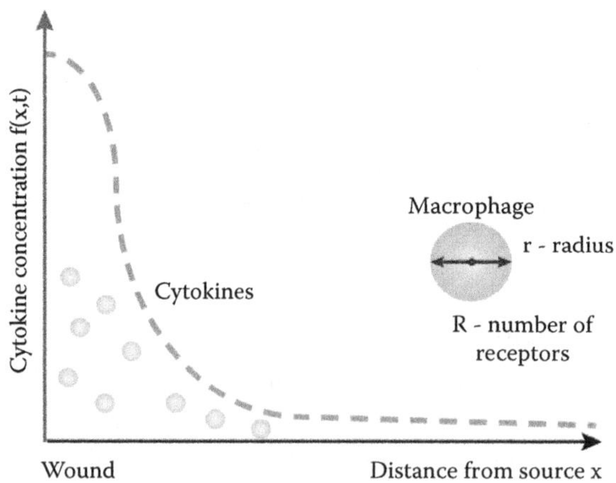

FIGURE 17.6
Cell migration model. Cytokines emenating from a wound are sensed by receptors, R, on the surface of macrophages (modelled as spheres with radius r, the concentration of the cytokines changes over time and with the distance to the wound, macrophages need to detect the gradient in order to find their way to the cytokine source.

faster than the migration of the macrophage itself; we can therefore apply the steady state description of the bound ligand (the cytokine) to the receptors with:

$$C(f(x)) = \frac{1}{2}\left(R + f(x,t) + K_d\right) - \sqrt{\frac{1}{4}(R + f(x,t) + K_d)^2 - Rf(x,t)},$$

where C is the concentration of cytokine bound to receptors, R is the concentration of receptors at the front and at the rear of the cell, K_d is the diffusion constant, and $f(x,t)$ describes the cytokine concentration at position x and time t.

The migration behaviour of the macrophage can be modelled as a biased persistent random walk [61], where for simplicity we describe a cell trajectory as a sequence of α_t, which is the angle between a motion vector of the cell and the negative x-axis (pointing towards the site of injury):

$$\alpha_t = \underbrace{wN_c(\alpha_{t-1}, \sigma_p)}_{\text{persistence}} + \underbrace{(1-w)N_c(0, \sigma_b)}_{\text{bias}}.$$

Here, w denotes the weight between the wrapped normal distributions (N_c) describing the biased and the persistent motion. The strengths of the bias (σ_b) and the persistence (σ_p) depend on the detected differences in cytokine concentration at the front and at the back of the macrophage:

$$\triangle C_{\max} = \underset{\mathrm{y}}{\operatorname{argmax}}(C(f(x,t)) - C(f(x+2r,t))).$$

Furthermore we define:

$$\rho_p = p_{\max}\left(1 + d_p\left(\frac{C(f(x-r,t)) - C(f(x+r,t))}{\triangle C_{\max}} - 1\right)\right),$$

$$\rho_b = b_{\max}\left(1 + d_b\left(\frac{C(f(x-r,t)) - C(f(x+r,t))}{\triangle C_{\max}} - 1\right)\right),$$

where $p_{\max}$ and $b_{\max}$ describe the maximum possible persistence and bias, respectively. Finally, we obtain the strength of the bias and persistence via:

$$\sigma_p = -2log(\rho_p) \text{ and } \sigma_b = -2log(\rho_b).$$

From previous studies [13,59], we can know that we can assume the cytokine distribution follows a diffusion type gradient:

$$f(x) = \frac{A}{\sqrt{4\pi D_c}} e^{-x^2/4\pi D_c}, \tag{17.16}$$

where A is the strength of the cytokine source, and D_c is the diffusion coefficient of the cytokine. We now defined a cell migration model described by a sequence of α_t for each cell. From this sequence, one can easily compute the cell trajectories with coordinates (x, y) over time given the initial coordinates at time $t = 0$ and the distribution of the step length, which we define as:

$$s_t \sim \sqrt{dt} * N^+(1,1), \tag{17.17}$$

where $N^+(1,1)$ is a truncated normal distribution, truncated at 0, and $dt = 0.001$.

The resulting model contains a set of parameters that we aim to infer based on cell migration data. In the last decade, a genetically modified zebrafish has been developed that contains green fluorescent protein (GFP) labelled macrophages. This allows us to track macrophages over time after wounding the zebrafish tail *in vivo*. The cells tracks can be extracted and summarised using the above mentioned statistics. For this example, we chose to compute the observed distributions of the directionality coefficient D in dependence of the distance from the wound (three spatial clusters) and time after wounding (five temporal clusters) (see data in Figure 17.7). For each spatio-temporal cluster we compute the distribution $\mathcal{S}$ of the straightness indices $S_D^{(i)}$ for the extracted trajectories:

$$S_D^{(i)} = \frac{d_i}{l_i}, \tag{17.18}$$

where l_i is the total length of the trajectory and:

$$d_i = |x_0, x_{end}|,$$

is the Euclidian distance between start and end point of each trajectory i, equation (17.18) can thus be written as:

$$S_D^{(i)} = \frac{|x_0, x_{end}|}{l_i}. \tag{17.19}$$

In order to compare the distributions we use the Kolmogorov–Smirnov distance between their respective histograms, the resulting distance function is:

$$d = \sum_{t=1}^{N_t} \sum_{s=1}^{N_s} K(\mathcal{S}^{s,t}, \mathcal{S}^{*s,t}),$$

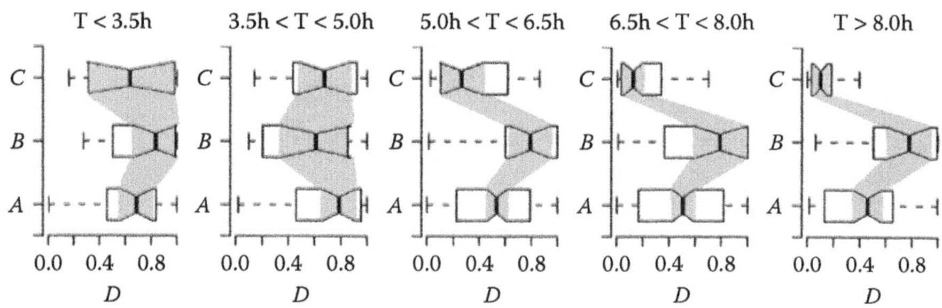

FIGURE 17.7
Representation of the cell migration data. The cell tracks are summarised using the directionality coefficient. Shown are the distributions of the directionality coefficient, D, divided into three different clusters according to the distance from the wound (A: $x < 250$ μm, B: 250 $\mu m < x < 500$ μm, and C: $x > 500$ μm) and furthermore clustered according to time, T, post-wounding.

where $\mathcal{S}$ and $\mathcal{S}^*$ are the distributions of S for the experimental data and the simulated data, respectively, N_s and N_t are the number of spatial and temporal groups (here three and five), and K is the Kolmogorov–Smirnov distance for pairs of empirical distribution functions. The Kolmogorov–Smirnov distance is defined as:

$$K(\mathcal{S}, \mathcal{S}^*) = \sup_x |\mathcal{S}(x) - \mathcal{S}^*(x)|,$$

with $\mathcal{S}$ and $\mathcal{S}^*$ are the two empirical distribution functions that we aim to compare.

We now have: (i) a model to describe macrophage migration in response to wounding, (ii) a dataset described by the distributions of directionality coefficients, and (iii) a distance function (Kolmogorov–Smirnov distance). Application of ABC-SMC allows us to estimate the parameters that define the cell migration, but also the parameters that describe the cytokine gradient. Sampling from the posterior parameter distribution we can then, for example, visualise the inferred cytokine gradients for the five temporal clusters as shown in Figure 17.8, from this analysis we can learn, among other things, that the cytokine signals are produced at the wound site for up to 7 hours. Such insights provide clues as to how wound signalling gradients are established and may be managed. Here the analysis was carried out in embryos (as these are optically transparent), but the inflammatory response to, for example, wounding will also need to be understood in older individuals, where it can be linked to ageing [62]. Mechanistic models will help to understand better the inflammatory response to a wound, which can have undesirable side-effects, for example, scaring, currently ABC seems to be the only statistical framework capable of dealing with these problems.

One modelling approach that is gaining popularity in this domain is *agent-based model* [63]. They consist of agents – in our context, for example, individual organisms or cells – that interact with their environment and each other.

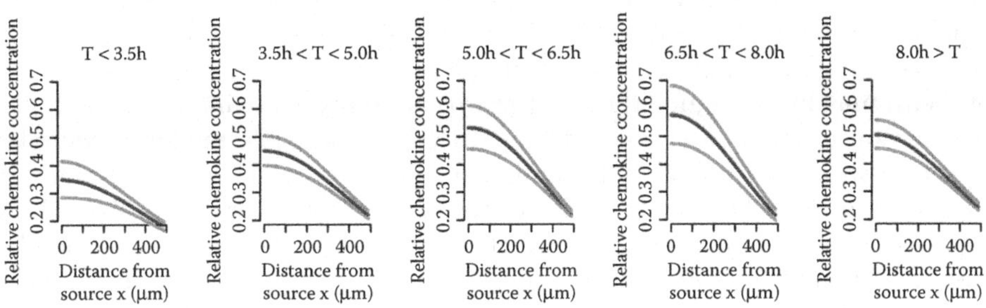

FIGURE 17.8
Spatio-temporal characteristics of cytokine gradients. The estimated cytokine gradients are shown for the five time intervals post wounding. Dark grey lines indicate the means of simulations from the posterior parameter distributions, and light grey lines are the respective 5%-ile and 95%-ile.

Interactions can be mediated by direct contact or via the exchange of (e.g. biochemical) signals that determine how agents behave next. ABC methods have already been employed to study models of stem cell dynamics, tumour growth models of stem cells, and cancer dynamics in intestinal crypts (which are the sites from which intestinal lining tissue is being renewed) [64], as well as tumour evolution [71]. Determination of a likelihood is typically impossible for these models and, despite the great cost of simulating such models [65], ABC approaches are beginning to bear fruit. The main challenge of applying ABC in this context seems to be the lack of suitably fast simulation software: mathematical and computational modelling typically involve small to moderate numbers of simulations, whereas simulation-based inference requires orders of magnitude larger numbers of simulation runs for many different parameter combinations.

17.6 ABC and Beyond

The virtue of ABC is to open up Bayesian inference to problems that defy conventional, exact inference. Where statistical inference that accounts for uncertainty and confidence appropriately [6] is otherwise not possible, optimisation is often used. ABC is perhaps best used as a stop-gap for problems for which exact Bayesian inference is (at least for now) computationally too expensive or not yet applicable. Parameter estimation and especially model selection, which is possible for dynamical systems as no summary statistics are used, are central to many problems in systems biology. And for many of these problems, ABC offers the most direct and convenient way of applying the Bayesian framework.

The central aim of statistical inference in systems biology is to get a better understanding of how biological systems – cells, tissues or even whole organisms—work, through the development of mechanistic models, from these models testable hypotheses can be distilled that probe our understanding further [66]. Even preliminary understanding of these systems (or of our models of them) may in turn guide the way towards applying exact inference methods, perhaps on approximate, computationally more amenable models. Examples of such approximations include the linear noise approximation [23], moment expansion approaches [26], and finite state projection approaches [67], which all provide approximations to the stochastic dynamics normally described by the chemical master equation [70]. These approaches allow us to simulate the dynamics at reduced computational cost, but often they also allow us to derive expressions of tractable likelihoods, that can be used in Bayesian inference. Now, however, instead of performing approximate inference on 'exact' models, we can employ exact inference on approximate models. Given that these models are already oversimplifications or abstractions of a much more complicated reality, it is hard to say which approximation is more problematic.

Tools to assess how much the assumption of a given model structure affects further analyses are only now emerging [37], and we can integrate these into our statistical analysis workflows. Nevertheless, there remains the question of how bad the ABC approximation using the full model is compared to approximating the model to yield a tractable likelihood.

While ABC may only be a stop-gap for many dynamical systems, for multi-scale problems – at least those that are of biological interest – ABC offers perhaps the only viable inferential framework that is currently available [72]. Here, however, the question as to whether the available or observable data are sufficient for parameter inference, let alone model selection, remains open, the answer will probably be very problem specific. For some migration processes, it will generally be possible to apply ABC and obtain robust estimates efficiently. For some more complicated multi-scale systems, including many agent-based models, this problem needs to be carefully assessed.

Multi-scale systems are gaining prominence also outside of systems biology and provide rich pickings for statisticians, especially those interested in ABC inference. Whether we can use ABC as a true approximation to conventional inference, or as a base for an alternative inferential framework, as has sometimes been suggested, remains to be seen. And addressing the lack of powerful simulation approaches – or side-stepping simulation in favour of emulation [68,69] – will be a first necessary step in establishing ABC in these contexts.

References

[1] J Ross and A P Arkin. Complex systems: From chemistry to systems biology. *Proceedings of the National Academy of Sciences of the United States of America*, 106(16):6433–6434, 2009.

[2] D S Goodsell. *The Machinery of Life.* New York: Springer Science & Business Media, 2009.

[3] C J Oates and S Mukherjee. Network inference and biological dynamics. *The Annals of Applied Statistics*, 6(3):1209–1235, 2012.

[4] T W Thorne, P Fratta, M G Hanna, A Cortese, V Plagnol, E M Fisher, and M P H Stumpf. Graphical modelling of molecular networks underlying sporadic inclusion body myositis. *Molecular Biosystems*, 9(7):1736–1742, 2013.

[5] J Gunawardena. Models in biology: 'Accurate descriptions of our pathetic thinking'. *BMC Biology*, 12(1):29, 2014.

[6] P D W Kirk, A C Babtie, and M P H Stumpf. Systems biology (un)certainties. *Science*, 350(6259):386–388, 2015.

[7] M P H Stumpf. Approximate Bayesian inference for complex ecosystems. *F1000Prime Reports*, 6(60):60, 2014.

[8] P Fearnhead, D Prangle, M P Cox, P J Biggs, and N P French. Semi-automatic selection of summary statistics for ABC model choice. *Statistical Applications in Genetics and Molecular Biology*, 13(1):67–82, 2014.

[9] C P Robert, J M Cornuet, J M Marin, and N S Pillai. Lack of confidence in approximate Bayesian computation model choice. *Proceedings of the National Academy of Sciences*, 108(37):15112–15117, 2011.

[10] R D Wilkinson. Approximate Bayesian computation (ABC) gives exact results under the assumption of model error. *Statistical Applications in Genetics and Molecular Biology*, 12(2):129–141, 2013.

[11] M Cohen, A Kicheva, A Ribeiro, R Blassberg, K M Page, C P Barnes, and J Briscoe. Ptch1 and Gli regulate Shh signalling dynamics via multiple mechanisms. *Nature Communications*, 6:6709, 2015.

[12] J Liepe, H G Holzhütter, E Bellavista, P M Kloetzel, M P H Stumpf, and M Mishto. Quantitative time-resolved analysis reveals intricate, differential regulation of standard- and immuno-proteasomes. *eLife*, 4:e07545, 2015.

[13] J Liepe, H Taylor, C P Barnes, M Huvet, L Bugeon, T W Thorne, J R Lamb, M J Dallman, and M P H Stumpf. Calibrating spatio-temporal models of leukocyte dynamics against in vivo live-imaging data using approximate Bayesian computation. *Integrative Biology: Quantitative Biosciences from Nano to Macro*, 4(3):335–345, 2012.

[14] J Liepe, S Filippi, M Komorowski, and M P H Stumpf. Maximizing the information content of experiments in systems biology. *PLoS Computational Biology*, 9(1):e1002888, 2013.

[15] G R Holmes, S R Anderson, G Dixon, A L Robertson, C C Reyes-Aldasoro, S A Billings, S A Renshaw, and V Kadirkamanathan. Repelled from the wound, or randomly dispersed? Reverse migration behaviour of neutrophils characterized by dynamic modelling. *Journal of The Royal Society Interface*, 9(77):3229–3239, 2012.

[16] T Toni and M P H Stumpf. Simulation-based model selection for dynamical systems in systems and population biology. *Bioinformatics (Oxford, England)*, 26(1):104–110, 2010.

[17] A L Maclean, C Lo Celso, and M P H Stumpf. Population dynamics of normal and leukaemia stem cells in the haematopoietic stem cell niche show distinct regimes where leukaemia will be controlled. *Journal of the Royal Society Interface*, 10(81):20120968–20120968, 2013.

[18] N Domedel-Puig, I Pournara, and L Wernisch. Statistical model comparison applied to common network motifs. *BMC Systems Biology*, 4(1):18, 2010.

[19] P Kirk, T Toni, and M P H Stumpf. Parameter inference for biochemical systems that undergo a Hopf bifurcation. *Biophysical Journal*, 95(2):540–549, 2008.

[20] S Rogers and M Girolami. A Bayesian regression approach to the inference of regulatory networks from gene expression data. *Bioinformatics*, 21(14):3131–3137, 2005.

[21] T Toni, D Welch, N Strelkowa, A Ipsen, and M P H Stumpf. Approximate Bayesian computation scheme for parameter inference and model selection in dynamical systems. *Journal of the Royal Society Interface*, 6(31):187–202, 2009.

[22] D J Wilkinson. *Stochastic Modelling for Systems Biology*. Boca Raton, FL: CRC Press, 2011.

[23] E W J Wallace, D T Gillespie, K R Sanft, and L R Petzold. Linear noise approximation is valid over limited times for any chemical system that is sufficiently large. *IET Systems Biology*, 6(4):102–115, 2012.

[24] A Golightly and D Wilkinson. Bayesian sequential inference for stochastic kinetic biochemical network models. *Journal of Computational Biology*, 13(3):838–851, 2006.

[25] A Golightly and D J Wilkinson. Bayesian parameter inference for stochastic biochemical network models using particle Markov chain Monte Carlo. *Interface Focus*, 1(6):807–820, 2011.

[26] A Ale, M P H Stumpf, and P Kirk. A general moment expansion method for stochastic kinetic models. *The Journal of Chemical Physics*, 138(17):174101, 2013.

[27] A Golightly and D J Wilkinson. Bayesian inference for stochastic kinetic models using a diffusion approximation. *Biometrics*, 61(3):781–788, 2005.

[28] M Secrier, T Toni, and M P H Stumpf. The ABC of reverse engineering biological signalling systems. *Molecular Biosystems*, 5(12):1925–1935, 2009.

[29] S A Sisson, Y Fan, and M M Tanaka. Sequential Monte Carlo without likelihoods. *Proceedings of the National Academy of Sciences*, 104(6):1760–1765, 2007.

[30] J F Apgar, D K Witmer, F M White, and B Tidor. Sloppy models, parameter uncertainty, and the role of experimental design. *Molecular BioSystems*, 6(10):1890–1900, 2010.

[31] K Erguler and M P H Stumpf. Practical limits for reverse engineering of dynamical systems: A statistical analysis of sensitivity and parameter inferability in systems biology models. *Molecular Biosystems*, 7(5):1593–1602, 2011.

[32] R N Gutenkunst, J J Waterfall, F P Casey, K S Brown, C R Myers, and J P Sethna. Universally sloppy parameter sensitivities in systems biology models. *PLoS Computational Biology*, 3(10):1871–1878, 2007.

[33] M Komorowski, M J Costa, D A Rand, and M P H Stumpf. Sensitivity, robustness, and identifiability in stochastic chemical kinetics models. *Proceedings of the National Academy of Sciences of the United States of America*, 108(21):8645–8650, 2011.

[34] J Liepe, C P Barnes, E Cule, K Erguler, P Kirk, T Toni, and M P H Stumpf. ABC-SysBio–approximate Bayesian computation in Python with GPU support. *Bioinformatics (Oxford, England)*, 26(14):1797–1799, 2010.

[35] J Liepe, C P Barnes, P Kirk, S Filippi, T Toni, and M P H Stumpf. A framework for parameter estimation and model selection from experimental data in systems biology using approximate Bayesian computation. *Nature Protocols*, 9(2):439–456, 2014.

[36] P Kirk, T W Thorne, and M P H Stumpf. Model selection in systems and synthetic biology. *Current Opinion in Biotechnology*, 24(4):767–774, 2013.

[37] A C Babtie, P D W Kirk, and M P H Stumpf. Topological sensitivity analysis for systems biology. *Proceedings of the National Academy of Sciences of the United States of America*, 111(52):18507–18512, 2014.

[38] O Ratmann, C Andrieu, C Wiuf, and S Richardson. Model criticism based on likelihood-free inference, with an application to protein network evolution. *Proceedings of the National Academy of Sciences*, 106(26):10576–10581, 2009.

[39] C P Barnes, S Filippi, M P H Stumpf, and T W Thorne. Considerate approaches to constructing summary statistics for ABC model selection. *Statistics and Computing*, 22(6):1181–1197, 2012.

[40] A Grelaud, C P Robert, and J M Marin. ABC methods for model choice in Gibbs random fields. *Comptes Rendus Mathematique*, 347:205–210, 2009.

[41] T Toni, Y Ozaki, P Kirk, S Kuroda, and M P H Stumpf. Elucidating the in vivo phosphorylation dynamics of the ERK MAP kinase using quantitative proteomics data and Bayesian model selection. *Molecular Biosystems*, 8(7):1921–1929, 2012.

[42] N Nandagopal and M B Elowitz. Synthetic biology: Integrated gene circuits. *Science*, 333(6047):1244–1248, 2011.

[43] C P Barnes, D Silk, and M P H Stumpf. Bayesian design strategies for synthetic biology. *Interface Focus*, 1(6):895–908, 2011.

[44] C P Barnes, D Silk, X Sheng, and M P H Stumpf. Bayesian design of synthetic biological systems. *Proceedings of the National Academy of Sciences of the United States of America*, 108(37):15190–15195, 2011.

[45] C J Ryan, P Cimermančič, Z A Szpiech, A Sali, R D Hernandez, and N J Krogan. High-resolution network biology: Connecting sequence with function. *Nature Reviews Genetics*, 14(12):865–879, 2013.

[46] E D Kolaczyk. *Statistical Analysis of Network Data: Methods and Models.* New York: Springer Science & Business Media, 2009.

[47] O Ratmann, O Jørgensen, T Hinkley, M P H Stumpf, S Richardson, and C Wiuf. Using likelihood-free inference to compare evolutionary dynamics of the protein networks of *H. pylori* and *P. falciparum*. *PLoS Computational Biology*, 3(11):e230, 2007.

[48] T A Gibson and D S Goldberg. Improving evolutionary models of protein interaction networks. *Bioinformatics* (*Oxford, England*), 27(3):376–382, 2011.

[49] M P H Stumpf, W P Kelly, T W Thorne, and C Wiuf. Evolution at the system level: The natural history of protein interaction networks. *Trends in Ecology & Evolution*, 22(7):366–373, 2007.

[50] S N Dorogovtsev and J F F Mendes. *Evolution of Networks.* From Biological Nets to the Internet and WWW. Oxford, UK: Oxford University Press, 2013.

[51] C Wiuf, M Brameier, O Hagberg, and M P H Stumpf. A likelihood approach to analysis of network data. *Proceedings of the National Academy of Sciences*, 103(20):7566–7570, 2006.

[52] T W Thorne and M P H Stumpf. Graph spectral analysis of protein interaction network evolution. *Journal of the Royal Society, Interface*, 9(75):2653–2666, 2012.

[53] N M Rashidi, M K Scott, N Scherf, A Krinner, J S Kalchschmidt, K Gounaris, M E Selkirk, I Roeder, and C L Celso. In vivo time-lapse

imaging of mouse bone marrow reveals differential niche engagement by quiescent and naturally activated hematopoietic stem cells. *Blood*, 124(1):79–83, 2014.

[54] M A R Ferreira and H K H Lee. *Multiscale Modeling.* A Bayesian Perspective. London, UK: Springer Science & Business Media, 2007.

[55] S Aeschbacher, M A Beaumont, and A Futschik. A novel approach for choosing summary statistics in approximate Bayesian computation. *Genetics*, 192(3):1027–1047, 2012.

[56] P Fearnhead and D Prangle. Constructing summary statistics for approximate Bayesian computation: Semi-automatic approximate Bayesian computation. *Journal of the Royal Statistical Society: Series B* (*Statistical Methodology*), 74:419–474, 2012.

[57] M A Nunes and D J Balding. On optimal selection of summary statistics for approximate Bayesian computation. *Statistical Applications in Genetics and Molecular Biology*, 9(1):34, 2010.

[58] G R Holmes, S R Anderson, G Dixon, S A Renshaw, and V Kadirkamanathan. A Bayesian framework for identifying cell migration dynamics. *Conference Proceedings: ... Annual International Conference of the IEEE Engineering in Medicine and Biology Society. IEEE Engineering in Medicine and Biology Society. Conference*, 2013:3455–3458, 2013.

[59] H B Taylor, J Liepe, C Barthen, L Bugeon, M Huvet, P Kirk, S B Brown, J R Lamb, M P H Stumpf, and M J Dallman. P38 and JNK have opposing effects on persistence of in vivo leukocyte migration in zebrafish. *Immunology and Cell Biology*, 91(1):60–69, 2013.

[60] J B Beltman, A F M Marée, and R J de Boer. Analysing immune cell migration. *Nature Reviews Immunology*, 9(11):789–798, 2009.

[61] E A Codling, M J Plank, and S Benhamou. Random walk models in biology. *Journal of the Royal Society Interface*, 5(25):813–834, 2008.

[62] A M Valdes, D Glass, and T D Spector. Omics technologies and the study of human ageing. *Nature Reviews Genetics*, 14(9):601–607, 2013.

[63] H Kaul and Y Ventikos. Investigating biocomplexity through the agent-based paradigm. *Briefings in Bioinformatics*, 16(1):137–152, 2013.

[64] A Sottoriva and S Tavaré. Integrating approximate Bayesian computation with complex agent-based models for cancer research. *Proceedings of COMPSTAT'2010* (Chapter 5):57–66, 2010.

[65] M Scianna and L Preziosi. *Cellular Potts Models: Multiscale Extensions and Biological Applications.* Boca Raton, FL: CRC Press, 2013.

[66] T Toni, G Jovanovic, M Huvet, M Buck, and M P H Stumpf. From qualitative data to quantitative models: Analysis of the phage shock protein stress response in Escherichia coli. *BMC Systems Biology*, 5(1):69, 2011.

[67] B Munsky and M Khammash. The finite state projection algorithm for the solution of the chemical master equation. *The Journal of Chemical Physics*, 124(4):044104, 2006.

[68] S Conti and A O'Hagan. Bayesian emulation of complex multi-output and dynamic computer models. *Journal of Statistical Planning and Inference*, 140(3):640–651, 2010.

[69] C C Drovandi, A N Pettitt, and M J Faddy. Approximate Bayesian computation using indirect inference. *Journal of the Royal Statistical Society: Series C (Applied Statistics)*, 60(3):317–337, 2011.

[70] C Gardiner. Stochastic Methods: A Handbook For The Natural And Social Sciences. Springer, 2009.

[71] A Sottoriva, H Kang, Z Ma, T A Graham, M P Salomon, J Zhao, P Marjoram, K Siegmund, M F Press, D Shibata, and C Curtis. A Big Bang model of human colorectal tumor growth. *Nature Genetics*, 47(3):209–216, 2015.

[72] P J M Jones, A Sim, H B Taylor, L Bugeon, M J Dallman, B Pereira, M P H Stumpf, and J Liepe. Inference of random walk models to describe leukocyte migration. *Physical biology*, 12(6):066001, 2015.

18

Application of ABC to Infer the Genetic History of Pygmy Hunter-Gatherer Populations from Western Central Africa

Arnaud Estoup, Paul Verdu, Jean-Michel Marin, Christian Robert, Alex Dehne-Garcia, Jean-Marie Cornuet, and Pierre Pudlo

CONTENTS

18.1 Introduction 541
18.2 Simulation of Datasets 544
18.2.1 Observed dataset 544
18.2.2 Models and parameters 545
18.2.3 Computer programs 548
18.3 Results and Discussions 549
18.3.1 Model choice: ABC random forest 549
18.3.2 Parameter estimation 555
18.3.3 Model-posterior checking 559
18.4 Conclusions and Perspectives 562
Acknowledgements 563
References 563

18.1 Introduction

Approximate Bayesian computation (ABC; Beaumont et al., 2002) represents an elaborate approach to model-based inference in a Bayesian setting in which model likelihoods are difficult to calculate and must be estimated by massive simulations. We will not detail here the general statistical features of

ABC, as they have been reviewed in previous publications (Beaumont, 2010; Bertorelle et al., 2010; Csillėry et al., 2010; Marin et al., 2012; Sunnaker et al., 2013) and in other chapters of this book. ABC methods undoubtedly widen the realm of models for which statistical inference can be considered. The method arose in population genetics (Tavaré, 1997; Beaumont et al., 2002) and is increasingly used in other fields, including epidemiology, system biology, ecology, and agent-based modelling (reviewed in Beaumont, 2010).

Although full-likelihood methods have been developed in evolutionary and population genetics for some models, they typically rely on Markov chain Monte Carlo that has problems converging on large datasets and there is a limited range of model structures that can be analysed (Beaumont, 2010). By contrast, ABC is very flexible and is, hence, well adapted to investigate complex models of species and population history often involving serial or independent divergence events, changes of population sizes, and genetic admixture or migration events (Fagundes et al., 2007; Lombaert et al., 2010). In an ABC framework, such events can be easily simulated and, hence, incorporated into different historical and demographic evolutionary models (often called 'scenarios') that can be formally tested against each other with respect to the observed data. The method can then be used to estimate the posterior distributions of demographic parameters of interest, such as divergence times, admixture rates, and effective population sizes, in a given scenario (usually the most likely one). In practice, ABC users in the field of population and evolutionary biology can base their analysis on simulation programs, such as SIMCOAL (Laval and Excoffier, 2004), ms (Hudson, 2002), or MaCS (Chen et al., 2009) and then use various statistical software to post-process their simulation outputs. Several ABC programs have recently been developed to provide non-specialist users with more integrated computational solutions varying in user-friendliness (see the list of ABC packages and toolboxes in Beaumont, 2010; Csillėry et al., 2010).

In this chapter, we used a set of recent ABC-based methods to thoroughly analyse a human microsatellite genetic dataset from Western Central African Pygmy and non-Pygmy populations. Central Africa and the Congo Basin are currently peopled by the largest group of forest hunter-gatherer populations worldwide, which have been historically called 'Pygmies' in reference to the mythical population of short stature described by the ancient Greek poet Homer (Hewlett, 2014). Each Central African Pygmy group is in the neighbourhood of several sedentary agricultural populations (hereafter called 'non-Pygmies') with whom they share complex sociocultural and economic relationships, including social rules regulating intermarriages between communities (Verdu et al., 2013; Hewlett, 2014). Due to the lack of ancient human remains in the equatorial forest, the origins of Pygmies and neighbouring non-Pygmies, including population divergence times, remains largely unknown (Cavalli-Sforza et al., 1994; Cavalli-Sforza and Feldman, 2003).

Moreover, Western colonisers from the nineteenth century somewhat arbitrarily collapsed into a single 'Pygmy' group more than 20 populations that were, and still are, culturally and geographically isolated in reality, which further clouded our understanding of evolutionary relationships among these populations. Thus, whether all Central African Pygmy populations shared a common or an independent origin, and when during history did populations diverge from one another and from neighbouring non-Pygmies, was still largely debated in the anthropology and ethnology communities (Cavalli-Sforza, 1986; Verdu et al., 2009; Hewlett, 2014). To address these questions, Verdu et al. (2009) genotyped strongly variable genetic markers (namely, microsatellite loci) in a dense sample of non-Pygmy and neighbouring Pygmy populations from Western Central Africa. The resulting genetic dataset was analysed using ABC techniques to compare a set of possible evolutionary scenarios and estimate key parameters under the most likely historical model identified.

In the application of ABC methods presented here, we have extended a subset of the Verdu et al. (2009) genetic dataset by adding a European population sample previously genotyped at the same microsatellite loci and for which the divergence time was fixed during the inferential process according to previous estimates (Fagundes et al., 2007; Gravel et al., 2011). By considering a dataset including such a so-called 'scaling population', we hoped to obtain more precise inferences regarding several original key population parameters of Central African populations' history, such as the divergence times between Pygmies and non-Pygmies (see Excoffier et al., 2013 for an illustration of inferences using a scaling human population in a different historical and genetic context). Additional noticeable novelties of the ABC analysis presented here compared to Verdu et al. (2009) include the application of ABC random forest algorithms to make model choice (Pudlo et al., 2016) and ABC model-posterior checking methods to evaluate the goodness of fit between the inferred genetic history and the observed dataset.

In the following, we first describe the observed genetic dataset for which one wants to make inferences, the set of evolutionary models we compared, with their respective historical, demographic, and mutational parameters, their associated prior distributions, and the way datasets were simulated for ABC analyses. Second, we present the model choice analyses we carried out using ABC random forest algorithms. Third, we present the estimation of historical and demographic parameters we carried out under the most likely of the compared models for the peopling of Central Africa, with a particular interest in evaluating the effect (or lack of effect) of using a scaling population to improve inferences. Finally, we report the model-posterior checking analyses we carried out to evaluate the goodness of fit between the final inferred genetic history and the observed dataset. Each section includes methodological aspects, results, and elements of discussion.

18.2 Simulation of Datasets

18.2.1 Observed dataset

The dataset included the genotyping at 26 microsatellite loci of 183 unrelated individuals from four Pygmy groups (i.e. the Baka, Bezan, Kola, and Koya; 29–32 individuals per group), neighbouring non-Pygmy individuals (33 individuals) from Cameroon and Gabon (Western Central Africa), and a European French population (29 individuals) for which genotype data at the same microsatellite loci was available (Rosenberg et al., 2002). This dataset corresponds to a subset of the African dataset Verdu et al. (2009) used in their initial ABC treatments, which originally included the genotyping at 28 microsatellite loci of 400 Pygmy and non-Pygmy individuals from Cameroon and Gabon (see Figure 18.1; and see Verdu et al., 2009's Table S1 for details about geographic location of population samples and their genetic grouping). In the present study, we considered a reduced sample set as compared to Verdu et al. (2009) for both computational efficiency and to homogenise sample sizes across Pygmy and non-Pygmy populations. We kept the Bezan, Koya, and Kola Pygmy samples identical to those used in Verdu et al. (2009) (they had roughly the same sample sizes i.e. about 30 individuals). We re-sampled 32 Baka Pygmy individuals and 33 non-Pygmy African individuals among the 117 Baka individuals and 194 non-Pygmy Africans considered originally in Verdu et al. (2009), respectively.

We added European individuals from France as a scaling population sample, with the hope to obtain more precise ABC inferences about original parameters associated to the Central African peopling history, especially for Pygmy population history, our main topic of interest here. We built a correspondence table for the allele calls between the datasets of Verdu et al. (2009) and Rosenberg et al. (2002) by genotyping the 28 microsatellite loci in ten individuals from the Human Genome Diversity Project (HGDP)-Centre d'Etude du Polymorphisme Humain (CEPH) panel (Cann et al., 2002). Allele calls for two loci (D14S1280 and D21S1432) could not be unambiguously normalised between the two datasets, and these two loci were thus discarded reducing the marker dataset from 28 microsatellites used in Verdu et al. (2009), to 26 here. All Central African microsatellite data originally used in Verdu et al. (2009) are available upon request at the European Genome-phenome Archive (EGA, http://www.ebi.ac.uk/ega/) under accession number EGAS00001000652, and all HGDP-CEPH data at the Marshfield institute (http://research.marshfieldclinic.org/genetics/genotypingData_Statistics/humanDiversityPanel.asp). The exact dataset used in the present study is available upon request to some authors (PV or AE), following ethical appropriateness.

18.2.2 Models and parameters

We considered the same set of eight complex evolutionary models, hereafter, referred to as scenarios, as in Verdu et al. (2009). These scenarios with their historical and demographic parameters are represented in Figure 18.1, following the notation of Verdu et al. (2009).

These eight scenarios formalise two main types of evolutionary histories debated in the anthropology community. Class 1 scenarios (scenarios 1a, 1b, 1c, and 1d) correspond to a common evolutionary history of the four Pygmy groups, where they all originate from the same ancestral Pygmy population, which initially diverged from the non-Pygmy population in a more remote past. Class 2 scenarios (scenarios 2a, 2b, 2c, and 2d) correspond to an independent evolutionary history of the four Pygmy groups, where they each originate independently from the non-Pygmy population. Within each scenario class, the scenarios differ by two types of events: (1) the possibility or not of recent and ancient asymmetrical admixture events between each Pygmy population and the non-Pygmy one, as was already suggested by previous anthropological and genetic studies (Cavalli-Sforza, 1986; Hewlett, 1996; Destro-Bisol et al., 2004), and (2) the possibility or not of a change of population size in the non-Pygmy population. The European French population, used for scaling purposes, first underwent a demographic bottleneck of duration DB_{ooa} (*ooa* for 'Out Of Africa') during which its size is NF_{ooa} and then reached a stable population size N_{ooa}. Following Excoffier et al. (2013, and see reference therein), this scaling population was assumed to diverge from the African non-Pygmy population at a time fixed at 50,000 years (i.e. 2,000 generations assuming a generation time of 25 years).

We chose flat prior distributions as in Verdu et al. (2009) for all demographic parameters: uniform distributions bounded between 100 and 10,000 diploid individuals for all Pygmy populations and ancestral population sizes (N_i, N_{ap}, N_{ai}, and N_A, with i between 1 and 4), between 1,000 and 100,000 for the non-Pygmy (N_{np}) and the European French (N_{ooa}) populations, and between 10 and 1,000 for the number of European founders (NF_{ooa}) for a bottleneck duration (DB_{ooa}) between 1 and 30 generations. Priors were drawn from uniform distributions between 1 and 5,000 generations for all divergence times (t_p, t_{pnp}, t_{pnpi}, with i between 1 and 4), for the population size variation times (t_{nei}, with i between 1 and 4), and for the times of 'ancient' introgression of non-Pygmy genes into ancestral Pygmy lineages (tr_a, and tr_{ai}, with i between 1 and 4). For the time of change in effective population size (t_A), considered only in scenarios 1a/1b and 2a/2b, we drew our priors from a uniform distribution bounded between 1 and 10,000 generations. For the 'recent' introgression times from non-Pygmies into the Pygmy lineages (tr, and tr_{ri}, with i between 1 and 4), we chose loguniform prior distributions

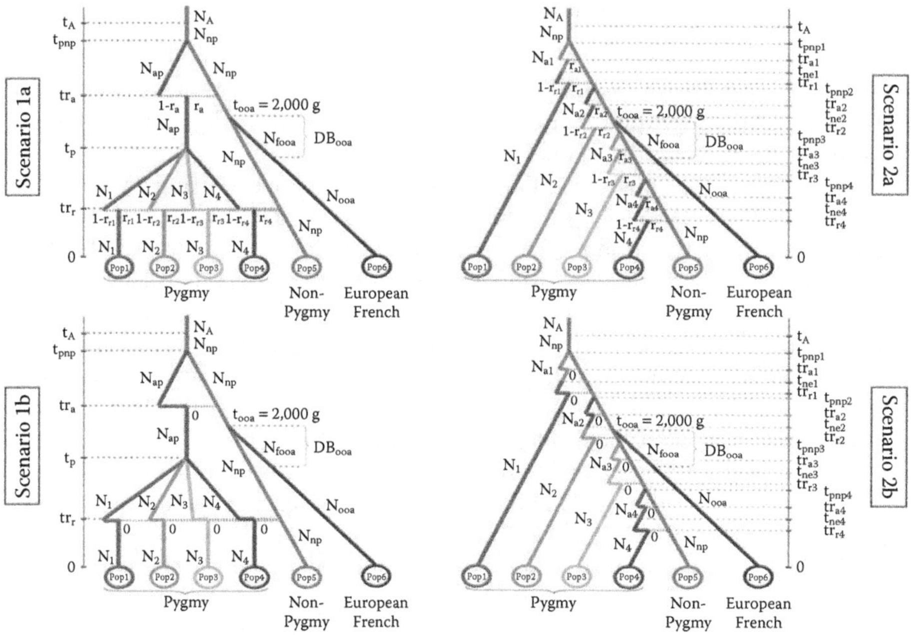

FIGURE 18.1
Eight complex competing scenarios of origin and diversification of Pygmy populations from Western Central Africa. The scenarios and parameters are similar to those in Verdu et al. (2009) and follow the same notation (Figure in color available by request from AE). The only exception is the addition of European individuals from France providing a scaling population sample which diverged from African non-Pygmies at a time fixed at 50,000 years (i.e. $t_{ooa} = 2{,}000$ generations). Scenarios 1a–d correspond to a common origin of Pygmy populations that diversified from a single ancestral Pygmy population at time t_p. The ancestral Pygmy population itself diverged at time t_{pnp} from the non-Pygmy population. Scenarios 2a–d correspond to an independent origin of Pygmy groups that independently diverged from the non-Pygmy population at times t_i. We simulated in the eight scenarios, two potential events of introgression (cf. parameters t_{ri} and r_i) from the non-Pygmy lineage into each Pygmy lineage independently. Finally, Scenarios 1a, 1b, 2a, and 2b include a potential stepwise change of effective population size that occurred in the non-Pygmy lineage at time t_A. Scenarios 1b/2b were identical to scenarios 1a/2a, except that all introgression rates (r_i) were set to zero. Scenarios 1c/2c were identical to scenarios 1a/2a, but did not consider the potential change in non-Pygmy effective size. Finally, in scenarios 1d/2d, neither introgression events nor change in effective size were considered. For all scenarios, N_i indicates the effective population size of population i. Note that for scenarios 2a–d, split times were drawn independently for each Pygmy lineage, and, thus, the order

(Continued)

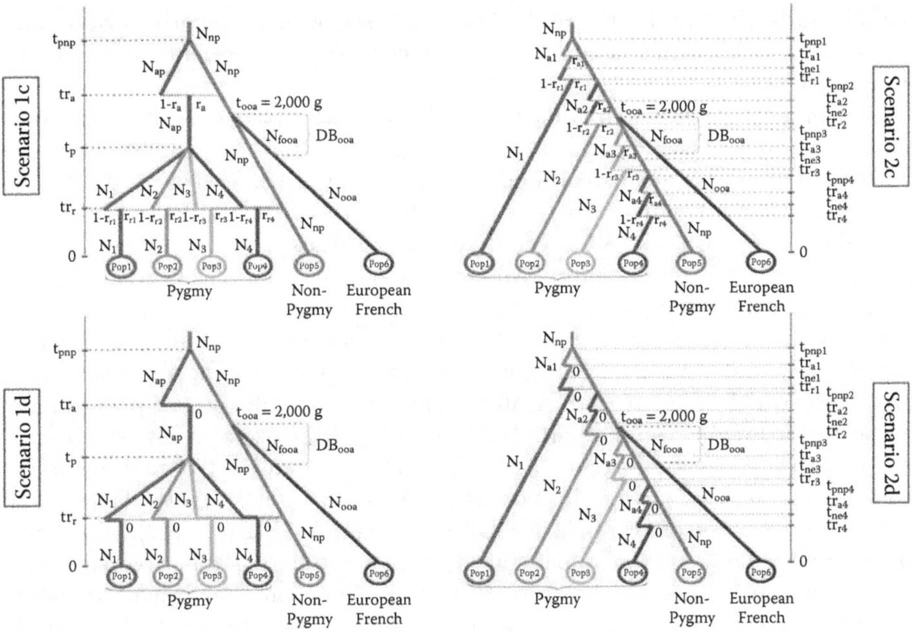

FIGURE 18.1 (Continued)
in which these lineages split is not pre-defined. See main text for details regarding prior distributions of parameters.

bounded between 1 and 5,000 generations. For the class 1 scenarios, we set the conditions $tr_r < t_p < tr_a < t_{pnp}$. For the class 2 scenarios, we set the conditions $tr_{ri} < t_{nei} < tr_{ai} < t_{pnpi}$ with i between 1 and 4.

For the parameterisation of microsatellite mutation rates, each locus was assumed to follow a generalised stepwise mutation model with a possible range of 40 contiguous allelic states (Estoup et al., 2002). The mean mutation rate ($\bar{\mu}$) was drawn from a uniform distribution bounded between 10^{-4} and 10^{-3} (Dib et al., 1996; Estoup et al., 2002), and mutation rates for each locus were drawn independently from a gamma distribution (mean $= \bar{\mu}$ and shape $= 2$). The mean parameter of the geometric distribution ($\bar{p}$) of the length in number of repeats of mutation events was drawn from a uniform distribution bounded between 0.1 and 0.3 (Dib et al., 1996; Excoffier et al., 2005), and the parameters for individual loci were drawn from a gamma distribution (mean $= \bar{p}$ and shape $= 2$). Some allele lengths, measured in base pairs, were odd and some were even, a pattern implying that there has been an indel mutation, giving a length that is not a multiple of the motif length. In order to simulate uneven insertion/deletion events that could be suspected for several of our microsatellite loci based on observed allele sizes, we draw the mutation parameter μSNI (for single nucleotide instability) from a loguniform distribution with bounds 10^{-9} and 10^{-4} and a gamma distribution (mean = μSNI and shape = 2) for drawing rates of single nucleotide insertion-deletion events at individual loci.

As for any Bayesian inference, the shape of the priors used for dataset simulations may affect both the posterior probabilities of scenarios and the posterior parameter estimation under ABC inference (Sunnaker et al., 2013). Therefore, in the original Verdu et al. (2009) study, we conducted all ABC procedures assuming a set of alternative non-flat priors for the simulations (cf. prior set 2 in Verdu et al., 2009). We found that the posterior probabilities and distributions were moderately affected by the shape of the prior distribution in our specific case (see Verdu et al., 2009 Supplementary Material). For sake of concision, we did not repeat here this time-consuming, but otherwise necessary procedure which aims at empirically evaluating the influence of prior shape on posterior inferences for each ABC case-study (Bertorelle et al., 2010).

A major interest of ABC methods is their potential to treat complex and, hence, relatively realistic models. This is why we straightly formalised and analysed a set of complex evolutionary models (Figure 18.1). However, the verbally stated main evolutionary question (i.e. independent versus non-independent histories of the four Pygmy groups) can be drawn, at least roughly, from a comparison of a lower number of less parameterised (though less realistic) historical and demographic models than those described earlier and in Figure 18.1. To illustrate the impact of assuming substantially simplified models, we considered and analysed a set of two simplified versions of our complex scenarios 1a and 2a, in which we removed all introgression events and assumed that all Pygmy and non-Pygmy populations had the same effective population size (drawn into a uniform distribution bounded between 100 and 30,000 diploid individuals). Prior distributions for divergence times and mutational parameters remained the same as for previous complex scenarios. In the following, we will refer to this pair of simplified scenarios as 'the simple scenario 1' and 'the simple scenario 2' throughout.

18.2.3 Computer programs

For both the simulation of data following the aforementioned model-prior design and all post-process statistical treatments (with the exception of the ABC random forest treatments applied for model choice; Pudlo et al., 2016), we used the package Do It Yourself Approximate Bayesian Computation (DIYABC) v2.1.0 (Cornuet et al., 2014), freely available with a detailed user-manual and example projects for academic and teaching purposes at http://www1.montpellier.inra.fr/CBGP/diyabc. Briefly, Cornuet et al. (2008, 2010, 2014) developed DIYABC to provide a user-friendly interface, which allows biologists with little background in programming to perform inferences via ABC. DIYABC is a coalescent-based program (Nordborg, 2001), which can consider complex population histories, including any number of divergence (without migration), admixture, and population size variation events, for population samples that may have been collected at different times. The package accepts various types of molecular data (microsatellites, DNA sequences, and SNP), evolving under various mutation models, and located on various chromosome types (autosomal, X or Y chromosomes, and mitochondrial DNA). A text file including the instructions based on the simple DIYABC coding language (cf. DIYABC user) to make simulations for the eight complex scenarios of Figure 18.1 (and the simple scenarios 1 and 2) is available upon request from AE.

Regarding ABC random forest treatments, computations were performed with the random forest statistical framework implemented in the randomForest package of R (Liaw and Wiener, 2002), with all methodologies detailed in Pudlo et al. (2016) implemented in the R package abcrf available on the CRAN. We are currently implementing the ABC-random forest (RF) algorithms of Pudlo et al. (2016) in the next version of the software DIYABC to provide a user-friendly interface to implement this promising statistical method.

18.3 Results and Discussions

As a pre-amble, it is worth stressing that we willingly biased our methodological choices towards ABC methods already implemented in the package DIYABC and the recently developed ABC random forest methodologies. This bias is explained by the fact that most of the authors of this chapter are directly involved into the development of such methods and computer program and are therefore optimally positioned to use, explain, and critically discuss them. Nevertheless, a number of similar and alternative ABC methods and computer programs have been developed by other groups and could certainly have been successfully applied to the presently studied dataset. Useful references to alternative ABC methods and programs can be found in recent reviews on ABC, such as those by Beaumont (2010) and Sunnaker et al. (2013). For instance, the computer packages *ABCtoolbox* (Wegmann et al., 2010) and *abc* (Csillėry et al., 2012) provide useful operational and alternative ABC algorithms (e.g. Markov chain Monte Carlo without likelihood, a particle-based sampler and ABC-generalized linear model (GLM) for ABCtoolbox, neural network regression-based methods for *abc*) for making ABC inferences about complex evolutionary scenarios using various types of molecular data.

18.3.1 Model choice: ABC random forest

Choosing among a finite set of models (scenarios) is a crucial inferential issue, as it allows the identification of major historical and evolutionary features formalised into the set of compared scenarios. In the eight complex scenarios we studied, such features include: whether there is an independent or non-independent origin of Pygmy groups; the possibility of introgression events between Pygmy and non-Pygmy populations; and the possibility of a change in size of the non-Pygmy population. Both theoretical arguments and simulation experiments indicate that model posterior probabilities are poorly evaluated by ABC, even though the numerical approximations of such probabilities can preserve the proper ordering of the compared models (Robert et al., 2011; Pudlo et al., 2016). Pudlo et al. (2016) have recently proposed a novel approach based on a machine learning tool named 'random forests' (Breiman, 2001) to conduct selection among the highly complex models covered by ABC algorithms. In their approach named 'ABC-RF', Pudlo et al. (2016) proposed both to step away from selecting the most probable model from estimated posterior probabilities and to reconsider the very problem of constructing efficient summary statistics. First, given an arbitrary pool of available statistics, they completely bypass selecting among these. This new perspective directly proceeds from machine learning methodology (Breiman, 2001). Second, because posterior probabilities are poorly evaluated by ABC, ABC-RF postpones the approximation of model posterior probabilities to a second stage. ABC-RF analyses, hence, include two successive steps. As a first step, ABC-RF predicts the model that best fits the observed dataset by constructing a (machine learning) classifier from simulations from the prior predictive distribution, known as the reference table in ABC (i.e. records of a given number of datasets simulated from the priors under different models and summarised with an extensive pool of statistics). As a second step, Pudlo et al. (2016) demonstrated that a reliable estimation of posterior probability of the previous selected model can be obtained through a secondary random forest that regresses the model selection error in the first step over the available summary statistics.

As compared to past ABC implementations for model choice, ABC-RF offers improvements at least at four levels: (1) on all experiments we studied (including those detailed in Pudlo et al., 2016), it significantly reduces the classification error as measured by the probability to choose a wrong model when drawing model index and parameter values into priors (a quantity, hereafter, named the prior error rate); (2) it is robust to the number and choice of summary statistics, as RF can handle many superfluous and/or strongly correlated statistics with no impact on the performance of the method; (3) the computing effort is considerably reduced, as RF requires a much smaller reference table compared with alternative methods (i.e. a few thousands of simulated datasets versus hundred thousand to millions of simulations per compared model); and (4) it provides a reliable estimate of posterior probability of the selected model (i.e. the model that best fits the observed dataset).

We first used ABC-RF to discriminate among the eight complex scenarios presented in Figure 18.1 and then computed the posterior probability of the best supported model. Following Pudlo et al. (2016), ABC-RF analyses were processed on 80,000 simulated datasets (10,000 per scenario), drawing parameter values into the prior distributions detailed in the previous Section 18.2.2. Since the summary statistics proposed by DIYABC (Cornuet et al., 2014) describe genetic variation per population (e.g. number of alleles), per pair (e.g. genetic distance), or per triplet (e.g. admixture rate) of populations, averaged over the 26 loci (see the DIYABC 2.1.0 user-manual page 16 for details about such statistics), we have included 204 of those statistics plus the seven linear discriminant analysis axes as additional summary statistics in our ABC-RF treatments. For sake of comparison, we also processed standard ABC treatments (using a logistic regression approach modified following Estoup et al., 2012) for model choice on the present dataset using a substantially larger number of simulated datasets (i.e. 8×10^6). It should be noted that the standard ABC treatments for model choice processed in Verdu et al. (2009) relied on a subset of 48 summary statistics describing genetic variation within and between populations which were selected based on the expertise of the authors in population genetics, as well as on a large number of simulated datasets (i.e. 4×10^6).

The projection of the reference table on the first four linear discriminant analysis axes provides a first visual indication about our capacity to discriminate among the eight compared scenarios (Figure 18.2). Simulations under the different scenarios moderately overlapped, suggesting a rather good power to discriminate scenarios. In agreement with this, we obtained a probability to choose a wrong model when drawing model index and parameter values into priors (i.e. a prior error rate) equal to 0.211. As a first inferential clue, one can note that the location of the observed dataset (indicated by a star symbol in Figure 18.2) suggests, albeit without formal quantification, a marked association with the scenario 1a and, to a lesser extent, with the scenario 2a.

Figure 18.3 shows that RFs are able to automatically determine the (most) relevant statistics for model comparison. Interestingly, many of them were not selected by the experts in Verdu et al. (2009), especially some crude estimates of admixture rates based on population triplets (i.e. maximum likelihood coefficient of admixture (AML) statistics; Choisy et al., 2004). A possible explanation is that experts in population genetics are biased towards choosing summary statistics that are informative for parameter estimation under a given model. However, according to our own experience on this issue, the most informative statistics for model choice might not be the same as those that are informative for parameter estimates (see also Robert et al., 2011). Hence, the set of best statistics found with ABC-RF should not be considered as an optimal set for further parameter estimates under a given model with standard ABC techniques (see Section 18.3.2).

The outcome of the first step of the ABC-RF statistical treatment applied to a given target dataset is a classification vote for each model which represents the number of times a model is selected in a forest of n trees. The model with the highest classification vote corresponds to the model best suited to the target dataset among the set of compared models. Note that there are obviously a number of other possible models that might fit the data just as well, if not better than the 'best' model found among the present finite set of models. In our case study, the classification vote estimated for the observed human

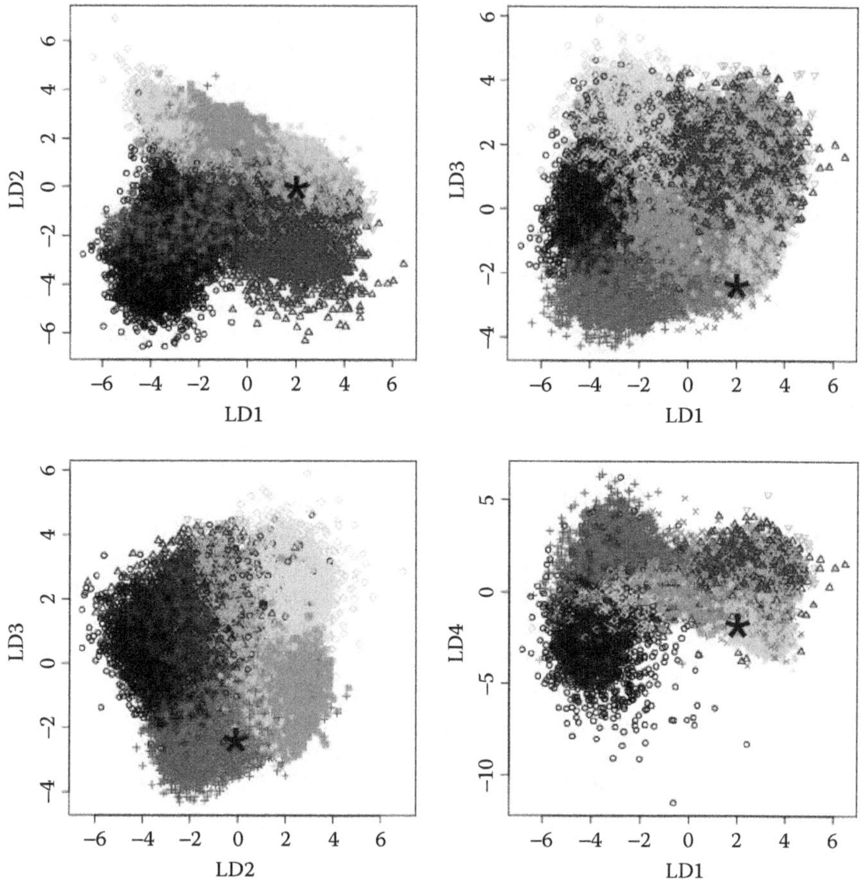

FIGURE 18.2
Projection on the first four linear discriminant analysis axes of the microsatellite population datasets simulated under the eight complex scenarios we compared. Colours correspond to the indices of the scenarios presented in Figure 18.1. Scenario 1a: chocolate, 1b: red, 1c: blue, 1d: black, 2a: pink, 2b: green, 2c: orange, and 2d: gold. The location of the additional (observed) dataset is indicated by a large black star. (Figure in color available by request from AE).

microsatellite dataset was by far the highest for the scenario 1a (i.e. 430 of the $n = 500$ RF-trees selected scenario 1a; see Pudlo et al., 2016 and Breiman, 2001 for a justification of considering a forest of $n = 500$ trees). The evolutionary scenario selected by our RF method fully agrees with the earlier conclusion of Verdu et al. (2009), based on approximations of posterior probabilities with local logistic regression using a substantially higher number of simulated datasets and a different set of summary statistics. As a reminder, scenario 1a corresponds to a common origin of all Western Central African Pygmy

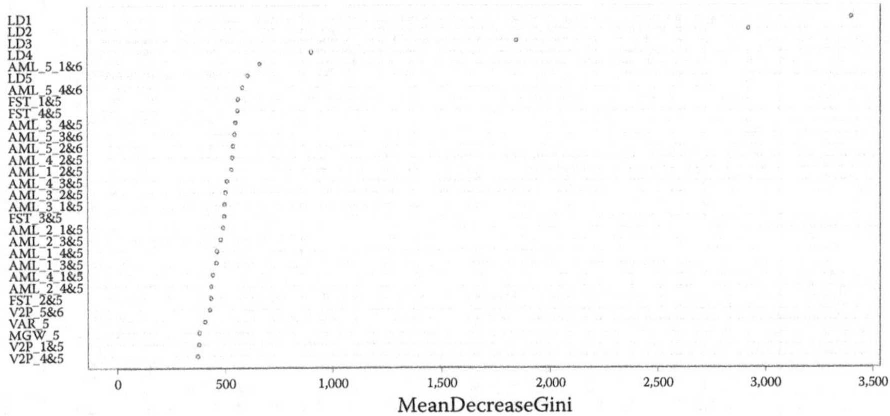

FIGURE 18.3
Contributions of the 30 most informative statistics to the random forests. The contribution of a statistic is evaluated with the mean decrease in node impurity (MeanDecreaseGini) in the trees of the random forest. LD_x = linear discriminant axe x. The meaning of the acronyms of other statistics can be found in the DIYABC 2.1.0 user-manual.

groups considered, with the ancestral Pygmy populations having diverged from the non-Pygmy lineage in a more remote past. Furthermore, scenario 1a encompasses both recent and ancient asymmetrical introgression events from the non-Pygmy gene pool into each Pygmy population considered and a change of population size in the non-Pygmy lineage.

There is no direct connection between the frequencies of the model allocations of the data among the tree classifiers (i.e. the classification vote) and the posterior probabilities of the competing models (see Figure S2 in Pudlo et al., 2016). We therefore carried out the second RF analytical step corresponding to the algorithm 3 in Pudlo et al. (2016) to obtain a reliable estimation of posterior probability of scenario 1a equal to 0.898. This high posterior probability value provides a strong confidence in selecting the scenario 1A as the model best suited to the target dataset among the set of eight complex compared models. For comparison, more standard ABC treatments for model choice (using the 1% closest simulated datasets from a total of 8×10^6 datasets; Estoup et al., 2012) gave a higher, and probably somewhat overoptimistic, posterior probability of scenario 1a equal to 0.948 (a value close to that obtained in Verdu et al., 2009, i.e. prob. = 0.960). The latter statistical treatment required a ca. 100 times longer computational duration than when using the ABC-RF approach.

Table 18.1 shows that increasing the number of simulated datasets in the reference table decreases the mean and standard deviation values of the ABC-RF estimates of both the prior error rate and the posterior probability of the best supported model. The gain of increasing the number of recorded

TABLE 18.1
Estimation Using ABC-RF of Prior Error Rate and Posterior Probability of the Best Supported Model for Different Sizes of the Reference Table and Replicate Analyses

Reference Table Size[a]	**10,000**		**40,000**		**80,000**		**80,000**[c]		**100,000**	
Replicate Analysis	**PER**	**PPbm**[b]	**PER**	**PPbm**[b]	**PER**	**PPbm**[b]	**PER**	**PPbm**[b]	**PER**	**PPbm**[b]
# 1	*0.238*	*0.812*	*0.215*	*0.841*	*0.212*	0.904	0.211	*0.814*	*0.209*	0.882
# 2	0.238	0.859	0.218	0.874	0.210	0.920	*0.214*	0.875	0.209	0.903
# 3	0.231	0.817	0.218	0.866	0.211	0.911	0.212	0.908	*0.208*	0.908
# 4	0.235	0.882	0.216	0.847	0.210	0.870	0.212	0.854	0.209	*0.929*
# 5	0.236	0.816	0.219	0.912	0.212	0.911	0.214	0.844	0.208	0.925
# 6	0.233	0.826	0.218	*0.924*	0.212	0.900	0.212	0.879	0.209	0.895
# 7	*0.230*	*0.940*	0.218	0.877	0.211	*0.875*	*0.209*	0.870	0.209	*0.873*
# 8	0.233	0.831	0.217	0.892	0.211	0.878	0.210	0.885	0.209	0.884
# 9	0.235	0.843	*0.220*	0.844	0.211	0.891	0.213	0.900	0.209	0.908
# 10	0.233	0.934	0.217	0.924	*0.209*	*0.923*	0.211	*0.921*	0.209	0.907
Mean	**0.234**	**0.856**	**0.218**	**0.880**	**0.211**	**0.898**	**0.212**	**0.875**	**0.209**	**0.901**
SD	**0.004**	**0.048**	**0.002**	**0.032**	**0.001**	**0.019**	**0.002**	**0.032**	**0.001**	**0.018**

Notes: We here considered the complex modelling set up including eight evolutionary scenarios (Figure 18.1).

Abbreviations: Prior error rate (PER) and posterior probability of the best supported model (PPbm).

[a] Reference table size in number of records (i.e. datasets compiled under the form of the scenario identity and corresponding summary statistics) simulated from DIYABC.

[b] The same scenario (i.e. scenario 1a) was the best supported for all reference table sizes and for all replicate analyses. Replicate analyses have been processed on the same reference table except for those detailed in the column 80,000.

[c] Which were processed on different reference tables, each one including 80,000 records. Mean and standard deviation values are in bold characters. Min and max values are in italic characters.

simulated datasets to build the trees of the random forest is substantial between 10,000 and 40,000 records and then becomes limited for higher numbers of records, especially between 80,000 and 100,000 records. Although not sizeable, a non-negligible surplus of estimation variance is observed when replicate analyses are processed on different reference tables (see the two columns labelled 80,000 in Table 18.1). This is expected due to the substantial stochastic variation present among reference tables of relatively small size. A practical lesson of the results in Table 18.1 is that processing an ABC-RF analysis using 10,000 records per scenario (here 80,000 records) is a sensible choice, at least in the present case study. Pudlo et al. (2016) reached similar conclusions from a set of other real and controlled datasets analysed using the same ABC-RF methodology (see the section 'Practical recommendations regarding the implementation of the algorithms' in Pudlo et al., 2016).

The verbally stated main evolutionary question (i.e. independent vs. non-independent histories of the four Pygmy groups) can be drawn, at least roughly, from a comparison of only two simplified (and, hence, less realistic) models, named scenario 1 and 2 (see Section 18.2.2). The ABC-RF approach applied on this pair of simplified scenarios provides the highest classification vote for the scenario 1 (i.e. 328 of the $n = 500$ RF-trees selected scenario 1). The posterior probability of this best supported scenario was lower than that obtained when analysing the set of eight complex scenarios: it was equal to 0.733 and 0.761 for reference tables including 20,000 and 50,000 simulated datasets, respectively. Such lower posterior probabilities may reflect, at least partly, some noticeable difficulties of any of the two simplified models to fit with the real history and observed dataset. Nevertheless, in agreement with previous results, the scenario 1 corresponds to a non-independent history of the four Pygmy groups.

18.3.2 Parameter estimation

As it is in the case of model selection, the choice of summary statistics is crucial in parameter estimation by ABC. Standard ABC algorithms might indeed suffer from the curse of dimensionality and correlation among explanatory variables (i.e. multi-co-linearity) during the regression step, and, hence, yield poor results when the number of statistics is large. At least theoretically, the dimensionality issue might be offset by increasing the number of simulations, but the amount of time needed to do so might be unreasonable for most concrete applications. Note that it remains highly challenging to assess in a generic manner to which extent such dimensionality issues may be critical, as it depends on the analysed observed dataset, the summary statistics chosen, and/or the scenario settings. The statistical techniques recently developed to select summary statistics provide noticeable improvements, at least in some cases (reviewed in Blum et al., 2013). We are currently developing a method similar in essence to the ABC random forest algorithms developed for model choice, which would be applicable to the estimation of demographic, historical, and mutation parameters under a given model. We hope that such a method will allow (more) accurate estimation of parameters with a lower computing effort (i.e. using a much smaller reference table than for standard ABC inferences), without the need to choose a more or less arbitrary subset of summary statistics within a large pool of possible ones.

For the sake of simplicity and computation efficiency (i.e. statistical techniques for choosing summary statistics have not been implemented in DIYABC), we used a standard ABC methodological framework (Beaumont et al., 2002; Cornuet et al., 2008) applied to the same set of 'expert-chosen' statistics as in Verdu et al. (2009) to estimate posterior parameter distributions under the most likely complex scenario (i.e. scenario 1a). We considered 48 summary statistics in total: the mean number of alleles per locus and population sample, the mean genetic diversity, the mean allele size variance expressed in base pairs, and all pairwise F_{st}'s, and genetic distances $(\delta\mu)^2$ between population samples (see the DIYABC 2.1.0 user-manual page 16 for details about such statistics). Using parameter values drawn from the prior distributions detailed in Section 18.2.2, we produced a reference table containing one million simulated datasets. Following Beaumont et al. (2002), we then used a local linear regression to estimate the parameter posterior distributions. We took the 10,000 (1%) simulated datasets closest to our observed dataset for the regression, after applying a logit transformation to parameter values. Considering instead the 1‰ closest simulated datasets only slightly changed posterior distribution estimates (results not shown).

We were particularly interested in evaluating the usefulness of adding a scaling population (here, the French population) to obtain more precise inferences specifically regarding original divergence time parameters in scenario 1a: t_p (divergence time among Pygmy populations), and t_{pnp} (among the ancestral Pygmy population and non-Pygmy population). To tackle this question, we produced a second reference table for the scenario 1a containing one million simulated datasets, where the Out of Africa divergence time of the French population was drawn from a uniform distribution bounded between 1 and 5,000 generations instead of being fixed at 2,000 generations as previously (i.e. 50,000 years). Parameter estimation was carried out in the same way as earlier. Accuracies in parameter estimates were further compared for the calibrated and non-calibrated settings by computing the relative median absolute deviation (RMedAD) from a set of 10,000 test datasets (i.e. pseudo-observed datasets for which the true parameter values are known; hereafter, named 'pods'), with parameters drawn from prior distributions. The RMedAD measure corresponds to the 50% quantile (over the 10,000 pods) of the median (over each pod) of the absolute difference between each value of the pod posterior distribution sample (estimated as earlier) and the true value divided by the true value. Lower RMedAD values correspond to more precise estimates.

For most historical and demographic parameters of the complex scenario 1a, the posterior distributions do not differ strongly from the priors, except for the common separation time of the different Pygmy populations (t_p),

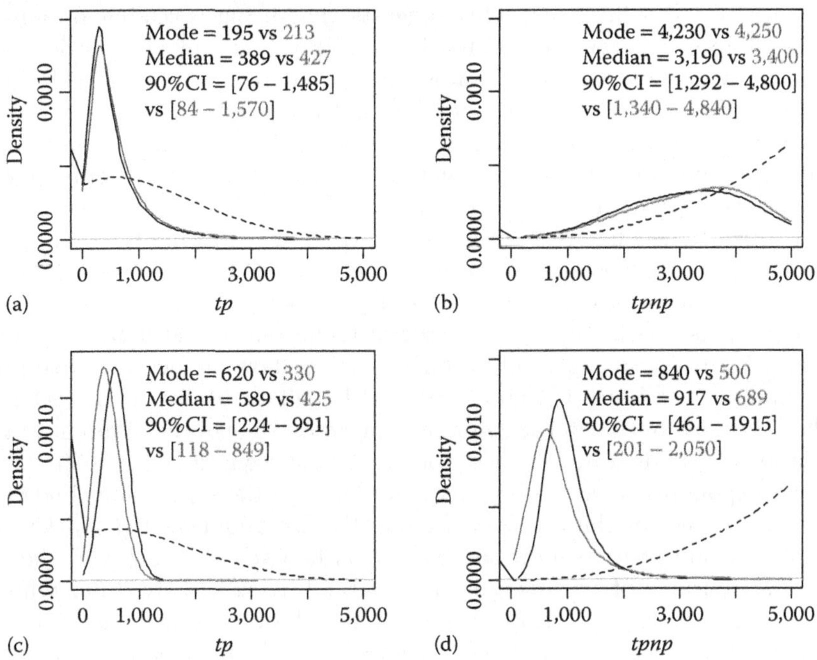

FIGURE 18.4
Posterior distributions of two historical parameters of interest for both a scaled and a non-scaled setting of the complex scenario 1a (panels a and b) and the simple scenario 1 (panels c and d). Plain dark grey curves = posterior distributions for the scaled settings; plain light grey curves = posterior distributions for the non-scaled settings; and dotted black curves = prior distributions. Point estimates and 90% credibility intervals (CI) are indicated for each parameter (in dark grey and light grey for the scaled and non-scaled settings, respectively). t_p = separation time of the different Pygmy populations from a single ancestral Pygmy population. t_{pnp} = separation time of the ancestral Pygmy population from the non-Pygmy population. Prior distributions of tp and tpnp are not flat due to conditions on time events (see Section 2.2 of the main text).

which shows a clear peak corresponding to a mode at 195 generations (i.e. ca. 4,900 years assuming a generation time of 25 years; see Figure 18.4). Moreover, the credibility intervals are large for all parameters. Overall, these results indicate that the genetic data contain relatively little information concerning most parameters in the model, with the notable exception of the divergence time among Pygmy populations, one of the key parameters of interest to anthropologists. Interestingly, we found that the posterior distributions of parameters are similar whether or not we resort to the European scaling population in the inferential process (see Figure 18.4 for an illustration on divergence times of interest). This suggests that using a scaling population does not improve parameter estimation under the complex scenario 4a. In agreement with this, RMedAD values computed from either a scaled or a non-scaled setting of scenario 1a indicate that the original Pygmy population parameters are not or only slightly more accurately estimated when using a scaling population. For all parameters, RMedAD values are only slightly smaller for the scaling setting (i.e. 0.38%–4.26% smaller). Note that we reached similar conclusions when computing RMedAD values drawing the pod parameter values from the posterior distributions estimated for the observed dataset rather than from prior distributions (results not shown).

Unlike the scenario 1a, the posterior distributions of historical and demographic parameters strongly differ from the priors showing nicely shaped peaks in the simplified scenario 1 (Figure 18.4). Although credibility intervals largely overlap, the distribution of the common separation time of the different Pygmy populations (t_p) has a higher mode in the simple scenario 1 (i.e. 620 generations corresponding to ca. 15,500 years) and that of the separation time of the ancestral Pygmy population from the non-Pygmy population (t_{pnp}) shows a clear peak around 840 generations (i.e. ca. 21,000 years). Moreover, the posterior distributions of parameters in scenario 1 more clearly differ whether or not we resort to the European scaling population in the inferential process, indicating a significant effect of considering a scaling population in this case (Figure 18.4). In agreement with this, RMedAD values measured from pods indicate a significant gain in precision overall when considering the simplified scenario 1 (i.e. 11.5%–16.5% smaller RMedAD values for the scaling setting). This suggests that the high level of complexity of scenario 1a is an important factor explaining the poor improvement of parameter estimation observed when using a scaling population. We have further confirmed this result by additional RMedAD analyses processed on different versions of the complex scenario 1a (results not shown).

It is worth stressing that the historical and demographic simplifications assumed in scenario 1 (i.e. absence of introgression events and same effective population size in all Pygmy and non-Pygmy populations) are to a large extent incompatible with previous knowledge regarding Pygmy and non-Pygmy populations and with some of the initial anthropological questioning that motivated our study. For instance, the assumption that all populations, Pygmy and non-Pygmy, had the same stable effective population size prevents us from evaluating possible (and actually likely) differences in the demographic histories experienced separately by each population. This is unfortunate, as it is of major interest to evolutionary biologists to understand whether past demographic changes, such as bottlenecks or expansions, may have significantly shaped the different genetic diversity patterns observed today across populations. Moreover, previous anthropologic, ethnographic, and genetic studies documented the occurrence of genetic introgression among Central African populations (Cavalli-Sforza, 1986; Destro-Bisol et al., 2004; Verdu et al., 2009, 2013; Hewlett et al., 2014), a feature consistent with the finding that the complex scenario 1a (which encompasses admixture events among populations) is much more likely than alternative complex scenarios without admixture events. Therefore, substantial improvement in the accuracy of parameter estimation through the using of a scaling population seems to be achievable, in our case study, only at the cost of unrealistic simplifications of our evolutionary models. Moreover, this casts serious doubts on the relevance in terms of historical interpretation of the outwardly precise estimates of the historical parameter obtained with scenario 1 (see also the following Section 18.3.3: Model-posterior checking).

18.3.3 Model-posterior checking

The ABC methods we considered so far allowed us to select the most likely model (scenario) among a finite set of scenarios and to estimate posterior distributions of parameters under this scenario. However, they do not provide any goodness-of-fit type information. Bayesian model choice can indeed be understood as a Bayesian testing procedure, but with parametric alternatives. Such alternatives do not answer the question whether we missed some important phenomenon in the set of eight complex (or two simple) scenarios we have considered in the present study. The Bayesian model choice procedure returns the best model among the set of compared scenarios, but does not assure that the observed data is typical of datasets produced by the corresponding stochastic model. In other words, how well the inferred scenario-posterior combination matches with the observed dataset remains to be evaluated. An established approach to model criticism in the Bayesian setting involves comparing some function of the observed data to a reference distribution, such as the posterior predictive distribution, the latter corresponding to the distribution of future observations conditional on the observed data (Rubin, 1984; Gelman et al., 2003).

Here, we used the ABC model-posterior checking method implemented in DIYABC (Cornuet et al., 2010; see also Csillèry et al., 2010), which was largely inspired by Gelman et al. (2003) and Cook et al. (2006). The principle is as follows: if a scenario-posterior combination fits the observed data correctly, then data simulated under this scenario with parameters drawn from associated posterior distributions should be close to the observed data. The lack of fit of the model to the data can be measured by determining the frequency at which test quantities measured on the observed dataset are extreme with respect to the distributions of the same test quantities computed from the simulated datasets (i.e. the posterior predictive distributions). In practice, the test quantities are chosen among the large set of ABC summary statistics proposed in the program DIYABC. For each test quantity (q corresponding to the chosen summary statistics), a lack of fit of the observed data with respect to the posterior predictive distribution can be measured by the cumulative distribution function values of each test quantity defined as $\text{Prob}(q_{\text{simulated}} < q_{\text{observed}})$. Tail-area probability can be easily computed for each test quantity as Prob($q_{\text{simulated}} < q_{\text{observed}}$) and 1.0 − Prob ($q_{\text{simulated}} < q_{\text{observed}}$) for Prob ($q_{\text{simulated}} < q_{\text{observed}}$) ≤ 0.5 and >0.5, respectively. Such tail-area probabilities, also named posterior predictive p-values (i.e. *ppp*-values), represent the probabilities that the replicated data (simulated ABC summary statistics) could be more extreme than the observed data (observed ABC summary statistics). *Ppp*-values should be taken with caution (Meng, 1994): they aim at checking whether the model, with posterior distribution adjusted on the data, can really fit the data, and therefore the data are used twice. In practice, *ppp*-values can be interpreted as a kind of guideline for tracking model-posterior misfit: a few *ppp*-values around 0.05 are not necessarily a big deal, whereas the presence of *ppp*-values around 0.001 is pretty suspect. Hence, too many observed summary statistics falling in the tails of distributions, especially if some of those *ppp*-values are pretty low, cast serious doubts on the adequacy of the model-posterior combination to the observed dataset. Finally, because *ppp*-values are computed for a number of (often non-independent) test statistics, a method such as that of Benjamini and Hochberg (1995) can be used to control the false discovery rate (Cornuet et al., 2010). Such multiple-test correction might not be optimal, however, essentially because *ppp*-values are not calibrated due to the data being used twice, even if one will preferentially avoid as test statistics those that have been used to adjust the parameter posterior distributions of the selected scenario (Cornuet et al., 2010, and see the following for an application).

We carried out the ABC model-posterior checking analysis on our observed microsatellite dataset as follows. From the 10^6 datasets simulated under the selected scenario (i.e. scenario 1a), we obtained a posterior sample of 10^4 values from the posterior distributions of parameters through a rejection step based on Euclidian distances and a linear regression post-treatment (Beaumont et al., 2002). We simulated 10^4 datasets with parameter values drawn with replacement from this posterior sample. There is a risk of over-estimating the quality of the fit by using the same statistics twice. This problem, which clearly arises within an ABC framework, is actually a general issue in statistical inference. As underlined in many text books in statistics (e.g. Gelman et al., 2003), it is advised not to perform model checking using information that has already been used for training (i.e. model fitting; see also Cornuet et al., 2010 for illustrations on simulated datasets). Optimally, model-posterior checking should be based on test quantities that do not correspond to the summary statistics which have been used for previous inferential steps. This is naturally possible with DIYABC, as the package proposes a large choice of summary statistics. In practice, one will avoid, as a priority, the statistics that have been used to adjust the parameter posterior distributions of the selected scenario (here, the 48 statistics of Verdu et al., 2009, described in the previous section). Our set of test statistics therefore included the 96 remaining single and pairwise summary statistics available in DIYABC that were not used for previous ABC parameter estimation (see the DIYABC 2.1.0 user-manual page 16). Note that the ABC random forest method we applied for model choice used information from all statistics available in DIYABC, which might be a problem if one also wants to exclude the statistics involved in this initial inferential step too. However, additional treatments on a subset of test quantities discarding the most informative statistics in the ABC random forest did not change our model-posterior checking results (results not shown).

When applying an ABC model checking analysis on the complex scenario 1a, we found that only two of the 96 test quantities had low *ppp*-values (i.e. $ppp = 0.028$ and 0.034), the latter values being not significant when applying the Benjamini and Hochberg 1995s method as an attempt to control the false discovery rate. This result supports that the inferred scenario-posterior combination does not obviously misfit the observed dataset. In agreement with this, the projections of the simulated datasets on the principal component axes from the scenario-posterior combination were relatively well grouped and centred on the target point corresponding to the observed dataset (Figure 18.5). It is worth stressing that, due to the modest size of the dataset (i.e. 26 microsatellite loci), only situations of major inadequacy of the model-posterior combination to the observed dataset are likely to be identified. In agreement with this, only three test quantities had low *ppp*-values (not significant after the Benjamini and Hochberg 1995s correction) when carrying out the same model-posterior checking analysis on the scenario 2a ranked second in model choice, although the latter was found to be far less likely than scenario 1a. By contrast, some major lack of fit was observed when considering the scenario 1, which corresponds to a simplified version of the complex scenarios 1a. We found that in this case, 31 of the 96 test quantities had

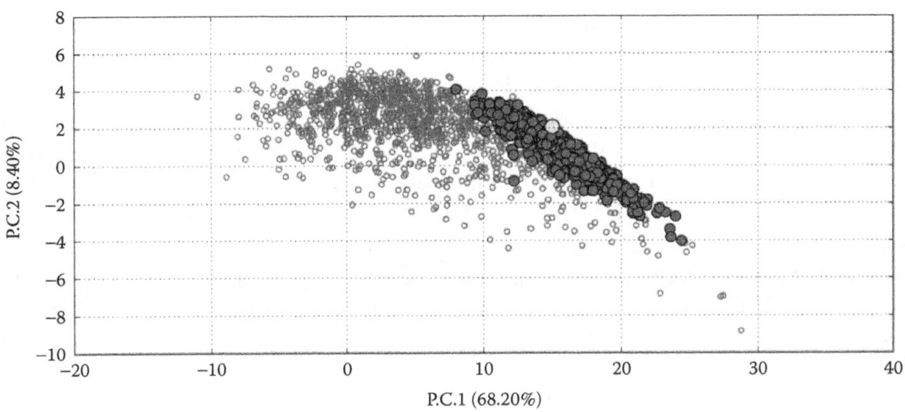

FIGURE 18.5
Principal component analysis of 96 test quantities when processing model checking for the scenario 1a. Only the two first axes are presented. Empty small light grey circles = datasets simulated from priors (subset of 2,000 plots) and plain large dark grey circles = datasets simulated from posteriors (subset of 2,000 plots). Plain large white circle = observed dataset.

low *ppp*-values (six of them being very low, i.e. $ppp < 0.001$), with 28 of them remaining significant after applying the Benjamini and Hochberg 1995s correction. Such a major lack of fit raises serious doubts on the relevance in terms of historical interpretation of the posterior distributions (albeit, the latter showed nicely shaped peaks) obtained for several historical parameters under the simple scenario 1 (Figure 18.4). It is worth noting that recent genetic analyses of human populations relying on substantially larger datasets often point favourably towards even more complex population models than the set of eight complex models considered in the present study (e.g. models including more refined patterns of migration/introgression and population size fluctuations; e.g. Excoffier et al., 2013).

18.4 Conclusions and Perspectives

Our analysis of a modified version of the genetic dataset of Verdu et al. (2009) confirmed the initial historical interpretation of the authors, even though this new dataset included smaller population sample sizes, a smaller number of genetic markers, and the addition of a European scaling population sample. We found a probable common origin of all Western Central African populations categorised as Pygmies by Western explorers, despite the vast cultural, morphological, and genetic diversity observed today among these populations (Hewlett, 2014). Moreover, we found a recent (about 4,900 years before present (YPB)) common origin for the Western Central African Pygmy populations, together with a more ancient (about 80,000 YBP) divergence between the ancestral Pygmy and non-Pygmy populations. This last result is consistent with Verdu et al. (2009) results, as well as with two other following studies that addressed similar questions using ABC approaches applied to autosomal and mitochondrial sequence data (Patin et al., 2009; Batini et al., 2011). We also confirmed recent asymmetrical and heterogeneous genetic introgressions from non-Pygmies into each Pygmy population. Altogether, these results are in agreement with the ethno-historical scenario proposed by Verdu et al. (2009), in which the relatively recent expansion of non-Pygmy agriculturalist populations in Western Central Africa which occurred 2,000–5,000 YBP may have modified the pre-existing social relationships in the ancestral Pygmy population, in turn resulting in its fragmentation into isolated groups. Since then, enhanced genetic drift in isolated populations with small effective sizes, and different levels of genetic introgression from non-Pygmies into each Pygmy population, led to the rapid genetic diversification of the various Western Central African Pygmy populations observed today.

Regarding methodological aspects, our analyses illustrate how the ABC random forest approach can be useful for picking through different highly parameterised models, how ABC can be used to make inferences about parameter values in particular models and how model-posterior checking approaches can be easily applied. Unfortunately, key divergence time parameters of Pygmy populations, and, in particular, the original divergence time of ancestral Pygmy and non-Pygmy populations, were not more precisely estimated than in Verdu et al. (2009), despite the addition of a European sample used as scaling population. We found that, at least in our case study, the limited benefit of using a scaling population was due, to some extent, to the low number of genotyped markers and, to a larger extent, to the complexity of the historical models considered. To our knowledge, this study is the first onc that quantitatively evaluates the actual gain of using a scaling population for parameter estimation.

Finally, we found that many historical and demographic parameters were overall poorly estimated. This result illustrates the general inferential issue that one may have, on the one hand, a good power to discriminate among models and, on the other hand, a poor precision in the estimation of many parameters under the selected model. This result also indicates that it is of perennial importance to consider the anthropological interpretations of our results for what they are: working hypotheses that will imperatively need to be further elucidated in future studies. Using a much larger number of molecular markers may provide new clues to reconstruct the widely unknown peopling history of Central Africa. For instance, Patin et al. (2014) investigated in deep the timing of the admixture onset between Pygmies and non-Pygmies using massive genome-wide datasets comprising more than a million markers genotyped for each individual. Based on such data, they inferred that complex admixture processes which started a few thousand years ago left significant signatures in the genome-wide diversity patterns observed in Central African populations today. This type of study highlights the major potential of increasingly available genome-wide datasets for both model and non-model organisms for future inference on complex population origins and demographic histories. More generally, we believe that ABC methods, especially ABC-RF algorithms, will be of considerable interest for the statistical processing of massive datasets whose availability rapidly increases in various fields of research including, but not limited to, population genetics.

Acknowledgements

We thank Mark Beaumont and one anonymous reviewer for providing helpful and cogent comments on a previous version of the manuscript.

References

Batini, C., Lopes, J., Behar, D.M. et al. 2011. Insights into the demographic history of African Pygmies from complete mitochondrial genomes. *Mol Biol Evol* 28:1099–1110.

Beaumont, M.A. 2010. Approximate Bayesian computation in evolution and ecology. *Annu Rev Ecol Evol Syst* 41:379–406.

Beaumont, M.A., Zhang, W.Y., and D.J. Balding. 2002. Approximate Bayesian computation in population genetics. *Genetics* 162:2025–2035.

Benjamini, Y. and Y. Hochberg. 1995. Controlling the false discovery rate: A practical and powerful approach to multiple testing. *J Roy Stat Soc B* 57:289–300.

Bertorelle, G., Benazzo, A., and S. Mona. 2010. ABC as a flexible framework to estimate demography over space and time: Some cons, many pros. *Mol Ecol* 19:2609–2625.

Blum, M., Nunes, M., Prangle, D., and S. Sisson. 2013. A comparative review of dimension reduction methods in Approximate Bayesian Computation. *Stat Sci* 28:189–208.

Breiman, L. 2001. Random forests. *Machine Learning* 45:5–32.

Cann, H.M., de Toma, C., Cazes, L. et al. 2002. A human genome diversity cell line panel. *Science* 296:261–262.

Cavalli-Sforza, L.L. 1986. *African Pygmies: An Evaluation of the State of Research.* Orlando, FL: Academic Press.

Cavalli-Sforza, L.L. and M.W. Feldman. 2003. The application of molecular genetic approaches to the study of human evolution. *Nat Genet* 33:266–275.

Cavalli-Sforza, L.L., Menozzi, P., and A. Piazza. 1994. *The History and Geography of Human Genes.* Princeton, NJ: Princeton University Press.

Chen, G.K., Marjoram, P., and J.D. Wall. 2009. Fast and flexible simulation of DNA sequence data. *Genome Res* 19:136–142.

Choisy, M., Franck, P., and J.-M. Cornuet. 2004. Estimating admixture proportions with microsatellites: Comparison of methods based on simulated data. *Mol Ecol* 13:955–968.

Cook, S., Gelman, A., and D.B. Rubin. 2006. Validation of software for Bayesian models using posterior quantiles. *J Comput Graph Stat* 15:675–692.

Cornuet, J.-M., Pudlo, P., Veyssier, J. et al. 2014. DIYABC v2.0: A software to make Approximate Bayesian Computation inferences about population history using single nucleotide polymorphism, DNA sequence and microsatellite data. doi:10.1093/bioinformatics/btt1763.

Cornuet, J.-M., Ravigne, V., and A. Estoup. 2010. Inference on population history and model checking using DNA sequence and microsatellite data with the software DIYABC (v1.0). *BMC Bioinformatics* 11:401.

Cornuet, J.-M., Santos, F., Beaumont, M.A. et al. 2008. Inferring population history with DIY ABC: A user-friendly approach to Approximate Bayesian Computation. *Bioinformatics* 24:2713–2719.

Csilléry, K., Blum, M.G.B., Gaggiotti, O., and O. François. 2010. Approximate Bayesian Computation (ABC) in practice. *TREE* 25:410–418.

Csilléry, K., François, O., and M.G.B. Blum. 2012. ABC: An R package for Approximate Bayesian Computation (ABC). *Methods Ecol Evol* 3:475–479.

Destro-Bisol, G., Donati, F., Coia, V. et al. 2004. Variation of female and male lineages in sub-Saharan populations: the importance of sociocultural factors. *Mol Biol Evol* 21:1673–1682.

Dib, C., Faure, S., Fizames, C. et al. 1996. A comprehensive genetic map of the human genome based on 5,264 microsatellites. *Nature* 380:152–154.

Estoup, A., Jarne, P., and J.-M. Cornuet. 2002. Homoplasy and mutation model at microsatellite loci and their consequences for population genetics analysis. *Mol Ecol* 11:1591–1604.

Estoup, A., Lombaert, E., Marin, J.-M. et al. 2012 Estimation of demo-genetic model probabilities with Approximate Bayesian Computation using linear discriminant analysis on summary statistics. *Mol Ecol Res:* 12:846–855.

Excoffier, L., Dupanloup, I., Huerta-Sanchez, E., Sousa, V., and M. Foll. 2013. Robust demographic inference from genomic and SNP data. *PLoS Genetics* 9:e1003905.

Excoffier, L., Estoup, A., and J.-M. Cornuet. 2005. Bayesian analysis of an admixture model with mutations and arbitrarily linked markers. *Genetics* 169:1727–1738.

Fagundes, N.J., Ray, N., Beaumont, M. et al. 2007. Statistical evaluation of alternative models of human evolution. *Proc Natl Acad Sci USA* 104:17614–17619.

Gelman, A., Carlin, J.B., Stern, H.S., and D.B. Rubin. 2003. *Bayesian Data Analysis.* London, UK: Chapman & Hall/CRC.

Gravel, S., Henn, B.M., Gutenkunst, R.N. et al. 2011. Demographic history and rare allele sharing among human populations. *Proc Natl Acad Sci USA* 108:11983–11988.

Hewlett, B. 1996. Cultural diversity among African Pygmies. In *Cultural Diversity among Twentieth-Century Foragers: An African Perspective*, S. Kent (Ed.), pp. 215–244. Cambridge, UK: Cambridge University Press.

Hewlett, B.S. 2014. *Hunter-Gatherers of the Congo Basin: Cultures, Histories and Biology of African Pygmies.* New Brunswick, NJ: Transactions Publishers, 353 p.

Hudson, R.R. 2002. Generating samples under a Wright-Fisher neutral model. *Bioinformatics* 18:337–338.

Laval, G., and L. Excoffier. 2004. SIMCOAL 2.0, a program to simulate genomic diversity over large recombining regions in a subdivided population with a complex history. *Bioinformatics* 20:2485–2487.

Liaw, A., and M. Wiener. 2002. Classification and regression by randomforest. *R News* 2:18–22.

Lombaert, E., Guillemaud, T., Cornuet, J.-M., Malausa, T., Facon, B., and A. Estoup. 2010. Bridgehead effect in the worldwide invasion of the biocontrol harlequin ladybird. *PLoS One* 5:e9743.

Marin, J.M., Pudlo, P., Robert, C.P., and R.J. Ryder. 2012. Approximate Bayesian Computational methods. *Stat Comput* 22:1167–1180.

Meng, X.L. 1994. Posterior predictive p-values. *Annals Stat* 22:1142–1160.

Nordborg, M. 2001. Coalescent theory. In: *Handbook of Statistical Genetics*, D.J. Balding, M. Bishop, and C. Cannings (Eds.), pp. 179–212. Chichester, UK: John Wiley & Sons.

Patin, E., Laval, G., Barreiro, L.B. et al. 2009. Inferring the demographic history of African farmers and Pygmy hunter-gatherers using a multilocus resequencing dataset. *PLoS Genet* 5:e1000448.

Patin, E., Siddle, K.J., Laval, G. et al. 2014. The impact of agricultural emergence on the genetic history of African rainforest hunter-gatherers and agriculturalists. *Nat Com* 5:3163.

Pudlo, P., Marin, J.M., Estoup, A., Cornuet, J.M., Gautier, M., and C.P. Robert. 2016. Reliable ABC model choice via random forests. *Bioinformatics* 32:859–866.

Robert, C.P., Cornuet, J.-M., Marin, J.-M., and N.S. Pillai. 2011. Lack of confidence in Approximate Bayesian Computation model choice. *Proc Natl Acad Sci USA* 108:15112–15117.

Rosenberg, N.A., Pritchard, J.K., Weber, J.L. et al. 2002. Genetic structure of human populations. *Science* 298, 2381–2385.

Rubin, D.B. 1984. Bayesianly justifable and relevant frequency calculations for the applied statistician. *Ann Stat* 12:1151–1172.

Sunnaker, M., Busetto, A.G., Numminen, E., Corander, J., Foll, M., and C. Dessimoz. 2013. Approximate Bayesian Computation. *PLoS Comput Biol* 9:e1002803. doi:10.1371/journal.pcbi.1002803.

Tavaré, S., Balding, D.J., Griffiths, R.C., and P. Donnelly. 1997. Inferring coalescence times from DNA sequence data. *Genetics* 145:505–518.

Verdu, P., Austerlitz, F., Estoup, A. et al. 2009. Origins and genetic diversity of Pygmy hunter-gatherers from Western Central Africa. *Curr Biol* 19:312–318.

Verdu, P., Becker, N.S., Froment, A. et al. 2013. Sociocultural behavior, sex-biased admixture, and effective population sizes in Central African Pygmies and non-Pygmies. *Mol Biol Evol* 30:918–937.

Wegmann, D., Leuenberger, C., Neuenschwander, S., and L. Excoffier. 2010. ABCtoolbox: A versatile toolkit for approximate Bayesian computations. *BMC Bioinformatics* 11:116.

19

ABC for Climate: Dealing with Expensive Simulators

Philip B. Holden, Neil R. Edwards, James Hensman, and Richard D. Wilkinson

CONTENTS

19.1	Introduction		570
19.2	History Matching and ABC		573
19.3	Emulation		576
	19.3.1	Sequential history matching	576
	19.3.2	A simple climate example	579
19.4	Climate Model Case Study		581
	19.4.1	The global carbon cycle	581
	19.4.2	Emulator-informed ABC design	582
	19.4.3	Applications	586
		19.4.3.1 Probabilistic simulation outputs: The uncertain response of the carbon cycle to anthropogenic CO_2 emissions	586
		19.4.3.2 Calibrating model parameters: The strength of the terrestrial carbon sink	587
		19.4.3.3 Model understanding: What determines the spatial distribution of dissolved carbon in the ocean?	587
		19.4.3.4 Coupling applications: Coupling climate models and climate change impact models	588
19.5	Future Applications		589
References			590

19.1 Introduction

One of the primary challenges faced when calibrating a simulator using approximate Bayesian computation (ABC) is overcoming the computational constraints posed by working with limited resource. The requirement to repeatedly simulate from a model can make inference extremely computationally expensive. Consequently, much of the methodological development in ABC has focused on improving computational efficiency, either through the use of more efficient Monte Carlo algorithms or through the use of statistical methods to ameliorate the effect of using a large tolerance.

The difficulty of dealing with limited computer power is felt more keenly in climate science than in most other disciplines. A major focus of climate research concerns the construction of ever more accurate and comprehensive simulators of the climate system. Since the 1970s, global climate models have evolved from representing only the large-scale circulation of the global atmosphere (Holloway Jr and Manabe, 1971) to models that incorporate complex dynamic representations of land surface, ocean, sea ice, atmospheric aerosols, ocean biogeochemistry, vegetation, soils, and atmospheric chemistry (Flato et al., 2013). Separate Earth system components are coupled through the exchange of fluxes, which describe the flow of some quantity between them (e.g. energy, moisture, CO_2) and bypassing any state variables that are needed to define boundary conditions (e.g. land surface albedo, sea surface temperature). 'Intermediate complexity' models (which use simplified model components and lower resolution in return for a more complete description of the Earth system and higher computational efficiency) may also include dynamic representations of other important elements, such as ice sheets, permafrost, ocean sediments, and weathering (Flato et al., 2013), but these additional, long-timescale components require orders of magnitude longer simulations to reach equilibrium. Modern climate models are generally, and more accurately, described as 'Earth system models' or ESMs.

This evolution in complexity has been accompanied by a 5-fold increase in spatial resolution, allowing the resolution of important finer scale processes. This increased resolution (combined with shorter time-steps that are required for numerical stability at higher spatial resolution) has *alone* led to an $O(1{,}000)$-fold increase in computational demands since the 1970s. In general, higher resolution allows more direct and more realistic representation of smaller-scale processes, although this does not guarantee better projections, in part because more complex models are more challenging to calibrate. A feature of climate modelling is that multi-decadal climate projections must be used before data are available to validate them, while past data give only approximate clues to the expected behaviour of model discrepancy because expected changes greatly exceed the range of variability in the instrumental period.

It is perhaps inevitable, given the continual striving for more complex models and the highest possible resolution, that state-of-the-art ESMs will always be at the limits of what is practicable with available computing power. The UK Met Office Hadley Centre's computer comprises eight 'supernodes' of IBM Power775 supercomputer servers, which were installed in 2012 at a cost of more than £11 million. The ESMs run at the Hadley Center and at equivalent climate modelling institutions in other countries are extremely computationally expensive, requiring months of such supercomputing to perform a single simulation of order 100 years. Even the intermediate complexity model GENIE-1 (Holden et al., 2013b) used in our case study (Section 19.4) requires several days (on a single central processing unit (CPU) node) to perform each O(10 kyear) 'spin-up' simulation to reach equilibrium, so that simulation ensembles require implementation on multi-node computing clusters. The simulated climates are large complex datasets which comprise temporally resolved three-dimensional spatial arrays of up to $\sim$100 state variables. These outputs, in particular the outputs of carefully designed model inter-comparison projects, are often analysed in great detail, in a comparable way to how scientists in other fields analyse the outputs from empirical studies; model projections are the best and only predictions we have of future climate.

An ESM configuration is determined by the settings of many 100s of model parameters. These include switches (which determine the precise numerical schemes applied), physical constants that are approximately known, but vary spatially in the real world (such as the reflectivity of ice), and parameterisations of 'sub-gridscale' processes such as cloud formations, which have 'tuned' values that are known to result in reasonable model behaviour. This complexity (many weakly constrained inputs, high dimensional outputs, and expensive simulators) has meant that careful statistical calibration (either with Bayesian or frequentist approaches) does not have a long history in climate science. Often different modules of an Earth system simulator are separately 'tuned' before being bolted together. For example, the atmospheric component can be tuned independently of the ocean component by prescribing sea-surface temperatures with observational values. The components may all be tuned independently before being coupled, with no guarantee that what was a good tuning in an isolated module will work well in the coupled model. After coupling, a small subset of model parameters are adjusted so that the coupled model is consistent with large-scale observational constraints. It has been shown, perhaps unsurprisingly, that such a tuning process does not produce a unique solution, so that different combinations of parameters can lead to equally plausible model realisations (Mauritsen et al., 2012).

As statistical methodology develops, scientists are beginning to perform more careful parameter estimation in their models. More rigorous parameter estimation methods are often developed with (relatively fast) intermediate complexity models, for example, Annan et al. (2005), thereby informing application to higher-complexity models, for example, Marquis et al. (2014).

We can view climate simulators as black boxes which map from parameter values $\theta \in \Theta$, to climate states $f(\theta) = \mathcal{C}_{\text{sim}}$. The aim of a Bayesian calibration, is to find the posterior distribution:

$$\pi(\theta|\mathcal{C}_{\text{obs}}) \propto \int \pi(\mathcal{C}_{\text{obs}}|\mathcal{C}_{\text{sim}})\pi(\mathcal{C}_{\text{sim}}|\theta)\mathrm{d}\mathcal{C}_{\text{sim}}\pi(\theta), \tag{19.1}$$

where $\mathcal{C}_{\text{obs}}$ is a set of observations of the climate system (Kennedy and O'Hagan, 2001; Rougier, 2007). Here, $\pi(\theta)$ is the prior distribution for θ, $\pi(\mathcal{C}_{\text{sim}}|\theta)$ is the simulator likelihood function (which is typically unknown), and $\pi(\mathcal{C}_{\text{obs}}|\mathcal{C}_{\text{sim}})$ is the statistical model relating the simulator to physical climate. This calculation, however, is typically far too ambitious to perform in practice. Computational restrictions generally limit us to an ensemble of N simulator runs $\{\theta^{(i)}, \mathcal{C}_{\text{sim}}^{(i)}\}_{i=1}^{N}$. Typically, N is small, ruling out most Monte Carlo-based calibration approaches. We are left needing to estimate $\pi(\theta|\mathcal{C}_{\text{obs}})$ as best we can, often by adding further approximation.

A further problem faced by climate scientists is that simulator discrepancy (often called 'model error') can be considerable (Murphy et al., 2004). And whilst the physical models of climate, $\pi(\mathcal{C}_{\text{sim}}|\theta)$, are well developed, statistical models of the simulator discrepancy relating simulated to observed climate, $\pi(\mathcal{C}_{\text{obs}}|\mathcal{C}_{\text{sim}})$, have only begun to be developed relatively recently (Rougier and Goldstein, 2014). The large simulator discrepancy makes most simulators incapable of reproducing all aspects of the climate record simultaneously and can mean that the simulator parameters are no longer directly comparable to their physical namesakes, making prior specification challenging.

So what is possible? We know that ABC, given infinite computational resources and a perfect simulator, can in theory produce arbitrarily accurate posteriors (i.e. the ABC posterior can be made arbitrarily close to the true posterior). But for many problems, computational resources are often severely constrained and simulator discrepancy can be significant and largely unmodelled. Climate science is interesting for the statistician, as it presents extreme cases of both these issues.

A key idea allowing calibration in many of these expensive simulators is the idea of replacing the simulator with an emulator (or meta-model), which is a cheap statistical surrogate used in place of the simulator (Sacks et al., 1989; Santner et al., 2003; O'Hagan, 2006). Emulation techniques are attracting considerable interest in the climate community. They are used, for instance, to approximate probabilistic model outputs (Sansó et al., 2008; Rougier et al., 2009; Harris et al., 2013), for parameter estimation (Olson et al., 2012; Sham Bhat et al., 2012), to facilitate model understanding (Lee et al., 2012; Holden et al., 2015), and to provide numerically efficient model surrogates for coupling applications (Castruccio et al., 2014; Holden et al., 2014; Oyebamiji et al., 2015). The application we will describe here is in the ABC design of 'plausible' simulation ensembles (Holden et al., 2010; Edwards et al., 2011), using emulation in order to overcome the prohibitive limitations imposed by simulator cost.

19.2 History Matching and ABC

Climate science presents the double whammy of computationally expensive simulators, and simulator discrepancy that is too large to ignore, but which is not well understood or modelled. Both of these issues make a careful Bayesian calibration (as described by Equation 19.1) difficult. What can be achieved? Our aim is to compare observations of Earth's climate $\mathcal{C}_{\text{obs}}$, with simulator predictions $\mathcal{C}_{\text{sim}} = f(\theta)$, in order to learn about the parameter θ. ABC is an approach for obtaining a probabilistic calibration and seeks to match simulator output to observations, approximating the distribution:

$$\pi_{ABC}(\theta|\mathcal{C}_{\text{obs}}) \propto \int \mathbb{I}(\rho(\mathcal{C}_{\text{obs}}, \mathcal{C}_{\text{sim}}) \leq \epsilon)\pi(\mathcal{C}_{\text{sim}}|\theta)\text{d}\mathcal{C}_{\text{sim}}\pi(\theta). \tag{19.2}$$

The acceptance kernel $\mathbb{I}(\rho(\mathcal{C}_{\text{sim}}, \mathcal{C}_{\text{obs}}) \leq \epsilon)$ implicitly implies a uniform distribution for the simulator discrepancy (Wilkinson, 2013), but this is usually viewed as a pragmatic compromise, rather than a modelling decision.

An alternative to a probabilistic calibration is to do a history match (Williamson et al., 2013), which has been used in studies involving complex computer models, such as oil reservoir modelling (Craig et al., 1997), cosmology (Vernon et al., 2010), epidemiology (Andrianakis et al., 2015), and climate science (Edwards et al., 2011). History matching, like calibration, seeks to identify regions of the input space that give acceptable matches between simulator output, $\mathcal{C}_{\text{sim}}$, and observed data, $\mathcal{C}_{\text{obs}}$. But instead of finding a probability distribution over Θ, we instead seek merely to rule out implausible regions of input space, for example, those θ that the simulator suggests could not have lead to $\mathcal{C}_{\text{obs}}$, even after having accounted for the simulator discrepancy. Often large parts of the input space give simulated climates that are very different from the observed data, and which can, hence, be ruled to be physically implausible and removed from further consideration.

We define $\mathcal{P}_{\mathcal{C}}$ to be a set of plausible climate states that represent an acceptable match between simulation and observation. We define $\mathcal{P}_\theta$ to be the subset of the parameter space that leads to plausible simulated climates, for example:

$$\mathcal{P}_\theta = \{\theta \in \Theta : f(\theta) \in \mathcal{P}_{\mathcal{C}}\}.$$

Often, the vast majority of the input space gives rise to unacceptable matches to the observed data (sometimes $\mathcal{P}_\theta = \varnothing$), and it is these regions that we are trying to rule out as implausible. For example, for an ESM, we might define $\mathcal{P}_{\mathcal{C}}$ to be any simulated climate that has global surface air temperature within 2°C of the observed value, the maximum value of Atlantic meridional overturning circulation, a measure of the large-scale circulation of the ocean, within 5 Sv (1 Sv $= 10^6 m^3 s^{-1}$) of observations, and the global mass of vegetation to be within 200 giga-tonnes carbon of observations, though clearly the choice of appropriate metrics and acceptance ranges is highly simulator-dependent. $\mathcal{P}_\theta$ is then the set of model parameters that would generate plausible climates for the ESM in question.

Note the similarity to ABC here. If the prior distribution for θ is uniform on Θ, for example, $\pi(\theta) \propto \mathbb{I}_{\theta \in \Theta}$, if $f(\theta)$ is deterministic (as is often, at least approximately, the case in climate science), and if we use $\mathbb{I}_{f(\theta) \in \mathcal{P}_\mathcal{C}}$ as the ABC acceptance kernel, then:

$$\pi_{ABC}(\theta|\mathcal{C}) \propto \begin{cases} 1 \text{ if } \theta \in \mathcal{P}_\theta \\ 0 \text{ otherwise.} \end{cases}$$

If we interpret a posterior probability of zero, as the statement that θ is implausible, then history matching and ABC are thus the same. Note also the direct relationship between the discrepancy considerations built into $\mathcal{P}_\mathcal{C}$, and the way ABC performs 'Monte Carlo' exact inference for the model that has a discrepancy defined by the acceptance kernel (Wilkinson, 2013).

History matching and ABC have in common that they do not use a detailed model of the discrepancy, but instead characterise it using simple criteria. A philosophical difference between the two approaches perhaps lies in the degree of thought given to the plausible set $\mathcal{P}_\mathcal{C}$. In history matching, the plausibility criteria are often based on measurement error variances and the expected magnitude of the simulator discrepancy (Vernon et al., 2010). Consequently, $\mathcal{P}_\theta$ consists of those parameter values θ that have not yet been ruled out as implausible by our knowledge of the simulator and its discrepancy, and the observed data and measurement error. The result is usually not interpreted probabilistically, but only as values that we can not yet rule to be implausible given our current state of knowledge. In contrast, in ABC, the choice of metric ρ and tolerance ϵ are usually based pragmatically on the characteristics of the algorithm, rather than on physical aspects of the underlying problem. Often, ϵ is chosen to generate a specified number of acceptances. For example, if the computational budget allows for 10^8 simulator runs, and we want 10^4 accepted values in order to approximate the posterior, we set ϵ to the value that leads to 0.01% of simulations being accepted (i.e. Biau et al., 2015, interpret ABC as a nearest neighbour algorithm).

A further difference lies in the choice of information to include in $\mathcal{P}_\mathcal{C}$ (i.e. what summary statistics to use). Climate simulators provide a large variety of outputs, and some of these are better able to reproduce observed climate than others. For example, temperatures are generally better reproduced than precipitation, consequently, it is more common to calibrate to the former than the latter. In contrast, ABC has its roots in genetics, where perhaps the simulator output is less varied and, consequently, more focus is given to the automatic selection of summary statistics, often chosen on the basis of what is most informative for θ (Blum et al., 2013). This approach is unlikely to be suitable in climate science. Some outputs for which the simulator discrepancy is particularly large (precipitation say) may well be very 'informative' about θ if we do not allow for discrepancy, but this would only misguide and may lead us to incorrectly rule out large swathes of parameter space as implausible. Variables which are not well simulated are often included in ESMs, either because they improve the overall simulation through the representation of important feedbacks, or because they are considered important outputs in their own right in spite of higher discrepancy associated with the outputs. Whether a weak calibration constraint on these outputs is appropriate will depend on the details of the discrepancy. Where a known missing process gives a significant contribution to regional error for instance, such as large precipitation errors in monsoon regions as a result of unresolved topographic variation, using a too precise calibration constraint (equivalently too small a model discrepancy) would distort the rest of the solution.

A key question for any simulator is whether given a set of plausibility conditions, the simulator is capable of producing any plausible simulated climates. That is, is $\mathcal{P}_\theta$ empty? If $\mathcal{P}_\theta$ is empty, it is an indication that we understand less than we thought about the simulator and system. Either there is an error in our implementation of the simulator, or we have under-estimated the magnitude of the simulator discrepancy or measurement error. The fact that the result of a history match can be to find there are no plausible parameter values should not be seen as a negative aspect of the approach, as it forces us to confront the cold reality that something is missing from our understanding of the system. In contrast, likelihood based techniques such as Markov chain Monte Carlo (and pragmatic ABC applications, where ϵ is chosen to guarantee a particular acceptance rate), result in a posterior distribution $\pi(\theta|\mathcal{C}_{\text{obs}})$ regardless of how close the simulated climates are to real climate. It is thus sensible when using these techniques to carefully check that the calibrated simulator does indeed produce acceptable fits. While it can often be useful to find the distribution $\pi(\theta|\mathcal{C}_{\text{obs}})$ (or an approximation to it) regardless of the simulator quality, note that if discrepancy is ignored, $\pi(\theta|\mathcal{C}_{\text{sim}})$ can often be more constrained, or equivalently $|\mathcal{P}_\theta|$ smaller, than is justified (Brynjarsdóttir and O'Hagan, 2014).

Note that even if a probabilistic calibration is required, a history match can be performed first in order to rule out regions of space which are clearly implausible. This can dramatically reduce the area needed to be explored during the more challenging probabilistic calibration. If using a stochastic simulator, for which θ may never be completely ruled as implausible (as $\pi(\theta|\mathcal{C}_{\text{sim}}) > 0$ for all θ say), this can still be advantageous. We can rule out parameter regions for which the likelihood is considerably smaller than at the maximum likelihood estimator (MLE), with only a small increase in the approximation error (Wilkinson, 2014), again making a subsequent probabilistic calibration easier.

Assuming that $\mathcal{P}_\theta$ is not empty, the question then becomes, can we find elements of $\mathcal{P}_\theta$ and, better still, can we characterise all of $\mathcal{P}_\theta$? The complexity of climate science is such that even incomplete specifications of $\mathcal{P}_\theta$ are useful, as discussed in Section 19.4. This is because interest lies not in $\mathcal{P}_\theta$, but in what it implies about future climate, for example, in the implied calibrated distribution for other aspects of the climate system:

$$\pi(\mathcal{C}_{\text{future}}|\mathcal{C}_{\text{obs}}) = \int \pi(\mathcal{C}_{\text{future}}|\theta)\pi(\theta|\mathcal{C}_{\text{obs}})d\theta,$$

and so even partial descriptions of $\mathcal{P}_\theta$ are useful in constraining our beliefs about future climate behaviour. Our aim is thus, given a limited computational budget of N simulator evaluations, can we find $\mathcal{P}_\theta$ and the corresponding set of plausible future climates? ABC applications usually use millions of simulator evaluations. What can we do if instead we can only afford 100 or 1,000 simulator evaluations? The answer is going to be even more approximate than in ABC, and furthermore, we will necessarily have to make some modelling assumptions if we wish to make progress. The key tool that has arisen for doing this is the emulator, or meta-model.

19.3 Emulation

If the simulator, $f(\theta)$, is expensive to evaluate, we can instead try to find an approximation, $\tilde{f}(\theta)$, called an emulator or meta-model, which provides a good approximation to $f(\theta)$, but which is computationally cheap (Sacks et al., 1989; O'Hagan, 2006). We can then either use $\tilde{f}$ to answer the question of interest (e.g. calibrating the simulator) or use it to guide the choice of the next parameter value at which, to evaluate f.

We start by generating an ensemble of simulator evaluations $\mathcal{D} = \{\theta_i, f(\theta_i)\}_{i=1}^N$, which we then use to build $\tilde{f}$. Building an emulator is a regression problem and, consequently, a myriad of different techniques have been used, including linear regression and its variants, neural networks, and Gaussian processes (GPs, also known as 'Kriging'), with GPs proving the most popular class of model thus far. The functional form of the simulator is not known a priori, and so neither is the best regression model, but a reasonable approximation can usually be found using GPs, as long as the response is a smooth continuous function of θ. For the purposes of calibration, the key properties of any emulator are predictive accuracy, quantification of uncertainty in the predictions, and speed of prediction. In climate science, where the output fields being modelled are often spatio-temporal fields, the regression model is usually combined with a dimension reduction technique, to project the output onto a lower dimensional manifold (Higdon et al., 2008; Holden and Edwards, 2010; Wilkinson et al., 2010).

19.3.1 Sequential history matching

For many problems, the plausible set $\mathcal{P}_\theta$ may constitute only a small fraction of the prior space Θ. Furthermore, $\mathcal{P}_\theta$ may consist of multiple disconnected regions. For Monte Carlo methods, this can make designing an effective sampler difficult, as MCMC chains (or particles) can fail to explore all plausible regions. For emulator methods, the difficulty lies in building a model that can approximate the simulator in all regions of space. For example, stationary covariance functions that assume a constant length-scale throughout space are commonly used in GPs and may be inappropriate. Other problems arise if $f(\theta)$ varies over too wide a range, which is common if $f(\theta)$ is a likelihood function (Wilkinson, 2014). If we need an emulator of $f(\theta)$ that is valid in all of Θ, then we can look to use a non-stationary covariance function or a more flexible model, such as a treed-GP (Gramacy and Lee, 2008). However, for calibration, we only need to approximate the simulator when $f(\theta)$ is close to being plausible. In other parts of parameter space, it is only necessary to say θ is implausible with a high degree of confidence. It does not matter if an estimate of $f(\theta)$ is poor, as long as we are correct in saying $f(\theta) \notin \mathcal{P}_C$.

GP predictions are more accurate in regions rich in data. Thus, the key issue when building a GP emulator is the choice of the design points, $\mathcal{D}_\theta = \{\theta_i\}_{i=1}^n$, at which we evaluate the simulator. Space filling designs, such as maximin Latin hypercubes (McKay et al., 2000) or low discrepancy sequences (such as Sobol sequences, Morokoff and Caflisch, 1994) are the default choice of design and usually lead to reasonable global approximations. But they are less well suited to calibration problems, in which we usually want to focus on just a small region of parameter space.

Instead of a space filling design, we can seek to build the design sequentially: given the current design, we build an emulator that describes our current knowledge of $f(\theta)$. We then use the emulator to decide where next to run the simulator and so on. The basic idea is as follows:

1. Start with an *a priori* plausible set $\mathcal{P}_\theta^{(0)} = \Theta$.
2. Choose design $\mathcal{D}_\theta^{(1)} = \{\theta_i \in \Theta : i = 1, \ldots, n_1\}$, and run the simulator to get ensemble $\mathcal{D}^{(1)} = \{(\theta_i, C_i = f(\theta_i) : \theta_i \in \mathcal{D}_\theta^{(1)}\}$.
3. Build emulator $\tilde{f}_{(1)}$ and use it to predict the plausible set $\tilde{\mathcal{P}}_\theta^{(1)}$.
4. Choose new design points $\mathcal{D}_\theta^{(2)}$, and run the simulator to get $\mathcal{D}^{(2)}$.
5. Build emulator $\tilde{f}^{(2)}$ and use it to predict the plausible set $\tilde{\mathcal{P}}_\theta^{(2)}$.
6. Etc.

The details of each step vary in each problem. The plausibility criteria are usually defined so that they become more stringent at each iteration. The first plausibility condition $\mathcal{P}_C^{(1)}$ may be relatively weak, with $\mathcal{P}_C^{(1)}, \mathcal{P}_C^{(2)}, \ldots, \mathcal{P}_C^{(W)}$ slowly approaching the final desired criterion $\mathcal{P}_C^{(W)}$. If the difference between $\mathcal{P}_\theta^{(i)}$ and $\mathcal{P}_\theta^{(i+1)}$ is too large, we may find the emulator accuracy is insufficient, causing us to incorrectly rule-out some regions of space (type-I errors).

The plausibility criteria can be relaxed by changing the number of measurements we need to match and the closeness of the required match. Note the superficial similarity to sequential Monte Carlo (SMC)-ABC approaches, in that the approximation is iteratively improved as we learn.

The emulator used at each stage may be based upon all the previous simulator runs, adding new data points in important regions (see the following for details), or it can be built from scratch. For example, in Vernon et al. (2010), they build an emulator, $\tilde{f}^{(i)}$, to predict $f(\theta)$ for $\theta \in \mathcal{P}_\theta^{(i-1)}$, the estimated plausible region from the previous iteration. The emulator is not required to predict for $\theta \notin \mathcal{P}_\theta^{(i-1)}$. The benefit of this is that the simulator response is likely to be less variable within $\mathcal{P}_\theta^{(i-1)}$ than in Θ, making it easier to model. The disadvantage is that if some regions are incorrectly ruled to be implausible in iteration $i-1$, this mistake can never be rectified.

The most important algorithmic decision is the choice of design, $\mathcal{D}_\theta^{(i)}$, at each iteration, for example, given an emulator, how should we choose locations θ at which to run the simulator? If we only wish the emulator to predict well in $\mathcal{P}_\theta^{(i-1)}$, then we only need a design in $\mathcal{P}_\theta^{(i-1)}$. Vernon et al. (2010) take the approach of seeking to use a space filling design on $\mathcal{P}_\theta^{(i-1)}$, such as a Latin hypercube. To do this, they create a large design on Θ and then reject any point not predicted to lie in $\mathcal{P}_\theta^{(i-1)}$ by $\tilde{f}^{(i)}$, which is also the approach we describe in Section 19.4. If we instead seek a global emulator valid for all $\theta \in \Theta$, but which is accurate in the important regions, then it can be beneficial to add simulator runs to the design one at a time. The critical regions are those where the emulator is most uncertain about whether $\theta \in \mathcal{P}_\theta$. This is typically either in regions for which we have no data or near the edge of the plausible region, where we are unsure if $\theta \in \mathcal{P}_\theta$ or not given the accuracy of the emulator. If we use a GP emulator, then our prediction of $f(\theta)$ is Gaussian:

$$f(\theta) \sim N(\mu^{(i)}(\theta), \Sigma^{(i)}(\theta)),$$

where $\mu^{(i)}$ and $\Sigma^{(i)}$ are the mean and covariance function of $\tilde{f}^{(i)}$. This allows us to calculate the probability that $\theta \in \mathcal{P}_\theta$. For example, if our criterion is that θ is plausible if $D_- \leq f(\theta) \leq D_+$, then:

$$p(\theta) = \mathbb{P}_{\tilde{f}^{(i)}}(\theta \in \mathcal{P}_\theta) = \Phi\left(\frac{D_+ - \mu(\theta)}{\Sigma(\theta)^{\frac{1}{2}}}\right) - \Phi\left(\frac{D_- - \mu(\theta)}{\Sigma(\theta)^{\frac{1}{2}}}\right).$$

In some regions $p(\theta)$ will be close to zero, indicating that we are confident that θ is implausible, and in others close to one, indicating the converse. It is regions in which we are most uncertain that we wish to target, as these represent parameter values that we can neither rule in nor out. One approach to selecting new design points is to choose points to minimise the entropy of this surface (Hennig and Schuler, 2012; Chevalier et al., 2014). The entropy represents how close to certain knowledge we are. If we let $\bar{H}$ be the average entropy of the emulator prediction of the plausibility surface:

$$\bar{H} = \int -p(\theta) \log p(\theta) - (1 - p(\theta)) \log(1 - p(\theta)) \mathrm{d}\theta,$$

then we can ask, if we were to add a simulator evaluation at θ, what is the expected value of $\bar{H}$ given the expected resulting information? We can then add θ to the design in order to minimise $\mathbb{E}(\bar{H}|\mathcal{D}^{(i-1)} \cup \{\theta\})$. This approach places new points in regions that most quickly reduce the uncertainty about the plausible region $\mathcal{P}_\theta$.

19.3.2 A simple climate example

As an illustration of the potential benefit of these techniques, we consider a relatively simple two-box climate simulator (Emanuel, 2002), which models atmospheric and ocean heat transport and storage, with water vapour as a positive feedback. The simulator is useful for the purposes of demonstration, as 10 years of model time takes approximately 5 seconds of CPU time, allowing a large number of model runs to be done. MATLAB® code for this simulator is available online.

We present the results of a simple history-matching task, calibrating two parameters: DTcrit_conv, the critical vertical temperature gradient that triggers convection, which we allowed to vary in the range [30, 50]; and gamma, the emissivity parameter for water vapour, which we varied in the range [1, 2]. We try to find the parameter values that give a global surface temperature between 294.5 K and 295.5 K once the model is in equilibrium. The CO_2 concentration was set to 560 ppm, and all other parameters were set to their default values (EPcm, 2010). These choices are arbitrary and only intended for illustration of the methodology.

Applying a simple ABC rejection algorithm and allowing for 1,000 simulator evaluations gave us 106 accepted parameter values, which are shown in the left-hand plot in Figure 19.1, with the light grey points showing the ten values accepted after only 100 simulator evaluations. In contrast, the middle and right-hand plots show the result of using a GP emulator with a maximin Latin hyper-cube (MLH) design of 10 and 30 simulator evaluations. After ten simulator evaluations, the emulator has some idea of where the plausible region is, but with errors, for example, in the bottom right hand corner. After 30 simulator evaluations, it has accurately found all of the plausible region, with just a little uncertainty at the edge of the region (shown by the grey shading).

This approach, however, relies upon finding a good design. If an accurate emulator results, then it will do well at predicting the plausible region. Here we can see, in the case where we had only ten design points, that no information is available about the bottom right hand corner of the parameter space, and consequently the model does less well there. As the design is chosen in advance of the simulations being run, finding a good design involves an element of luck.

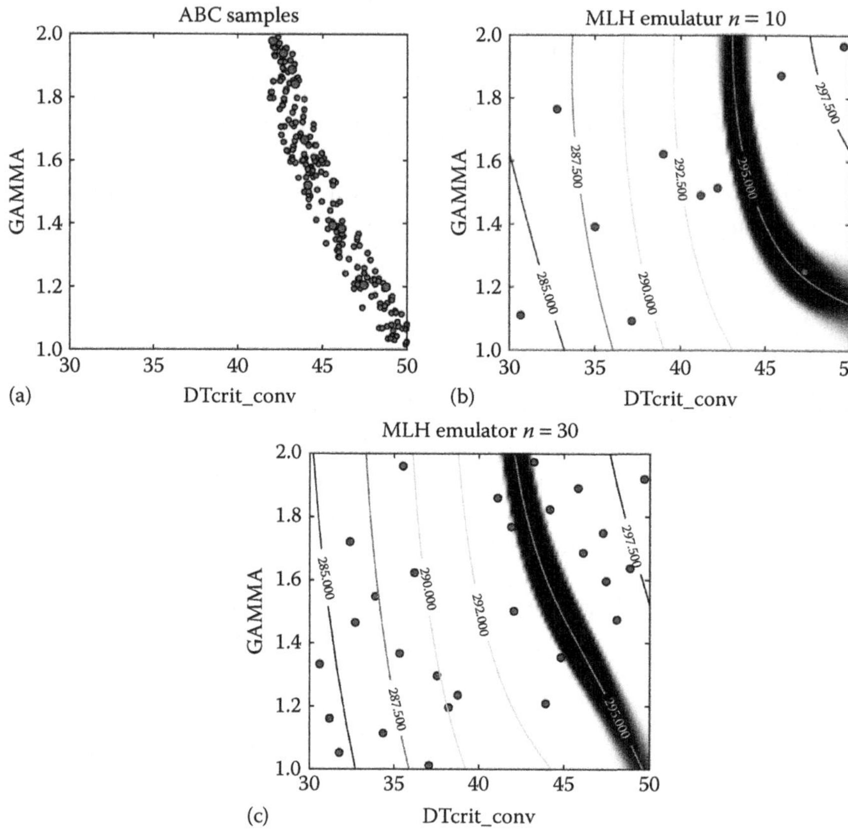

FIGURE 19.1
(a) Accepted samples from the rejection ABC algorithm after 100 (light grey) and 1,000 (dark grey) simulator evaluations. (b and c) The estimated plausible region using an emulator trained with a maximin Latin hypercube design (points shown in grey) with 10 (middle) and 30 (right) simulator evaluations. The shading indicates the estimated value of $\mathbb{P}(\theta \in \mathcal{P}_\theta)$. The contour lines are the estimated response surface $f(\theta)$.

If we instead use a sequential design and add design points one at a time in order to minimise the expected average entropy of the resulting history match, then we can significantly improve the speed with which we find $\mathcal{P}_\theta$. The two plots in Figure 19.2 show the resulting history match after four and ten simulator evaluations. After only ten simulator evaluations, we have found $\mathcal{P}_\theta$ with superior accuracy to that found after 30 simulator evaluations using the MLH design.

Note that the acceptance rate in the ABC algorithm was approximately 10%, considerably higher than in most problems (we had a 1% acceptance rate

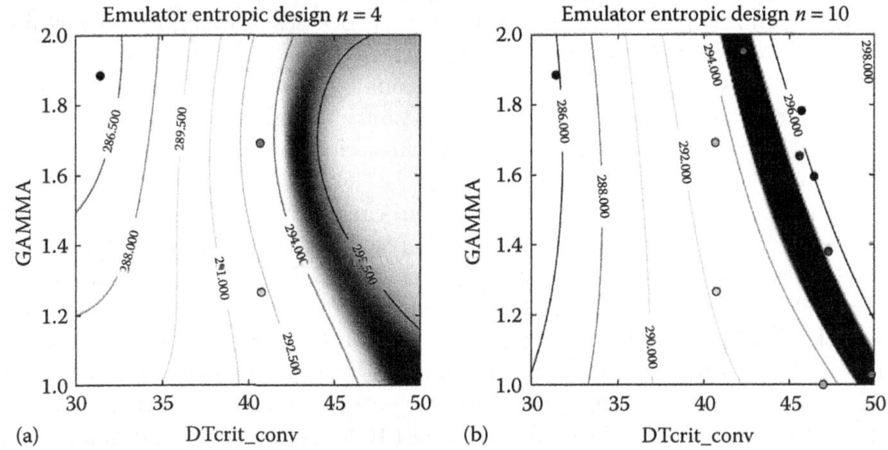

FIGURE 19.2
Results from using an entropy based sequential design. The left-hand column shows the estimated response surface (contours) and $\mathbb{P}(\theta \in \mathcal{P}_\theta)$ (shading), with the design points overlaid. The large dark grey point is the most recently added point. The right-hand column shows the entropy surface. The top row uses four simulator evaluations, and the bottom row uses ten simulator evaluations, all added according to the entropy criterion.

in the case study described in Section 19.4). As the acceptance rate decreases, the value of using an emulator to predict $\mathcal{P}_\theta$ increases, as the emulator is able to predict where the plausible region is, whereas ABC can only find the region by chance, as it uses no information about the shape of the underlying surface. In contrast, the major advantage of the Monte Carlo approach is that it is less prone to errors (although mixing errors commonly occur in practice), unlike the emulator approach, which can mislead if the fitted model is inaccurate, and thus requires careful supervision.

19.4 Climate Model Case Study

19.4.1 The global carbon cycle

Human emissions of CO_2 into the atmosphere are a principal cause of climate change. However, this 'anthropogenic' CO_2 does not remain in the atmosphere indefinitely. It is taken up by vegetation and by the oceans, and, eventually, (after many thousands of years) it is deposited as carbonate sediments at the ocean floor. Understanding these processes is crucial for future climate projections. Climate change is driven by changes in CO_2 concentration and it is therefore determined by the interplay between anthropogenic emissions and the carbon cycle. Many carbon cycle processes are highly uncertain. Projections of year 2100 CO_2 concentrations from different ESMs driven by the same assumption of future emissions typically vary by ± 100 ppm (Friedlingstein et al., 2006). This uncertainty range is greater than the total increase to date (2015) due to all historical anthropogenic emissions ($\sim$120 ppm).

To investigate uncertainties in the global carbon cycle, we need a model of appropriate complexity that is capable of resolving the important processes, but which is sufficiently numerically efficient. The GENIE-1 intermediate complexity ESM (Holden et al., 2013b) is one such model. The computational speed of GENIE-1 comes mainly from the use of a very simple two-dimensional (2D) model of the atmosphere and relatively coarse model resolution (grid cells of $\sim 1,000 \times 1,000$ km). The carbon cycle of GENIE-1 comprises a terrestrial carbon model, a three-dimensional (3D) dynamic ocean, dynamic sea ice, ocean biogeochemistry, and ocean sediments. Given appropriate model parameter choices, GENIE-1 simulates realistic spatial distributions of carbon storage in vegetation, soil, ocean, and carbonate sediment. However, the future response of the climate cycle to ongoing emissions depends upon the specific parameter choices and will vary even amongst parameter sets that have been constrained to produce similar (and reasonable) modern climate states. To quantify this uncertain response, we require an ensemble of simulations that samples widely from plausible input parameter space.

The timescales for different carbon cycle processes vary considerably. Equilibrium timescales are $\sim$10s years for vegetation, $\sim$100s years for soil, $\sim 1,000$s years for the ocean and $\sim 10,000$'s years for carbonate sediments. In order to simulate an Earth with a carbon cycle in approximate equilibrium (i.e. prior to human interference), a simulation of at least 10,000 years is required.* Although several orders of magnitude faster than than state-of-the-art ESMs, GENIE-1 requires $\sim$4 CPU days to simulate 10,000 real years. The exploration of high-dimensional input space and identification of plausible subspaces is therefore a highly demanding computational problem, which we address through emulator-informed ABC.

19.4.2 Emulator-informed ABC design

The philosophy of the design approach is to vary key model parameters over the entire range of plausible values and to accept those parameter combinations that lead to climate states that cannot be uncontroversially ruled out as implausible (Edwards et al., 2011). We are seeking to explore all plausible simulator realisations in order to capture the range of possible feedback strengths. The input ranges we apply, Θ, are generally broader than ranges that are applied in model tuning exercises. This is in part to enable us to fully quantify model behaviour over plausible parameter space, $\mathcal{P}_\theta$, and in part to

*Shorter spin-ups are sufficient for models that neglect sediments.

improve the validity of the ensemble for application to diverse climate states, such as the Last Glacial Maximum.

The experimental set-up is described in Holden et al. (2013a). We varied 24 model parameters in the ensemble. The choice of parameters was governed by consideration of the processes that are thought to contribute to the natural variability of atmospheric CO_2 on glacial-interglacial timescales (Kohfeld and Ridgwell, 2009) and, hence, to which the distribution of carbon may be sensitive in general. Five atmospheric parameters were varied. These parameters control the spatial distribution of simulated temperature and precipitation and, hence, drive changes in vegetation, sea-ice coverage, and ocean circulation. Five parameters were varied in the vegetation model, controlling photosynthesis and respiration rates. Five ocean parameters were varied. These control ocean circulation and, hence, the spatial distribution of carbon, alkalinity, dissolved oxygen, and nutrients in the ocean. Sea-ice diffusivity was varied, primarily because of its effect on ocean circulation by altering the transport of freshwater. Nine ocean biogeochemistry parameters were varied. These parameters drive changes in the rates of atmosphere-ocean gas exchange, plankton photosynthesis, and the remineralisation of the organic products of this photosynthesis. The rate of remineralisation controls the transport of carbon from the surface of the ocean to the deep.

A 500-member ensemble of 25,000-year simulations was first performed using a MLH design.* The plausibility of each simulator run was evaluated using eight different output quantities, usually termed *metrics* in the climate literature. These simple metrics impose no constraints on the spatial distribution of modelled outputs. They instead provide global-scale constraints on atmosphere (global average temperature), ocean (strength of North Atlantic overturning and Antarctic deep water formation), Antarctic sea-ice coverage, global vegetation carbon, global soil carbon, ocean biogeochemistry (average dissolved oxygen concentration in the global ocean), and ocean sediments (the average percentage of $CaCO_3$ in the surface sediment). Only four of the 500 MLH simulations were found to satisfy all eight plausibility constraints, which given that $\dim\theta = 24$, is insufficient for any meaningful statistical analysis. The MLH ensemble took more than ten years of computing to complete, demonstrating that a naive application of ABC is infeasible for this application.

As described in Section 19.3, we can use emulators to guide the search to find plausible regions of parameter space. Regression-based emulators, including linear and quadratic terms, were built for each of the eight metrics (outputs) specified earlier. Prior to fitting, variables were linearly mapped onto the range $[-1,1]$, so that odd and even terms were orthogonal, aiding variable selection. The models were built using a stepwise model selection

*We note that while efficiencies can be gained in certain applications by initialising each ensemble member with output from an existing equilibrium simulation, such an approach is not likely to be useful here as our approach is designed to sample widely differing Earth system states.

scheme, initially using the Akaike information criterion as the selection criterion and then subsequently shrunk further by applying the more stringent Bayes information criterion. This procedure of first growing the model beyond the Bayes information criterion constraint and then shrinking helps to avoid local minima in the stepwise search.

Parameters were then sampled uniformly from the a priori plausible region and the emulators used to predict if they would lead to plausible simulations. Parameters were accepted as potentially plausible when the emulators predicted plausible values for all eight metrics. The plausibility ranges used were based on the observed climate record, the simulator discrepancy, and the emulator accuracy. Each accepted parameter combination was then used as a design point in a further simulation.

As simulations completed, the emulators were rebuilt using the additionally available data. This process progressively improved the success rate of the emulator predictions (i.e. the percentage of emulator predicted plausible parameters that led to plausible simulations) from 24% to 65%. In total, the simulator was run for 1,000 parameter values predicted to be plausible by the emulator. This produced 885 completed simulations of which 471 were plausible (the remaining 115 simulations terminated before completion, a common occurrence with climate simulators). This 471-member plausible set forms the Emulator Filtered Plausibility Constrained (EFPC) ensemble. The generation of these simulations required a further 25 years of computing time. Without the ~50-fold increase in efficiency gained by using an emulator to predict the plausible region, this would have required an infeasible amount (more than 1,000 years) of CPU time.

While it is clear that ABC strongly constrains the outputs (metrics) that are explicitly filtered for, it is worth noting that it indirectly constrains all aspects of the Earth system and leads to improved simulated magnitudes *and spatial distributions* of state variables generally. Figure 19.3 provides an illustrative output of the EFPC ensemble and of the benefits of the ABC filtering. The figure illustrates cross-sections of ocean alkalinity through the Atlantic and Pacific Oceans, comparing ensemble means of the unfiltered MLH simulations (left) and filtered EFPC simulations (centre) with observational data (right). Ocean alkalinity exerts a strong control on atmospheric carbon dioxide by determining the degree to which dissolved carbon dioxide is dissociated into bicarbonate and carbonate ions, in turn determining the rates of exchange of dissolved carbon in the ocean with the atmosphere (carbon dioxide) and the sediments (calcium carbonate). Alkalinity is not directly constrained by the ABC metrics, but its distribution is influenced by them, for instance, through the constraints imposed on ocean circulation strength and the sediment carbonate concentration. Relative to the MLH ensemble, the EFPC ensemble shows elevated surface concentrations, decreased concentrations in the deep Atlantic (apparently associated with the Atlantic overturning circulation in the unfiltered ensemble), and increased penetration of high alkalinity towards southern latitudes in the deep Pacific. Although discrepancies with

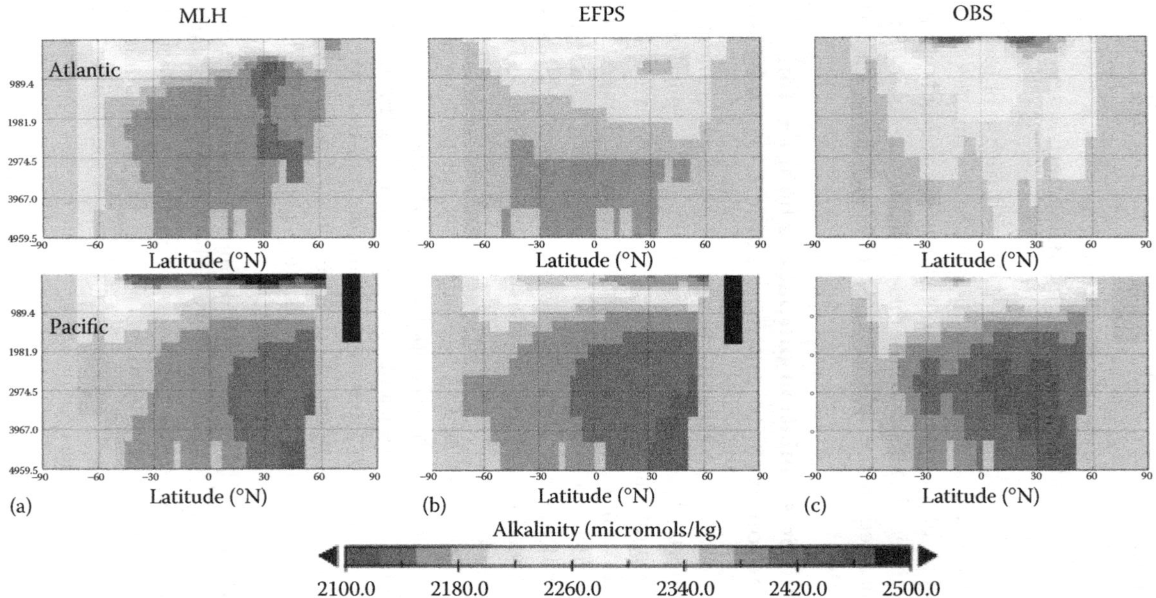

FIGURE 19.3
Cross-sections of ocean alkalinity through the Atlantic (25°W) and Pacific (155°W) Oceans. The figure compares the mean of the training MLH ensemble (a) and the plausibility filtered EFPC ensemble (b) with observations (c).

observations remain, which may reflect structural deficiencies in the simulator, each of these trends produces better ensemble-averaged agreement with the observed distribution.

19.4.3 Applications

Although the use of ABC to derive a posterior distribution is useful in itself, our primary motivation is to identify a set of plausible parameters for application to diverse simulation problems. The EFPC parameter set has been used in a range of experiments, considering both past and future climate change. For clarity, it is worth emphasising that while these experiments did not use ABC directly, they were all rendered tractable by the use of emulator-informed ABC to design the underlying simulation ensemble. A selection of these experiments are summarised in the following, each with a focus on a different category of application.

19.4.3.1 Probabilistic simulation outputs: The uncertain response of the carbon cycle to anthropogenic CO_2 emissions

We (PBH and NRE) contributed a suite of carbon cycle experiments for the Fifth Assessment Report (AR5) of the Intergovernmental Report on Climate Change (IPCC). Fifteen intermediate complexity ESMs from around the world performed these experiments. The focus was on historical change (Eby et al., 2013) and long-term future change (Zickfeld et al., 2013), considering long timescale problems that are not tractable by state-of-the-art ESMs, thus requiring the use of reduced complexity models such as GENIE-1. Forty seven experiments were performed.

We applied a subset of the EFPC parameter set, in part to aid computational tractability, in view of the 47 separate experiments required, and, in part, to eliminate a bias in the transient response of the ensemble. The EFPC parameter set is constrained to simulate a plausible pre-industrial climate, but no constraint was imposed upon the dynamic response to anthropogenic emissions. Four important model parameters were not constrained by preindustrial plausibility, two relating to cultivated vegetation (deforestation for agriculture was neglected in the preindustrial simulations), a parameter controlling the direct effect of CO_2 on photosynthesis ('CO_2 fertilisation', see following section), and a parameter controlling the uncertain effect of clouds on the Earth's radiation budget in a warmer planet. The dynamic response was therefore filtered through a historical forcing experiment, which imposed anthropogenic forcing, including CO_2 emissions, since pre-industrial times in an EFPC ensemble of transient simulations. Twenty parameter sets, selected at random from the EFPC parameter sets, but constrained to approximately reproduce the present day atmospheric CO_2 concentration, were accepted and applied to the IPCC experiments.

We do not attempt to summarise the results of these extensive multi-model comparisons here, but note that the GENIE-1 perturbed-parameter ensemble was found to provide an unbiased representation of the multi-model ensemble, being approximately centred on the mean of the 15 models and with comparable uncertainty. These uncertainties were presented in a related model intercomparison paper (Joos et al., 2013).

19.4.3.2 Calibrating model parameters: The strength of the terrestrial carbon sink

The IPCC experiments revealed a general tendency of intermediate complexity ESMs to understate the magnitude of the terrestrial carbon sink (the anthropogenic CO_2 taken up by vegetation on land). The major uncertainty in the terrestrial sink relates to CO_2 fertilisation. Experimental evidence almost without exception shows a stimulation of leaf photosynthesis when plants are exposed to elevated CO_2 (Körner, 2006). In addition to this direct affect on photosynthesis, the short timescale physiological effect of reduced stomatal opening increases water-use efficiency and additionally increases the efficiency of photosynthesis (Field et al., 1995). However, the strength of the fertilisation effect is poorly quantified, especially under natural conditions. Some studies have failed to detect a measurable effect in nature, while others suggest that any effects may be short term, as CO_2 is only one of a number of potentially limiting factors on plant growth (Körner, 2006).

We addressed this calibration problem in Holden et al. (2013b). Using output from a 671-member ensemble of transient GENIE-1 simulations derived from the EFPC parameter sets, we built an emulator of the change in atmospheric CO_2 concentration change since the pre-industrial period. We then applied this emulator to sample the 28-dimensional input parameter space. A Bayesian calibration suggests that the increase in gross primary productivity (GPP) in response to a doubling of CO_2 from pre-industrial values is very likely (90% confidence) to exceed 20%, with a most likely value of 40%–60%.

19.4.3.3 Model understanding: What determines the spatial distribution of dissolved carbon in the ocean?

In Holden et al. (2013a), we applied the EFPC ensemble to a transient experiment over the recent industrial era (1858–2008 AD). The temporal evolution of atmospheric CO_2 and its isotopic composition are known from observational data, and these simulated quantities were made to follow the observations through a relaxation term. The objective of the experiment was to better understand the mechanisms by which the anthropogenic CO_2 emissions are taken up by the ocean.

To achieve this, we analysed the change in distributions of ocean concentrations of dissolved inorganic carbon and its stable isotope $\delta^{13}C$, considering two-dimensional latitudinal-vertical transects through the Atlantic and Pacific. These two transects were combined into a single vector for each simulation (to ensure inter-basin effects were consistently represented), and these vectors were combined into an ensemble matrix. Singular vector decomposition was applied to the DIC and $\delta^{13}C$ matrices in order to extract the dominant modes of their spatial variability across the ensemble. Emulators of the component scores elicited further understanding of these modes by identifying which model parameters were driving each mode of variability.

This, together with physical interpretation of the spatial patterns of each mode, enabled us to identify the principal processes driving them, on the assumption that the dominant parameters governing uncertainty in the response of each mode could be identified with the most important parameterised processes controlling the respective modes. We showed that the main processes governing the uptake of anthropogenic CO_2 and $\delta^{13}C$ are quite distinct: an important conclusion because observations of the isotopic composition are used to infer rates of ocean CO_2 uptake. Uncertainty in anthropogenic $\delta^{13}C$ uptake is dominated by air-sea gas exchange, which explains 63% of modelled variance. This mode of variability is largely absent from the ensemble variability in CO_2 uptake, which is instead driven by uncertainties in mixing rates between the surface and deep oceans.

19.4.3.4 Coupling applications: Coupling climate models and climate change impact models

The evaluation of climate impacts requires coupling climate models, impact models, and economic models together within an 'integrated assessment model' (IAM) framework. In such couplings, climate data (e.g. regional temperature, precipitation) are passed to the IAM for computation of climate impact functions, and the IAM passes back anthropogenic forcing (such as CO_2 emissions or land use change). Computational demands mean that it is generally infeasible to couple complex climate models into IAMs. Various approaches are taken to address this, using either simplified models or statistical representations of more complex models. Recently, effort has focussed on the use of emulators of climate models as surrogates for the simulator in these coupling applications.

Economic models provide projections of CO_2 emissions. They typically convert emissions into concentrations through the use of simple 'box-models', describing rates of carbon transfer between the atmospheric, terrestrial and oceanic reservoirs. We have recently applied the EFPC parameter set to build an emulator of the GENIE-1 carbon cycle model for incorporation into integrated assessment models (Foley et al., 2016). An 86-member subset of the EFPC parameter set was used to generate an ensemble of future climate-carbon cycle experiments, with future emissions prescribed as modified Chebyshev polynomials.

The emulation approach followed the '1-step' dimensionally reduced emulation methodology of Holden et al. (2015), emulating a singular value decomposition of the ensemble outputs. Emulators of the first four component scores were derived as functions of the 28 model parameters and the 6 concentration profile coefficients. The emulator outputs are, unsurprisingly, dominated by the Chebyshev coefficients. However, uncertainty for a given forcing scenario is generated through emulator dependencies on GENIE-1 parameters. The resulting carbon cycle emulator has been coupled into an integrated assessment framework that also includes a macroeconometric model of the global economy E3MG (Mercure et al., 2014), an agent-based model of technology substitution dynamics *future technology transformations (FTT)-power* (Mercure, 2012) and a spatiotemporally resolved emulator of the climate system (Holden et al., 2014). We have applied the framework to assess the impact on the climate of emissions reduction policies in the electricity sector (Mercure et al., 2014), addressing the cascade of uncertainty through the coupled system.

19.5 Future Applications

It may never be possible to apply statistical approaches to robustly calibrate a truly state-of-the-art climate simulator. They are defined by the limits of available computing power, and, consequently, very few simulations are possible with these models. This begs the important question of how far could one go with simulator complexity and still be able apply these methods. We have demonstrated the application of emulator-informed ABC to generate a 471-member ensemble of a model that takes $\sim$10 days to perform each simulation. The computational constraints ultimately determined the number of parameters we could vary; a rule of thumb dictates that we use a minimum of ten ensemble members for each varied active input (Loeppky et al., 2009). It is worth noting that a useful ensemble varying only, say, 5 parameters would need $\sim$50 simulations and could have been achieved for a 10-fold slower model.

The improvements in methodology demonstrated in Section 19.3.2, suggest efficiencies that should significantly extend the applicability of the approach. The use of GP emulation generally allows a better statistical model than linear regression, and, therefore, would be expected to improve the success rate of the emulator filtering. This will certainly be the case when a parametric mean function is used and the GP is applied only to emulate the residual. The uncertainty estimates provided by the GP should also improve the success rate of the emulator filtering, for instance, by only accepting parameters for which there is a high probability of plausibility. Furthermore, a significant improvement arises from the use of a sequential design process, which was shown to yield a 3-fold increase in efficiency in our example. For more complex simulators, we will want to make use of parallel computation. The sequential approach then changes from adding one design point at a time, to adding d, where d is the number of available cores. Finding the d optimal points that minimise the expected entropy is difficult, and is an area of active research, but even suboptimal designs can give significant improvements over the default space-fillings designs. For stochastic simulators, many of the same techniques can be applied. The likelihood function now needs to be estimated, significantly increasing the difficulty, but progress is being made in this direction (Meeds and Welling, 2014; Oakley and Youngman, 2014; Wilkinson, 2014; Gutmann and Corander, 2015).

These improvements in efficiency should render application to 'previous-generation' ESMs, such as HadCM3* tractable on multi-node computing clusters, certainly so on distributed computing systems, such as climateprediction.net, which last year facilitated more than 7,500 years of climate modelling on the personal computers of the general public.

References

A box climate model: EPcm. Model documentation v4. www.sp.ph.ic.ac.uk/~aczaja/EP_ClimateModel.html, 2010.

Andrianakis, I., I. R. Vernon, N. McCreesh, T. J. McKinley, J. E. Oakley, R. N. Nsubuga, M. Goldstein, and R. G. White. Bayesian history matching of complex infectious disease models using emulation: A tutorial and a case study on hiv in uganda. *PLoS Computational Biology*, 11(1):e1003968, 2015.

Annan, J., J. Hargreaves, N. Edwards, and R. Marsh. Parameter estimation in an intermediate complexity earth system model using an ensemble Kalman filter. *Ocean Modelling*, 8(1):135–154, 2005.

Biau, G., F. Cérou, and A. Guyader. New insights into Approximate Bayesian Computation. *Annales de l'Institut Henri Poincaré, Probabilités et Statistiques*, 51:376–403, 2015.

Blum, M. G., M. A. Nunes, D. Prangle, and S. A. Sisson. A comparative review of dimension reduction methods in approximate Bayesian computation. *Statistical Science*, 28(2):189–208, 2013.

Brynjarsdóttir, J., and A. O'Hagan. Learning about physical parameters: The importance of model discrepancy. *Inverse Problems*, 30(11):114007, 2014.

Castruccio, S., D. J. McInerney, M. L. Stein, F. Liu Crouch, R. L. Jacob, and E. J. Moyer. Statistical emulation of climate model projections based on precomputed GCM runs. *Journal of Climate*, 27(5):1829–1844, 2014.

Chevalier, C., D. Ginsbourger, J. Bect, E. Vazquez, V. Picheny, and Y. Richet. Fast parallel kriging-based stepwise uncertainty reduction with application to the identification of an excursion set. *Technometrics*, 56(4):455–465, 2014.

*HadCM3 performs more than ten years per CPU day on eight nodes of a linux cluster.

Craig, P. S., M. Goldstein, A. H. Seheult, and J. A. Smith. Pressure matching for hydrocarbon reservoirs: A case study in the use of Bayes linear strategies for large computer experiments. In C. Gatsonis, J. S. Hodges, R. E. Kass, R. McCulloch, P. Rossi, and N. D. Singpurwalla (Eds.), *Case Studies in Bayesian Statistics*, pp. 37–93. New york: Springer, 1997.

Eby, M., A. J. Weaver, K. Alexander, K. Zickfeld, A. Abe-Ouchi, A. Cimatoribus, E. Crespin et al. Historical and idealized climate model experiments: an intercomparison of Earth system models of intermediate complexity. *Climate of the Past*, 9:1111–1140, 2013.

Edwards, N. R., D. Cameron, and J. Rougier. Precalibrating an intermediate complexity climate model. *Climate Dynamics*, 37(7–8):1469–1482, 2011.

Emanuel, K. A simple model of multiple climate regimes. *Journal of Geophysical Research*, 107(D9):ACL–4, 2002.

Field, C., R. Jackson, and H. Mooney. Stomatal responses to increased CO_2: Implications from the plant to the global scale. *Plant, Cell & Environment*, 18(10):1214–1225, 1995.

Flato, G., J. Marotzke, B. Abiodun, P. Braconnot, S. C. Chou, W. Collins, P. Cox et al. Evaluation of climate models. In *Climate Change 2013: The Physical Science Basis. Contribution of Working Group I to the Fifth Assessment Report of the Intergovernmental Panel on Climate Change*, pp. 741–866, 2013.

Foley, A.M., P. B. Holden, N. R. Edwards, J.-F. Mercure, P. Salas, H. Politt, and U. Chewpreecha. Climate model emulation in an integrated assessment framework: A case study for mitigation policies in the electricity sector. *Earth System Dynamics*, 7:119–132, 2016.

Friedlingstein, P., P. Cox, R. Betts, L. Bopp, W. Von Bloh, V. Brovkin, P. Cadule et al. Climate-carbon cycle feedback analysis: Results from the C4MIP model intercomparison. *Journal of Climate*, 19(14):3337–3353, 2006.

Gramacy, R. B., and H. K. Lee. Bayesian treed Gaussian process models with an application to computer modeling. *Journal of the American Statistical Association*, 103(483):1119–1130, 2008.

Gutmann M. U., and J. Corander. Bayesian optimization for likelihood-free inference of simulator-based statistical models. *arXiv preprint arXiv:1501.03291*, 2015.

Harris, G. R., D. M. Sexton, B. B. Booth, M. Collins, and J. M. Murphy. Probabilistic projections of transient climate change. *Climate Dynamics*, 40:2937–2972, 2013.

Hennig, P., and C. J. Schuler. Entropy search for information-efficient global optimization. *The Journal of Machine Learning Research*, 13(1):1809–1837, 2012.

Higdon, D., J. Gattiker, B. Williams, and M. Rightley. Computer model calibration using high-dimensional output. *Journal of the American Statistical Association*, 103(482):570–583, 2008.

Holden, P., and N. Edwards. Dimensionally reduced emulation of an AOGCM for application to integrated assessment modelling. *Geophysical Research Letters*, 37(21):L21707, 2010.

Holden, P., N. Edwards, P. Garthwaite, K. Fraedrich, F. Lunkeit, E. Kirk, M. Labriet, A. Kanudia, and F. Babonneau. PLASIM-ENTSem v1.0: A spatio-temporal emulator of future climate change for impacts assessment. *Geoscientific Model Development*, 7(1):433–451, 2014.

Holden, P., N. Edwards, P. Garthwaite, and R. Wilkinson. Emulation and interpretation of high-dimensional climate model output. *Journal of Applied Statistics*, 2015. doi:10.1080/02664763.2015.1016412.

Holden, P., N. Edwards, D. Gerten, and S. Schaphoff. A model-based constraint on CO_2 fertilisation. *Biogeosciences*, 10(1):339–355, 2013b.

Holden, P., N. Edwards, S. Müller, K. Oliver, R. Death, and A. Ridgwell. Controls on the spatial distribution of oceanic $\delta^{13}C_{DIC}$. *Biogeosciences*, 10:1815–1833, 2013a.

Holden, P. B., N. Edwards, K. Oliver, T. Lenton, and R. Wilkinson. A probabilistic calibration of climate sensitivity and terrestrial carbon change in GENIE-1. *Climate Dynamics*, 35(5):785–806, 2010.

Holloway Jr., J. L., and S. Manabe. Simulation of climate by a global general circulation model: I. hydrologic cycle and heat balance. *Monthly Weather Review*, 99(5):335–370, 1971.

Joos, F., R. Roth, J. Fuglestvedt, G. Peters, I. Enting, W. von Bloh, V. Brovkin et al. Carbon dioxide and climate impulse response functions for the computation of greenhouse gas metrics: A multi-model analysis. *Atmospheric Chemistry and Physics*, 13(5):2793–2825, 2013.

Kennedy, M., and A. O'Hagan. Bayesian calibration of computer models (with discussion). *Journal of the Royal Statistical Society: Series B (Statistical Methodology)*, 63:425–464, 2001.

Kohfeld, K. E., and A. Ridgwell. Glacial-Interglacial variability in atmospheric CO_2. In C. Quere and E. S. Saltzman (Eds.), *Surface Ocean-Lower Atmosphere Processes*, pp. 251–286. Washington, DC: American Geophysical Union, 2009.

Körner, C. Plant CO_2 responses: An issue of definition, time and resource supply. *New phytologist*, 172(3):393–411, 2006.

Lee, L., K. Carslaw, K. Pringle, and G. Mann. Mapping the uncertainty in global CCN using emulation. *Atmospheric Chemistry and Physics*, 12(20): 9739–9751, 2012.

Loeppky, J. L., J. Sacks, and W. J. Welch. Choosing the sample size of a computer experiment: A practical guide. *Technometrics*, 51(4):366–376, 2009.

Marquis, J., Y. Richardson, P. Markowski, D. Dowell, J. Wurman, K. Kosiba, P. Robinson, and G. Romine. An investigation of the Goshen County, Wyoming, tornadic supercell of 5 June 2009 using EnKF assimilation of mobile mesonet and radar observations collected during VORTEX2. Part I: Experiment design and verification of the EnKF analyses. *Monthly Weather Review*, 142(2):530–554, 2014.

Mauritsen, T., B. Stevens, E. Roeckner, T. Crueger, M. Esch, M. Giorgetta, H. Haak et al. Tuning the climate of a global model. *Journal of Advances in Modeling Earth Systems*, 4(3):M00A01, 2012.

McKay, M. D., R. J. Beckman, and W. J. Conover. A comparison of three methods for selecting values of input variables in the analysis of output from a computer code. *Technometrics*, 42(1):55–61, 2000.

Meeds, E., and M. Welling. GPS-ABC: Gaussian process surrogate approximate Bayesian computation. *arXiv preprint arXiv:1401.2838*, 2014.

Mercure, J.-F. FTT: Power: A global model of the power sector with induced technological change and natural resource depletion. *Energy Policy*, 48:799–811, 2012.

Mercure, J.-F, H. Pollitt, U. Chewpreecha, P. Salas, A. M. Foley, P. B. Holden, and N. R. Edwards. The dynamics of technology diffusion and the impacts of climate policy instruments in the decarbonisation of the global electricity sector. *Energy Policy*, 73:686–700, 2014.

Morokoff, W. J., and R. E. Caflisch. Quasi-random sequences and their discrepancies. *SIAM Journal on Scientific Computing*, 15:1251–1279, 1994.

Murphy, J. M., D. M. Sexton, D. N. Barnett, G. S. Jones, M. J. Webb, M. Collins, and D. A. Stainforth. Quantification of modelling uncertainties in a large ensemble of climate change simulations. *Nature*, 430(7001):768–772, 2004.

Oakley, J. E., and B. D. Youngman. Calibration of complex computer simulators using likelihood emulation. *arXiv preprint arXiv:1403.5196*, 2014.

O'Hagan, A. Bayesian analysis of computer code outputs: A tutorial. *Reliability Engineering & System Safety*, 91(10):1290–1300, 2006.

Olson, R., R. Sriver, M. Goes, N. M. Urban, H. D. Matthews, M. Haran, and K. Keller. A climate sensitivity estimate using Bayesian fusion of instrumental observations and an Earth System model. *Journal of Geophysical Research: Atmospheres (1984–2012)*, 117(D4):D04103, 2012.

Oyebamiji, O. K., N. R. Edwards, P. B. Holden, P. H. Garthwaite, S. Schaphoff, and D. Gerten. Emulating global climate change impacts on crop yields. *Statistical Modelling*, 15(6):499–525, 2015.

Rougier, J. Probabilistic inference for future climate using an ensemble of climate model evaluations. *Climatic Change*, 81(3–4):247–264, 2007.

Rougier, J. and M. Goldstein. Climate simulators and climate projections. *Annual Review of Statistics and Its Application*, 1:103–123, 2014.

Rougier, J., D. M. Sexton, J. M. Murphy, and D. Stainforth. Analyzing the climate sensitivity of the HadSM3 climate model using ensembles from different but related experiments. *Journal of Climate*, 22(13):3540–3557, 2009.

Sacks, J., W. J. Welch, T. J. Mitchell, and H. P. Wynn. Design and analysis of computer experiments. *Statistical Science*, 4:409–423, 1989.

Sansó, B., C. E. Forest, D. Zantedeschi. Inferring climate system properties using a computer model. *Bayesian Analysis*, 3(1):1–37, 2008.

Santner, T. J., B. J. Williams, and W. I. Notz. *The Design and Analysis of Computer Experiments*. New York: Springer Verlag, 2003.

Sham Bhat, K., M. Haran, R. Olson, and K. Keller. Inferring likelihoods and climate system characteristics from climate models and multiple tracers. *Environmetrics*, 23(4):345–362, 2012.

Vernon, I., M. Goldstein, and R. G. Bower. Galaxy formation: A Bayesian uncertainty analysis. *Bayesian Analysis*, 5:619–669, 2010.

Wilkinson, R. D. Bayesian calibration of expensive multivariate computer experiments. In L. Biegler, G. Biros, O. Ghattas, M. Heinkenschloss, D. Keyes, B. Mallick, Y. Marzouk, L. Tenorio, B. van Bloemen Waanders, and K. Wilcox (Eds.), *Large-Scale Inverse Problems and Quantification of Uncertainty*, pp. 195–215. Chichester, UK: John Wiley & Sons, 2010.

Wilkinson, R. D. Approximate Bayesian computation (ABC) gives exact results under the assumption of model error. *Statistical Applications in Genetics and Molecular Biology*, 12:129–141, 2013.

Wilkinson, R. D. Accelerating ABC methods using Gaussian processes. *JMLR Workshop and Conference Proceedings Volume 33: Proceedings of the Seventeenth International Conference on Artificial Intelligence and Statistics*, 33:1015–1023, 2014.

Williamson, D., M. Goldstein, L. Allison, A. Blaker, P. Challenor, L. Jackson, and K. Yamazaki. History matching for exploring and reducing climate model parameter space using observations and a large perturbed physics ensemble. *Climate Dynamics*, 41(7–8):1703–1729, 2013.

Zickfeld, K., M. Eby, A. J. Weaver, K. Alexander, E. Crespin, N. R. Edwards, A. V. Eliseev et al. Long-term climate change commitment and reversibility: An EMIC intercomparison. *Journal of Climate*, 26(16):5782–5809, 2013.

20

ABC in Ecological Modelling

Matteo Fasiolo and Simon N. Wood

CONTENTS

20.1 Simulation-Based Methods in Ecology 597
20.1.1 Intractable ecological models 597
20.1.2 Inference for state space models 600
20.1.3 SL versus tolerance-based ABC 604
20.2 Example: A Chaotic Prey–Predator Model 608
20.2.1 Description of data and priors 610
20.2.2 Comparison using simulated data 612
20.2.3 Results from the Kilpisjarvi dataset 613
20.3 Discussion 616
Acknowledgements 619
References 619

20.1 Simulation-Based Methods in Ecology

20.1.1 Intractable ecological models

Ecology aims to understand the abundance and distribution of organisms. This essentially quantitative task is made difficult by the complex web of interactions that exist between living things. In the face of such daunting ecological complexity, dynamic models play an important role in separating fundamental mechanisms from matters of detail. In particular, they allow theoretical ideas to be sharpened into well defined quantitative hypotheses, and this in turn opens up the possibility of testing these hypotheses using data.

But there is a catch. To be useful, ecological dynamic models must often resort to 'cartooning' of some ecological processes. Simplification is essential if the model is not to become a 'model-of-everything', hence, a reasonably parsimonious model may not be intended to reproduce the full data y_{obs} in all its features. For example, while the full data might be characterized by a spatial or temporal structure, it is often convenient to use a lumped model that ignores these dimensions. Similarly, when the data contain several classes of organisms, computational considerations might lead to a model that aggregates key statistics, such as population counts, over different classes. Under these circumstances, reducing the full data to a set of summary statistics, $s_{obs} = S(y_{obs})$, might not lead to any loss of information during parameter estimation or model selection (Hartig et al., 2011).

Basing statistical inference on aggregate summary statistics might be necessary even when working with individual based models, which are often used to understand ecological outcomes that depend intricately on the interactions of individuals within a population. Forest stand growth models are an example. In these models, individual trees of many species may be grown to maturity, all competing continuously for light and nutrients as they do so. Here, the mismatch between data and model is of a different kind. For example, in a real forest, we would obtain data consisting of measurements on individual trees. The same measurements can often be made on the model trees, but a particular model individual does not correspond to any real individual. We are left with no choice but to base inference on summary statistics, which suggests the use of ABC-type methods. One such example is Hartig et al. (2014) who uses synthetic likelihood (Wood, 2010), an approximate method closely related to ABC, to fit the FORest Mixture INDividual-based (FORMIND) individual-based forest model to Ecuadorian tropical forest field data. While the model deals with individual trees, its output is summarised using 112 statistics such as biomass, growth rate, and tree counts, obtained by aggregating trees over several diameter classes.

Other reasons for considering the use of summary statistics relate to highly non-linear dynamics, of the sort that are often found in populations of small animals, with high rates of fecundity and mortality. Indeed, even if our models are perfect descriptions of the driving ecological mechanisms, dynamic irregularity can make reliable inference very difficult to achieve by conventional means. If our models are less perfect, the interaction of such irregularity with small infelicities in the model's ability to match the data can lead to substantial inferential errors. Wood (2010) shows that these problems can arise in ecological systems as simple as the Ricker map (May, 1976) and illustrates how the extreme sensitivity of near chaotic systems to small changes in dynamically important parameters can cause minuscule moves in the parameter space to result in massive changes in likelihood values. In this circumstance, it is obviously appealing to base inference on summary statistics of the data that the model should be able to reproduce, rather than on the full data. Indeed, Wood (2010) and Fasiolo et al. (2014) argue that ABC-type methods can offer an appealing robustness here, provided that they are used in conjunction with appropriately robust statistics.

Even in the absence of the difficulties just discussed, ecological models can have tractability problems. Most of the conventional statistical tools used to find the parameter values or models that are most consistent with data (and possibly with prior knowledge), rely on the likelihood function, $p(y_{obs}|\theta)$. Unfortunately, for many models of ecological interest, $p(y_{obs}|\theta)$ is not available directly or is otherwise problematic, thus posing an obstacle to the whole inferential process. This difficulty can occur for several possible reasons, but one common problem is the presence of hidden or latent states. Specifically, we often know that the dynamics of an observed process y_{obs} are related to those of other processes n, which are hidden from us. In such cases, the likelihood could ideally be obtained by integrating the latent states out of the joint probability density of data and hidden states:

$$p(y_{obs}|\theta) = \int p(y_{obs}, n|\theta)\, dn. \tag{20.1}$$

In practice, this integration problem is usually analytically intractable, while the efficient implementation of numerical or Monte Carlo integration schemes often require additional assumptions, such as those detailed in Section 20.1.2.

Classical examples of partially observed systems of ecological interest are predator-prey systems, where the abundance of one of the two components is often completely unknown. In Section 20.2, we consider the prey–predator model proposed by Turchin and Ellner (2000), which has been used to describe the population dynamics of Fennoscandian voles. In that example trap data provide noisy estimates of voles abundance, but no such proxy is available for predatory weasels. A similar example is provided by Kendall et al. (2005), who evaluate alternative explanations for the regular oscillations in population density of insect pest pine looper moths. They consider, among others, a parasitoid and a food quality model, and they fit them using only data on moth population density. Given that ecological systems are observed with noise in most cases, the issue of hidden states is widespread, and it appears in studies concerned with animal movement (Langrock et al., 2012; Morales et al., 2004), population abundance estimation (Farnsworth et al., 2007), and essentially whenever remote tracking data are available (Jonsen et al., 2005).

The rapid growth in computational resources has supported the development of several approaches meant to tackle the issue of intractable likelihoods. Some of these approaches exploit the fact that faster computation makes forward model simulation, that is simulation of data y from $p(y|\theta)$, cheap enough that it can be repeated many thousands of times. In particular, it is possible to use forward simulations to find the set of parameter values or models that are able to closely reproduce the full data, y_{obs}, or more often some of its most informative features, s_{obs}. ABC represents one class of such methods which, being based on a Bayesian framework, generally try to address questions regarding parameter estimation or model selection by approximately sampling the corresponding posteriors $p(\theta|s_{obs})$ and $p(Mod|s_{obs})$. The rejection sampler described in Algorithm 20.1 is probably the simplest exponent of the ABC family.

Algorithm 20.1: An ABC Rejection Sampler

The most basic instance of an ABC sampler, targeting an approximation to $p(\theta|s_{obs})$, is the following rejection algorithm:

1. Sample M parameter vectors $\theta^1, \ldots, \theta^M$ from the prior $\pi(\theta)$.
2. For each parameter vector θ^i, with $i = 1, \ldots, M$, simulate a corresponding datasets Y^i from $p(y|\theta^i)$.
3. Transform the simulated datasets $Y^1, \ldots, Y^M$ to vectors of summary statistics $S^1 = S(Y^1), \ldots, S^M = S(Y^M)$.
4. Calculate the distances $d^i = d(s_{obs}, S^i)$, for $i = 1, \ldots, M$, using an appropriate distance measure $d(\cdot, \cdot)$.
5. Store the $N \leq M$ parameter vectors $\theta^{i_1}, \ldots, \theta^{i_N}$, whose corresponding distances $d^{i_1}, \ldots, d^{i_N}$ are all lower than a tolerance $h > 0$.

Notice that $\theta^{i_1}, \ldots, \theta^{i_N}$ are effectively a sample from:

$$\pi_{ABC}(\theta|s_{obs}) \propto p\{d(s_{obs}, s) < h|\theta\}\pi(\theta),$$

which should be a close approximation to $\pi(\theta|s_{obs})$, if h is sufficiently small.

In the remainder of this chapter, we focus on a particular family of intractable models: state space models. In Section 20.1.2, we briefly describe this class of partially observed models, which are very popular in the ecological literature, and we introduce two approaches that can be used to perform statistical inference for such models. In Section 20.1.3, we discuss how one of these approaches, synthetic likelihood (SL), differs from other ABC methods, while in Section 20.2, we consider the predator-prey model of Turchin and Ellner (2000) and compare the available methods using both simulated and field data. Finally, in Section 20.3, we conclude by making some practical considerations regarding the benefits and drawbacks of using ABC or SL, rather than less approximate methods, when working with state space models.

20.1.2 Inference for state space models

State space models (SSMs) represent a special class of models with hidden or partially observed states. In these models, the hidden states follow Markov processes, whose conditional pdf has the following property:

$$p(n_t|n_1, \ldots, n_{t-1}, \theta) = p(n_t|n_{t-1}, \theta), \tag{20.2}$$

where $t \in \{1, \ldots, T\}$ and θ is a vector of static parameters. Property (20.2) implies that the future states are statistically independent of the past, upon conditioning on the present. Generally, the hidden ecological processes are coupled with an observation process according to which observed data points are conditionally independent, given the underlying states (King, 2014)

$$p(y_t|n_t, y_1, \ldots, y_{t-1}, \theta) = p(y_t|n_t, \theta), \tag{20.3}$$

where we defined $y_t = y_{obs,t}$, to simplify the notation. Typically, the term SSMs is used to indicate partially observed Markov processes with continuous state spaces, while models with discretely valued states are called hidden Markov models (HMMs). In the following, we focus on SSMs, but most considerations apply also to hidden Markov models.

As for most partially observed systems, the likelihood of SSMs is generally not available directly. Indeed, for such models $p(y_{1:T}|\theta)$, where $y_{1:T} = \{y_1, \ldots, y_T\}$, is available analytically only if both $p(n_t|n_{t-1}, \theta)$ and $p(y_t|n_t, \theta)$ are linear and Gaussian (Kalman, 1960). Fortunately, the Markov property (20.2) mitigates the intractability of these models, because it allows estimation of the likelihood by performing the required T-dimensional integration efficiently. In particular, the Markov property is exploited by particle filters (Doucet and Johansen, 2009) to break down the integration problem into T sequential integration steps. These computational tools can be used to obtain Monte Carlo estimates $\hat{p}(y_{1:T}|\theta)$ of the full-likelihood function. We describe the Sequential Importance Re-Sampling (SIR) algorithm, which is the simplest instance of a particle filter, in Algorithm 20.2.

A more general solution to the problem of intractable likelihoods is offered by SL (Wood, 2010). This is a simulation-based and approximate approach, which is closely related to ABC methods. Rather than approximating the full likelihood function, SL transforms the data to a set of summary statistics s_{obs} and approximates $p(s_{obs}|\theta)$ parametrically. In particular, SL assumes that the summary statistics are approximately normally distributed, conditionally on the parameters:

$$S \sim \text{N}\{\mu(\theta), \Sigma(\theta)\}, \tag{20.4}$$

where the functions $\mu(\theta)$ and $\Sigma(\theta)$ are generally unknown. Given that the parametric density assumption does not hold exactly in general, the resulting synthetic likelihood, $p_{SL}(s_{obs}|\theta)$, should be considered an approximation to $p(s_{obs}|\theta)$. Point-wise estimates of the synthetic likelihood can be obtained by using the procedure described in Algorithm 20.3.

There exists a strong relationship between SL and the simulation-based approach of Diggle and Gratton (1984), who proposed to estimate the full likelihood $p(y_{obs}|\theta)$ pointwisely, by simulating data from the model and approximating its distribution using a non-parametric density estimator. Most ABC algorithms follow a less likelihood-centric approach, because they generally aim at sampling from $\pi(\theta|s_{obs})$ directly. This is the case, for instance, in ABC rejection, Markov chain Monte Carlo (MCMC) and sequential Monte Carlo (SMC) algorithms (Beaumont, 2010). Section 20.1.3 discusses how SL differ from other ABC methods in more details.

Algorithm 20.2: Sequential Importance Re-Sampling (SIR)

This algorithm was proposed by Gordon et al. (1993) and has been hugely successful in the context of SSMs. It implements a sequential importance sampling procedure, with a re-sampling step that is used to discard particles with low weights, thus mitigating the particle depletion problem (Doucet and Johansen, 2009). An estimate of the likelihood at θ can be obtained using the following steps:

1. Draw particles N_0^i, for $i = 1, \ldots, M$, from the prior distribution of the initial state $N_0^i \sim \pi(n_0)$.

2. For $t = 1$ to T:

   (a) *Prediction step*: propagate the particles forward in time:

   $$N_t^i \sim p(n_t | n_{t-1}^i, \theta), \quad \text{for} \quad i = 1, \ldots, M.$$

   (b) *Update step*: calculate the normalised weight of each particle:

   $$w^i = \frac{\tilde{w}^i}{\sum_{i=1}^{N} \tilde{w}^i}, \quad \tilde{w}^i = p(y_t | n_t^i, \theta), \quad \text{for} \quad i = 1, \ldots, M.$$

   (c) Estimate the current component of the likelihood:

   $$\hat{p}(y_t | y_{1:t-1}, \theta) = \frac{1}{M} \sum_{i=1}^{M} \tilde{w}^i.$$

   (d) Re-sample the particles multi-nomially with replacement, using probabilities equal to the normalised weights.

3. Estimate the likelihood using the decomposition:

$$\hat{p}(y_{1:T} | \theta) = \hat{p}(y_1 | \theta) \prod_{t=2}^{T} \hat{p}(y_t | y_{1:t-1}, \theta).$$

The point estimates $\hat{p}(y_{obs}|\theta)$ and $\hat{p}_{SL}(s_{obs}|\theta)$, obtained using, respectively, SIR and SL, can be used within a Metropolis–Hastings (MH) algorithm. Specifically, if SL is used, the MH acceptance probability is given by:

$$\alpha = \min\left\{1, \frac{\hat{p}_{SL}(s_{obs}|\theta^*) p(\theta|\theta^*) \pi(\theta^*)}{\hat{p}_{SL}(s_{obs}|\theta) p(\theta^*|\theta) \pi(\theta)}\right\}, \tag{20.5}$$

Algorithm 20.3: Evaluating the Synthetic Likelihood

Point-wise estimates of the synthetic likelihood, at an arbitrary position θ_p in the parameter space, can be obtained as follows:

1. Simulate M datasets $Y^1, \ldots, Y^M$ from the model $p(y|\theta_p)$ and transform them into d-dimensional summary statistic vectors $S^1 = S(Y^1), \ldots, S^M = (Y^M)$.

2. Estimate mean and covariance matrix of the summary statistics, using standard estimators:
$$\hat{\mu}(\theta_p) = \frac{1}{M}\sum_{i=1}^{M} S^i,$$
$$\hat{\Sigma}(\theta_p) = \frac{1}{M-1}\sum_{i=1}^{M}\{S^i - \hat{\mu}(\theta_p)\}\{S^i - \hat{\mu}(\theta_p)\}^T,$$
or possibly more robust alternatives.

3. Evaluate the corresponding Gaussian density at the observed statistics, that is:
$$\hat{p}_{SL}(s_{obs}|\theta_p) = (2\pi)^{-\frac{d}{2}}|\hat{\Sigma}(\theta_p)|^{-\frac{1}{d}} \times \exp\left[-\frac{1}{2}\{s_{obs} - \hat{\mu}(\theta_p)\}^T\hat{\Sigma}(\theta_p)^{-1}\{s_{obs} - \hat{\mu}(\theta_p)\}\right].$$

where $p(\theta^*|\theta)$ is the transition kernel and $\pi(\theta)$ is the prior density. When $\hat{p}(y_{obs}|\theta)$ is used in place of $\hat{p}_{SL}(s_{obs}|\theta)$ in (20.5), the resulting sampler is called a particle marginal Metropolis–Hastings (PMMH) algorithm (Andrieu et al., 2010). Under the assumptions detailed by Andrieu and Roberts (2009), this sampler targets $\pi(\theta|y_{obs})$, thus representing an exact-approximate algorithm. When SL is used, the situation is more complex because, unless the statistics are normally distributed across the parameter space, the resulting synthetic likelihood Metropolis-Hasting (SLMH) algorithm will target $\pi(\theta|s_{obs})$ only approximately.

The main drawback of using SLMH or PMMH is their high computational cost: the value of the (synthetic) likelihood function at the proposed parameters θ^* has to be estimated at each MH step, and this can be expensive for complex models. For this reason, Wilkinson (2014) and Gutmann and Corander (2015) avoid using SLMH, by explicitly approximating the synthetic likelihood function $\hat{p}(s_{obs}|\theta)$ using Gaussian processes. Their approaches clearly extend to situations where the likelihood is estimated using a particle filter. An additional complication of MH algorithms using noisy likelihood estimates is that they are often affected by poor mixing, because the sampler tends to get trapped when an unusually high estimate of the likelihood is reached (an ad hoc solution is to simply re-estimate the value of the (synthetic) likelihood at latest accepted position, θ, at every MH step). This problem is discussed by Doucet et al. (2015) and Sherlock et al. (2014), who study how to tune MH algorithms which make use of noisy and unbiased likelihood estimates.

Given the earlier issues, ABC methods might appear to be more efficient than SLMH or PMMH, because at each iteration they typically simulate only a single summary statistics vector from $p(s|\theta)$. However, the accuracy and the acceptance ratio of ABC samplers are, respectively, inversely and directly proportional to the tolerance h. This trade-off makes it is difficult to formulate a clear statement about the computational efficiency of ABC methods, relative to SLMH and PMMH.

While in Section 20.1.3, we discuss the merits and drawbacks of SL relative to other ABC methods, we come back to SLMH and PMMH in Section 20.2, where we use them to fit the SSM of Turchin and Ellner (2000) to ecological data.

20.1.3 SL versus tolerance-based ABC

The choice of summary statistics is crucial for the performance of ABC methods, hence, the topic has been the subject of much research. See Blum et al. (2013) for a comprehensive review of methods for dimension reduction or statistics selection. SL and ABC methods share some requirements regarding the choice of summary statistics. More specifically, in parameter, estimation problems, the summary statistics should contain as much information as possible about the parameters, so that $\pi(\theta|y_{obs})$ will be approximately proportional to $\pi(\theta|s_{obs})$.

Beside this common ground, SL differs from ABC methods in several ways, and this entails some diverging requirements on the summary statistics. In particular, reducing the number summary statistics is more critical to ABC methods than to SL. In fact, the non-parametric approach followed by most ABC methods, implies that the convergence rate of the resulting posterior distributions slows down rapidly as the dimension of the statistics vector increases (Blum, 2010). On the other hand, the parametric likelihood estimator used by SL ensures that this method is much less sensitive to the number of summary statistics used. This difference in scalability has important practical implications. In particular, SL allows practitioners to focus on the challenging task of identifying informative summary statistics, without having to worry too much about keeping their number low. Obviously SL's scalability in the number of statistics does not come without a cost, but it has to be paid for in parametric assumptions, whose effect might be hard to quantify.

Another potential issue with ABC algorithms, such as the rejection sampler in Algorithm 20.1, is that they often measure the distance between the observed and simulated statistics using a squared Mahalanobis distance:

$$d(s_{obs}, S) = ||s_{obs}, S||_A^2 = (s_{obs} - S)^T A(s_{obs} - S),$$

where A is a scaling matrix. The choice of A is fundamental when the summary statistics have very different scales or when there are subsets of highly correlated statistics. A possible solution is to simulate N vectors of summary statistic at some location θ_p in the parameters space and use the inverse of the empirical covariance matrix of the simulated summary statistics as scaling matrix $A = \hat{\Sigma}(\theta_p)^{-1}$. This simple choice works well in many cases, but it can lead to unsatisfactory results when the covariance of the summary statistics varies strongly with model parameters.

As an illustration of this problem, we consider a stochastic version of the Ricker map:

$$Y_t \sim \text{Pois}(\phi X_t), \quad N_t = rN_{t-1}e^{-N_{t-1}+Z_t}, \quad Z_t \sim N(0, \sigma^2),$$

where N_t is the population size at time t, r is the intrinsic growth rate of the population, ϕ is a scaling parameter, and Z_t can be interpreted as environmental noise. In the following, we employ the set of 13 summary statistics proposed by Wood (2010), who used them to fit this model with SL.

In order to quantify the importance of the scaling matrix A in this setting, we performed the following simulation experiment:

- Define a sequence of equally spaced values v_k, for $k = 1, 2, \ldots, 50$, ranging from 2.8 to 3.8.

- For each value v_k:

  1. Simulate a path $Y_{1:T}$ from the Ricker map, using $T = 50$ and parameter values $\log r = 3.8$, $\sigma^2 = 0.3$, and $\phi = 10$. Define $s_{obs} = S(Y_{1:T})$.
  2. Set the initial parameter vector θ_p to $\log r = v_k$, $\sigma^2 = 0.3$, and $\phi = 10$.
  3. Simulate 10^4 paths from the model using parameters θ_p, transform each of them into a vector summary statistics, and calculate their empirical covariance $\hat{\Sigma}(\theta_p)$.
  4. Sample $\pi_{ABC}(\theta|s_{obs})$ using the SMC-ABC routine proposed by Toni et al. (2009), where $\hat{\Sigma}(\theta_p)^{-1}$ is used as scaling matrix. We refer the reader to Toni et al. (2009) for details about this algorithm, but point out that this is a sequential scheme where the tolerance h is reduced at each step and that we terminated the algorithm when the acceptance ratio of the most recent iteration was below 1%.

We repeated the whole experiment seven times, and the results are illustrated in Figure 20.1. Here, the x-axis represents the value of $\log(r)$ at which the scaling matrix was estimated, while the y-axis represents the lowest tolerance h achieved before the termination of the SMC-ABC algorithm. This plot shows how crucial is the choice of scaling matrix in situations where $\Sigma(\theta)$ varies widely with θ: if the scaling matrix is not adequate, the tolerance cannot be reduced enough. In an applied ecological setting, where the true parameters

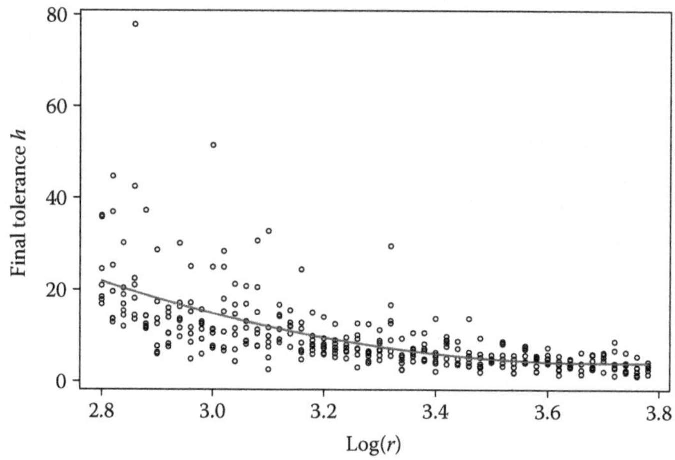

FIGURE 20.1
Lowest achievable tolerance h versus value of $\log r$ at which the scaling matrix is estimated. The grey line is a quadratic regression fit.

are unknown and the model of interest is more complex than the one used here, this means that a practitioner might struggle to find either a reasonable guess for the scaling matrix or a set of summary statistics whose covariance is not strongly dependent on θ.

Another choice that has to be made, in order to use tolerance-based ABC procedures, is the selection of h. The tolerance can be a small scalar constant, as in the MCMC-ABC algorithm of Marjoram et al. (2003), or it can be a vector of decreasing tolerances, as in the SMC-ABC algorithm of Toni et al. (2009). In order to obtain a better approximation to $\pi(\theta|s_{obs})$, h should be chosen to be as small as possible, but the acceptance probability will decrease with the tolerance. A common choice is to select a tolerance that allows a predetermined acceptance ratio to be achieved, but in some cases this strategy can lead to invalid results, as detailed in Silk et al. (2013).

The regression adjustment of Beaumont et al. (2002) can be used to mitigate the discrepancy between the observed and the simulated statistics, which is proportional to the tolerance h. However, the result of this correction is generally still dependent on h, which controls the bias-variance trade-off of the regression (Beaumont et al., 2002). Hence, using this procedure does not necessarily lead to higher accuracy in parameter estimation. For example, Fearnhead and Prangle (2012) obtained worse results with the regression correction than from the raw ABC output, using the Ricker model and the same summary statistics considered here.

SL is not afflicted by the difficulties just described, because it is tolerance-free, and the summary statistics are scaled automatically and dynamically by the empirical covariance matrix $\hat{\Sigma}(\theta)$. Obviously, this robustness comes at a cost: a single point-wise synthetic likelihood estimate requires a number of simulations sufficient to estimate the covariance matrix. In addition, even though for many commonly used statistics the central limit theorem (CLT) assures asymptotic normality, in small samples the normal approximation might be crude, while in some contexts it might be difficult to devise asymptotically normal statistics.

As a simple example of the former problem, let us consider a sample of size N from an exponential distribution with rate α. Here the maximum likelihood (ML) estimator of α is given by the reciprocal of the sample average:

$$s = \frac{1}{\bar{x}} = \left(\frac{\sum_{i=1}^{N} x_i}{N}\right)^{-1}.$$

Given that s is a sufficient statistic for α, the likelihood function can be factorised as follows:

$$p(x|\alpha) = h(x)f(s,\alpha) \propto f(s,\alpha),$$

hence, the likelihood is proportional to a function of only s and α. By the CLT, the distribution of s is asymptotically normal, but we want to verify how well we can approximate the likelihood using SL when $N = 10$. Figure 20.2 shows the log-likelihood (dashed) and the estimated synthetic log-likelihood (black) for $\alpha \in [0.5, 2]$. The true value of α is 1. With such a small sample size, the distribution of the simulated statistic is far from normal, and in fact, the synthetic log-likelihood is quite off target. In cases such as this, where the number of summary statistics is low, it is straightforward to use

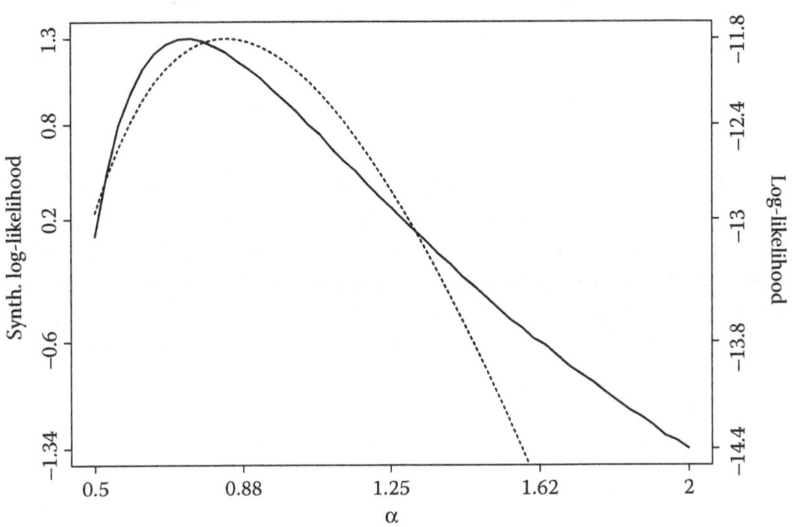

FIGURE 20.2
Synthetic log-likelihood function (black line) versus true log-likelihood function (broken line) for a Exp($\alpha = 1$) distribution.

transformations to improve to normality assumption, as proposed by Wood (2010). However, in an higher-dimensional setting, approximate multi-variate normality might be difficult to assess or improve. More importantly, achieving multi-variate normality for a certain set of parameters does not assure that this approximation will hold elsewhere in the parameter space.

20.2 Example: A Chaotic Prey–Predator Model

In order to illustrate the performance of SLMH and PMMH, we consider a modified version of the prey–predator model proposed by Turchin and Ellner (2000), which has been used to describe the dynamics of Fennoscandian voles (Microtus and Clethrionomys). More specifically, the model was an attempt at explaining the shift in voles abundance dynamics from low-amplitude oscillations in central Europe and southern Fennoscandia to high-amplitude fluctuations in the north. One of the possible drivers of this shift is the absence of generalist predators in the north, where voles are hunted primarily by weasels (Mustela nivalis) (Turchin and Ellner, 2000). According to this hypothesis, the lack of the stabilising effect of generalist predators is the main factor determining the observed instability of vole abundances in the north.

The predator-prey dynamics are given by the following system of differential equations (Turchin and Ellner, 2000):

$$\frac{dN}{dt} = r(1 - e\sin 2\pi t)N - \frac{r}{K}N^2 - \frac{GN^2}{N^2 + H^2} - \frac{CNP}{N + D} + \frac{N}{K}\frac{dw}{dt},$$

$$\frac{dP}{dt} = s(1 - e\sin 2\pi t)P - sQ\frac{P^2}{N}, \tag{20.6}$$

where $dw(t_2) - dw(t_1) \sim N[0, \sigma^2(t_2 - t_1)]$, with $t_2 > t_1$, is a Brownian motion process with constant volatility σ. The model is formulated in continuous time, because voles do not reproduce in discrete generations (Turchin and Hanski, 1997). Here N and P indicate vole and weasel abundances, respectively. In the absence of predators, voles abundance grows at a seasonal logistic rate. Parameters r and s represents the intrinsic population growth rates of voles and weasels, while K is the carrying capacity of the former. These parameters are averaged over the seasonal component, which is modelled through a sine function with amplitude e and period equal to one year, with peak growth achieved in the summer. Generalist predation is modelled through a type III functional response, under which generalists progressively switch from alternative prey to hunting voles, as vole density increases. The maximal rate of mortality inflicted by generalists is G, while H is the half saturation parameter.

Predation by weasels follows a type II response, where C is the maximal predation rate of individual weasels and D is the half saturation prey density.

No prey-switching behaviour occurs under this functional response, which is consistent with weasels being specialist predators. Weasel abundance grows at a seasonal logistic rate, where the carrying capacity depends on prey density. Parameter Q specifies the number of voles needed to support and replace an individual weasel, and it determines the ratio of prey to predator densities at equilibrium.

Differently from Turchin and Ellner (2000), who include environmental stochasticity in the system by randomly perturbing all model parameters using Gaussian noise with pre-specified volatility, we choose to explicitly perturb the prey equation using a Brownian motion process and to include its volatility σ in the vector of unknown parameters.

Vole abundance is not observed directly, but a proxy is provided by trapping data. We assume that the number of trapped voles is Poisson distributed:

$$Y_t \sim \text{Pois}(\Phi N_t),$$

where $t \in \{1, \dots, T\}$ is the set of discrete times when trapping took place. No such proxy is available for weasels density, hence, predator abundance represents a completely hidden state.

Following Turchin and Ellner (2000), model (20.6) is not fitted directly to data, but it is rescaled to a dimensionless form first. In particular, if we define:

$$n = \frac{N}{K}, \quad p = \frac{QP}{K}, \quad d = \frac{D}{K}, \quad a = \frac{C}{K}, \quad g = \frac{G}{K}, \quad h = \frac{H}{K}, \quad \text{and} \quad \phi = \Phi K,$$

the reduced system is given by:

$$\frac{dn}{dt} = r(1 - e\sin 2\pi t)n - rn^2 - \frac{gn^2}{n^2 + h^2} - \frac{anp}{n+d} + n\frac{dw}{dt},$$

$$\frac{dp}{dt} = s(1 - e\sin 2\pi t)p - s\frac{p^2}{N},$$

$$Y_t \sim \text{Pois}(\phi n_t). \tag{20.7}$$

While Turchin and Ellner (2000) implicitly re-scaled the simulations from the model, in order to match their means with that of the observed data, we formally estimate the scaling parameter ϕ.

Turchin and Ellner (2000) fitted the model by using a method which they call non-linear forecasting, which is an instance of simulated quasi-maximum likelihood method (Smith, 1993). One of the drawbacks of their estimation procedure is that it does not take into account the fact that trapping data provides noisy estimates of vole density. Another issue is that their method could not be used to estimate parameters that affect the variance of conditional distributions $p(n_t|n_{t-1}, n_{t-2}, \dots)$, but not their mean (Turchin and Ellner, 2000).

20.2.1 Description of data and priors

While Turchin and Ellner (2000) consider several datasets, here we focus on the time series concerning vole abundance (mainly Clethrionomys rufocanus) in Kilpisjarvi, Finland. The data, shown in Figure 20.3, consist of 90 data points collected during the springs (mid-June) (triangles) and autumns (September) (stars) of each year, between 1952 and 1997. Each data point represents the number of voles trapped in a specific trapping season, divided by the number of hundred trap-nights used in that season. After 1980, the number of trap-nights was fixed to around 1000, but in earlier years this number is not available: it varied from a minimum of 500 to more than 1000 (Perry, 2000). This correction for the sampling effort implies that, if the number of the trapped voles in each season is approximately Poisson distributed, the trapping index is not.

We have dealt with this problem by multiplying the data in Figure 20.3 by 10 and by rounding each data point to the nearest integer. This solution should give near-exact results for data collected after 1980, and a good

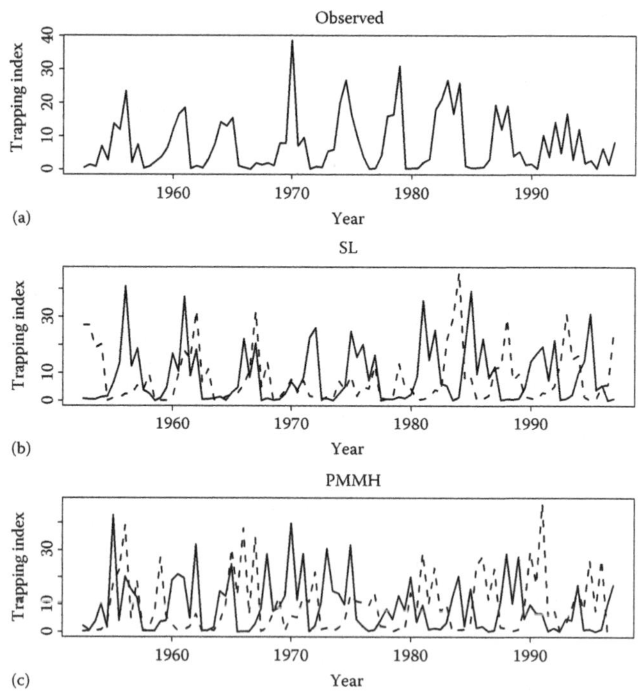

FIGURE 20.3
(a) Observed voles trapping index in Kilpisjarvi, between 1952 and 1997. (b and c) Two realisations (solid and dashed) of model 20.6, using parameters equal to the posterior means given by SLMH and PMMH.

TABLE 20.1
Priors Used for the Voles–Weasels Model

Parameter	Prior distribution
r	$N(\mu = 5, \sigma = 1)$
e	$N(\mu = 1, \sigma = 1)$
g	$\text{Exp}(\lambda = 7)$
h	$\text{Gamma}(\kappa = 4, \theta = 40)$
a	$N(\mu = 15, \sigma = 15)$
d	$N(\mu = 0.04, \sigma = 0.04)$
s	$N(\mu = 1.25, \sigma = 0.5)$
σ	$\text{Unif}(0.5, \infty)$
ϕ	$\text{Unif}(0, \infty)$

approximation for all data points representing a considerable population, thanks to the normal approximation to the Poisson distribution.

A useful source of prior information is represented by Turchin and Hanski (1997), where life history and data from short experiments were used to estimate the parameters of model (20.7). We report the prior distributions for each parameter in Table 20.1. The expected values of the prior distributions have been chosen on the basis of the remarks of Turchin and Hanski (1997), and we refer the reader to this reference for further details. The specific distributions and variabilities used for the priors have been chosen based on an attempt at quantifying the remarks of Turchin and Hanski (1997) regarding their confidence in their independently derived estimates. Admittedly, this process entails a certain degree of arbitrariness. No prior information was available for ϕ and σ, hence, we have used improper uniform priors for both parameters.

For SL, we used the following set of 17 summary statistics:

- Autocovariances of $n_1, \ldots, n_T$ up to lag 5.
- Mean population $\bar{n}$.
- Difference between mean and median population $\bar{n} - \tilde{n}$.
- Coefficients $\beta_1, \ldots, \beta_5$ of the regression $n_{t+1} = \beta_1 n_t + \beta_2 n_t^2 + \beta_3 n_{t-6} + \beta_4 n_{t-6}^2 + \beta_5 n_{t-6}^3 + z_t$.
- Coefficients of a cubic regression of the ordered differences $n_t - n_{t-1}$ on their observed values.
- Number of turning points, $\#n$.

This choice of statistics deserves some comments. Notice that, under suitable assumptions, all the earlier statistics are asymptotically normal as $T \to \infty$, due to the CLT. This provides some asymptotic justification to the Gaussian approximation used by SL. The autocovariances and the coefficient of the polynomial autoregressive model were meant to capture the dynamics of prey abundance on a short ($\beta_{1,2}$) and long ($\beta_{3,4,5}$) term basis. The degrees of the polynomials were choosed visually, by plotting n_t against n_{t-1} and n_{t-6}. Intermediate lags, such as n_{t-3}, were excluded, because they would have led to very strong correlations between the regression coefficients. The marginal distribution of n_t is summarised by $\bar{n}$ and $\bar{n} - \tilde{n}$, while the cubic regression coefficients aim at capturing the marginal structure of $n_t - n_{t-1}$. The number of turning points was introduced with the intention of capturing the volatility σ^2. This is because increasing σ^2 generally leads $\#n$ closer to 1/2, which is typical of random walk behaviour.

20.2.2 Comparison using simulated data

In order to verify the accuracy of SLMH and PMMH for this prey–predator model, we have simulated 24 datasets of length $T = 90$, using parameters values $r = 4.5$, $e = 0.8$, $g = 0.2$, $h = 0.15$, $a = 8$, $d = 0.06$, $s = 1$, $\sigma = 1.5$, and $\phi = 100$. We have then estimated the parameters with both methods, using 2.5×10^4 MCMC iteration, the first 5×10^3 of which was discarded as burn-in period, and 10^3 simulation from the model at each step. All the chains were initialised at the same parameter values. The resulting root mean squared errors (RMSEs) and variance-to-squared-bias ratios are reported in Table 20.2. While the RMSEs are quite similar for most parameters, the Table suggests that PMMH gives more accurate estimates for the scaling parameter ϕ and possibly for the generalist predation rate g. Indeed, SLMH estimates of ϕ are biased downward and are around ten times more variable than the estimates obtained with PMMH. In the case of g, the significance of the t-test should not be over-interpreted, given that it is attributable to PMMH achieving almost zero error on a single run.

TABLE 20.2
RMSEs and Variance-to-Squared-Bias Ratios (in Parentheses) for SLMH and PMMH. P-values for Differences in Log-Squared Errors Have Been Calculated Using t-Tests

Parameter	RMSE SLMH	RMSE PMMH	P-Value	Best
r	0.33(3.3)	0.25(9.9)	0.49	PMMH
e	0.19(0.1)	0.2(0.1)	0.78	SLMH
g	0.09(0.2)	0.08(0.5)	0.05	PMMH
h	0.04(0.2)	0.03(0.4)	0.15	PMMH
a	2.12(1.3)	1.97(1)	0.48	PMMH
d	0.02(0.5)	0.02(0.6)	0.57	SLMH
s	0.07(18.6)	0.08(10.9)	0.22	SLMH
σ	1.97(2.5)	0.71(2.1)	0.36	PMMH
ϕ	16.04(3.9)	4.85(7.4)	< 0.001	PMMH

From a computational point of view, the two algorithms performed similarly. In particular, on a single 2.50 GHz Intel i7-4710MQ CPU, point-wise estimates of $p(y_{obs}|\theta)$ or $p_{SL}(y_{obs}|\theta)$ cost around 1.55 and 1.35 seconds, when 10^3 particles or simulated statistics are used. This time difference is marginal, and probably highly dependent on implementation details. However, it is worth pointing out that it is much easier to parallelise the computation of $\hat{p}_{SL}(s_{obs}|\theta)$ than that of $\hat{p}(y_{obs}|\theta)$. This is because of SIR's re-sampling step, which breaks the parallelisms at each time-step t (see Algorithm 20.1). For a review of parallelisation strategies for the re-sampling step, see Li et al. (2015). A possibly simpler solution is to compute several estimates $\hat{p}_1(y_{obs}|\theta)$, $\dots$, $\hat{p}_C(y_{obs}|\theta)$ in parallel, by running SIR with a fraction of the total number of particles M on each of the C cores, and then averaging them at each PMMH step to obtain a single estimate of $p(y_{obs}|\theta)$.

20.2.3 Results from the Kilpisjarvi dataset

We fitted the Kilpisjarvi dataset using 1.5×10^5 MCMC iteration, of which the first 10^4 were discarded as burn-in period. At each step, we used 10^3 simulations from the model (SLMH) or particles (PMMH). The resulting posterior means are reported in Table 20.3, while the marginal posterior densities of the parameters as shown in Figure 20.4.

SLMH and PMMH give similar estimates for most parameters, with substantial differences only for σ and ϕ. Indeed, PMMH's estimate of the former parameter is much higher than that obtained using SL. Interestingly, Fasiolo et al. (2014) encountered a similar pattern when fitting the blowfly model of Wood (2010) to Nicholson's experimental datasets (Nicholson, 1954, 1957). In that context, the process noise estimates were much higher under PMMH than under SL, on all datasets. This biased PMMH's estimates of the remaining parameters towards stability, particularly on two of the datasets. As we will show later in this section, this stabilising effect of high process noise estimates on the dynamics is less noticeable here.

Figure 20.3 compares the observed data with trajectories simulated from model (20.6), using parameters equal to the posterior means given by SLMH

TABLE 20.3
Estimated Posterior Means (Standard Deviations) for Model 20.6

	r	e	g	h	a
SLMH	4.85(0.63)	0.78(0.12)	0.11(0.11)	0.1(0.05)	8.0(3.3)
PMMH	5.11(0.7)	0.84(0.14)	0.14(0.11)	0.1(0.05)	6.3(2.1)
	d	s	σ	ϕ	
SLMH	0.07(0.03)	1.04(0.21)	8.4(2.3)	270.5(63.5)	
PMMH	0.08(0.03)	1.04(0.23)	14.8(1.7)	184.2(26.9)	

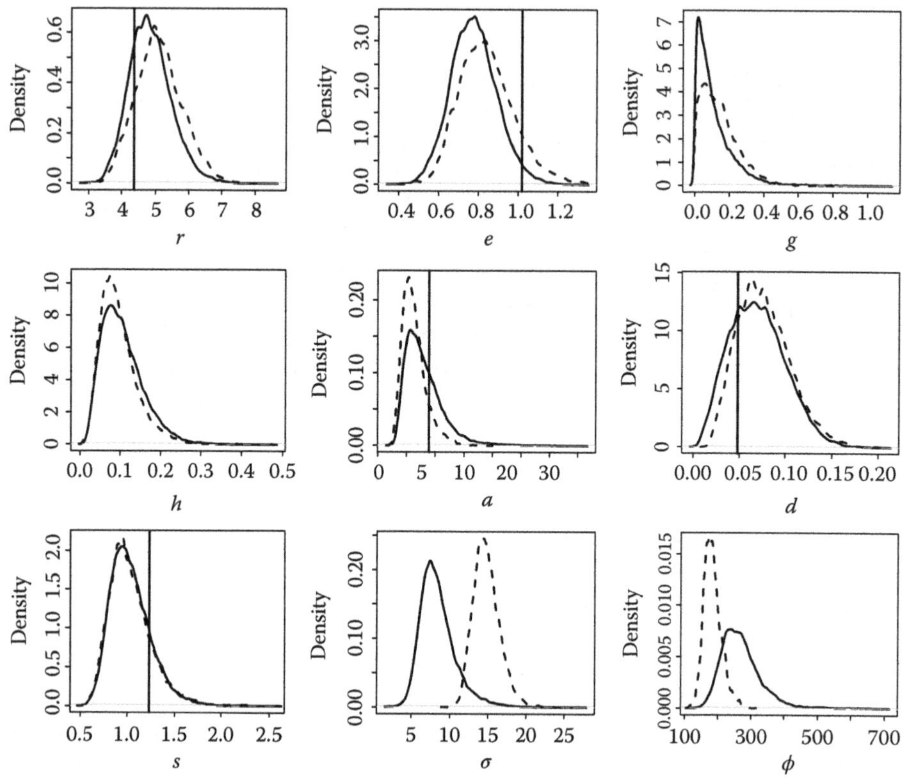

FIGURE 20.4
Marginal posterior densities for voles model using SLMH (black) and PMMH (broken). The vertical lines correspond to estimates reported by Turchin and Ellner (2000), obtained using NLF (available only for five parameters).

and PMMH. Both methods seem to produce dynamics that are qualitatively similar to the observed ones, with the paths simulated using PMMH's estimates being slightly more irregular, which is attributable to the higher process noise estimate.

Besides comparing observed and simulated trajectories, in the context of SL, it is also advisable to check whether the summary statistics are indeed approximately normally distributed. In particular, it is important to verify whether this assumption holds within the highest posterior density region. For this reason, we simulated $M = 10^4$ summary statistics, $S^{1:M} = \{S^1, \ldots, S^M\}$, from the model, using parameters equal to the estimated posterior mean. Then we used the methods of Krzanowski (1988) to produce the normality plots shown in Figure 20.5. In particular, we transformed $S^{1:M}$ to a variable that should be $\chi^2(17)$, under normality of S. Figure 20.5a compares observed and theoretical log quantiles. Departures on the right end of the plot indicate that

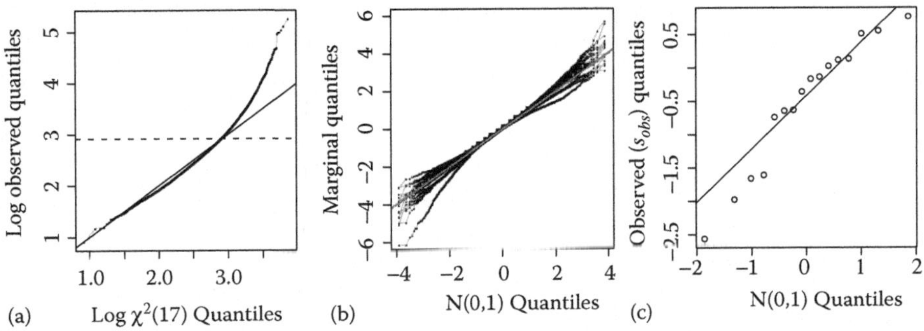

FIGURE 20.5
Normality plots for the simulated summary statistics. See main text for details.

the normal approximation is poor in the tails. In this case this is not much of a problem, as the dashed line, which indicates the Mahalonobis distance between s_{obs} and the sample mean of $S^{1:M}$, falls within the region where the normal approximation is adequate. Figure 20.5b shows marginal normal q-q plots for the simulated statistics. Marginal normality seems to hold reasonably well for most statistics, with the exception of β_5, whose distribution is skewed to the left. An analogous qq-plot for normalised observed statistics s_{obs} is shown in Figure 20.5c. This plot is suggestive of departures from normality, but this approach does not have much power, unless the dimension of s_{obs} is fairly large.

One of the main scientific questions model (20.6) was meant to address was whether the observed dynamic in vole densities can be classified as chaotic. To answer this question, we have randomly sampled 10^3 parameter sets from posterior samples obtained by SLMH and PMMH. We have then used each parameter set to simulate a trajectory from the deterministic skeleton of model (20.6) for 10^5 months, which were discarded in order to let the system leave the transient, and used additional 10^4 months of simulation to estimate the maximal Lyapunov exponent as in Wolf et al. (1985). By doing this, we obtained the two approximate posterior densities of the Lyapunov exponent shown in Figure 20.6. Notice that the posterior produced by PMMH is slightly more skewed to the left relatively to that obtained with SL, which suggests that the system dynamics are estimated to be more stable under the former methods. Together with the high estimate of σ^2, this confirms the tendency of PMMH to inflate the noise and to bias the estimated dynamics towards stability. While this effect was very pronounced under the blowfly model studied by Fasiolo et al. (2014), in this case it is very mild. Indeed, the median Lyapunov exponent is equal to -6×10^{-4} for SLMH and -0.015 for PMMH. These estimates are very close to each other and to the one (-0.02) reported by Turchin and Ellner (2000) for this dataset and provide more model-based evidence supporting the hypothesis that this system lives on the edge of chaos.

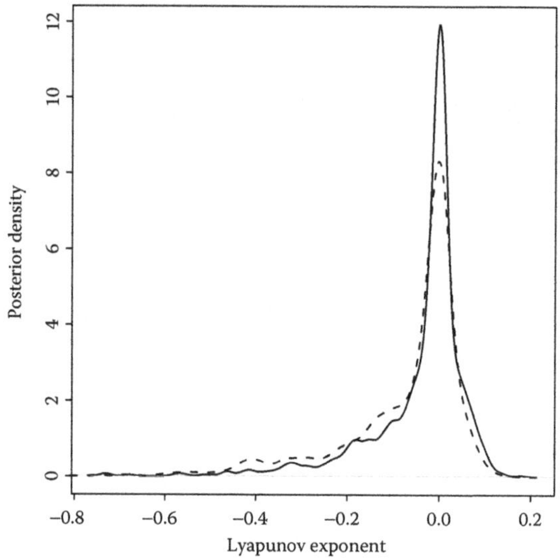

FIGURE 20.6
Approximate posterior densities of Lyapunov exponents for SLMH (black) and PMMH (broken).

20.3 Discussion

The example presented in this work gives the flavour of what can be accomplished using SL or particle filters, in the context of ecological SSMs. Both approaches provided a sample from the parameters' posterior distribution, which is the result of a full Bayesian analysis that incorporates both prior and likelihood-based information. While Table 20.2 suggests that SL might have lost information regarding some of the parameters, in Section 20.2.3, we point out that the estimates provided by PMMH might be slightly biased towards stability. In essence, PMMH estimates the process noise σ^2 to be quite high, which leads PMMH to explain the observed dynamics using noise, rather than by moving the system away from stability, using other dynamically important parameters. When dealing with highly non-linear models, it is worth being aware of this tendency, because it can lead to system dynamics being classified as stable even when they are not, as shown by Fasiolo et al. (2014), using the blowfly model of Wood (2010). Fortunately, SLMH and PMMH strongly agree in classifying the dynamics of the prey–predator system considered in this work as near-chaotic, hence, either approach could have been used to answer the main scientific question underlying model (20.6).

From the point of view of applied ecologists, ease of use and automation are arguably as important as statistical and computational efficiency.

In Section 20.2, we have shown that the choice of scaling matrix can be very important for ABC methods. Selecting this parameter correctly can be particularly difficult when little or no prior knowledge about model parameters is available. From this point of view, SL is at an advantage with respect to other ABC methods because, once the summary statistics have been selected, there is very little tuning to do. Obviously, SL pays for this tuning-free property with a normality assumption, which might result in lower accuracy.

The summary statistics selection process, which SL cannot escape, can be the most time consuming and arbitrary step of the inferential process. In the example presented in Section 20.2, we obtained good results, in terms of parameter accuracy, by using the statistics of Wood (2010) with some modifications. In our experience, this is the exception, rather than the rule. In fact, even though Blum et al. (2013) offer several systematic approaches for statistics selection, studying the model output by visualising characteristics such as empirical transition densities, periodicity, and dependencies between states is still indispensable for most models of reasonable complexity.

ABC methods have become popular tools for dealing with complex phylogeographic (Hickerson et al., 2010), phylogenetic (Rabosky, 2009), and individual based (Hartig et al., 2014) models, but they do not seem to have been equally successful for dynamical SSMs of ecological interest. The main reasons for this might be that particle filters represent an obvious alternative, and that at the moment it is not clear whether ABC methods can outperform them along any dimension of the inferential process. In fact, particle filters have the important advantage of using the full data, y_{obs}, thus, avoiding both the information loss and the issue of choosing the summary statistics. On the other hand, this use of all the data makes filtering more susceptible to model mis-specification problems, in which failure, to capture the data generating mechanism exactly can have a substantial negative impact on inference.

The robustness properties of methods based on summary or 'intermediate' statistics, in particularly the protection they can offer against model misspecification and outliers, has been widely recognised and exploited in econometrics, but it seems to have attracted less attention in the wider statistical community (Jiang and Turnbull, 2004). Hence, it would be interesting to verify whether ABC methods share any of the robustness properties of more traditional statistics-based approaches. If this turns out to be the case, one possibility is that ABC methods will be used in support of more accurate, but possibly less robust, methods based on the full likelihood, such as particle filters. This was suggested by Fasiolo et al. (2014), in the context of highly non-linear ecological and epidemiological models, and by Owen et al. (2014), who propose a hybrid procedure where an ABC sampler is used in support of a PMMH algorithm. While both works have suggested that ABC methods are more robust than particle filters to bad initialisations, the first one has also found that they are less affected by outliers and that they can provide reliable parameter estimates when dealing with highly non-linear models characterized by extremely multi-modal full likelihoods.

Although the use of summary statistics wastes information and requires an often time-consuming statistics selection process, ABC methods have some features that are very appealing from a practical perspective. In fact, they are purely simulation-based or 'plug-and-play' (Bhadra et al., 2011), because they only require simulation of data from the model and transformation to summary statistics. This property makes these methods general-purpose, because they can be used to fit any model for which a simulator is available, with little or no assumptions required. Hence, ABC methods can potentially accelerate the model development process: once the summary statistics have been chosen, testing new model versions requires only updating the simulator. In addition, this generality allows practitioners to explore models that violate the assumptions necessary for particle filters to work, such as Markovian dynamics or the tractability of the observational density $p(y_t|x_t)$.

Similar practical considerations hold also in regard to the programming effort necessary to implement each method. For models of moderate complexity, no ABC or particle filtering method can be entirely implemented in a traditional interpreted language (such as *R*). In fact, any of these methods requires at least part of the code to be written in a compiled language (such as *C/C++*). In the case of ABC methods, this is often simple to do, because the largest share of the computational time is spent simulating data and transforming it to summary statistics, so it is often sufficient to write only these steps in a compiled language. On the other hand, particle filters generally do not simulate whole datasets, but work in sequential steps, so it is difficult to isolate the parts of these algorithms that have to be implemented efficiently. This means that it might be necessary to write these procedures entirely in a compiled language, which slows down the model development and evaluation process.

For these reasons, software tools providing frameworks and algorithms for doing inference for SSMs are very useful to statistical ecologists. One such example is the *pomp* *R* package (King et al., 2014), which we used to set up the model described in Section 20.2. This package focuses mainly on tools based on particle filtering, but it offers also several approximate approaches, and it can greatly reduce the programming effort, if the model of interest fits the framework provided by the package. While statistical suites are available for tolerance-based ABC methods and for SL, such as the *EasyABC* (Jabot et al., 2014) and the *synlik* (Fasiolo and Wood, 2014) *R* packages, these do not focus on SSMs in particular, thus, reflecting the wide range of application of the underlying statistical methodologies.

In conclusion, ABC methods offer an approach to intractable ecological models that forgo information in exchange for generality and, possibly, robustness. While this trade-off has shown to be fruitful in many branches of ecological modelling (Hartig et al., 2011), particularly when the model is not intended to reproduce the data exactly, future work will determine whether ABC methods will play a major role in the context of SSMs, possibly alongside less approximate approaches.

Acknowledgements

This work was performed under partial support of the EPSRC grant EP/I000917/1 and EP/K005251/1.

References

Andrieu, C., A. Doucet, and R. Holenstein (2010). Particle Markov chain Monte Carlo methods. *Journal of the Royal Statistical Society: Series B (Statistical Methodology) 72*(3), 269–342.

Andrieu, C. and G. O. Roberts (2009). The pseudo-marginal approach for efficient Monte Carlo computations. *The Annals of Statistics 37*(2), 697–725.

Andrieu, C. and J. Thoms (2008). A tutorial on adaptive MCMC. *Statistics and Computing 18*(4), 343–373.

Beaumont, M. A. (2010). Approximate Bayesian computation in evolution and ecology. *Annual Review of Ecology, Evolution, and Systematics 41*, 379–406.

Beaumont, M. A., W. Zhang, and D. J. Balding (2002). Approximate Bayesian computation in population genetics. *Genetics 162*(4), 2025–2035.

Bhadra, A., E. L. Ionides, K. Laneri, M. Pascual, M. Bouma, and R. C. Dhiman (2011). Malaria in northwest india: Data analysis via partially observed stochastic differential equation models driven by lévy noise. *Journal of the American Statistical Association 106*(494), 440–451.

Blum, M. G. B. (2010). Approximate Bayesian computation: A nonparametric perspective. *Journal of the American Statistical Association 105*(491), 1178–1187.

Blum, M. G. B., M. A. Nunes, D. Prangle, and S. A. Sisson (2013). A comparative review of dimension reduction methods in approximate Bayesian computation. *Statistical Science 28*(2), 189–208.

Diggle, P. J. and R. J. Gratton (1984). Monte Carlo methods of inference for implicit statistical models. *Journal of the Royal Statistical Society. Series B (Statistical Methodology) 46*, 193–227.

Doucet, A. and A. M. Johansen (2009). A tutorial on particle filtering and smoothing: Fifteen years later. *Handbook of Nonlinear Filtering 12*, 656–704.

Doucet, A., M. Pitt, G. Deligiannidis, and R. Kohn (2015). Efficient implementation of Markov chain Monte Carlo when using an unbiased likelihood estimator. *Biometrika 102*, 295–313.

Farnsworth, K. D., U. H. Thygesen, S. Ditlevsen, and N. J. King (2007). How to estimate scavenger fish abundance using baited camera data. *Marine Ecology Progress Series 350*, 223.

Fasiolo, M., N. Pya, and S. N. Wood (2014). Statistical inference for highly non-linear dynamical models in ecology and epidemiology. *arXiv preprint arXiv:1411.4564*.

Fasiolo, M. and S. N. Wood (2014). *An Introduction to Synlik (2014). R Package Version 0.1.1.*

Fearnhead, P. and D. Prangle (2012). Constructing summary statistics for approximate Bayesian computation: Semi-automatic approximate Bayesian computation. *Journal of the Royal Statistical Society: Series B (Statistical Methodology) 74*(3), 419–474.

Gordon, N. J., D. J. Salmond, and A. F. M. Smith (1993). Novel approach to nonlinear/non-Gaussian Bayesian state estimation. *IEE Proceedings F (Radar and Signal Processing) 140*, 107–113.

Gutmann, M. U. and J. Corander (2015). Bayesian optimization for likelihood-free inference of simulator-based statistical models. *arXiv preprint arXiv:1501.03291*.

Hartig, F., J. M. Calabrese, B. Reineking, T. Wiegand, and A. Huth (2011). Statistical inference for stochastic simulation models – theory and application. *Ecology Letters 14*(8), 816–827.

Hartig, F., C. Dislich, T. Wiegand, and A. Huth (2014). Technical note: Approximate Bayesian parameterization of a process-based tropical forest model. *Biogeosciences 11*, 1261–1272.

Hickerson, M. J., B. C. Carstens, J. Cavender-Bares, K. A. Crandall, C. H. Graham, J. B. Johnson, L. Rissler, P. F. Victoriano, and A. D. Yoder (2010). Phylogeography's past, present, and future: 10 years after. *Molecular Phylogenetics and Evolution 54*(1), 291–301.

Jabot, F., T. Faure, and N. Dumoullin (2014). *EasyABC: Performing Efficient Approximate Bayesian Computation Sampling Schemes.* R package version 1.3.1.

Jiang, W. and B. Turnbull (2004). The indirect method: Inference based on intermediate statistics – a synthesis and examples. *Statistical Science 19*(2), 239–263.

Jonsen, I. D., J. M. Flemming, and R. A. Myers (2005). Robust state-space modeling of animal movement data. *Ecology 86*(11), 2874–2880.

Kalman, R. E. (1960). A new approach to linear filtering and prediction problems. *Journal of basic Engineering 82*(1), 35–45.

Kendall, B. E., S. P. Ellner, E. McCauley, S. N. Wood, C. J. Briggs, W. W. Murdoch, and P. Turchin (2005). Population cycles in the pine looper moth: Dynamical tests of mechanistic hypotheses. *Ecological Monographs 75*(2), 259–276.

King, A. A., E. L. Ionides, C. M. Bretó, S. P. Ellner, M. J. Ferrari, B. E. Kendall, M. Lavine et al. (2014). *pomp: Statistical Inference for Partially Observed Markov Processes* (*R Package*).

King, R. (2014). Statistical ecology. *Annual Review of Statistics and Its Application 1*(1), 401–426.

Krzanowski, W. (1988). *Principles of Multivariate Analysis.* New York: Oxford University Press.

Langrock, R., R. King, J. Matthiopoulos, L. Thomas, D. Fortin, and J. M. Morales (2012). Flexible and practical modeling of animal telemetry data: Hidden Markov models and extensions. *Ecology 93*(11), 2336–2342.

Li, T., M. Bolic, and P. M. Djuric (2015). Resampling methods for particle filtering: Classification, implementation, and strategies. *Signal Processing Magazine, IEEE 32*(3), 70–86.

Marjoram, P., J. Molitor, V. Plagnol, and S. Tavaré (2003). Markov chain Monte Carlo without likelihoods. *Proceedings of the National Academy of Sciences 100*(26), 15324–15328.

May, R. M. (1976). Simple mathematical models with very complicated dynamics. *Nature 261*(5560), 459–467.

Morales, J. M., D. T. Haydon, J. Frair, K. E. Holsinger, and J. M. Fryxell (2004). Extracting more out of relocation data: Building movement models as mixtures of random walks. *Ecology 85*(9), 2436–2445.

Nicholson, A. J. (1954). An outline of the dynamics of animal populations. *Australian Journal of Zoology 2*(1), 9–65.

Nicholson, A. J. (1957). The self-adjustment of populations to change. In *Cold Spring Harbor Symposia on Quantitative Biology*, Volume 22, pp. 153–173. Cold Spring Harbor, NY: Cold Spring Harbor Laboratory Press.

Owen, J., D. J. Wilkinson, and C. S. Gillespie (2014). Scalable inference for Markov processes with intractable likelihoods. *arXiv preprint arXiv:1403.6886*.

Perry, J. N. (2000). *Chaos in Real Data: The Analysis of Non-linear Dynamics from Short Ecological Time Series*, Volume 27. Dordrecht, the Netherlands: Springer.

Rabosky, D. L. (2009). Heritability of extinction rates links diversification patterns in molecular phylogenies and fossils. *Systematic Biology 58*(6), 629–640.

Sherlock, C., A. H. Thiery, G. O. Roberts, J. S. Rosenthal (2014). On the efficiency of pseudo-marginal random walk metropolis algorithms. *The Annals of Statistics 43*(1), 238–275.

Silk, D., S. Filippi, and M. P. H. Stumpf (2013). Optimizing threshold-schedules for sequential approximate Bayesian computation: Applications to molecular systems. *Statistical Applications in Genetics and Molecular Biology 12*(5), 603–618.

Smith, A. A. (1993). Estimating nonlinear time-series models using simulated vector autoregressions. *Journal of Applied Econometrics 8*(S1), S63–S84.

Toni, T., D. Welch, N. Strelkowa, A. Ipsen, and M. P. Stumpf (2009). Approximate Bayesian computation scheme for parameter inference and model selection in dynamical systems. *Journal of the Royal Society Interface 6*(31), 187–202.

Turchin, P. and S. P. Ellner (2000). Living on the edge of chaos: Population dynamics of fennoscandian voles. *Ecology 81*(11), 3099–3116.

Turchin, P. and I. Hanski (1997). An empirically based model for latitudinal gradient in vole population dynamics. *The American Naturalist 149*(5), 842–874.

Wilkinson, R. D. (2014). Accelerating ABC methods using Gaussian processes. *arXiv preprint arXiv:1401.1436*.

Wolf, A., J. B. Swift, H. L. Swinney, and J. A. Vastano (1985). Determining Lyapunov exponents from a time series. *Physica D: Nonlinear Phenomena 16*(3), 285–317.

Wood, S. N. (2010). Statistical inference for noisy nonlinear ecological dynamic systems. *Nature 466*(7310), 1102–1104.

21

ABC in Nuclear Imaging

Y. Fan, Steven R. Meikle, Georgios I. Angelis, and Arkadiusz Sitek

CONTENTS

21.1 Introduction to Nuclear Imaging 623
21.2 Compartmental Models in PET 627
21.3 Parameter Estimation in Compartmental Models 629
21.4 A Simple ABC Algorithm for Kinetic Models 631
21.5 Application to a Neurotransmitter Response Model 632
21.5.1 Prior and sampling distributions 634
21.5.2 Summary statistics selection 636
21.5.3 Tolerance level determination 638
21.5.4 Comparisons of different estimation methods 638
21.6 Conclusions and Discussions 643
References 644

21.1 Introduction to Nuclear Imaging

Nuclear imaging technologies produce non-invasive measures of a broad range of physiological functions, using externally detected electromagnetic radiation originating from radiopharmaceuticals administered to the subject. The main nuclear imaging modalities PET (positron emission tomography) and SPECT (single photon emission computed tomography) are the backbones of the field of molecular imaging, where they are used extensively in both clinical settings and pre-clinical research with animals to study disease mechanisms and test effectiveness of new therapies. Typical applications include glucose metabolism studies for cancer detection and evaluation and cardiac imaging, imaging of blood flow and volume, and it is one of the few methods available to neuroscientists to non-invasively study biochemical processes within the living brain, such as receptor binding, drug occupancy, or neurotransmitter release.

In nuclear imaging, a subject is administered with a small amount of radiopharmaceutical called a *tracer*. Radiation is created when the nuclei of the tracer decay and produce photons in the range 35–511 keV that are then detected by radiation sensitive detectors external to the subject. The energy of the photons must be high enough to allow the photon to leave the subject's body, but low enough to allow absorption in the detector. Photons can be produced either as a direct product of the nuclear reaction that occurred or indirectly. For example, SPECT is based on single photon detection produced by the decay of the radioisotopes. A gamma camera (Anger, 1964) is rotated around the subject and acquires photon counts at different projection angles, typically a full 180 degree set of projections are needed. Figure 21.1 shows two of the positions during data acquisition. On the contrary, PET is based on indirect photon detection produced by positron annihilation, where a radioactive decay produces a positron which annihilates with an electron and produces a pair of photons travelling in opposite directions along a straight line path. The photons are detected by a large number of detectors surrounding the subject, forming a ring. The PET detector ring is stationary, see Figure 21.2. Basic physical principles underlying PET and SPECT imaging and instrumentation for data acquisition can be found in the reviews of Cherry and Dahlbom (2004) and Wernick and Aarsvold (2004).

The goal of imaging is to study the concentrations of radioactive nuclei that are in the imaged object, assuming that they are attached to the

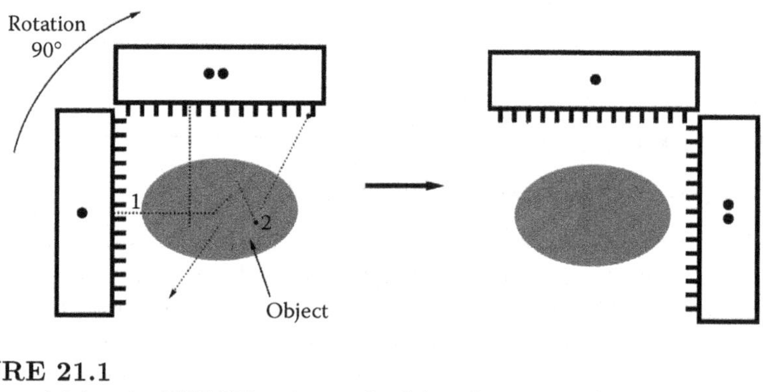

FIGURE 21.1
Data acquisition in SPECT using a dual head scanner. Each of two heads is an independent gamma camera (Anger, 1964). With this acquisition setup, the system needs only 90 degree rotation to acquire counts from all directions in a 2D plane around the object. There are many intermediate steps (in the order of 64) between configuration on the left and right at which the data are acquired. Dotted lines illustrate hypothetical paths of gamma photons which can be absorbed by the object (attenuated) (#2) or scattered and then detected (#1). These are two examples of many possible interactions. (Reproduced from Sitek, A., *Statistical Computing in Nuclear Imaging*, CRC Press, Boca Raton, FL, 2014. With permission.)

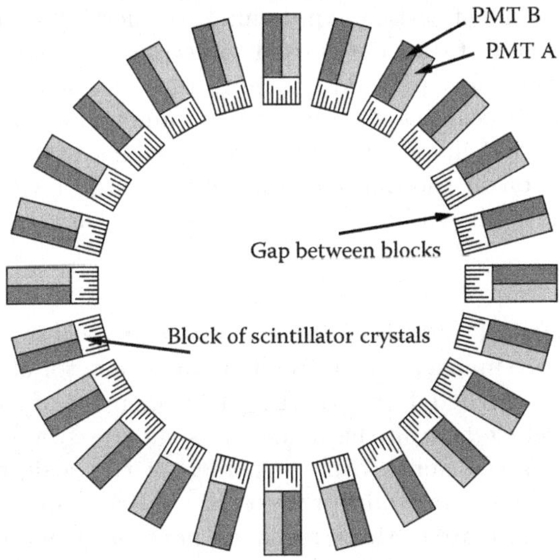

FIGURE 21.2
PET camera comprising 24 PET block detectors. Photomultiplier tubes (PMTs) are attached to scintillation crystals for signal amplification and interaction localisation. Two of the four PMTs per detector block needed to localise a gamma ray interaction are shown as PMT A and PMT B. (Reproduced from Sitek, A., *Statistical Computing in Nuclear Imaging*, CRC Press, Boca Raton, FL, 2014. With permission.)

tracer molecules. The description is simplified by subdividing the imaged volume into non-overlapping equal sized volume elements called *voxels*, and only the numbers of nuclei inside each voxel are considered. In a typical setup of the inverse problem, the quantity of interest is the expected number of decays per voxel, which leads to the data being modelled as Poisson distribution. Nuclear imaging can be extended to so called *dynamic imaging*, where changes in distribution of the tracer are investigated over time. This is done by dividing the time in which the changes are observed (typically in the order of 30 minutes to an hour) into time frames (about 20 to 60) and reconstruct voxelised time frames independently. The reconstructed time frames are then analysed by algorithms, such as the one presented in this chapter.

The voxelised *image* data are reconstructed from acquired data (counts) by any one of a number of reconstruction algorithms, see for example Qi and Leahy (2006). This is called the *tomographic reconstruction*. The primary factors limiting reconstructed image quality are detector resolution, which determines the maximum resolution of the reconstructed images; and the total number of detected counts, which determines the minimum noise level that can be achieved at the maximum resolution. In the following discussion, we restrict our attention to model-based reconstruction algorithms, which are of more interest to the statistical community. In model-based reconstruction approaches, a probabilistic model is used to account for the physical and geometric factors that affect photon detection. In its simplest form, the image reconstruction can be seen as a problem of parameter estimation, where the acquired data (counts) are Poisson random variables with mean equal to a linear transformation of the parameters. Let $\mathbf{y}$ be the measured projection data and $\mathbf{x}$ be the unknown image; $\mathbf{x}$ is related to $\mathbf{y}$ via:

$$E(\mathbf{y}) = \mathbf{P}\mathbf{x}.$$

The projection matrix $\mathbf{P}$ models the probability of an emission from each voxel element in the source image being detected at each detector element. In simple terms, $\mathbf{x}$ is the reconstructed image, where each element of the matrix $\mathbf{x}$ corresponds to a voxel element in the image. $\mathbf{y}$ is the measured count, assumed to have Poisson distribution. To give an idea of the scale of the problem, a single 3-dimensional scan could produce $10^7 - 10^8$ counts, with 10^6 image parameters to be estimated. Many methods have been proposed for solving the inverse problem, starting with the expectation maximisation (EM) algorithm of Shepp and Vardi (1982), to the ordered subsets algorithm of Hudson and Larkin (1994), to many more sophisticated algorithms which deal with the problem of ill-conditioning often arising in PET applications (where the solutions to the inverse problem are sensitive to small changes in the data). Fessler (1996), Leahy and Qi (2000), and Qi and Leahy (2006) provide detailed reviews on the statistical challenges in model-based reconstruction methods. Sitek (2014) discusses in depth the statistics of detected counts.

The 3-dimensional images of the radiotracer distributions, when monitored over time, provide insights into the physiological state of the organism in vivo. This dynamic imaging, often referred to as *functional imaging*, focuses on how tracers accumulate and clear from the tissue, enabling the physiological function associated with that tissue to be measured. Typically, the changes are characterised by using models of biological processes occurring in the voxel or the region of interest (ROI), which is a group of voxels corresponding to a particular anatomical region. As explained earlier, dynamic data are obtained by dividing the total acquisition time into intervals or *time frames*, and the data acquired in each time frame are reconstructed independently, representing the average concentration of the tracer in a voxel over the time interval. It is possible to process the data so that the correlation between time frames is taken into account, but in practice simpler approaches are used because of the ease of processing. Analyses of the 4-dimensional spatio-temporal dataset often proceeds by modelling the changes in concentration of the tracer using appropriate compartmental models of temporal data at each voxel or averaged groups of voxels (ROIs). These temporal data are termed the time activity curve (TAC). Compartmental models provide estimates of biologically meaningful parameters. The parameters of the models can be estimated for each voxel separately (as opposed to a group of voxels, ROI) to produce parametric images (one 3D image for each parameter) that describe important physiological information about the subject. An important consideration in parametric image estimation is robustness to noise, as noise in the voxel TAC can be high. Additionally, any estimation has to be very fast due to the large number of voxels associated with each image and large numbers of TACs to process.

In this chapter, we first briefly describe compartmental models in PET in Section 21.2, we then introduce a simple ABC algorithm in the context of PET kinetic modelling in Section 21.4. Section 21.5 provides a detailed example of ABC implementation for a neurotransmitter response model, and in Section 21.6, we conclude with some discussions about the potential of ABC in medical imaging.

21.2 Compartmental Models in PET

As discussed, PET is a technique so that given a time sequence of images, one can monitor the interaction of a particular radiotracer molecule with the body's physiological processes. For instance, blood flow can be measured by using radioactive water (with ^{15}O replacing ^{16}O in water $H_2{}^{16}O$ molecules, by bombarding them with protons) as a tracer, and metabolism can be measured with a radioactive glucose analog.

Kinetic models for PET typically derive from the one-, two-, or three-compartment model with a model input function. In PET, one normally assumes that all tissues in the body see the same input function, and this is typically a measured concentration of radioactivity in the blood plasma during the experiment. In compartmental modelling, it is assumed that within a voxel, whatever radioactive species contribute to the radioactive signal are in uniform concentration and can be characterised as being in one or more unique states. Assuming the system is in steady state, each of these states is assigned a compartment, which in turn is described by the rates of a change in concentration within a single ordinary differential equation. The coefficients of the differential equations or the kinetic parameters are reflective of inherent properties of the particular radiotracer molecule in the system, providing information about any hypothesised processes.

As an illustration of the compartmental model, consider the example given in Sitek (2014), Chapter 5. Figure 21.3 illustrates the possible physiological states of the tracer compound ^{18}FDG, a glucose analog. The compound is delivered to the blood, and transported into the cells. Three possible states can be identified: (1)^{18}F-Fluorodeoxyglucose (^{18}FDG, analog of glucose) within the plasma, (2) unmetabolised ^{18}FDG present in the cells or the interstitial spaces between cells, and (3) phosphorylated ^{18}FDG which is trapped in the cell (Figure 21.3). Compartmental models are then built by describing the connections between the states of the molecules, describing the influx to, and efflux from each compartment, in the form of ordinary differential equations.

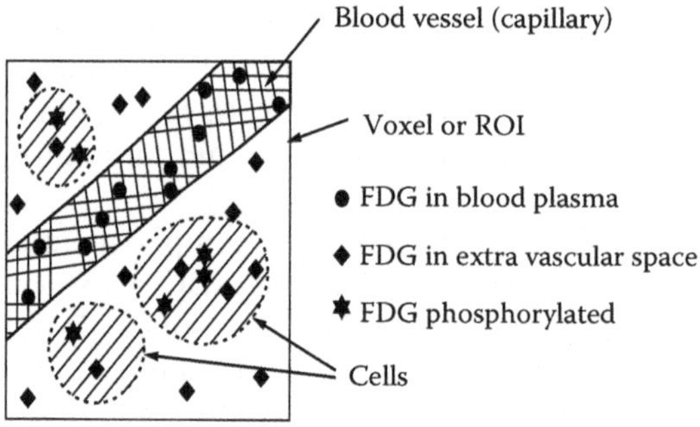

FIGURE 21.3
Representation of the voxel or ROI. The tracer (in this example FDG) is assumed to be in either of three states: in blood plasma, in the extra vascular space, or in a phosphorylated state within the cell. (Reproduced from Sitek, A., *Statistical Computing in Nuclear Imaging*, CRC Press, Boca Raton, FL, 2014. With permission.)

The one-tissue compartmental model is the simplest model that frequently arises in PET applications, describing the bi-directional flux of tracer between blood and tissue. See Figure 21.4 for a pictorial depiction of the model. The one-tissue compartment model is characterised by the tracer concentration in the tissue over time $C_t(t)$, the arterial blood (or blood plasma input function) $C_a(t)$, and two first-order kinetic rate constants (K_1, k_2). The tracer flux from blood to tissue is $K_1C_a(t)$ and the flux from tissue to blood is $k_2C_t(t)$, so the net tracer flux into tissue is given by the ordinary differential equation as:

$$\frac{dC_t(t)}{dt} = K_1C_a(t) - k_2C_t(t),$$

which is solved to obtain:

$$C_t(t) = K_1C_a(t) \otimes \exp(-k_2t), \tag{21.1}$$

where the symbol $\otimes$ denotes the 1-dimensional convolution. For a PET image, $C_t(t)$ is the measured radioactivity concentration in a voxel or ROI and $C_a(t)$

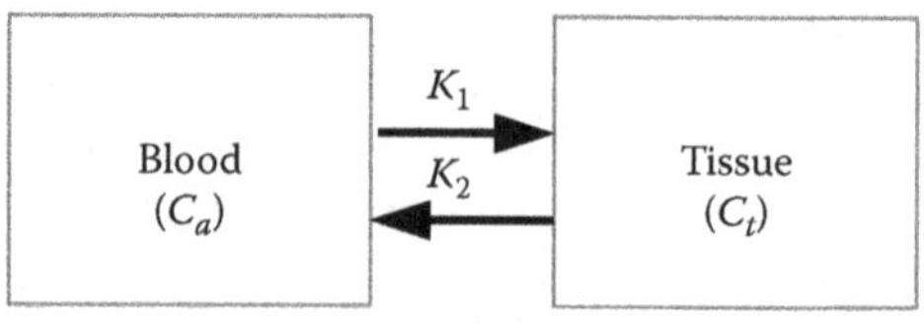

FIGURE 21.4
One-tissue compartment model describing the flow of the tracer between blood (C_a) and tissue (C_t). K_1 and k_2 are the kinetic rate constants, see equation 21.1.

is the arterial blood concentration of the tracer measured in a sample drawn during a scan. If the PET data are not corrected for physical decay, the parameter k_2 includes a component of radioactive decay. For further interpretation of kinetic rate parameters, see Morris et al. (2004).

More complex compartmental models distinguish different biochemical or physiological states of the tracer in tissue. After entering a cell, the tracer is available for binding in a free form at the concentration $C_1(t)$. Free tracer can specifically be bound to its target molecule, with concentration $C_2(t)$, but it may also specifically bind to some cell components that are not known in detail, $C_3(t)$. The system of differential equations can be derived analogously to the one-tissue compartment model, but is much more complex, with six unknown parameters that may be difficult to estimate. In practice the system is often reduced to a two-compartment model by treating free and non-specifically bound tracer as a single compartment, provided that the rates of exchange between the free and bound states are sufficiently rapid compared with the net influx into the combined compartment. Authoritative reviews on the subject can be found in Morris et al. (2004), Innis et al. (2007), and Gunn et al. (2015).

21.3 Parameter Estimation in Compartmental Models

The amount of data available to fit the model is relatively small, typically around 20–40 time points per voxel or ROI. The estimation of parameters based on these data, sometimes ten or more of them, can be non-trivial. In more realistic and complex models, parameter identifiability becomes an issue due to the sparsity of data. Therefore, the adoption of a particular model is by necessity a simplification of the truth (Gunn et al., 2002). Robustness of parameter estimation in the presence of high level of noise, particularly in voxel-wise estimations, where the noise-to-signal ratio can be high, poses another difficulty. In addition, since a separate estimation procedure has to be performed for each voxel, this might typically be around one million voxels, computational speed needs to be taken into consideration.

A typical approach to parameter estimation in kinetic modelling proceeds via a variety of least squares fitting procedures (Carson, 1986; Feng et al., 1996), weighted integration (Carson et al., 1986), or basis function techniques (Gunn et al., 1997). Many authors have commented on the difficulties with using non-linear least squares methods, particularly with noisy data, often failing to converge, producing estimates with large variances (which can be the case even in noiseless data) (Gunn et al., 2002; Alpert and Yuan, 2009). This has led to methods that employ penalised optimisation to stablise parameter estimates (Zhou et al., 2001; Gunn et al., 2002).

Whilst the limitations of the basis function technique of Gunn et al. (1997) are well known, its simplicity and ease of implementation has made it a preferred method for parameter estimation of kinetic models for PET data. The basic idea is to linearise the kinetic equation, and then use (weighted) least squares methods to obtain parameter estimates. Consider, for example, the one-tissue compartment model (21.1), the parameter K_1 is linear, whilst the parameter k_2 is non-linear. The non-linear term is then dealt with by choosing a discrete spectrum of parameter values for k_2 and forming the corresponding basis functions:

$$B_i(t) = C_a(t) \otimes \exp(-k_2^i t),$$

for $i = 1, \dots, n$, where the values of k_2^i are taken from a physiologically plausible range of values for k_2. Equation (21.1) then becomes linear in K_1, where:

$$C_t^i(t) = K_1^i B_i(t).$$

The parameters K_1^i can now be solved for each basis function $B_i(t)$ using linear least squares, and the parameter set (K_1^i, k_2^i) that produces the minimum residual sum of squares is taken as the optimal solution, (Cunningham and Jones, 1993; Meikle et al., 1998). Gunn et al. (1997) reported that in their experimentation, only 100 basis functions were needed to obtain good results, making the method very time efficient. It is interesting to note that the idea of fitting a spectrum of values of k_2^i and then choosing the most likely value according to some goodness of fit criterion is very similar to ABC, where ABC formalises the selection of the candidate parameter set with a prior distribution. Whilst the method of Gunn et al. (1997) is not formally Bayesian, the authors note the superior performance of the estimation when a constraint or a bounded region is placed on the non-linear parameters, thus implicitly placing a prior distribution on the unknown parameters.

The scarcity of data in kinetic modelling lends itself naturally to Bayesian modelling, where inclusion of priors can provide better estimates. This approach has been advocated more recently by several authors (Alpert and Yuan, 2009; Zhou et al., 2013; Malave and Sitek, 2015). Most applications of Bayesian modelling in medical imaging proceed in a frequentist fashion, that is, one often simply finds the maximum a posteriori estimate of the posterior using any number of optimisation tools, see for example Lin et al. (2014). Recently, Malave and Sitek (2015) and Sitek (2014) have advocated a proper treatment of Bayesian inference in the medical imaging community, given that uncertainty quantification is particularly relevant when the observational data has a very low signal-to-noise ratio.

Typically, the full Bayesian inference proceeds by assuming an error model for the time activity curve. The most common model is the independent Gaussian error model, with the variance at each time point assumed to be proportional to the observed data point. Markov chain Monte Carlo (MCMC) is the default posterior sampling method. However, despite its wide usage, the Gaussian error model is often not appropriate. Zhou et al. (2013) found that a t-distribution worked better for the examples they studied. In reality, the error distribution is highly positively skewed at time points with low activity if a non-negativity constraint is used with reconstruction and more symmetric at higher activity time points. In simulation studies, Poisson error is often introduced to the deterministic data. A second difficulty is that MCMC itself requires tuning and convergence assessment. While the former can be automated to some extent by automatic tuning algorithms (Garthwaite et al., 2016), the latter would ideally require repeat analyses at dispersed starting points. This can be computationally infeasible when the analyses involves hundreds of thousands of repeat simulations.

21.4 A Simple ABC Algorithm for Kinetic Models

ABC offers an alternative to MCMC. Traditionally, ABC is used when the likelihood function is not tractable. In the current setting, ABC offers a way of computing full Bayesian analyses without the need to specify an exact error distribution: we only require the ability to simulate summary statistics. The most obvious advantage is its ease of interpretation and application, which makes fully Bayesian inference easily achievable for practical users of Bayesian methodology. In this chapter, we will restrict our attention to the simplest of ABC algorithms, the standard rejection sampling method. For the parameter vector $\boldsymbol{\theta} = (\theta_1, \ldots, \theta_p)'$, this is achieved by the following three steps:

1. Sample parameters $\theta_i, i = 1, \ldots, p$ from the sampling distribution, $\text{Uniform}(a_i, b_i)$
2. Compute $\hat{C}_t(t)$ using $\boldsymbol{\theta}$, and the corresponding S_{sim}
3. Retain $\boldsymbol{\theta}$ if $\sum_t |S_{sim}^t - S_{obs}^t| < \epsilon$

The sampling distributions $\text{Uniform}(a_i, b_i)$ are proportional to the prior distributions for each parameter, $\text{Uniform}(a_i^*, b_i^*)$, we will discuss how to obtain a good sampling distribution in Section 21.5.1. $\hat{C}_t(t)$ is the estimated activity concentration, using the trial value of $\boldsymbol{\theta}$. For example, $\boldsymbol{\theta} = (K_1, k_2)$, if using Equation (21.1); S_{sim}^t and S_{obs}^t are the simulated and observed summary statistics, respectively, at the t-th time point. ϵ is a pre-determined error tolerance value. The choice of summary statistics will be discussed in Section 21.5.

It is clear from the above, that in repeated estimations for different voxels, steps 1 and 2 do not need to be repeated. This is because the values $\hat{C}_t(t)$ computed for one voxel can be re-used for others and the additional computational cost in step 3 is relatively small.

In this algorithm, we have not replicated the noise in the data. Since we are not interested in estimating the parameters in the error distribution, those are considered nuisance parameters. What we assume here is that there exist summary statistics that are (nearly) sufficient for the kinetic parameters. We will discuss the selection of summary statistics in more detail in the example section.

This simple form of ABC is similar to the popular basis function approach of Gunn et al. (1997), where the summary statistics are just taken as the original data. ABC formalises the constraints on the parameters in the form of a prior and, instead of using least squares for some of the parameters, ABC samples all parameters. In addition, the ABC method provides parameter uncertainty estimation by probabilistically retaining some of the sampled parameters.

21.5 Application to a Neurotransmitter Response Model

Development of neurochemical assays that capture temporal signatures is critical because the neurotransmitter dynamics may encode both normal and abnormal cognitive or behavioural functions in the brain. The elucidation of specific patterns of neurotransmitter fluctuations are beneficial to the study of a wide range of neuropsychiatric diseases, including alcohol and substance abuse disorders (Morris et al., 2005; Normandin et al., 2012).

Morris et al. (2005) developed a new model, called ntPET, for quantifying time-varying neurotransmitter concentrations. The new model enhances the standard tracer kinetic model, accounting for both time-varying dynamics of the radiotracer [^{11}C]raclopride and the endogenous neurotransmitter dopamine that competes with it for the same D2 receptor binding sites. For the input function, a reference region approach is used instead of arterial sampling, where the activity concentration measurements in the reference region of tissue are assumed to contain negligible specific binding signal (Morris et al., 2004). Experimental data are acquired in two separate PET scans, one conducted with the subject at rest and the other immediately following a stimulus. Normandin et al. (2012) further developed this model to be used with a single scan session and proposed a basis function approach for the simplification of computation; they call the method lp-ntPET (linear parametric-neurotransmitter PET). In our simulation studies, we will generate simulated data using ntPET, and fit the model lp-ntPET to the simulated data, since the latter is a simplification of the former.

The operational equation for the lp-ntPET model takes the form:

$$C_t(t) = R_1 C_R(t) + k_2 \int_0^t C_R(u)du - k_{2a} \int_0^t C_t(u)du - \gamma \int_0^t C_t(u)h(u)du, \tag{21.2}$$

where $C_t(t)$ and $C_R(t)$ are the concentration of the tracer in the target tissue and reference regions, respectively. The parameters R_1, k_2 and k_{2a} describe the kinetics of tracer uptake and retention in the tissue. The parameter γ describes the neurotransmitter response magnitude.

The function $h(t)$ describes the non-steady state component of the kinetic model (with γ encoding the magnitude), given by:

$$h(t) = \left(\frac{t - t_D}{t_P - t_D}\right)^{\alpha} \exp\left(\alpha\left[1 - \frac{t - t_D}{t_P - t_D}\right]\right) u(t - t_D),$$

where $u(t)$ is the unit step function. The variable t_D is the delay time at which the response starts relative to the start of scan, t_P is the peak time of maximal response magnitude, and α is the sharpness of the function. The lp-ntPET model has seven parameters, four that describe tracer kinetics and response magnitude $(R_1, k_2, k_{2a}, \gamma)$, and three describing the time course of the neurotransmitter/activation response (t_D, t_P, α). This formulation is a simplification of the ntPET model which has eleven parameters.

Equation (21.2) can be expressed in matrix form $y = Ax$, as:

$$\begin{bmatrix} C_t(t_1) \\ \vdots \\ C_t(t_m) \end{bmatrix} = \begin{bmatrix} C_R(t_1) & \int_0^{t_1} C_R(u)du & -\int_0^{t_1} C_t(u)du & -\int_0^{t_1} C_t(u)h(u)du \\ \vdots & \vdots & \vdots & \vdots \\ C_R(t_m) & \int_0^{t_m} C_R(u)du & -\int_0^{t_m} C_t(u)du & -\int_0^{t_m} C_t(u)h(u)du \end{bmatrix} \times \begin{bmatrix} R_1 \\ k_2 \\ k_{2a} \\ \gamma \end{bmatrix}. \tag{21.3}$$

So, for fixed values of t_P, t_D, and α in the function $h(t)$, the earlier representation can be solved using linear least squares.

Normandin et al. (2012) propose an efficient computational algorithm for parameter estimation for lp-ntPET. The idea is similar to the basis function method of Gunn et al. (1997). Setting the basis function to be $B_i(t) = \int_0^t C_t(u)h_i(u)du$ (this corresponds to the last column entry of the matrix A), then for basis function $B_i(t)$, a solution is obtained for equation (21.3), where $\hat{x} = (A^T WA)^{-1} A^T Wy$ with the weight matrix having diagonal elements inversely proportional to the variance of the PET measurement of C_t in the matching row of the matrix equation, since it is commonly assumed that the variance of the tracer concentration is proportional to the observed value. A similar assumption is made in most Bayesian models using Gaussian error assumption, see for example Zhou et al. (2013). Clearly, in the presence of high noise, such an assumption can lead to poor parameter estimation. Finally, a large library of basis functions are calculated over different combinations of t_D, t_P, and α, and the parameter set that minimises the residual sum of squares is then chosen as the final estimate. If the non-negativity constraint is to be used, for example, for the parameter γ, then an iterative weighted least squares approach is adopted.

In the next section, we consider the application of ABC to the problem of neurotransmitter response modelling described earlier. We obtain simulated data, using the nt-PET model, and use ABC to fit the simpler lp-ntPET model to the data at varying levels of noise. The noise is Poisson with a mean proportional to the simulated activity concentration. Simulation data are obtained over 60 time frames each with one minute duration.

21.5.1 Prior and sampling distributions

For simplicity, we use the Uniform distributions $U(a_i^*, b_i^*), i = 1, \ldots, 7$ as the prior distributions for the seven unknown parameters $(R_1, k_2, k_{2a}, \gamma, t_D, t_P, \alpha)$, all of which are non-negative. In practice, the investigator may have a rough idea of the range of plausible values for the parameters. In this example, we set the priors as $U(0, 20), U(0, 10), U(0, 10), U(0, 5)$ for the first four parameters. For parameters t_D, t_P, α, Normandin et al. (2012) discussed the choice of priors for these parameters and found that the response to a stimulus at 20 minutes should occur before 25 minutes. Here, we use for t_D a flat prior around the value 20, so $t_D \sim U(15, 25)$. This is reasonable to do in most cases because the displacement modelled by $h(t)$ is caused by an external stimulus that the experimenter controls and commences at a known time, for example, a drug injection at 20 minutes. We set the priors for t_P as $U(t_D + 1, 35)$ and $\alpha \sim U(0, 25)$; these are essentially the largest numerical ranges that produce sensible simulated data.

For an efficient ABC algorithm, we require a good sampling distribution $U(a_i, b_i)$. A good starting point for the sampling distribution is to use the prior distributions, for example, set $a_i = a_i^*, b_i = b_i^*$. This is typically too diffuse for the algorithm to work efficiently, unless the prior happens to concentrate around the highest density regions of the posterior. Here, we employ a sequential method of narrowing down the range, for example, finding values $a_i \geq a_i^*$ and $b_i \leq b_i^*$. We begin by applying the ABC algorithm of Section 21.4, starting with $a_i = a_i^*$ and $b_i = b_i^*$, and a large initial tolerance level of $\epsilon = 200$. The tolerance is gradually reduced to around ten, over several intermediate steps. With each reduction in the ϵ value, we use the parameter range obtained from the ABC algorithm at the previous iteration to define new a_i and b_i. The samples after each of the first three iterations are plotted in Figure 21.5, for R_1, k_2, and k_{2a}. For example, for k_{2a} shown in the right panel, the first iteration used $U(0, 10)$ as the sampling distribution, with a tolerance of $\epsilon = 200$. Applying the algorithm of Section 21.4, the range for this parameter has reduced to between 0 and 0.8, as indicated by the solid line. At the next iteration, we use $U(0, 0.8)$ as the new sampling distribution, with a tolerance of $\epsilon = 50$; the dotted line indicates the range for this parameter after the second iteration, which will then form the sampling distribution for the next iteration, and so on. The process is then continued until we obtain a

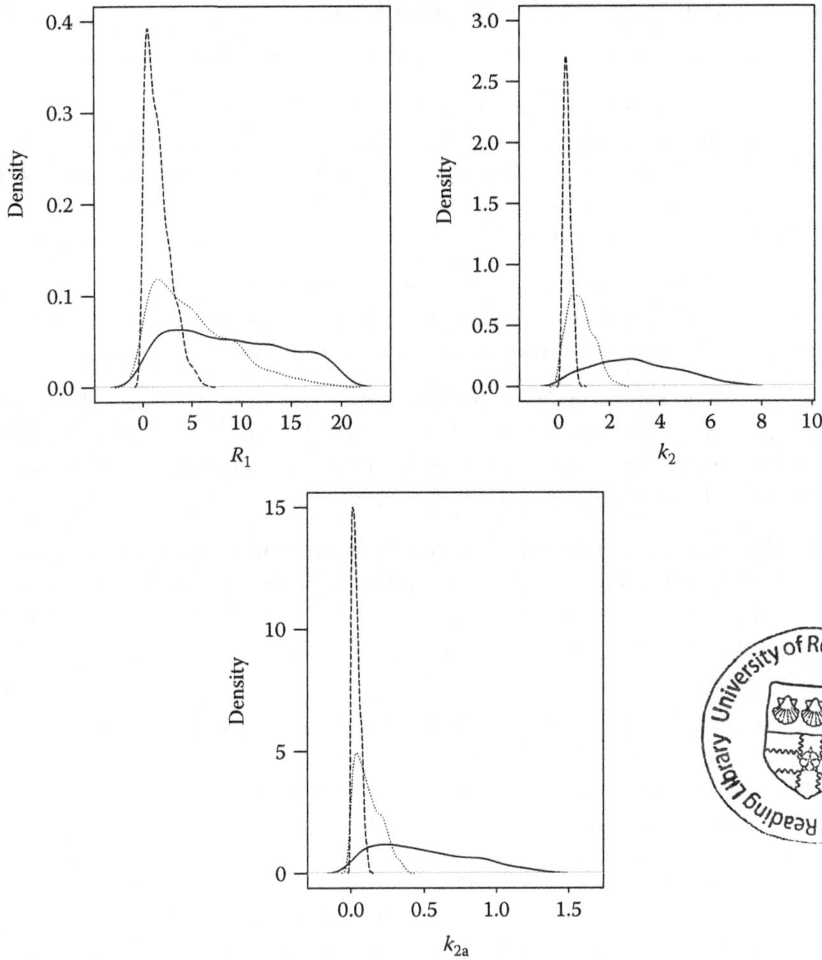

FIGURE 21.5
Samples for R_1, k_2, and k_{2a} after each of the first three iterations, indicated by solid, dotted, and dashed lines, respectively. Their respective tolerance levels are $\epsilon = 200, 50, 10$.

reasonably informative range for $U(a_i, b_i)$. In our simulated dataset, the final sampling distributions were $R_1 \sim U(0,5)$, $k_2 \sim U(0,1)$, $k_{2a} \sim U(0,0.2)$, $\gamma \sim U(0,2)$. Note that this sequential procedure is valid with the algorithm in Section 21.4, as long as the sampling distribution is proportional to the prior. A more elaborate sequential sampling scheme can be found in Sisson et al. (2007).

21.5.2 Summary statistics selection

We consider four different summary statistics, $S_1, \ldots, S_4$:

- S_1: Spline smoothed data. This is obtained by using the `R` package's `smooth.spline` function, using cross validation. The discrepancy between observed and simulated data is taken as the sum of the absolute differences between the smoothed observed data and the smoothed simulated data over each time point.

- S_2: The full dataset. The discrepancy between observed and simulated data is taken as the sum of the absolute differences between the raw observed data and the simulated data over each time point.

- S_3: The scaled dataset. The discrepancy is the sum of the absolute differences between the raw observed data and the simulated data, where the error at each time point is now scaled by the empirical estimate of the standard deviation of the raw difference.

- S_4: The weighted least squares. For each simulated sample of t_D, t_P, and α, the weighted least squares estimate of R_1, k_2, k_{2a}, and γ is estimated for the observed data and simulated data, the discrepancy is taken as the sum of the absolute difference between the four weighed least squares estimates.

The spline smoothed data can be considered as sample means at each data point and should be nearly sufficient for the parameters of interest. Figure 21.6 (top two rows) shows the TACs for two different activation levels (200% of baseline activation in the top row and 100% in the second row, over three different noise levels, ranging from high to low, shown from left to right). The dotted lines in the figures indicate the raw data, the dashed lines are the spline smoothed estimates of the raw data, and the solid lines are the true (noiseless) curves. These plots indicate that the spline estimate is very close to the true curve, particularly in low noise level cases, and even in the case of very high noise, it still provides very good estimate of the true TAC.

Similarly Figure 21.6 (bottom two rows) shows the simulated full dataset indicated by dashed lines. The plotted simulated dataset is estimated at a given set of parameter values (not necessarily optimal for the datasets plotted). We can see that at large noise levels, the simulated dataset cannot expect to fully replicate the original dataset, as we do not simulate noise here. Therefore in any ABC applications, when the raw data are used in this way, we do not expect the tolerance to be able to go to zero. In the lower noise levels, the discrepancy between the simulated and observed data is less marked, as would be expected.

In order to assess which summary statistics performed best, we considered the use of posterior predictive distributions, Gelman et al. (2004). For each summary statistic, we obtain 1 million samples from the sampling distribution and for each statistic, retain the 1,000 samples with the smallest error

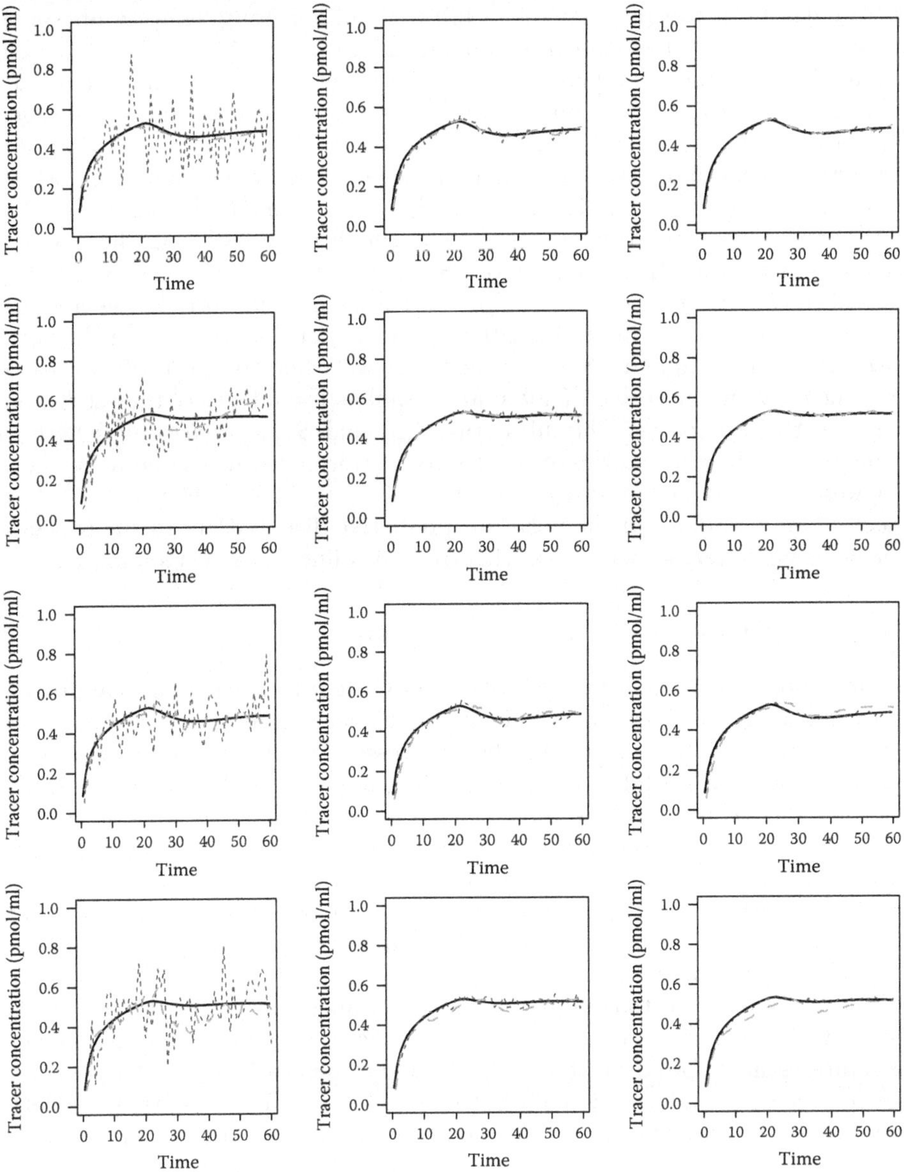

FIGURE 21.6
Rows top to bottom correspond to simulated TAC using the model with 200% and 100% activation. Columns from left to right correspond to high to low noise levels. True mean curves (solid line), observed noisy data (dotted line), and smoothed data (dashed line). S_1: top two rows. S_2: bottom two rows.

as samples from the posterior. The initial 1 million samples were obtained by setting $\epsilon = 10$, based on the Euclidean distance between observed and simulated data. The nominal value of 10 was used because simulation was fast at this value of ϵ, while minimising the burden on computational storage. For each posterior sample, we generated a dataset and plotted the posterior predictive mean and credibility intervals together with the spline smoothed observed data in Figure 21.7. The plots show data generated from the model with 200% activation. Top two rows have a high noise level and the bottom two rows have a moderate noise level. Solid lines indicate the observed data, dashed lines indicate the posterior predictive mean, and dotted lines are the corresponding interval limits for a 95% posterior predictive interval. In both cases, the spline summary S_1 performed very well, both in terms of capturing the true curve within the 95% interval, as well as the fidelity of the estimated curve to the true curve. The full dataset, S_2 and S_3, showed similar performances to each other and gave reasonable performance when the noise level is lower. The weighted least squares estimate S_4 performed the worst and has much more variability in the posterior predictive distribution. In the remainder of this chapter, we will work with S_1, the spline smoothed summary.

21.5.3 Tolerance level determination

For the determination of ϵ in step 3 of the ABC algorithm, the typical approach is to gradually decrease the value of ϵ until no further improvements can be made. Figure 21.8 illustrates the progression of the estimated marginal posteriors at $\epsilon \approx 7.8, 2.6, 1.7$, corresponding approximately to the 0.8, 0.02, and 0.001 percentiles of the sampled errors in our initial simulation of the one million samples. The solid line corresponds to the largest error, and the dotted line is the one with the smallest error. Note that the figures show marginal posteriors beyond the range of the prior distributions. This is due to the effect of smoothing for the purpose of visualisation, the true samples should not go beyond the prior distributions.

It is evident here that while at larger ϵ values the posterior variance is inflated, the posterior means do not change too much between varying values of ϵ. Interestingly for parameters γ, t_D, t_P and α, decreasing the values of ϵ did not produce more information about the parameters, suggesting that the data are fairly uninformative about these parameters. In our MCMC simulations, we observed similar behaviour with these parameters, suggesting that these parameters of the lp-ntPET model may not be estimable from the data.

21.5.4 Comparisons of different estimation methods

In this section, we compare the performances of ABC, WLS, and MCMC on simulation datasets. Figure 21.9 shows the posterior distribution obtained from ABC using the smallest ϵ value of 1.7, for a single set of simulated data. The model used for the simulation has 200% activation and a very high

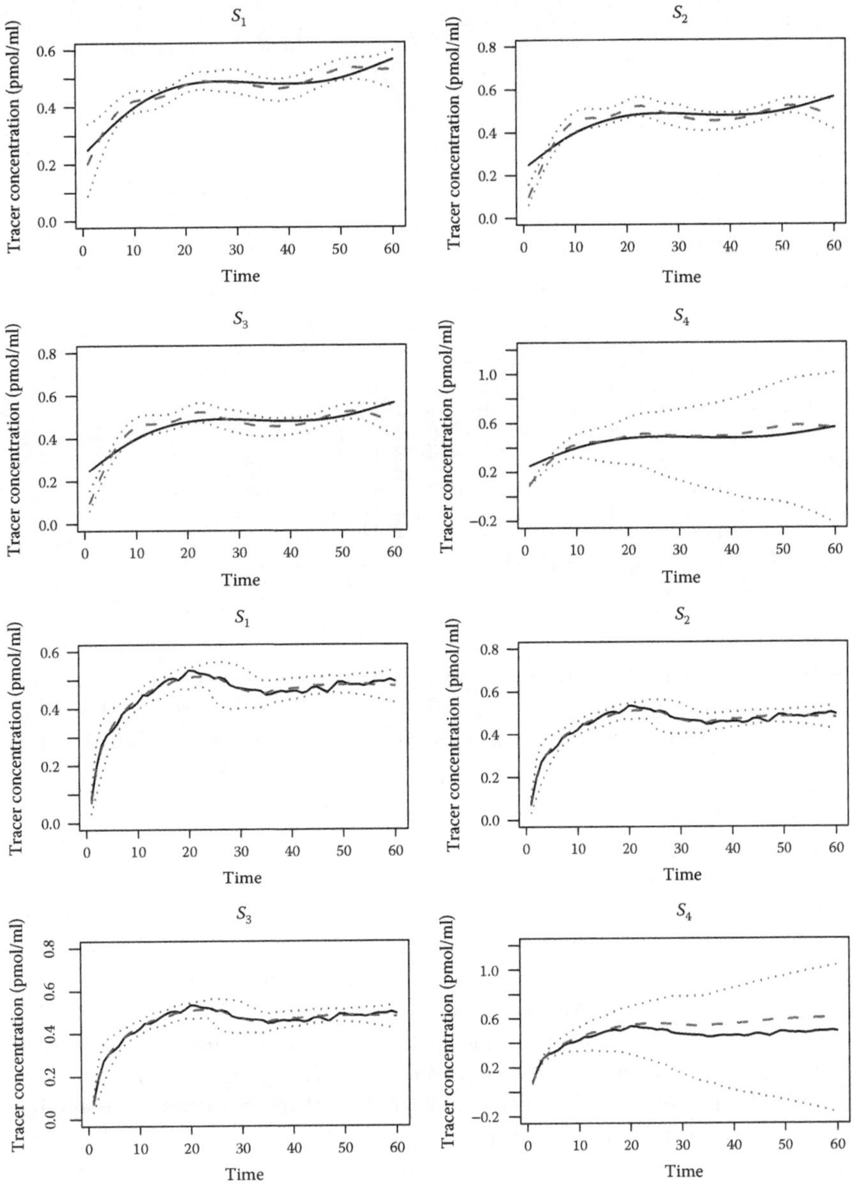

FIGURE 21.7
Posterior predictive plots for model with 200% activation; top two rows at high noise level and bottom two rows at moderate noise level. Results shown for the four different summary statistics S_1 to S_4. Solid lines indicate smoothed observed data, and dashed and dotted lines are mean, 0.025 and 0.975 percentiles, respectively, of the posterior predictive distribution.

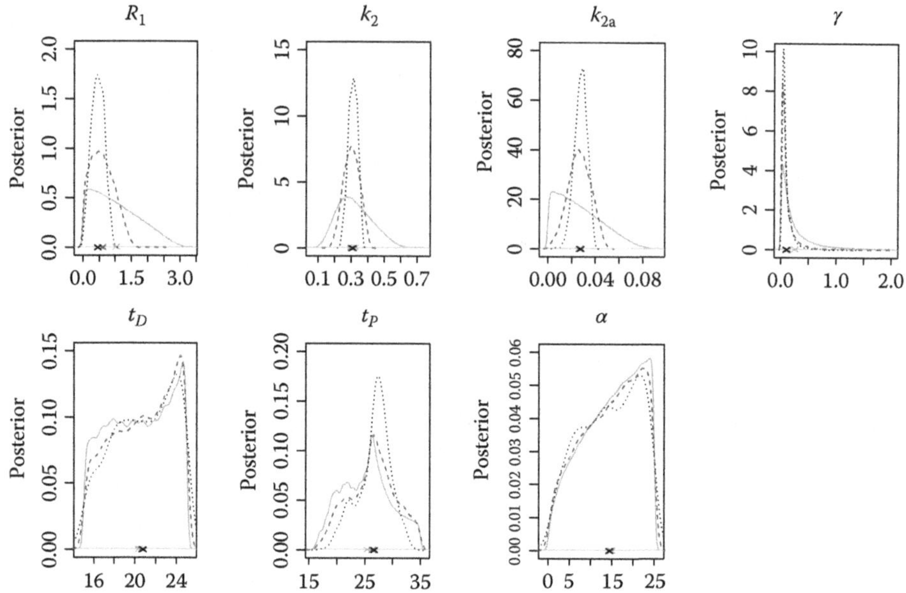

FIGURE 21.8
Evolution of the estimated marginal posterior distribution for the seven parameters, at different values of $\epsilon \approx 7.8, 2.6, 1.7$ (corresponding to the 0.8, 0.02, and 0.001 percentiles of the one million samples). Indicated by solid, dashed, and dotted lines, respectively. $\times$ indicates the posterior means.

noise level. Circles indicate the true parameter value used to obtain simulated data, triangles indicate the posterior mean, and pluses are the weighted least squares (WLS) estimate of Normandin et al. (2012). For WLS, we have simulated 100,000 values of t_D, t_P, and α from the same prior used for ABC and computed the estimate following Normandin et al. (2012). We have also implemented MCMC assuming an independent Gaussian error distribution with variances proportional to the observed TAC. However, it turns out that the MCMC algorithm is highly sensitive to the starting values, and chains can get stuck easily for many starting points, including those based on the true values. In addition, the trace plots indicate that the MCMC sampler has bad mixing behaviour, and these appear to be difficult to overcome using the standard MCMC sampler. Most of our MCMC samplers were unable to converge within a reasonable amount of computational time. This may have been caused by the mis-specification of the error model, since the errors in these data are known to be more complicated than Gaussian. The assumption that the variance of the error is proportional to the observed TAC, would likely induce a highly non-smooth likelihood surface, particularly when data are noisy. The behaviour of the MCMC output for parameters t_D, t_P, and α

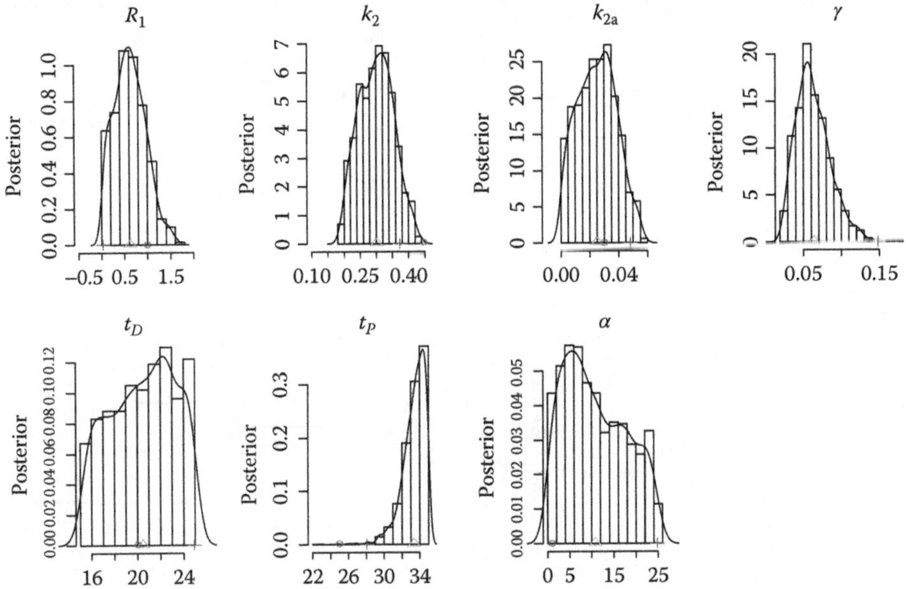

FIGURE 21.9
Final estimates of the marginal posterior distributions using ABC, at a value of $\epsilon \approx 1.7$. Circles indicate the true parameter value, triangles indicate the ABC posterior mean, and plusses indicates the weighted least squares estimate using Normandin et al. (2012).

are erratic; these parameters are essentially un-estimable. Indeed, the posterior distribution of these parameters suggests that the data indeed have very little information about the values of t_D and α, as the posteriors are largely unchanged from our prior distribution. This is also seen in the results from ABC, shown in Figure 21.9. We found that MCMC tended to over-estimate the R_1 parameter, while this is underestimated by ABC and WLS in some cases. MCMC was able to give very precise estimates of t_D close to the true value, while it found it difficult to estimate t_P. The situation is reversed for ABC, which found t_D difficult to estimate, while t_P was relatively straightforward. It is difficult to know the exact reason for these discrepancies, a lack of convergence in the MCMC sampler could partially explain some of the differences, model mis-specification is another possibility.

Figure 21.10 shows the posterior mean estimates from ABC (top row) and least squares estimates for 100 noise realisations (at 100% activation and highest noise level). We have excluded results from MCMC simulations due to the unreliable results obtained. At these noise and activation specifications, the parameter estimations were the most problematic. Results in Figure 21.10 demonstrate that ABC estimates are much less variable than WLS, although for both algorithms, the parameter R_1 is largely underestimated.

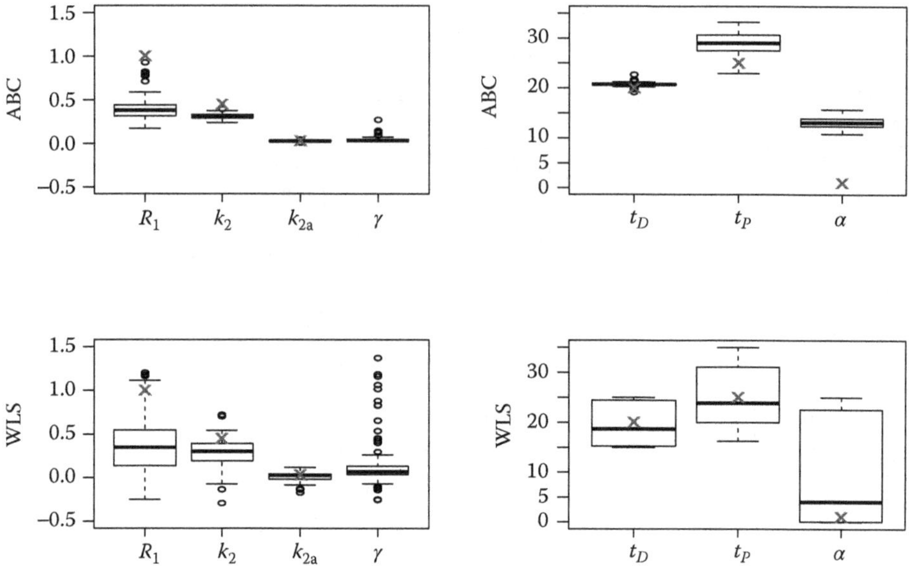

FIGURE 21.10
Box plots of posterior mean estimates over 100 noise realisations using ABC (top row), the corresponding WLS estimates (bottom row). Crosses indicate the true values.

The ABC estimator is more robust for all the parameters, but particularly so for R_1 and the time course response parameters t_D, t_P, and α, where the variability of the estimates as demonstrated by the box plots are much smaller.

We further investigated the cause for the apparently large bias in the R_1 estimation in both ABC and WLS. We found that parameter estimates are somewhat sensitive to the prior specification of the parameter α, and using a smaller range of $U(0, 3)$, we were able to obtain better estimates for both algorithms. However, this still did not provide a substantial improvement to the bias in the R_1 estimates. Figure 21.11 shows the comparative box plots for the R_1 parameter as estimated by ABC and WLS, over four noise levels and two different activation levels (200% and 100%). While it can be seen that the estimates of the 200% activation model are generally better than the 100% activation model, in both cases, the estimates worsen with noise, exhibiting high bias and high variance. The performance of the ABC estimator in the higher noise cases are generally superior to WLS. In low noise cases, WLS are often similar or even better than ABC, suggesting that the benefit of a Bayesian analysis lies in the more noisy problems.

In terms of the large bias in R_1, one possibility is that it could be an inherent bias of the lp-ntPET model, but this seems unlikely to explain away all the

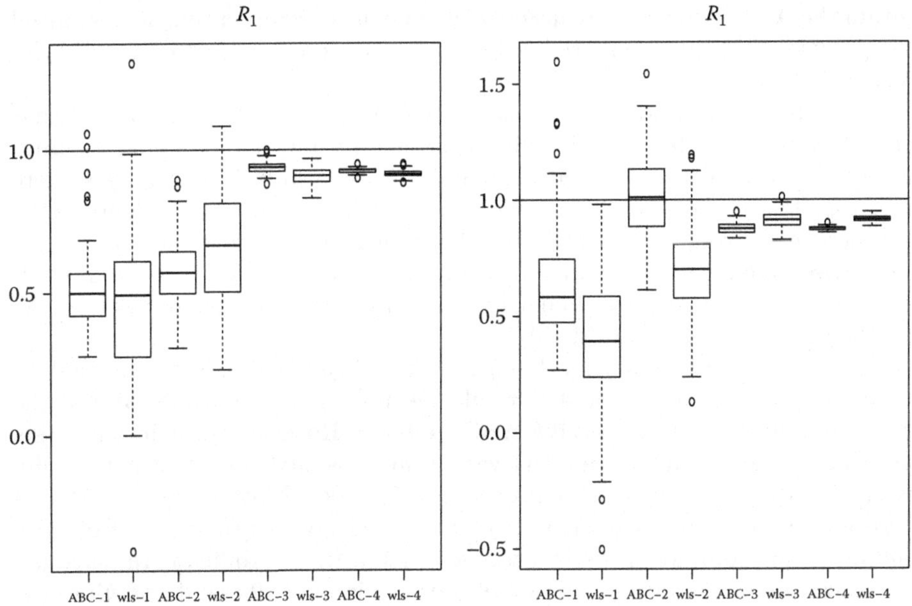

FIGURE 21.11
Box plots of posterior mean estimates over 100 noise realisations based on 100% activation model (left) and 200% activation model (right). X-axis correspond to results from ABC and WLS alternately, for four different noise levels (1–4) from highest to lowest noise levels. Horizontal line indicates the true value at 1.

bias. Figure 21.11 suggests that for lower noise levels, the results are close to the true value. This suggests that the biases maybe due to the way we handle the noise. A closer look at Figures 21.6, first column of top two rows, suggests that the spline-based summary deviates from the true TAC, while in lower noise data, there is much better accordance between the summaries and the true curve. This suggests that a more robust spline estimator, less sensitive to the distribution of the noise, may yield better results.

21.6 Conclusions and Discussions

This chapter examined the use of ABC for medical imaging data. In these types of data, it is often necessary to perform parameter estimation for multiple datasets, sometimes in the order of tens of thousands. A computational advantage of ABC in this scenario is that simulation of synthetic datasets within the ABC step will only need to be done once, representing a substantial computational saving compared with more traditional estimation procedures such as MCMC.

Our simulation studies comparing ABC, MCMC, and WLS showed that MCMC was unstable, and difficult to implement under our model assumptions. ABC and WLS obtained comparable results and, in most cases, were able to retrieve the true parameter values. In higher noise problems, ABC produced more robust estimates than WLS, which will prove more useful for voxel-wise estimations. In terms of computational time, WLS is the fastest. Both ABC and MCMC are time consuming, but over multiple datasets, ABC is substantially faster than MCMC.

We expect that in less noisy datasets, with relatively simple kinetic models, WLS would perform well, and it would be difficult to justify the use of the more computationally expensive ABC method. However, even in this case, there are added benefits from a Bayesian analysis that are often not readily available from the frequentist approach. For instance, Normandin et al. (2012) were interested in the significance of the magnitude parameter γ. However, finding an appropriate statistical test for such a task is difficult. In Bayesian inference, the posterior distribution of γ from Figure 21.9 readily provides the credibility interval for the parameter and allows us to assess the significance of a parameter immediately. Alternatively, posterior model comparison can be carried out relatively straightforwardly; see Chapter 6, in this handbook for more details on ABC model choice. Finally, the posterior distribution provides some information on how well the data are able to estimate certain parameters in a given model, see, for example, parameters t_D and α, and this could serve as an exploratory tool for the development of new models.

The ABC algorithm described in this chapter shares some similarity to the WLS approach of Normandin et al. (2012), where a basis function approach is used to estimate the time course response curve. Their WLS can be seen as a hybrid of Bayesian and frequentist methods. The main differences are that ABC requires the selection of a summary statistic and WLS searches for the modal estimate, while ABC computes the full posterior. In both algorithms, parameters become harder to estimate when the noise level is high. One possibility in ABC is to consider summaries which are more robust to noise. Another possible direction is to extend the analysis, currently assuming voxel independence, to allow borrowing of information from nearby voxels. It would be interesting to see how this can be performed efficiently within the ABC setting.

References

Alpert, N. M. and F. Yuan (2009). A general method of Bayesian estimation for parametric imaging of the brain. *Neuroimage 45*, 1183–1189.

Anger, H. O. (1964). Scintillation camera with multichannel collimators. *Journal of Nuclear Medicine 65*, 515–531.

Carson, R. E. (1986). *Positron Emission Tomography and Autoradiography: Principles and Applications for the Positron Emission Tomography and Autoradiography: Principles and Applications for the Brain and Heart*, Chapter Parameter estimation in positron emission tomography, pp. 347–390. New York: Raven Press.

Carson, R. E., S. C. Huang, and M. E. Green (1986). Weighted integration method for local cerebral blood flow measurements with positron emission tomography. *Journal of Cerebral Blood Flow and Metabolism 6*, 245–258.

Cherry, S. R. and M. Dahlbom (2004). *PET: Molecular Imaging and its Biological Applications*, Chapter PET: Physics, instrumentation and scanners, pp. 1–124. Berlin, Germany: Springer.

Cunningham, V. J. and T. Jones (1993). Spectral analysis of dynamic PET studies. *Journal of Cerebral Blood Flow and Metabolism 13*(1), 15–23.

Feng, D., S. C. Huang, Z. Wang, and D. Ho (1996). An unbiased parametric imaging algorithm for uniformly sampled biomedical system parameter estimation. *IEEE Transactions on Medical Imaging 15*, 521–518.

Fessler, J. A. (1996). Mean and variance of implicitly defined biased estimators (such as penalized maximum likelihood): Applications to tomography. *IEEE Transactions on Image Processing 5*, 493–506.

Garthwaite, P. H., Y. Fan, and S. A. Sisson (2016). Adaptive optimal scaling of Metropolis–Hastings algorithms using the Robbins–Monro process. *Communications in Statistics: Theory and Methods 45*, 5098–5111.

Gelman, A., J. B. Carlin, H. S. Stern, and D. B. Rubin (2004). *Bayesian Data Analysis*. Texts in Statistical Science. Boca Raton, FL: Chapman & Hall/CRC Press.

Gunn, R. N., S. R. Gunn, F. E. Turkheimer, J. A. D. Aston, and V. J. Cunningham (2002). Positron Emission Tomography compartmental models; A basis pursuit strategy for kinetic modeling. *Journal of Cerebral Blood Flow and Metabolism 22*, 1425–1439.

Gunn, R. N., A. A. Lammertsma, S. P. Hume, and V. J. Cunningham (1997). Parametric imaging of ligand-receptor binding in PET using a simplified reference region model. *Neroimage 6*(4), 279–287.

Gunn, R. N., M. Slifstein, G. E. Searle, and J. C. Price (2015). Quantitative imaging of protein targets in the human brain with PET. *Physics in Medicine and Biology 60*, R363–R411.

Hudson, H. M. and R. S. Larkin (1994). Accelerated image reconstruction using ordered subsets of projection data. *IEEE Transactions on medical imaging 13*(4), 601–609.

Innis, R.B., V. J. Cunningham, J. Delforge, M. Fujita, A. Gjedde, R. N. Gunn, J. Holden et al (2007). Consensus nomenclature for in vivo imaging of reversibly binding radioligands. *Journal of Cerebral Blood Flow and Metabolism 27*, 1533–1539.

Leahy, R. M. and J. Qi (2000). Statistical approaches in quantitative positron emission tomography. *Statistics and Computing 10*, 147–165.

Lin, Y., J. Haldar, Q. Li, P. Conti, and R. Leahy (2014). Sparsity constrained mixture modeling for the estimation of kinetic parameters in dynamic PET. *IEEE Transactions on Medical Imaging 33*, 173–185.

Malave, P. and A. Sitek (2015). Bayesian analysis of a one-compartment kinetic model used in medical imaging. *Journal of Applied Statistics 42*(1), 98–113.

Meikle, S. R., J. C. Matthews, V. J. Cunningham, D. L. Bailey, L. Livieratos, T. Jones, and P. Price (1998). Parametric image reconstruction using spectral analysis of PET projection data. *Physics in Medicine and Biology 43*, 651–666.

Morris, E. D., C. J. Enders, K. Schmidt, B. T. Christian, R. F. Muzic, and R. E. Fisher (2004). *Emission Tomography: The Fundamentals of PET and SPECT*, Chapter Kinetic modeling in PET, pp. 499–540. Emission Tomography: The Fundamentals of PET and SPECT. Amsterdam, the Netherlands: Academic Press.

Morris, E. D., K. K. Yoder, C. Wang, M. Normandin, Q.-H. Zheng, B. Mock, R. F. M. Raymond Jr., and J. C. Froehlich (2005). ntPET: A new application of PET imaging for characterizing the kinetics of endogenous neurotransmitter release. *Molevular Imaging 4*(4), 473–489.

Normandin, M. D., W. K. Schiffer, and E. D. Morris (2012). A linear model for estimation of neurotransmitter response profiles from dynamic PET data. *Neuroimage 59*, 2689–2699.

Qi, J. and R. M. Leahy (2006). Iterative reconstruction techniques in emission computed tomography. *Physics in Medicine and Biology 51*, 541–578.

Shepp, L. A. and Y. Vardi (1982). Maximum likelihood reconstruction for emission tomography. *IEEE Transactions on medical imaging MI-1*(2), 113–122.

Sisson, S. A., Y. Fan, and M. M. Tanaka (2007). Sequential Monte Carlo without likelihoods. *Proceedings of the National Academy of Sciences 104*, 1760–1765.

Sitek, A. (2014). *Statistical Computing in Nuclear Imaging.* Series in Medical Physics and Biomedical Engineering. Boca Raton, FL: CRC Press.

Wernick, M. N. and J. N. Aarsvold (2004). *Emission Tomography: The Fundamentals of PET and SPECT.* New York: Academic Press.

Zhou, Y., J. A. D. Aston, and A. M. Johansen (2013). Bayesian model comparison for compartmental models with applications in positron emission tomography. *Journal of Applied Statistics 40*, 993–1016.

Zhou, Y., S. C. Huang, and M. Bergsneider (2001). Linear ridge regression with spatial constraint for generation of parameter images in dynamic positron emission tomography studies. *IEEE Transactions on Nuclear Science 48*, 125–130.

Index

Note: Page numbers followed by f and t refer to figures and tables respectively.

A

ABC algorithm, 308. *See also* Approximate Bayesian computation (ABC)
- box plots, 642f
- computational advantage, 643–644
- for expensive simulators, 99
- for kinetic models, 631–632
- marginal posterior distribution, 641f
- MCMC, 199, 203
- to perform site update, 425

ABC approximation methods
- flexible regression density estimator, 230–232
- Gaussian copula, 219–222, 221f
- marginal adjustment strategy, 214–215
- multivariate g-and-k model, 222, 224–225, 225f
- non-linear multiple quantile regression analysis, 226–229, 229f
- posterior distribution, 213–299
- stereological extreme analysis, 232–235
- summary statistics, distribution, 229–230
- toy model, 215–219, 216f, 217f, 218f

ABC composite parameter (ABC-cp) approach, 185t, 202–203

ABC importance/rejection sampler variants, 95
- importance sampling, 96–98
- marginal ABC samplers, 100–101
- rejection-based ABC algorithms, 99
- rejection control importance sampling, 95–96
- stopping rule, ABC rejection sampling with, 99

ABC indirect inference (ABC II) methods, 180, 184t, 185–189, 191

ABC indirect parameter (ABC IP) method, 138, 184t, 186–187

ABC indirect score (ABC IS) approach, 139, 180, 184t, 187

ABC Markov chain Monte Carlo (ABC-MCMC) algorithm, 246, 399–401
- 1-hit algorithm, 262
- *ABCtoolbox*, 402
- adaptive, 262–263
- averaging estimators, 255–257
- *EasyABC*, 401–402
- inexact algorithms, 263–264
- noiseless algorithm, 250–251
- performance measures, 249–250
- rates of convergence, 251–253, 253t
- rejuvenation, 257–260
- samplers, 102–109
- stratification, 260

ABC posterior approximations, 22–23, 271–274
  binding function effect, 275–276, 275f
  checking method, 559
  concentration, 272–274, 273f
  versus full-likelihood posterior, 40f
  for *Gamma* target distribution, 29f
  mean, 279–281
  model error, 276, 277f
  overview, 277–279
  rate of concentration, 274–275, 274f
  target distribution, 23f
  using summary statistics, 38f
abc R packages, 370–371, 373
  cross-validation process, 385
  inferring BF, 388–389
  methods, 379
  multi-dimensional posteriors, 380–381
  post-sampling adjustments, 379–380
  rejection algorithm, 374
  validation, model choice, 390–391
`abc.star` R package, 300
  dispersions testing, 303–305, 304f
  locations testing, 300–303
  rates testing, 305–306
ABCtoolbox R packages, 370–371, 373–374
  ABC-MCMC, 402
  arguments in, 375, 377, 384
  biased posteriors, checking for, 386–387
  cross-validation process, 385–386
  estimation settings for, 376t–377t
  with fastsimcoal2, 405–406
  inferring BF, 389–390
  multi-dimensional posteriors, 381–382
  normal/uniform distribution model by, 392f
  output files, 378t
  posterior densities, 375f
  post-sampling adjustments, 380
  rejection algorithm, 374–378
  simulations for, 397–399, 398t, 400t
  statistical tests, 383
  validation, model choice, 391–392
Accelerating SL, 327–328
Adaptive multiple importance sampling (AMIS), 347, 350
  BC_{el} algorithm, 352–353
Agent-based model, 328, 443–444, 531
Approximate Bayesian computation (ABC), 14–15, 88, 243, 322, 514, 541–542
  in 11 Biaka pygmies, 39t
  accept/reject decisions in, 30f
  approximate posterior distribution, 15–23, 21f
  asymptotics, 269–286
  Bayesian approach, 62–64
  benefit, 403
  BF, 158, 161
  calibrations, 289, 291–292
  cell biology application, 335f, 336f
  cell migration, 527–532
  checking analysis, 561
  coalescent trees and mutation, 55–59, 57f, 58f
  computationally intensive methods, 60–62
  for design, 522
  emulator-informed, 582–586
  expensive simulators, rejection-based, 99
  g-and-*k* distribution analysis, 11–14, 12f

GLM algorithm, 379
hypothesis testing approach, 313f
importance sampling algorithm, 91–98, 111, 496–497
inference with, 496–501
interpretations, 43–44
iSIR with, 259
k-nn importance sampling algorithm, 98
levels in approximation, 41–43
likelihood-free intuition, 6–7
marginal posterior densities, 502f, 503t
MCMC algorithm, 350
overview, 3–5
parameter estimation by, 555–559, 557f
in population genetics analysis, 34–41, 35t
posterior. *See* ABC posterior approximations
quantile regression forest in, 82
regression approaches, 71–82
rejection sampling algorithm, 17, 25–26, 32–33, 90, 94, 580f, 600
RF model, 549–555, 554t
samplers, 88–117
SMC framework and algorithm, 114–115, 521f
software, 370t
standard *versus* calibration, 310f
statistical inference, 59–60
stereological extremes analysis, 7–11, 9f
stopping rule, rejection sampling with, 99
summary statistics in, 23–34, 329–330, 337f
takes off, 64–65
variants, 395
Z-test for, 294, 296, 297f, 299
Approximate sufficiency methods, 130
Augmented space ABC-MCMC samplers, 105–107
Autoregressive (AR) model, 183
Autoregressive moving average (ARMA) model, 183
Auxiliary likelihood methods, 129, 137, 191
advantages/disadvantages, 140
likelihood distance, 138
MLE, 138
properties, 139–140
scores, 139

B

Bayes Factors (BF), 351
ABC algorithm, 158, 161
boxplots, 352f
inferring, 388–390
Bayesian approach, 62–64
Bayesian calibration, 572–573
Bayesian computational methods, 14
Bayesian computation for copulas (BCOP) algorithm, 355–356
Bayesian EL, 338–339
Bayesian framework, 322
Bayesian indirect likelihood (BIL), 184t
with parametric auxiliary model, 189–192
Bayesian inference method, 3, 72
Bayesian likelihood-free methods, 184t–185t
Bayesian modelling, applications, 630
Bayesian SL, 325–327
Bayes sufficiency, 128
BC_{el} algorithm, 321, 347–352
AMIS, 352–353
extensions, 352–358, 354f
histogram, 349f
R code for, 348–349
BCOP (Bayesian computation for copulas) algorithm, 355–356
Benchmark model assumptions, 465–466

BF. *See* Bayes Factors (BF)
Biased posteriors, checking for, 386–387
BIL. *See* Bayesian indirect likelihood (BIL)
Binary networks, 523
Binding function, 181–182, 275–276, 312
Biosensors, 522
Block-parallel EP, 420–423
Bolivian dataset, 485
Boosting methods, 135
BSL approach, 322, 325
  MCMC, 326–327
  for posterior inference, 326–327
  uBSL, sensitivity, 332t, 334t
Built-in function `hist()`, 379

C

Carbon cycle processes, 581–582
CARTs (classification and regression trees) algorithm, 164
CDF (Cumulative Density Function), 428
Cell migration, ABC models, 527
  cytokines, 527–528
  data, 530f
  macrophages, 528–529
  model, 528f
  spatio-temporal characteristics, 531f
Chaotic prey–predator model, 608–609
  data and priors, 610–612
  from Kilpisjarvi dataset, 613–616, 613t, 614f
  using simulated data, 612–613, 612t
Chi-X Europe, 440f, 464
Classical SL, 323–324
Classification and regression trees (CARTs) algorithm, 164
Climate models, ABC for
  case study, 581–589
  emulation, 576–581
  features, 570
  history matching and, 573–576
  overview, 570–572
Climate simulators, 574
Coalescent model, 76
Coalescent trees and mutation, 55–59, 57f, 58f
Compartmental models
  one-tissue, 628, 628f, 630
  parameter estimation in, 629–631
  in PET, 627–629
Computational cost, 133
Computationally intensive methods, 60–62
Copula models, 353
Coupling applications, 588–589
Covariance mutation operator, 463
Cross-validation process, 384–386
  abc, 385
  ABCtoolbox, 385–386
Cumulative Density Function (CDF), 428
Cumulative distributions function, 351, 351f
Curse of dimensionality, 127, 141

D

Data generation process, 5
Datasets, simulation
  computer programs, 548–549
  models and parameters, 545–548
  observed, 544
Dimension reduction techniques, 134–135
Dirac measure, 4, 88
Direct conversion, 507
Divide and conquer in ABC
  EP algorithms, 417–428
  overview, 416–417
  spatial extremes application, 428–431
Do It Yourself Approximate Bayesian Computation (DIYABC) program, 159, 548–549, 560

Double exponential distribution, 160
Drug resistance. *See also* Multidrug resistance (MDR)
  acquisition, 493–494
  transmission and, 483
Dynamical systems, parameter estimation for, 515
  inference for, 515–516
  and model selection, 520–522
  posterior analysis for, 517–519
Dynamic imaging, 625–626

E

Earth system models (ESMs), 570–571
EasyABC R packages, 370–371
  ABC-MCMC, 401–402
  simulations for rejection, 396–397
Ecological modelling, ABC in
  chaotic prey–predator model, 608–616
  intractable, 597–600
  simulation-based methods in, 597–608
Effective sample size (ESS), 93, 330, 349
Efficient method of moments (EMM) approach, 183
EFPC. *See* Emulator Filtered Plausibility Constrained (EFPC) ensemble
EGA (European Genome-phenome), 544
EL. *See* Empirical likelihood (EL)
Ellipsoidal inclusions model, 233f, 234f, 235f
`el.test` library, 344–345
EM (expectation maximisation) algorithm, 626
EMM (efficient method of moments) approach, 183
Empirical likelihood (EL), 322, 339–341
  Bayesian, 338–339
  BC_{el} algorithm, 347–358
  estimation, 343–345
  features, 341–343
  in practice, 345–347
`emplik` library, 344–345
Emulation techniques, 572, 576
  methods, 577
  sequential history matching, 576–579
  two-box climate simulator, 579–581, 581f
Emulator Filtered Plausibility Constrained (EFPC)
  ensemble, 584
  parameter set, 586–588
Emulator-informed ABC design, 582–586
Entropy/loss minimisation approaches, 131
EP-ABC (expectation propagation ABC) algorithm, 156–157
EP algorithms. *See* Expectation-propagation (EP) algorithms
Equivalence hypothesis tests, 292–294
  within ABC, 294–299
ESMs (Earth system models), 570–571
ESS (effective sample size), 93, 330, 349
European Genome-phenome Archive (EGA), 544
Evolutionary models, scenarios, 545–548
Ewens, W., 59
Expectation maximisation (EM) algorithm, 626
Expectation propagation ABC (EP-ABC) algorithm, 156–157
Expectation-propagation (EP) algorithms, 417
  block-parallel, 422–423
  convergence, 431

Expectation-propagation (EP) algorithms (*Continued*)
- exponential properties, 418–419
- feasibility, 419
- flavours, 424
- Gaussian sites, 420
- iid case, 426–428
- parallel, 422
- practical considerations, 423
- presentation, 417–418, 426
- principle, 424–425
- sequential, 421
- site updates, 419–423
- theoretical properties, 423–424

F

Fay–Herriot (FH) model, 347
Finite sites models, 59–60
Fitting regression models, 74–76
Flexible regression density estimator, 230–232
Forest stand growth models, 598
Functional imaging, 626

G

g-and-*k* distribution analysis, 11–14, 12f
GARCH (generalized autoregressive conditional heteroskedasticity) model, 356
Gaussian copula ABC method, 219–222, 221f
Gaussian processes (GPs), 327–328, 576
- emulation, 589
- predictions, 577

Gaussian sites, 420
Generalized autoregressive conditional heteroskedasticity (GARCH) model, 356
General Linear Model (GLM) algorithm, 379
Generative model, 137
Gene regulatory network, 517, 517f
Generic pseudo-marginal algorithm, 247
Genetic dataset, 542–543
Genetic mutation/cross-over operators, 461–462
GFP (green fluorescent protein), 530
GLM (General Linear Model) algorithm, 379
Glucose metabolism, 623
GPs. *See* Gaussian processes (GPs)
Green fluorescent protein (GFP), 530
Gross primary productivity (GPP), 587

H

H3N2's disease, ABC calibrations in, 307–308, 309f
HDI (high posterior density intervals), 386
Hidden Markov models (HMMs), 526, 526f, 601
High-dimensional ABC approaches, 235–236
High posterior density intervals (HDI), 386
History matching and ABC, 573–574
HMMs (hidden Markov models), 526, 526f, 601
Human population genetics model, 170–172, 171f, 172f, 173f
Hypothesis tests, 312

I

IAM (integrated assessment model), 588
IFM (inference from the margins) approach, 354–355
IgG (immunoglobulin-G) dataset, 226, 227f
iid. *See* Independent and identically distributed (iid)
II method. *See* Indirect inference (II) method
ILI (influenza-like-illness), 306, 306f
Immunoglobulin-G (IgG) dataset, 226, 227f

Importance sampling, ABC, 91–95
Independent and identically distributed (iid), 155–156
  datasets, 181, 325
  parallel expectation propagation-ABC in, 426–428
Indirect inference (II) method, 179–180, 470
  auxiliary model, 181
  developments, 181–185
  mapping/binding function, 181
Inexact algorithms, 263–264
Infectious disease model, 196–199, 200f
Inference for SSMs, 600–604
Inference from the margins (IFM) approach, 354–355
Inference with ABC, 496–498
  parameter specifications and prior distributions, 499–501
  summary statistics, 498–499
Inferring BF, 388
  abc, 388–389
  ABCtoolbox, 389–390
Infinitely many alleles and sites model, 34, 56
Influenza-like-illness (ILI), 306, 306f
Integrated assessment model (IAM), 588
Intergovernmental Report on Climate Change (IPCC), 586
Intermediate complexity models, 570–571
Intractable ecological models, 597–600
Intractable likelihood stochastic model, 443
IPCC (Intergovernmental Report on Climate Change), 586
iSIR algorithm, 259

K

Kaplan–Meier curve, 346
Kernel functions, 16, 16t, 17f
Kingman's coalescent model, 34
k-nearest neighbours (k-nn) method, 155
Kolmogorov–Smirnov distance, 530–531
Kullback–Leibler (KL) divergence, 222, 223t

L

Laplace distribution, 160
Lasso regression approach, 76
Lazy summary statistic, MCMC ABC algorithm, 198–199
LDA (linear discriminant analysis), 394, 552f
Likelihood-free methods, 14–15, 452
  intuition level, 6–7
  rejection sampling algorithm, 7, 10, 13
Likelihood function, 5, 157
  synthetic log *versus* true log, 607, 607f
Limit order book (LOB), 438, 444
  agent-based model, 443–451
  Bayesian model formulation, 451–454
  complex model structure, 455
  data description application, 464–465, 465f
  and multi-queue simulation models, 439–441
  simulated intra-day representations, 470f, 471f
  stochastic processes, 441–442
  summary statistics in Bayesian, 454–455
Limit order cancellation/submission process, 445–449
Linear algebra notation, 489t
Linear discriminant analysis (LDA), 394, 552f
Linear noise approximation (LNA), 196
Linear parametric-neurotransmitter PET (lp-ntPET), 632–633

Linear regression methods, 134–135
Liquidity, 441–442
 demander agent, 449–451
 providers/demanders, 445–451
LNA (linear noise approximation), 196
LOB. *See* Limit order book (LOB)
Local error rates, 144–145
Local regression method, 156
Loclinear method, 379
Logit transformation, 76–77
Log return, 466–467
lp-ntPET (linear parametric-neurotransmitter PET), 632–633
Lyapunov exponent, posterior densities, 615, 616f

M

Mahalanobis distance, 13, 31, 88, 203
MAP. *See* Maximum a posteriori (MAP) model
Mapping/binding function, 181
Marginal ABC sampler, 100–101
Marginal adjustment approach, 214–219, 216f, 217f, 218f
Market orders, 439, 443–444
Markov chain Monte Carlo (MCMC) methods, 102–109, 324, 601, 630
 ABC algorithm, 350
 ABCtoolbox, 402
 BSL, 326–327, 331f, 333f
 EasyABC, 401–402
 Metropolis–Hastings algorithm in, 4, 338
 performing, 399–402
 samplers, 640
 simulations, 638
 uBSL, 327, 332f, 334f
Mass action, new infections by, 490–491, 491t
Maximum a posteriori (MAP) model, 163, 384
 approximating posterior probability, 165–166
 posterior probability estimation, 163–164
 RF construction, 164–165
 via machine learning, 162–166
Maximum likelihood estimator (MLE), 138, 184, 187, 575
Max-stable processes, 200–201
MCMC. *See* Markov chain Monte Carlo (MCMC) methods
MCWM (Monte Carlo within Metropolis), 263
MDR. *See* Multidrug resistance (MDR)
Meta-Gaussian distribution, 219–220
Metropolis–Hastings (MH) algorithm, 102, 245, 602
MEVDs (multivariate extreme value distributions), 199
Microsatellite loci, 544
MLE. *See* Maximum likelihood estimator (MLE)
Model choice, ABC, 140–141, 153–154, 387–388
 algorithm, 155
 curse of dimensionality, 141
 curse of insufficiency, 157–162, 160f, 161f, 162f
 encompassing model, 143
 inferring BF, 388–390
 local error rates, 144–145
 local logistic regression algorithm, 156
 mutual information, 143
 prior error rate, 169t, 170t
 projection/classification methods, 143–144
 RF algorithm, 165
 simulate only simulate, 155–157
 sufficiency/consistency, 141–142
 summary statistics, 392–395, 395t
 toy models, 166–170, 168f, 169f, 389t
 validation, 390–392

`modelChoiceValidation` argument, 391
Model-posterior checking, 559–562
Molecular dataset, 486t–487t
Moment matching, 419
Monte Carlo error, 42–43, 281–282
Monte Carlo within Metropolis (MCWM), 263
Most recent common ancestor (MRCA), 56
ms program, simulations, 36–37
Multi-dimensional posteriors
  abc, 380–381
  ABCtoolbox, 381–382
Multidrug resistance (MDR), 482
  acquisition, 501f
  strains, 482
  transmission and treatment failure to TB, 505–506, 506t
Multi-locus VNTR analysis, 483, 496
Multi-objective-II estimation framework, 470
Multi-scale models, 525–526
  inference for, 526–527
Multivariate extreme value distributions (MEVDs), 199
Multivariate g-and-k model, 222, 224–225, 225f
Mutual information method, 131–132, 143
Mycobacterium tuberculosis, 482, 500

N

Network analysis and inference, 522–523
  distances and summaries, 523–525
  evolution models, 523, 525f
  summary statistics for, 524
`neuralnet` method, 389
Neurotransmitter response model, 632–634
  estimation methods, 638–643, 639f, 640f, 643f
  prior and sampling distributions, 634–635, 635f
  summary statistics selection, 636–638
  tolerance level determination, 638
Newton–Lagrange algorithm, 344
Noiseless algorithm approximation, 250–251
Noisy likelihood perspective, 244–246
Non-linear multiple quantile regression analysis, 226–229, 229f
Non-linear regression method, 135
Non-parametric regression approaches, 212, 328
Non-Pygmies populations, 542–543, 545
  divergence time parameters, 556, 563
  introgression events, 550
Nuclear imaging, ABC in, 623–627

O

Observed datasets, 544
Ocean alkalinity, 584, 585f
One-tissue compartmental model, 628, 628f, 630
Ordinary differential equations (ODE), 515–516

P

Parallel EP, 422, 427
Parameter estimation
  in compartmental models, 629–631
  in kinetic modelling, 629
  model choice, 555–559, 557f
  robustness, 629
Parameter inference, 373–374
  estimation validation, 382–387
  multi-dimensional posteriors, 380–382
  post-sampling adjustments, 378–380
  rejection algorithm, 374–378
  statistics for, 392–394

Parameter validation
estimation, 382–387
testing uniformity, 387f
using methods, 385f
Parametric Bayesian indirect inference (pBII) methods, 180
Parametric Bayesian indirect likelihood (pBIL), 180, 325
Parametric BIL on the full data level (pdBIL) method, 185t, 189–190
Parametric BIL on the summary statistic level (psBIL) method, 185t, 191
Partial least squares (PLS) methods, 134, 392
analysis, 393
components, 405
versus RMSEP, 393f
transformations, 402
Partial posterior distribution, 72
Particle degeneracy, 109–110
Particle marginal Metropolis–Hastings (PMMH) algorithm, 603–604
Lyapunov exponents for, 616f
pBII (parametric Bayesian indirect inference) methods, 180
pBIL (parametric Bayesian indirect likelihood), 180, 325
pdBIL (parametric BIL on the full data level) method, 185t, 189–190
PET. *See* Positron emission tomography (PET)
Photons, 624
Phylodynamics model, 76
Pitman–Koopman–Darmois theorem, 128
PLS. *See* Partial least squares (PLS) methods
PMMH. *See* Particle marginal Metropolis–Hastings (PMMH) algorithm
Polynomial mutation, 462–463
Positron emission tomography (PET), 623
compartmental models in, 627–629
detector ring, 624, 625f
kinetic models for, 627
Posterior distributions, 72
analysis, 519f
estimators, 77
Spearman's ρ, 357, 357f
Posterior predictive p-values (ppp-values), 560–562
Post-processing techniques, 42
Post-sampling adjustments, 378–379
abc, 379–380
ABCtoolbox, 380
Potts model, Bayesian inference for, 328
ppp-values (posterior predictive p-values), 560–562
Predator-prey systems, 599
Principal component analysis, 561, 561f
Probabilistic calibration, 573, 575
Probabilistic simulation outputs, 586–587
Projection methods, 129, 134
advantages/disadvantages, 137
boosting, 135
classification methods, 143–144
dimension reduction techniques, 134–135
linear regression, 134–135
PLS, 134
training data, 136
Protein-protein interaction networks, 523
psBIL (parametric BIL on the summary statistic level) method, 185t, 191
Pseudo-marginal algorithms, 246–248

Pseudo-marginal Monte Carlo methods, 101
Pseudo-observed datasets, 386
Pygmies populations, 542–543, 545
 divergence time parameters, 556, 563
 introgression events, 550
 origin and diversification, 546f–547f

Q

Quantile regression analysis, 226–229

R

Radiation, 624
Radiotracer distributions, 3D image, 626
Random forests (RFs) technique, 163–165, 549
 ABC model, 549–555, 554t
 informative statistics to, 553f
Region of interest (ROI), 626, 628f
Regression adjustment techniques, 71–72
 application, 79–80, 79f, 80f, 81f
 benefits, 283–285, 283f
 fitting regression models, 74–76
 homoscedasticity, 74
 parameter transformations, 76–77
 partial posterior distribution, 72
 principle, 73–74, 75f
 regression methods, 81–82
 rejection algorithm by adjustment, 73
 shrinkage, 77
 theoretical results, 77–78
Regression-based emulators, 583
Regularisation approaches, 132
Rejection algorithm
 abc, 374
 ABCtoolbox, 374–378
 by adjustment, 73
Rejection control method, 95–96
Rejection sampling algorithm, 6, 89–91
Relative median absolute deviation (RMedAD), 556, 558
Resistance acquisition, models, 501–503
 drugs at equal rates, 504–505, 505f
 posterior predictive distribution, 504f
 resistance to drugs, 503–504
Reverse sampler (RS) approach, 192, 194, 195f
RFs. *See* Random forests (RFs) technique
Ricker map, 598, 605
RMedAD (relative median absolute deviation), 556, 558
ROI (region of interest), 626, 628f
Root mean squared errors (RMSEs), 405, 612, 612t
Root mean square error of prediction (RMSEP), 393f, 406f
R packages, 370–371, 395
 abc, 370–371. *See also abc* R packages
 ABCtoolbox, 370–371. *See also ABCtoolbox* R packages
 `density()`, 374, 378, 382
 EasyABC, 370–371
 `hist()`, 374
 `runif()`, 373
 summary statistics in, 372
RS (reverse sampler) approach, 192, 194, 195f

S

Sample degeneracy, 93
Sea-ice diffusivity, 583
Semi-automatic ABC, 135
Semi-parametric hierarchical EL (SHEL) model, 347
Sequential EP, 421
Sequential history matching, 576–579
Sequential Importance Re-Sampling (SIR) algorithm, 601–602

Sequential importance sampling (SIS), 110–113
Sequential Monte Carlo (SMC)
algorithm, 4, 109–110, 416, 516, 601
samplers, 113–116. *See also* SMC samplers
SIS, 110–113
SFS. *See* Site-frequency spectrum (SFS)
SHEL (semi-parametric hierarchical EL) model, 347
SI (susceptible-infected) model, 196
Simplification process, 597–598
Simulated datasets, 133, 159, 636
Simulated quasi-maximum likelihood (SQML) approach, 182, 324
Simulated *versus* observed statistics, 383f
Simulation-based methods
inference for SSMs, 600–604
intractable, 597–600
SL *versus* tolerance-based ABC, 604–608, 606f
Simulations, 395–396
MCMC, 399–402
mechanism, 72
population genetics, 402–407, 403t, 404t
for rejection, 396–399
toy models, ABC, 372–373
Single Nucleotide Polymorphisms (SNPs), 79, 170–172
Single photon emission computed tomography (SPECT), 623–624, 624f
SIR (Sequential Importance Re-Sampling) algorithm, 601–602
SIS (sequential importance sampling), 110–113
Site-frequency spectrum (SFS), 402–403
Perl script to, 408–409
synonymous, 404t
SL method. *See* Synthetic likelihood (SL) method
SLMH. *See* Synthetic likelihood Metropolis–Hastings (SLMH) algorithm
SMC samplers
adaptive schedules, 460–461
algorithm, 458
covariance mutation operator, 463
estimation algorithm configuration, 467–468, 468f
genetic mutation/cross-over operators, 461–462
intractable likelihood Bayesian models, 458–463
mutation kernel choice, 461–463
overview, 456–457
particle fitness/parameters distributions, 468–470, 469f, 472f
polynomial mutation, 462–463
simulated binary crossover, 462
stages, 457
SNPs (Single Nucleotide Polymorphisms), 79, 170–172
Spatial extremes application, 199–203, 204f
numerical results on, 429–431, 430f, 431f
overview, 428–429
summary statistics, 429
SPECT (single photon emission computed tomography), 623–624, 624f
SQML (simulated quasi-maximum likelihood) approach, 182, 324
State space models (SSMs), 600–601
Statistical inference process, 59–60
Stereological extremes analysis, 7–11, 9f, 232–235
Stochastic agent representation, 445–451

Stratification strategy, 260
Streptomycin drug, 482
Subset selection methods, 129–130
  advantages/disadvantages, 133
  approximate sufficiency, 130
  entropy/loss minimisation, 131
  mutual information, 131–132
  regularisation approaches, 132
  related methods, 132
Sufficient statistics concept, 128
Summary statistics, 429. *See also* Summary statistics selection methods
  in ABC, 23–34, 329–330, 337f
  ABC posterior approximations, 38f
  in Bayesian LOB, 454–455
  data and, 371–372, 372t
  inference with ABC, 498–499
  mean number, 33, 33t
  model choice, 392–395, 395t
  for network analysis and inference, 524
  normality plots for, 615f
  practical issues with, 28–34
  principles, 23–28
  in R packages, 372
  sampling distribution, ABC, 229–230
  selection, 636–638
  spatial extremes application, 429
Summary statistics selection methods, 126–127
  ABC model choice, 141–145
  asymptotic theory, 132
  auxiliary likelihood methods, 137–140
  curse of dimensionality, 127
  empirical performance, 145–146
  projection methods, 134–137
  strategies, 129
  subset selection methods, 128–133
  sufficient statistics concept, 128
Susceptible-infected (SI) model, 196
Sweeting, T. J., 65
Synthetic likelihood (SL) method, 230, 322, 600
  accelerating, 327–328
  Bayesian, 325–327
  classical, 323–324
  evaluating, 603
  technique, 191–192
  versus tolerance-based ABC, 604–608, 606f
Synthetic likelihood Metropolis-Hasting (SLMH) algorithm, 603–604
  Lyapunov exponents for, 616f
  posterior densities, 614f
Synthetic log-likelihood function, 607, 607f
Systems biology, ABC in, 532–533
  for dynamical systems, 515–519
  multi-scale models, 525–532
  network analysis and inference, 522–525
  overview, 513–514
  selection in, 519–522
  statistical inference in, 532

T

TAC. *See* Time activity curve (TAC)
Tail-area probability, 559–560
Terrestrial carbon sink strength parameters, 587
Threshold parameter, 73, 130
Time activity curve (TAC), 626–627, 636, 637f
Tomographic reconstruction, 625
Toy models, ABC, 371
  approximation methods, 215–219, 216f, 217f, 218f
  benefits, 371
  data and summary statistics, 371–372, 372t
  model choice by, 166–170, 168f, 169f, 389t
  simulations, 372–373

Tracer, nuclear imaging, 624, 628
Tripletwise extremal coefficients, 202–203
True log-likelihood function, 607, 607f
Tuberculosis (TB), acquisition of MDR in, 482
 antibiotic drugs for, 482
 conditions, 495–496
 cure, recovery, and death, 492
 dataset in Bolivia, 484–485
 detection and treatment, 492–493
 drug resistance, acquisition, 493–494
 MDR, 483
 mutation of marker, 494–495
 patterns, 483
 transmission and treatment failure to MDR, 505–506, 506t
 treatment, 482
Tukey P-value, 383–384, 384t
Two one-sided T-test and Z-test, 293–294, 302f

V

Validation, model choice, 390
 abc, 390–391
 ABCtoolbox, 391–392
Variable numbers of tandem repeats (VNTRs), 483, 483f
 genotypes, 485, 494
 loci, 500
 model, 485, 488–490, 488f
Voles trapping index, 610, 610f
Voles–weasels model, 611t
Voxel elements, 625–626, 628f

W

Weighted least squares (WLS), 640, 642, 642f, 644
Wilks' theorem, 341

X

Xetra trading system, 442, 476